International
REVIEW OF
Neurobiology
Volume 87

International

REVIEW OF

Neurobiology

Volume 87

SERIES EDITORS

RONALD J. BRADLEY

Department of Psychiatry, College of Medicine
The University of Tennessee Health Science Center
Memphis, Tennessee, USA

R. ADRON HARRIS

Waggoner Center for Alcohol and Drug Addiction Research
The University of Texas at Austin
Austin, Texas, USA

PETER JENNER

Division of Pharmacology and Therapeutics
GKT School of Biomedical Sciences
King's College, London, UK

Essays on Peripheral Nerve Repair and Regeneration

EDITED BY

STEFANO GEUNA
Department of Clinical and Biological Sciences
University of Turin
Turin, Italy

PIERLUIGI TOS
Reconstructive Microsurgery Unit
Department of Orthopedics
C.T.O. Hospital
Turin, Italy

BRUNO BATTISTON
Reconstructive Microsurgery Unit
Department of Orthopedics
C.T.O. Hospital
Turin, Italy

ELSEVIER

AMSTERDAM • BOSTON • HEIDELBERG • LONDON
NEW YORK • OXFORD • PARIS • SAN DIEGO
SAN FRANCISCO • SINGAPORE • SYDNEY • TOKYO
Academic Press is an imprint of Elsevier

Academic Press is an imprint of Elsevier
360 Park Avenue South, New York, NY 10010-1700
525 B Street, Suite 1900, San Diego, California 92101-4495, USA
32 Jamestown Road, London NW1 7BY, UK

This book is printed on acid-free paper.

For information on all Academic Press publications visit our Web site at www.elsevierdirect.com

ISBN-13: 978-0-12-375084-6

Printed and bound by CPI Group (UK) Ltd, Croydon, CR0 4YY

Transferred to Digital Print 2011

CONTENTS

Methods and Protocols in Peripheral Nerve Regeneration Experimental Research: Part I—Experimental Models

Pierluigi Tos, Giulia Ronchi, Igor Papalia, Vera Sallen, Josette Legagneux, Stefano Geuna, and Maria G. Giacobini-Robecchi

Methods and Protocols in Peripheral Nerve Regeneration Experimental Research: Part II—Morphological Techniques

Stefania Raimóndo, Michele Fornaro, Federica Di Scipio, Giulia Ronchi, Maria G. Giacobini-Robecchi, and Stefano Geuna

Methods and Protocols in Peripheral Nerve Regeneration Experimental Research: Part III—Electrophysiological Evaluation

Xavier Navarro and Esther Udina

Methods and Protocols in Peripheral Nerve Regeneration Experimental Research: Part IV—Kinematic Gait Analysis to Quantify Peripheral Nerve Regeneration in the Rat

Luís M. Costa, Maria J. Simões, Ana C. Maurício, and Artur S. P. Varejão

Current Techniques and Concepts in Peripheral Nerve Repair

Maria Siemionow and Grzegorz Brzezicki

Artificial Scaffolds for Peripheral Nerve Reconstruction

Valeria Chiono, Chiara Tonda-Turo, and Gianluca Ciardelli

Conduit Luminal Additives for Peripheral Nerve Repair

Hede Yan, Feng Zhang, Michael B. Chen, and William C. Lineaweaver

Tissue Engineering of Peripheral Nerves

Bruno Battiston, Stefania Raimondo, Pierluigi Tos, Valentina Gaidano, Chiara Audisio, Anna Scevola, Isabelle Perroteau, and Stefano Geuna

Mechanisms Underlying the End-to-Side Nerve Regeneration

Eleana Bontioti and Lars B. Dahlin

Experimental Results in End-To-Side Neurorrhaphy

Alexandros E. Beris and Marios G. Lykissas

End-to-Side Nerve Regeneration: From the Laboratory Bench to Clinical Applications

Pierluigi Tos, Stefano Artiaco, Igor Papalia, Ignazio Marcoccio, Stefano Geuna, and Bruno Battiston

Novel Pharmacological Approaches to Schwann Cells as Neuroprotective Agents for Peripheral Nerve Regeneration

Valerio Magnaghi, Patrizia Procacci, and Ada Maria Tata

Melatonin and Nerve Regeneration

Ersan Odaci and Suleyman Kaplan

Transthyretin: An Enhancer of Nerve Regeneration

Carolina E. Fleming, Fernando Milhazes Mar, Filipa Franquinho, and Mónica M. Sousa

Enhancement of Nerve Regeneration and Recovery by Immunosuppressive Agents

Damien P. Kuffler

The Role of Collagen in Peripheral Nerve Repair

Guido Koopmans, Birgit Hasse, and Nektarios Sinis

Gene Therapy Perspectives for Nerve Repair

Serena Zacchigna and Mauro Giacca

Use of Stem Cells for Improving Nerve Regeneration

Giorgio Terenghi, Mikael Wiberg, and Paul J. Kingham

Transplantation of Olfactory Ensheathing Cells for Peripheral Nerve Regeneration

Christine Radtke, Jeffery D. Kocsis, and Peter M. Vogt

Manual Stimulation of Target Muscles has Different Impact on Functional Recovery after Injury of Pure Motor or Mixed Nerves

Nektarios Sinis, Thodora Manoli, Frank Werdin, Armin Kraus, Hans E. Schaller, Orlando Guntinas-Lichius, Maria Grosheva, Andrey Irintchev, Emanouil Skouras, Sarah Dunlop, and Doychin N. Angelov

Electrical Stimulation for Improving Nerve Regeneration: Where do we Stand?

Tessa Gordon, Olewale A. R. Sulaiman, and Adil Ladak

Phototherapy in Peripheral Nerve Injury: Effects on Muscle Preservation and Nerve Regeneration

SHIMON ROCHKIND, STEFANO GEUNA, AND ASHER SHAINBERG

Age-Related Differences in the Reinnervation after Peripheral Nerve Injury

UROŠ KOVAČIČ, JANEZ SKETELJ, AND FAJKO F. BAJROVIĆ

Neural Plasticity After Nerve Injury and Regeneration

XAVIER NAVARRO

Future Perspective in Peripheral Nerve Reconstruction

LARS DAHLIN, FREDRIK JOHANSSON, CHARLOTTA LINDWALL, AND MARTIN KANJE

CONTRIBUTORS

Numbers in parentheses indicate the pages on which the authors' contributions begin.

Doychin N. Angelov (417), Anatomical Institute I, University of Cologne, Germany

Stefano Artiaco (281), Department of Orthopaedics, Traumatology, Rehabilitation, Plastic and Reconstructive Sciences, Second University of Naples, Naples, Italy

Chiara Audisio (227), Department of Animal and Human Biology, University of Turin, Turin, Italy

Fajko F. Bajrović (465), Institute of Pathophysiology, Faculty of Medicine, University of Ljubljana, SI-1000 Ljubljana, Slovenia

Bruno Battiston (1, 227, 281), Reconstructive Microsurgery Unit, Department of Orthopedics, C.T.O. Hospital, Turin 10126, Italy

Alexandros E. Beris (269), Department of Orthopaedic Surgery, University of Ioannina, School of Medicine, Ioannina, Greece

Eleana Bontioti (251), Department of Hand Surgery, Evgenidio Hospital, Athens, Greece

Grzegorz Brzezicki (141), Cleveland Clinic, Department of Plastic Surgery, 9500 Euclid Avenue, Cleveland, Ohio 44195, USA

Michael B. Chen (199), Division of Plastic Surgery, University of Mississippi Medical Center, Jackson, Mississippi, USA

Valeria Chiono (173), Department of Mechanics, Politecnico di Torino, Corso Duca Degli Abruzzi 24, 10129 Torino, Italy

Gianluca Ciardelli (173), Department of Mechanics, Politecnico di Torino, Corso Duca Degli Abruzzi 24, 10129 Torino, Italy

Luís M. Costa (127), Department of Veterinary Sciences, CITAB, University of Trás-os-Montes e Alto Douro, P.O. Box 1013, 5001-801 Vila Real, Portugal

Krzysztof Czaja (27), Department of Veterinary, Comparative Anatomy, Pharmacology, and Physiology, College of Veterinary Medicine, Washington State University, Pullman, Washington 99164, USA

Lars B. Dahlin (251, 507), Department of Hand Surgery, Malmö University Hospital, SE-205 02 Malmö, Sweden

Sarah Dunlop (417), School of Animal Biology and Western Australian Institute for Medical Research, The University of Western Australia, Perth, Australia

Carolina E. Fleming (337), Nerve Regeneration Group, Instituto de Biologia Molecular e Celular—IBMC, R. Campo Alegre 823, 4150-180 Porto, Portugal

Michele Fornaro (9, 27, 81), Department of Clinical and Biological Sciences, San Luigi Gonzaga School of Medicine, University of Turin, 10043 Turin, Italy; and Cavalieri Ottolenghi Scientific Institute of Neurobiology, University of Turin, 10043 Turin, Italy

Filipa Franquinho (337), Nerve Regeneration Group, Instituto de Biologia Molecular e Celular—IBMC, R. Campo Alegre 823, 4150-180 Porto, Portugal

Valentina Gaidano (227), Department of Clinical and Biological Sciences, San Luigi Gonzaga School of Medicine, University of Turin, 10043 Turin, Italy

Stefano Geuna (1, 27, 47, 81, 227, 281, 445), Cavalieri Ottolenghi Scientific Institute of Neurobiology, University of Turin, Turin 10043, Italy; and Department of Clinical and Biological Sciences, San Luigi Gonzaga School of Medicine, University of Turin, Turin 10043, Italy

Mauro Giacca (381), Molecular Medicine Laboratory, International Centre for Genetic Engineering and Biotechnology (ICGEB), Trieste 34149, Italy

Maria G. Giacobini-Robecchi (47, 81), Department of Clinical and Biological Sciences, San Luigi Gonzaga School of Medicine, University of Turin, Turin 10043, Italy

Tessa Gordon (433), Division of Neuroscience, University of Alberta, Edmonton, Alberta, Canada T6G 2S2

Maria Grosheva (417), ENT-Department, University of Cologne, Germany

Orlando Guntinas-Lichius (417), ENT-Department, Friedrich-Schiller University, Jena, Germany

Birgit Hasse (363), SCT Spinal Cord Therapeutics GmbH, Max-Planck-Str. 15a, 40699 Erkrath, Germany

Andrey Irintchev (417), ENT-Department, Friedrich-Schiller University, Jena, Germany

Fredrik Johansson (507), Department of Cell and Organism Biology, Lund University, SE-223 62 Lund, Sweden

Martin Kanje (507), Department of Cell and Organism Biology, Lund University, SE-223 62 Lund, Sweden

Suleyman Kaplan (9, 317), Department of Histology and Embryology, Ondokuz Mayis University School of Medicine, 55139 Samsun, Turkey

Paul J. Kingham (393), Section of Anatomy, Department of Integrative Medical Biology, Umeå University, Umeå, Sweden; and Blond McIndoe Laboratories, Tissue Injury and Repair Group, School of Medicine, University of Manchester, Manchester, United Kingdom

Jeffery D. Kocsis (405), Rehabilitation Research Center, Veterans Affairs Connecticut Healthcare System, West Haven, Connecticut 06516, USA; and Department of Neurology and Center for Neuroscience and Regeneration

Research, Yale University School of Medicine, New Haven, Connecticut 06510, USA

Guido Koopmans (363), SCT Spinal Cord Therapeutics GmbH, Max-Planck-Str. 15a, 40699 Erkrath, Germany

Uroš Kovačič (465), Institute of Pathophysiology, Faculty of Medicine, University of Ljubljana, SI-1000 Ljubljana, Slovenia

Armin Kraus (417), Department of Hand, Plastic, Reconstructive Surgery and Burn Unit, Eberhard-Karls-University of Tuebingen, BG Trauma Center Tuebingen, Germany

Damien P. Kuffler (347), Institute of Neurobiology, University of Puerto Rico Medical Sciences Campus, San Juan, Puerto Rico 00901, USA

Adil Ladak (433), Division of Plastic Surgery, University of Alberta, Edmonton, Alberta, Canada T6G 2S2

Josette Legagneux (47), Laboratory of Microsurgery, School of Surgery, Assistance Publique, Hôpitaux de Paris, France

Charlotta Lindwall (507), Institute of Neuroscience and Physiology, Gothenburg University, SE-413 90, Gothenburg, Sweden

William C. Lineaweaver (199), Rankin Plastic Surgery Center, Brandon, Mississippi, USA

Marios G. Lykissas (269), Department of Orthopaedic Surgery, University of Ioannina, School of Medicine, Ioannina, Greece

Valerio Magnaghi (295), C.I.Ma.I.Na., Interdisciplinary Centre for Nanostructured Materials and Interfaces, University of Milan, 20133 Milan, Italy; and Department of Endocrinology, Physiopathology, Applied Biology, Universita degli Studi di Milano, 20133 Milan, Italy

Thodora Manoli (417), Department of Hand, Plastic, Reconstructive Surgery and Burn Unit, Eberhard-Karls-University of Tuebingen, BG Trauma Center Tuebingen, Germany

Fernando Milhazes Mar (337), Nerve Regeneration Group, Instituto de Biologia Molecular e Celular—IBMC, R. Campo Alegre 823, 4150-180 Porto, Portugal

Ignazio Marcoccio (281), Hand and Microsurgery Unit, Istituto Clinico Città di Brescia, Italy

Ana C. Maurício (127), UMIB, Porto University, 4099-003 Porto, Portugal; and Department of Veterinary Clinics, Biomedics Sciences Institute of Abel Salazar, Porto University, 4099-003 Porto, Portugal

Xavier Navarro (105, 483), Institute of Neurosciences and Department of Cell Biology, Physiology and Immunology, Universitat Autònoma de Barcelona, E-08193 Bellaterra, Spain; and Centro de Investigación en Red sobre Enfermedades Neurodegenerativas (CIBERNED), Spain

Ersan Odaci (9, 317), Department of Histology and Embryology, Karadeniz Technical University School of Medicine, 61080 Trabzon, Turkey

Igor Papalia (1, 47, 281), Department of Surgical Disciplines, University of Messina, Messina, Italy

Isabelle Perroteau (227), Department of Animal and Human Biology, University of Turin, 10043 Turin, Italy

Patrizia Procacci (295), Department of Human Morphology and Biomedical Sciences-Città Studi, Università delgi Studi Milano, 20133 Milan, Italy

Christine Radtke (405), Department of Plastic, Hand and Reconstructive Surgery, Hannover Medical School, Hannover, Germany

Stefania Raimondo (27, 81, 227), Cavalieri Ottolenghi Scientific Institute of Neurobiology, University of Turin, Turin 10043, Italy; and Department of Clinical and Biological Sciences, San Luigi Gonzaga School of Medicine, University of Turin, Turin 10043, Italy

Shimon Rochkind (445), Division of Peripheral Nerve Reconstruction, Department of Neurosurgery, Tel Aviv Sourasky Medical Center, Tel Aviv University, Israel

Giulia Ronchi (27, 47, 81), Cavalieri Ottolenghi Scientific Institute of Neurobiology, University of Turin, Turin 10043, Italy; and Department of Clinical and Biological Sciences, San Luigi Gonzaga School of Medicine, University of Turin, Turin 10043, Italy

Bunyamin Sahin (9), Department of Anatomy, Ondokuz Mayis University School of Medicine, 55139 Samsun, Turkey

Vera Sallen (47), Institut de la Main, Clinique Jouvenet, Paris, France

Anna Scevola (227), Reconstructive Microsurgery Unit, Department of Orthopedics, C.T.O. Hospital, Turin, Italy

Hans E. Schaller (417), Department of Hand, Plastic, Reconstructive Surgery and Burn Unit, Eberhard-Karls-University of Tuebingen, BG Trauma Center Tuebingen, Germany

Federica Di Scipio (27, 81), Department of Clinical and Biological Sciences, San Luigi Gonzaga School of Medicine, University of Turin, Turin 10043, Italy

Asher Shainberg (445), Faculty of Life Science, Bar-Ilan University, Israel

Maria Siemionow (141), Cleveland Clinic, Department of Plastic Surgery, 9500 Euclid Avenue, Cleveland, Ohio 44195, USA

Maria J. Simões (127), UMIB, Porto University, 4099-003 Porto, Portugal; and Department of Veterinary Clinics, Biomedics Sciences Institute of Abel Salazar, Porto University, 4099-003 Porto, Portugal

Nektarios Sinis (363, 417), Klinik für Hand-, Plastische-, Rekonstruktive und Verbrennungschirurgie, Eberhard-Karls-Universität Tübingen, BG-Unfallklinik, Schnarrenbergstr. 95, 72076 Tübingen, Germany; and Department of Hand, Plastic, Reconstructive Surgery and Burn Unit, Eberhard-Karls-University of Tuebingen, BG Trauma Center Tuebingen, Germany

Janez Sketelj (465), Institute of Pathophysiology, Faculty of Medicine, University of Ljubljana, SI-1000 Ljubljana, Slovenia

Emanouil Skouras (417), Department of Trauma, Hand and Reconstructive Surgery, University of Cologne, Germany

Mónica M. Sousa (337), Nerve Regeneration Group, Instituto de Biologia Molecular e Celular—IBMC, R. Campo Alegre 823, 4150-180 Porto, Portugal

Olewale A. R. Sulaiman (433), Department of Neurosurgery, Ochsner Clinic Foundation, New Orleans, Louisiana 70131, USA

Ada Maria Tata (295), Department of Cell and Developmental Biology, Neurobiology Center "Daniel Bovet", "La Sapienza" University, Rome, Italy

Giorgio Terenghi (393), Blond McIndoe Laboratories, Tissue Injury and Repair Group, School of Medicine, University of Manchester, Manchester, United Kingdom

Chiara Tonda-Turo (173), Department of Mechanics, Politecnico di Torino, Corso Duca Degli Abruzzi 24, 10129 Torino, Italy

Pierluigi Tos (1, 27, 47, 227, 281), Reconstructive Microsurgery Unit, Department of Orthopedics, C.T.O. Hospital, Turin 10126, Italy

Esther Udina (105), Institute of Neurosciences and Department of Cell Biology, Physiology and Immunology, Universitat Autònoma de Barcelona, E-08193 Bellaterra, Spain; and Centro de Investigación en Red sobre Enfermedades Neurodegenerativas (CIBERNED), Spain

Bunyami Unal (9), Department of Histology and Embryology, Ataturk University School of Medicine, 25100 Erzurum, Turkey

Artur S. P. Varejão (127), Department of Veterinary Sciences, CITAB, University of Trás-os-Montes e Alto Douro, P.O. Box 1013, 5001-801 Vila Real, Portugal

Peter M. Vogt (405), Department of Plastic, Hand and Reconstructive Surgery, Hannover Medical School, Hannover, Germany

Frank Werdin (417), Department of Hand, Plastic, Reconstructive Surgery and Burn Unit, Eberhard-Karls-University of Tuebingen, BG Trauma Center Tuebingen, Germany

Mikael Wiberg (393), Section of Hand and Plastic Surgery, Department of Surgical and Perioperative Science, University Hospital, Umeå, Sweden; and Section of Anatomy, Department of Integrative Medical Biology, Umeå University, Umeå, Sweden

Hede Yan (199), Division of Plastic Surgery, University of Mississippi Medical Center, Jackson, Mississippi, USA

Serena Zacchigna (381), Molecular Medicine Laboratory, International Centre for Genetic Engineering and Biotechnology (ICGEB), Trieste 34149, Italy

Feng Zhang (199), Division of Plastic Surgery, University of Mississippi Medical Center, Jackson, Mississippi, USA

PREFACE

ESSAYS ON PERIPHERAL NERVE REPAIR AND REGENERATION

Nerve traumas and diseases are very frequent and, although these clinical conditions do not usually represent a threat to patient survival, their consequences on the quality of life and the related socioeconomic costs are relevant (Evans, 2001; Midha, 2006; Ruijs *et al.*, 2005). Interest in the study of peripheral nerve repair and regeneration has increased significantly over the last 20 years since, while in the past most nerve traumas and diseases were not surgically treated, today the number of nerve reconstructions performed is progressively increasing due to the continuous improvement in surgical technology and the spread of microsurgical skills among surgeons worldwide. Microsurgery is slowly emerging from a pioneering age when nerve reconstruction was performed only in few highly specialized centers.

Unfortunately, in spite of the impressive technical advancements in nerve reconstruction, complete recovery and normalization of nerve function almost never occur and the clinical outcome is often poor (Battiston *et al.*, 2005; Casha *et al.*, 2008; Gordon *et al.*, 2003; Höke, 2006; Lundborg, 2002). Therefore, we expect that the increasing number of patients receiving nerve surgery will represent an enormous stimulus for more research in peripheral nerve repair and regeneration with the goal of gaining knowledge on the basic mechanisms and, eventually, defining innovative strategies for improving nerve recovery in human and veterinary medicine.

In line with this growing interest, this special issue of the *International Review of Neurobiology* is aimed at providing an overview of the state-of-the-art knowledge in peripheral nerve repair and regeneration by bringing together a number of reviews that critically address some of the most important issues in this biomedical field.

After the first three review articles, which give a background overview on the scientific context, this special issue includes four methodology-oriented articles that are not usually included in review collections. They are aimed at providing the reader with practical methodological information which could be helpful for the design of an adequate experimental set up for peripheral nerve regeneration studies as well as to facilitate the interpretation of the results. The other review articles of this special issue address some of today's hot topics in nerve regeneration research from reconstructive and tissue engineering techniques to some of the most promising biomolecular and pharmacological approaches for promoting nerve regeneration and functional recovery. Contributors to this special issue have different and interdisciplinary backgrounds from experimental microsurgery to molecular biology and from neurobiology and neuroanatomy to biochemistry and material sciences.

We hope that this special issue will help, not only experienced nerve researchers in addressing the number of challenging scientific questions that still need answers in this intriguing research field, but also to attract young scientists and clinicians to form a new generation of peripheral nerve regeneration researchers that will be able to face tomorrow's scientific challenges using an integrated, interdisciplinary, and multitranslational approach.

Stefano Geuna
Pierluigi Tos
Bruno Battiston

References

Battiston, B., Geuna, S., Ferrero, M., and Tos, P. (2005). Nerve repair by means of tubulization: Literature review and personal clinical experience comparing biological and synthetic conduits for sensory nerve repair. *Microsurgery* **25,** 258–267.

Casha, S., Yong, V. W., and Midha, R. (2008). Minocycline for axonal regeneration after nerve injury: A double-edged sword. *Exp. Neurol.* **213,** 245–248.

Evans, G. R. (2001). Challenges to nerve regeneration. *Semin. Surg. Oncol.* **19,** 312–318.

Gordon, T., Sulaiman, O., and Boyd, J. G. (2003). Experimental strategies to promote functional recovery after peripheral nerve injuries. *J. Peripher. Nerv. Syst.* **8,** 236–250.

Höke, A. (2006). Mechanisms of disease: What factors limit the success of peripheral nerve regeneration in humans? *Nat. Clin. Pract. Neurol.* **2,** 448–454.

Lundborg, G. (2002). Enhancing posttraumatic nerve regeneration. *J. Peripher. Nerv. Syst.* **7,** 139–140.

Midha, R. (2006). Emerging techniques for nerve repair: Nerve transfers and nerve guidance tubes. *Clin. Neurosurg.* **53,** 185–190.

Ruijs, A. C., Jaquet, J. B., Kalmijn, S., Giele, H., and Hovius, S. E. (2005). Median and ulnar nerve injuries: A meta-analysis of predictors of motor and sensory recovery after modern microsurgical nerve repair. *Plast. Reconstr. Surg.* **116,** 484–494.

PERIPHERAL NERVE REPAIR AND REGENERATION RESEARCH: A HISTORICAL NOTE

Bruno Battiston,* Igor Papalia,† Pierluigi Tos,* and Stefano Geuna‡

*Reconstructive Microsurgery Unit, Department of Orthopedics, C.T.O. Hospital, Turin 10126, Italy

†Department of Surgical Disciplines, University of Messina, Messina, Italy

‡Department of Clinical and Biological Sciences, San Luigi Gonzaga School of Medicine, University of Turin, Turin 10043, Italy

Although the most significant advances in nerve repair and regeneration have been acquired over the last few decades, the study of nerve repair and regeneration potential dates back to ancient times namely to Galen in the second century A.D. This brief historical note outlines the milestones which have guided us to our present knowledge. In particular, we focus on the nineteenth century and the first decades of the twentieth century, an age in which the fathers of neurosurgery and neurobiology established the basis for most of the nerve repair and regeneration concepts used today. Finally, we shine a light on the most current history to show how recent pressure to use modern interdisciplinary and translational approach represents a sort of rediscovery of the scientific habits of the fathers of modern biomedicine, who used to carry out research from an integrated and broad point of view rather than from a super-specialized and specific one as it is often used today.

I. Introduction

Although the study of peripheral nerve regeneration potential dates back to ancient times (Naff and Ecklund, 2001; Terzis *et al.*, 1997), it is only since the second half of nineteenth century that a body of literature on nerve regeneration and nerve repair strategies began to accumulate, starting with the milestone observations of Augustus Waller (1850) (reprinted in Stoll *et al.*, 2002).

INTERNATIONAL REVIEW OF NEUROBIOLOGY, VOL. 87
DOI: 10.1016/S0074-7742(09)87001-3

0074-7742/09 $35.00

The aim of this historical note is to quickly trace the long history from the ancient anecdotal evidence to last century's scientific advancements which has led to the present state of the art knowledge in nerve regeneration research. Several historical articles have already reviewed various aspects of peripheral nerve regeneration and reconstruction and should be referred to further information (Ijpma *et al.*, 2008; Naff and Ecklund, 2001; Papalia *et al.*, 2007; Terzis *et al.*, 1997).

While the first written descriptions of peripheral nerves date back to the fourth century B.C. in Hippocrates' writings (Adams, 1868), the first descriptions of nerve repair and regeneration potential can be found in Galen's writings in the second century A.D. (Terzis *et al.*, 1997). Further descriptions of nerve sutures were reported by Paul von Aegina in the seventh century (Streppel *et al.*, 2000) and the Persian physicians Rahzes and Avicenna in the last years of the first millennium (Sunderland, 1981).

During the first half of the second millennium, further reports on nerve regeneration potential can be found in the work of several distinguished surgeons such as Guglielmo di Saliceto, Guido Lanfranchi, Guy de Chauliac, and Leonardo di Bertapaglia (Ladenheim, 1989; Terzis *et al.*, 1997). In spite of the above-mentioned references to nerve sutures, it has been proposed that the birth date of nerve reconstruction should be dated in the sixteenth century and attributed to Gabriele Ferrara who was the first to provide a detailed and clear description of a technique for suturing a severed nerve (Artico *et al.*, 1996).

II. The 19th Century

Although the regenerative potential of peripheral nerves after surgical repair was supported by works published in the seventeenth and eighteenth centuries, including the notable work of Cruikshank (1795) on the physiological aspects of nerve regeneration, it was during the nineteenth century that the study of neural repair significantly increased. The nineteenth century was the beginning of the new age of Life Sciences due to the introduction of the new histological techniques which would shed light on the fine structure of tissues and cells. The study of the nervous system too saw an impressive surge during this century and interest in the potential of the peripheral nerve to regenerate and the possible strategies to repair it grew from the milestone observations of Augustus Waller (1850) (reprinted in Stoll *et al.*, 2002). In his seminal paper entitled *Experiments on the section of the glossopharyngeal and hypoglossal nerves of the frog, and observations of the alterations produced thereby in the structure of their primitive fibres* (Waller, 1850), Waller described for the first time the progressive disorganization of the medulla of the nerve (i.e., the axons) which occurs downstream to nerve transection and which also involved the white substance of Schwann (i.e., the myelin sheaths). Since then, the study of the surgical repair of severed nerves has grown together with the study of the mechanisms of nerve regeneration, as carried out by neuroscientists, including some of the fathers of modern neurobiology such as Camillo Golgi, the discoverer of the black reaction, and Santiago Ramon y Cajal (Guillery, 2005).

Noteworthy is the work of Paget who, in 1947, reported functional recovery after median nerve primary repair in a young patient (Koopmans *et al.*, 2009, this issue) and of Létiévant (1873) who published the *Traité des Sections Nerveuses* where he provided a comprehensive overview of the different surgical approaches for repairing a complete nerve transection ("*synthése du nerf*") (p. 427) also including the first description of end-to-side nerve repair (Papalia *et al.*, 2007).

The first description of an autograft nerve reconstruction was reported by Philipeaux and Vulpian (1870) followed by the work of Albert (1878, 1885) who also was the first to perform an allograft nerve repair.

The first described attempt to bridge a nerve defect using a tube was made by Glück (1880), who employed a piece of bone to bridge a nerve gap based on the study carried out by Neuber one year before using reabsorbable decalcified bone tubes (Ijpma *et al.*, 2008; Neuber, 1879). Though unsuccessful, this attempt was followed by experiments by Vanlair, who obtained successful nerve fiber regeneration across a 3-cm-long tube made of decalcified bone (Vanlair, 1882, 1885).

III. The 20th Century

Two very nice overviews of the works of peripheral nerve repair and regeneration carried out throughout the nineteenth century can be found in the comprehensive papers by Powers (1904) and Sherren (1906).

In the first paper, Powers (1904) summarized the work carried out during the previous century on the bridging of nerve defects, including a paper written in Russian by Spijarny where nearly 200 cases of nerve suture were reported, and concluded that although it hardly seems possible to say definitively what is the best form for bridging nerve defects, nerve anastomosis "... for the present it would seem that they should be preferred."

In his work, Sherren (1906) tabled a number of previous studies including 8 experiments on human nerve grafts, 22 on animal nerve grafts, 73 on nerve anastomosis, and 8 on nerve crossing. For each work, the author noted the conditions of employment, the method and the results and concluded by emphasizing the importance of "operations upon peripheral nerves" thus trying to point out the limitation of the knowledge and to outline the direction in which he believed it must proceed in order to obtain still greater success.

Another nice paper is that published by Sachs and Malone (1922), reporting on their experimental studies on nerve regeneration across nerve gaps. One of the strongest points of this paper is the use of histology to support the surgical finding as a demonstration of the growing synergies that used to take place between basic and clinical science in nerve regeneration research in the first half of the twentieth century.

Noteworthy is also the paper by Ballance *et al.* (1926), where these authors reported on their experience on nerve anastomosis and nerve crossing in monkeys and cats. In particular, they reported various types of end-to-side and side-to-end including double lateral anastomosis and they documented their findings with nice histological drawings too.

In the same years, some negative results were also reported including the papers by Stookey (1922) and Babcock (1927). While the former paper was specifically negative towards the possibility that nerve flaps can be successful in repairing nerve defects ("On the futility of bridging nerve defects by means of nerve flaps"), the second paper even raised doubts also about the usefulness of various nerve repair techniques including also some techniques that are currently used today, namely nerve grafting and tubulization (Fig. 1).

During the first three decades of the twentieth century, the interest in peripheral nerve repair and regeneration was lively and saw a synergism between basic and clinical scientists; however, in the second half of the past century this trend had decreased probably because of the arising criticisms about the real usefulness of the nerve reconstruction techniques in promoting nerve regeneration. Although very important works were carried out by many surgeons worldwide (including very famous surgeons such as Herbert Seddon and Sydney Sunderland), research along most of the remaining years of the twentieth century was mainly dedicated to optimization of the surgical techniques for nerve reconstruction. The observation that peripheral nerve axons retain a capability for spontaneous regeneration after trauma led researchers to focus on how to repair the nerves and not on how to improve nerve regeneration. On the other hand, basic neurosciences saw a great expansion towards neurochemistry and molecular neurobiology and the interest towards the study of the regeneration of peripheral nerves decreased.

IV. Conclusions

It is just over the last years that research synergy between surgical science and the new tendencies of molecular neurobiology began to rise again. The increasing awareness that, although possible, peripheral regeneration is far from being optimal (Battiston *et al.*, 2009, this issue; Lundborg, 2002) led to the awareness among surgeons that the next advancements in peripheral nerve reconstruction

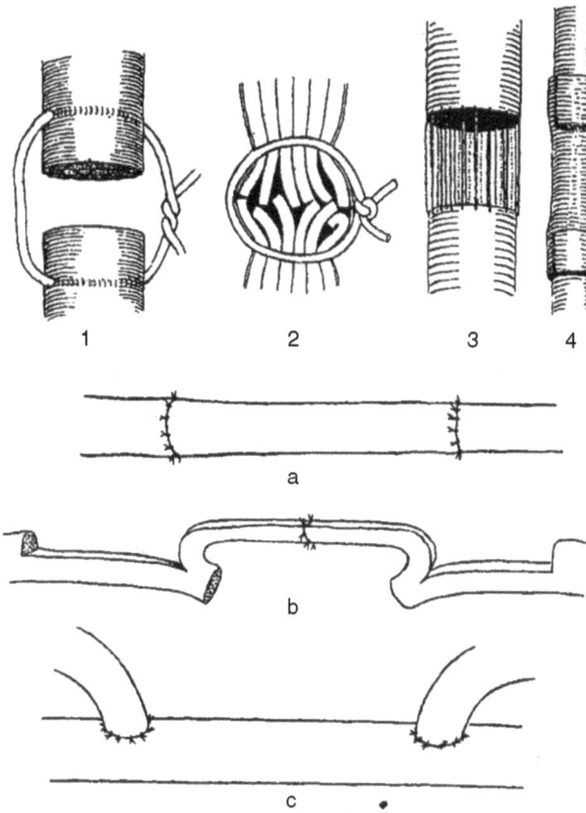

FIG. 1. Drawing from Babcock's 1927 paper illustrating the faulty methods for nerve reconstruction which include some techniques that are currently used today, namely nerve grafting and tubulization.

would need a stronger biological basis and, on the other hand, and continuous increase in basic scientists' commitment to peripheral nerve regeneration research occurred, as shown by the dedication of special issues of important international neuroscience journals over the last years. This new trend towards interdisciplinary and multitranslational research opens several new scientific fields and makes it possible to foresee that the next decades will see significant scientific advancements in nerve repair and regeneration.

In addition, revisiting history of nerve regeneration can be important not only to understand how we arrived to the state of the art scientific knowledge but also to rediscover some old ideas that, although innovative, have not been expanded adequately because of the technical limitations but might become innovative when reappraised today. What happened with end-to-side nerve repair is paradigmatic of how an innovation can be reported and investigated for many years, then forgotten for a long lapse of time, and eventually rediscovered concurrently with the advances in the scientific technology and knowledge (Papalia *et al.*, 2007). Perhaps, a careful revisiting of the long history of nerve repair and regeneration research may reveal other old discoveries that are worthy of being reappraised today.

Acknowledgments

This work was supported by grants from the MUR (Italian Ministry of University and Research), the Compagnia di San Paolo (Bando Programma Neuroscienze), and the Regione Piemonte (Bando Ricerca Sanitaria Finalizzata).

References

Adams, F. (1868). "The Genuine Works of Hippocrates." William Wood & Co., New York. http://www.chlt.org/sandbox/dh/Adams.

Albert, E. (1878). *Verhandlung des Naturwissenschaftlichen und Medizinischen Vereins in Innsbruck* **9,** 97–121.

Albert, E. (1885). Einige operationem an nerven. *Wien. Med.* **26,** 1222.

Artico, M., Cervoni, L., Nucci, F., and Giuffrè, R. (1996). Birthday of peripheral nervous system surgery: The contribution of Gabriele Ferrara (1543–1627). *Neurosurgery* **39,** 380–382.

Babcock, W. W. (1927). A standard technique for operations on peripheral nerves with special reference to closure of large gaps. *Surgery Gynec.Obstet.* **45,** 364.

Ballance, C., Colledge, L., and Bailey, L. (1926). Further results of nerve anastomosis. An illustrated record of some experiments in which: 1. The central and peripheral ends of a divided nerve were implanted at varying distances apart into a neighbouring normal nerve. 2. Certain nerve-trunks of the limbs were divided and anastomosed by suture in cross-wise fashion. *Br. J. Surg.* **13,** 533–558.

Battiston, B., Raimondo, S., Tos, P., Gaidano, V., Audisio, C., Scevola, A., Perroteau, I., and Geuna, S. (2009). Tissue engineering of peripheral nerves. *Int. Rev. Neurobiol.* **87,** 225–249.

Cruikshank, W. (1795). Experiments on the nerves, particular on their reproduction; and on the spinal marrow of living animals. *Philos. Trans. R. Soc. London* **85,** 177–189.

Glück, T. (1880). Ueber Neuroplastic Auf dem Wege der transplantation. *Arch. Klin. Chir.* **25,** 606–616.

Guillery, R. W. (2005). Observations of synaptic structures: Origins of the neuron doctrine and its current status. *Philos. Trans. R. Soc. Lond. B Biol. Sci.* **360,** 1281–1307.

Ijpma, F. F., Van De Graaf, R. C., and Meek, M. F. (2008). The early history of tubulation in nerve repair. *J. Hand. Surg. Eur.* **33,** 581–586.

Koopmans, G., Hasse, B., and Sinis, N. (2009). The role of collagen in peripheral nerve repair. *Int. Rev. Neurobiol.* **87,** 363–379.

Ladenheim, J. C. (1989). "Leonard of Bertapaglia: On Nerve Injuries and Skull Fracture." Futura, New York.

Létiévant, E. (1873). "Traité des Sections Nerveuses." J.B. Baillière et Fils, Paris.

Lundborg, G. (2002). Enhancing posttraumatic nerve regeneration. *J. Peripher. Nerv. Syst.* **7,** 139–140.

Naff, N. J., and Ecklund, J. M. (2001). History of peripheral nerve surgery techniques. *Neurosurg. Clin. N. Am.* **12,** 197–209.

Neuber, G. (1879). Ein antiseptischer Dauerverband nach gründlicher Blutstillung. *Arch. Klin. Chir.* **24,** 314–330.

Papalia, I., Geuna, S., D'Alcontres, F. S., and Tos, P. (2007). Origin and history of end-to-side neurorrhaphy. *Microsurgery* **27,** 56–61.

Philipeaux, J. M., and Vulpian, A. (1870). Note sur des essays de greffe d'u troncon de nerf lingualentre les deux bouts du nerf hypoglosse. Apres excision d'un segment du dernier nerf. *Arch. Phys. Norm. Pathol.* **3,** 618.

Powers, C. A. (1904). The bridging of nerve defects. A contribution to the surgery of nerves. *Ann. Surg.* **40,** 632–643.

Sachs, E., and Malone, J. Y. (1922). An experimental study on the methods for bridging nerve defects. *Arch. Surg.* **5,** 314–333.

Sherren, J. (1906). Some points in the surgery of the peripheral nerves. *Edinb. Med. J.* **20,** 297–316.

Stoll, G., Jander, S., and Myers, R. R. (2002). Degeneration and regeneration of the peripheral nervous system: From Augustus Waller's observations to neuroinflammation. *J. Peripher. Nerv. Syst.* **7,** 13–27.

Stookey, B. (1922). "Surgical and Mechanical Treatment of Peripheral Nerves." Saunders, Philadelphia.

Streppel, M., Heiser, T., and Stennert, E. (2000). Historical development of facial nerve surgery with special reference to hypoglossal-facial nerve anastomosis. *HNO* **48,** 801–808.

Sunderland, S. (1981). The anatomic foundation of peripheral nerve repair techniques. *Orthop. Clin. N. Am.* **12,** 245–266.

Terzis, J. K., Sun, D. D., and Thanos, P. K. (1997). Historical and basic science review: Past, present and future of nerve repair. *J. Reconstr. Microsurg.* **13,** 215–225.

Vanlair, C. (1882). De la régénération des nerfs périphériques par le procédé de la suture tubulaire. *Arch. Biol. (Paris)* **3,** 379–496.

Vanlair, C. (1885). Nouvelles recherches expérimentales sur la régénération des nerfs. *Arch. Biol. (Paris)* **6,** 127–235.

Waller, A. (1850). Experiments on the section of the glossopharyngeal and hypoglossal nerves of the frog, and observations of the alterations produced thereby in the structure of their primitive fibers. *Philos. Trans. R. Soc. Lond.* **140,** 423–429 (reprinted in Stoll *et al.*, 2002).

DEVELOPMENT OF THE PERIPHERAL NERVE

Suleyman Kaplan,* Ersan Odaci,† Bunyami Unal,‡ Bunyamin Sahin,§ and Michele Fornaro¶

*Department of Histology and Embryology, Ondokuz Mayis University School of Medicine, 55139 Samsun, Turkey

†Department of Histology and Embryology, Karadeniz Technical University School of Medicine, 61080 Trabzon, Turkey

‡Department of Histology and Embryology, Ataturk University School of Medicine, 25100 Erzurum, Turkey

§Department of Anatomy, Ondokuz Mayis University School of Medicine, 55139 Samsun, Turkey

¶Department of Clinical and Biological Sciences, San Luigi Gonzaga School of Medicine, University of Turin, 10043 Turin, Italy

Normal function of the peripheral nerve (PN) is based on morphological integrity and relationship between axons, Schwann cells, and connective sheaths, which depends on the correct development of all these components. Most of the relevant studies in this field were carried out using animal models, since reports on the development of the human PNs from the time of prenatal formation to postnatal development are limited as it is quite difficult to find many nerves in fetuses. In this review paper, we will address the main developmental stages of axons, Schwann cells, and connective tissue sheaths in PNs. Knowledge on the development of PNs and their main components is important for the study of nerve repair and regeneration. This knowledge can be helpful for designing innovative treatment strategies since, like with other organs, the development and regeneration processes share many biological features.

INTERNATIONAL REVIEW OF
NEUROBIOLOGY, VOL. 87
DOI: 10.1016/S0074-7742(09)87002-5

0074-7742/09 $35.00

I. Introduction

A peripheral nerve (PN) is composed of connective tissue and neural components. A nerve is also described as a discrete organ of the PNS since the nervous fibers' components are completed by a surrounding connective tissue stroma and addition to a nutritive blood supply (Kerns, 2008). Nerve fibers can be either unmyelinated or myelinated (Flores *et al.*, 2000; Geuna *et al.*, 2009, Landon, 1976, this issue). While unmyelinated fibers are composed of several axons, enveloped as a group by a single Schwann cell, myelinated fibers consist of a single axon, enveloped individually by a single Schwann cell. A multilaminated myelin sheath is formed by Schwann cells' membrane wrapping around the myelinated nerve fibers (Fig. 1) (Flores *et al.*, 2000).

The connective tissue structures of the PN consist of three distinct sheaths: endoneurium, perineurium, and epineurium, from innermost to outermost, respectively. These structures form a framework that organizes and protects the nerve fibers and axons (Flores *et al.*, 2000). The epineurium made up the connective tissue that surrounds the entire nerve trunk and blends with the connective tissue of nearby parts. The perineurium is the middle-level connective tissue sheath around the nerve fibers, and it made up of concentrically arranged, more compact cellular layers. The perineurium encloses individual fascicles of longitudinally running nerve fibers. It is the innermost sheath surrounding the Schwann cells and the nerve fibers within (Landon, 1976).

The normal development of the PNs and their morphological structures (axon, Schwann cell, and the components of epineurium, perineurium, and endoneurium) are of crucial importance for the integrity of body function since they control many tissues, organs, and systems associated with different homeostatic functions (Kerns, 2008). In this paper, studies on the development of parenchyma (axon, Schwann cell) and stromal (connective tissue sheaths—epineurium, perineurium, and endoneurium) components of the PN are reviewed, while another paper in this special issue of the *International Review of Neurobiology* (Geuna *et al.*, 2009, this issue) will address, more in-depth, the normal structure of the adult PN. Our knowledge in this field is principally based on studies performed using experimental animal models (i.e., rat, mice, cat, and chick). In fact, because of ethical considerations and difficulty in finding human samples from prenatal to postnatal periods, there are no sufficient studies on the development of the morphological features of human PNs from prenatal to postnatal stages.

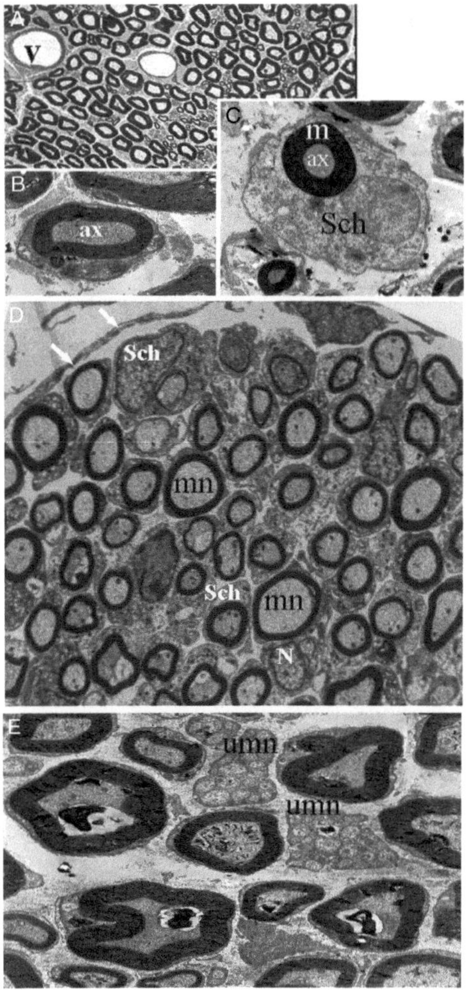

FIG. 1. Light and electron microscopy of sciatic nerve in adult rats (A–E). Schwann cells are not easily detectable in light microscopy (A), but clearly seen in electron microscopy (B–E). A myelinating Schwann cell envelopes one-to-one a small axon (C). Arrows point to the border of the perineurium (D). At higher magnification, unmyelinated fibers within myelinated fibers are detectable (E). v, vessel; ax, axon; Sch, Schwann cell; mn, myelinated nerve fiber; unm, unmyelinated nerve fiber; N, nucleus of Schwann cell. (See Color Insert.)

II. Development of the Neural Components of the Peripheral Nerve

A. Developmental Properties of the Schwann Cells

In vertebrates, the most represented glial cell components, which house PNs are the Schwann cells (Jessen, 2004). Schwann cells not only build a protective sheath around neuronal processes and myelinate large-caliber axons in the adult, but they are also quite related to neurons during the earliest stages of their common development. Moreover, later development and maintenance of PN morphology and function is crucially dependent on the controlled and bidirectional cell cross talking between Schwann cells and neurons (Lobsiger *et al.*, 2002). The normal development of PNs requires an appropriate number of Schwann cells (Atanasoski *et al.*, 2008). However, growth factors and signaling pathways that control Schwann cell proliferation and differentiation have not been clearly described yet (Jessen and Mirsky, 2002; Lobsiger *et al.*, 2002). The proliferating activity of Schwann cells transiently changes in the prenatal and postnatal periods and in particular, an increase of proliferation is seen throughout postnatal time points. Prenatal and postnatal Schwann cell proliferation is differently regulated. The postnatal wave of proliferation is dispensable for generating the number of Schwann cells required for correct myelination of axons. These conclusions are based on findings that CDk4-deficient Schwann cells divide normally before birth and virtually stop proliferating after birth (Atanasoski *et al.*, 2008).

Schwann cell development, including its unique molecular markers, signaling responses and tissue relationships during the PN development, from the initial stages of gliogenesis to myelinization has been widely investigated in rodents (Woodhoo and Sommer, 2008). It is well known that the cell precursors of the central nervous system (CNS) begin to develop in the early stages of embryogenesis through a series of processes called neurulation (Moore and Persaud, 1993). The cellular rod that determines the primitive axis of the embryo, the notochord, is incorporated into the vertebral system. This structure induces the overlying ectodermic tissue to form the neural plate at approximately 2 weeks of gestation in humans (Moore and Persaud, 1993; Rice and Barone, 2000). In humans, the neural plate invaginates along its central axis to form the neural groove with neural folds on each side on approximately gestational day 18, and these folds have begun to close up and fuse by the end of the third week of gestation. Therefore, the neural tube is formed near the anterior end of the notochord. This fusion progresses from cranial towards caudal direction. During the neural tube formation from the overlying ectoderm, a population of cells diverges from the surface ectoderm at the apex of the neural folds to form the neural crest (Moore and Persaud, 1993; Rice and Barone, 2000). Schwann cells derive from the portion of the neural epithelium that gives rise to the neural crest as described above (Jessen, 2004; Le Douarin and Smith, 1988).

Schwann cells share their origin from the neural crest with the spinal ganglia sensory neurons (Moore and Persaud, 1993) as well as several non-neuronal cell types such as pigmented melanocytes in the skin, smooth muscle cells in the outflow tract of the heart, craniofacial bones, cartilage, and connective tissue (Le Douarin and Smith, 1988; Rice and Barone, 2000; Woodhoo and Sommer, 2008). The major role of Schwann cells is the formation and maintenance of myelin of the PNs and ensheathing the unmyelinated axons. We thus describe two subpopulations of Schwann cells: myelinating Schwann cells which are involved in myelin formation, and unmyelinating Schwann cells that surround unmyelinated axons (Arai *et al.*, 1998; Jessen, 2004). The myelinating Schwann cells form insulating sheaths around axons that have similar structure and function of oligodendrocytes in the CNS. The unmyelinating Schwann cells are likely to have metabolic and mechanical support functions comparable to astrocytes in the CNS. It has been known that Schwann cells are essential for neuronal survival during development, and control the successful regeneration and functional recovery in damaged nerves (Jessen, 2004).

1. *The Differentiation of the Schwann Cell: From the Neural Crest to Mature Schwann Cell*

It has been described that neural crest-derived stem cells (NCSCs) and the migratory neural crest cells were found in various organs that originate from the neural crest (Bixby *et al.*, 2002; Delfino-Machin *et al.*, 2007; Fernandes *et al.*, 2004; Sieber-Blum and Grim, 2004; Trentin *et al.*, 2004) such as the PNs (i.e., sciatic nerve) and the DRGs (Hagedorn *et al.*, 1999; Morrison *et al.*, 1999; Woodhoo and Sommer, 2008). The neural crest gives rise to a Schwann cell precursor (SCPs) population (Feltri *et al.*, 2008; Jessen and Mirsky, 2005) that successively differentiates into immature Schwann cells that reach the PN as final target. (Jessen and Mirsky, 1997, 2005; Ndubaku and De Bellard, 2008; Rummler *et al.*, 2004). However, Schwann cells of dorsal and ventral roots and satellite glial cells of the ganglia also partially develop from another early cell pool such as boundary cap cells (Maro *et al.*, 2004; Mirsky *et al.*, 2008). Finally, the immature Schwann cells differentiate into mature Schwann cells (Ndubaku and De Bellard, 2008).

Studies in mouse embryos showed that the SCPs colonize nerves directly through the ventro-lateral migratory stream, and spinal roots after becoming boundary cap cells, a transient population that occupies the boundary between CNS and PNS from embryonic day (ED)-10 to postnatal day (PND)-5 (Feltri *et al.*, 2008; Le Douarin and Smith, 1988; Maro *et al.*, 2004). Then SCPs migrate along outgrowing axons at the ED-12/13.5 although it has been described that they can also anticipate neuronal growth cones on their path to peripheral targets (Feltri *et al.*, 2008; Wanner *et al.*, 2006). Therefore, the differentiation of the Schwann cell population destined to populate spinal PNs occurs in three stages

(Woodhoo and Sommer, 2008) during embryonic development: (1) the neural crest cells by means of the gliogenesis process are specified to become SCPs; (2) they are specialized into immature Schwann cells (Dong *et al.*, 1999; Jessen *et al.*, 1994); (3) around birth, immature Schwann cells first start to differentiate into mature myelinating and subsequently the nonmyelinating glial cells localized in adult mammalian nerves (Mirsky *et al.*, 2008).

It is well known that the development of immature Schwann cells from the neural crest is different in mammalian species. For example in rat, SCPs are present in the spinal nerves at ED-14/15 and differentiate into immature Schwann cells around ED-17/18 (Jessen *et al.*, 1994; Reynolds *et al.*, 1991; Woodhoo and Sommer, 2008). In mouse, this developmental process occurs 2 days earlier than in rats so that the SCPs are already seen in PNs around ED-12/13, and they give rise to Schwann cells at ED-15/16 (Dong *et al.*, 1999; Feltri *et al.*, 2008; Woodhoo and Sommer, 2008). In humans, this process begins at the 12th week *in utero* in the sciatic nerves and, in general, in the nerves that belong to the brachial plexus (Cravioto, 1965).

2. *The Interaction of the Schwann Cell and Axon in Developing PNs*

Many studies have suggested that axons play a major role in determining whether a Schwann cell will display a myelinating or unmyelinating phenotype (Aguayo *et al.*, 1976a,b; Weinberg and Spencer, 1976). Therefore, there are a number of potentially distinct physical interactions between Schwann cells and axons during the development of PN as Schwann cells proliferate, migrate, surround axons, and form myelin sheaths (Feltri *et al.*, 1994).

The first evidence of cross talking between immature Schwann cells and neuronal axons occurs around ED-17/18 for rat nerves while in mice has been seen at ED-15/16 (Webster and Favilla, 1984). Thereafter, Schwann cells send cytoplasmic processes to groups of axons and progressively fasciculate them, and the process of radial sorting begins. This process depends on interactions between $\beta 1$ integrin receptors located on the Schwann cell membranes and laminins located in the basal lamina of Schwann cells (Feltri *et al.*, 2008; Woodhoo and Sommer, 2008). In this process, axons, that are greater than 1 μm in diameter, are selectively wrapped by immature Schwann cells and form one-to-one relationships with them (Fig. 2). This is a necessary process in the formation of myelination of large diameter axon, which starts around birth in rodents (Feltri *et al.*, 2008; Woodhoo and Sommer, 2008). This prerequisite for peripheral myelination in mouse continues until PND-10, and includes the one-to-one relationship between Schwann cell and axons as well as the insertion of Schwann cell processes within axons to wrap them. In this process, a myelinating Schwann cell surrounds the axons with several layers of membrane to form myelin, according to the rule of one-to-one per myelin segment (Fig. 3) (Feltri *et al.*, 2008).

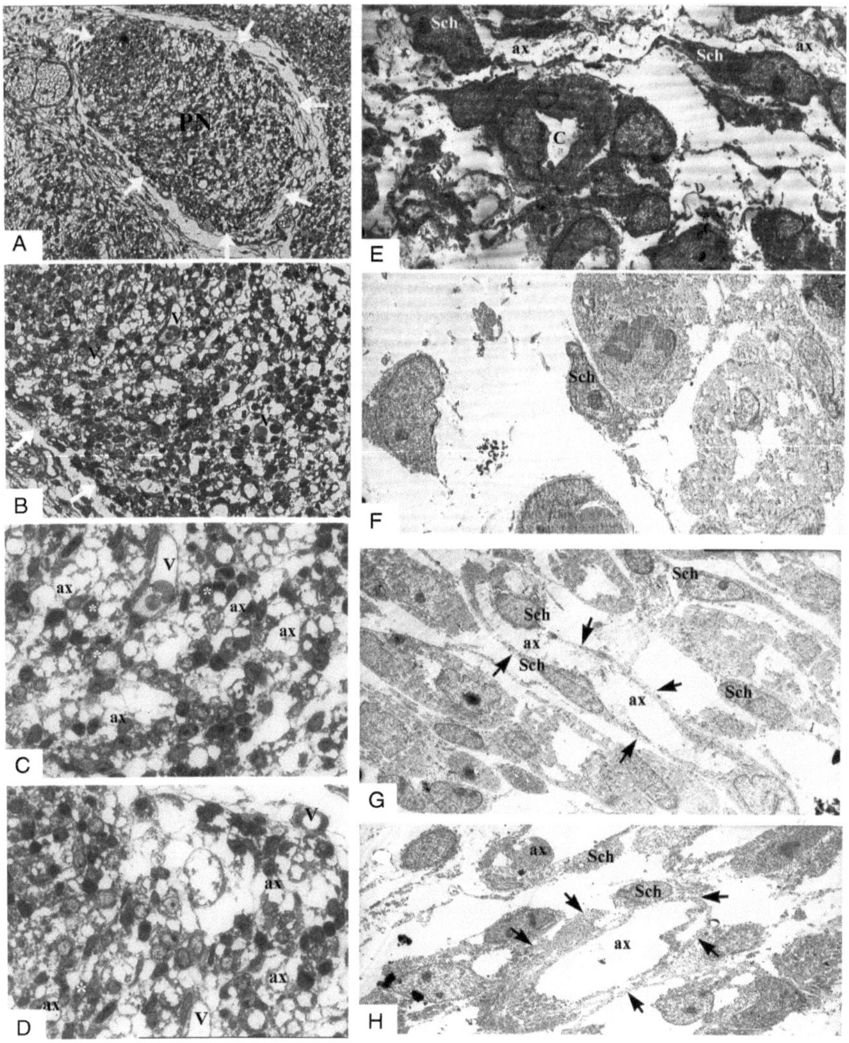

FIG. 2. Light (A–D) and electron microscopy (E–H) of sciatic nerve 20 ED rats in transversal and longitudinal section, respectively. During nerve development, Schwann cells (Sch) appear first followed by axonal growth cones. Light microscopy shows the structure of the sciatic nerve (A–D) while the relationship between Schwann cell and axons is shown in electron microscopy. Schwann cells are not well organized surrounding axons (E–H). PN, peripheral nerve; v, vessel; ax, axon; c, capillary; Sch, Schwann cell; Arrows (in the left picture) shows the borders of a developing sciatic nerve. Arrows (in the right pictures) point the borders of a nerve fiber. (See Color Insert.)

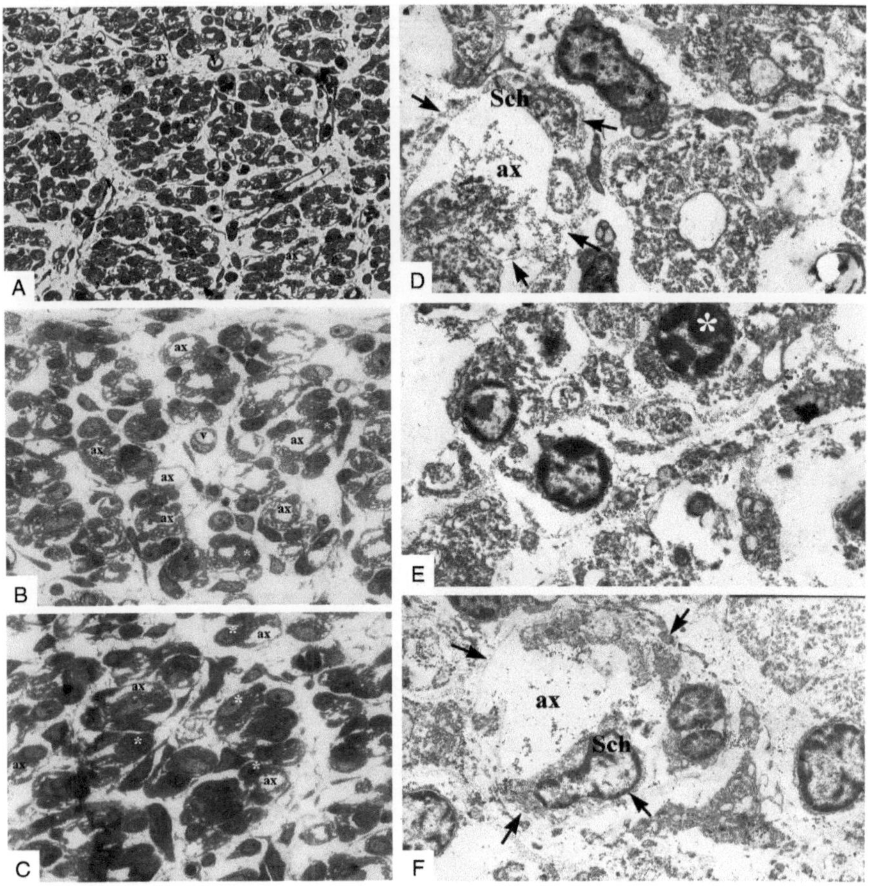

FIG. 3. Light (A–C) and electron microscopy (D–F) of sciatic nerve of newborn rats in a transversal section. At this stage, compared to 20ED, Schwann cells (Sch) are better organized around axons (D–F). * Indicates an apoptotic cell. Arrows (D, F) point to the nerve fiber borders. v, vessel; ax, axon; Sch, Schwann cell. (See Color Insert.)

Instead, unmyelinating Schwann cells surround axons smaller than 1 μm in diameter (Feltri *et al.*, 2008; Griffin and Thompson, 2008; Jessen and Mirsky, 2008; Mirsky *et al.*, 2008; Woodhoo and Sommer, 2008). During this stage of nerve development, axon and Schwann cell numbers need to match and if this that does not happen, developmental neuronal death is largely seen at these late embryonic stages (Davies, 1996) thus suggesting that Schwann cells control the survival of neurons (Woodhoo and Sommer, 2008).

B. Developmental Properties of the Axon in Peripheral Nerves

Developing peripheral axon properties such as axonal growth, axonal navigation, axonal maturation, and their myelination have been previously investigated and valuable findings have been reported in the following studies (Hockfield and McKay, 1985; Song *et al.*, 1999). The growth of axonal size and the development of myelination, axonal number, and classes were investigated by means of electron microscopy during fetal and early postnatal life in the rat phrenic nerve (Song *et al.*, 1999). It was found that the formation of fascicles in the rat phrenic nerve begins at ED-15 while Schwann cells penetrated the nerve from ED-17. Myelination process begins at PND-0, and the total number of axons decreased from ED-15 to ED-19; thereafter the number does not change until PND-0 before rising to almost adult values by PND-7 (Song *et al.*, 1999).

The postnatal increasing in number of axons may be due to a large influx of unmyelinated axons. Neurons in the DRG from C2 to C6 contributed peripheral processes to the phrenic nerve at ED-13. Phrenic afferents arrived in the spinal cord by ED-13, penetrated the dorsal horn at ED-14 and terminal fields for phrenic afferents became apparent by ED-17. It was suggested that phrenic afferent differentiation is largely complete by birth (Song *et al.*, 1999). The same authors showed that the myelination process of the rat phrenic nerve began between ED-21 and birth, and about 10% of axons developed as myelinated fibers around birth. At PND-7, about 32% of axons were already myelinated as compared to a value of 40% in the adult. Their results concerning the myelination are in agreement with Fraher's (1976) study that showed a 15% of myelinated axons around birth counted in the peripheral side of the rat cervical ventral roots.

1. *The Development of the PNs Network*

It has been suggested that normal development of the PN network is based on incredible ability of axon wiring to locate and recognize their appropriate synaptic patterns. The axonal wiring occurs in two steps: the "pathfinding" and the "target selection" (Zou and Lyuksyutova, 2007). Developing axons choose specific routes in the embryos during pathfinding stage followed by growth cones navigating towards their targets (Hockfield and McKay, 1985; Tessier-Lavigne and Goodman, 1996). Axons' targets sometimes may be far away from their soma and they generally pass through intermediate targets to form axon networks. In a previous study, it was suggested that outgrowth and termination of nerve fibers must be guided to their respective end-organs and other connection sites by selective chemical or electrical forces (Sperry, 1963). A hypothesis called "chemoaffinity" was formulated, suggesting that the chemical differences among axons mediate route and target specificity (Sperry, 1963). McKay and coworkers (1983) obtained some direct evidences about developing axons expressing different molecules on their surfaces in the adult compare to embryo confirming the chemoaffinity hypotheses. After selection of their postsynaptic targets within similar groups of cells, fibers converge to their target area. It was suggested that synaptic connections are organized in a point-to-point topographic or converging patterns and frequently in a lamina-specific manner (Zou and Lyuksyutova, 2007).

The terminal enlargement of a growing axon, called the growth cone, arises from filopodia and leads the growth of an axon along its route. The growth cone navigates over long distances along specific pathways to find the correct targets (Hockfield and McKay, 1985; Tessier-Lavigne and Goodman, 1996). The guiding role of the growth cone seems to be dependent on at least four different mechanisms: contact attraction, chemoattraction, contact repulsion, and chemorepulsion. All these mechanisms act simultaneously and in a coordinated manner for a proper pathfinding. Each of these mechanisms is mediated by mechanistically and evolutionarily conserved ligand–receptor systems (Tessier-Lavigne and Goodman, 1996). While a range of permissive and attractive effects is called attraction, repulsion refers to a range of inhibitory and repulsive effects in these mechanisms (Baier and Bonhoeffer, 1992; McKenna and Raper, 1988; Tessier-Lavigne and Goodman, 1996).

Many axons grow to reach and innervate their targets; two features simplify this task (Tessier-Lavigne and Goodman, 1996): (1) the axon trajectories divided into short and individual segments are called the first feature. These segments may be a few hundred micrometers long and generally terminate at specialized cells that are called intermediate targets for the axons. Already existing guidance information provides the axons to select and to initiate growth along the next segment of the trajectory (Tessier-Lavigne and Goodman, 1996). (2) The second feature, known as selective fasciculation strategy, simplifies the wiring of the nervous system in a stepwise manner. According to this strategy, in early developmental stages, the pioneer axons develop through an axon-free environment thus tracing the path for later axons to follow; after selective fasciculation strategy, many developing axons use these preexisting tracks for at least some of their path. They may switch from one fascicle to another at specific choice points (Bastiani *et al.*, 1984; Raper *et al.*, 1983, 1984; Tessier-Lavigne and Goodman, 1996). This strategy explains fiber assembly of large nerves (Tessier-Lavigne and Goodman, 1996). In rat, early in the development (ED-10), axons are observed in the spinal cord at rostral cervical levels. As axons grow into the periphery from ED-11 onward, their growth cones and filopodia contact other axons to form fascicles and begin to innervate peripheral structures (Hockfield and McKay, 1985). The distinct stages of axon maturation were showed in rat embryo by means of the expression of different antigens. Three different monoclonal antibodies such as Rat-202, Cat-l01, and Cat-201, specifically expressed at different time point in developing rat axons, were used. The order of antibody detection during development of the axons is Rat-202, followed by Cat-101 and Cat-201 (ED-11, ED-17, and in the third postnatal week, respectively) (Hockfield and McKay, 1985).

2. *The Development of the Axon Bundles*

Axons with a diameter of 0.1–1.5 μm that complement a definitive PN trunk arise from the spinal cord within large bundles surrounded by two cellular layers. While the inner layer is made of mesenchymal precursors of the connective tissue elements that will surround the mature nerve, the outer layer is made of migrating Schwann cells that will proliferate and, crossing the mesenchymal layer, will populate the axons (Landon, 1976). In rodents (mice and rats), this process as been described around the ED-12 (Asbury, 1967), while in human it begins at week 12 of gestation in the fetal sciatic nerves and in nerves that belong to the brachial plexus (Cravioto, 1965). The author investigated the Schwann cells of the sciatic nerve and brachial plexuses of human fetuses at 12, 14, 16, and 22 weeks of intrauterine life. He described four successive stages in the development of these cells called: pseudosyncytial, migration, cell division, axonal separation, and myelinization.

The main function of Schwann cells during the axonal separation is to penetrate the axon bundles and separate each single axon. The accomplishment of this stage would be achieved in association with a continued and massive division of the Schwann cells (Cravioto, 1965). After axon isolation has been accomplished, the Schwann cells start the myelinization process, which in humans occurs around the 14–16 weeks of intrauterine life. In summary, Schwann cells' task is to establish both the geometry and physiological characteristics of the mature nerve bundles. This task occurs in a relatively short time in human fetus (Landon, 1976). Finally, it has been suggested that Schwann cells, surrounding axons, guarantee a layer between nerve tissue and mesenchymal tissues thus playing both a morphological and perhaps physiological role during all stages of human PNs development (Cravioto, 1965).

III. Development of the Nonneural Components of the Peripheral Nerve

Biochemical and morphological structure of the mature connective or supportive tissue that constitute the PN sheath have been well described using different techniques (Burkel, 1967; Gamble and Eames, 1964; Jessen and Mirsky, 1999; Kerns, 2008; Landon, 1976; Thomas, 1963; Waggener and Beggs, 1976). The PN sheath is made of three components called endoneurium, perineurium, and epineurium. Its organization and structure guarantees flexibility and robustness to the PN and also supports and protects the peripheral axons from mechanical and chemical attacks (Flores *et al.*, 2000; Geuna *et al.*, 2009, this issue; Jessen and Mirsky, 1999; Kerns, 2008; Parmantier *et al.*, 1999). The innermost part of the sheath is called the endoneurium that surrounds the Schwann cell and axon unit; it consists of fibroblasts and their products: the collagen fibers and extracellular matrix (Jessen and Mirsky, 1999; Kerns, 2008).

At a different level, many fibers together form a nerve fascicle, which is surrounded by the perineurium. Each nerve is generally made of more than one fascicle. The perineurium is described as a cellular tube since its wall is composed of several layers of flattened epithelial-like perineurial cells that are covered by a basal lamina. Each cell of the perineurium is joined by gap junctions and to form tight, impermeable junctions with each other (Jessen and Mirsky, 1999; Kerns, 2008). For this reason, the perineurium is also known as one site for the blood–nerve barrier, which prevents large or unwanted molecules and also cellular infiltration into the endoneurium (Kerns, 2008; Parmantier *et al.*, 1999). The outermost connective tissue layer is named epineurium. It lies immediately outside the perineurium and surrounds all the fascicles defining the nerve; it contains large amounts of loose arrangement of collagen fibers and includes adipose tissue (Jessen and Mirsky, 1999; Kerns, 2008).

A. The Embryonic Origin of Cell Types of Nerve Sheath

The embryonic origin of all the different cell types that constitute the nerve sheath has been investigated for long time, although the results are still debated. These studies are considered milestones in this field since they suggested the answers to many important questions. For example, does the neural crest gives rise to the perineurial cells as well as Schwann cells or do perineurial cells share their origin from mesenchyme with fibroblast? (Bunge *et al.*, 1980; Cravioto, 1965; Jessen and Mirsky, 1999; Parmantier *et al.*, 1999).

Recent findings suggested that the majority of perineurial cells do not originate from neural crest thus suggesting that they are not lineally related to Schwann cells nor endoneurial fibroblasts (Bunge *et al.*, 1980). It is not excluded, though, that a small percent of perineurial cells on the endoneurial side of the perineurium could be neural crest-derived.

On the other end, the origin of nerve pericytes is still not clear: some authors suggested an origin from neural crest while other thought they originated from blood vessels, that is, mesoderm origin. These data, all together, suggest that Schwann cells are the only population originating from the neural crest although it was proved that NCSCs cultured from sciatic nerve are able to generate *in vitro* glia and myofibroblasts, in addition to neurons. Those endoneurial fibroblasts in addition to myelinating and unmyelinating Schwann cells derive from the neural crest (Joseph *et al.*, 2004).

Schwann cells through their secreted signals regulate the formation of perineurium (Parmantier *et al.*, 1999) and the development of the PN. Moreover, they lead to complicated interaction between neural crest-derived and non-neuronal crest-derived progenitors (Joseph *et al.*, 2004). We know that during the earliest stages of PN development complex, Schwann cells organize themselves as single cells and cords around bundles of axons (Du Plessis *et al.*, 1996) thus separating the embryonic nerve from the surrounding mesenchyme (Cravioto, 1965). After this, Schwann cells proliferate and invade the nerve, subdividing the bundles of axons into increasingly smaller units with their cell processes (Cravioto, 1965; Du Plessis *et al.*, 1996; Peters and Muir, 1959). Finally, Schwann cells isolating single axons start the process of myelination (Allt, 1969; Du Plessis *et al.*, 1996; Gamble, 1966; Ochoa, 1971).

B. Development of the Peripheral Nerve Sheath

In a study performed in chick sciatic nerve, three phases were determined for the development of the perineurium. First phase is described as an early primitive phase during which the embryonic perineurium can be distinguished from the surrounding mesenchyme. Second phase is explained as an intermediate phase of differentiation with the formation of a multilayered cellular network around the Schwann cell–axon complexes. Third phase consists of a final phase of maturation during which the perineurial sheath showed features correlating with those of a functional barrier (Du Plessis *et al.*, 1996). In this study, the earliest phase of development is showed as a mesenchymal origin for the perineurium and its cells initially appear as fibroblast-like cells. Perineurial differentiation was closely connected to the developmental events in the Schwann cell–axon complexes during the most active Schwann cell proliferation period (Du Plessis *et al.*, 1996). In fact, it has been proved that factors released by the Schwann cell–axon complexes may be responsible for perineurial differentiation and organization of the surrounding mesenchyme (Du Plessis *et al.*, 1996; Jessen and Mirsky, 1999; Parmantier *et al.*, 1999).

Moreover, many studies also showed that Schwann cell-derived signals are required for the development of the PN sheath and the transition of mesenchymal cells to form the epithelium-like structure of the perineurial tube (Jessen and Mirsky, 1999; Parmantier *et al.*, 1999; Sharghi-Namini *et al.*, 2006). As a support to this hypothesis, the morphology of the adult nerve is grossly abnormal if the protein Desert Hedgehog (Dhh), a member of the Hedgehog family secreted by Schwann cells, is lacking. In absence of Dhh protein the perineurial cells fail to express the gap junction protein connexin 43 (Jessen and Mirsky, 1999; Mirsky *et al.*, 2002; Parmantier *et al.*, 1999). Dhh deficiency affects the blood–nerve barrier in the perineurium. The consequences are not only seen in an increasing of the endoneurium permeability to large proteins but also in unwanted cells passing into the endoneurium (Parmantier *et al.*, 1999). Those data, also supported by the finding that a Dhh deficiency causes a reduction of collagen synthesis, suggested that Dhh is involved in the formation of the perineurium and epineurial connective tissue (Bunge *et al.*, 1989; Olsson, 1990). On the other hand, the lack of Dhh causes many and more complex effects for the perineurium and the endoneurium (Bunge *et al.*, 1989; Jessen and Mirsky, 1999; Parmantier *et al.*, 1999).

The regular development of the perineurium in the embryonic nerves of mice occurs in two major stages: (1) mesenchymal cells generate a thin, loose, and permeable sheath (Parmantier *et al.*, 1999); (2) the primitive sheath is arranged as an ordered multilayered structure. This stage includes the elaboration of a mature basal lamina and the expression of connexin 43 by perineurial cells. This step does not occur in nerves of Dhh-deficient mice thus suggesting that the second stage and not the first one of perineurial development seems to be depending on Dhh signaling from Schwann cells (Parmantier *et al.*, 1999). Dhh protein can be detected by means of *in situ* hybridization in developing normal nerves as early as ED-11.5 (Jessen and Mirsky, 1999). The structure and function of the perineurium are severely altered in mice with Dhh deficiency (Jessen and Mirsky, 1999; Mirsky *et al.*, 2002). In these animals, the perineurium is remarkably thin with one–three cell layers instead of five–eight. Finally, the perineurial cells fail to express connexin 43 and the collagen sheath is scanty or absent in some places of the epineurium (Mirsky *et al.*, 2002). Therefore, the mentioned data, all together, strongly support the hypothesis that Dhh is involved in the formation of not only the perineurium, but also of the endoneurium and epineurium connective tissue (Bunge *et al.*, 1989; Mirsky *et al.*, 2002; Parmantier *et al.*, 1999; Sharghi-Namini *et al.*, 2006). In conclusion, many data also suggested that Schwann cells and their precursors are involved in fashioning the connective tissue sheaths of the nerves (Jessen and Mirsky, 1999).

IV. Conclusion

The studies on the nervous system development are helpful in our understanding of cellular cross talking, pathogenic mechanisms, and their developmental anomalies. Among these studies, the development of the PN occupied a prominent place. This can be explained by the tight correlations between nerve development and nerve regeneration, a field of research that in the past 50 years made extraordinary improvement opening clinical application for reparative medicine.

The study of PN development has been facilitated by an easy availability of samples (i.e., sciatic nerves) from many animal experimental models as well as by an improvement of experimental approaches as morphological analysis, staining methods, electrophysiological techniques, and experimental manipulation (Atanasoski *et al.*, 2008; Canan *et al.*, 2008; Jessen and Mirsky, 2002; Keskin *et al.*, 2004; Landon, 1976; Lobsiger *et al.*, 2002). Developmental studies of the PN have been mostly conducted in experimental animals due to ethical reasons and difficulty of finding human samples from prenatal to postnatal stages (Jenq *et al.*, 1986; Marlot and Duron, 1979; Schäfer and Friede, 1988; Song *et al.*, 1999). However, in literature a complete description of the development of a single nerve from prenatal to postnatal stages is still lacking mostly because of the difficulty of working with the most common rodents used as experimental model (Song *et al.*, 1999). For this reason, further investigations conducted on large animal will probably be useful to fill up the gaps in understanding the development of the PN.

Acknowledgments

This work was supported by grants from the MUR (Italian Ministry of University and Research), ex-60% fund, and the Compagnia di San Paolo (Bando Programma Neuroscienze).

References

Aguayo, A. J., Charron, L., and Bray, G. M. (1976a). Potential of Schwann cells from unmyelinated nerves to produce myelin: A quantitative ultrastructural and radiographic study. *J. Neurocytol.* **5,** 565–573.

Aguayo, A. J., Epps, J., Charron, J., and Bray, G. M. (1976b). Multipotentiality of Schwann cells in cross-anastomosed and grafted unmyelinated nerves—Quantitative microscopy and radioautography. *Brain Res.* **104,** 1–20.

Allt, G. (1969). Ultrastructural features of the immature peripheral nerve. *J. Anat.* **105,** 283–293.

Arai, H., Hirato, J., and Nakazato, Y. (1998). A novel marker of Schwann cells and myelin of the peripheral nervous system. *Pathol. Int.* **48,** 206–214.

Asbury, A. K. (1967). Schwann cell proliferation in developing mouse sciatic nerve. A radioautographic study. *J. Cell Biol.* **34,** 735–743.

Atanasoski, S., Boentert, M., De Ventura, L., Pohl, H., Baranek, C., Beier, K., Young, P., Barbacid, M., and Suter, U. (2008). Postnatal Schwann cell proliferation but not myelination is strictly and uniquely dependent on cyclin-dependent kinase 4 (cdk4). *Mol. Cell Neurosci.* **37,** 519–527.

Baier, H., and Bonhoeffer, F. (1992). Axon guidance by gradients of a target-derived component. *Science* **255,** 472–475.

Bastiani, M. J., Raper, J. A., and Goodman, C. S. (1984). Pathfinding by neuronal growth cones in grasshopper embryos. III. Selective affinity of the G growth cone for the P cells within the A/P fascicle. *J. Neurosci.* **4,** 2311–2328.

Bixby, S., Kruger, G. M., Mosher, J. T., Joseph, N. M., and Morrison, S. J. (2002). Cell-intrinsic differences between stem cells from different regions of the peripheral nervous system regulate the generation of neural diversity. *Neuron* **35,** 643–656.

Bunge, M. B., Williams, A. K., Wood, P. M., Uitto, J., and Jeffrey, J. J. (1980). Comparison of nerve cell and nerve cell plus Schwann cell cultures, with particular emphasis on basal lamina and collagen formation. *J. Cell Biol.* **84,** 184–202.

Bunge, M. B., Wood, P. M., Tynan, L. B., Bates, M. L., and Sanes, J. R. (1989). Perineurium originates from fibroblasts: Demonstration *in vitro* with a retroviral marker. *Science* **243,** 229–231.

Burkel, W. E. (1967). The histological fine structure of perineurium. *Anat. Rec.* **158,** 177–190.

Canan, S., Aktas, A., Ulkay, B., Colakoglu, S., Ragbetli, M. C., Ayyildiz, M., Geuna, S., and Kaplan, S. (2008). Prenatal exposure to a non-steroidal anti-inflammatory drug or saline solution impairs sciatic nerve morphology: A stereological and histological study. *Int. J. Dev. Neurosci.* **26,** 733–738.

Cravioto, H. (1965). The role of Schwann cells in the development of human peripheral nerves. An electron microscopic study. *J. Ultrastruct. Res.* **12,** 634–651.

Davies, A. M. (1996). Paracrine and autocrine actions of neurotrophic factors. *Neurochem. Res.* **21,** 749–753.

Delfino-Machin, M., Chipperfield, T. R., Rodrigues, F. S., and Kelsh, R. N. (2007). The proliferating field of neural crest stem cells. *Dev. Dyn.* **236,** 3242–3254.

Dong, Z., Sinanan, A., Parkinson, D., Parmantier, E., Mirsky, R., and Jessen, K. R. (1999). Schwann cell development in embryonic mouse nerves. *J. Neurosci. Res.* **56,** 334–348.

Du Plessis, D. G., Mouton, Y. M., Muller, C. J., and Geiger, D. H. (1996). An ultrastructural study of the development of the chicken perineurial sheath. *J. Anat.* **189,** 631–641.

Feltri, M. L., Scherer, S. S., Nemni, R., Kamholz, J., Vogelbacker, H., Scott, M. O., Canal, N., Quaranta, V., and Wrabetz, L. (1994). Beta 4 integrin expression in myelinating Schwann cells is polarized, developmentally regulated and axonally dependent. *Development* **120,** 1287–1301.

Feltri, M. L., Suter, U., and Relvas, J. B. (2008). The function of RhoGTPases in axon ensheathment and myelination. *Glia* **56,** 1508–1517.

Fernandes, K. J., McKenzie, I. A., Mill, P., Smith, K. M., Akhavan, M., Barnabe-Heider, F., Biernaskie, J., Junek, A., Kobayashi, N. R., Toma, J. G., Kaplan, D. R., Labosky, P. A., *et al.* (2004). A dermal niche for multipotent adult skin-derived precursor cells. *Nat. Cell Biol.* **6,** 1082–1093.

Flores, A. J., Lavernia, C. J., and Owens, P. W. (2000). Anatomy and physiology of peripheral nerve injury and repair. *Am. J. Orthop.* **29,** 167–173.

Fraher, J. P. (1976). The growth and myelination of central and peripheral segments of ventral motoneurone axons. A quantitative ultrastructural study. *Brain Res.* **105,** 193–211.

Gamble, H. J. (1966). Further electron microscope studies of human foetal peripheral nerves. *J. Anat.* **100,** 487–502.

Gamble, H. J., and Eames, R. A. (1964). An electron microscope study of the connective tissues ofhuman peripheral nerve. *J. Anat.* **98,** 655–663.

Geuna, S., Raimondo, S., Ronchi, G., Di Scipio, F., Tos, P., and Fornaro, M. (2009). Histology of the peripheral nerve and changes occurring during nerve regeneration. *Int. Rev. Neurobiol.* **87,** 27–46.

Griffin, J. W., and Thompson, W. J. (2008). Biology and pathology of nonmyelinating Schwann cells. *Glia* **56,** 1518–1531.

Hagedorn, L., Suter, U., and Sommer, L. (1999). P0 and PMP22 mark a multipotent neural crest-derived cell type that displays community effects in response to TGF-beta family factors. *Development* **126,** 3781–3794.

Hockfield, S., and McKay, R. D. (1985). Identification of major cell classes in the developing mammalian nervous system. *J. Neurosci.* **5,** 3310–3328.

Jenq, C. B., Chung, K., and Coggeshall, R. E. (1986). Postnatal loss of axons in normal rat sciatic nerve. *J. Comp. Neurol.* **244,** 445–450.

Jessen, K. R. (2004). Glial cells. *Int. J. Biochem. Cell Biol.* **36,** 1861–1867.

Jessen, K. R., and Mirsky, R. (1997). Embryonic Schwann cell development: The biology of Schwann cell precursors and early Schwann cells. *J. Anat.* **191,** 501–505.

Jessen, K. R., and Mirsky, R. (1999). Schwann cells and their precursors emerge as major regulators of nerve development. *Trends Neurosci.* **22,** 402–410.

Jessen, K. R., and Mirsky, R. (2002). Signals that determine Schwann cell identity. *J. Anat.* **200,** 367–376.

Jessen, K. R., and Mirsky, R. (2005). The origin and development of glial cells in peripheral nerves. *Nat. Rev. Neurosci.* **6,** 671–682.

Jessen, K. R., and Mirsky, R. (2008). Negative regulation of myelination: Relevance for development, injury, and demyelinating disease. *Glia* **56,** 1552–1565.

Jessen, K. R., Brennan, A., Morgan, L., Mirsky, R., Kent, A., Hashimoto, Y., and Gavrilovic, J. (1994). The Schwann cell precursor and its fate: A study of cell death and differentiation during gliogenesis in rat embryonic nerves. *Neuron* **12,** 509–527.

Joseph, N. M., Mukouyama, Y. S., Mosher, J. T., Jaegle, M., Crone, S. A., Dormand, E. L., Lee, K. F., Meijer, D., Anderson, D. J., and Morrison, S. J. (2004). Neural crest stem cells undergo multi-lineage differentiation in developing peripheral nerves to generate endoneurial fibroblasts in addition to Schwann cells. *Development* **131,** 5599–5612.

Kerns, J. M. (2008). The microstructure of peripheral nerves. *Tech. Reg. Anesth. Pain Manag.* **12,** 127–133.

Keskin, M., Akbas, H., Uysal, A. O., Canan, S., Ayyildiz, M., Agar, E., and Kaplan, S. (2004). Enhancement of nerve regeneration and orientation across a gap by using the nerve graft within the nerve vein conduit graft: A functional, stereological, and electrophysiological study. *Plastic Reconstr. Surg.* **113,** 1372–1379.

Landon, D. N. (1976). The PN. Chapman and Hall, London.

Le Douarin, N. M., and Smith, J. (1988). Development of the peripheral nervous system from the neural crest. *Annu. Rev. Cell Biol.* **4,** 375–404.

Lobsiger, C. S., Taylor, V., and Suter, U. (2002). The early life of a Schwann cell. *Biol. Chem.* **383,** 245–253.

Marlot, D., and Duron, B. (1979). Postnatal maturation of phrenic, vagus and intercostal nerves in the kitten. *Biol. Neonate* **36,** 264–272.

Maro, G. S., Vermeren, M., Voiculescu, O., Melton, L., Cohen, J., Charnay, P., and Topilko, P. (2004). Neural crest boundary cap cells constitute a source of neuronal and glial cells of the PNS. *Nat. Neurosci.* **7,** 930–938.

McKay, R. D., Hockfield, S., Johansen, J., Thompson, I., and Frederiksen, K. (1983). Surface molecules identify groups of growing axons. *Science* **222,** 788–794.

McKenna, M. P., and Raper, J. A. (1988). Growth cone behavior on gradients of substratum bound laminin. *Dev. Biol.* **130,** 232–236.

Mirsky, R., Jessen, K. R., Brennan, A., Parkinson, D., Dong, Z., Meier, C., Parmantier, E., and Lawson, D. (2002). Schwann cells as regulators of nerve development. *J. Physiol. (Paris)* **96,** 17–24.

Mirsky, R., Woodhoo, A., Parkinson, D. B., Arthur-Farraj, P., Bhaskaran, A., and Jessen, K. R. (2008). Novel signals controlling embryonic Schwann cell development, myelination and dedifferentiation. *J. Peripher. Nerv. Syst.* **13,** 122–135.

Moore, K. L., and Persaud, T. V. N. (1993). The nervous system. *In* "The Developing Human: Clinically Oriented Embryology," 5th Ed., pp. 385–422. W.B. Saunders Company, Philadelphia.

Morrison, S. J., White, P. M., Zock, C., and Anderson, D. J. (1999). Prospective identification, isolation by flow cytometry, and *in vivo* self-renewal of multipotent mammalian neural crest stem cells. *Cell* **96,** 737–749.

Ndubaku, U., and De Bellard, M. E. (2008). Glial cells: Old cells with new twists. *Acta Histochem.* **110,** 182–195.

Ochoa, J. (1971). The sural nerve of the human foetus: Electron microscope observation and counts of axons. *J. Anat.* **108,** 231–245.

Olsson, Y. (1990). Microenvironment of the peripheral nervous system under normal and pathological conditions. *Crit. Rev. biol.* **5,** 265–311.

Parmantier, E., Lynn, B., Lawson, D., Turmaine, M., Namini, S. S., Chakrabarti, L., McMahon, A. P., Jessen, K. R., and Mirsky, R. (1999). Schwann cell-derived desert hedgehog controls the development of peripheral nerve sheaths. *Neuron* **23,** 713–724.

Peters, A., and Muir, A. R. (1959). The relationship between axons and Schwann cells during development of peripheral nerves in the rat. *Q. J. Exp. Physiol.* **44,** 117–130.

Raper, J. A., Bastiani, M. J., and Goodman, C. S. (1983). Guidance of neuronal growth cones: Selective fasciculation in the grasshopper embryo. *Cold Spring Harb. Symp. Quant. Biol.* **2,** 587–598.

Raper, J. A., Bastiani, M. J., and Goodman, C. S. (1984). Pathfinding by neuronal growth cones in grasshopper embryos. IV. The effects of ablating the A and P axons upon the behavior of the G growth cone. *J. Neurosci.* **4,** 2329–2345.

Reynolds, M. L., Fitzgerald, M., and Benowitz, L. I. (1991). GAP-43 expression in developing cutaneous and muscle nerves in the rat hindlimb. *Neuroscience* **41,** 201–211.

Rice, D., and Barone, S. Jr (2000). Critical periods of vulnerability for the developing nervous system: Evidence from humans and animal models. *Environ. Health Perspect.* **3,** 511–533.

Rummler, L. S., Dinh, P. T., and Gupta, R. (2004). The anatomy and biochemistry of myelin and myelination. *Oper. Tech. Orthop.* **14,** 146–152.

Schäfer, K., and Friede, R. L. (1988). The onset and rate of myelination in six peripheral and autonomic nerves of the rat. *J. Anat.* **159,** 181–195.

Sharghi-Namini, S., Turmaine, M., Meier, C., Sahni, V., Umehara, F., Jessen, K. R., and Mirsky, R. (2006). The structural and functional integrity of peripheral nerves depends on the glial-derived signal desert hedgehog. *J. Neurosci.* **26,** 6364–6376.

Sieber-Blum, M., and Grim, M. (2004). The adult hair follicle: Cradle for pluripotent neural crest stem cells. *Birth Defects Res. C Embryo Today* **72,** 162–172.

Song, A., Tracey, D. J., and Ashwell, K. W. (1999). Development of the rat phrenic nerve and the terminal distribution of phrenic afferents in the cervical cord. *Anat. Embryol.* **200,** 625–643.

Sperry, R. W. (1963). Chemoaffinity in the orderly growth of nerve fiber patterns and connections. *Proc. Natl. Acad. Sci. USA* **50,** 703–710.

Tessier-Lavigne, M., and Goodman, C. S. (1996). The molecular biology of axon guidance. *Science* **274,** 1123–1133.

Thomas, P. K. (1963). The connective tissue of peripheral nerve: An electron microscope study. *J. Anat.* **97,** 35–44.

Trentin, A., Glavieux-Pardanaud, C., Le Douarin, N. M., and Dupin, E. (2004). Self-renewal capacity is a widespread property of various types of neural crest precursor cells. *Proc. Natl. Acad. Sci. USA* **101,** 4495–4500.

Waggener, J. D., and Beggs, J. (1976). The membranous covering of neural tissues: An electron microscopy study. *J. Neuropathol. Exp. Neurol.* **26,** 412–426.

Wanner, I. B., Guerra, N. K., Mahoney, J., Kumar, A., Wood, P. M., Mirsky, R., and Jessen, K. R. (2006). Role of N-cadherin in Schwann cell precursors of growing nerves. *Glia* **54,** 439–459.

Webster, H. deF., and Favilla, J. T. (1984). Development of peripheral nerve fibers. *In* "Peripheral Neuropathy" (P. J. Dyck, P. K. Thomas, E. H. Lambert, and R. P. Bunge, Eds.), 2nd Ed. pp. 329–359. WB Saunders, Philadelphia.

Weinberg, H. J., and Spencer, P. S. (1976). Studies on the control of myelinogenesis. II. Evidence for neuronal regulation of myelin production. *Brain Res.* **113,** 363–378.

Woodhoo, A., and Sommer, L. (2008). Development of the Schwann cell lineage: From the neural crest to the myelinated nerve. *Glia* **56,** 1481–1490.

Zou, Y., and Lyuksyutova, A. I. (2007). Morphogens as conserved axon guidance cues. *Curr. Opin. Neurobiol.* **17,** 22–28.

HISTOLOGY OF THE PERIPHERAL NERVE AND CHANGES OCCURRING DURING NERVE REGENERATION

Stefano Geuna,*,† Stefania Raimondo,*,† Giulia Ronchi,*,† Federica Di Scipio,* Pierluigi Tos,‡ Krzysztof Czaja,§ and Michele Fornaro*,†

*Department of Clinical and Biological Sciences, San Luigi Gonzaga School of Medicine, University of Turin, Turin 10043, Italy

†Cavalieri Ottolenghi Scientific Institute of Neurobiology, University of Turin, Turin 10043, Italy

‡Reconstructive Microsurgery Unit, Department of Orthopedics, C.T.O. Hospital, Turin 10126, Italy

§Department of Veterinary, Comparative Anatomy, Pharmacology, and Physiology, College of Veterinary Medicine, Washington State University, Pullman, Washington 99164, USA

Peripheral nerves are complex organs that can be found throughout the body reaching almost all tissues and organs to provide motor and/or sensory innervation. A parenchyma (the *noble* component made by the nerve fibers, i.e., axons and Schwann cells) and a stroma (the *scaffold* made of various connective elements) can be recognized.

Although morphological analysis is the most common approach for studying peripheral nerve regeneration, researchers are not always aware of several histological peculiarities of these organs. Therefore, the aim of this review is to describe, at a structural and ultrastructural level, the main features of the parenchyma and the stroma of the normal undamaged nerve as well as the most important morphological changes that occur after nerve damage and during posttraumatic nerve regeneration. The paper is aimed at providing the reader with the basic framework information on nerve morphology. This would enable the correct interpretation of morphological data obtained by many experimental studies on peripheral nerve repair and regeneration such as those outlined in

INTERNATIONAL REVIEW OF NEUROBIOLOGY, VOL. 87
DOI: 10.1016/S0074-7742(09)87003-7

0074-7742/09 $35.00

several other papers included in this special issue of the *International Review of Neurobiology*.

I. Introduction

Peripheral nerves are organs that expand throughout the body, forming a complex arborization that very much resembles that found in blood vessels (Fig. 1), sharing with it developmental pathways (Zacchigna *et al.*, 2008). The peripheral nerves emerging from the central nervous system (CNS) are divided into two

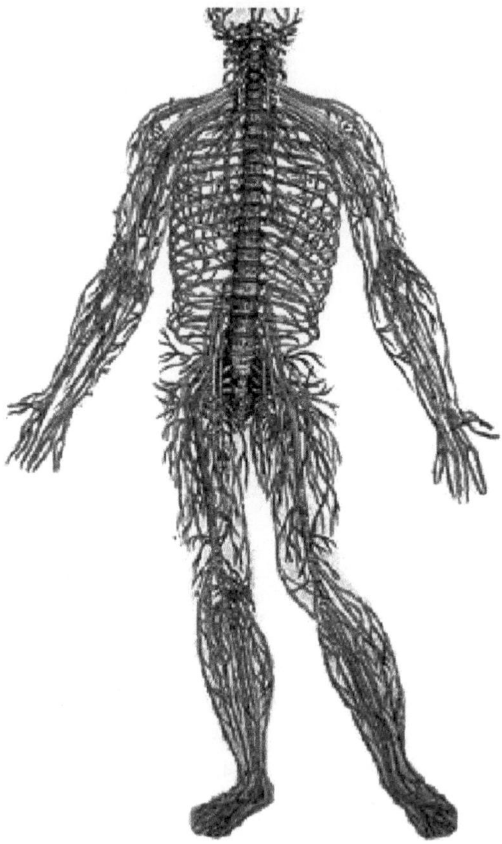

FIG. 1. Peripheral nerves expand throughout the body, forming a complex arborization that very much resembles that found in blood vessels. Taken from Vesalius (1514–1564).

categories: the cranial and the spinal nerves. Multiple branches originate from these main stems, and terminals reach all body districts. Although nerve trunks located in the various parts of the body differ with respect to the fiber-type composition (and thus functional significance) and the presence and number of fascicles (Sunderland, 1978; Sunderland and Bradley, 1949), the morphology of these nerve trunks is relatively similar in all districts (Lundborg, 2004) with the only exception being the first two cranial nerves, namely the olfactory and optic nerves.

Peripheral nerves are usually classified into three main categories, depending on fiber-type composition: (i) sensory, (ii) motor, and (iii) mixed nerves (Williams, 1999). With only few exceptions (VIII cranial nerve and the mesencephalic root of the V cranial nerve), sensory nerve fibers originate from pseudounipolar neurons located in the sensory ganglia. On the other hand, motor nerve fibers originate from somatic and autonomic motor neurons located in the CNS. While somatic motor fibers directly reach the target skeletal muscle fibers, autonomic motor fibers create synapses in an ortho- or parasympathetic ganglion where the second-order autonomic neuron is located and the axon of which eventually reaches the target visceral organs (Williams, 1999).

The aim of this paper is to describe and illustrate the main structural and ultrastructural features of the peripheral nerve. In addition, different diseases can affect peripheral nerves which spread their branches and endings throughout the whole body. This makes these organs particularly vulnerable to traumatic damage and thus, the second part of this article, we will point out the most important changes that occur during posttraumatic nerve regeneration. The technical issues concerning morphological analysis of nerves will not be addressed in this review because they are described in detail in an accompanying methodology-oriented paper (Raimondo *et al.*, 2009, this issue).

II. Structure and Ultrastructure of the Peripheral Nerve

It is possible to speculate that a nerve morphologically recall the same organization of a parenchymatous organ since, like other parenchymatous organs, a parenchyma and a stroma can be distinguished in the peripheral nerve. The former is represented by nerve fibers made by axons and the surrounding Schwann cells, the *noble* component of the nerve, while the stroma is composed by several connective elements some of which (the perineurial cells) are peculiar of the nerve.

In the remainder of this chapter, we will discuss the nerve morphology in normal conditions, while in chapter III the changes occurring during regeneration will be addressed. All descriptions refer to the adult since the developmental aspects of nerves are addressed in a dedicated article (Kaplan *et al.*, 2009, this issue).

A. The "Parenchyma" of the Nerve

The smallest functional unit of a peripheral nerve is the nerve fiber. Several schemes of classification of peripheral nerve fibers have been used, based on various parameters such as conduction velocity, function, fiber diameter, and other attributes. Anatomically, the strategy adopted from Schwann cells to enclose axons allows us to distinguish two subgroups of fibers: myelinated and unmyelinated nerve fibers (Fig. 2A–D). All larger mammalian axons are myelinated; myelin is responsible for the glistening whiteness of peripheral nerves and central white matter. Axons smaller than 1 μm in diameter are usually unmyelinated.

1. *Myelinated Nerve Fibers*

Myelinated nerve fibers consist of a single axon that is enveloped individually by a single Schwann cell. The membrane of this Schwann cell wraps around the nerve fiber to form a multilaminated myelin sheath.

Within the peripheral nervous system (PNS), myelin is produced by the Schwann cells. The myelin sheath can be thought of as a flat glial process that spirally wraps around the axon (Fig. 2A). The intracellular and extracellular spaces of the glial process are lost as the external and internal faces of the membrane become tightly apposed. In electron microscopy, the compacted external surfaces of myelin are seen as minor dense lines that alternate with the compacted inner cytoplasmic surfaces corresponding to the major dense lines (Fig. 2E and F). The inner and outer zones of occlusion of the spiral process are continuous with the minor dense line and are called the inner and outer mesaxons (Fig. 2G and H, arrows). The major dense line is continuous with the cytoplasmic face of the membrane at all regions where compaction is lost and appears to be quite stable. In contrast, the minor dense line appears to be labile (Blaurock *et al.*, 1986; Napolitano and Scallen, 1969; Williams and Hall, 1971).

In myelinated fibers, the territory of the Schwann cell defines an internode, the interval between internodes being the Ranvier's node (Williams, 1999). The internodal length varies directly with the diameter of the fibers, from 150 to 1500 μm (Kashef, 1966). In the PNS, the myelin sheaths on both side of a node terminate in paranodal bulbs, which often show an asymmetry related to growth. The surface of the bulbs is fluted as they approach the nodes. The grooves in the external surface of the myelin sheath that are produced by fluting are filled by Schwann cell cytoplasm, which is rich in mitochondria (Berthold, 1968; Landon and Williams, 1963). Each myelinated segment is separated from the enclosed axon by a narrow periaxonal space (15–20 nm), which, although nominally part of the extracellular space, is functionally isolated from the extracellular space at the paranodes.

Along the interparanodal myelin in normal myelinated fibers, we see oblique interruptions in which the membrane compaction is lost. These oblique

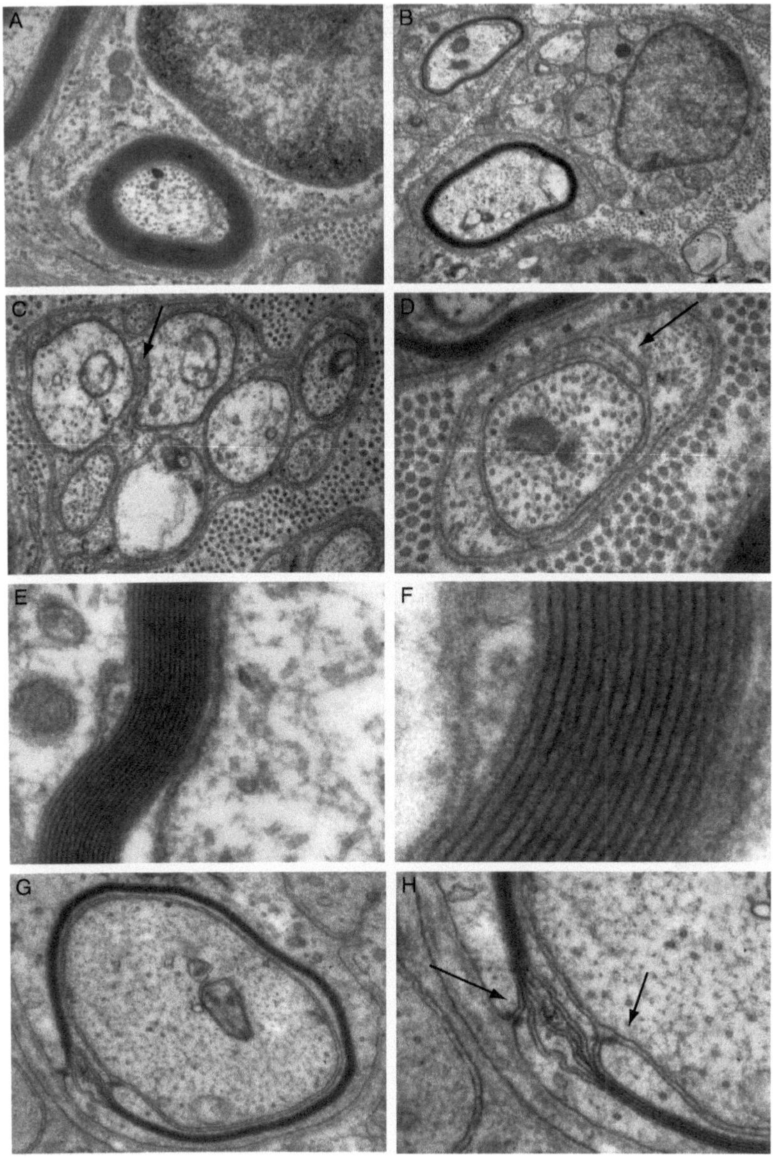

FIG. 2. Myelinated and unmyelinated fibers are shown in electron microscopy. The membrane of a Schwann cell wraps around a single axon forming a multilaminated myelin sheath (A). Unmyelinated fibers are shown in (C). The arrows point to tongues of Schwann cell cytoplasm that separate the axons from each other and ends forming mesoaxons (D, arrow). At higher magnification, the compacted minor dense lines alternated with the mayor dense lines forming the myelin are detectable (E, F). The inner and outer mesoaxons corresponding to the inner and outer zones of occlusion of the spiral process are shown (G, H, arrows).

interruptions are called Schmidt-Lanterman incisures (Williams, 1999). The dimensions and relationships of the myelinated segment features are altered to varying degrees in pathological conditions. Thus, following nerve crush or the induction of primary demyelination the paranodal myelin loses contact with the axon and the Schmidt-Lanterman incisures dilate as the adjacent minor dense line opens. This causes an irreversible collapse of the myelin periodicity (Hall and Gregson, 1971; Williams and Hall, 1971).

In general, myelination is seen only in axons above a certain diameter, about 1.5 μm in the PNS and 1 μm in the CNS (Matthews, 1968). Axonal diameter was thought to be critical in determining myelination; however, since there is considerable overlap between the size of the smallest myelinated and the largest unmyelinated axons, axonal caliber is unlikely to be the only factor.

2. *Unmyelinated Nerve Fibers*

Unmyelinated nerve fibers are composed of several nerve axons enveloped as a group by a single Schwann cell (Fig. 2B and C).

In cutaneous nerves and dorsal spinal roots, about 75% of mammalian axons are unmyelinated. They structure about 50% of the fibers of nerves projecting to muscles and 30% of the nerve bundles in ventral spinal roots. Autonomic postganglionic axons are almost exclusively unmyelinated. Unmyelinated axons are small (0.15–2.00 μm in diameter) and grouped within a sequential series of Schwann cells. In mature nerves, the mode of enclosure of each group of axons shows inter- and intraspecific variation. Axons are usually separated from each other by tongues of Schwann cell cytoplasm (Fig. 2C, arrow), but these axons are sometimes further isolated by separate processes of cytoplasm, that converge in the perinuclear region (Gamble and Eames, 1964). The line of invagination during development is marked by a mesoaxon (Fig. 2D, arrow), a double layer of Schwann cell plasma membrane. At the exterior of the Schwann cell, the layers separate and are continuous with the plasma membrane. Because of this arrangement, endoneurial tissue fluid reaches the periaxonal spaces between the mesoaxonal membranes. These intercellular spaces allow the movement of ions when action potentials are conducted along the enclosed axon. In the absence of a myelin sheath and nodes, salutatory conduction does not occur, and the interrupted passage of impulses is very slow, with velocity about 0.5–4.0 m/s.

A three-dimensional reconstruction from sections of somatic autonomic nerves revealed that the spatial relationships between axons and Schwann cells alter continuously within each cell (Aguayo *et al.*, 1973). The transfer of axons between Schwann cells usually occurs at the extremities of adjacent glial cells, where their cytoplasmic processes interdigitate (Gamble *et al.*, 1978).

B. The "Stroma" of the Nerve: Nerve Fibers

Unlike the CNS where connective tissue is mostly localized at the meningeal level all around the nervous tissue, in the PNS neurons, axons and glial cells are surrounded and supported by a reach connective scaffold as a support for an adequate resistance to stretch and compression forces applied during body movements.

Classically, nerve trunks, whether uni- or multifascicular, are surrounded by an epineurium; individual fasciculi are enclosed by a multilayered perineurium, which in turn surrounds the endoneurium or intrafascicular connective tissue (Fig. 3).

The epineurium is a supporting and protective connective tissue carrying the main supply channels of the intraneural vascular system: the vasa nervorum, which pass across the perineurium to communicate with the network of arterioles and venules within the endoneurium. Embriologically, the epineurium is derived from mesoderm. In human, the epineurium normally constitutes the 30–70% of the total cross-sectional area of the nerve bundle. As a general rule, the more fasciculi present in a peripheral nerve, the thicker the epineurium. The relative amount of epineurium varies among nerves, levels, and individuals (Sunderland, 1978; Sunderland and Bradley, 1949). Around the joints epineurium is often more abundant than elsewhere.

This connective tissue contains fibroblasts, collagen (types I and II), and variable amounts of fat, which seems to have a role in protecting the nerve this tissue surrounds.

The perineurium is a dense and mechanically strong sheath that surrounds each fascicle (Key and Retzius, 1876).

This sheath extends from the CNS–PNS transitional zone to the periphery, where it continues with the capsules of muscle spindles and the encapsulated sensory endings. At unencapsulated endings and neuromuscular junctions the perineurium ends open. This may be a critical point for the entry into the endoneurial space of substances that otherwise could not penetrate the perineurium along the course of the nerve. The perineurium consists of alternating layers of flattened polygonal cells and collagen: up to 15 layers are present around the fascicles of mammalian nerve trunks (Akert *et al.*, 1976; Thomas and Jones, 1967; Thomas and Olsson, 1984). Each cell layer is enclosed by a basal lamina. The cell layers are separated by spaces containing longitudinally oriented capillaries. Collagen fibrils and elastic fibers are located in the same spaces (Thomas and Jones, 1967). According to many studies, the epithelium-like flattened cells represent only an inner part of the true perineurium, whereas this cellular part of the perineurial sheath is encircled by an outer layer containing fibrous tissue gradually merging onto the connective tissue of the epineurium (Millesi and Terzis, 1984; Sunderland, 1978). This distinction is important from the surgical point of view, because it should be possible to place sutures in the perineurial membrane

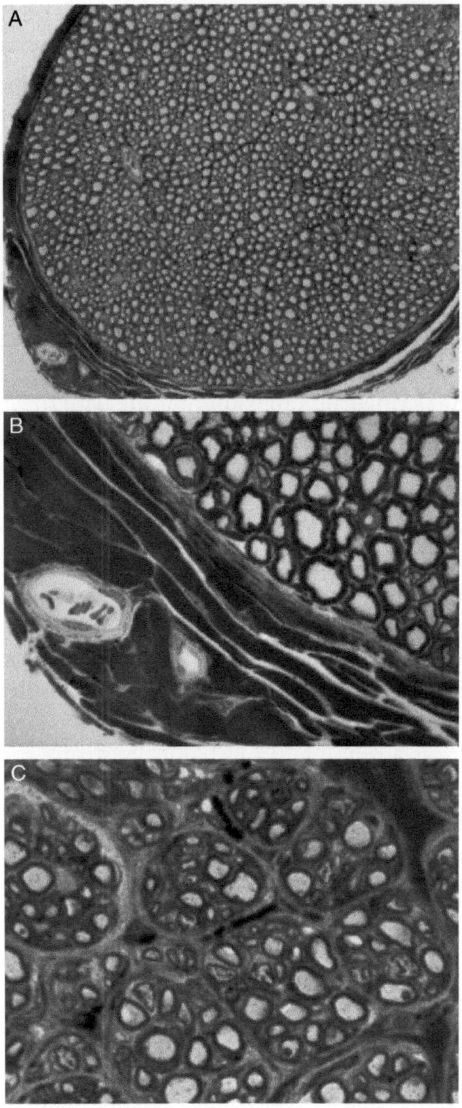

FIG. 3. Transverse sections of a rat peripheral nerve stained with toluidine blue. The nerve fasciculi, the epi-, peri-, and endoneurial connective tissue sheaths are shown at lower (A) and higher magnification (B, C). The epineurium supports and contains all the nerves carrying the main intraneural vascular system: the vasa nervorum (B). The perineurium and endoneurium are particularly evident in a distal stump of a regenerated nerve where compartmentation occurs (C). (See Color Insert.)

without penetrating the layer. These cells characteristically contain numerous pinocytotic vesicles and often bundles of microfilaments. This finding, associated with the fact that perineurium cells are rich in phosphorylating enzymes, underlies the fact that perineurium functions as a metabolically active diffusion barrier. It is probable that the perineurium together with the blood–nerve barrier plays an essential role in maintaining the osmotic milieu and the fluid pressure within the endoneurium (Williams, 1999).

The mechanical strength of the perineurium is impressive. The intrafascicular pressure can be experimentally raised 300–750 mm Hg before rupture of the perineurial membrane occurs (Selander and Sjöstrand, 1978).

The endoneurium represents a loose, soft, connective tissue that embeds and protects the fascicles, cushioning them during the movements of an extremity, and protecting them against external trauma (Lundborg, 2004). The endoneurium is a loose collagenous matrix with large extracellular spaces. The matrix contains fibroblasts, macrophages, mast cells, extracellular matrix components (collagen fiber, mucopolysaccharide ground substance), and a capillary network (Thomas *et al.*, 1993). The fibrous and cellular components of the endoneurium are bathed in endoneurial fluid (Low, 1984). Endoneurial fluid pressure is slightly higher than that of the surrounding epineurium. It is believed that the resulting pressure gradients function to minimize endoneurial contamination by toxic substances external to the nerve bundle (Powell *et al.*, 1979).

Most of the cell population in the endoneurium consists of Schwann cells and endothelial cells, while fibroblasts make up only 4% of the total (Causey and Barton, 1959). In the endoneurium, the collagen fibrils are closely packed around each nerve fiber to form the supporting walls of the "endoneurial tubes."

III. Morphological Changes after Nerve Damage and Regeneration

Trauma to peripheral nerve trunks may result in various extents of nerve fiber injury. The axonal fate is a critical factor in determining the extent, time course, and recovery following nerve injury.

After a peripheral nerve sustains a traumatic injury, complex pathophysiologic changes, including morphologic and metabolic changes, occur at the injury site. These complex changes also occur in the nerve cell body, in the segments proximal and distal to the injury site, and in the distal endings of both muscle end-plates and sensory receptors. Changes in the nerve at the site of injury begin almost immediately. With crushing or transection of a nerve trunk, significant changes take place in normal morphology and tissue organization proximally and distally to the lesion. In the following sections, the main changes occurring in the segments proximal and distal to the injury site will be separately analyzed.

A. The Proximal Nerve Segment

Transection of an axon means amputation of a major part of the axoplasmic volume from the cell. It is therefore not surprising that such a traumatic event may lead not only to profound changes in cell body structure and function but also to cell death (Purves and Nja, 1978). These changes occur in both the dorsal root ganglia sensory neurons and in the motor neurons of the spinal cord anterior horn.

Changes can be seen in the nerve cell body as early as several hours after the injury. The series of morphologic changes that ensue in the cell body after injury are known as chromatolysis, and they entail cell body and nucleolar swelling, and nuclear eccentricity. All of these changes involve an alteration of the metabolic machinery from being primarily concerned with transmitting nerve impulses to fabricating structural components for reconstruction of the injured nerve (Ducker *et al.*, 1969; Lieberman, 1971). The neurons switch from a "signaling mode" to a "growing mode" (Fu and Gordon, 1997), and protein synthesis switches from neurotransmitter-related substances to those required for axonal reconstruction (Müller and Stoll, 1998; Terzis and Smith, 1990). Metabolic changes include altered synthesis of many neuropeptides (Hökfelt *et al.*, 1994) and changes in synthesis of cytoskeletal proteins (Fornaro *et al.*, 2008; Tetzlaff *et al.*, 1988) and growth-associated proteins (Schreyer and Skene, 1991; Tetzlaff *et al.*, 1991).

In the proximal segment, axons degenerate for some distance back from the site of injury, leaving the corresponding endoneurial tubes (the basal laminae of the Schwann cell) behind as empty cylinders. This retrograde degeneration may extend over one or several internodal segments, the length depending on the severity of the lesion (Cajal, 1928).

Within hours after injury, the axon in the proximal segment produces a great number of collateral and terminal sprouts that advance distally along the tube on the inside of the basal lamina (Fawcett and Keynes, 1990; Mira, 1984). The terminal sprouts arise from the tip of the remaining axon. Within hours of axotomy, small axoplasmic outgrowths have been observed from axoplasmic tips (Zelená *et al.*, 1968). This first wave of sprouts is followed by a second wave, appearing within the first 2 days (Cajal, 1928; Grafstein and McQuarrie, 1978; Mira, 1984). Early sprouts can apparently degenerate before the definitive sprouting phase occurs. The time required for the definite sprouts to appear has been called the "initial delay" (Sunderland, 1978). A recent study on rat regenerating sciatic nerve (Witzel and Brushart, 2003) showed that sprouts have great variability in their behavior. There were "direct" projections (i.e., single sprouts crossing the gap), often traveling laterally in the interstump gap before entering a distal Schwann cell tube. "Arborizing" projections, in contrast, sampled 5–10 distal tubes from among more than 100 within their 50- to 100-μm spread. A single axon traveling within distal Schwann cell tubes continued to sprout collaterals, suggesting that the process of sprouting is a natural concomitant of regeneration. Schwann cell tubes in the distal segment were sometimes reinnervated by sprouts from several different parent axons.

Recent research shows that Schwann cells play an important role in nerve regeneration at the site of injury. Schwann cells elaborate processes that include physical conduits that guide axons to their targets. The rate of axon regeneration is limited by the extension of these Schwann cell processes rather than by axonal growth (Son and Thompson, 1995). The regenerating units will initially lack myelin even when the parent axon is a myelinated fiber. With time, these unmyelinated fibers will become myelinated (Flores *et al.*, 2000).

To reach the distal segment, the advancing sprouts have to pass a critical area between the proximal and distal stumps of the cut nerve: the interstump zone. The final success of the nerve regeneration is, to a great extent, dependent on what happens at this level and in what way local chemical and cellular reaction can influence the growth of sprouts toward their peripheral pathways.

1. *Perikaryal Phenotype Following Nerve Damage and Regeneration*

Axonal injury exposes the intracellular compartment to the extracellular environment, triggering ion fluxes and antidromic electrical activity that initiate pathways for neuronal death (Nadeau *et al.*, 2005; Navarro *et al.*, 2007; Zhang and Yannas, 2005). Moreover, damage to neurons or their axons induces phenotypic changes as indicated by alterations in mRNA transcription (Krekoski *et al.*, 1996; Salis *et al.*, 2007; Sebert and Shooter, 1993), protein synthesis (Ji *et al.*, 2007; Lundstrom *et al.*, 2005; Navarro *et al.*, 2007; Roglio *et al.*, 2008; Weragoda and Walters, 2007), and membrane receptor profiles (Karchewski *et al.*, 2004; Obata *et al.*, 2006; Oyelese *et al.*, 1995; Seniuk, 1992; Terenghi, 1999; Tonra *et al.*, 1998). Likewise, axonal transport (Hoffman and Luduena, 1996; Stone *et al.*, 2004), the secretion of neuropeptides and neurotrophic factors (Guseva and Chelyshev, 2006; Mulderry, 1994; Wang *et al.*, 2008; White and Mansfield, 1996) also are changed following injury to neurons or their axons. Finally, it is now apparent that an end result of injury is that a considerable proportion of all primary afferent neurons contributing to an injured nerve will die, with estimates ranging from 7% to 50%, depending upon the exact nature of the experimental model (Hiura, 2000; Hiura *et al.*, 1999; Navarro *et al.*, 2007). Hence, in addition to axonal regeneration, the potential for functional recovery after injury depends on restoration of neuronal numbers, and on development of appropriate neuronal phenotypes.

Previously reported studies reveal that peripheral nerve injuries induce a cascade of events progressing throughout the plastic changes to restoration of the damaged connections. In damaged neurons, axons begin to sprout after a delay of 3–42 days (Czaja *et al.*, 2008; Su and Cho, 2003). Nerve fibers grow by sprouting neurites that advance through the repair site only to be pruned down when the endoneurial tubes of the distal stump are reached (Donnerer, 2003). Although neurite growth is facilitated by contact guidance from neurite outgrowth-promoting factors (Yoshii *et al.*, 2004), it also is dependent upon the neurons' inherent regenerative capacity. This is enhanced by adoption of the regenerative phenotype, partly in response to injury factors (Navarro *et al.*, 2007). As a result, axons preferentially reinnervate the distal stump over neighboring tissues, and display preferential reinnervation in the selection of endoneurial tubes (Brushart *et al.*, 1998; Kovacic *et al.*, 2007; Rajan *et al.*, 2003; Redett *et al.*, 2005). Moreover, several studies show that damage to the adult nervous system induces factors and mechanisms that control neuronal proliferation, migration, differentiation, and connectivity during development (Ghashghaei *et al.*, 2007; Navarro *et al.*, 2007; Taupin, 2006). The rate at which new neurons appear is not constant but can be increased or decreased in response to stress (Mirescu and Gould, 2006), activity (Bordey, 2006), drugs (Huang and Herbert, 2006; Perera *et al.*, 2007), or type of neuronal injury (Groves *et al.*, 2003; Kokaia and Lindvall, 2003; Zhang *et al.*, 2006).

B. The Distal Nerve Segment

After nerve transection, the distal segment undergoes a slow process of degeneration known as Wallerian degeneration (Fig. 4A). This process starts immediately after injury and involves myelin breakdown and proliferation of Schwann cells. Schwann cells and macrophages are recruited to the injury site, and over a period of 3–6 weeks they phagocytize all the myelin and cellular debris.

Within hours after transection, the axon membrane fuses and seals the ends. Disintegration of the axons starts within the first days. The first stages of this process are characterized by a granular disintegration of axoplasmic microtubules and neurofilaments due to proteolysis (Lubińska, 1982; Schlaepfer, 1977; Vial, 1958).

The loss of axon–Schwann cell contact is a signal that causes the Schwann cell proliferation. Schwann cells upregulate the synthesis of several types of neurotrophic factors as NGF (Heumann, 1987; Thoenen *et al.*, 1988). In addition to NGF, Schwann cells also produce and present the neurotrophins BDNF, NT-3, NT-4/5, and NT-6 to the outgrowing axons (Funakoshi *et al.*, 1993) and the glial growth factor neuregulin (Geuna *et al.*, 2007). Proliferating Schwann cells organize themselves into columns (named bands of Büngner) and the regenerating axons associate with them by growing distally between their basal membranes.

The advancement of regenerating axons in the distal segment is promoted by neurite outgrowth-promoting factors, such as laminin and fibronectin (Baron-Van Evercooren *et al.*, 1982; Hall, 1997; Liu, 1996). A number of cell adhesion molecules such as N-CAM, L1, the myelin-associated glycoprotein, and

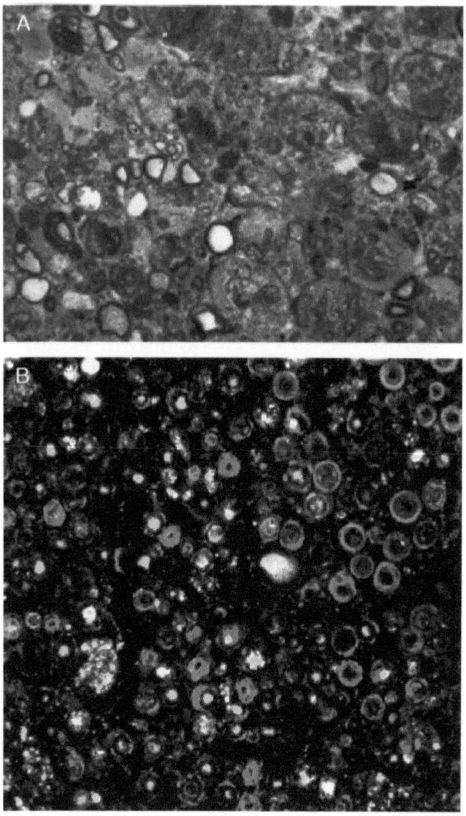

FIG. 4. After nerve transection, the distal segment undergoes a slow process of degeneration known as Wallerian degeneration (A). Signs of degeneration regarding axons and myelin disintegration are shown. After few days, few new regenerated fibers surrounded by new-formed thin myelin sheath are detectable (A, B). A double immunofluorescence shows different caliber of myelinated regenerated axons neurofilament-positive (green) surrounded by Schwann cells S100-immunopositive (red) (B). (See Color Insert.)

tumor-associated glycoprotein (TAG)-1, also play an important role (Daniloff *et al.*, 1986; Walsh and Doherty, 1996). In the distal segment, axon sprouts (which do not take an extraneural course) either approach a Schwann cell column or may grow at random into the connective tissue of the nerve. The Schwann cell columns are invaded by axon sprouts arising from parent axons in the proximal segment (Fig. 4B).

Since an excess number of sprouts invade the distal Schwann cell columns (Aguayo *et al.*, 1973; Sanders and Young, 1946), the initial number of axons present in the distal nerve segment may considerably exceed the number in the same nerve proximal to the lesion (Povlsen and Hildebrand, 1993). With time, some of the regenerated axons, which have reached appropriate distal targets, enlarge, mature, and regain a close-to-normal diameter (Sanders and Young, 1946) as result of a trophic supply from the target organs. Other branches that do not reach the target are pruned away and disappear (Griffin and Hoffman, 1993). After a few months of nerve regeneration, we will see a reorganization of the nerve trunk into a large number of miniature compartments, each surrounded by a new perineurium. Cajal (1928) described a process in which the distal stump of a divided nerve became separated into numerous nerve bundles, or "minifascicles," to replace the original large fascicle (Fig. 3C). This phenomenon is known as "compartmentation" (Morris *et al.*, 1972). Initially, it occurs also in the proximal stump of a cut nerve and in the gap between the two ends as the axons advance. The stimulus to compartmentation is probably a disturbance of the endoneurial environment resulting from damage to the perineurium. The formation of numerous miniature fascicles expresses the need for restitution of the normal endoneurium environment around the nerve fibers as quickly as possible by restoring the perineurial barrier (Lundborg, 2004).

Prolonged denervation of the distal segment results in a progressive increase in collagen content and extensive changes in the distribution of collagen types have been observed in the endoneurium and perineurium (Salonen *et al.*, 1985). Collagen production in the endoneurium may result from fibroblast activity but it may also be a result of Schwann cell activity (Barton, 1962; Thomas, 1964).

When assessing the rate of axonal outgrowth in experimental animals, several factors seem to play a role, such as the nature of the lesion, the species and the method of assessment. The quality of outgrowth obtained after transection and suture is always worse than that obtained after a crush injury. The regeneration rate in rat and rabbit nerves falls within the range of 2.0–3.5 mm/day after transection and repair and 3.0–4.4 mm/day after a crush lesion (Lundborg, 2004).

IV. Conclusions

Histological parameters are the far most used predictors of peripheral nerve damage and regeneration (Castro *et al.*, 2008; Vleggeert-Lankamp, 2007). Therefore, adequate knowledge on nerve histology is a prerequisite for peripheral nerve research. We have focused our attention on traumatic injury and regeneration of a "normal" nerve without addressing the neuropathological changes occurring as a consequence of various nerve diseases since this article is included in a special issue of the *International Review of Neurobiology* dedicated to peripheral nerve repair and regeneration and not to neuropathology of nerves.

This paper is aimed at providing the peripheral nerve researcher with the basic framework information on nerve morphology that can facilitate the correct interpretation of the morphological data obtained in experimental studies. Yet, it may help researchers in selecting the best morphological technique for reaching their scientific goals. Finally, this work will hopefully lead the reader to appreciate how histology, carried out by both traditional and modern methods, can be a valuable tool for the scientific advancement in nerve repair and regeneration.

Acknowledgments

This work was supported by grants from the MUR (Italian Ministry of University and Research), ex-60% fund, FIRB fund (code: RBAU01BJ95), PRIN2005 fund (code: 2005057088), the Compagnia di San Paolo (Bando Programma Neuroscienze), and the Regione Piemonte (Bando Ricerca Sanitaria Finalizzata). Stefania Raimondo is recipient of a PostDoc grant partially supported by the Regione Piemonte (Azione Contenimento del Brain Drain).

References

Aguayo, A. J., Peyronnard, J. M., and Bray, G. M. (1973). A quantitative ultrastructural study of regeneration from isolated proximal stumps of transected unmyelinated nerves. *J. Neuropathol. Exp. Neurol.* **32,** 256–270.

Akert, K., Sandri, C., Weibel, E. R., Peper, K., and Moor, H. (1976). The fine structure of the perineural endothelium. *Cell Tissue Res.* **165,** 281–295.

Baron-Van Evercooren, A., Kleinman, H. K., Ohno, S., Marangos, P., Schwartz, J. P., and Dubois-Dalcq, M. E. (1982). Nerve growth factor, laminin, and fibronectin promote neurite growth in human fetal sensory ganglia cultures. *J. Neurosi. Res.* **8,** 179–193.

Barton, A. A. (1962). An electron microscope study of degeneration and regeneration of nerve. *Brain* **85,** 799–808.

Berthold, C. H. (1968). Ultrastructure of the node-paranode region of mature feline ventral lumbar spinal-root fibres. *Acta Soc. Med.* **73**(Suppl. 9), 37–70.

Blaurock, A. E., Yale, J. L., and Roots, B. I. (1986). Ca-controlled, reversible structural transition in myelin. *Neurochem. Res.* **11,** 1103–1129.

Bordey, A. (2006). Adult neurogenesis: Basic concepts of signaling. *Cell Cycle* **5,** 722–728.

Brushart, T. M., Gerber, J., Kessens, P., Chen, Y. G., and Royall, R. M. (1998). Contributions of pathway and neuron to preferential motor reinnervation. *J. Neurosci.* **18,** 8674–8681.

Cajal, R. S. (1928). "Degeneration and Regeneration of the Nervous System." Oxford University Press, London.

Castro, J., Negredo, P., and Avendaño, C. (2008). Fiber composition of the rat sciatic nerve and its modification during regeneration through a sieve electrode. *Brain Res.* **1190,** 65–77.

Causey, G., and Barton, A. A. (1959). The cellular content of the endoneurium of peripheral nerve. *Brain* **82,** 594–598.

Czaja, K., Burns, G. A., and Ritter, R. C. (2008). Capsaicin-induced neuronal death and proliferation of the primary sensory neurons located in the nodose ganglia of adult rats. *Neuroscience* **154,** 621–630.

Daniloff, J. K., Levi, G., Grumet, M., Rieger, F., and Edelman, G. M. (1986). Altered expression of neuronal cell adhesion molecules induced by nerve injury and repair. *J. Cell Biol.* **103,** 929–945.

Donnerer, J. (2003). Regeneration of primary sensory neurons. *Pharmacology* **67,** 169–181.

Ducker, T. B., Kempe, L. G., and Hayes, G. J. (1969). The metabolic background for peripheral nerve surgery. *J. Neurosurg.* **30,** 270–280.

Fawcett, J. W., and Keynes, R. J. (1990). Peripheral nerve regeneration. *Annu. Rev. Neurosci.* **13,** 43–60.

Flores, A. J., Lavernia, C. J., and Owens, P. W. (2000). Anatomy and physiology of peripheral nerve injury and repair. *Am. J. Orthop.* **29,** 167–173.

Fornaro, M., Lee, J. M., Raimondo, S., Nicolino, S., Geuna, S., and Giacobini-Robecchi, M. G. (2008). Neuronal intermediate filament expression in rat dorsal root ganglia sensory neurons: An *in vivo* and *in vitro* study. *Neuroscience* **153,** 1153–1163.

Fu, S. Y., and Gordon, T. (1997). The cellular and molecular basis of peripheral nerve regeneration. *Mol. Neurobiol.* **14,** 67–116.

Funakoshi, H., Frisén, J., Barbany, G., Timmusk, T., Zachrisson, O., Verge, V. M., and Persson, H. (1993). Differential expression of mrnas for neurotrophins and their receptors after axotomy of the sciatic nerve. *J. Cell Biol.* **123,** 455–465.

Gamble, H. J., and Eames, R. A. (1964). An electron microscope study of the connective tissues of human peripheral nerve. *J. Anat.* **98,** 655–663.

Gamble, H. J., Fenton, J., and Allsopp, G. (1978). Electron microscope observations on the changing relationships between unmyelinated axons and Schwann cells in human fetal nerves. *J. Anat.* **127,** 363–378.

Geuna, S., Nicolino, S., Raimondo, S., Gambarotta, G., Battiston, B., Tos, P., and Perroteau, I. (2007). Nerve regeneration along bioengineered scaffolds. *Microsurgery* **27,** 429–438.

Ghashghaei, H. T., Lai, C., and Anton, E. S. (2007). Neuronal migration in the adult brain: Are we there yet? *Nat. Rev. Neurosci.* **8,** 141–151.

Grafstein, B., and McQuarrie, I. G. (1978). Role of the nerve cell body in axonal regeneration. *In* "Neuronal Plasticity." (C. W. Cotman, Ed.), pp. 155–195. Raven Press, New York.

Griffin, J. W., and Hoffman, P. N. (1993). Degeneration and regeneration in the peripheral nervous system. *In* "Peripheral Neuropathy." (P. J. Dyck and P. Thomas, Eds.), WB Saunders, Philadelphia.

Groves, M. J., Schanzer, A., Simpson, A. J., An, S. F., Kuo, L. T., and Scaravilli, F. (2003). Profile of adult rat sensory neuron loss, apoptosis and replacement after sciatic nerve crush. *J. Neurocytol.* **32,** 113–122.

Guseva, D., and Chelyshev, Y. (2006). The plasticity of the DRG neurons belonging to different subpopulations after dorsal rhizotomy. *Cell Mol. Neurobiol.* **26,** 1225–1234.

Hall, S. (1997). Axonal regeneration through acellular muscle grafts. *J. Anat.* **190,** 57–71.

Hall, S. M., and Gregson, N. A. (1971). The *in vivo* and ultrastructural effects of injection of lysophosphatidyl choline into myelinated peripheral nerve fibres of the adult mouse. *J. Cell Sci.* **9,** 769–789.

Heumann, R. (1987). Regulation of the synthesis of nerve growth factor. *J. Exp. Biol.* **132,** 133–150.

Hiura, A. (2000). Neuroanatomical effects of capsaicin on the primary afferent neurons. *Arch. Histol. Cytol.* **63,** 199–215.

Hiura, A., Nakagawa, H., Koshigae, Y., Yoshizako, A., Kubo, Y., and Ishizuka, H. (1999). Age-related changes in the response to thermal noxious heat and reduction of C-fibers by neonatal treatment with capsaicin. *Somatosens. Mot. Res.* **16,** 115–121.

Hoffman, P. N., and Luduena, R. F. (1996). The axonal transport of beta III-tubulin is altered in both branches of sensory axons after injury of the rat sciatic nerve. *Brain Res.* **708,** 182–184.

Hökfelt, T., Zhang, X., and Wiesenfeld-Hallin, Z. (1994). Messenger plasticity in primary sensory neurons following axotomy and its functional implications. *Trends Neurosci.* **17,** 22–30.

Huang, G. J., and Herbert, J. (2006). Stimulation of neurogenesis in the hippocampus of the adult rat by fluoxetine requires rhythmic change in corticosterone. *Biol. Psychiatry* **59,** 619–624.

Ji, R. R., Kawasaki, Y., Zhuang, Z. Y., Wen, Y. R., and Zhang, Y. Q. (2007). Protein kinases as potential targets for the treatment of pathological pain. *Handb. Exp. Pharmacol.* **177,** 359–389.

Kaplan, S., Odaci, E., Unal, B., Sahin, B., and Fornaro, M. (2009). Development of the peripheral nerve. *Int. Rev. Neurobiol.* **87,** 9–26.

Karchewski, L. A., Bloechlinger, S., and Woolf, C. J. (2004). Axonal injury-dependent induction of the peripheral benzodiazepine receptor in small-diameter adult rat primary sensory neurons. *Eur. J. Neurosci.* **20,** 671–683.

Kashef, R. (1966). The node of Ranvier. PhD thesis. University of London.

Key, A., and Retzius, G. (1876). "Studien in der anatomie des nerven systems und des bindegewebes." Samson & Wallin, Stockholm.

Kokaia, Z., and Lindvall, O. (2003). Neurogenesis after ischaemic brain insults. *Curr. Opin. Neurobiol.* **13,** 127–132.

Kovacic, U., Tomsic, M., Sketelj, J., and Bajrovic, F. (2007). Collateral sprouting of sensory axons after end-to-side nerve coaptation—A longitudinal study in the rat. *Exp. Neurol.* **203,** 358–369.

Krekoski, C. A., Parhad, M., and Clark, A. W. (1996). Attenuation and recovery of nerve growth factor receptor mrna in dorsal root ganglion neurons following axotomy. *J. Neurosci. Res.* **43,** 1–11.

Landon, D. N., and Williams, P. L. (1963). Ultrastructure of the node of Ranvier. *Nature* **199,** 575–577.

Lieberman, A. R. (1971). The axon reaction: A review of the principal features of perikaryal responses to axon injury. *Int. Rev. Neurobiol.* **14,** 49–124.

Liu, H. M. (1996). Growth factors and extracellular matrix in peripheral nerve regeneration, studied with a nerve chamber. *J. Peripher. Nerv. Syst.* **1,** 97–110.

Low, P. A. (1984). Endoneural fluid pressure and microenviroment of the nerve. *In* "Peripheral Neuropathy." (P. J. Dyck, P. K. Thomas, E. H. Lambert, and R. Bunge, Eds.), pp. 599–617. WP Saunders, Philadelphia.

Lubińska, L. (1982). Patterns of Wallerian degeneration of myelinated fibres in short and long peripheral stumps and in isolated segments of rat phrenic nerve. Interpretation of the role of axoplasmic flow of the trophic factor. *Brain Res.* **233,** 227–240.

Lundborg, G. (2004). "Nerve Injury and Repair." Elsevier, Philadelphia, PA.

Lundstrom, L., Elmquist, A., Bartfai, T., and Langel, U. (2005). Galanin and its receptors in neurological disorders. *Neuromol. Med.* **7,** 157–180.

Matthews, M. A. (1968). An electron microscopic study of the relationship between axon diameter and the initiation of myelin production in the peripheral nervous system. *Anat. Rec.* **161,** 337–351.

Millesi, H., and Terzis, J. K. (1984). Nomenclature in peripheral nerve surgery. Committee report of the International Society of Reconstructive Microsurgery. *Clin. Plast. Surg.* **11,** 3–8.

Mira, J. C. (1984). Effects of repeated denervation on muscle reinnervation. *Clin. Plast. Surg.* **11,** 31–38.

Mirescu, C., and Gould, E. (2006). Stress and adult neurogenesis. *Hippocampus* **16,** 233–238.

Morris, J. H., Hudson, A. F., and andWeddell, G. (1972). A study of degeneration and regeneration in the rat sciatic nerve based on electron microscopy. II. The development of the "regenerating unit". *Z. Zellforsch.* **124,** 103–130.

Mulderry, P. K. (1994). Neuropeptide expression by newborn and adult rat sensory neurons in culture: Effects of nerve growth factor and other neurotrophic factors. *Neuroscience* **59,** 673–688.

Müller, H. W., and Stoll, G. (1998). Nerve injury and regeneration: Basic insights and therapeutic interventions. *Curr. Opin. Neurol.* **11,** 557–562.

Nadeau, S., Hein, P., Fernandes, K. J., Peterson, A. C., and Miller, F. D. (2005). A transcriptional role for C/EBP beta in the neuronal response to axonal injury. *Mol. Cell Neurosci.* **29,** 525–535.

Napolitano, L. M., and Scallen, T. J. (1969). Observations on the fine structure of peripheral nerve myelin. *Anat. Rec.* **163,** 1–6.

Navarro, X., Vivo, M., and Valero-Cabre, A. (2007). Neural plasticity after peripheral nerve injury and regeneration. *Prog. Neurobiol.* **82,** 163–201.

Obata, K., Katsura, H., Sakurai, J., Kobayashi, K., Yamanaka, H., Dai, Y., Fukuoka, T., and Noguchi, K. (2006). Suppression of the p75 neurotrophin receptor in uninjured sensory neurons reduces neuropathic pain after nerve injury. *J. Neurosci.* **26,** 11974–11986.

Oyelese, A. A., Eng, D. L., Richerson, G. B., and Kocsis, J. D. (1995). Enhancement of GABAA receptor-mediated conductances induced by nerve injury in a subclass of sensory neurons. *J. Neurophysiol.* **74,** 673–683.

Perera, T. D., Coplan, J. D., Lisanby, S. H., Lipira, C. M., Arif, M., Carpio, C., Spitzer, G., Santarelli, L., Scharf, B., Hen, R., Rosoklija, G., Sackeim, H. A., *et al.* (2007). Antidepressant-induced neurogenesis in the hippocampus of adult nonhuman primates. *J. Neurosci.* **27,** 4894–4901.

Povlsen, B., and Hildebrand, C. (1993). Axonal regeneration in an articular branch following rat sciatic nerve lesions. *J. Neurol. Sci.* **120,** 153–158.

Powell, H. C., Myers, R. R., Costello, M. L., and Lampert, P. W. (1979). Endoneurial fluid pressure in Wallerian degeneration. *Ann. Neurol.* **5,** 550–573.

Purves, D., and Nja, A. (1978). Trophic maintenance of synaptic connections in autonomic ganglia. *In* "Neuronal Plasticity." (C. W. Cotman, Ed.), pp. 27–47. Raven Press, New York.

Raimondo, S., Fornaro, M., Di Scipio, F., Ronchi, G., Giacobini-Robecchi, M. G., and Geuna, S. (2009). Methods and protocols in peripheral nerve regeneration experimental research. *Part II*—Morphological techniques **87,** 81–103.

Rajan, B., Polydefkis, M., Hauer, P., Griffin, J. W., and McArthur, J. C. (2003). Epidermal reinnervation after intracutaneous axotomy in man. *J. Comp. Neurol.* **457,** 24–36.

Redett, R., Jari, R., Crawford, T., Chen, Y. G., Rohde, C., and Brushart, T. M. (2005). Peripheral pathways regulate motoneuron collateral dynamics. *J. Neurosci.* **25,** 9406–9412.

Roglio, I., Giatti, S., Pesaresi, M., Bianchi, R., Cavaletti, G., Lauria, G., Garcia-Segura, L. M., and Melcangi, R. C. (2008). Neuroactive steroids and peripheral neuropathy. *Brain Res. Rev.* **57,** 460–469.

Salis, C., Setton, C. P., Soto, E. F., and Pasquini, J. M. (2007). The mRNA of transferrin is expressed in Schwann cells during their maturation and after nerve injury. *Exp. Neurol.* **207,** 85–94.

Salonen, V., Lehto, M., Vaheri, A., Aro, H., and Peltonen, J. (1985). Endoneurial fibrosis following nerve transection. An immunohistological study of collagen types and fibronectin in the rat. *Acta Neuropathol.* **67,** 315–321.

Sanders, F. K., and Young, J. Z. (1946). The influence of peripheral connection on the diameter of regenerating nerve fibers. *J. Exp. Biol.* **22,** 203–212.

Schlaepfer, W. W. (1977). Structural alterations of peripheral nerve induced by the calcium ionophore A23187. *Brain Res.* **136,** 1–9.

Schreyer, D. J., and Skene, J. H. (1991). Fate of GAP-43 in ascending spinal axons of DRG neurons after peripheral nerve injury: Delayed accumulation and correlation with regenerative potential. *J. Neurosci.* **11,** 3738–3751.

Sebert, M. E., and Shooter, E. M. (1993). Expression of mRNA for neurotrophic factors and their receptors in the rat dorsal root ganglion and sciatic nerve following nerve injury. *J. Neurosci. Res.* **36,** 357–367.

Selander, D., and Sjöstrand, J. (1978). Longitudinal spread of intraneurally injected local anesthetics. An experimental study of the initial neural distribution following intraneural injections. *Acta Anaesthesiol. Scand.* **22,** 622–634.

Seniuk, N. A. (1992). Neurotrophic factors: Role in peripheral neuron survival and axonal repair. *J. Reconstr. Microsurg.* **8,** 399–404.

Son, Y. J., and Thompson, W. J. (1995). Schwann cell processes guide regeneration of peripheral axons. *Neuron* **14,** 125–132.

Stone, J. R., Okonkwo, D. O., Dialo, A. O., Rubin, D. G., Mutlu, L. K., Povlishock, J. T., and Helm, G. A. (2004). Impaired axonal transport and altered axolemmal permeability occur in distinct populations of damaged axons following traumatic brain injury. *Exp. Neurol.* **190,** 59–69.

Su, H. X., and Cho, E. Y. (2003). Sprouting of axon-like processes from axotomized retinal ganglion cells induced by normal and preinjured intravitreal optic nerve grafts. *Brain Res.* **991,** 150–162.

Sunderland, S. S. (1978). "Nerves and Nerve Injuries," 2nd Ed. Churchill Livingstone, Edinburgh.

Sunderland, S., and Bradley, K. C. (1949). The cross-sectional area of peripheral nerve trunks devoted to nerve fibers. *Brain* **72,** 428–449.

Taupin, P. (2006). Adult neurogenesis and neuroplasticity. *Restor. Neurol. Neurosci.* **24,** 9–15.

Terenghi, G. (1999). Peripheral nerve regeneration and neurotrophic factors. *J. Anat.* **194**(Pt. 1), 1–14.

Terzis, J., and Smith, K. (1990). "The Peripheral Nerve. Structure, Function and Reconstruction." Raven Press, New York.

Tetzlaff, W., Bisby, M. A., and Kreutzberg, G. W. (1988). Changes in cytoskeletal proteins in the rat facial nucleus following axotomy. *J. Neurosci.* **8,** 3181–3189.

Tetzlaff, W., Alexander, S. W., Miller, F. D., and Bisby, M. A. (1991). Response of facial and rubrospinal neurons to axotomy: Changes in mrna expression for cytoskeletal proteins and GAP-43. *J. Neurosci.* **11,** 2528–2544.

Thoenen, H., Bandtlow, C., Heumann, R., Lindholm, D., Meyer, M., and Rohrer, H. (1988). Nerve growth factor: Cellular localization and regulation of synthesis. *Cell Mol. Neurobiol.* **8,** 35–40.

Thomas, P. K. (1964). Changes in the endoneurial sheaths of peripheral myelinated nerve fibres during wallerian degeneration. *J. Anat.* **98,** 175–182.

Thomas, P. K., and Jones, D. G. (1967). The cellular response to nerve injury. II. Regeneration of the perineurium after nerve section. *J. Anat.* **101,** 45–55.

Thomas, P., and Olsson, Y. (1984). "Microscopia Anatomy and Junction of the Connective Tissue Components of Peripheral Nerve." Peripheral neuropathy, Philadelphia.

Thomas, P., Berthold, C., and Ochoa, J. (1993). "Microscopic Anatomy of the PNS." Peripheral neuropathy, Philadelphia.

Tonra, J. R., Curtis, R., Wong, V., Cliffer, K. D., Park, J. S., Timmes, A., Nguyen, T., Lindsay, R. M., Acheson, A., and Di Stefano, P. S. (1998). Axotomy upregulates the anterograde transport and expression of brain-derived neurotrophic factor by sensory neurons. *J. Neurosci.* **18,** 4374–4383.

Vial, J. D. (1958). The early changes in the axoplasm during wallerian degeneration. *J. Biophys. Biochem. Cytol.* **4,** 551–555.

Vleggeert-Lankamp, C. L. (2007). The role of evaluation methods in the assessment of peripheral nerve regeneration through synthetic conduits: A systematic review. *Laboratory investigation. J. Neurosurg.* **107,** 1168–1189.

Walsh, F. S., and Doherty, P. (1996). Cell adhesion molecules and neuronal regeneration. *Curr. Opin. Cell Biol.* **8,** 707–713.

Wang, T. H., Meng, Q. S., Qi, J. G., Zhang, W. M., Chen, J., and Wu, L. F. (2008). NT-3 expression in spared DRG and the associated spinal laminae as well as its anterograde transport in sensory neurons following removal of adjacent DRG in cats. *Neurochem. Res.* **33,** 1–7.

Weragoda, R. M., and Walters, E. T. (2007). Serotonin induces memory-like, rapamycin-sensitive hyperexcitability in sensory axons of aplysia that contributes to injury responses. *J. Neurophysiol.* **98,** 1231–1239.

White, D. M., and Mansfield, K. (1996). Vasoactive intestinal polypeptide and neuropeptide Y act indirectly to increase neurite outgrowth of dissociated dorsal root ganglion cells. *Neuroscience* **73,** 881–887.

Williams, P. L. (1999). "Gray's Anatomy." Churchill livingstone, London.

Williams, P. L., and Hall, S. M. (1971). Chronic Wallerian degeneration—An *in vivo* and ultrastructural study. *J. Anat.* **109,** 487–503.

Witzel, C., and Brushart, T. (2003). Morphology of peripheral axon regeneration. *J. Peripher. Nerv. Syst.* **8,** 75–76.

Yoshii, S., Shima, M., Oka, M., Taniguchi, A., Taki, Y., and Akagi, M. (2004). Nerve regeneration along collagen filament and the presence of distal nerve stump. *Neurol. Res.* **26,** 145–150.

Zacchigna, S., Ruiz de Almodovar, C., and Carmeliet, P. (2008). Similarities between angiogenesis and neural development: What small animal models can tell us. *Curr. Top. Dev. Biol.* **80,** 1–55.

Zelená, J., Lubińska, L., and Gutmann, E. (1968). Accumulation of organelles at the ends of interrupted axons. *Z. Zellforsch. Mikrosk. Anat.* **91,** 200–219.

Zhang, M., and Yannas, I. V. (2005). Peripheral nerve regeneration. *Adv. Biochem. Eng. Biotechnol.* **94,** 67–89.

Zhang, Y. L., Qiu, S. D., Zhang, P. B., and Shi, W. (2006). Brdu-labelled neurons regeneration after cerebral cortex injury in rats. *Chin. Med. J. (Engl.)* **119,** 1026–1029.

METHODS AND PROTOCOLS IN PERIPHERAL NERVE REGENERATION EXPERIMENTAL RESEARCH: PART I—EXPERIMENTAL MODELS

Pierluigi Tos,* Giulia Ronchi,[‡] Igor Papalia,[†] Vera Sallen,[§] Josette Legagneux,[¶] Stefano Geuna,[‡] and Maria G. Giacobini-Robecchi[‡]

*Reconstructive Microsurgery Unit, Department of Orthopedics, C.T.O. Hospital, Turin 10126, Italy

[†]Department of Surgical Disciplines, University of Messina, Messina, Italy

[‡]Department of Clinical and Biological Sciences, San Luigi Gonzaga School of Medicine, University of Turin, Turin 10043, Italy

[§]Institut de la Main, Clinique Jouvenet, Paris, France

[¶]Laboratory of Microsurgery, School of Surgery, Assistance Publique, Hôpitaux de Paris, France

This paper addresses several basic issues that are important for the experimental model design to investigate peripheral nerve regeneration. First, the importance of carrying out adequate preliminary *in vitro* investigation is emphasized in light of the ethical issues and with particular emphasis on the concept of the Three Rs (Replacement, Reduction, and Refinement) for limiting *in vivo* animal studies. Second, the various options for the selection of the animal species for nerve regeneration research are reviewed. Third, the two main experimental paradigms of nerve lesion (axonotmesis vs. neurotmesis followed by microsurgical reconstruction) are critically outlined and compared. Fourth, the various nerve

DOI: 10.1016/S0074-7742(09)87004-9

0074-7742/09 $35.00

models that have most commonly been employed are overviewed focusing in particular on forearm mixed nerves and on behavioural tests for assessing their function: the ulnar test and the grasping test which is useful for assessing both median and radial nerves in the rat. Finally, the importance of considering the influence of various factors and diseases which could interfere with the nerve regeneration process is emphasized in the perspective of a wider adoption of experimental models which more closely mimic the environmental and clinical conditions found in patients.

I. Introduction

Adequate methodological adoption should be the basis for reaching any scientific goal, but unfortunately this requirement isn't always met in nerve regeneration research (Geuna and Varejão, 2008; Geuna *et al.*, 2004; Vleggeert-Lankamp, 2007). One of the reasons might be the difficulty in obtaining the important methodological information from published research papers, the methods sections of which are usually very synthetic, due to page limit constraints, and often incomplete.

This review is the first of a series of four methodology-oriented papers that have been included in this special issue on nerve regeneration of the *International Review of Neurobiology* with the aim of providing the reader with an up-to-date critical overview on the important elements that should be considered for designing and carrying out a successful study. While this first paper will address the selection of the experimental models and the study design, the other three reviews will focus on techniques for evaluating study results, namely morphological (Raimondo *et al.*, 2009, this issue), electrophysiological (Navarro and Udina, 2009, this issue), and behavioural (Costa *et al.*, 2009, this issue).

II. *In Vitro* Models of Axonal Elongation

The attention given to ethical issues in biomedical research involving animals has greatly increased over the last years. One of the most important achievements is the progressive spread among the scientific community of the "Three Rs" (replacement, reduction, and refinement of animal studies) concept put forward by Russell and Burch (1992).

As far as the first principle, replacement, is concerned, the selection of *in vitro* models of axon elongation should always be considered for nerve regeneration research and can go in three directions: immortalized cell lines (neuronal and glial), primary cultures (neuronal and glial), and organotypic and 3D cultures.

A. Immortalized Neuronal and Glial Cell Lines

A number of immortalized neuronal and glial cell lines (Hara *et al.*, 2008; Sak and Illes, 2005; Shastry *et al.*, 2001; Trotter, 1993) have been obtained either from neoplastic nervous tissue or by genetic manipulation of neuronal and glial precursors. These lines represent stem/precursor cells that can differentiate into neurons under adequate medium conditions. The main advantage of cell lines compared to primary cultures is the availability of a large and unlimited amount of cells without requiring the sacrifice of animals and with limited costs. Yet, primary culture preparation is labour-intensive, the cell population is heterogeneous, often containing contaminating cells, and survives only few weeks in culture (Moreno-Flores *et al.*, 2006).

Cell lines are thus particularly adequate for large-scale studies on basic mechanisms, at cellular and molecular level, of neuronal and glial functions where a number of *in vitro* assays are required. Yet, cell lines can be used for the preliminary comparative screening of new approaches for promoting cell differentiation, including axonal elongation. Finally, 2D cocultures of neuronal and glial cell lines are used to investigate the basic mechanisms of neuro–glial and axo–glial interactions under well defined and reproducible conditions.

The main disadvantage of cell lines is related to the possibility that they can react differently from animal tissue cells to environmental conditions, including treatments and manipulations that are investigated *in vitro*. In fact, it should always be taken into consideration that the neoplastic origin (or the genetic manipulation to induce immortalization) may have altered the biological properties of cells (Falkenburger and Schulz, 2006) and therefore the translation of the results obtained from cell line studies must be interpreted with great caution.

B. Primary Neuronal and Glial Cultures

The relevant biological differences that may occur between cell lines and cells from living tissues provide the justification for the employment of primary cell cultures in nerve regeneration studies, in spite of the above-mentioned shortcomings (Moreno-Flores *et al.*, 2006). In fact, effective techniques for obtaining primary cultures from most neuronal and glial cell populations are available today. Regarding nerve cells, the most used for investigating nerve regeneration are motor neurons (De Paola *et al.*, 2007) and primary sensory neurons (Scanlin *et al.*, 2008) since most peripheral nerve axons come from these neurons.

Regarding glial cells, Schwann cells are the most used since they are the glial cells of peripheral nerves (Geuna *et al.*, 2009, this issue). In addition, primary cultures of olfactory ensheathing cells also deserve mention because of the versatility of their employment which has raised great expectations for their use for grafting purposes not only in the peripheral but also in the central nervous system (Pellitteri *et al.*, 2007; Radtke *et al.*, 2009; Raisman, 2007).

C. 3D and Organotypic Cocultures

The usefulness of the *in vitro* study of nerve regeneration can be improved if the culturing conditions mimic the 3D organization of the nerve tissue. This can be created either by 3D cocultures, where the spatial organization of neuronal and glial cells is maintained by synthetic scaffolds, or by organotypic cultures of full tissue. The former approach is very promising though requires complex matrices (Bozkurt *et al.*, 2007; Gingras *et al.*, 2008). On the other hand, organotypic cultures, especially obtained from dorsal root ganglia explants (Fornaro *et al.*, 2008), are much easier to be obtained and provide a very good model for peripheral nerve *in vitro* regeneration research (Fig. 1). Thus, though still poorly known, in our opinion, their employment should be promoted among peripheral

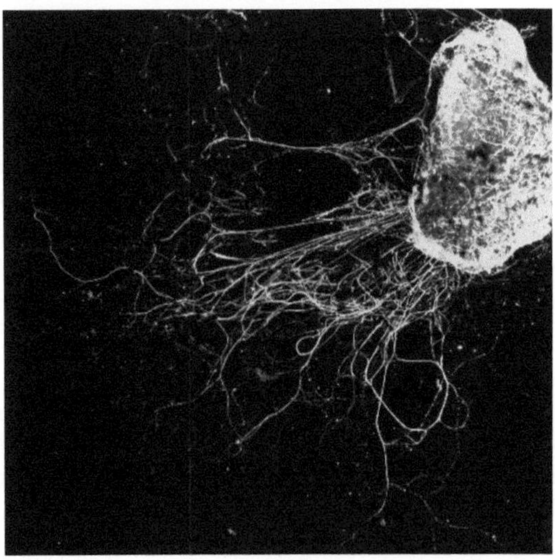

Fig. 1. (A) Organotypic culture of primary sensory neurons from dorsal root ganglion explants labeled by antineurofilament-200kD (green) and antiperipherin (red). Magnification: 40×. (See Color Insert.)

nerve researchers. To obtain this type of organotypic culture, the DRGs are removed, reduced, and maintained in Leibovitz's medium for 1 h under sterile conditions. The connective-tissue capsules are reduced using fine forceps and then ganglia are divided in half. The halves of ganglia are adhered onto matrigel-coated (diluted 1:1 in the culture medium) coverslips and incubated at 37 °C for 1 h. Explants are maintained for several days in defined serum-free medium at 37 °C with 5% CO_2 (Fornaro *et al.*, 2008).

Finally, recently published studies using this type of *in vitro* model are refining the techniques used in order to increase the culture's potential. Using a genetic algorithm, which had been optimized to promote growth, axons showed improved growth rate (Tse *et al.*, 2007). With such mathematical modeling to explore and predict axon regeneration mechanisms, these culturing protocols have become even more intriguing.

III. *In Vivo* Animal Models for the Study of Nerve Repair and Regeneration

When an investigator wants to move from an *in vitro* to an *in vivo* experimental model, it is important to choose the animal model which best fits with the study goals, while taking into consideration the pros and cons of the different options available.

While in most biomedical application rats and mice are by far the two most employed laboratory animals, in nerve regeneration studies there is a clear prevalence of rat use. A PubMed analysis of a random sample of 1500 research papers on nerve regeneration showed that more than 90% of them adopted the rat animal model. The main reason appears to be the larger physical size of rat nerves which reduces the complexity of the microsurgical procedures (Tos *et al.*, 2008), the possibility to have standardized and comparable functional tests and the fact that rats are more resilient than mice.

The anatomy of rat nerves is well established (Greene, 1963) and, in general, very similar to human anatomy. However, it should be noted that differences in both anatomical organization and function of nerve between rat and human have been described, especially regarding the forelimb (Bertelli and Mira, 1995; Papalia *et al.*, 2003, 2006; Ronchi *et al.*, 2009). On the other hand, in mice, the small nerve size and consequently the advanced microsurgical skills required for performing epineurial suturing without causing any epineurial damage, have certainly represented an important limitation to the employment of mice for nerve regeneration research (Tos *et al.*, 2008). However, the recent worldwide progressive spread of microsurgical skills in the medical community (Chan *et al.*, 2007), which goes hand in hand with the continuously increasing number of medical centers carrying out reconstructive nerve surgery, makes it possible to foresee that these technical difficulties will not represent an important limitation to the future employment of mouse nerve models. Notably, it has recently been shown that at least two of the most commonly used immunomarkers for rat axons are also effective for investigating mouse (Tos *et al.*, 2008).

In addition, the availability of a number of genetically modified mouse colonies will most probably increase mouse employment since transgenic models will allow the carrying out of biological "dissection" of various components and mechanisms of the nerve regeneration process. Table I lists examples of the transgenic mouse studies on nerve regeneration illustrating the potential results that could be achieved using this experimental approach. It should be noticed that the PubMed accurate data mining carried out to obtain this list revealed that the majority of papers were published over the last year.

Whereas rat and mouse studies in most cases represent the first choice for nerve regeneration studies, several authors believe that the translation to clinical application may benefit from a preclinical study on large animal nerves since, as for many other organs, the regeneration process of nerves in large animals is more similar to humans (Fullarton *et al.*, 2000; Lawson and Glasby, 1995). In addition, most studies on brachial plexus reconstruction are not possible in small newborn animals and to asses long distances, large animals are the only possibility (Hems *et al.*, 1994; Hess *et al.*, 2007).

Various large animal models have been employed for nerve regeneration (Table II) including rabbits, sheep, pigs, and primates. The use of cats and dogs has been progressively reduced over the last years because of more restrictive laws on the employment of these animals for experimental surgery. While it is beyond the aim of this paper to revise the literature in detail about animal models used for nerve regeneration, we have decided to include only a brief synoptic table which summarizes, for a randomly selected small sample of research papers, the study goals and the indications of the protocols for anesthesia that, from a practical point of view, are critical issues in animal research (Table II).

IV. Experimental Lesion Paradigms for Nerve Regeneration Research

Two main experimental lesion paradigms can be adopted for nerve regeneration studies: (1) axonotmesis, that is, induction of nerve fiber interruption by crush injury without discontinuing the nerve; (2) neurotmesis, that is, complete transection of the whole nerve, followed by microsurgical nerve reconstruction. The two lesion paradigms are strictly related to the corresponding clinical conditions observable in patients (Table III) with the only difference that, unlike man, in the rat the crush injury does not lead to neuroma formation.

TABLE I

SYNOPTIC TABLE OF NERVE REGENERATION STUDIES IN TRANSGENIC MOUSE EXPERIMENTAL MODELS

References	Type of transgene	Type of study	Results
(Hirota *et al.* 1996)	Double transgenic expressing IL-6 and IL-6 receptor	Hypoglossal nerve ligation	Transgenic mice showed improved regeneration. These results suggest that IL-6 signal may play an important role in nerve regeneration
(Gondré *et al.* 1998)	ΔSCIP transgene	Sciatic nerve crush injury	The transgenic mice showed markedly accelerated regeneration and hypertrophy of both myelin and axons
(Inserra *et al.* 2000)	IL-6-null mice	Sciatic nerve crush injury and end-to-end neurorrhaphy	The absence of IL-6 does not impair peripheral nerve recovery after injury. The histomorphometric findings were consistent with the functional results, suggesting that IL-6 does not have a significant effect on nerve regeneration
(Kim *et al.* 2003)	Transgenic mice expressing Nogo-C in peripheral Schwann cells	Sciatic nerve crush injury	The transgenic mice regenerate axons less rapidly than do wild-type (WT) mice. This is associated with a decreased recovery rate for motor function after sciatic nerve injury. Thus, expression of the Nogo-66 domain by otherwise permissive myelinating cells is sufficient to hinder axonal reextension after trauma
(Rong *et al.* 2004)	Transgenic mice expressing DN RAGE in mononuclear phagocytes and/or peripheral neurons	Sciatic nerve crush injury	After lesion, transgenic mice displayed decreased functional and morphological recovery, and myelinated fiber density. In double transgenic mice, regeneration was even further impaired, suggesting the critical interplay between RAGE-modulated inflammation and neurite outgrowth in nerve repair
(Triolo *et al.* 2006)	GFAP-null mice	Sciatic nerve crush injury	Without lesion, peripheral nerves develop and function normally in GFAP-null mice; no significant differences in axonal sorting, Schwann-cell axon relationship, and myelination were observed. Axonal regeneration after damage was delayed. Mutant Schwann cells maintained the ability to dedifferentiate but showed defective proliferation

(*continued*)

TABLE I (*continued*)

References	Type of transgene	Type of study	Results
(Jungnickel *et al.* 2006)	Heterozygous FGF-2 mice	Sciatic nerve crush injury	Overexpression of FGF-2 has no influence on axonal growth, maturation,or myelination during development. After lesion, in transgenic mice, the number of proliferating Schwann cells was significantly increased compared to WTs, suggesting that endogenously synthesized FGF-2 influences early peripheral nerve regeneration by regulating Schwann cell proliferation, axonal regrowth, and remyelination
(Kotulska *et al.* 2006)	Mice that overexpress SOD1	End-to-end (sciatic nerve)	SOD1 overexpression is deleterious for nerve regeneration processes and aggravates neuropathic pain-like state in mice. This can be at least partially ascribed to disturbed inflammatory reactions at the injury site
(Hayashi *et al.* 2007)	Double transgenic thy1-CFP and S100-GFP mice	Nerve allograft on sciatic nerve	Double transgenic mice whose Schwann cells constitutively express green fluorescent protein (GFP) and axons express cyan fluorescent protein (CFP) can be used to serially evaluate the temporal relationship between nerve regeneration and Schwann cell migration through acellular nerve grafts
(Lorenzetto *et al.* 2008)	Mice deficient in Reelin	Saphenous nerve crush injury	The regeneration process takes place with apparently the same modality as in control nerves, but with an impairment of axonal growth. This is due to a lower growth rate of axons. The hypothesis is that Reelin intervenes in the early phases after nerve damage
(Bannerman *et al.* 2008)	Neuropilin2 deficient mice	Sciatic nerve crush injury	Prior to the sciatic nerve crush, transgenic mice, although slightly smaller than adult WT mice, have a normal gait and normal numbers of myelinated axons in sciatic nerve. After lesion, axonal regeneration, remyelination of the regenerating axons, and recovery of normal gait were all significantly slower in transgenic mice than in the control mice. Thus, neuropilin-2 appear to facilitate peripheral nerve axonal regeneration
(López-Vales *et al.* 2008)	cPLA2 null mice	Sciatic nerve crush injury	The delayed myelin clearance and Wallerian degeneration after sciatic nerve crush injury in mice lacking cPLA2 and iPLA2 activities is accompanied by a delay in axon regeneration, target reinnervation, and functional recovery. These results indicate that the intracellular PLA2s contribute significantly to various aspects of Wallerian degeneration

(Sahenk *et al.* 2008)	Heterozygous NT-$3^{+/-}$ mice	Sciatic nerve crush injury	Without lesion, myelinated fiber density and size distribution in the transgenic mice did not differ from the WT. After lesion, there is an impairment in nerve regeneration in transgenic mice with a retardation of the myelination process. These observations indicate that NT-$3^{+/-}$ status of the SCs, but not of the axons, is responsible for impaired nerve regeneration and that NT-3 is essential for SC survival in early stages of regeneration-associated myelination in the adult peripheral nerve
(Kittaka *et al.* 2008)	knockout of the GM2/GD2 synthase gene	Hypoglossal nerve crush injury	Transgenic mice exhibited marked impairment of regenerative activity both in the number of surviving neurons and in the number of peroxidase-positive neurons. It might seem possible that the neurodegeneration in ganglioside-lacking mutant mice is due to toxic effects of accumulated glycolipids in the individual KO mice
(Hu *et al.* 2008)	BACE1-null mice	Sciatic nerve crush injury	Prior to the sciatic nerve crush, myelin sheath is thinner and the *g* ratio is higher in BACE-null mice than in WT mice. After lesion, genetic deletion of BACE1 affects sciatic nerve remyelination. The impaired remyelination appears to stem from the loss of neuregulin-1 cleavage by BACE1. The hypothesis is that the BACE1-cleaved extracellular domain of axonal neuregulin-1 binds to Schwann cell ErbB receptors, which in turn regulate remyelination
(Lee *et al.* 2009)	IL-6-null mice	Sciatic nerve crush injury	In a nerve crush model of IL-6-null mice, the functional recovery index of the sciatic nerve after injury was significantly lower only at early postoperative days, compared to WT mice. Thus, it may be possible that WT mice achieve a more rapid recovery by the IL-6/STAT3/GFAP pathway

TABLE II

Examples of Nerve Regeneration Studies in Different Animal Models

Animal	Gender/Weight	Type of study	Drug (dose)	References
Rat	Male/300–350 g	Sciatic nerve crush injury	Nembutal (60 mg/kg of body weight) i.p.	(Chen *et al.* 1992)
	Female/200–225 g	End-to-side neurorrhaphy (peroneal nerve on tibial nerve)	Nembutal (50 mg/kg body weight) i.p.	(Liu *et al.* 1998)
	Male/250–350 g	Sciatic nerve crush injury	Ketamine 9 mg:100 g, Rompun 1.25 mg:100 g, Atropine 0.025 mg:100 g body weight i.p.	(Bervar 2000)
	Male/180–220 g	End-to-end neurorrhaphy (sciatic nerve) and sciatic nerve crush injury	Ketamine (100 mg/kg), xylazine (5.2 mg/kg), and acepromazine (1 mg/kg)	(Madison *et al.* 2000)
	Male/300–350 g	End-to-side neurorrhaphy (peroneal nerve on tibial nerve)	Sodium pentobarbiturate (60 mg/kg body weight) i.p.	(De Sá *et al.* 2004)
	–	End-to-side neurorrhaphy (peroneal nerve on tibial nerve)	Medetomidine hydrochloride (0.5 mg/kg) and ketamine (75 mg/kg) s.c.	(Hess *et al.* 2006)
	Female/250–300 g	End-to-side neurorrhaphy (median merve on radial nerve)	Ketamine (40 mg/250 g) and cloropromazine (3.75 mg/250 g) i.p.	(Papalia *et al.* 2007)
Mouse	–	Sciatic nerve crush injury	2.5% Avertin i.p.	(Kim *et al.* 2003)
	–	Sciatic nerve crush injury	Avertin (trichloroethanol, 0.02 ml/g of body weight)	(Triolo *et al.* 2006)
	–	Allograft (sciatic nerve)	Ketamine (75 mg/kg) and medetomidine (100 mg/kg) s.c.	(Hayashi *et al.* 2007)
	–	Saphenous nerve crush injury	Solution of 23% Domitor (1 mg/ml) and 4% Ketavet 50 in sterile saline (25 ml/kg)	(Lorenzetto *et al.* 2008)
	Male/30 g	End-to-end neurorrhaphy (median nerve)	Ketamine (9 mg/100 g-body weight), xylazine (1.25 mg/100 g-body weight), and atropine (0.025 mg/100 g body weight) i.m.	(Tos *et al.* 2008)
	Female/20–22 g	Allografts and isografts (sciatic nerve)	0.24 ml Hypnorm (fentanyl citrate 0.135 mg/ml and fluanisone 10 kg/ml) and Midazolam 5 mg/ml	(Kvist *et al.* 2008)

Rabbit	3.5–4 kg	End-to-side neurorrhaphy (motor nerve branch of the rectus femoris on the motor branch of the vastus medialis)	Rompun (2% 0.2 ml/kg) and Narketan 10 (0.65 ml/kg) s.c., then intubated and kept under general anesthesia by using halothane, nitrous oxide, and oxygen	(Giovanoli *et al.* 2000)
	Male	Nerve transfer to the median nerve using parts of the ulnar and radial nerves	Rompun1 (1 mg/kg) and ketamine (1 mg/kg) i.m.	(Lutz *et al.* 2000)
	Male	End-to-end neurorrhaphy (median nerve)	Ketamine (35 mg/kg) and xylazine (5 mg/kg) with maintenance doses administered as needed	(Ruch *et al.* 2004)
	2500 g	End-to-side neurorrhaphy (ulnar nerve on median nerve)	Ketamine (40 mg/kg), dormicum (40 mg/kg), and atropine (0.2 mg/kg) i.m. and maintained as required	(Zhang *et al.* 2006)
	Female/2500–3500 g	Practical nerve morphometry	0.7 ml Hypnorm i.m. (fentanyl citrate 0.15 mg/ml; fluanisone 10 mg/ml), 0.4 ml diazepam i.v. and maintenance anaesthesia with fentanyl and fluanisone (“Hypnorm”) i.v. as required	(Urso-Baiarda and Grobbelaar 2006)
	Female/2–2.5 kg	End-to-end neurorrhaphy (peroneal nerve)	Isoflurane (2.5–3.5% by mask)	(Henry *et al.* 2009)
Sheep	Female/60 kg	Muscle grafts on median nerve	Midazolam (0.5 mg/kg) and etomidate (0.5 mg/kg) i.v. The sheep were intubated with a cuffed endotracheal tube and ventilated with a fresh gas flow of 21 mm^{-1} oxygen and 41 mm^{-1} nitrous oxide. Anaesthesia was maintained with 1–2% halothane. The sheep were then paralyzed with mivacurium (200 µg/kg) and neuromuscular blockade monitored with a nerve stimulator at the facial nerve	(Lawson and Glasby 1995)
	–	Median nerve repair by entubulation with a biodegradable glass tube	Thiopentone (bolus dose) i.v. Anesthesia is maintained by administering a mixture of oxygen, nitrous oxide, and vaporized halothane	(Kelleher *et al.* 2006)

(continued)

TABLE II (*continued*)

Animal	Gender/Weight	Type of study	Drug (dose)	References
	Male	Multiple neurotizations of the lumbar roots with lower intercostal nerves	Thiopental (bolus of 1.0 g/sheep) and maintained with halothane (1%–2%) in a mixture of nitrous oxide (0.5–1 l/min) and oxygen (1.5 l/min)	(Vialle *et al.* 2008)
Minipig	65–75 kg	Autograft and allograft (ulnar nerve)	Ketamine (2 mg/kg), xylazine (2 mg/kg), and zolazepam mixture tiletamine (4 mg/kg) i.m. Anesthesia was maintained by inhalational isoflurane	(Atchabahian *et al.* 1998)
	–	Allograft (ulnar nerve)	Acepromazine maleate (0.1 mg/kg) and atropine sulfate (0.2 mg/kg) preanesthetics and ketamine hydrochloride (2.2 mg/kg), tiletamine hydrochloride/zolazepam hydrochloride (4.4 mg/kg), and xylazine (2.2 mg/kg) induction agents were administered via i.m. injection and used for all interventional procedures. Animals then underwent endotracheal intubation for mechanical ventilation and administration of isoflurane to achieve an adequate plane of anesthesia	(Brenner *et al.* 2005)
	Male	Autograft and allograft (ulnar nerve)	Acepromazine (0.1 mg/kg) and atropine sulfate (0.2 mg/kg) i.m. were used for preanesthesia. Ketamine (2.2 mg/kg) i.m., Telazol (4.4 mg/kg i.m.), and Xylazine (2.2 mg/kg i.m.) were subsequently used for induction of anesthesia. Animals then underwent endotracheal intubation for mechanical ventilation, and isoflurane was administered to maintain an adequate plane of anesthesia	(Jensen *et al.* 2005)
Monkey	Male/3–4 kg	Autograft and allograft (ulnar nerve)	Ketamine (12 mg/kg) and midazolam (1 mg/kg) given by i.m., and repeated as needed	(Aubà *et al.* 2006)

TABLE III

Classification of Nerve Lesions According to Sunderland (Left Column)

Interrruption of axon conduction of the action potential	Axonotmesis	MILD nerve lesion (it does not require surgical repair)
Loss of axon continuity		
Loss of fiber continuity		
Loss of perineurium continuity	Neurotmesis	SEVERE nerve lesion (it requires surgical repair)
Loss of epineurium continuity		

In the other two columns a simplified classification which is used in experimental nerve regeneration research is illustrated.

A. Axonotmesis

Experimental axonotmesis is usually induced by means of a crush lesion which interrupts nerve fibers without severing the connective tissue of the nerve trunk (Sarikcioglu *et al.*, 2007; Varejão *et al.*, 2004) (Fig. 2). In this way, the injured axons are provided with an optimal regeneration pathway, represented by the nerve segment distal to the injury, which undergoes Wallerian degeneration (Fig. 2F and G), without the need for microsurgical repair by epineurial suture.

Most of the methods that have been reported in the literature to administer the crush injury were not standardized in terms of force and pressure administered and thus reproducible (reviewed in Ronchi *et al.*, 2009; Varejão *et al.*, 2004). In 2001, Beer *et al.* devised a standardized and reproducible clamp, in terms of force and pressure exerted as well as duration of the compression (Beer *et al.*, 2001). This method was then successfully used in the rat sciatic (Amado *et al.*, 2008; Luís *et al.*, 2007, 2008; Varejão *et al.*, 2004) and median (Ronchi *et al.*, 2009) nerve models. This standardized clamp device (Fig. 2A) is manufactured and commercially available by the Institute of Industrial Electronic and Material Sciences, University of Technology, Vienna, Austria. The clamp is equipped with three different springs (41/43/49) and two washers, which can be used in different combinations in order to exert different forces to the nerve according to a table provided by the manufacturer. In our laboratory, we use springs no. 43 with both washers, a combination which exerts a force to the median nerve of 61.3 N and a final pressure of 17.02 MPa (Ronchi *et al.*, 2009). Figures 2B–E describes the surgical steps for median nerve crush injury. Immediately after a 30-seconds injury (Fig. 2D), the crushed area of all median nerves appears flattened although nerve continuity is preserved (Fig. 2E).

The axonotmesis lesion paradigm has two main advantages in comparison to neurotmesis. First, it is less technically challenging, a great advantage for all peripheral nerve researchers not trained in microsurgery. Second, interanimal variability in the postoperative outcome is rather low (Ronchi *et al.*, 2009; Varejao *et al.*, 2004), and much lower than after neurotmesis followed by microsurgical

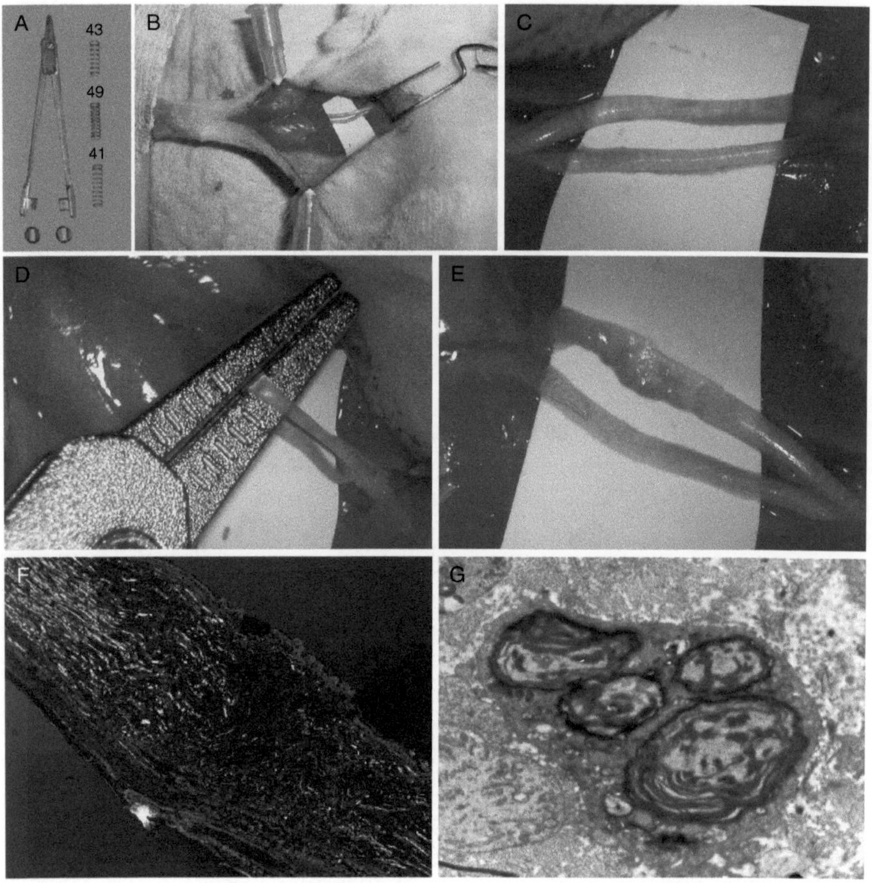

FIG. 2. (A) The device used to produce the crush injury is equipped with three different springs (41/43/49) and two washers, which can be used in different combinations in order to exert different forces to the nerve according to a table provided by the manufacturer. (B) The median and ulnar nerves are isolated. (C) At higher magnification, the different morphology of the median (larger) and ulnar (smaller) nerves can be appreciated. (D) The median nerve is clamped and compression time can be decided by the operator. (E) Immediately after the acute compression injury, the crushed area of the median nerve appears flattened although nerve continuity is preserved. (F) Double staining with antibodies against neurofilament 200kD and S100 that shows the interruption of nerve fibers at crush site. (G) Wallerian degeneration is shown by electron microscopy. Magnifications: F = 400×; G = 10,000×.

neurorrhaphy, thus making this procedure particularly adequate when a reproducible regeneration process is required, such as for the study of biological mechanisms of peripheral nerve fiber regeneration or rationale development for new therapeutic agents for promoting posttraumatic nerve repair.

The main disadvantages of the axonotmesis model is represented by the fastness of the regeneration process in basal conditions (i.e., without any treatment), which might make it difficult to disclose differences between experimental groups.

B. Neurotmesis

Figure 3 illustrates the comparison of neurotmesis model in rat and mouse at gross anatomy as well as at light and electron microscopy. The complete nerve transection (with or without removal of a nerve segment) requires surgical repair to reestablish epineurial continuity (Fig. 3A and F). This experimental paradigm not only provides the model for the comparative investigation of new types of

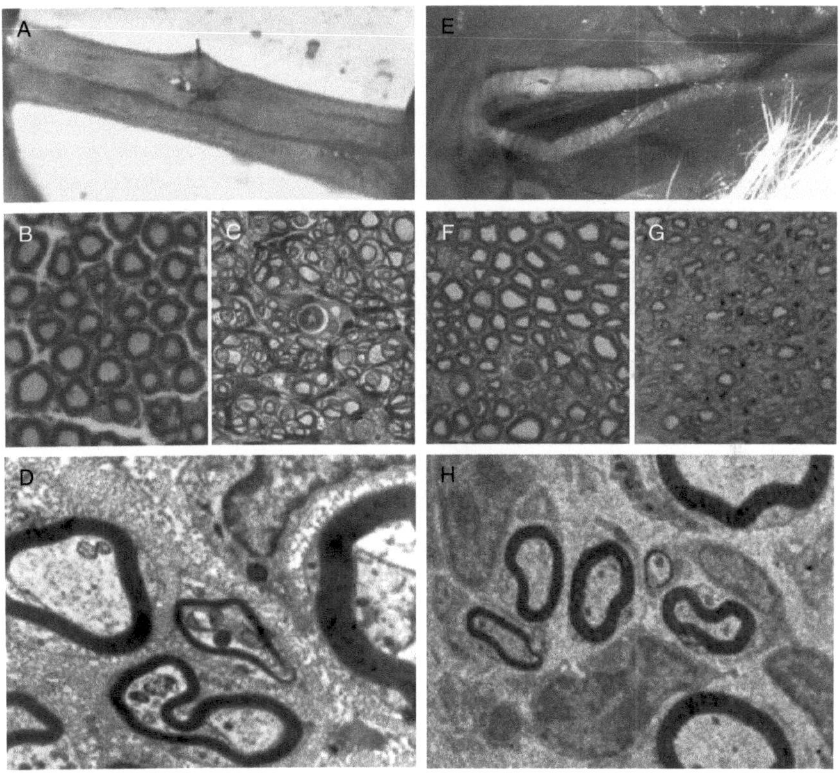

Fig. 3. (A–D) Rat and (E–H) Mouse. (A) and (E) show end-to-end neurorrhaphy of median nerve in rat and mouse, respectively. (B) and (F) show Toluidine blue stained microsections of control nerves in rat and mouse, respectively. (C) and (G) show Toluidine blue stained microsections of regenerated nerves in rat and mouse, respectively. (D) and (H) show electron microscopy of regenerated rat and mouse nerves, respectively. Magnifications: B, C, F, and G = 600×; D and H = 10,000×.

microsurgical and tissue engineering approaches for nerve reconstruction but it also provides a good model for assessing the effectiveness of various postoperative treatments (drugs, physical therapy, diet, etc.).

In rats, microsurgical epineurial suturing reconstruction is carried out by 2–3 stitches with nonabsorbable 7/0 to 11/0 monofilaments depending on the size of the nerve (Fig. 3A). In the mouse, 11/0 or even 12/0 are required and the weakness of the nerve and its connective tissue makes this surgery very complex as suturing must be done taking care that forceps never touch the nerve itself (Fig. 3E) (Tos *et al.*, 2008).

In comparison to axonotmesis, axonal regeneration is much slower, in terms of both morphological and functional predictors, thus making it easier to disclose differences between experimental groups. For example, after rat median nerve axonotmesis, functional recovery begins after 12 days and reaches the plateau after 28 days, while after neurotmesis the motor recovery starts at day 30, and reaches the plateau at day 120. This point is a critical one in terms of clinical translation of experimental results on nerve regeneration promotion to patients. Thus, it can be even suggested that preclinical studies on new therapeutic agents for improving nerve regeneration should be carried out preferentially on experimental models based on neurotmesis followed by complex nerve reconstruction for which poor outcome is expected (e.g., end-to-side neurorrhaphy). In this way, if a new therapeutic approach is effective, significant differences in terms of morphological, electrophysiological, and functional predictors will be more easily detected in the statistical comparison among experimental groups.

V. Selection of the Nerve Model

The animal body contains many nerves and, although the structure of peripheral nerves is similar (Geuna *et al.*, 2009, this issue), several factors can guide the choice of the nerve model for an experimental study. The main factor is certainly the size of the nerve and, in fact, the large size appears to be the main reason why the sciatic nerve is the most frequently used nerve model (Varejao *et al.*, 2004). In addition, the presence and number of collateral branches should also be considered since availability of a nerve segment with no (or few) collaterals is fundamental for avoiding excessive intersample variability.

Another important factor is certainly the actual clinical translational aim, that is, if a study is carried out with a perspective of clinical translation to maxillo-facial surgery, selection of the facial or hypoglossal nerve might be more reasonable than sciatic or median nerve. The contrary is true, if the clinical translational aim is focused on limb surgery. In this view, availability of an adequate behavioral test for motor and/or sensory function assessment which might facilitate the interpretation of the study results is also very important.

A. Hindlimb Nerves

The rat sciatic nerve has been, and still continues to be, the most used model to assess motor function recovery (Bervar, 2000; Brown *et al.*, 1991; Varejão *et al.*, 2001; Yu *et al.*, 2001). It can be certainly sustained that the main advances in peripheral nerve regeneration research have been based on the employment of the sciatic nerve model which still represents a valid experimental approach due to the several behavioral functional tests available (Nichols *et al.*, 2005; Varejão *et al.*, 2004), such as computerized gait analysis (Bozkurt *et al.*, 2008; Deumens *et al.*, 2007; Luis *et al.*, 2007).

What are the reasons for the supremacy of the sciatic nerve model? Certainly one reason is its large size (larger than all other nerves) that facilitates experimental microsurgery. Yet, the easy surgical access and the few collateral branches given before its division at the knee are important points in favor of this experimental model. Another explanation can be the willingness of researchers, who are selecting the experimental model for a new study, to get their data comparable to previous similar studies.

Besides the sciatic nerve, other hindlimb nerves have been used in many important studies, including the femoral (i.e., Huang *et al.*, 2009; Robinson and Madison, 2009), tibial (i.e., Apel *et al.*, 2009; Moradzadeh *et al.*, 2008), and peroneal (i.e., Alluin *et al.*, 2009; Chabas *et al.*, 2008) nerves. The last two nerves are often studied together (i.e., De Sá *et al.*, 2004; Hess *et al.*, 2006). Notably, these nerves are used in large animal models probably because of their size. For the contrary reason, in the mouse almost all studies have been carried out in the sciatic nerve both in wild-type (Baptista *et al.*, 2007; Islamov *et al.*, 2002; Pereira Lopes *et al.*, 2006; Shao *et al.*, 2007) and transgenic animals (Table I).

Although availability of a reliable functional test is the key element in the selection of the nerve model, in this article, we will not review the behavioral methods for hindlimb functional analysis since this issue is addressed in details in another paper of this special issue of the *International Review of Neurobiology* (Costa *et al.*, 2009, this issue).

B. Forelimb Nerves

While in the twentieth century forelimb nerves have been used only occasionally (Bertelli and Mira, 1995), over the last decade their use (especially the median nerve) has progressively increased. One of the main reasons is that animal welfare is more preserved (Papalia *et al.*, 2003, 2006). Other reasons include that experimental results are more likely to be translated to the clinical practice since the surgical interventions for repairing a damaged human nerve are usually performed at the upper limb level and that hand functions require fine and skilled finger movement and a behavior that is quite similar between rodents and humans (Whishaw *et al.*, 1992).

On the other hand, one of the main limiting factors of forelimb-based nerve models is that the smaller nerve size requires advanced microsurgical skills especially when surgery is carried out in mice (Tos *et al.*, 2008).

Since, as already reported, availability of a reliable functional test is the key element in the selection of the nerve model, this paragraph will focus and address functional analysis of the three mixed nerves of the rat forearm using two tests that we have developed and already used in various studies on nerve regeneration, namely the grasping test and the ulnar test. For review of other functional tests of rat forelimb nerves, readers can refer to previous works (Bontioti *et al.*, 2003; Galtrey and Fawcett, 2007; Nichols *et al.*, 2005; Sinis *et al.*, 2006).

1. *Functional Anatomy of Finger Movements*

Like in man, three terminal mixed nerves of the brachial plexus reach the forearm: median, ulnar, and radial (Greene, 1963) (Fig. 4). The flexor movements of the rat finger are controlled by the median and ulnar nerves while finger extensor muscles are innervated by the radial nerve (Greene, 1963). A difference between rodents and humans exists, however, with respect to flexion since, while in man both median and ulnar nerves contribute to innervate both extrinsic and

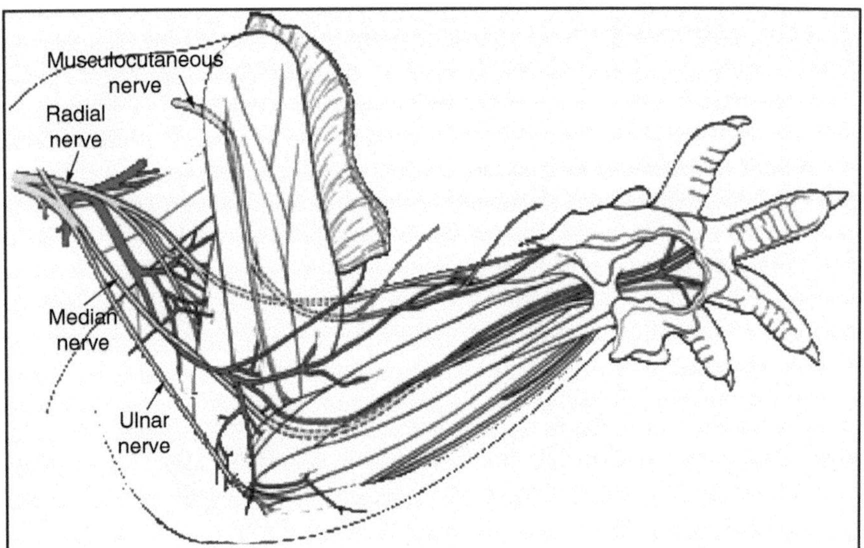

FIG. 4. Rat forelimb nerve anatomy (modified from Greene, 1963).

intrinsic muscles (Kozin *et al.*, 1999), in the rat extrinsic finger flexor muscles are prominently innervated by median nerve (Bertelli and Mira, 1995; Papalia *et al.*, 2003) while ulnar nerve controls most intrinsic muscles (lumbrical and interossei) with the only exception of the flexor pollicis brevis (Greene, 1963). This difference in the innervation of extrinsic and intrinsic muscles controlling finger flexion allows us to differentially evaluate the function of these two nerves on the basis of animal's prehensile activities since the strength of the grip is predominantly controlled by the median nerve through its action on the flexion of the distal phalanges, while the coordination of the grip is predominantly controlled by the ulnar nerve through its action on fine phalangeal and metacarpal movements of laterality, rotation, and opposition, which optimize application of the strength on the object (Papalia *et al.*, 2006).

As a consequence, when the median nerve alone is transected, performance is null both in the grasping (i.e., prehension of an object that is very easy to be taken, namely a bar) and ulnar test (i.e., prehension of an object that is very difficult to be taken, namely a sphere) because the animal loses the ability to bend the distal phalanges (Papalia *et al.*, 2003, 2006; Wang *et al.*, 2008). By contrast, when the ulnar nerve alone is impaired, the animal can still bend the distal phalanges of the fingers and thus its performance in the grasping test is only slightly impaired (Wang *et al.*, 2008) while its performance in the ulnar test is dramatically reduced (Papalia *et al.*, 2006) because of the absence of fine phalangeal and metacarpal movements (notably, after ulnar nerve impairment, the performance in the ulnar test never completely disappears since the preserved ability to bend the distal phalanges of the fingers allows a partial prehension of spherical objects).

2. *The Grasping Test*

The grasping test was first introduced by Bertelli and Mira (1995) as a simple method for assessing the flexor function in rat median nerve model. In 2003, we proposed a modified procedure for carrying out this behavioral test which coped with some limitations that we experienced using the original method, namely the tendency to walk on the grid and the possible employment of the wrist flexion to hold the grid bars (Fig. 5). The small tower with only three bars forming a triangle on its top that we have devised avoids rat walking on it while the band put just under the three "grasping" bars avoids that the rat introduces the entire paw under the bar to hold it with the wrist (Papalia *et al.*, 2003).

The device is connected to a precision dynamometer (BS-GRIP Grip Meter, 2Biological Instruments, Varese, Italy) and the test is carried out by holding the rat by its tail and lowering it towards the device and then, when the animal grips the grid, pulling it upward until it loses its grip (Fig. 5B and C). Each animal is tested three times and either the maximum or the average weight that the animal

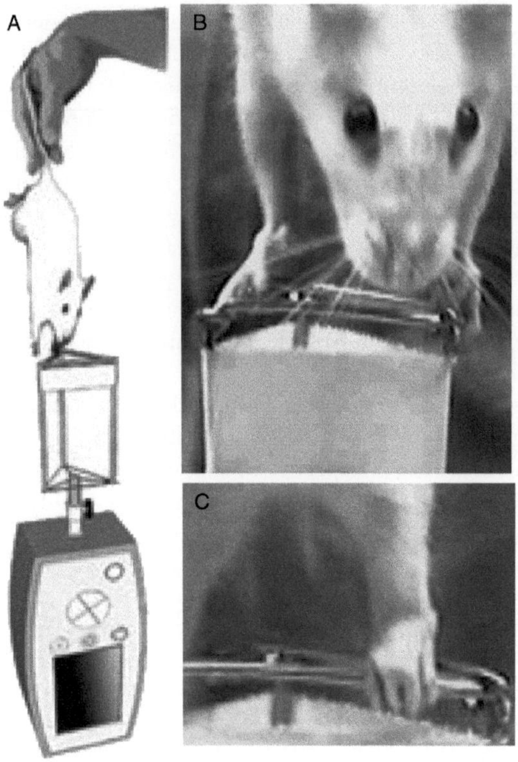

FIG. 5. The grasping test device adapted for rats. (B) and (C) A rat grasping the bar with its fingers.

manages to hold before losing its grip is recorded. The presence of two investigators is recommended: one for holding the rat and verifying correctness of the grip and the other for recording the numerical data.

From a practical viewpoint, the way to lift animals is particularly relevant because differences in how investigator pulls the rat by its tail can influence measurements. Since strength and quickness of animal lifting cannot be standardized, it is very important it is performed always by the same person who shall try to reproduce the same strength and quickness for each test and, if possible, also blindness of the investigator who lifts animals should be sought.

The grasping test can be also adapted for mice (Tos *et al.*, 2008) changing the grip device, namely using a grid instead of bars (Fig. 6), and by pulling the animals horizontally rather than vertically, while no attempt has been made to adapt the ulnar test (see next paragraph) since the very small size makes mouse ulnar nerve surgery very difficult.

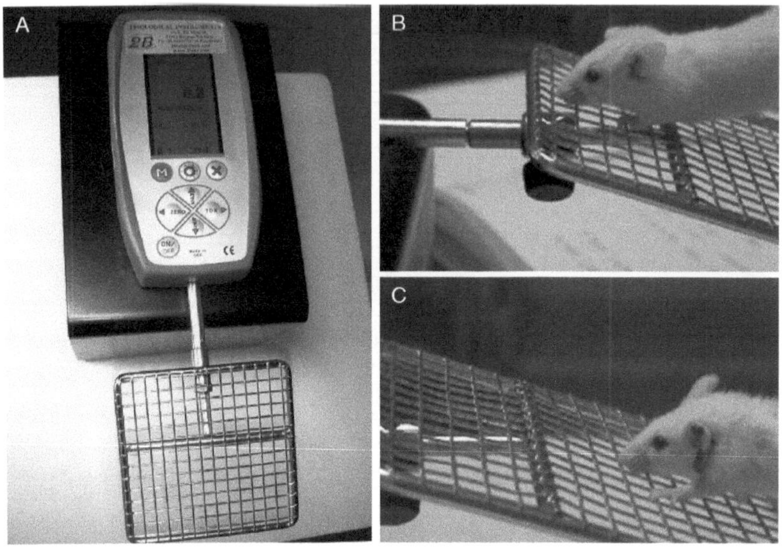

FIG. 6. (A) The grasping test device adapted for mice by means of a grid instead of bars. (B) and (C) A mouse grasping the grid with its fingers.

3. *The Ulnar Test*

The ulnar test is similar to the grasping test since it also aims to measure the force exerted by the rat paw while pulling it up holding it by its tail (Papalia *et al.*, 2006). The difference is represented by the device (Fig. 7) that is made by a 15-cm squared wooden board to which 19 iron nails are inserted being at a distance of 1.25 cm one from the other. Each nail holds a plastic sphere the size of which (5 mm) was chosen to fit with the size of the paw of the rats. The distance between the spheres and the board was 10 mm. The board is covered by a white plastic round plate in order to focus animal attention on the spheres and the entire device is fixed to a precision balance on time of testing.

Similarly to the grasping test, the rat is approached to the device holding it by its tail (Fig. 7) and when the animal grips one sphere with its paw it is gently pulled until it loses the grip and the maximum (or average) weight that the rat manages to hold up before loosing the grip is recorded (the rat is approached to the device for three times).

Careful animal surveillance showed that both behavioral tests provoke minimal distress to the animal and no painful sensation. Animal testing is simple and quick and is effective in detecting the date on which recovery starts after nerve impairment and in following its improvement with time (Papalia *et al.*, 2003, 2006; Ronchi *et al.*, 2009). The availability of this couple of tests can further promote the

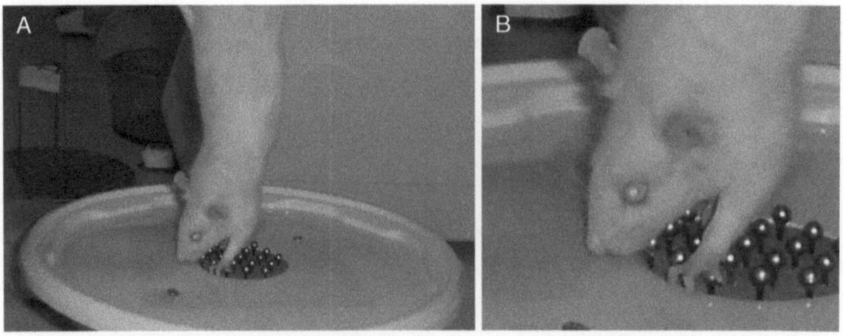

FIG. 7. (A) The ulnar grasping test device. (B) Higher magnification showing a rat gripping one sphere.

use of forelimb experimental nerve models allowing to independently assess the function of the two nerves such as for the study of multiple Y-shaped-tubulization nerve repair. A recent study on combined median and ulnar nerve repair by means of Y-shape muscle-vein-combined conduits permitted to demonstrate that functional recovery of both nerves can be obtained, independently from the proximal donor nerve employed, and that tissue, and not topographic, specificity guides nerve fiber regeneration (Geuna *et al.*, 2007; Lee *et al.*, 2007).

4. *Functional Assessment of the Radial Nerve by Means of the Grasping Test*

In the attempt to define a behavioral test useful for measuring function of the third mixed nerve of the rat forearm and for improving the battery of evaluation tools available to the nerve researcher, we have investigated the possibility that grasping test performance could also be influenced by radial nerve impairment. In fact, wrist extension, controlled by the radial nerve, is agonistic to finger flexion, and thus grip strength reduction would be expected after radial nerve functional impairment.

In five adult female rats under general anesthesia by ketamine (40 mg/250 g) and cloropromazine (3.75 mg/250 g) and clean conditions, we have performed end-to-end reconstruction of the left radial nerve. Experimental procedures were carried out in the Laboratory of Microsurgery of the Ecole de Chirurgie de Paris. Approval for this study was obtained from the local Institution's Animal Care and Ethics Committee, and in accordance with the European Communities Council Directive of 24 November 1986 (86/609/EEC). To prevent interferences with the grasping test device during testing because of the use of the contra-lateral forepaw (Papalia, *et al.*, 2003, 2006), the contra-lateral median nerve was transected at the middle third of the brachium and its proximal stump was sutured in the pectoralis major muscle to avoid spontaneous reinnervation. In five other animals, right median nerve resection only was performed and was used as control. Postoperative follow up was 12 months and animals were tested by grasping test as described above at four time points: month 1, 4, 8, and at time of euthanasia. From each animal, a 1-cm-long segment of the radial nerve distal to the injury site was removed fixed in 2.5% glutaraldehyde and 0.5% saccharose in 0.1 M Sorensen phosphate buffer for 6–8 h and then postfixed for 2 h in 2% osmium tetroxide in order to stain myelin sheaths (Di Scipio *et al.*, 2008). The nerves were then dehydrated and embedded in paraffin. Series of 8-μm thick transverse sections were cut starting from the distal stump of each radial nerve specimen and quantitatively examined by design-based quantitative morphology (Geuna *et al.*, 2004). Statistical analysis was performed using the one-way repeated measures analysis of variance (RM-ANOVA) test applied on the values from the different time-point assessments followed by post-hoc multiple pair-wise comparisons using the Student-Neuman-Keuls (SNK) test. Statistical significance was established as $p < 0.05$.

Results of the stereological analysis of myelinated nerve fibers showed, as expected, that regenerated nerve fibers have a significant ($p < 0.05$) increase in the total number and mean density of myelinated nerve fibers and a significant decrease in the mean fiber size. Results of the behavioral assessment showed that, radial nerve lesion induced a significant ($p < 0.05$) decrease in grasping test performance which dropped from an average control value of 254 $\pm$ 13, to 129 $\pm$ 31. Then at month-4 postoperative it returned to control values (240 $\pm$ 10) remaining not significantly different form controls until the end point of this experiment (month-12 postoperative).

It should be noted that a recent study by Windebank's group (Wang *et al.*, 2008a) led, unexpectedly, to opposite results namely the absence of obvious decrease in grip strength after radial nerve lesion. These authors interpreted the unexpected piece of result on the basis of the observation, obtained by two-dimensional digital video motion analysis (Wang *et al.*, 2008b), that although wrist extension decreased after radial nerve lesions the position of the wrist did not fall below neutral (180°). The reason may be that a different line of pull of the tendons in rats keeps tension on the wrist and prevents it from dropping. However, in the light of our present results the possibility that the higher sensibility of our testing device may explain the discrepancy between the results of two studies might be also taken into consideration and the potential occurrence of grip strength impairment also after radial nerve injury should be considered when designing an experimental nerve regeneration study.

Although our results suggest that the grasping test can be used for assessing not only median but also radial nerve function, the variability in radial nerve fiber's composition along brachium due to several collateral branches (Santos *et al.*, 2007) points to median and ulnar nerves as preferable models for nerve regeneration research in the forelimb.

C. Other Nerve Models

Several peripheral nerves not belonging to limbs have been used for regeneration and repair studies but it is beyond our goal to address this here. However, we wish to just mention the facial nerve, for which a nice functional test is available (i.e., Hadlock *et al.*, 2008), the hypoglossal nerve (i.e., González-Forero *et al.*, 2004), the pure sensory inferior alveolar nerve (i.e., Atsumi *et al.*, 2000) in the head, and the vagus (i.e., Bregeon *et al.*, 2007), and cavernous (i.e., Ding *et al.*, 2009), among the autonomic nerves. We wish also to emphasize the importance of investigating nerve regeneration not only on mixed somatic nerve models of the limbs and head but also in autonomic and sensory nerve models too. In fact, while the high clinical relevance of mixed nerve lesions justify the prevailing use of somatic mixed nerve models, the possibility that sensory nerves can have different regeneration patterns (Moradzadeh *et al.*, 2008) should be taken into consideration in light of clinical translation of the experimental results. The same is true for autonomic nerve regeneration that is acquiring increasing importance in relation to urologic surgery (May *et al.*, 2005).

VI. Interfering Conditions and Disease Models

Usually, experimental nerve regeneration studies are carried out on ideal subjects, that is, young and healthy animals. This might represent a problem when researchers seek to translate experimental results to the clinics since patients often do not match these two characteristics (i.e., they are not young and/or concurring diseases are present). Since this discrepancy is likely to represent one of the causes of the failure in translating laboratory bench results to the patient bed, the employment experimental models with old animals (Geuna and Tos, 2008; Kovačič *et al.*, 2009, this issue) and/or concurring diseases (such as infections, diabetes, etc.) (Jolivalt *et al.*, 2008; Zochodne *et al.*, 2007) should be adopted to verify the effectiveness of a new technique for improving nerve regeneration.

A further factor that deserves mention is delayed nerve regeneration since it has been shown that a delay in surgical nerve repair results in impaired nerve regeneration and functional recovery both in rodents and humans (Richardson, 1997; Saito and Dahlin, 2008).

Finally, sexual dimorphism should also been taken into consideration since there is evidence that nerve regeneration is more pronounced in females because of the neuroprotective effects of sex hormones (Kovačič *et al.*, 2003; Roglio *et al.*, 2008).

Interstrain variability deserves a final mention. While it has been shown that few differences in peripheral nerve recovery appear to exist between rat strains and that uniform conclusions may be drawn regardless of strain used (Strasberg *et al.*, 1999), definite strain differences were observed in the degree of autotomy following nerve lesion both in rat (Carr *et al.*, 1992) and mouse (Rubinstein *et al.*, 2003). Lewis rats and C57BL/6J and C57BL/10J mouse strains appear to be the most appropriate for nerve regeneration research especially with nerve models of the hindlimb where autotomy is more suitable to occur.

VII. Conclusions

This methodology-oriented paper is expected to provide some elements which might facilitate researchers in choosing the best experimental model for their nerve regeneration research. Although we wish to emphasize that there is no absolutely best experimental protocol and thus the choice should be left to each researcher after accurate consideration of many different factors, we would also like to try to put forward some personal recommendations that, of course, do not claim to be the last word but rather aim to represent a further step towards shared criteria for selecting the most appropriate nerve regeneration experimental model for any given study.

1. As for other biomedical fields, nerve regeneration studies should be driven by the "Three Rs" concept (Russell and Burch, 1992). This concept is based on three principles that can be outlined as follows (Robinson, 2005): (1) *Replacement* which means the use of nonanimal methods such as cell cultures, human volunteers, and computer modeling instead of animals to achieve a scientific aim; (2) *Reduction* which means the use of methods that enable researchers to obtain comparable amounts of information from fewer animals, or more information from the same number of animals. (3) *Refinement* which means the use of methods that alleviate or minimize potential pain, suffering or distress, and that enhance animal welfare for those animals that cannot be replaced. All three rules must be considered while designing *in vivo* nerve regeneration studies by careful preliminary evaluation of *in vitro* models in substitution and/or preparation of an *in vivo* investigation (replace), adoption of adequate data collection systems to optimize data collection and analysis (reduce), and finally by taking all possible measures to reduce animal pain and distress (refine).
2. The choice of the animal models should be driven by several factors which take into consideration various elements related to the resources available as well as to the study goals. As a general rule, the rat can be considered the standard "first choice" model because of the larger nerve size compared to

the mouse. Even if small size makes mouse nerve manipulation difficult, this can be a good option when the molecular mechanisms of nerve regeneration are investigated because of the availability of a large number of transgenic strains. Finally, the selection of larger animals can be considered, in the view of clinical translation, as a second investigation step for validating rodents' results.

3. Selection of the lesion paradigm should be directed by the study goals. *Axonotmesis* is particularly suitable when a reproducible regeneration process is required, such as for the study of the biological mechanisms of regeneration or rationale development for new therapeutic agents. *Neurotmesis* should be preferred for preclinical studies on new therapeutic agents since significant differences in terms of morphological, electrophysiological, and functional predictors will be more easily detected among experimental groups.
4. Selection of the nerve models can be guided by several factors that have been outlined in this review. In particular, one of the emerging issues is the contrast between traditional hindlimb vs forelimb nerve models (Bontioti *et al.*, 2003; Nichols *et al.*, 2005; Ronchi *et al.*, 2009; Sinis *et al.*, 2006). Although the debate is still open, our present knowledge does not allow us to conclude that one of these two models is superior to the other. Researchers must choose the experimental model based on their specific requirements and expertise, knowing each model's limitations and using the results within those limitations, rather than hewing to a more rigid point of view about which model is best. As a general rule, it can be recommended that nerve models are selected on the basis of the translational goals considering both the anatomical location as well as the prevailing nerve fiber composition (mixed, sensory, or autonomic). Selection of autonomic and sensory nerve models deserves mention since most nerve regeneration studies are carried out in mixed somatic nerve models of the limbs and head. While this prevalence is justified by the prevailing clinical relevance of mixed nerve lesions, the possibility that sensory and especially autonomic nerves can have different regeneration patterns should be taken into consideration in light of clinical translation of the experimental results.
5. The presence of physiological factors that can influence nerve regeneration (e.g., sexual dimorphism and/or aging) should be taken into consideration when evaluating the results and yet the deliberate introduction of interfering pathological conditions in the experimental model (e.g., infection or diabetes) should be adopted for a comprehensive assessment of new treatment protocols.

Acknowledgments

This work was supported by grants from the MUR (Italian Ministry of University and Research), ex-60% fund, FIRB fund (code: RBAU01BJ95), PRIN2005 fund (code: 2005057088), the Compagnia di San Paolo (Bando Programma Neuroscienze), and the Regione Piemonte (Progetto Ricerca Sanitaria Finalizzata).

References

Alluin, O., Wittmann, C., Marqueste, T., Chabas, J. F., Garcia, S., Lavaut, M. N., Guinard, D., Feron, F., and Decherchi, P. (2009). Functional recovery after peripheral nerve injury and implantation of a collagen guide. *Biomaterials* **30,** 363–373.

Amado, S., Simões, M. J., Armada da Silva, P. A. S., Luís, A. L., Shirosaki, Y., Lopes, M. A., Santos, J. D., Fregnan, F., Gambarotta, G., Raimondo, S., Fornaro, M., Veloso, A. P., *et al.* (2008). Use of hybrid chitosan membranes and N1E-115 cells for promoting nerve regeneration in an axonotmesis rat model. *Biomaterials* **29,** 4409–4419.

Apel, P. J., Alton, T., Northam, C., Ma, J., Callahan, M., Sonntag, W. E., and Li, Z. (2009). How age impairs the response of the neuromuscular junction to nerve transection and repair: An experimental study in rats. *J. Orthop. Res.* **27,** 385–393.

Atchabahian, A., Genden, E. M., MacKinnon, S. E., Doolabh, V. B., and Hunter, D. A. (1998). Regeneration through long nerve grafts in the swine model. *Microsurgery* **18,** 379–382.

Atsumi, Y., Imai, T., Matsumoto, K., Sakuda, M., Kurisu, K., and Wakisaka, S. (2000). Effects of neonatal injury of the inferior alveolar nerve on the development and regeneration of periodontal nerve fibers in the rat incisor. *Brain Res.* **21,** 201–209.

Aubá, C., Hontanilla, B., Arcocha, J., and Gorría, O. (2006). Peripheral nerve regeneration through allografts compared with autografts in FK506-treated monkeys. *J. Neurosurg.* **105,** 602–609.

Bannerman, P., Ara, J., Hahn, A., Hong, L., McCauley, E., Friesen, K., and Pleasure, D. (2008). Peripheral nerve regeneration is delayed in neuropilin 2-deficient mice. *J. Neurosci. Res.* **86,** 3163–3169.

Baptista, A. F., Gomes, J. R., Oliveira, J. T., Santos, S. M., Vannier-Santos, M. A., and Martinez, A. M. (2007). A new approach to assess function after sciatic nerve lesion in the mouse—Adaptation of the sciatic static index. *J. Neurosci. Methods* **161,** 259–264.

Beer, G. M., Steurer, J., and Meyer, V. E. (2001). Standardizing nerve crushes with a non-serrated clamp. *J. Reconstr. Microsurg.* **17,** 531–534.

Bertelli, J. A., and Mira, J. C. (1995). The grasping test: A simple behavioral method for objective quantitative assessment of peripheral nerve regeneration in the rat. *J. Neurosci. Methods* **58,** 151–155.

Bervar, M. (2000). Video analysis of standing—An alternative footprint analysis to assess functional loss following injury to the rat sciatic nerve. *J. Neurosci. Methods* **102,** 109–116.

Bontioti, E., Kanje, M., and Dahlin, L. B. (2003). Regeneration and functional recovery in the upper extremity of rats after various types of nerve injuries. *J. Peripher. Nerv. Syst.* **8,** 159–168.

Bozkurt, A., Brook, G. A., Moellers, S., Lassner, F., Sellhaus, B., Weis, J., Woeltje, M., Tank, J., Beckmann, C., Fuchs, P., Damink, L. O., Schügner, F., Heschel, I., and Pallua, N. (2007). *In vitro* assessment of axonal growth using dorsal root ganglia explants in a novel three-dimensional collagen matrix. *Tissue Eng.* **13,** 2971–2979.

Bozkurt, A., Tholl, S., Wehner, S., Tank, J., Cortese, M., O'Dey, D., Deumens, R., Lassner, F., Schügner, F., Gröger, A., Smeets, R., Brook, G., *et al.* (2008). Evaluation of functional nerve recovery with Visual-SSI-A novel computerized approach for the assessment of the static sciatic index (SSI). *J. Neurosci. Methods* **170,** 117–122.

Bregeon, F., Alliez, J. R., Héry, G., Marqueste, T., Ravailhe, S., and Jammes, Y. (2007). Motor and sensory re-innervation of the lung and heart after re-anastomosis of the cervical vagus nerve in rats. *J. Physiol.* **15,** 1333–1340.

Brenner, M. J., Lowe, J. B., III, Fox, I. K., Mackinnon, S. E., Hunter, D. A., Darcy, M. D., Duncan, J. R., Wood, P., and Mohanakumar, T. (2005). Effects of Schwann cells and donor antigen on long-nerve allograft regeneration. *Microsurgery* **25,** 61–70.

Brown, C. J., Evans, P. J., Mackinnon, S. E., Bain, J. R., Makino, A. P., Hunter, D. A., and Hare, G. (1991). Inter- and intraobserver reliability of walking-track analysis used to assess sciatic nerve function in rats. *Microsurgery* **12,** 76–79.

Carr, M. M., Best, T. J., Mackinnon, S. E., and Evans, P. J. (1992). Strain differences in autotomy in rats undergoing sciatic nerve transection or repair. *Ann. Plast. Surg.* **28,** 538–544.

Chabas, J. F., Alluin, O., Rao, G., Garcia, S., Lavaut, M. N., Risso, J. J., Legre, R., Magalon, G., Khrestchatisky, M., Marqueste, T., Decherchi, P., and Feron, F. (2008). Vitamin D2 potentiates axon regeneration. *J. Neurotrauma* **25,** 1247–1256.

Chan, W. Y., Matteucci, P., and Southern, S. J. (2007). Validation of microsurgical models in microsurgery training and competence: A review. *Microsurgery* **27,** 494–499.

Chen, L. E., Seaber, A. V., Glisson, R. R., Davies, H., Murrell, G. A., Anthony, D. C., and Urbaniak, J. R. (1992). The functional recovery of peripheral nerves following defined acute crush injuries. *J. Orthop. Res.* **10,** 657–664.

Costa, L. M., Maria, J., Simões, M. J., Maurício, A. C., and Varejão, A. S. P. (2009). Methods and protocols in peripheral nerve regeneration experimental research. Part IV—Kinematic gait analysis to quantify peripheral nerve regeneration in the rat. *Int. Rev. Neurobiol.* **87,** 127–139.

De Paola, M., Buanne, P., Biordi, L., Bertini, R., Ghezzi, P., and Mennini, T. (2007). Chemokine MIP-2/CXCL2, acting on CXCR2, induces motor neuron death in primary cultures. *Neuroimmunomodulation* **14,** 310–316.

De Sá, J. M., Mazzer, N., Barbieri, C. H., and Barreira, A. A. (2004). The end-to-side peripheral nerve repair. Functional and morphometric study using the peroneal nerve of rats. *J. Neurosci. Methods* **136,** 45–53.

Deumens, R., Jaken, R. J., Marcus, M. A., and Joosten, E. A. (2007). The catwalk gait analysis in assessment of both dynamic and static gait changes after adult rat sciatic nerve resection. *J. Neurosci. Methods* **164,** 120–130.

Ding, X. G., Li, S. W., Zheng, X. M., Hu, L. Q., Hu, W. L., and Luo, Y. (2009). The effect of platelet-rich plasma on cavernous nerve regeneration in a rat model. *Asian J. Androl.* **11,** 215–221.

Di Scipio, F., Raimondo, S., Tos, P., and Geuna, S. (2008). A simple protocol for paraffin embedded myelin sheath staining with osmium tetroxide for light microscope observation. *Microsc. Res. Tech.* **71,** 497–502.

Falkenburger, B. H., and Schulz, J. B. (2006). Limitations of cellular models in Parkinson's disease research. *J. Neural. Transm. Suppl.* **70,** 261–268.

Fornaro, M., Lee, J. M., Raimondo, S., Nicolino, S., Geuna, S., and Giacobini-Robecchi, M. (2008). Neuronal intermediate filament expression in rat dorsal root ganglia sensory neurons: An *in vivo* and *in vitro* study. *Neuroscience* **153,** 1153–1163.

Fullarton, A. C., Lenihan, D. V., Myles, L. M., and Glasby, M. A. (2000). Obstetric brachial plexus palsy: A large animal model for traction injury and its repair. Part 1: Age of the recipient. *J. Hand Surg.* **25,** 52–57.

Galtrey, C. M., and Fawcett, J. W. (2007). Characterization of tests of functional recovery after median and ulnar nerve injury and repair in the rat forelimb. *J. Peripher. Nerv. Syst.* **12,** 11–27.

Geuna, S., and Tos, P. (2008). Comparison of biodegradable conduits within aged rat sciatic nerve defects. *Plast. Reconstr. Surg.* **121,** 705–706.

Geuna, S., and Varejão, A. S. (2008). Evaluation methods in the assessment of peripheral nerve regeneration. *J. Neurosurg.* **109,** 360–362.

Geuna, S., Gigo-Benato, D., Rodrigues, C., and Ade, C. (2004). On sampling and sampling errors in histomorphometry of peripheral nerve fibers. *Microsurgery* **24,** 72–76.

Geuna, S., Nicolino, S., Raimondo, S., Gambarotta, G., Battiston, B., Tos, P., and Perroteau, I. (2007). Nerve regeneration along bioengineered scaffolds. *Microsurgery* **27,** 429–438.

Geuna, S., Raimondo, S., Ronchi, G., Di Scipio, F., Tos, P., and Fornaro, M. (2009). Histology of the peripheral nerve and changes occurring during nerve regeneration. *Int. Rev. Neurobiol.* **87,** 27–46.

Gingras, M., Beaulieu, M. M., Gagnon, V., Durham, H. D., and Berthod, F. (2008). *In vitro* study of axonal migration and myelination of motor neurons in a three-dimensional tissue-engineered model. *Glia* **56,** 354–364.

Giovanoli, P., Koller, R., Meuli-Simmen, C., Rab, M., Haslik, W., Mittlböck, M., Meyer, V. E., and Frey, M. (2000). Functional and morphometric evaluation of end-to-side neurorrhaphy for muscle reinnervation. *Plast. Reconstr. Surg.* **106,** 383–392.

Gondré, M., Burrola, P., and Weinstein, D. E. (1998). Accelerated nerve regeneration mediated by Schwann cells expressing a mutant form of the POU protein SCIP. *J. Cell Biol.* **141,** 493–501.

González-Forero, D., Portillo, F., Sunico, C. R., and Moreno-López, B. (2004). Nerve injury reduces responses of hypoglossal motoneurones to baseline and chemoreceptor-modulated inspiratory drive in the adult rat. *J. Physiol.* **15,** 991–1011.

Greene, E. C. (1963). "Anatomy of the Rat." Hafner Publishing Company, New York, NY. A re-edition of this book has been recently published by Braintree Scientific Inc., Braintree, MA. http://www.braintreesci.com.

Hadlock, T., Kowaleski, J., Lo, D., Bermejo, R., Zeigler, H. P., Mackinnon, S., and Heaton, J. T. (2008). Functional assessments of the rodent facial nerve: A synkinesis model. *Laryngoscope* **118,** 1744–1749.

Hara, K., Yasuhara, T., Maki, M., Matsukawa, N., Masuda, T., Yu, S. J., Ali, M., Yu, G., Xu, L., Kim, S. U., Hess, D. C., and Borlongan, C. V. (2008). Neural progenitor NT2N cell lines from teratocarcinoma for transplantation therapy in stroke. *Prog. Neurobiol.* **85,** 318–334.

Hayashi, A., Koob, J. W., Liu, D. Z., Tong, A. Y., Hunter, D. A., Parsadanian, A., Mackinnon, S. E., and Myckatyn, T. M. (2007). A double-transgenic mouse used to track migrating Schwann cells and regenerating axons following engraftment of injured nerves. *Exp. Neurol.* **207,** 128–138.

Hems, T. E., Clutton, R. E., and Glasby, M. A. (1994). Repair of avulsed cervical nerve roots. An experimental study in sheep. *J. Bone Joint Surg.* **76,** 818–823.

Henry, F. P., Goyal, N. A., David, W. S., Wes, D., Bujold, K. E., Randolph, M. A., Winograd, J. M., Kochevar, I. E., and Redmond, R. W. (2009). Improving electrophysiologic and histologic outcomes by photochemically sealing amnion to the peripheral nerve repair site. *Surgery* **145,** 313–321.

Hess, J. R., Brenner, M. J., Myckatyn, T. M., Hunter, D. A., and Mackinnon, S. E. (2006). Influence of aging on regeneration in end-to-side neurorrhaphy. *Ann. Plast. Surg.* **57,** 217–222.

Hess, J. R., Brenner, M. J., Fox, I. K., Nichols, C. M., Myckatyn, T. M., Hunter, D. A., Rickman, S. R., and Mackinnon, S. E. (2007). Use of cold-preserved allografts seeded with autologous Schwann cells in the treatment of a long-gap peripheral nerve injury. *Plast. Reconstr. Surg.* **119,** 246–259.

Hirota, H., Kiyama, H., Kishimoto, T., and Taga, T. (1996). Accelerated Nerve Regeneration in Mice by upregulated expression of interleukin (IL) 6 and IL-6 receptor after trauma. *J. Exp. Med.* **183,** 2417–2419.

Hu, X., He, W., Diaconu, C., Tang, X., Kidd, G. J., Macklin, W. B., Trapp, B. D., and Yan, R. (2008). Genetic deletion of BACE1 in mice affects remyelination of sciatic nerves. *FASEB J.* **22,** 2970–2980.

Huang, J., Hu, X., Lu, L., Ye, Z., Wang, Y., and Luo, Z. (2009). Electrical stimulation accelerates motor functional recovery in autograft-repaired 10 mm femoral nerve gap in rats. *J. Neurotrauma* (Epub ahead of print).

Inserra, M. M., Yao, M., Murray, R., and Terris, D. J. (2000). Peripheral nerve regeneration in interleukin 6-deficient mice. *Arch. Otolaryngol. Head Neck Surg.* **126,** 1112–1116.

Islamov, R. R., Hendricks, W. A., Jones, R. J., Lyall, G. J., Spanier, N. S., and Murashov, A. K. (2002). 17-Beta-estradiol stimulates regeneration of sciatic nerve in female mice. *Brain Res.* **943,** 283–286.

Jensen, J. N., Brenner, M. J., Tung, T. H., Hunter, D. A., and Mackinnon, S. E. (2005). Effect of FK506 on peripheral nerve regeneration through long grafts in inbred swine. *Ann. Plast. Surg.* **54,** 420–427.

Jolivalt, C. G., Vu, Y., Mizisin, L. M., Mizisin, A. P., and Calcutt, N. A. (2008). Impaired prosaposin secretion during nerve regeneration in diabetic rats and protection of nerve regeneration by a prosaposin-derived peptide. *J. Neuropathol. Exp. Neurol.* **67,** 702–710.

Jungnickel, J., Haase, K., Konitzer, J., Timmer, M., and Grothe, C. (2006). Faster nerve regeneration after sciatic nerve injury in mice over-expressing basic fibroblast growth factor. *J. Neurobiol.* **66,** 940–948.

Kelleher, M. O., Al-Abri, R. K., Lenihan, D. V., and Glasby, M. A. (2006). Use of a static magnetic field to promote recovery after peripheral nerve injury. *J. Neurosurg.* **105,** 610–615.

Kim, J. E., Bonilla, I. E., Qiu, D., and Strittmatter, S. M. (2003). Nogo-C is sufficient to delay nerve regeneration. *Mol. Cell Neurosci.* **23,** 451–459.

Kittaka, D., Itoh, M., Ohmi, Y., Kondo, Y., Fukumoto, S., Urano, T., Tajima, O., Furukawa, K., and Furukawa, K. (2008). Impaired hypoglossal nerve regeneration in mutant mice lacking complex gangliosides: Down-regulation of neurotrophic factors and receptors as possible mechanisms. *Glycobiology* **18,** 509–516.

Kotulska, K., LePecheur, M., Marcol, W., Lewin-Kowalik, J., Larysz-Brysz, M., Paly, E., Matuszek, I., and London, J. (2006). Overexpression of copper/zinc-superoxide dismutase in transgenic mice markedly impairs regeneration and increases development of neuropathic pain after sciatic nerve injury. *J. Neurosci. Res.* **84,** 1091–1097.

Kovačič, U., Sketelj, J., and Bajrovic, F. F. (2003). Sex-related difference in collateral sprouting of nociceptive axons after peripheral nerve injury in the rat. *Exp. Neurol.* **184,** 479–488.

Kovačič, U., Sketelj, J., and Bajrović, F. F. (2009). Age-related differences in the reinnervation after peripheral nerve injury. *Int. Rev. Neurobiol.* **87,** 465–482.

Kozin, S. H., Porter, S., Clark, P., and Thoder, J. J. (1999). The contribution of the intrinsic muscles to grip and pinch strength. *J. Hand Surg. (Am.)* **24,** 64–72.

Kvist, M., Kanje, M., Ekberg, H., Corbascio, M., and Dahlin, L. B. (2008). Costimulation blockade in transplantation of nerve allografts: Long-term effects. *J. Peripher. Nerv. Syst.* **13,** 200–207.

Lawson, G. M., and Glasby, M. A. (1995). A comparison of immediate and delayed nerve repair using autologous freeze-thawed muscle grafts in a large animal model. The simple injury. *J. Hand Surg. (Br.)* **20,** 663–700.

Lee, J. M., Tos, P., Raimondo, S., Fornaro, M., Papalia, I., Geuna, S., and Giacobini-Robecchi, M. G. (2007). Lack of topographic specificity in nerve fiber regeneration of rat forelimb mixed nerves. *Neuroscience* **144,** 985–990.

Lee, H. K., Seo, I. A., Suh, D. J., Hong, J. I., Yoo, Y. H., and Park, H. T. (2009). Interleukin-6 is required for the early induction of glial fibrillary acidic protein in Schwann cells during Wallerian degeneration. *J. Neurochem.* **108,** 776–786.

Liu, H. G., Hong, G. X., Wang, F. B., and Chen, F. (1998). Motoneurotrophins derived from limb buds protect the motoneurons in anterior spinal cord after nerve injury and promote nerve regeneration. *Brain Res.* **800,** 216–226.

López-Vales, R., Navarro, X., Shimizu, T., Baskakis, C., Kokotos, G., Constantinou-Kokotou, V., Stephens, D., Dennis, E. A., and David, S. (2008). Intracellular phospholipase A(2) group IVA and group VIA play important roles in Wallerian degeneration and axon regeneration after peripheral nerve injury. *Brain* **131,** 2620–2631.

Lorenzetto, E., Panteri, R., Marino, R., Keller, F., and Buffelli, M. (2008). Impaired nerve regeneration in reeler mice after peripheral nerve injury. *Eur. J. Neurosci.* **27,** 12–19.

Luis, A. L., Amado, S., Geuna, S., Rodrigues, J. M., Simoes, M. J., Santos, J. D., Fregnan, F., Raimondo, S., Veloso, A. P., Ferreira, A. J., Armada-da-Silva, P. A., Varejao, A. S., *et al.* (2007). Long-term functional and morphological assessment of a standardized rat sciatic nerve crush injury with a non-serrated clamp. *J. Neurosci. Methods* **163,** 92–104.

Luís, A. L., Rodrigues, J. M., Geuna, S., Amado, S., Simões, M. J., Fregnan, F., Ferreira, A. J., Veloso, A. P., Armada-Da-Silva, P. A., Varejão, A. S., and Maurício, A. C. (2008). Neural cell transplantation effects on sciatic nerve regeneration after a standardized crush injury in the rat. *Microsurgery* **28,** 458–470.

Lutz, B. S., Chuang, D. C., Chuang, S. S., Hsu, J. C., Ma, S. F., and Wei, F. C. (2000). Nerve transfer to the median nerve using parts of the ulnar and radial nerves in the rabbit—Effects on motor recovery of the median nerve and donor nerve morbidity. *J. Hand Surg. (Br.)* **25,** 323–324.

Madison, R. D., Zomorodi, A., and Robinson, G. A. (2000). Netrin-1 and peripheral nerve regeneration in the adult rat. *Exp. Neurol.* **161,** 563–570.

May, F., Vroemen, M., Matiasek, K., Henke, J., Brill, T., Lehmer, A., Apprich, M., Erhardt, W., Schoeler, S., Paul, R., Blesch, A., Hartung, R., *et al.* (2005). Nerve replacement strategies for cavernous nerves. *Eur. Urol.* **48,** 372–378.

Moradzadeh, A., Borschel, G. H., Luciano, J. P., Whitlock, E. L., Hayashi, A., Hunter, D. A., and Mackinnon, S. E. (2008). The impact of motor and sensory nerve architecture on nerve regeneration. *Exp. Neurol.* **212,** 370–376.

Moreno-Flores, M. T., Bradbury, E. J., Martín-Bermejo, M. J., Agudo, M., Lim, F., Pastrana, E., Avila, J., Diaz-Nido, J., McMahon, S. B., and Wandosell, F. (2006). A clonal cell line from immortalized olfactory ensheathing glia promotes functional recovery in the injured spinal cord. *Mol. Ther.* **13,** 598–608.

Navarro, X., and Udina, E. (2009). Methods and protocols in peripheral nerve regeneration experimental research. Part III—Electrophysiological evaluation. *Int. Rev. Neurobiol.* **87,** 103–124.

Nichols, C. M., Myckatyn, T. M., Rickman, S. R., Fox, I. K., Hadlock, T., and Mackinnon, S. E. (2005). Choosing the correct functional assay: A comprehensive assessment of functional tests in the rat. *Behav. Brain Res.* **163,** 143–158.

Papalia, I., Tos, P., Stagno d'Alcontres, F., Battiston, B., and Geuna, S. (2003). On the use of the grasping test in the rat median nerve model: A re-appraisal of its efficacy for quantitative assessment of motor function recovery. *J. Neurosci. Methods* **127,** 43–74.

Papalia, I., Tos, P., Scevola, A., Raimondo, S., and Geuna, S. (2006). The ulnar test: A method for the quantitative functional assessment of posttraumatic ulnar nerve recovery in the rat. *J. Neurosci. Methods* **154,** 198–203.

Papalia, I., Cardaci, A., d'Alcontres, F. S., Lee, J. M., Tos, P., and Geuna, S. (2007). Selection of the donor nerve for end-to-side neurorrhaphy. *J. Neurosurg.* **107,** 378–382.

Pellitteri, R., Spatuzza, M., Russo, A., and Stanzani, S. (2007). Olfactory ensheathing cells exert a trophic effect on the hypothalamic neurons *in vitro*. *Neurosci. Lett.* **417,** 24–29.

Pereira Lopes, F. R., Camargo de Moura Campos, L., Dias Correa, J., Jr., Balduino, A., Lora, S., Langone, F., Borojevic, R., and Blanco Martinez, A. M. (2006). Bone marrow stromal cells and resorbable collagen guidance tubes enhance sciatic nerve regeneration in mice. *Exp. Neurol.* **198,** 457–468.

Radtke, C., Aizer, A. A., Agulian, S. K., Lankford, K. L., Vogt, P. M., and Kocsis, J. D. (2009). Transplantation of olfactory ensheathing cells enhances peripheral nerve regeneration after microsurgical nerve repair. *Brain Res.* **1254,** 10–17.

Raimondo, S., Fornaro, M., Di Scipio, F., Ronchi, G., Giacobini-Robecchi, M. G., and Geuna, S. (2009). Methods and protocols in peripheral nerve regeneration experimental research. Part II—Morphological techniques. *Int. Rev. Neurobiol.* **87,** 79–101.

Raisman, G. (2007). Repair of spinal cord injury by transplantation of olfactory ensheathing cells. *C. R. Biol.* **330,** 557–560.

Richardson, P. M. (1997). Recovery of biceps function after delayed repair for brachial plexus injury. *J. Trauma* **42,** 791–792.

Robinson, V. (2005). Finding alternatives: An overview of the 3Rs and the use of animals in research. *School Sci. Rev.* **87,** 1–4.

Robinson, G. A., and Madison, R. D. (2009). Influence of terminal nerve branch size on motor neuron regeneration accuracy. *Exp. Neurol.* **215,** 228–235.

Roglio, I., Bianchi, R., Gotti, S., Scurati, S., Giatti, S., Pesaresi, M., Caruso, D., Panzica, G. C., and Melcangi, R. C. (2008). Neuroprotective effects of dihydroprogesterone and progesterone in an experimental model of nerve crush injury. *Neuroscience* **155,** 673–685.

Ronchi, G., Nicolino, S., Raimondo, S., Tos, P., Battiston, B., Papalia, I., Varejão, A., Giacobini-Robecchi, M. G., Perroteau, I., and Geuna, S. (2009). Functional and morphological assessment of a standardized crush injury of the rat median nerve. *J. Neurosci. Methods* **179,** 51–57.

Rong, L. L., Yan, S. F., Wendt, T., Hans, D., Pachydaki, S., Bucciarelli, L. G., Adebayo, A., Qu, W., Lu, Y., Kostov, K., Lalla, E., Yan, S. D., *et al.* (2004). RAGE modulates peripheral nerve regeneration via recruitment of both inflammatory and axonal outgrowth pathways. *FASEB J.* **18,** 1818–1825.

Rubinstein, R. E., Deem, K. C., Jensen, J., MacKinnon, S. E., and Tung, T. H. (2003). Strain differences in autotomy in mice after peripheral nerve transection or repair. *Microsurgery* **23,** 363–368.

Ruch, D. S., Deal, D. N., Ma, J., Smith, A. M., Castle, J. A., Walker, F. O., Martin, E. V., Yoder, J. S., Rushing, J. T., Smith, T. L., and Koman, L. A. (2004). Management of peripheral nerve defects: External fixator-assisted primary neurorrhaphy. *J. Bone Joint Surg. Am.* **86,** 1405–1413.

Russell, W. M., and Burch, R. L. (1992). "The Principles of Humane Experimental Technique." Universities Federation for Animal Welfare (UFAW) Blackwell Publishing, Hertfordshire.

Sahenk, Z., Oblinger, J., and Edwards, C. (2008). Neurotrophin-3 deficient Schwann cells impair nerve regeneration. *Exp. Neurol.* **212,** 552–556.

Saito, H., and Dahlin, L. B. (2008). Expression of ATF3 and axonal outgrowth are impaired after delayed nerve repair. *BMC Neurosci.* **9,** 88.

Sak, K., and Illes, P. (2005). Neuronal and glial cell lines as model systems for studying P2Y receptor pharmacology. *Neurochem. Int.* **47,** 401–412.

Santos, A. P., Suaid, C. A., Fazan, V. P., and Barreira, A. A. (2007). Microscopic anatomy of brachial plexus branches in Wistar rats. *Anat. Rec. (Hoboken)* **290,** 477–485.

Sarikcioglu, L., Yaba, A., Tanriover, G., Demirtop, A., Demir, N., and Ozkan, O. (2007). Effect of severe crush injury on axonal regeneration: A functional and ultrastructural study. *J. Reconstr. Microsurg.* **23,** 143–149.

Scanlin, H. L., Carroll, E. A., Jenkins, V. K., and Balkowiec, A. (2008). Endomorphin-2 is released from newborn rat primary sensory neurons in a frequency- and calcium-dependent manner. *Eur. J. Neurosci.* **27,** 2629–2642.

Shao, C., Liu, M., Wu, X., and Ding, F. (2007). Time-dependent expression of myostatin RNA transcript and protein in gastrocnemius muscle of mice after sciatic nerve resection. *Microsurgery* **27,** 487–493.

Shastry, P., Basu, A., and Rajadhyaksha, M. S. (2001). Neuroblastoma cell lines—A versatile *in vitro* model in neurobiology. *Int. J. Neurosci.* **108,** 109–126.

Sinis, N., Schaller, H. E., Becker, S. T., Lanaras, T., Schulte-Eversum, C., Muller, H. W., Vonthein, R., Rosner, H., and Haerle, M. (2006). Cross-chest median nerve transfer: A new model for the evaluation of nerve regeneration across a 40 mm gap in the rat. *J. Neurosci. Methods* **156,** 166–172.

Strasberg, J. E., Strasberg, S., Mackinnon, S. E., Watanabe, O., Hunter, D. A., and Tarasidis, G. (1999). Strain differences in peripheral-nerve regeneration in rats. *J. Reconstr. Microsurg.* **15,** 287–293.

Tos, P., Ronchi, G., Nicolino, S., Audisio, C., Raimondo, S., Fornaro, M., Battiston, B., Graziani, A., Perroteau, I., and Geuna, S. (2008). Employment of the mouse median nerve model for the experimental assessment of peripheral nerve regeneration. *J. Neurosci. Methods* **169,** 119–127.

Triolo, D., Dina, G., Lorenzetti, I., Malaguti, M., Morana, P., Del Carro, U., Comi, G., Messing, A., Quattrini, A., and Previtali, S. C. (2006). Loss of glial fibrillary acidic protein (GFAP) impairs Schwann cell proliferation and delays nerve regeneration after damage. *J. Cell Sci.* **119,** 3981–3993.

Trotter, J. (1993). The development of myelin-forming glia: Studies with primary cell cultures and immortalized cell lines. *Perspect. Dev. Neurobiol.* **1,** 149–154.

Tse, T. H., Chan, B. P., Chan, C. M., and Lam, J. (2007). Mathematical modeling of guided neurite extension in an engineered conduit with multiple concentration gradients of nerve growth factor (NGF). *Ann. Biomed. Eng.* **35,** 1561–1572.

Urso-Baiarda, F., and Grobbelaar, A. O. (2006). Practical nerve morphometry. *J. Neurosci. Methods* **156,** 333–341.

Varejão, A. S., Meek, M. F., Ferreira, A. J., Patrício, J. A., and Cabrita, A. M. (2001). Functional evaluation of peripheral nerve regeneration in the rat: Walking track analysis. *J. Neurosci. Methods* **108,** 1–9.

Varejão, A. S. P., Cabrita, A. M., Meek, M. F., Bulas-Cruz, J., Melo-Pinto, P., Raimondo, S., Geuna, S., and Giacobini-Robecchi, M. G. (2004). Functional and morphological assessment of a standardized rat sciatic nerve crush injury with a non-serrated clamp. *J. Neurotrauma* **21,** 1652–1670.

Vialle, R., Lozeron, P., Loureiro, M. C., and Tadié, M. (2008). Multiple lumbar roots neurotizations with the lower intercostal nerves. Preliminary clinical and electrophysiological results in a sheep model. *J. Surg. Res.* **149,** 199–205.

Vleggeert-Lankamp, C. L. (2007). The role of evaluation methods in the assessment of peripheral nerve regeneration through synthetic conduits: A systematic review. *J. Neurosurg.* **107,** 1168–1189.

Wang, H., Sorenson, E. J., Spinner, R. J., and Windebank, A. J. (2008a). Electrophysiologic findings and grip strength after nerve injuries in the rat forelimb. *Muscle Nerve* **38,** 1254–1265.

Wang, H., Spinner, R. J., Sorenson, E. J., Windebank, A. J. (2008b). Measurement of forelimb function by digital video motion analysis in rat nerve transection models. *J. Peripher. Nerv. Syst.* **13,** 92–102.

Whishaw, I. Q., Pellis, S. M., and Gorny, B. P. (1992). Skilled reaching in rats and humans: Evidence for parallel development or homology. *Behav. Brain Res.* **47,** 59–70.

Yu, P., Matloub, H. S., Sanger, J. R., and Narini, P. (2001). Gait analysis in rats with peripheral nerve injury. *Muscle Nerve* **24,** 231–239.

Zhang, Z., Johnson, E. O., Vekris, M. D., Zoubos, A. B., Bo, J., Beris, A. E., and Soucacos, P. N. (2006). Repair of the main nerve trunk of the upper limb with end-to-side neurorrhaphy: An experimental study in rabbits. *Microsurgery* **26,** 245–252.

Zochodne, D. W., Guo, G. F., Magnowski, B., and Bangash, M. (2007). Regenerative failure of diabetic nerves bridging transection injuries. *Diabetes Metab. Res. Rev.* **23,** 490–496.

METHODS AND PROTOCOLS IN PERIPHERAL NERVE REGENERATION EXPERIMENTAL RESEARCH: PART II—MORPHOLOGICAL TECHNIQUES

Stefania Raimondo,[1] Michele Fornaro,[1] Federica Di Scipio, Giulia Ronchi, Maria G. Giacobini-Robecchi, and Stefano Geuna

Department of Clinical and Biological Sciences, San Luigi Gonzaga School of Medicine, University of Turin, Turin 10043, Italy

[1]These authors contributed equally to this work

This paper critically overviews the main procedures used for carrying out morphological analysis of peripheral nerve fibers in light, confocal, and electron microscopy. In particular, this paper emphasizes the importance of osmium tetroxide post-fixation as a useful procedure to be adopted independently from the embedding medium. In order to facilitate the use of any described techniques, all protocols are presented in full details. The pros and cons for each method are critically addressed and practical indications on the different imaging approaches are reported.

Moreover, the basic rules of morpho-quantitative stereological analysis of nerve fibers are described addressing the important concepts of design-based sampling and the disector. Finally, a comparison of stereological analysis on

DOI: 10.1016/S0074-7742(09)87005-0

0074-7742/09 $35.00

myelinated nerve fibers between paraffin- and resin-embedded rat radial nerves is reported showing that different embedding procedures might influence the distribution of size parameters.

I. Introduction

Morphological analysis is the far most common method for the study of peripheral nerve regeneration (Castro *et al.*, 2008; Vleggeert-Lankamp, 2007). In fact, although in the clinical perspective functional assessment is the key element for the assessment of the nervous system, the investigation of nerve morphology can give us important information on various aspects of the regeneration processes (Hall, 2005; Geuna *et al.*, 2009, this volume) which relates with nerve function (Kanaya *et al.*, 1996).

The aim of this methodology-oriented paper is to describe the main morphological techniques for investigating the structure and ultrastructure of peripheral nerves with particular emphasis on the methods for the quantitative assessment of the morphological indicators of nerve function loss and recovery by design-based 2D stereology.

II. Light Microscopy

A. Fixation Procedures

Although different types of fixatives can be used for peripheral nerve histology, including Carnoy's fixative and Bouin's fluid fixation, we use 4% paraformaldehyde (Fluka, Buchs, Switzerland) in PBS (Phosphate Buffered Saline) for 2–4 h, followed by washing in 0.2% glycine in PBS. To obtain good histological quality, perfusion is not required and it is enough to fix the nerve specimens by immersion in the fixative solution. During the first few seconds of fixation, the nerve segment has to be maintained straight in a small fixative drop in order to facilitate specimen's orientation and cutting.

B. Embedding Procedures

The two most commonly embedding procedures for light microscopy are paraffin or cryo-embedding. The two techniques have both advantages and disadvantages and they can be alternatively chosen depending on the type of analysis that must be done.

Paraffin embedding provides a stronger support for the tissue and, in general, guaranties a better histology compared to cryo-embedding. On the other hand, the main limitation of paraffin is that antigenic sites are less exposed reducing the efficiency of an immunohistochemical analysis; moreover, the risk of tissue autofluorescence is higher. To overcome the latter limitation, prior to immunolabeling, sections can be processed with methods that facilitate antigen–antibody binding, including: (a) three microwaves cycles of 5 min in EDTA solution (100 mM); (b) incubation in NH_4Cl for 10 min.

With cryo-embedding, tissue quality is less maintained compared to paraffin because the sudden change from liquid to solid phase of the tissue fluids. To overcome this problem, it is recommended to carry out sample cryo-protection with subsequent passages in increasing solutions of sucrose before the freezing step. The main advantage of cryo-embedding is that antigenic sites are less masked thus facilitating immunohistochemistry.

1. *Paraffin Embedding Protocol*

Specimens undergo a dehydration procedure in ethanol from 50% to 100%. Dehydration is followed by a diaphanization step in xylol or a substitute such as Bioclear (Bio-Optica, Milano, Italy). Specimens are then maintained in liquid paraffin at 60 °C over night (step 1) and then passed to a second passage in liquid paraffin at 60 °C (step 2) before polymerization at room temperature.

Nerve sections are usually cut in a thickness range of 5–10 μm. Before staining, slides need to be deparaffinated and rehydrated with decreasing ethanol passages.

2. *Cryo-embedding Protocol*

The specimens are rehydrated with PBS and cryo-protected with three passages in increasing solutions of sucrose (7.5% for 1 h, 15% for 1h, 30% overnight) in 0.1 M PBS. Thereafter, specimens are maintained in a 1:1 solution of sucrose 30% and optimal cutting temperature medium (OCT) for 30 min and then embedded in 100% OCT. Specimens must then be store at −80 °C.

Nerve sections are usually cut in a thickness range of 10–15 μm and must then be stored at −20 °C. For staining, sections are taken out of freezer to room temperature and, as soon as they are acclimatized, they can be further processed.

C. Staining Procedures

1. *Hematoxylin and Eosin Staining*

Hematoxylin and eosin is the most commonly used stain for light microscopy observation in histology and histopathology. Hematoxylin labels nuclei in blue while eosin is detectable as a pink stain in cell cytoplasm. The slides are immersed in 0.1% hematoxylin (we use the product from Ciba, Basel, Switzerland) for

10 min, washed in tap water for 15 min, then immersed in 0.1% eosin (we use the product from Ciba) for 5 min and washed in distilled water. The sections are finally dehydrated in ethanol and mounted in DPX (we use the product from Fluka).

Although very popular, it must be emphasized that hematoxylin and eosin is not an adequate method for nerve tissue staining because the myelin sheaths are not labeled and they are thus difficult to be detected (Fig. 1A).

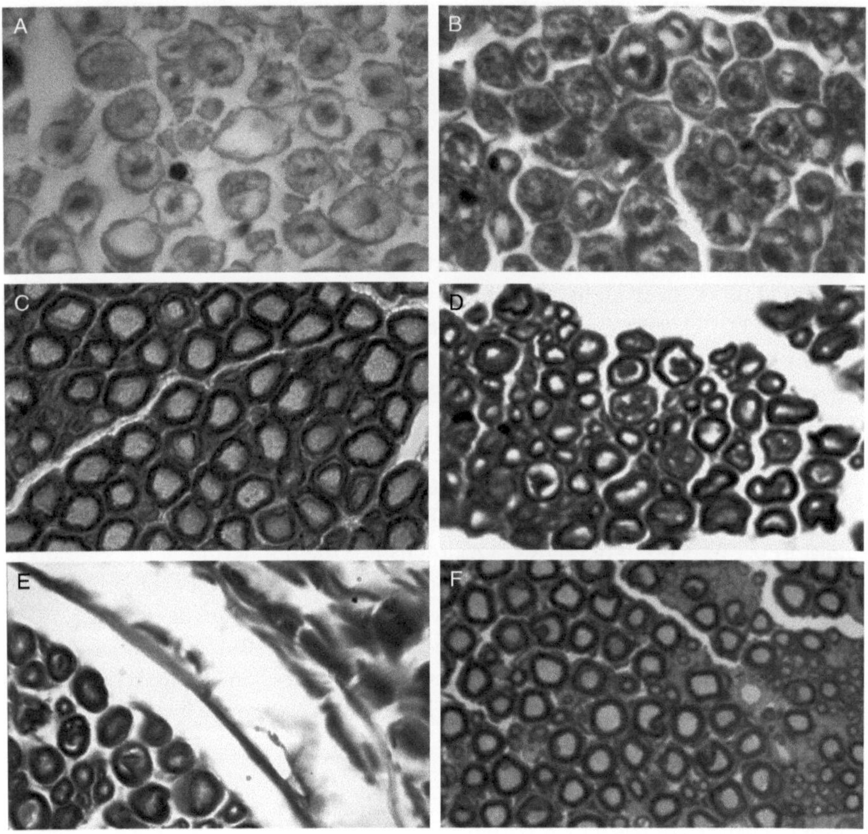

FIG. 1. High resolution light photomicrographs of cross sections of rat median nerve specimens processed by different methods. (A) Paraffin embedding and hematoxylin and eosin staining. (B) Paraffin embedding and Masson's trichrome staining. (C) Sections stained with osmium tetroxide before paraffin embedding. (D, E) Pre-embedding osmium tetroxide stained section counterstained with Masson's trichrome. (F) Resin embedding (with osmium tetroxide pre-embedding staining) and toludine blue staining. Magnification 600×. (See Color Insert.)

2. *Masson's Trichrome Staining*

The quality of the histology of nerve sections stained with Masson's trichrome is higher compared to hematoxylin and eosin because it highlights also the connective tissue. However, unless osmium tetroxide postfixation is carried out, myelin sheaths are not labeled with this method too (Fig. 1B).

For Masson's trichrome staining, in our laboratory we use a *Masson trichrome with aniline blue* kit (Bio-Optica): six drops of Weigert's iron hematoxylin (solution A) and six drops of Weigert's iron hematoxylin (solution B) are combined together and used to stain slides for 10 min. Without washing, the slides are then drained and incubated with ten drops of alcoholic picric acid solution for 4 min. After washing in distilled water, sections are stained with ten drops of Ponceau acid fuchsin for 4 min and washed again in distilled water. Further on, ten drops of phosphomolybdic acid solution are added to the section for 10 min. Without washing, the slides are drained and 10 drops of aniline blue are added to the section for 5 min. Finally, after washing in distilled water, dehydrating rapidly in ethanol and clearing in xylol/Bioclear (Bio-Optica), the slides are mounted in DPX (Fluka).

3. *Pre-embedding Myelin Sheath Stain with Osmium Tetroxide before Paraffin Embedding*

The rationale for this procedure is to introduce osmium tetroxide's immersion prior to the embedding procedure also in case of paraffin embedding. This technique allows a better fixation of the myelin resulting in a better quality of the imaging. In fact, due to its action as a lipid fixative, post-fixation in osmium prevents myelin sheath swelling, which usually occurs during paraffin embedding, and provides the typical dark and sharp myelin stain, which greatly facilitates the identification of nerve fibers (Fig. 1C).

After fixation in 4% paraformaldehyde and washing in 0.2% glycine in PBS for few minutes, specimens are immersed for 2 h in 2% osmium tetroxide (Sigma, St. Louis, MO) in Soerensen phosphate buffer (see Section IV.A). The nerves are then dehydrated in numerous passages in ethanol as described in the procedure for resin embedding (see Section IV.B) in order to completely remove excess of osmium from tissue. The specimens are then embedded in paraffin, cut and counter-stained with either hematoxylin and eosin or Masson's trichrome. Whereas myelin sheaths can be sharply detected right after applying the osmium post-fixation, (Fig. 1C), a very good histological quality can be obtained by Masson's trichrome counterstaining, which in particular allows a clear imaging of the nerve's connective structures (Fig. 1D, E).

4. *Toluidine Blue Staining of Semithin Sections from Resin-embedded Blocks*

The best quality for nerve analysis in light microscopy is obtainable after resin embedding (see Section IV.B) and toluidine blue staining (Fig. 1F). With this procedure, most of the myelinated axons can be clearly identified and myelin sheaths are sharply delimited due to lipid staining of osmium tetroxide post-fixation.

Semi-thin sections of nerve samples are usually cut in a thickness range of 1–3 μm with an ultramicrotome (we use a Ultracut UCT, Leica Microsystems, Wetzlar, Germany) and stained with 1% Toluidine blue (Fluka) in 1% borax on a 80 °C hot plate for 30–45 s.

5. *Polychrome Staining of Semithin Sections from Resin-embedded Blocks*

This method serves the same purpose as the Toluidine blue procedure for staining semithin sections, but provides with red and blue colors (Hoffman *et al.*, 1983). After staining with 1% Toluidine blue (Fluka) in 1% borax on a 80 °C hot plate for 30–45 s, sections are incubated with a 1:1 solution of 0.1% basic fuchsin and in 1% borax on a 80 °C hot plate for few seconds.

III. Immunohistochemistry and Confocal Microscopy

A. Fixation Procedures

The most used fixation solution for immunohistochemistry and confocal microscopy is 4% paraformaldehyde as described for light microscopy (Section II.A). However, it is important to emphasize that since sample fixation can compromise immunolabeling by covering the antigenic sites, nerve segments intended for immunohistochemistry should be kept in fixative for less than 2 h depending on specimen's size.

B. Embedding Procedures

For immunohistochemistry, tissue samples can be embedded in paraffin or ice as described above (Section II.B). Yet, the strategies for unmasking antigen sites can be applied as recommended in Section II.B. For nerve immunohistochemistry, we usually prefer embedding in paraffin. Cryo-embedding procedure must be used on GFP-autofluorescent samples since paraffin embedding, because of the ethanol passages, would delete GFP-autofluorescence.

1. *"Etching" Procedure for Immunohistochemistry after Pre-embedding Osmium Tetroxide Staining*

For immunohistochemistry and confocal laser microscopy on sections obtained from nerve specimens post-fixed in osmium tetroxide, the slides must be etched, after deparaffination, by incubating them in 3% H_2O_2 (Sigma) for 10 min. This technique allows to use the same sample for both stereological and immunohistochemical analysis (Di Scipio *et al.*, 2008).

C. Antibodies and Immunostaining Procedures

Both axon and glia can be detected by immunohistochemistry using specific antibodies. In particular, the most used antibodies as axon markers are those against neurofilament (NF) subunits. In our laboratory, we have used both anti-NF 200 kDa (monoclonal, mouse, Sigma) and anti-PAN-NF (polyclonal, rabbit, Biomol). A-PAN-NF reacts with all three NF proteins (68 kDa, 150 kDa, and 200 kDa) and therefore it allows staining almost all myelinated nerve fibers. Figure 2 shows sciatic nerve of monkey (Fig. 2A), rat (Fig. 2B), mouse (Fig. 2C), stained with NF-200 kDa. For mouse nerve tissue, a better result has been obtained using a-PAN-NF (Fig. 2D).

Another useful axonal marker is anti-peripherin (polyclonal, rabbit, Chemicon, Billerica, MA, USA) that predominantly labels unmyelinated axons. Double labeling with anti-peripherin and anti-NF 200 kDa (Fig. 3A, B), which predominantly labels myelinated axons, permits to distinguish between the two types of fibers (Fornaro *et al.*, 2008). Other axonal markers that we commonly use are the anti-PGP 9.5 (polyclonal, rabbit, Biogenesis), that is found specifically in the PNS (Fig. 2E), and anti-GAP43 (growth associated protein 43)(polyclonal, goat, Santa Cruz Biotechnologies, USA) that is expressed at high levels during development and axonal regeneration. Finally, a marker selectively specific for motor axons is the anti-ChAT (choline acetyltransferase) (polyclonal, goat, Chemicon) (Fig. 2G).

As far as Schwann cell recognition is concerned, they can be detected by immunohistochemistry using specific glial markers, such as GFAP and S100. Anti-GFAP antibody (in our lab we use both monoclonal, mouse, Dako, Denmark and polyclonal, rabbit, Sigma) is the commonly used marker for immature and un-myelinating Schwann cells. Anti-S100 antibody (polyclonal, rabbit, Sigma or Dako) labels the cytoplasm and nucleus of Schwann's cells (Fig. 2F) and has been shown to be a very good marker of human peripheral nerves (Gonzalez-Martinez *et al.*, 2003).

Glial markers can be associated with neuronal markers in double immunostaining providing useful information on the relationship between axons and glial cells (Fig. 3C).

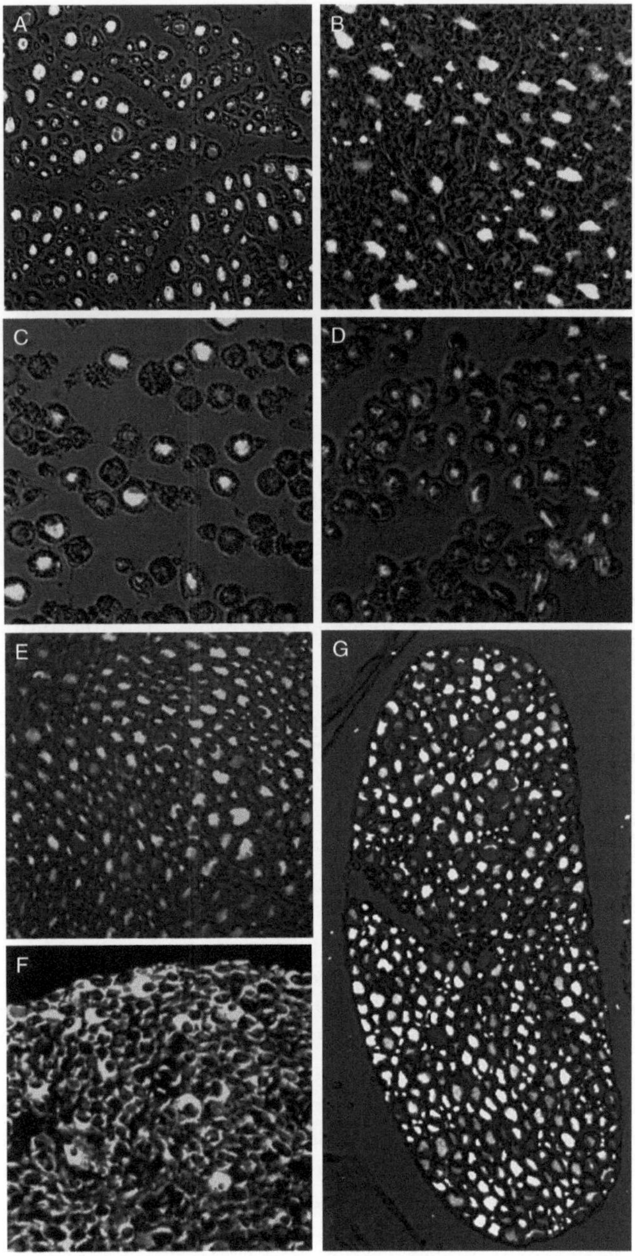

FIG. 2. Confocal images of different animal species normal nerves. (A–C) Immuno-staining with anti-NF 200 kDa of monkey (A), rat (B), and mouse (C) sciatic nerve. (D) Immuno-staining with

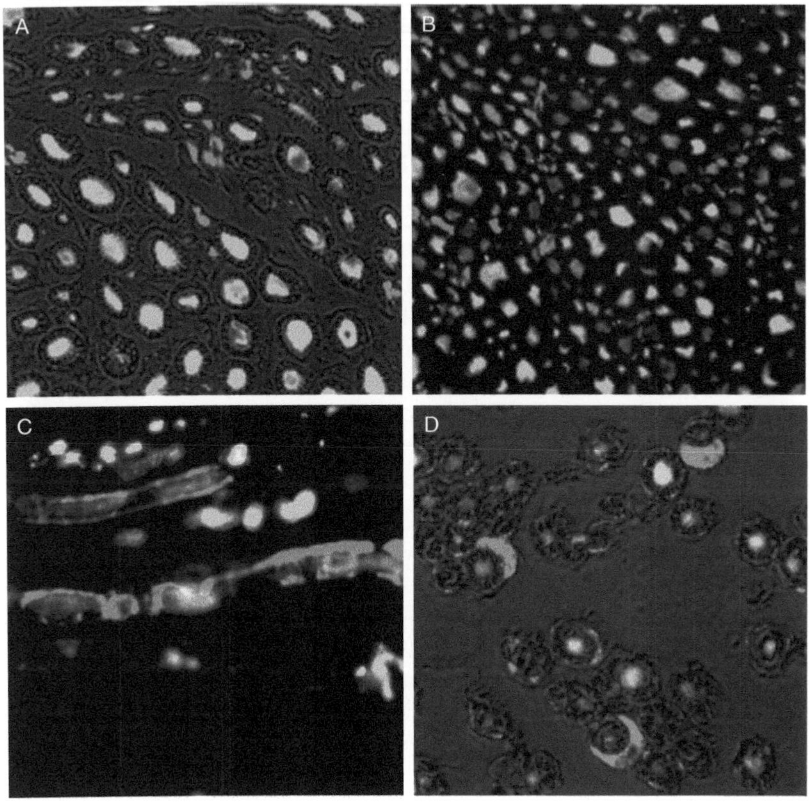

FIG. 3. Confocal images of monkey normal nerve (A) and rat normal sciatic nerve (B) double labeled with anti-NF 200 kDa and anti-peripherin. (C). Regenerating rat fibers double stained with anti-NF 200 kDa and S100. (D). Mouse median nerve double stained with anti-NF 200 kDa and erbB2. Magnifications: A–C = 600×; D = 1000×. (See Color Insert.)

Beside their use as markers of different axons and glia, immunohistochemical analysis is also a useful tool to investigate cell function and molecular activity, for example the cellular signaling pathways. Particularly interesting for nerve regeneration is the NRG/erbB pathway system (Audisio *et al.*, 2008; Casha *et al.*, 2008). In several experimental studies we specifically focused on erbB2 expression in Schwann cells, testing different antibodies. The best results were obtained with the polyclonal antibody from Genetex (Fig. 3D).

anti-PAN NF mouse sciatic nerve. (E,F) PGP9.5 (E) and S100 (F) immunolabeling of a rat sciatic nerve. (G) Spinal cord ventral root immunostained with anti-Choline-Acetyltransferase (ChAT). Magnifications: A = 400×; B = 600×; C,D = 900×; E,F = 500×; G = 300×. (See Color Insert.)

1. *Immunofluorescence*

For immunofluorescence, the sections are rinsed in PBS, blocked with normal serum (1%), (the use of a normal serum made in the same species of the secondary antibody is recommended), for 1 h and then incubated overnight with the primary antibody. For double labeling, different primary antibodies can be used contemporarily as long as they are made in different animal species. If both antibodies are made in the same species, an "unconjugated affinity Fab fragment IgG" protocol (Jackson Immunoresearch Laboratories, Baltimore, MD, USA) can be used (Fornaro *et al.*, 2003). After primary antibody(ies) incubation, sections are washed three times in PBS and incubated for 1 h in a solution containing the secondary antibody(ies) conjugated with a fluorofore and selected in order to recognize the species of primary antibodies. After three washes in PBS, sections are finally mounted with a Dako fluorescent mounting medium and stored at 4 °C before being analyzed.

2. *Immunoperoxidase*

For immunoperoxidase staining the sections are rinsed in PBS and the endogenous peroxidase is inhibited with an incubation of 10 minutes in a solution of methanol (50%) and H_2O_2 (1%) in PBS. Sections are then blocked with normal serum (1%), made in the same species of the secondary antibody, for 1 h and then incubated overnight with a primary antibody. The sections are washed three times in PBS and incubated for 1 h in a solution containing a biotinylated secondary antibody against the same species of the primary antibody. After three washes in PBS samples are then processed with peroxidase-conjugated Vectastain ABC kit (Vector, Burlingame, CA, USA) and revealed with diaminobenzidine (Sigma). For double immuno-staining, the two immunolabeling must be carried out separately and revealed using different enzyme-systems, such as peroxidase/phosfatase. The peroxidase protocol can also be used as a pre-embedding stain technique for electron microscopy immunolabeling.

IV. Electron Microscopy

A. Fixation Procedures

We fix nerve samples in a solution of 2.5% purified glutaraldehyde (Histo-line Laboratories s.r.l., Milano, Italy) and 0.5% saccarose (Merck, Darmstadt, Germany) in 0.1 M Sörensen phosphate buffer, pH 7.4, for 6–8 h, then wash and store them in 0.1 M Sörensen phosphate buffer added with 1.5% saccarose at 4–6 °C prior to embedding (in our experience the nerves can be stored for several days or even weeks in buffer at 4–6 °C with no problem). During the first few seconds of fixation, the nerve segment has to be maintained straight in a small fixative drop in order to facilitate specimen's orientation and cutting.

Sörensen phosphate buffer is made with 56 g di-potassium hydrogen phosphate 3-hydrate ($K_2HPO_4 \cdot 3H_2O$) (Fluka) and 10.6 g sodium di-hydrogen phosphate 1-hydrate ($NaH_2PO_4 \cdot H_2O$) (Merck) in 1 l of doubly-distilled water.

Just before the embedding, nerves are washed for few minutes in the storage solution and then immersed for 2 h in 2% osmium tetroxide (Sigma) in the same buffer solution.

B. Embedding Procedures

The specimens are carefully dehydrated in passages in ethanol from 30% to 100% with at least five passages of 5 min each. After two passages of 7 min each in propylene oxide (Sigma) and 2 h in a 1:1 mixture of propylene oxide and Glauerts' mixture of resins, specimens are embedded in Glauerts' mixture of resins, which is made of equal parts of Araldite M and the Araldite Härter, HY 964 (Merck). At the resin mixture, 2% of accelerator 964, DY 064 is added (Merck). For the final step a plasticizer (0.5% of dibutylphthalate) is added to the resin in order to promote the polymerization of the embedding mixture.

C. Cutting and Staining Procedures

In our laboratory, thin sections of nerve samples are usually cut in a thickness range of 50–70 nm with an ultramicrotome (we use a Ultracut UCT, Leica Microsystems). Sections are collected and placed on grids previously coated with pioloform film. For transmission electron microscope, grids are usually stained with uranyl acetate (sature solution) for 15 min and lead citrate for 7 min, washed and dried. As alternative to uranyl acetate it's possible to use Platinum blue (Inaga *et al.*, 2007).

In the nerve, transmission electron microscopy analysis allows to investigate various ultrastructural features, including the organization of unmyelinated (Fig. 4A) and myelinated (Fig. 4B) axons. Figure 4C shows a typical artifact of myelin sheaths, namely small swelling areas (arrow), that is commonly detected in peripheral nerves and that can be misinterpreted as a pathological sign.

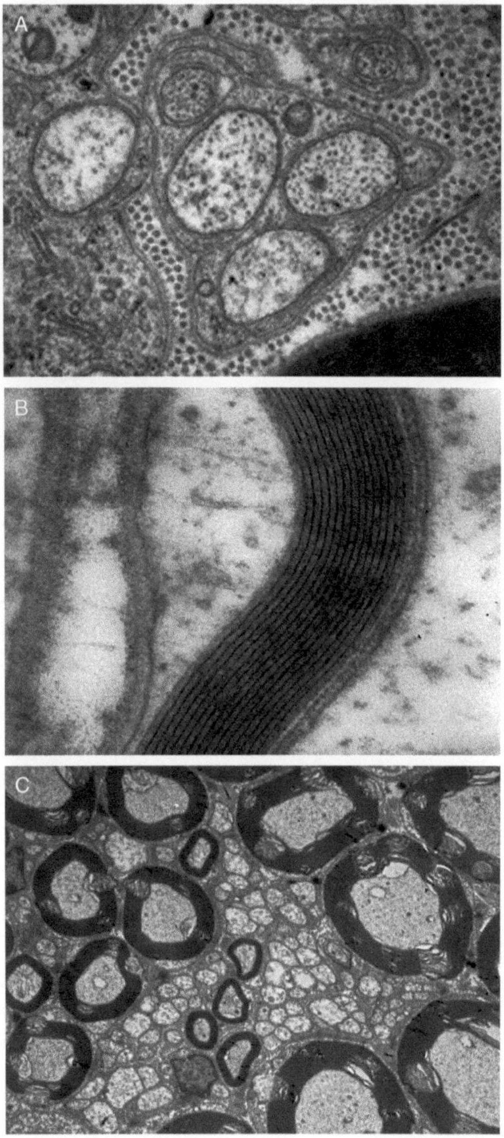

FIG. 4. Electron micrographs of nerve fibers. (A) Unmyelinated nerve fibers. (B) High magnification of a myelinated sheath. (C) Myelin sheath swelling, a typical artifact in large myelinated nerve fibers. Magnifications: A = 80,000×; B = 150,000×; C = 5000×.

V. Histomorphometry (Stereology)

Quantitative estimation of nerve fiber morphology (especially myelinated ones) is, together with functional assessment, a key investigation tool in nerve regeneration research (Geuna *et al.*, 2004; Kanaya *et al.*, 1996; Vleggeert-Lankamp, 2007). The most important geometrical parameters that can be used for the assessment nerve fibers are: (1) Number of fibers, (2) Density of fibers, (3) Diameter of fibers and axons (Maximum, Minimum, Circle-equivalent), (4) Cross-sectional area of fibers and axons, (5) Perimeter of fibers and axons, (6) Myelin thickness, (7) Myelin-thickness/axon-diameter ratio, and (8) Fiber-diameter/axon-diameter ratio or axon-diameter/fiber-diameter (*g*-ratio).

Although number and density of nerve fibers are the most used indicators of nerve regeneration, both parameters need to be carefully interpreted. In fact, a high number of regenerated nerve fibers can not only indicate a good regeneration, but also aberrant sprouting (in this case the contemporary assessment of fiber size can provide additional information). Data on fiber density are even more difficult to be interpreted since a high fiber density not always reflects good nerve regeneration, but can also reflect the presence of small regenerated axons. On the other hand, a low fiber density can reflect either larger axons (that is a good predictor) or also the presence of oedema in the regenerated nerve (that is a bad predictor). Again in this case, the contemporary assessment of fiber size can facilitate interpretation of density data.

Fiber and axon diameter are the classical parameter for nerve type identification since they have proven to be the main determinant of conduction velocity (Hoffman, 1995). Various types of diameters of nerve fibers and/or axons can be used to assess their size (Geuna *et al.*, 2001): the maximum diameter (which is strongly biased by obliquity of cross-sectional fiber profiles), the minimum diameter (which is strongly biased by fiber shrinkage), and the circle-equivalent diameter (which represents the diameter of a circle the area of which corresponds to the cross-sectional area of the fiber and/or axon). Cross-sectional area is another commonly used size estimation parameter for myelinated nerve fibers that, however, is not easy to be interpreted by readers since the diameter is the classical parameter used to classify nerve fibers (Geuna *et al.*, 2001; Hoffman, 1995).

Starting from rough data on the diameter of the fiber (D) and the axon (d), several other size parameters can be calculated by simple mathematical formulas: myelin thickness $[(D-d)/2]$, the myelin-thickness/axon-diameter ratio $[(D-d)/2d]$, the fiber-diameter/axon-diameter ratio (D/d), and its opposite the axon-diameter/fiber-diameter ratio, also called *g-ratio* (d/D).

These additional parameters are particularly important for the investigation of nerve development (Fraher *et al.*, 1990) as well as nerve regeneration since they better correlate with the functional outcome of nerve recovery (Kanaya *et al.*, 1996).

Since quantification of size parameters is not always easy from a technical viewpoint, the selection of the indicators to be used in a given nerve regeneration study should be also done on the basis of the quality of the histological material and the equipment available.

The quantitative assessment of tissue and organs on histological sections has been the subject of heated scientific debate over the last years. In particular, the emergence of an new approach to cope with bias in morphometrical analysis, namely *stereology*, has represented a significant advancement in neuromorphology (for literature review see: Baryshnikova *et al.*, 2006; Benes and Lange, 2001; Canan *et al.*, 2008; Coggeshall, 1992; Geuna, 2000, 2005; Guillery, 2002; Mayhew and Gundersen, 1996; Reed and Howard, 1998; Schmitz and Hof, 2005; von Bartheld, 2002; West, 1999).

Independently, of the parameters under investigation, there are at least five different sources of bias in the quantitative assessment of nerve fibers (Geuna *et al.*, 2001). First, the strain, gender and age of experimental animals (strain-related, gender-related, age-related foundations of bias). Second, the point (level) along the nerve axis where sections are cut (section-related foundations of bias). Third, the location of the sampling fields within the nerve cross-section profile (location-related foundations of bias). Fourth, the inclusion-exclusion rules for sampling fiber profiles within the sampling fields (morphology-related foundations of bias). Fifth, the method for measuring the selected size parameters (measurement-related foundations bias). The first two potential sources of bias are related to the study design. The other three sources are related to the sampling procedure and the method used for quantitative nerve fiber assessment and will thus be treated in this section focusing, in particular, on the basic principles and methods for design-based sampling and for nerve fiber stereology.

The 'golden rule' of sampling for any tissue and organ is the equal opportunity rule (Cruz-Orive and Weibel, 1981) which means that all objects must have the same opportunity of being included in the sample. The sampling paradigm that allows meeting the equal opportunity rule is called design-based sampling (Geuna, 2000). The term design refers to a system of sampling rules designed such that all objects in the sampling space have the same probability of being sampled. Design-based sampling can also be referred to as random sampling since its goal is to reach randomness.

Simple random sampling is the most basic random-based design and provides that all possible combinations of n sampling units have the same probability of being selected from among the total N sampling units in the population. However, this sampling approach requires a high amount of sampling to obtain a sufficient estimate precision and is impossible in histology as it would require the specimen to be glued together and resectioned after each section was selected (Geuna, 2000). Other random sampling designs include systematic, multistage, and stratified random sampling (Cochran, 1977). The most used approach in neuromorphology is systematic random sampling that is based on the systematic selection of every *n*th unit of the population from one randomly selected starting unit (where *n* is the distance between units that is decided in relation to the amount of sampling required). In case of nerve fiber stereology, the units are the single sampling boxes on a given nerve cross section, and, after the starting box is selected by chance, the following boxes are selected by systematically jumping at a given distance from the former box (Fig. 5A).

Once the sampling fields are selected within the nerve cross section by systematic random sampling, it is necessary to define a set of inclusion/exclusion rules for clearly determining which nerve fiber falls inside the sampling field, and which other does not (Geuna *et al.*, 2004). The bias originating from an unclear determination of inclusion/exclusion rules depends on the 'edge effect' (Gundersen, 1977) that is due to variability in the size and morphology of fiber profiles which may cause significant differences in the probability of each profile being intersected by the frame edges: larger fibers will have a higher probability of intersecting the frame edges and thus of partially falling into more than one sampling field than smaller fibers. If all edging fiber profiles are excluded, quantitative estimations will be biased towards a systematic underestimation of number and size of fibers, while if all edging fiber profiles are included, quantitative estimations will be biased toward a systematic overestimation. In must be noted that many papers reporting data on nerve histomorphometry do not provide any information on 'what happens' when a fiber profile intersects the histologic field edges.

To cope with the *edge effect*, the equal opportunity rule should be respected by adopting a set of inclusion/exclusion rules that assures that any fiber profile has the same chance of being sampled, irrespective of its morphological features. In other words, all fiber profiles must have the possibility of being selected in one histologic field only, irrespective of the number of edge intersections (Geuna *et al.*, 2004). For nerve fiber quantitative assessment, two stereological methods have been the most employed so far for coping with the edge effect: the unbiased sampling frame (Acar *et al.*, 2008; Canan *et al.*, 2008; Gundersen, 1978; Keskin *et al.*, 2004; Larsen, 1998) and the 2D disector (Gundersen, 1986; Geuna *et al.*, 2000, 2001). We have specific experience with the latter method that represents an adaptation of the disector principle (that is used for sampling object in 3D) and that it is basically an associated-point method, i.e., a method based on the identification of an "univocal" reference point in each particle (the "top"): the particle is then included in the sampling frame only if this point falls inside the frame independently from what happens to the rest of the particle (Geuna, 2000, 2005). In the 2D disector, the "top" is identified as the "higher" edge of a fiber profile and thus nerve fibers are considered inside the frame, and thus counted only when their "top" falls inside the sampling field borders (Fig. 5B). Whereas, the first description of the 2D-disector (Geuna *et al.*, 2000) was based on the

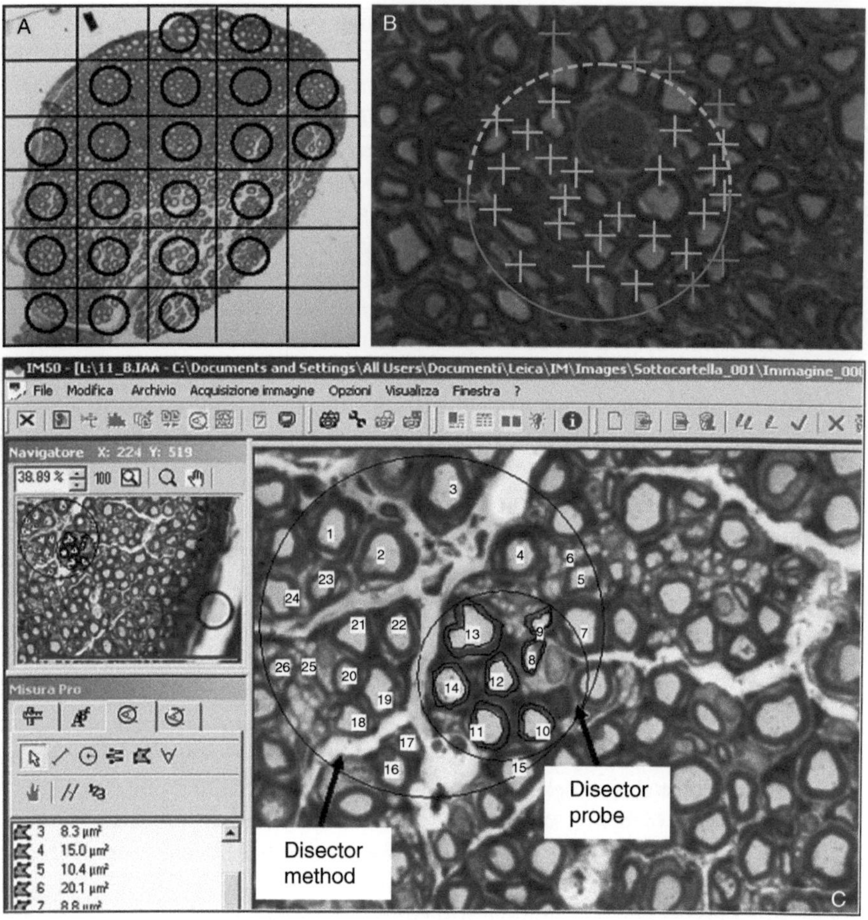

FIG. 5. (A) Systematic random sampling adapted for locating the sampling fields (the circles) all over the nerve profile. (B) Application of the 2D disector method for selecting fibers inside the circular sampling frame. Only fiber tops which fall inside the circle are selected. In case a fiber top falls exactly on the circle border an inclusion (green) and exclusion (red) half circle is preliminarily determined. (C) The selected fibers can be counted to estimate the total fiber number in that nerve (disector method). In the other case the disector is used as a probe to produce an unbiased sample of fibers respecting the equal opportunity rule (disector probe). Usually, the disector used as a probe is smaller than the disector for counting in order to avoid excessive workload in fiber measurement.

employment of a squared frame, we currently prefer to use a circular frame (Fig. 5) in order to reduce the probability of nerve fibers (that have a circular shape) to hit the frame border. To make the decision also when a fiber's top exactly falls on the line, an inclusion hemi circle (the higher dashed green one in the example of Fig. 5B) and an exclusion hemi circle (the lower red solid one) can be identified and the fiber top is excluded from counting when it touches the lower hemi circle and vice versa.

From a practical viewpoint, in our laboratory we use a DM4000B microscope equipped with a DFC320 digital camera and an IM50 image manager system (Leica Microsystems) (Fig. 5C). This system reproduces microscopic images (for quantitative morphology of myelinated nerve fibers, images should be captured through a 100× oil-immersion objective) on the computer monitor at a magnification adjusted by a digital zoom. A final magnification higher than 6000× enables accurate identification and morphometry analysis of myelinated nerve fibers. Figure 5C also shows the difference between the *disector (counting) method*, which allows obtaining an estimation of the number of objects, and the *disector probe* that allows selecting a random sample of objects for further carrying out measurements on them.

As most other authors, we carry out measurements just on one randomly selected section from each nerve. However, the use of a single section deserves mention since the quantitative parameters of nerve fibers can vary significantly depending on the nerve level and on the distance from the point of lesion (Santos *et al.*, 2007). Two methodological strategies can be adopted to avoid source of variability. The first and more laborious one is based on the calculation of mean values from data obtained on multiple sections taken at different levels of the nerve. A simpler alternative, is based on the use of a single section provided that a cutting procedure that assures that the section used for the quantitative assessment is taken at the same location along all nerves is adopted (e.g., 5 mm distal to the site of lesion site in a nerve regeneration study). If adequate sampling techniques are employed (e.g., the 2D disector), this approach provides unbiased data (Geuna *et al.*, 2000; Larsen, 1998).

Once the section is randomly selected, the total cross-sectional area of the nerve is measured and the sampling fields are then randomly selected using a simple procedure that we have described in details previously (Geuna *et al.*, 2000). Mean fiber density is then calculated by dividing the total number of nerve fibers within the sampling field by its area (N/mm^2). Total fibers number (N) is finally estimated by multiplying the mean fiber density by the total cross-sectional area of the whole nerve cross section.

Two-dimensional disector probes are then also used for the unbiased selection of a representative sample of myelinated nerve fibers in each of which both fiber and axon area are measured (Fig. 5C). From these two data, circle-fitting diameter of fiber (D) and axon (d) are calculated as well as myelin thickness $[(D-d)/2]$, myelin thickness/axon diameter ratio $[(D-d)/2d]$, and axon/fiber diameter ratio (d/D), the g-ratio (D/d).

Once a data set is obtained, the precision of the estimates is evaluated by calculating the coefficient of error (CE). Regarding quantitative estimates of fiber number, the CE*(n)* is obtained as follows: (Schmitz, 1998)

$$\mathrm{CE}(n)\frac{1}{\sqrt{\Sigma Q'}}$$

where $\Sigma Q'$ is the number of counted fibers in all disectors.

For size estimates, the coefficient of error is estimated as: (Geuna *et al.*, 2001)

$$\mathrm{CE}(z)\frac{\mathrm{SEM}}{\mathrm{Mean}}$$

where SEM = standard error of the mean.

The sampling scheme is usually designed in order to keep the CE below 0.10, which assures enough accuracy for neuromorphological studies (Pakkenberg and Gundersen, 1997).

Finally, numerical data are statistically analyzed by ANOVA. Statistical significance is established as $P < 0.05$. We perform statistical tests using the software "Statistica per discipline bio-mediche" (McGraw-Hill, Milano, Italia).

A. Comparison of Quantitative Estimates Between Resin- and Paraffin-Embedded Nerve Specimens

Although the gold standard for tissue processing is represented by toluidine blue staining of resin embedded semi-thin sections (Fig. 1F), we have recently described a simple protocol for pre-embedding staining of myelin sheath with osmium tetroxide on paraffin embedded sections (Fig. 1C) (Di Scipio *et al.*, 2008). Pre-embedding osmium tetroxide fixation avoids myelin sheath swelling (Fig. 1C) and provides a sharp myelin staining that makes it possible the clear recognition of most myelinated nerve fibers and the measurement of their main quantitative parameters (axon and fiber diameter and myelin thickness) both on resin and paraffin-embedded specimens. This method represents a valid alternative to the conventional resin embedding-based protocol in comparison to which is much cheaper and can be carried out in any histological laboratory.

Since it has been shown that variable tissue shrinkage can occur due to embedding procedures (Onishi *et al.*, 1974; Ward *et al.*, 2008), a question arises regarding the possibility to directly compare quantitative data on myelinated nerve fibers obtained on nerve samples embedded in paraffin vs resin.

Therefore, we have carried out a comparative stereological analysis on paraffin- and resin-embedded rat radial nerves in order to verify whether the different embedding procedures might influence the quantitative estimates of size parameters of the myelinated axons. Four adult female Wistar rats, weighing approximately 250 g, were used for the present study. Adequate measures were taken to minimize pain and discomfort taking into account human endpoints for animal suffering and distress. All procedures were performed with the approval of the Local Ethical Committee and in accordance with the European Communities Council Directive of 24 November 1986 (86/609/EEC). Under deep anesthesia by ketamine (40 mg/250 g) and cloropromazine (3.75 mg/250 g) and clean conditions, the left radial nerve was exposed at the middle third of the brachium and a 10-mm long segment withdrawn under operative microscope. Immediately after withdrawal, the nerve samples were divided into two segments of equally length. In order to facilitate the correct orientation for cutting, proximal specimen was marked by a 9-0 stitch on the proximal stump while in the distal specimen a 9-0 stitch was used to mark its distal stump.

The proximal specimens were fixed in 2.5% purified glutaraldehyde (Histoline Laboratories s.r.l.) and 0.5% saccarose in 0.1 M Sorensen phosphate buffer, post-fixed in 2% osmium tetroxide and processed for resin embedding; the distal specimens were fixed in 4% paraformaldehyde (Fluka) in PBS (Phosphate buffered saline), post-fixed in 2% osmium tetroxide and processed for paraffin embedding (see for detailed protocols: Di Scipio *et al.*, 2008). From the resin-embedded proximal specimen, a series of ten 2-μm-thick sections was cut starting from its distal stump, while from the paraffin-embedded specimen a series of ten 8-μm-thick sections was cut starting from its proximal stump. In this way, all sections in each specimen were taken within a 100-μm-long radial nerve segment. Resin sections were finally stained by toluidine blue for 2 min while no counter-stain was used for paraffin sections since myelin fibers are easily recognizable (Fig. 1C).

Statistical analysis was performed using the one-way analysis of variance (ANOVA) test for comparing each parameter's mean values and the Wilcoxon Rank Sum nonparametric test for comparison of fiber size distribution. Statistical significance was established as $P < 0.05$.

Results of the statistical comparison showed that when mean values are considered, no significant difference ($P > 0.05$) was detected between resin-embedded and paraffin-embedded myelinated nerves (Table I). On the other hand, when the Wilcoxon rank sum nonparametric statistical analysis was applied to fiber diameter and myelin thickness distribution histograms (Fig. 6), significant ($P < 0.05$) differences between the two types of tissue processing were observed.

VI. Conclusions

The experience of many years tell us that there is no single morphological technique which is intrinsically superior to the other and should thus be indicated as the gold standard for peripheral nerve regeneration research

TABLE I

COMPARISON OF STEREOLOGICAL ESTIMATES OF MYELINATED NERVE FIBERS IN RESIN-EMBEDDED AND PARAFFIN-EMBEDDED NORMAL RAT RADIAL NERVES (VALUES ARE MEANS ± S.D)

	Resin-embedded	Paraffin-embedded
Total number of myelinated fibers	4041 ± 327	3898 ± 851
Density of fibers (#/mm^2)	11,541 ± 857	11,718 ± 3.287
Fibers diameter (μm)	7.80 ± 1.74	7.71 ± 0.93
Axons diameter (μm)	5.61 ± 0.73	5.52 ± 0.37
Myelin thickness (μm)	1.09 ± 0.51	1.10 ± 0.33
M/d	0.20 ± 0.06	0.21 ± 0.05
D/d	1.40 ± 0.12	1.42 ± 0.11
g-ratio	0.72 ± 0.06	0.71 ± 0.05

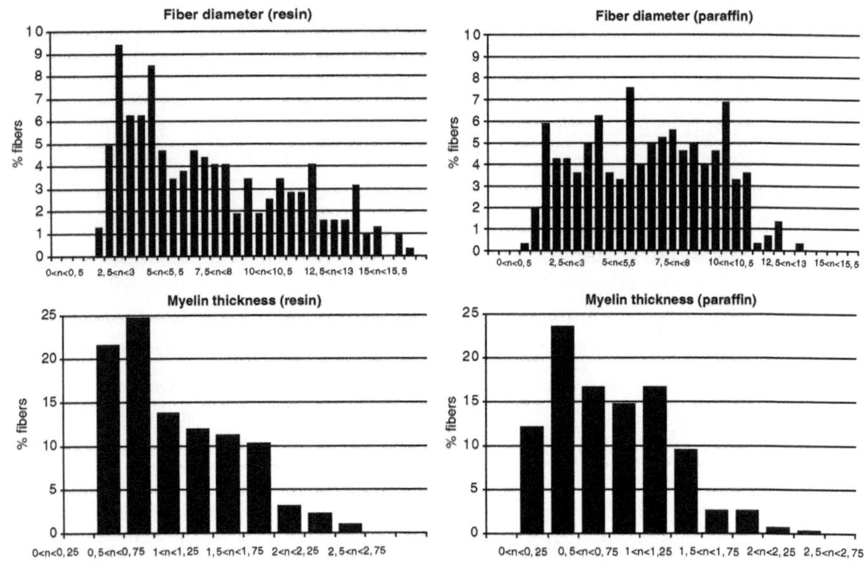

FIG. 6. Size distribution of fiber diameter and myelin thickness measurements comparing resin- and paraffin-embedded nerves. While mean values are not significantly different (Table I), nonparametric statistical analysis of size distribution ranks shows significant ($P < 0.05$) differences between the two types of embedding mediums.

(Vleggeert-Lankamp, 2007). On the other hand, the best methodology for histological assessment for any given study should be carefully selected on the basis of the study goals as well as of the available resources. Specifically, light microscopy should always be carried out and the simple introduction of osmium tetroxide pre-embedding staining can make light microscopy observation much more valuable independently from the embedding medium used (Di Scipio *et al.*, 2008). Immunohistochemistry is a powerful technique which has a wide range of applications in peripheral nerve regeneration research both as a mean to mark and thus recognize the various elements of the nerve and as a tool for exploring the biological mechanisms at a molecular level. Electron microscopy too is a powerful tool and, although its employment should be limited to selected research goals because of its high costs, this technique is particularly useful for investigating the early stages of nerve damage and regeneration at a subcellular level.

Histomorphometry is often the final step of morphological investigation since, to obtain an answer about any scientific question, a statistical comparison of numerical data between the different experimental groups must be sought. Although apparently simple, quantitative morphology is tricky and should be carried out carefully in order to avoid that bias creeps into the estimates (Geuna, 2000). Nowadays, stereology provides us a number of reliable methods for the quantitative assessment of peripheral nerve predictors during damage and regeneration. Original data reported in this paper suggest also that the direct comparison of quantitative results obtained from nerve specimens embedded in different mediums shall be evaluated carefully because of the influence of the embedding procedure on nerve fiber size distribution.

A final mention deserves the debate about the correlation between morphological and functional predictors of nerve regeneration. While it has been shown that some morphological parameters relate significantly with the functional predictor of nerve recovery (Kanaya *et al.*, 1996; Prodanov and Feirabend, 2007), in most cases morpho-functional correlation is poor and thus the value of the combined use of both types of methodological approaches in nerve repair and regeneration research should be emphasized (Castro *et al.*, 2008; Geuna and Varejao, 2008; Varejao *et al.*, 2004).

Acknowledgments

This work was supported by grants from the Regione Piemonte (Progetto Ricerca Sanitaria Finalizzata) and the Italian Ministry of University and Research. Stefania Raimondo is recipient of a PostDoc grant partially supported by the Regione Piemonte (Azione Contenimento del Brain Drain).

References

Acar, M., Karacalar, A., Ayyildiz, M., Unal, B., Canan, S., Agar, E., and Kaplan, S. (2008). The effect of autogenous vein grafts on nerve repair with size discrepancy in rats: An electrophysiological and stereological analysis. *Brain Res.* **1198,** 171–181.

Audisio, C., Nicolino, S., Scevola, A., Tos, P., Geuna, S., Battiston, B., and Perroteau, I. (2008). ErbB receptors modulation in different types of peripheral nerve regeneration. *Neuroreport* **19,** 1605–1609.

Baryshnikova, L. M., Von Bohlen Und Halbach, O., Kaplan, S., and Von Bartheld, C. S. (2006). Two distinct events, section compression and loss of particles ("lost caps"), contribute to *z*-axis distortion and bias in optical disector counting. *Microsc. Res. Tech.* **69,** 738–756.

Benes, F. M., and Lange, N. (2001). Two-dimensional versus three-dimensional cell counting: A practical perspective. *Trends Neurosci.* **24,** 11–17.

Canan, S., Bozkurt, H. H., Acar, M., Vlamings, R., Aktas, A., Sahin, B., Temel, Y., and Kaplan, S. (2008). An efficient stereological sampling approach for quantitative assessment of nerve regeneration. *Neuropathol. Appl. Neurobiol.* **34,** 638–649.

Casha, S., Yong, V. W., and Midha, R. (2008). Minocycline for axonal regeneration after nerve injury: A double-edged sword. *Exp. Neurol.* **213,** 245–248.

Castro, J., Negredo, P., and Avendaño, C. (2008). Fiber composition of the rat sciatic nerve and its modification during regeneration through a sieve electrode. *Brain Res.* **1190,** 65–77.

Cochran, W. G. (1977). Sampling Techniques, 3rd ed., John Wiley & Sons, Inc., New York.

Coggeshall, R. E. (1992). A consideration of neural counting methods. *Trends Neurosci.* **15,** 9–13.

Cruz-Orive, L. M., and Weibel, E. R. (1981). Sampling designs for stereology. *J. Microsc.* **122,** 235–257.

Di Scipio, F., Raimondo, S., Tos, P., and Geuna, S. (2008). A simple protocol for paraffin-embedded myelin sheath staining with osmium tetroxide for light microscope observation. *Microsc. Res. Tech.* **71,** 497–502.

Fornaro, M., Geuna, S., Fasolo, A., and Giacobini-Robecchi, M. G. (2003). HuC/D confocal imaging points to olfactory migratory cells as the first cell population that expresses a post-mitotic neuronal phenotype in the chick embryo. *Neuroscience* **122,** 123–128.

Fornaro, M., Lee, J. M., Raimondo, S., Nicolino, S., Geuna, S., and Giacobini-Robecchi, M. (2008). Neuronal intermediate filament expression in rat dorsal root ganglia sensory neurons: An *in vivo* and *in vitro* study. *Neuroscience* **153,** 1153–1163.

Fraher, J. P., O'Leary, D., Moran, M. A., Cole, M., King, R. H., and Thomas, P. K. (1990). Relative growth and maturation of axon size and myelin thickness in the tibial nerve of the rat. 1. Normal animals. *Acta Neuropathol.* **79,** 364–374.

Geuna, S. (2005). The revolution of counting "tops": Two decades of the disector principle in morphological research. *Microsc. Res. Tech.* **66,** 270–274.

Geuna, S., and Varejão, A. S. (2008). Evaluation methods in the assessment of peripheral nerve regeneration. *J. Neurosurg.* **109,** 360–362.

Geuna, S., Gigo-Benato, D., and Rodrigues Ade, C. (2004). On sampling and sampling errors in histomorphometry of peripheral nerve fibers. *Microsurgery* **24,** 72–76.

Geuna, S., Tos, P., Battiston, B., and Guglielmone, R. (2000). Verification of the two-dimensional disector, a method for the unbiased estimation of density and number of myelinated nerve fibers in peripheral nerves. *Ann. Anat.* **182,** 23–34.

Geuna, S., Tos, P., Guglielmone, R., Battiston, B., and Giacobini-Robecchi, M. G. (2001). Methodological issues in size estimation of myelinated nerve fibers in peripheral nerves. *Anat. Embryol.* **204,** 1–10.

Geuna, S., Raimondo, S., Ronchi, G., Di Scipio, F., Tos, P., and Fornaro, M. (2009). Histology of the peripheral nerve and changes occurring during nerve regeneration. *Int. Rev. Neurobiol.* This volume.

Gonzalez-Martinez, T., Perez-Piñera, P., Díaz-Esnal, B., and Vega, J. A. (2003). S-100 proteins in the human peripheral nervous system. *Microsc. Res. Tech.* **60,** 633–638.

Guillery, R. W. (2002). On counting and counting errors. *J. Comp. Neurol.* **447,** 1–7.

Gundersen, H. J. G. (1977). Notes on the estimation of the numerical density of arbitrary profiles: The edge effect. *J. Microsc.* **111,** 219–223.

Gundersen, H. J. G. (1978). Estimators of the number of objects per area unbiased by edge effects. *Microsc. Acta* **81,** 107–117.

Gundersen, H. J. G. (1986). Stereology of arbitrary particles. A review of unbiased number and size estimators and the presentation of some new ones, in memory of William R. Thompson. *J. Microsc.* **143,** 3–45.

Hall, S. (2005). The response to injury in the peripheral nervous system. *J. Bone Joint Surg.* **87,** 1309–1319.

Hoffman, E. O., Flores, T. R., Coover, J., and Garrett, II H. B. (1983). Polychrome stains for high resolution light microscopy. *Lab. Med.* **14,** 779–781.

Hoffman, P. N. (1995). The synthesis, axonal transport, and phosphorylation of neurofilaments determine axonal caliber in myelinated nerve fibers. *Neuroscientist* **1,** 76–84.

Inaga, S., Katsumoto, T., Tanaka, K., Kameie, T., Nakane, H., and Naguro, T. (2007). Platinum blue as an alternative to uranyl acetate for staining in transmission electron microscopy. *Arch. Histol. Cytol.* **70,** 43–49.

Kanaya, F., Firrell, J. C., and Breidenbach, W. C. (1996). Sciatic function index, nerve conduction tests, muscle contraction, and axon morphometry as indicators of regeneration. *Plast. Reconstr. Surg.* **98,** 1264–1271.

Keskin, M., Akbaş, H., Uysal, O. A., Canan, S., Ayyldz, M., Ağar, E., and Kaplan, S. (2004). Enhancement of nerve regeneration and orientation across a gap with a nerve graft within a vein conduit graft: A functional, stereological, and electrophysiological study. *Plast. Reconstr. Surg.* **113,** 1372–1379.

Larsen, J. O. (1998). Stereology of nerve cross sections. *J. Neurosci. Methods* **85,** 107–118.

Mayhew, T. M., and Gundersen, H. J. (1996). If you assume, you can make an ass out of u and me': A decade of the disector for stereological counting of particles in 3D space. *J. Anat.* **188,** 1–15.

Onishi, A., Offord, K., and Dyck, P. J. (1974). Studies to improve fixation of human nerves. 1. Effect of duration of glutaraldehyde fixation on peripheral nerve morphometry. *J. Neurol. Sci.* **23,** 223–226.

Pakkenberg, B., and Gundersen, H. J. (1997). Neocortical neuron number in humans: Effect of sex and age. *J. Comp. Neurol.* **384,** 312–320.

Prodanov, D., and Feirabend, H. K. (2007). Morphometric analysis of the fiber populations of the rat sciatic nerve, its spinal roots, and its major branches. *J. Comp. Neurol.* **503,** 85–100.

Reed, M. G., and Howard, C. V. (1998). Surface-weighted star volume: Concept and estimation. *J. Microsc.* **190,** 350–356.

Santos, A. P., Suaid, C. A., Fazan, V. P., and Barreira, A. A. (2007). Microscopic anatomy of brachial plexus branches in Wistar rats. *Anat. Rec.* **290,** 477–485.

Schmitz, C. (1998). Variation of fractionator estimates and its prediction. *Anat. Embryol.* **198,** 371–397.

Schmitz, C., and Hof, P. R. (2005). Design-based stereology in neuroscience. *Neuroscience* **130,** 813–831.

Varejão, A. S., Melo-Pinto, P., Meek, M. F., Filipe, V. M., and Bulas-Cruz, J. (2004). Methods for the experimental functional assessment of rat sciatic nerve regeneration. *Neurol. Res.* **26,** 186–194.

Vleggeert-Lankamp, C. L. (2007). The role of evaluation methods in the assessment of peripheral nerve regeneration through synthetic conduits: A systematic review. Laboratory investigation. *J. Neurosurg.* **107,** 1168–1189.

von Bartheld, C. (2002). Counting particles in tissue sections: Choices of methods and importance of calibration to minimize biases. *Histol. Histopathol.* **17,** 639–648.

Ward, T. S., Rosen, G. D., and von Bartheld, C. S. (2008). Optical disector counting in cryosections and vibratome sections underestimates particle numbers: Effects of tissue quality. *Microsc. Res. Tech.* **71,** 60–88.

West, M. J. (1999). Stereological methods for estimating the total number of neurons and synapses: Issues of precision and bias. *Trends Neurosci.* **22,** 51–61.

METHODS AND PROTOCOLS IN PERIPHERAL NERVE REGENERATION EXPERIMENTAL RESEARCH: PART III—ELECTROPHYSIOLOGICAL EVALUATION

Xavier Navarro and Esther Udina

Institute of Neurosciences and Department of Cell Biology, Physiology and Immunology, Universitat Autònoma de Barcelona, E-08193 Bellaterra, Spain
Centro de Investigación en Red sobre Enfermedades Neurodegenerativas (CIBERNED), Spain

Peripheral nerve injuries usually lead to devastating loss of motor, sensory, and autonomic functions in the patients. Due to the complex requirements for adequate axonal regeneration, functional recovery is often poorly achieved. Experimental models are a useful tool to investigate the mechanisms related to axonal regeneration and reinnervation and to test new strategies to improve functional recovery. Therefore, objective and reliable evaluation methods should be applied for the assessment of regeneration and function restitution after nerve injury in animal models. Electrophysiological tests are commonly used in clinical practice and can be also performed in animal models to determine the nature of peripheral nerve disorders, their severity, and their evolution. These tests provide an integrated approach using sensory and motor nerve conduction studies and electromyography, spinal reflex tests, and motor and sensory evoked potentials if appropriate. The low-invasiveness of several electrophysiological methods allows serial evaluation of sensory and motor reinnervation distal to the injury site at desired intervals without killing the animals or disturbing the regeneration process. This chapter gives a brief review of the most useful electrophysiological methods, their values, and limitations.

INTERNATIONAL REVIEW OF
NEUROBIOLOGY, VOL. 87
DOI: 10.1016/S0074-7742(09)87006-2

I. Introduction

Peripheral nerve injuries result in loss of neural control of motor, sensory, and autonomic functions in the denervated territory. These functions can be recovered if axotomized neurons survive, regrow the sectioned axons through the distal stump, and reestablish functional connections with appropriate peripheral target organs. Surgical repair has a great impact on the prognosis of traumatic nerve injuries by attempting to reestablish the continuity of the injured nerve. However, despite the refinement in microsurgical repair techniques, the results after nerve lesions are usually unsatisfactory, possibly due to the complex requirements for functional recovery. Compared with the clinical studies, animal experimentation allows better control of the variables influencing regeneration and recovery, so different variables can be studied separately. Animal models provide the application of objective methods to evaluate the early phases and the course of regeneration and to investigate molecular and cellular events in a manner not possible in clinical studies.

An understanding of the phases of functional recovery is necessary to know the value and the limitations of experimental methods for evaluation of nerve injuries. Three general phases can be considered in recovery: axonal regeneration, target reinnervation, and complex function recovery. Each phase is dependent upon the preceding one for a successful outcome. The first phase comprises two steps, the axonal growth across the injury and repair site, called the latency period, and the elongation of axons along the distal nerve segment. Severed axons readily emit regenerative sprouts that may cross the injury site and find an endoneurial tube, along which they regenerate supported by the reactive Schwann cells and the basal membrane. However, other axons can grow blindly into the connective tissue outside the endoneurium, where they may form a neuroma in continuity. The axons that elongate within the distal endoneurium will be useful only if they reach an appropriate target tissue, where they can establish functional connections, that is, synapses with muscle cells or reinnervation of sensory receptors. Axonal regeneration does not always result in effective reinnervation, since after nerve transection, regenerating axons have a probability of misdirecting regrowth, through a foreign branch or fascicle, to mismatching targets that they do not innervate (Madison *et al.*, 1996; Valero-Cabré and Navarro, 2002b). Following appropriate target reinnervation, regenerated axons mature with an increase of size and myelination that helps to restore normal conduction properties. Finally, the restitution of complex functions, such as fine motor control and sensory discrimination, requires also that the central connections return to a normal pattern after being lost or reorganized due to the nerve injury (Navarro *et al.*, 2007). Objective evaluation methods should be applied for the quantitative assessment of each of these phases of recovery after nerve injury, and also for the investigation of the different types of functions (motor, sensory, autonomic) conveyed by the injured nerves. This chapter gives a brief review of the most useful electrophysiological methods, their values, and limitations.

Elecrophysiological tests provide a quantitative measure of nerve activity in the normal and pathological states. Neurophysiological studies of patients with peripheral nerve injuries involve an integrated approach using sensory and motor nerve conduction studies and electromyography (EMG), as well as motor and sensory evoked potentials in some instances, as helpful adjuncts to the clinical assessment (Dorfman, 1990). EMG examination is mainly useful to detect fibrillation potentials or positive waves, which are indicative of acute or subacute muscle denervation due to axonal loss, and also to evaluate the motor unit action potentials (MUAPs) and their recruitment pattern. Nerve conduction studies allow to separate the degree of involvement of axons and myelin, to obtain differential information on motor and sensory nerve fibers, and can be used to test practically any peripheral nerve in the body. Clinical electrophysiological examination is addressed to obtain information on localization, pathophysiology, and severity of the injury, and to help in diagnosis and therapeutic decision. Serial electrophysiological investigations yield important information about the time-course of regeneration and reinnervation, and are able to reveal temporal differences with treatments.

II. Nerve Conduction Tests: Technical Bases

The electrical stimulation of a nerve results in the depolarization of the axons and the propagation of an action potential in both directions along the axons. The action potential is originated at the cathode level, referred to an anode placed at a short distance. Electrical stimuli of $\leq$0.1 ms duration are used to activate myelinated nerve fibers, with stimulus strength adjusted to about 20% above the intensity needed to obtain a maximal response. Further increase in stimulus current is unwarranted and may activate the nerve at a distance from the cathode, making measurement of conduction distance inaccurate (Krarup, 2004). The evoked action potential of individual axons has a characteristic spike waveform and conduction velocity (CV), which are dependent on the type of nerve fiber (diameter and myelination) and the physiological state (temperature, metabolism) (Gasser and Grundfest, 1939). However, these relationships may be altered in injured and regenerated fibers. A nerve may be studied by stimulating at one end and recording at a distance. When the whole nerve is stimulated at an intensity that activates all the contained axons, the recorded wave is a compound nerve action potential (CNAP), that is, the algebraic sum of the individual action potentials of all the axons stimulated and able to propagate the action potential. The size and shape of the CNAP is determined by the number of functional axons, the amplitude of single action potentials, the range of CVs, and the distance impulses travel along the nerve.

Stimulation of a mixed or a motor nerve elicits a twitch in the innervated muscles. The impulse conducted by α-motor axons causes synaptic transmission at the neuromuscular junction and the production of an action potential in the membrane of the skeletal muscle fibers, that can be recorded with electrodes placed on the muscle. The compound muscle action potential (CMAP) represents the summation of the action potentials of all the excited muscle fibers that respond to the nerve stimulation. Therefore, motor nerve conduction tests provide selective information on the function and regeneration of α-motor axons.

The essential components of an electrophysiological system are the electrodes, a stimulator, an amplifier, an oscilloscope or computer acquisition system, and a loudspeaker (Fig. 1) (for detailed information see Gitter and Stolov, 1995; Loeb and Gans, 1986). Electromyographic apparatus used in the clinic integrate all these components, but the system can also be made by separate components with

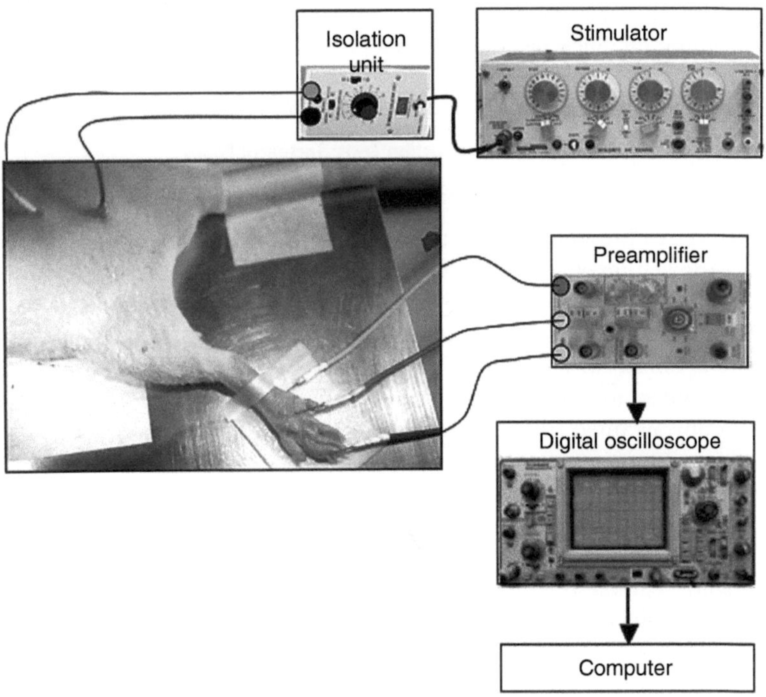

FIG. 1. Schematic diagram of the components of an electrophysiological system for evaluation of nerve regeneration. The photograph depicts the placement of stimulation electrodes at the sciatic notch, and the recording and ground electrodes at the hindpaw of a rat for nerve conduction studies.

more versatility. A biological electrical stimulator can give finely controlled pulses of varying duration and voltage, at different rates. The use of a stimulus isolation unit is recommended for safety and for reducing the stimulus artifact that may compromise the detection of nerve signals along short distances. In contrast to clinical techniques in which surface electrodes are commonly used to avoid the distress of needle insertion, in the experimental setting needle or wire electrodes are more appropriate for both stimulation and recording. Small needle electrodes as well as thin insulated wires are commercially available, whereas smaller or special types of electrodes may be manufactured in the laboratory. For example, needle electrodes of 27–28 gauge are adequate for nerve conduction tests in rats, whereas small needles of 50–100 μm diameter can be soldered to wire to be used in mice. For stimulation, two electrodes, one acting as cathode and the other as anode, are placed near the nerve of study. For recording, the active electrode is placed near the nerve site or on the muscle selected, and the reference electrode is placed distally near the nerve or at the tendon of the muscle; a ground electrode is placed in the same limb, between stimulating and recording electrodes. Amplifiers used for electrophysiology usually work in the differential mode and should be able to amplify the recorded signals from $\times 100$ to at least $\times 10{,}000$. Many kinds of extraneous electrical signals can contaminate the recording and render it useless; therefore, low- and high-pass filters are needed, and for many applications the isolation of the animal or the preparation in a grounded Faraday cage is required. Since the fluctuations of voltage in nerves and muscles occur at frequencies in the audible spectrum, they may be translated to sound waves by connecting the output of the amplifier to a loudspeaker as well as to the monitor. The characteristic sound produced by nerve and muscle action potentials is of value for identification of the potentials' origin and of artifact noises. An oscilloscope or a digitizing system is used for displaying the action potentials in real time. The polarity of the oscilloscope is set for the negative wave to appear upwards. Since the conduction of the nerve action potentials is very fast, visualization requires a fast sweep in the oscilloscope, between 10 and 100 ms. Signals are visualized in the screen at amplifications of 5–100 μV/square for nerve potentials, and of 0.1–10 mV/square for muscle potentials. Signals can be digitized and stored in the computer disk for quantitative measurements. While in the past recordings were stored on paper or magnetic media, the present trend is to directly digitize the data using an A/D board and a computer.

III. Electrophysiological Evaluation of Axonal Regeneration

The length and degree of axonal regeneration in the distal nerve can be evaluated by stimulating the nerve at one side, proximal to the injury and repair site, and recording at a distance on the regenerating nerve, or vice versa.

Depending on the length of the nerve one or more stimulating or recording pairs of electrodes may be placed at different positions to follow longitudinally the regenerated fibers. The technique is also applied during peroperative assessment of nerve injuries in the clinical setting (Brown and Veitch, 1994; Rosen and Jewett, 1980). Since CNAPs are of very small amplitude and usually dispersed in regenerated nerves at early phases, this method is frequently performed in an invasive manner, to avoid the skin resistance, by exposing the nerve in the animal and arranging a small pool with the skin around the exposed nerve or taking the nerve to an special chamber for *in vitro* studies. It is important to maintain the nerve moisturized, warmed and nourished during the full testing period. Electrodes used for stimulation and recording are placed in intimate contact with the exposed nerve, by using hook wires or cuff electrodes around the nerve. In this way, it is possible to evaluate different populations of nerve fibers that when propagating the action potential give origin to distinct compound waves according to their CV range.

By measuring the latency and the distance between stimulating and recording electrodes the CV can be calculated for each wave. In regenerating nerves the CV is altered by changes in axon diameter and myelination, and remains below normal values for long time after successful regeneration (Cragg and Thomas, 1964; Fugleholm *et al.*, 2000). The amplitude of the CNAP is related to the number and size of regenerated axons that are able to propagate the impulse. However, the CNAP amplitude is also affected by CV and distance; in normal nerves the wave of the CNAP increases in duration and declines in height with increasing conduction distance, due to temporal dispersion and phase cancellation phenomena. In regenerated nerves the reduction in axon diameter and internodal length alters the CV and thus influence the wave amplitude, introducing a technical problem for comparisons along time and with normal nerves. Nevertheless, the amplitude has been traditionally used because it is easy to measure and gives a reliable indication of the function of the axons in the nerve. The area under the curve of the CNAP does not change so significantly with the distance.

The invasive approach certainly precludes serial investigation in the same animal, but the electrophysiological results may be correlated with the histological investigation (mainly number and size of regenerated axons) of the same nerve processed thereafter (Fugleholm *et al.*, 2000; Vleggeert-Lankamp *et al.*, 2004), provided that damage is not induced during the testing procedure. The CNAP amplitude does not correlate with the total number of regenerated myelinated fibers, but has a highly significant relation, following an exponential function, with the number of large myelinated fibers (Fugleholm *et al.*, 2000), indicating that they are the ones mostly contributing in the electrophysiological recordings. An interesting application has been the chronic implantation of nerve electrodes, mainly of the cuff type around the nerve of study. The method provided recordings of whole nerve compound potentials as well as single fiber potentials with reproducible temporal features, although variability in the amplitudes and to a lower degree in the velocity of the CNAP was found in long-term observations, and slight nerve damage may be induced by the electrodes (Krarup *et al.*, 1988; Krarup and Loeb, 1988). With this approach, investigators could examine the time course of elongation and maturation of peripheral nerve fibers in the cat after different lesions of the tibial nerve, allowing detection of the earliest regeneration phase, comparisons of different types of nerve fibers, and study of the maturation of regenerated axons (Krarup *et al.*, 1988; Fugleholm *et al.*, 1994, 2000).

IV. Electrophysiological Evaluation of Regeneration and Reinnervation

Nerve conduction tests can be reliably used to assess reinnervation of nerves and muscles distal to the injury site by large myelinated nerve fibers in the whole animal. These techniques are minimally invasive, and thus have the advantage of allowing serial evaluation of reinnervation at desired intervals without sacrificing the animals or disturbing the regeneration process. Another advantage is that similar electrophysiological tests can be applied in experimental animals of different sizes (mouse, rat, rabbit, cat) to those regularly used in the evaluation of human patients, thus allowing to correlate findings from basic to clinical studies. The aim is to determine involvement and regeneration of sensory and motor fibers, and to characterize the pattern of denervation and reinnervation.

Experimental tests are generally performed under anesthesia, with the animal body and limbs fixed with tape on a flat surface. During the tests, the body temperature has to be maintained constant between 34 and 36 °C by means of thermostated heating pads or light bulbs. To ensure reproducibility the recording needles are placed under magnification to secure the same location guided by anatomical landmarks. For longitudinal studies encompassing several months, it is important to take into consideration the physiological changes that occur with maturation and aging of the animals, which are reflected by an increase in amplitude of action potentials and CV during the first 6 months of age in rodents, and a progressive later decline. Such age-related changes have more impact on regenerating nerves (Choi *et al.*, 1995; Verdú *et al.*, 1996, 2000). Comparisons of the values of corresponding nerves and muscles with those of control age-matched animals or to the contralateral side to nerve injury help to determine the degree of abnormalities.

A. Nerve Conduction Tests

Conduction tests can be applied to the injured nerve by stimulating at one site and recording at another distant site. The normal direction of propagation in an axon is called the orthodromic direction; an action potential propagating in the opposite direction follows an antidromic direction. In nerve conduction tests performed in the whole animal, antidromic potentials are often larger than orthodromic potentials, in particular in distal nerves, and therefore preferred by some electrophysiologists; however, the antidromic CNAP recorded in limb nerves may be disturbed by artifacts from near muscles activity evoked by stimulating the mixed nerve. The nerve is directly stimulated with short duration (50–100 μs) electrical pulses at sites proximal and distal to the injury, and the evoked CNAP may be recorded with needle electrodes placed near the distal nerve segment (Fig. 2B). The triphasic potential recorded represents the CNAP of large myelinated fibers. Components resulting from small Aδ and C fibers are usually not detected *in vivo.* Recording needle electrodes are preferred to surface electrodes because they allow for higher resolution to detect the small amplitude CNAPs, in the range of microvolts (Krarup, 2002). Electronic averaging of several responses improves detection of small CNAPs over the background noise. Signal acquisition with high sampling rates significantly improves the amplitude of the nerve potentials, which have more high frequency components than muscle potentials.

Nerve conduction tests used to evaluate regeneration in rats and mice have been described for a few nerves, mainly for the sciatic and caudal nerves (Table I). Nerve conduction tests applied to pure sensory nerves, such as saphenous, sural (although it has a small motor component in rodents), and digital nerves, give selective information on the functional state of myelinated cutaneous afferent fibers. However, in regenerated nerves the random nature of axonal growth when the injured parent nerve (femoral or sciatic) is a mixed nerve makes this assessment less perfect because motor axons can also grow through the normally sensory nerve (Madison *et al.*, 1996) and contribute to the CNAP.

The amplitude of the CNAP is measured from onset to the negative peak or from the negative peak to the positive peak. The first method is better for short nerves, where the CNAP may be followed or even partially overlapped by muscle action potentials. Indeed, muscle artifacts have to be avoided and discriminated for reliable measurements. The CNAP amplitude decreases with the length of the distance between stimulation and recording; therefore, the amplitude is larger with distal than with more proximal stimulation of the same nerve. At initial stages of regeneration, the action potentials may not be identified due to temporal dispersion of the regenerating axons, which conduct at variable low speeds (Krarup, 2002). The conduction time from the stimulus artifact to the onset of the CNAP, measured in milliseconds, is termed the latency. The latency is used to derive the CV of the fastest axons in each nerve segment tested.

B. Motor Nerve Conduction Tests

Motor nerve conduction is the most widely used electrophysiological technique for assessment of nerve regeneration and neuropathies in experimental models, since seminal reports in the 1960s (Fullerton and Barnes, 1966;

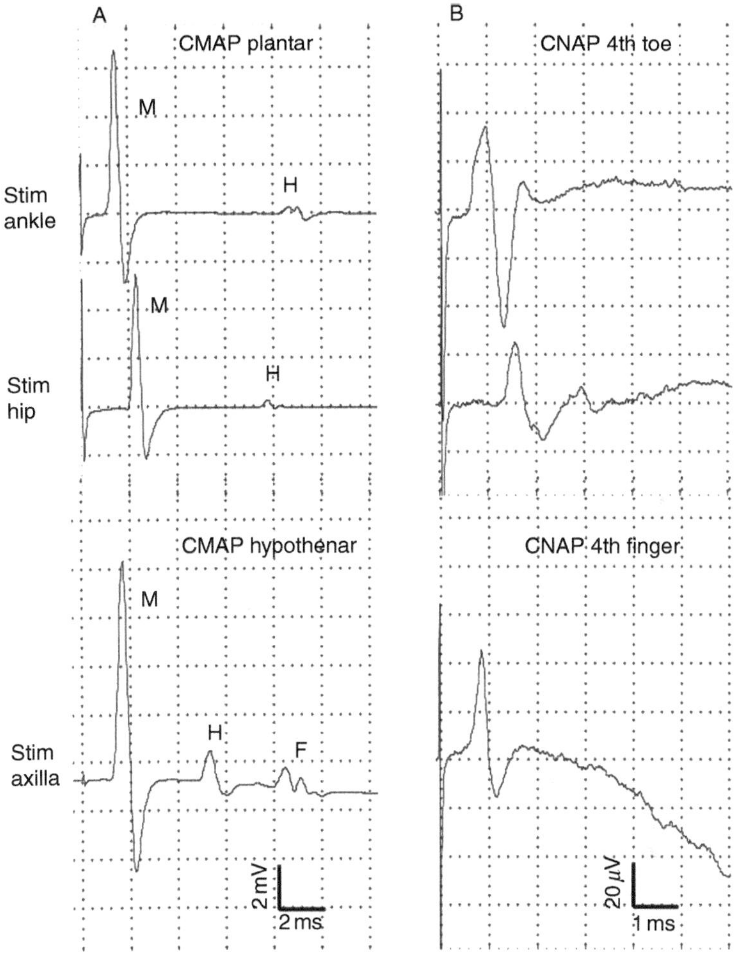

FIG. 2. Motor and sensory nerve conduction tests in the mouse. (A) CMAPs recorded in distal muscles of the hindpaw (sciatic nerve) and forepaw (ulnar nerve). Top: recording from the plantar muscle, after stimulation at the ankle and at the sciatic notch. Bottom: recording from the hypothenar muscles after stimulation at the axilla. In the recordings, the M wave (M), the H wave (H), and the more unstable F wave (F) can be observed. (B) CNAPs recorded from the pure sensory digital nerves of the toes (sciatic nerve) and fingers (ulnar nerve). Top: recording from the fourth digit of the hindpaw, after stimulation at the ankle and at the sciatic notch. Bottom: recording from the fourth digit of the forepaw after stimulation at the axilla.

Kaeser and Lambert, 1962). Motor nerve conduction tests are easier to perform, because of the larger amplitude of the potentials, than direct nerve conduction tests, and therefore have been used to assess axonal regeneration in different

TABLE I

SENSORY AND MOTOR NERVE CONDUCTION TESTS USED TO EVALUATE NERVE REGENERATION IN EXPERIMENTAL RESEARCH IN MICE, RATS, AND RABBITS

Nerve conduction tests			
Nerve	Stimulation	Recording	Key references
Sciatic	Sciatic notch	Tibial n. (ankle)	Navarro *et al.* (1994), Verdú *et al.* (1999), Udina *et al.* (2003), Valero-Cabré and Navarro (2002a), Kurokawa *et al.* (2004)
	Knee	Digital n	
	Ankle	Medial plantar n	
Median	Shoulder	Digital n.	Udina and Navarro (unpublished)
Ulnar	Elbow		
Caudal	Tail distal	Tail proximal	Apfel *et al.* (1992), Verdú *et al.* (1996)
Motor nerve conduction tests		*Muscle*	
Sciatic	Sciatic notch	Tibialis anterior	Navarro *et al.* (1994) Udina *et al.* (2003), Valero-Cabré and Navarro (2002b), Wolthers *et al.* (2005), Zeng *et al.* (1994), Beer *et al.* (2004)
	Knee	Gastrocnemius	
	Ankle	Plantar	
Median	Shoulder	Thenar	Wang *et al.* (2008), Hosoido *et al.* (2009)
Ulnar	Elbow	Hypothenar	
Radial	Arm	Extensor digit	
Caudal	Tail proximal	Tail distal	Verdú *et al.* (1996)
Facial	Facial n.	Vibrissal	Angelov *et al.* (1999), Byrne *et al.* (2005)

peripheral nerve models (Table I). The nerve is stimulated at sites proximal and distal to the injury with electrical pulses of increasing intensity until a maximal response is recorded, and the evoked CMAP is recorded with electrodes placed on the target muscles (Fig. 2A). The active recording electrode is placed subcutaneously on the belly of the muscle, at the endplate region, and the reference one at the muscle distal insertion.

The amplitude of the CMAP is determined by the number of muscle fibers innervated. The amplitude of the CMAP is measured from onset to the maximal negative peak, and in normal rats and mice is in the range of a few to tens millivolts. For early phases of regeneration, when evoked CMAPs are usually dispersed in several waves, the summed area under the waves may yield more accurate measurements. In contrast to the CNAP, the CMAP amplitude is little affected by the distance of conduction. The latency to the CMAP includes the conduction time of the impulse along the nerve and also the transmission time across the neuromuscular junction. Thus, the CV in the segment between one stimulation site and the recording electrode cannot be determined, since the slow conduction at terminal branches and the time of synaptic transmission would lead to erroneous values. The motor CV can be calculated in the nerve segment between two sites of stimulation, by dividing the difference of the latencies by the distance between the stimulation points.

C. Usefulness of Nerve Conduction Tests for Assessment of Nerve Regeneration

Combined application of sensory and motor nerve conduction tests allows for a meaningful evaluation of effective target reinnervation over time and the comparison of regeneration of sensory and motor axons. Severed axons retain their electrical excitability for up to 4–5 days, though neuromuscular transmission may fail earlier; therefore, electrophysiological examination may be started about one week after injury to assess the completeness of the nerve lesion. Thereafter, the first recorded responses give an estimate of the arrival of the fastest axons to the target nerves or muscles. The time-course of regeneration is of importance for the final recovery after peripheral nerve injuries. A study in monkeys showed that the time to muscle reinnervation is one of the best indicators of the levels of recovery attained months to years later (Krarup *et al.*, 2002). Electrophysiologically, reinnervation occurs slowly at first, then more rapidly, and finally taper off. For example, following sciatic nerve lesions in the rat, reinnervation of muscles in the leg, such as gastrocnemius and tibialis anterior, starts by the third week and the CMAP amplitude reaches a plateau by 3–4 months (Fig. 3), whereas reinnervation of plantar muscles occurs later, by the fourth week and increases until 5 months (Valero-Cabré and Navarro, 2001; Wolthers *et al.*, 2005). Reinnervation

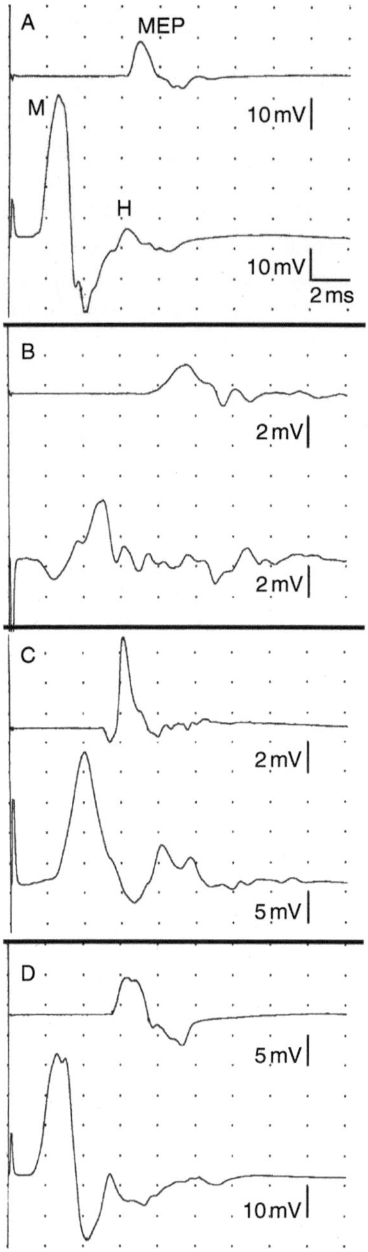

FIG. 3. Representative recordings of M and H waves (bottom recordings) and MEPs (top recordings) from the tibialis anterior muscle of a rat, in the intact hindlimb (A) and following sciatic nerve

of distal targets is delayed and does not achieve the same recovery levels than in more proximal muscles, probably due to the insufficient number of regenerating axons that reach long distances and the protracting effects of chronic denervation (Gordon *et al.*, 2003). Thus, the most sensitive tests to detect differences in final outcome are the recordings from distal targets, such as foot muscles and digital nerves (Navarro *et al.*, 1994; Valero-Cabré and Navarro, 2002b).

During the course of regeneration, the amplitude of CNAPs and CMAPs are the most useful parameter for evaluating the degree of denervation and reinnervation of nerves and muscles, respectively, distal to the injury. The CNAP amplitude is dependent on the number and size of myelinated $A\alpha\beta$ regenerated axons that reached the site of recording. The CMAP amplitude is proportional to the number of regenerated motor axons and the size of the corresponding motor units in the recorded muscle. The duration of the potentials reflects the homogeneity of the degree of myelination among all the axons in the nerve (i.e., maturity of the regenerated nerve). Nevertheless, a complete recovery of the amplitude does not mean that all the original axons in the injured nerve regenerated successfully. Failure of certain number of axons to reach their end-organ permits regenerating axons to express their capacity for collateral branching, increasing the ratio of target organ to axon several times the normal ratio (Brown *et al.*, 1980; Rafuse and Gordon, 1996). The degree of reinnervation, as expressed by the amplitude of compound action potentials, is thereby improved. Failure to normalize this parameter in experimental groups indicates that the capacity for regeneration and for collateral reinnervation is exceeded. An accurate estimation of the number of motor units, and thus of reinnervating motor axons, can be obtained by using the incremental stimulation technique with the motor nerve conduction setting (Lago *et al.*, 2007; Shefner, 2001).

The latency to the onset of the potentials and the derived CV of the fastest regenerated fibers reflect the degree of myelination and the size of those axons. In regenerated nerves the latency is prolonged and the CV decreased for long time after the lesion, and often does not show marked differences between treatment groups. The CV is less useful than the amplitude in evaluating nerve regeneration. Significant variations are frequently observed between small laboratory animals and on different days in the same animal, which are influenced by errors in measurement of nerve length (unless done under dissection) and to variations in body temperature.

---

section and repair at 21 days (B), 49 days (C), and 60 days (D). Note changes in amplitude scales in the recordings. The polyphasic M wave has delayed latency and low amplitude at the early phase of reinnervation (B), and tends to return to a monophasic shape, of increasing amplitude and near normal latency with time (C, D). The H wave and the MEP are relatively increased in amplitude during early (B) and mid (C) phases of reinnervation, resulting in an increase of the H/M ratio and the MEP/M ratio. Modified from Vivó *et al.*, (2008).

Combining electrophysiological and morphological analyses provides good insight in the changes that occur during regeneration (Udina *et al.*, 2003; Vleggeert-Lankamp *et al.*, 2004; Wolthers *et al.*, 2005). Both electrophysiological and histological methods were demonstrated to have resolving power to detect differences between treatments applied to severe nerve injuries (Vleggeert-Lankamp, 2007). Still, as they do not comprehensively reflect functional recovery, various tests of locomotion (walking track, grasping, etc.) and sensory responses (algesimetry, withdrawal response, etc.) have been devised as indicators of functional recovery after nerve lesions in experimental animals. They can be added in the research protocol to electrophysiology and histological methods, depending on the outcome objectives. It has to be taken into account that no correlations between electrophysiological results and functional tests of locomotion have been found in most experimental studies (Munro *et al.*, 1998; Valero-Cabré and Navarro, 2002b). The walking pattern remains severely impaired after nerve section despite of significant amount of muscle reinnervation, due to the impact of misdirected axonal growth and reinnervation of muscles by inappropriate motoneurons. Electrophysiological methods are useful for the objective assessment of selectivity of regeneration (Hou and Zhu, 1998; Lago *et al.*, 2007; Politis and Steiss, 1985; Valero-Cabré and Navarro, 2002b).

V. Electrophysiological Evaluation of Spinal Reflexes and Central Connectivity

A. Spinal Reflexes

Spinal reflexes are involved in the segmentary control of motor function. The stretch reflex is important for the control of muscle contraction and interjoint coordination during movement, whereas polysynaptic ipsilateral and contralateral reflexes allow withdrawal of the limb in front of a threatening stimulus and have a role as inductors of the central pattern generator of locomotion. Electrophysiological methods are useful to quantitatively evaluate the spinal reflexes and to assess dysfunctions secondary to peripheral nerve injuries, such as hyperreflexia, spasticity, and neuropathic pain.

When a peripheral nerve is electrically stimulated, both motor and sensory fibers are excited. The impulses orthodromically conducted by Ia afferents to the spinal cord cause the monosynaptic activation of homonymous and synergistic motoneurons receiving the afferent projections, and impulses are conducted by motor axons to the muscle. Because motoneurons fire nearly synchronously, a large wave can be recorded from the homonymous muscle. In motor nerve conduction tests, the direct CMAP or M wave is followed by a late response, called the H (for Hoffmann) wave (Fig. 3A) that is the electrical counterpart of the stretch reflex. The usual method for testing the H wave is alike that described for motor nerve conduction tests (Hosoido *et al.*, 2009; Meinck, 1976; Navarro *et al.*, 1996; Valero-Cabré and Navarro, 2001). In rodents, the H wave is evoked at a slightly lower threshold than the M wave and, unlike in humans, is consistently maintained with high stimulus intensity, although it may decrease in amplitude at supramaximal stimulus for the M wave (Meinck, 1976). The H wave can be easily distinguished from the late F waves, because the latter appear only at high intensity of stimulation and are more variable in shape and latency (Gozariu *et al.*, 1998). The amplitude of the H wave provides an indirect measure of the efficacy of transmission between Ia sensory fibers and α-motoneurons and of the activity of the motoneuron pool. The latency may be used to assess the central conduction time (H latency–M latency) and to derive the CV of proprioceptive sensory fibers (using two sites of stimulation).

Electrical stimulation of nerve sensory afferents induces polysynaptic withdrawal reflexes, in the pattern of an ipsilateral flexion response and a crossed extension response. Electrophysiologically, these responses are recordable in the ipsilateral hamstring and in the contralateral quadriceps and tibialis anterior muscles, respectively (Clare and Landau, 1975; Meyerson *et al.*, 1995; Navarro *et al.*, 1996; Valero-Cabré and Navarro, 2002a), following stimulation of sciatic or tibial nerves. By stimulating the sensory fibers of the nerve with electrical pulses of high intensity a multiburst reflex response with three differentiated components of activity is recordable. Differences of latency and stimulus threshold indicate that the three reflex components are conveyed by Aβ, Aδ, and C afferent fibers, respectively.

Studies after nerve injuries demonstrated that the H wave reappears early after the muscles become reinnervated, indicating that central sensory-motor connections are functional. The spinal H reflex is highly facilitated during the early phase of regeneration (Fig. 3) (Valero-Cabré and Navarro, 2001; Vivó *et al.*, 2008). Advancing muscle reinnervation reverts this facilitation, and as the M wave amplitude increases with time the H/M ratio declines. Similarly, withdrawal reflexes mediated by myelinated afferents (Aβ and Aδ) show a marked increase in relative amplitude during the first stages of regeneration, and tend to return down to normal values as reinnervation progresses (Valero-Cabré and Navarro, 2002a). However, even several months postlesion, reflex amplitudes remain at significantly higher levels than in control animals, particularly if there is only partial recovery.

B. Sensory and Motor Evoked Responses

Techniques for sensory and motor evoked potentials (MEPs) allow evaluating the whole neural pathway, from brain to periphery. In laboratory animals, somatosensory evoked potentials (SEPs) are typically recorded from the sensorimotor cortex following stimulation of peripheral nerves, while MEPs are elicited by stimulation of the brain and recorded from muscles.

After stimulation of a peripheral nerve, SEPs can be registered from the contralateral neck and scalp, with the greatest amplitude in the postcentral brain region. In order to identify these small potentials from background noise, SEP recording requires applying multiple stimuli and averaging tens to hundreds responses. Usually, electrical stimuli of 0.1 ms duration and intensity above threshold for myelinated fibers are applied at 1–6 Hz. Recordings can be made invasively with small ball electrodes lying on the cortex through craniotomy, with small screws fixed on the skull, or with subcutaneous needle electrodes. SEPs are less suitable for the determination of axonal regeneration than nerve conduction tests, since they are more difficult to record and more subjected to variations due to anesthesia levels and central synaptic efficacy. However, the method can be valuable for the examination of short nerves (Wang *et al.*, 2008), proximal lesions of the peripheral nerve (root and plexus), sensory or motor central disturbances, and changes in central somatotopic maps (Barbay *et al.*, 2002). SEPs have proven most useful in the objective evaluation of injury and diseases of the spinal cord (Nashmi *et al.*, 1997; García-Alías *et al.*, 2003).

Transcranial stimulation applied over the skull activates cortical and subcortical (brainstem) descending motor pathways (Adamson *et al.*, 1989; Schlag *et al.*, 2001) that connect with spinal interneurons, which in turn activate the motoneurons. Magnetic stimuli are too broad for small laboratory animals and can affect a too large area of the nervous system, so electrical pulses of supramaximal intensity are preferred for stimulation. Recordings are made as for motor nerve conduction tests. In rats and mice, a single shock elicits a brainstem-origin MEP of short latency (Fig. 3), whereas cortical MEPs require repeated stimulation and have a long latency. MEP evaluation is a sensitive method to study the function of motor descending pathways, particularly in animals with spinal cord injuries in which the MEP amplitude correlates with the amount of spared/regenerated tissue (García-Alías *et al.*, 2003; Nashmi *et al.*, 1997). MEPs have been rarely studied after peripheral nerve injury. In a recent study, facilitation of descending motor pathways was shown by an increased MEP/M amplitude ratio after sciatic nerve lesion (Vivó *et al.*, 2008). Thus, MEPs may be useful for investigating changes in motoneuron properties and in spinal and descending modulatory actions following injuries and recovery.

VI. EMG: Evaluation of Muscle Reinnervation

The normal muscle at rest is electrically silent. The insertion of a needle electrode mechanically excites some muscle fibers, causing a short burst of insertional activity. When the muscle under examination is voluntarily contracted, a number of motor units are activated, resulting in the recording of MUAPs. As the contraction increases, each motor unit fires at an increasing rate and more motor units are activated following a size-dependent recruitment pattern. During strong contractions, the number of recorded MUAPs is so numerous that they cannot be distinguished (interference pattern).

After nerve injury, there is absence of motor unit activity during volitional effort and in response to electrical stimulation in denervated muscles. The first abnormality of the EMG appears 1–3 weeks after injury and consists of the appearance of fibrillation potentials, due to spontaneous excitation of single, denervated muscle fibers. Fibrillations can be found until reinnervation occurs or until muscle fibers become markedly atrophic. There is no accurate method for quantitation of the number of different fibrillation potentials in a muscle and for comparison over time (Dorfman, 1990). Therefore, despite the relevance for clinical diagnosis of nerve lesions, detection of fibrillations has a limited role in experimental studies. The earliest positive evidence of muscle reinnervation is the appearance during voluntary effort or with nerve stimulation of MUAPs, characteristically of small amplitude and polyphasic. With time MUAPs increase in size and may become larger than in normal muscles, because of the higher compliment of muscle fibers that can be reinnervated by one regenerated motor axon. The EMG recording of muscle activation has, however, constrains in experimental animal studies, since it is hard to obtain graded voluntary muscle contraction in the target muscles.

Nevertheless, EMG can be used to record activity from sets of muscles using implanted intramuscular electrodes in awake, freely moving animals (Biedermann *et al.*, 2000; Scholle *et al.*, 2005; Whelan, 2003). To record EMG activity, pairs of fine wire electrodes stripped of their final insulation are implanted in the muscles of interest. The coated wires are passed subcutaneously to a connector secured to the skull with screws and dental cement. EMG signals permit an evaluation of the extent of myo-electrical activation in large regions of muscles and co-ordination between flexor and extensor muscles. With adequate techniques for electrode implantation and for signal analysis, the cross talk between neighbor muscles in the same limb is small. Simultaneous video-recording of the animal movement allows a temporal correlation of kinematic and EMG parameters (Scholle *et al.*, 2005). The amplitude of the rectified and filtered electromyogram can be used as an indirect measure of muscle activity, although the EMG signal, especially when muscles are highly activated, underestimates the activity of the muscle (Day and Hulliger, 2001). Chronic implantation of electrodes in the rat hindlimb muscles have allowed to obtain serial recordings and analyze the temporal course of reinnervation (English *et al.*, 2007), as well as the activation pattern during locomotion (Gramsbergen *et al.*, 2000). After sciatic nerve section and repair, EMG patterns in the operated hindlimb were irregular, coactivation of antagonistic muscles was present, and burst activity was badly adjusted to the phases of the step-cycle. Such abnormalities are attributed to the nonselective reinnervation of muscles by regenerated motor axons, whereas the spinal motoneurons maintain their original central activation pattern (Gramsbergen *et al.*, 2000).

VII. Electrophysiological Characterization of Electrical Properties of Regenerated Nerves

In peripheral nerve lesions, the amplitude and the CV of evoked responses are sensitive indices of the number of regenerated axons and their maturation, but provide little information about the axonal membrane properties. Indeed, axotomized neurons undergo plastic changes in their expression of ion channels, transducers, and receptors (Navarro *et al.*, 2007). The altered gene expression is related with modifications of neuronal electrical properties, including hyperexcitability, although electrophysiological changes vary among types of neurons and with the type of injury.

To provide insight into axonal membrane and ion channel functions, several methods have been developed to test the effect of polarization on the membrane potential, the recovery of excitability, and the specific properties of different types of nerve fibers, which can be applied in experimental animals as well as in human patients (Burke *et al.*, 2001). Threshold tracking methods (Bostock *et al.*, 1998) have been used as research tools for investigating physiological and pathophysiological mechanisms, such as ectopic discharges, differences between motor and sensory axons, and between normal and regenerated axons. Nerve excitability testing is a noninvasive approach that determines the electrical properties of the nerve membrane at the site of stimulation, and their changes caused by activation of ion channels and electrogenic ion pumps. In regenerating nerves, such electrophysiological studies have shown developmental and persistent changes in membrane function and excitability (Moldovan and Krarup, 2004, 2007; Sawai *et al.*, 2008). Recordings from splitted nerve filaments or by using microneurography have also allowed studying ectopic spontaneous activity and abnormalities in mechanical and thermal sensitivity in injured and regenerated nerves (Gorodetskaya *et al.*, 2003; Michaelis *et al.*, 1999; Serra *et al.*, 2009). More details on these methods are out of the scope of this review, although they are bringing important new information for the understanding of pathophysiological mechanisms involved in axonal injury and regeneration.

Acknowledgments

The authors' research was supported by grants from the Fondo de Investigación Sanitaria (PI060201 and PI080598) of Spain, the European Commission (TIME project, ICT-224012), and FEDER funds.

References

Adamson, J., Zapulla, R. A., Fraser, A., Ryder, J., and Malis, L. I. (1989). Effects of selective spinal cord lesions on the spinal motor evoked potentials (MEP) in the rat. *Electroencephalogr. Clin. Neurophysiol.* **74,** 469–480.

Angelov, D. N., Skouras, E., Guntinas-Lichius, O., Streppel, M., Popratiloff, A., Walther, M., Klein, J., Stennert, E., and Neiss, W. F. (1999). Contralateral trigeminal nerve lesion reduces polyneuronal muscle innervation after facial nerve repair in rats. *Eur. J. Neurosci.* **11,** 1369–1378.

Apfel, S. C., Arezzo, J. C., Lipson, L., and Kessler, J. A. (1992). Nerve growth factor prevents experimental cisplatin neuropathy. *Ann. Neurol.* **31,** 76–80.

Barbay, S., Peden, E. K., Falchook, G., and Nudo, R. J. (2002). An index of topographic normality in rat somatosensory cortex: Application to a sciatic nerve crush model. *J. Neurophysiol.* **88,** 1339–1351.

Beer, G. M., Burg, D., Zehnder, A., Seifert, B., Steurer, M., Grimaldi, H., and Meyer, E. (2004). Functional, electrophysiologic, and morphometric evaluation of nerve regeneration from coaptation on regenerated nerve fibers: Experimental study in rabbits. *J. Reconstr. Microsurg.* **20,** 159–166.

Biedermann, F., Schumann, N. P., Fischer, M. S., and Scholle, H. C. (2000). Surface EMG recordings by using a miniaturised matrix electrode: A new technique for small animals. *J. Neurosci. Methods* **97,** 69–75.

Bostock, H., Cikurel, K., and Burke, D. (1998). Threshold tracking techniques in the study of human peripheral nerve. *Muscle Nerve* **21,** 137–158.

Brown, M. C., Holland, R. L., Hopkins, W. G., and Keynes, R. J. (1980). An assessment of the spread of the signal for terminal sprouting within and between muscles. *Brain Res.* **210,** 145–151.

Brown, W. F., and Veitch, J. (1994). AAEM minimonograph #42: Intraoperative monitoring of peripheral and cranial nerves. *Muscle Nerve* **17,** 371–377.

Burke, D., Kiernan, M. C., and Bostock, H. (2001). Excitability of human axons. *Clin. Neurophysiol.* **112,** 1575–1585.

Byrne, P. J., Stuart, R. M., Fakhry, C., Lehar, M., and Flint, P. W. (2005). An electrophysiologic model for functional assessment of effects of neurotrophic factors on facial nerve reinnervation. *Arch. Facial Plast. Surg.* **7,** 114–118.

Choi, S. J., Harii, K., Lee, M. J., Furuya, F., and Ueda, K. (1995). Electrophysiological, morphological, and morphometric effects of aging on nerve regeneration in rats. *Scand. J. Plast. Reconstr. Surg. Hand Surg.* **29,** 133–140.

Clare, M. H., and Landau, W. (1975). Distal hind-limb reflex responses in cats with largely intact spinal reflex circuits. *Exp. Neurol.* **46,** 470–495.

Cragg, B. G., and Thomas, P. K. (1964). The conduction velocity of regenerated peripheral nerve fibres. *J. Physiol.* **171,** 164–175.

Day, S. J., and Hulliger, M. (2001). Experimental simulation of cat electromyogram: Evidence for algebraic summation of motor-unit action-potential trains. *J. Neurophysiol.* **86,** 2144–2158.

Dorfman, L. J. (1990). Quantitative clinical electrophysiology in the evaluation of nerve injury and regeneration. *Muscle Nerve* **13,** 822–828.

English, A. W., Chen, Y., Carp, J. S., Wolpaw, J. R., and Chen, X. Y. (2007). Recovery of electromyographic activity after transection and surgical repair of the rat sciatic nerve. *J. Neurophysiol.* **97,** 1127–1134.

Fugleholm, K., Schmalbruch, H., and Krarup, C. (1994). Early peripheral nerve regeneration after crushing, sectioning, and freeze studied by implanted electrodes in the cat. *J. Neurosci.* **14,** 2659–2673.

Fugleholm, K., Schmalbruch, H., and Krarup, C. (2000). Post reinnervation maturation of myelinated nerve fibers in the cat tibial nerve: Chronic electrophysiological and morphometric studies. *J. Peripher. Nerv. Syst.* **5,** 82–95.

Fullerton, P. M., and Barnes, J. M. (1966). Peripheral neuropathy in rats produced by acrylamide. *Br. J. Industr. Med.* **23,** 210–221.

García-Alías, G., Verdú, E., Forés, J., López-Vales, R., and Navarro, X. (2003). Functional and electrophysiological characterization of photochemical graded spinal cord injury in the rat. *J. Neurotrauma* **20,** 501–510.

Gasser, H. S., and Grundfest, H. (1939). Axon diameters in relation to the spike dimensions and the conduction velocity in mammalian A fibers. *Am. J. Physiol.* **127,** 393–399.

Gitter, A. J., and Stolov, W. C. (1995). AAEM minimonograph #16: Instrumentation and measurement in electrodiagnostic medicine—Part I and Part II. *Muscle Nerve* **18,** 799–811(see also pp. 812–824).

Gordon, T., Sulaiman, O., and Boyd, J. G. (2003). Experimental strategies to promote functional recovery after peripheral nerve injuries. *J. Peripher. Nerv. Syst.* **8,** 236–250.

Gorodetskaya, N., Constantin, C., and Jänig, W. (2003). Ectopic activity in cutaneous regenerating afferent nerve fibers following nerve lesion in the rat. *Eur. J. Neurosci.* **18,** 2487–2497.

Gozariu, M., Roth, V., Keime, F., Le Bars, D., and Willer, J. C. (1998). An electrophysiological investigation into the monosynaptic H-reflex in the rat. *Brain Res.* **782,** 343–347.

Gramsbergen, A., IJkema-Paassen, J., and Meek, M. F. (2000). Sciatic nerve transection in the adult rat: Abnormal EMG patterns during locomotion by aberrant innervation of hindleg muscles. *Exp. Neurol.* **161,** 183–193.

Hosoido, T., Motoyama, S., Goto, M., Mori, F., Tajima, T., Hirata, H., and Wada, N. (2009). Characteristics of H and M-waves recorded from rat forelimbs. *Neurosci. Lett.* **450,** 239–241.

Hou, Z., and Zhu, J. (1998). An experimental study about the incorrect electrophysiological evaluation following peripheral nerve injury and repair. *Electromyogr. Clin. Neurophysiol.* **38,** 301–304.

Kaeser, H. E., and Lambert, E. H. (1962). Nerve function studies in experimental polyneuritis. *Electroencephalogr. Clin. Neurophysiol.* **22,** 29–35.

Krarup, C. (2002). Nerve conduction studies in selected peripheral nerve disorders. *Curr. Opin. Neurol.* **15,** 579–593.

Krarup, C. (2004). Compound sensory action potential in normal and pathological human nerves. *Muscle Nerve* **29,** 465–483.

Krarup, C., Archibald, S. J., and Madison, R. D. (2002). Factors that influence peripheral nerve regeneration: An electrophysiological study of the monkey median nerve. *Ann. Neurol.* **51,** 69–81.

Krarup, C., and Loeb, G. E. (1988). Conduction studies in peripheral cat nerve using implanted electrodes: I. Methods and findings in controls. *Muscle Nerve* **11,** 922–932.

Krarup, C., Loeb, G. E., and Pezeshkpour, G. H. (1988). Conduction studies in peripheral cat nerve using implanted electrodes: II. The effects of prolonged constriction on regeneration of crushed nerve fibers. *Muscle Nerve* **11,** 933–944.

Kurokawa, K., de Almeida, D. F., Zhang, Y., Hébert, C. D., Page, J. G., Schweikart, K. M., and Oh, S. J. (2004). Sensory nerve conduction of the plantar nerve compared with other nerve conduction tests in rats. *Clin. Neurophysiol.* **115,** 1677–1682.

Lago, N., Rodríguez, F. J., Guzmán, M. S., Jaramillo, J., and Navarro, X. (2007). Effects of motor and sensory nerve transplants on amount and specificity of sciatic nerve regeneration. *J. Neurosci. Res.* **85,** 2800–2812.

Loeb, G. E., and Gans, C. (1986). "Electromyography for experimentalists." University of Chicago Press, Chicago, IL.

Madison, R. D., Archibald, S. J., and Brushart, T. M. E. (1996). Reinnervation accuracy of the rat femoral nerve by motor and sensory neurons. *J. Neurosci.* **16,** 5698–5703.

Meinck, H. M. (1976). Occurrence of the H reflex and the F wave in the rat. *Electroencephalogr. Clin. Neurophysiol.* **41,** 530–533.

Meyerson, B. A., Ren, B., Herregodts, P., and Linderoth, B. (1995). Spinal cord stimulation in animal models of mononeuropathy: Effects on the withdrawal response and the flexor reflex. *Pain* **61,** 229–243.

Michaelis, M., Blenk, K. H., Vogel, C., and Jänig, W. (1999). Distribution of sensory properties among axotomized cutaneous C-fibres in adult rats. *Neuroscience* **94,** 7–10.

Moldovan, M., and Krarup, C. (2004). Persistent abnormalities of membrane excitability in regenerated mature motor axons in cat. *J. Physiol.* **560,** 795–806.

Moldovan, M., and Krarup, C. (2007). Internodal function in normal and regenerated mammalian axons. *Acta. Physiol. (Oxf.)* **189,** 191–200.

Munro, C. A., Szalai, J. P., Mackinnon, S. E., and Midha, R. (1998). Lack of association between outcome measures of nerve regeneration. *Muscle Nerve* **21,** 1095–1097.

Nashmi, R., Imamura, H., Tator, C. H., and Fehlings, M. G. (1997). Serial recording of somatosensory and myoelectric motor evoked potentials: Role in assessing functional recovery after graded spinal cord injury in the rat. *J. Neurotrauma* **14,** 151–159.

Navarro, X., Lázaro, J. J., Butí, M., Verdú, E., Fábregas, P. J., and Castellano, B. (1996). Electrophysiological evaluation of spinal reflexes during epidural anesthesia in an experimental model. *Muscle Nerve* **19,** 29–36.

Navarro, X., Verdú, E., and Butí, M. (1994). Comparison of regenerative and reinnervating capabilities of different functional types of nerve fibers. *Exp. Neurol.* **129,** 217–224.

Navarro, X., Vivó, M., and Valero-Cabré, A. (2007). Neural plasticity after peripheral nerve injury and regeneration. *Progr. Neurobiol.* **82,** 163–201.

Politis, M. J., and Steiss, J. E. (1985). Electromyographic evaluation of a novel surgical preparation to enhance nervemuscle specificity that follows mammalian peripheral nerve trunk transection. *Exp. Neurol.* **87,** 326–333.

Rafuse, V. F., and Gordon, T. (1996). Self-reinnervated cat medial gastrocnemius muscles. I. Comparisons of the capacity for regenerating nerves to form enlarged motor units after extensive peripheral nerve injuries. *J. Neurophysiol.* **75,** 268–281.

Rosen, J. M., and Jewett, D. L. (1980). Physiological methods of evaluating experimental nerve repairs. *In* "Nerve repair and regeneration. Its clinical and experimental basis" (D. L. Jewett and H. R. McCarroll, Eds.), pp. 150–161. Mosby Co, St Louis, MO.

Sawai, S., Kanai, K., Nakata, M., Hiraga, A., Misawa, S., Isose, S., Hattori, T., and Kuwabara, S. (2008). Changes in excitability properties associated with axonal regeneration in human neuropathy and mouse Wallerian degeneration. *Clin. Neurophysiol.* **119,** 1097–1105.

Schlag, M. G., Hopf, R., and Redl, H. (2001). Serial recording of sensory, corticomotor, and brainstem-derived motor evoked potentials in the rat. *Somatosensory Motor Res.* **18,** 106–116.

Scholle, H. C., Biedermann, F., Arnold, D., Jinnah, H. A., Grassme, R., and Schumann, N. P. (2005). A surface EMG multi-electrode technique for characterizing muscle activation patterns in mice during treadmill locomotion. *J. Neurosci. Methods* **146,** 174–182.

Serra, J., Bostock, H., and Navarro, X (2009). *Microneurography in rats: a minimally invasive method to record single C fiber action potentials from peripheral nerves in vivo. Neurosci. Lett.* (in press).

Shefner, J. M. (2001). Motor unit number estimation in human neurological diseases and animal models. *Clin. Neurophysiol.* **112,** 955–964.

Udina, E., Ceballos, D., Gold, B. G., and Navarro, X. (2003). FK506 enhances reinnervation by regeneration and by collateral sprouting of nerve fibers. *Exp. Neurol.* **183,** 220–231.

Valero-Cabré, A., and Navarro, X. (2001). H reflex restitution and facilitation after different types of peripheral nerve injury and repair. *Brain Res.* **919,** 302–312.

Valero-Cabré, A., and Navarro, X. (2002a). Changes in crossed spinal reflexes after peripheral nerve injury and repair. *J. Neurophysiol.* **87,** 1763–1771.

Valero-Cabré, A., and Navarro, X. (2002b). Functional impact of axonal misdirection after peripheral nerve injuries followed by graft or tube repair. *J. Neurotrauma* **19,** 1475–1485.

Verdú, E., Butí, M., and Navarro, X. (1996). Functional changes of the peripheral nervous system with aging in the mouse. *Neurobiol. Aging* **17,** 73–77.

Verdú, E., Ceballos, D., Vilches, J. J., and Navarro, X. (2000). Influence of aging on peripheral nerve function and regeneration. *J. Peripher. Nerv. Syst.* **5,** 191–208.

Verdú, E., Vilches, J. J., Rodríguez, F. J., Ceballos, D., Valero, A., Navarro, X. (1999). Physiological and immunohistochemical characterization of cisplatin-induced neuropathy in mice. *Muscle Nerve* **22,** 329–340.

Vivó, M., Puigdemasa, A., Casals, L., Asensio, E., Udina, E., and Navarro, X. (2008). Immediate electrical stimulation enhances regeneration and reinnervation and modulates spinal plastic changes after sciatic nerve injury and repair. *Exp. Neurol.* **211,** 180–193.

Vleggeert-Lankamp, C. L. (2007). The role of evaluation methods in the assessment of peripheral nerve regeneration through synthetic conduits: a systematic review. Laboratory investigation. *J. Neurosurg.* **107,** 1168–1189.

Vleggeert-Lankamp, C. L., van den Berg, R. J., Feirabend, H. K., Lakke, E. A., Malessy, M. J., and Thomeer, R. T. (2004). Electrophysiology and morphometry of the Aalpha- and Abeta-fiber populations in the normal and regenerating rat sciatic nerve. *Exp. Neurol.* **187,** 337–349.

Wang, H., Sorenson, E. J., Spinner, R. J., and Windebank, A. J. (2008). Electrophysiologic findings and grip strength after nerve injuries in the rat forelimb. *Muscle Nerve* **38,** 1254–1265.

Whelan, P. J. (2003). Electromyogram recordings from freely moving animals. *Methods* **30,** 127–141.

Wolthers, M., Moldovan, M., Binderup, T., Schmalbruch, H., and Krarup, C. (2005). Comparative electrophysiological, functional, and histological studies of nerve lesions in rats. *Microsurgery* **25,** 508–519.

Zeng, L., Worseg, A., Albrecht, G., Grisold, W., Hopf, R., Redl, H., and Schlag, G. (1994). A noninvasive functional evaluation following peripheral nerve repair with electromyography in a rat model. *Plast. Reconstr. Surg.* **94,** 146–151.

METHODS AND PROTOCOLS IN PERIPHERAL NERVE REGENERATION EXPERIMENTAL RESEARCH: PART IV—KINEMATIC GAIT ANALYSIS TO QUANTIFY PERIPHERAL NERVE REGENERATION IN THE RAT

Luís M. Costa,* Maria J. Simões,[†,‡] Ana C. Maurício,[†,‡] and Artur S. P. Varejão*

*Department of Veterinary Sciences, CITAB, University of Trás-os-Montes e Alto Douro, P.O. Box 1013, 5001-801 Vila Real, Portugal

[†]Department of Veterinary Clinics, Biomedics Sciences Institute of Abel Salazar, Porto University, 4099-003 Porto, Portugal

[‡]UMIB, Porto University, 4099-003 Porto, Portugal

Functional recovery is one of the primary goals of therapeutic intervention in peripheral nerve research. The number and diversity of tests which have been used to assess functional recovery after experimental interventions often makes it difficult to recommend any particular indicator of nerve regeneration. Functional assessment after sciatic nerve lesion has long been focused on walking track analysis; however, it is important to note that the validity of the sciatic functional index has been questioned by several researchers. In the last decade, several authors have designed a series of sensitive quantitative methods to assess the recovery of locomotor function using computerized rat gait analysis. The objective of the present review is to provide a helpful tool for the peripheral nerve investigator, by integrating the most important gait kinematic measures described in the literature that can be gathered with this technology.

INTERNATIONAL REVIEW OF NEUROBIOLOGY, VOL. 87
DOI: 10.1016/S0074-7742(09)87007-4

I. Introduction

In the introduction to one of the most cited article on functional assessment in peripheral nerve research, de Medinaceli *et al.* (1982) wrote that "Recovery from peripheral nerve injury can be studied by a wide variety of techniques" but they also added that "Although easily assessed in man, in animals functional recovery has been difficult to measure." Despite remarkable advances in understanding nerve regeneration over the last two decades, selection of appropriate evaluation methods to quantify functional recovery after nerve lesion and repair is still a challenging problem faced by investigators. In experiments, the most often used experimental animal is the rat (*Rattus norvegicus*), in which several models of nerve injury can be performed.

Experiments on peripheral nerve regeneration are often performed on the rat sciatic model (Dellon and Mackinnon, 1989; Shen and Zhu, 1995). Additionally, other experimental models involving the brachial plexus (Bertelli and Mira, 1995; Papalia *et al.*, 2003a,b), facial nerve (Jergovic *et al.*, 2001), and caudal nerve (Back *et al.*, 2002) have been proposed to quantify recovery of function after peripheral nerve lesion.

In animal experiments, nerve regeneration can be assessed on the basis of electrophysiologic and histomorphometric changes; unfortunately, these two commonly employed classes of measures do not necessarily correlate with recovery of motor and sensory functions (Munro *et al.*, 1998). It is not generally agreed which type of assessment tool is the most useful descriptor of functional recovery; for this reason, the use of different methods for an overall assessment of nerve function has been recommended by several authors (for review, see Nichols *et al.*, 2005).

In order to overcome the limitations of the methods available to date, exciting work is being carried out on rat gait analysis which may significantly alter the future of peripheral nerve research. Gait analysis can accurately assess normal and abnormal gait, identify characteristics features of nerve damage, and quantify gait so that numeric comparisons can be made between experimental groups. Gait analysis is defined as the systematic measurement, description, and assessment of quantities that characterize locomotion, whereas kinematics is the term used to describe the movement itself, independent of the forces, both internal and external, that cause the movement (Gage *et al.*, 1995). Hruska *et al.* (1979) using a combination of techniques including video and footprint analysis, and Parker and Clarke (1990) with a study of gait topography concluded that free, spontaneous locomotion in rats is very consistent, symmetric, and replicable. The purpose of this article is to review the main parameters of rat gait kinematics that have been developed for functional evaluation of peripheral nerve regeneration.

II. Walking Track Analysis

In 1982, the pioneering work of de Medinaceli's laboratory led to a quantitative method of analyzing the sciatic nerve function in rats, known as the sciatic functional index (SFI), based on measurements of footprints from walking rats (de Medinaceli *et al.*, 1982). Since then, SFI has been employed extensively in studies of peripheral nerve injury and repair because it is inexpensive, requires no special equipment, and has generally been regarded as very accurate and reliable (Nichols *et al.*, 2005: Varejão *et al.*, 2001a).

A. Walking Tracks

Animals are tested in a confined walkway 8.2 cm wide by 42 cm long with a dark shelter at the end. After two or three conditioning trials during which rats often stop to explore the corridor, rats walk then steadily to the darkened cage at the end of the corridor. An overview of the literature shows that in different experiments the bottom of the track has been lined with various kinds of materials, and the plantar surface of the rat hind feet have been dipped or painted with several substances, in order to refine and improve the prints for walking track analysis (Brown *et al.*, 1989; Lowdon *et al.*, 1988; Monte-Raso *et al.*, 2008). The animals are immediately allowed to walk along the corridor leaving their footprints on the paper.

B. Analysis of Walking Tracks

Footprints are evaluated for three different parameters: (1) distance from the heel to the third toe, print length (PL); (2) distance from the first to the fifth toe, toe spread (TS); and (3) distance from the second to the fourth toe, intermediary toe spread (ITS). All three measurements are taken from the experimental (E) and normal (N) sides (Fig. 1). Using the following formula derived by Bain *et al.* (1989), SFI is calculated as follows:

$$\mathrm{SFI} = -38.3\left(\frac{\mathrm{EPL}-\mathrm{NPL}}{\mathrm{NPL}}\right) + 109.5\left(\frac{\mathrm{ETS}-\mathrm{NTS}}{\mathrm{NTS}}\right) + 13.3\left(\frac{\mathrm{EIT}-\mathrm{NIT}}{\mathrm{NIT}}\right) - 8.8$$

SFI is 0 for noninjured animals and −100 after complete transection of the sciatic nerve. Footprints of both the experimental side and the normal side are measured

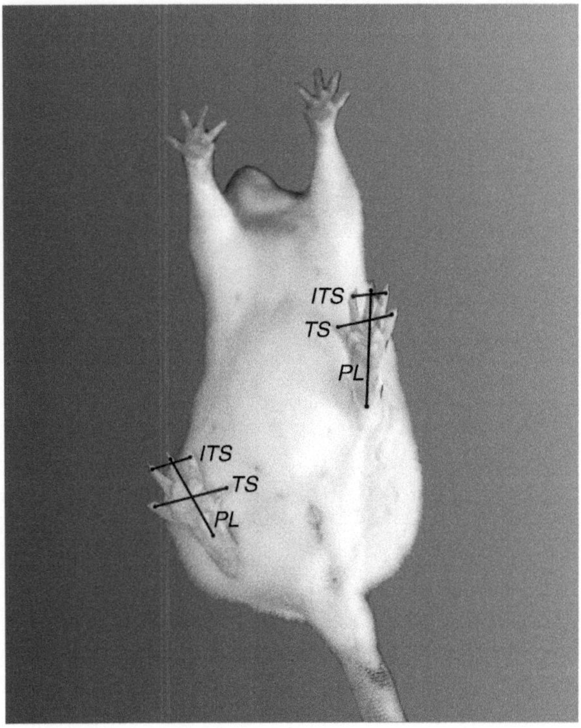

FIG. 1. Plantar surface of rat hind foot. PL = Print Length; TS = Toe Spread; ITS = Intermediary Toe Spread. The affected left hindlimb shows a reduction in the TS and ITS, while the PL is increased.

in millimeters either manually or using a scanner. The digitized images are stored and analyzed with appropriate software in order to increase the speed of analysis and simultaneously minimizing the probability of error in handling voluminous amounts of data (Monte-Raso *et al.*, 2008).

C. Limitations of the SFI

Despite the fact that walking track analysis is universally employed in nerve research, there are a number of problems that have limited its experimental use. Often, several walks are required to obtain visible print marks of both feet. Therefore, it is important to let the animals walk preoperatively, so that they become acquainted to the walking track apparatus (Bain *et al.*, 1989).

Some prints are not measurable owing to the development of flexion contractures, autotomy, smearing of the print, dragging of the tail across the print, or contamination with footprints. To prevent the development of fibrous tissue contractures secondary to neurological loss, one of the most common strategies is to manually exercise the injured leg on a weekly or biweekly basis or, by using a wire mesh, where the animal is allowed to play freely to provide continuous physiotherapy throughout the period of denervation (Strasberg *et al.*, 1996; Varejão *et al.*, 2004).

A peripheral nerve trauma causing axotomy frequently results in selfmutilation of the dennervated part in rodents. This behavior is called autotomy and may be due to a painful dysesthesia which is projected to the three lateral toes (Den Dunnen and Meek, 2001). For the prevention of this problem several deterrent taste substances have been used, with beneficial results (Sporel-Ozakat *et al.*, 1991). Administration of amitriptyline, a tricyclic antidepressant, may be of benefit as an autotomy preventive agent (Navarro *et al.*, 1994). This behavior can be significantly limited when housing male rats with females (Zellem *et al.*, 1989). Other experiments found an absence of selfmutilation in the Lewis strain of rat, indicating a strain dependency (Carr *et al.*, 1992).

III. Computerized Gait Analysis

It is important to realize that the number of kinematic variables (positions, velocities, and accelerations) required to describe one-step cycle is very high. It is only through imaging systems (cameras connected to computers) that we can achieve a full kinematic description during gait. Reflective markers are currently the most successful technique for obtaining joint motion data. Reflective joint markers are placed in the skin over specific bony landmarks. Generally, the kinematic data is gathered using a high-speed digital image camera, which is directly connected to a computer and positioned at right angles to a Perspex corridor or to a motorized treadmill (Fig. 2). The motorized treadmill is frequently used in experimental studies as it offers a controlled and convenient environment for testing and training. In a recent study we suggested that reliable kinematic measurements can be obtained from treadmill analysis in rats (Pereira *et al.*, 2006).

Recently, other methods have been developed to analyze locomotor behavior in the rat following nerve injury. Rat gait was analyzed using a tunnel positioned over a Tekscan pressure sensor grid (Boyd *et al.*, 2007). Following sciatic nerve injury this pressure mat measured timing variables and contact variables, including contact area, contact force, and contact pressure. A novel automated gait analysis system, the CatWalk, allows rapid and objective quantification of a large number of gait parameters, both static and dynamic (Bozkurt *et al.*, 2008). Using the sciatic nerve crush injury model, they showed that CatWalk can measure behavioral recovery examining deficits in dynamic gait parameters, coordination measures, and the intensity of paw prints. Furthermore, marked

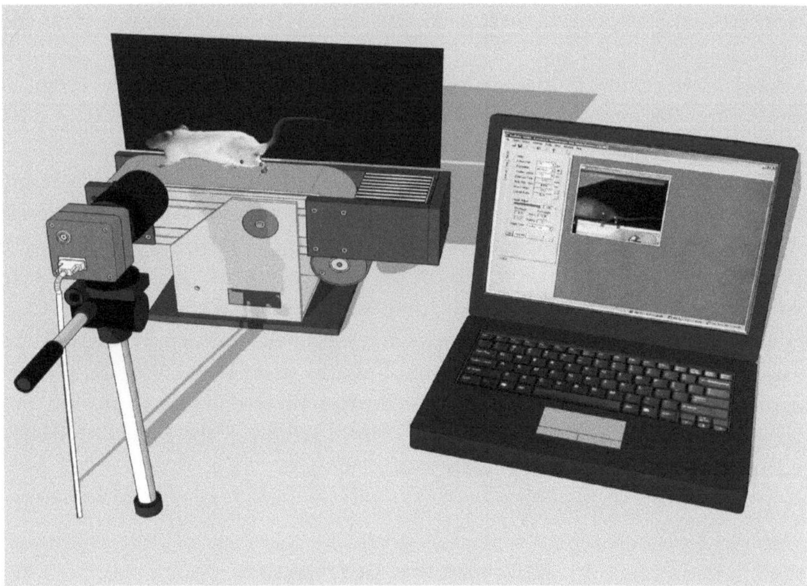

FIG. 2. The ankle angle is defined by two rigid segments: foot and leg.

strain differences were noted on the static paw parameters; base-of-support and relative paw position (Koopmans *et al.*, 2007).

A. Calculation of SFI

Studying footprint analysis, Walker *et al.* (1994) observed a significant variation in the PL with gait velocity; therefore, the SFI could not be reliably calculated. The experimental setup developed earlier by Westerga and Gramsbergen (1990) allowed a correct visualization and analysis of the foot placing and toe positions. The lateral and ventral views of the animal are visualized and the walking movements of the rat are recorded with a camera. Later, by using this method, the anatomic landmarks of the rat sole were easily measured, from consecutive and nonhesitant step cycles, and therefore SFI was promptly calculated (Dijkstra *et al.*, 2000).

B. Gait-Stance Duration

The gait cycle, the basic unit of measurement in gait analysis, is divided into two periods, stance and swing. These often are called gait phases. The stance phase begins when the foot comes in contact with the ground and ends when the foot is no longer in contact with the ground. Toe-off marks the beginning of the swing phase of the gait cycle (Varejão *et al.*, 2001b).

Measurement of the gait–stance duration, using the video technique, was first reported by Walker *et al.* (1994). Through single-frame analysis of videotaped recordings, they calculated the ratio of time of floor contact between the injured and uninjured hind paw (stance factor). In a noninjured rat the left and right foot will have a similar stance time, so the stance factor is approximately 1. After sciatic nerve injury, these authors found that the stance duration of the injured limb is significantly less than the opposite limb, reflecting nerve dysfunction alone, and not pain. The same investigators claim another advantage of this method, the possibility of carrying out a longitudinal study, even in the presence of autotomy.

C. Ankle Kinematics

During gait, deviations of individual joints such as the ankle, can and do occur in all three planes of motion, particularly the sagittal one. The system of markers used to define the ankle angle is similar between the different investigators; two rigid body segments are defined as vectors: leg and foot (Fig. 3). The ankle function is studied in the stance and swing phases.

In 1995, a new rat gait analysis parameter was developed to measure recovery of function after a peroneal nerve crush injury, the ankle angle, evaluated in the mid-swing and terminal swing (Santos *et al.*, 1995). These authors proposed the determination of a 2D geometrical model of the ankle joint by intersecting lines passing through the knee to the ankle and from metatarsal head to ankle.

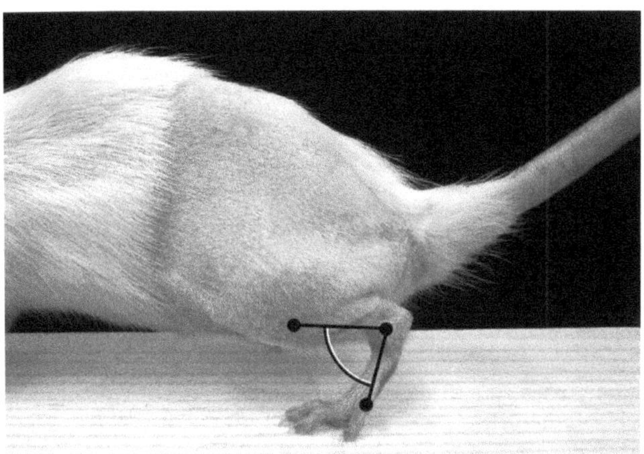

Fig. 3. Experimental setup for 2D gait analysis of treadmill walking.

Their postoperative kinematic data displayed an increased ankle angle related to the loss of dorsiflexion. Later, Yu *et al.* (2001) proposed the same joint angle orientation to calculate ankle kinematic variables at the mid-swing and terminal stance of the gait cycle. They demonstrated that sciatic and tibial lesions were best reflected in the ankle angle at terminal stance, and ankle angle at mid-swing contributed significantly to peroneal scores. Recently, video gait analysis was used to study ankle motion during treadmill walking. The ankle angle was assessed at mid-stance and terminal stance for evaluation of a new artificial nerve guide (Patel *et al.*, 2007). Recovery of 2D angular motion of the ankle joint was compared to scores for the SFI and was found to be more sensitive than the SFI (de Ruiter *et al.*, 2007).

With different marker locations as proposed by Lin *et al.* (1996), Santos *et al.* (1995) presented a new functional index, the ankle stance angle measured in the mid-stance, after sciatic nerve transection. For this study, the ankle angle was defined by two segments, one referring to the (i) shank and another connecting the (ii) lateral malleolus and the head of the fourth metatarsal. In this study, the kinematic plots of the ankle represented in the sagittal plane showed that the angle of the injured limb was close to normal at the beginning of the stance phase, but continued to decrease until the end of the stance phase.

In 2002, we developed a new two-segment biomechanical model to describe the magnitude and polarity of the ankle joint during the stance phase of walking (Varejão *et al.*, 2002). In this mechanical analysis, we recorded the coordinates of four markers which defined the leg and foot segments at the following landmarks: (i) the proximal point of the lower third of the tibia, (ii) lateral malleolus, (iii) calcaneus, and (iv) fifth metatarsal head. We made a clear division of the ankle motion in terms of dorsiflexion and plantarflexion using the following formula:

$$\theta\,\text{Ankle} = \theta\,\text{foot} - \theta\,\text{leg} - 90^\circ$$

If θAnkle was positive, the foot was considered dorsiflexed; if negative, the foot was considered plantarflexed; and if 0°, the foot was neither dorsiflexed nor plantarflexed. Immediately after sciatic nerve injury the role of the gastrocnemius muscles in controlling the forward moving leg is seriously affected, with the collapse of the tibia over the foot. In this pathological state the rats maintain flat-foot contact during the stance phase, with heel-rise merely being part of the total limb lift for swing (Varejão *et al.*, 2003a).

A long-term 2D biomechanical analysis (sagittal plan) was carried out applying a two-segment model of the ankle joint, adopted from our previously developed model. This study using a standardized rat sciatic nerve crush injury with a non-serrated clamp revealed that ankle kinematic parameters were still recovering their original values at week 12 postoperatively (Luís *et al*,., 2007).

D. Toe out Angle

In rodents, biomechanical behavior of the foot and ankle unit has been described with 2D models, assuming that motion is planar thus reducing the complexity of the data collection. Our lab proposed a new functional index, the toe out angle (TOA), defined as the degree of the angle between the direction of progression and a reference line in the sole of the foot (from the calcaneus to the tip of the third digit) (Varejão *et al.*, 2003b). We noted that the extent of external rotation of the foot correlated well with the functional recovery predicted by the SFI. Very recently, we included the transverse plane in order to describe the foot rotation during the stance and swing phases in normal and spinal injured rodents (Couto *et al.*, 2008). This kinematic parameter can be easily applied to peripheral nerve studies, providing a continuous presentation of the entire time plot of the foot rotation angle instead of a selected point of the stance phase of walking, which is critical for a detailed gait assessment.

IV. Gait Analysis in the Forelimb Nerve Injury Models

Although the large majority of the surgical interventions for repairing damaged human nerves are performed at the upper limb level, rat sciatic nerve injury is still the dominant model used in experimental setting. Over the years, most of the methods that have been developed to assess nerve regeneration are related to the fact that the rat forelimb has greater dexterity compared to the hindlimb (Bertelli and Mira, 1995; Papalia *et al.*, 2003a,b; Pagnussat *et al.*, 2009). Only a few articles have described new methods to assess functional recovery after median and ulnar injury during locomotor activity. Toe spread, intermediary toe spread, and print length have been assessed using traditional prints on the paper (Bontioti *et al.*, 2003) and digital video motion analysis (Wang *et al.*, 2008). Interestingly, it was found that injury to the median or ulnar nerve alone had no effect on the toe spread. However, a combined crush of both median and ulnar nerves produced a decrease in the toe spread. In contrast, the combined median and ulnar injury did not alter the print length of the pawprint (Bontioti *et al.*, 2003; Galtrey and Fawcett, 2007). A skilled walking test that is sensitive to placing impairments of forelimbs was developed by Metz and Whishaw (2002). This test requires that rats walk across the rungs of a horizontal ladder. Galtrey and Fawcett (2007) have demonstrated that the horizontal ladder is suitable for the measurement of behavioral recovery after median and ulnar injury.

Recently, a 2D video motion analysis was found to be a sensitive method to assess function in the rat forelimb nerve models (Wang *et al.*, 2008). A permanent marker was tattooed over five anatomic landmarks on the lateral side of the forelimb: the distal third of the ulna, the ulnar head, the fifth metacarpal base, the metacarpophalangeal (MP) joint, and the proximal interphalangeal joint. With this biomechanical model they collected the MP and wrist kinematic data at a sampling rate of 60 Hz. These authors found a profound decrease of wrist and MP flexion angles following median nerve injury or combined median and ulnar nerve injury. In contrast, ulnar nerve injury alone had no or little effect on wrist and MP joint motion.

V. Conclusions and Future Perspectives

The ultimate goal of the treatment of peripheral nerve injury is to restore nerve function. Therefore, much research is currently devoted to the creation of the optimal method to assess functional recovery associated with the treatment in experimental models. Certainly, one of the most significant challenges facing the peripheral nerve researcher is finding new ways to more accurately follow the course of recovery in nerve-injured rodents.

Despite the widespread use of traditional methods as indicators of functional outcome in peripheral nerve research, all have limitations and do not always correspond to the restoration of function. It is important to develop novel assessment procedures by means researchers are able to measure small changes in the course of functional recovery.

In conclusion, we recommend the kinematic gait analysis in order to quantify small biomechanical changes, previously inaccessible, during assessment of peripheral nerve regeneration.

Acknowledgment

This research was supported by the Luso-American Foundation (grant #L-V-315/2006) and by the Operational Programme for Science and Innovation 2010 (Portuguese Ministry of Science, Technology and Higher Education).

References

Back, S. K., Sung, B., Hong, S. K., and Na, H. S. (2002). A mouse model for peripheral neuropathy produced by a partial injury of the nerve supplying the tail. *Neurosci. Lett.* **322,** 153–156.

Bain, J. R., Mackinnon, S. E., and Hunter, D. A. (1989). Functional evaluation of complete sciatic, peroneal, and posterior tibial nerve lesions in the rat. *Plast. Reconstr. Surg.* **83,** 129–136.

Bertelli, J. A., and Mira, J. C. (1995). The grasping test: A simple behavioral method for objective quantitative assessment of peripheral nerve regeneration in the rat. *J. Neurosci. Methods* **58,** 151–155.

Bontioti, E., Kanje, M., and Dahlin, L. B. (2003). Regeneration and functional recovery in the upper extremity of rats after various types of nerve injuries. *J. Peripher. Nerv. Syst.* **8,** 159–168.

Boyd, B. S., Puttlitz, C., Noble-Haeusslein, L. J., John, C. M., Trivedi, A., and Topp, K. S. (2007). Deviations in gait pattern in experimental models of hindlimb paresis shown by a novel pressure mapping system. *J. Neurosci. Res.* **85,** 2272–2283.

Bozkurt, A., Deumens, R., Scheffel, J., O'Dey, D. M., Weis, J., Joosten, E. A., Führmann, T., Brook, G. A., and Pallua, N. (2008). CatWalk gait analysis in assessment of functional recovery after sciatic nerve injury. *J. Neurosci. Methods* **173,** 91–98.

Brown, C. J., Mackinnon, S. E., Evans, P. J., Bain, J. R., Makino, A. P., Hunter, D. A., and Hare, G. M. T. (1989). Self-evaluation of walking-track measurement using a sciatic function index. *Microsurgery* **10,** 226–235.

Carr, M. M., Best, T. J., Mackinnon, S. E., and Evans, P. J. (1992). Strain differences in autotomy in rats undergoing sciatic nerve transection or repair. *Ann. Plast. Surg.* **28,** 538–544.

Couto, P. A., Filipe, M., Magalhães, L. G., Pereira, J. E., Costa, L. M., Melo-Pinto, P., Bulas-Cruz, J., Maurício, A. C., Geuna, S., and Varejão, A. S. P. (2008). A comparison of two-dimensional and three-dimensional techniques for the determination of hindlimb kinematics during treadmill locomotion in rats following spinal cord injury. *J. Neurosci. Methods* **173,** 193–200.

de Medinaceli, L., Freed, W. J., and Wyatt, R. J. (1982). An index of the functional condition of rat sciatic nerve based on measurements made from walking tracks. *Exp. Neurol.* **77,** 634–643.

de Ruiter, G. C., Spinner, R. J., Alaid, A. O., Koch, A. J., Wang, H., Malessy, M. J., Currier, B. L., Yaszemski, M. J., Kaufman, K. R., and Windebank, A. J. (2007). Two-dimensional digital video ankle motion analysis for assessment of function in the rat sciatic nerve model. *J. Peripher. Nerv. Syst.* **12,** 216–222.

Dellon, A. L., and Mackinnon, S. E. (1989). Selection of the appropriate parameter to measure neural regeneration. *Ann. Plast. Surg.* **23,** 197–202.

Den Dunnen, W. F. A., and Meek, M. F. (2001). Sensory nerve function and auto-mutilation after reconstruction of various gap lengths with nerve guides and autologous nerve grafts. *Biomaterials* **22,** 1171–1176.

Dijkstra, J. R., Meek, M. F., Robinson, P. H., and Gramsbergen, A. (2000). Methods to evaluate functional nerve recovery in adult rats: walking track analysis, video analysis and the withdrawal reflex. *J. Neurosci. Methods* **96,** 89–96.

Gage, J. R., Deluca, P. A., and Renshaw, T. S. (1995). Gait analysis: Principles and applications. *J. Bone Joint Surg.* **77,** 1607–1623.

Galtrey, C. M., and Fawcett, J. W. (2007). Characterization of tests of functional recovery after median and ulnar nerve injury and repair in the rat forelimb. *J. Peripher. Nerv. Syst.* **12,** 11–27.

Hruska, R. E., Kennedy, S., and Silbergeld, E. K. (1979). Quantitative aspects of normal locomotion in rats. *Life Sci.* **25,** 171–180.

Jergovic, D., Stal, P., Lidman, D., Lindvall, B., and Hildbrand, C. (2001). Changes in a rat facial muscle after facial nerve injury and repair. *Muscle Nerve* **24,** 1202–1212.

Koopmans, G. C., Deumens, R., Brook, G., Gerver, J., Honig, W. M., Hamers, F. P., and Joosten, E. A. (2007). Strain and locomotor speed affect over-ground locomotion in intact rats. *Physiol. Behav.* **92,** 993–1001.

Lin, F. M., Pan, Y. C., Hom, C., Sabbahi, M., and Shenaq, S. (1996). Ankle stance angle: A functional index for the evaluation of sciatic nerve recovery after complete transection. *J. Reconstr. Microsurg.* **12,** 173–177.

Lowdon, M. R., Seaber, A. V., and Urbaniak, J. R. (1988). An improved method of recording rat tracks for measurement of the sciatic functional index of de Medinaceli. *J. Neurosci. Methods* **24,** 279–281.

Luís, A. L., Amado, S., Geuna, S., Rodrigues, J. M., Simões, M. J., Santos, J. D., Fregman, F., Raimondo, S., Veloso, A. P., Ferreira, A., Armada-da-Silva, P. A. S., Varejão, A. S. P., *et al.* (2007).

Long-term functional and morphological assessment of a standardized rat sciatic nerve crush injury with a non-serrated clamp. *J. Neurosci. Methods* **163,** 92–104.

Metz, G. A., and Whishaw, Q. (2002). Cortical and subcortical lesions impair skilled walking in the ladder rung walking test: A new task to evaluate fore- and hindlimb stepping, placing, and co-ordination. *J. Neurosci. Methods* **115,** 169–179.

Monte-Raso, V., Barbieri, C. H., Mazzer, N., Yamasita, A. C., and Barbieri, G. (2008). Is the sciatic functional index always reliable and reproducible? *J. Neurosci. Methods* **170,** 255–261.

Munro, C. A., Szalai, J. P., Mackinnon, S. E., and Midha, R. (1998). Lack of association between outcome measures of nerve regeneration. *Muscle Nerve* **21,** 1095–1097.

Navarro, X., Butí, M., and Verdú, E. (1994). Autotomy prevention by amitriptyline after peripheral nerve section in different strains of mice. *Rest. Neurol. Neurosci.* **6,** 151–157.

Nichols, C. M., Myckatyn, T. M., Rickman, S. R., Fox, K., Hadlock, T., and Mackinnon, S. E. (2005). Choosing the correct functional assay: A comprehensive assessment of functional tests in the rat. *Behav. Brain. Res.* **163,** 143–158.

Pagnussat Ade, S, Michaelsen, S. M., Achaval, M., and Netto, C. A. (2009). Skilled forelimb reaching in Wistar rats: Evaluation by means of Montoya staircase test. *J. Neurosci. Methods* **177,** 115–121.

Papalia, I., Tos, P., d'Alcontres, F. S., Battiston, B., and Geuna, S. (2003a). On the use of the grasping test in the rat median nerve model: A re-appraisal of its efficacy for quantitative assessment of motor function recovery. *J. Neurosci. Methods* **127,** 43–47.

Papalia, I., Tos, P., Stagno d'Alcontres, F., Battiston, B., and Geuna, S. (2003b). On the use of the grasping test in the rat median nerve model: A re-appraisal of its efficacy for quantitative assessment of motor function recovery. *J. Neurosci. Methods* **127,** 43–47.

Parker, A. J., and Clarke, K. A. (1990). Gait topography in rat locomotion. *Physiol. Behav.* **48,** 41–47.

Patel, M., Mao, L., Wu, B., and Vandevord, P. J. (2007). GDNF-chitosan blended nerve guides: A functional study. *J. Tissue Eng. Regen. Med.* **1,** 360–367.

Pereira, J. E., Cabrita, A. M., Filipe, V. M., Bulas-Cruz, J., Couto, P. A., Melo-Pinto, P., Costa, L. M., Geuna, S., Maurício, A. C., and Varejão, A. S. P. (2006). A comparison analysis of hindlimb kinematics during overground and treadmill locomotion in rats. *Behav. Brain Res.* **172,** 212–218.

Santos, P. M., Williams, S. L., and Thomas, S. S. (1995). Neuromuscular evaluation using rat gait analysis. *J. Neurosci. Methods* **61,** 79–84.

Shen, N., and Zhu, J. (1995). Application of sciatic functional index in nerve functional assessment. *Microsurgery* **16,** 552–555.

Sporel-Ozakat, R. E., Edwards, P. M., Hepgul, K. T., Savas, A., and Gispen, W. H. (1991). A simple method for reducing autotomy in rats after peripheral nerve lesions. *J. Neurosci. Methods* **36,** 263–265.

Strasberg, S. R., Watanabe, O., Mackinnon, S. E., Tarasidis, G., Hertl, M. C., and Wells, M. R. (1996). Wire mesh as a post-operative physiotherapy assistive device following peripheral nerve graft repair in the rat. *J. Periph. Nerv. Syst.* **1,** 73–76.

Varejão, A. S. P., Meek, M. F., Ferreira, A. J. A., Patrício, J. A. B., and Cabrita, A. M. S. (2001a). Functional evaluation of peripheral nerve regeneration in the rat: Walking track analysis. *J. Neurosci. Meth.* **108,** 1–9.

Varejão, A. S. P., Cabrita, A. M., Patrício, J. A., Bulas-Cruz, J., Gabriel, R. C., Melo-Pinto, P., Couto, P. A., and Meek, M. F. (2001b). Functional assessment of peripheral nerve recovery in the rat: gait kinematics. *Microsurgery* **21,** 383–388.

Varejão, A. S. P., Cabrita, A. M., Meek, M. F., Bulas-Cruz, J., Gabriel, R. C., Felipe, M., Melo-Pinto, P., and Winter, D. A. (2002). Motion of the foot and ankle during the stance phase in rats. *Muscle Nerve* **26,** 630–635.

Varejão, A. S. P., Cabrita, A. M., Meek, M. F., Bulas-Cruz, J., Filipe, M., Gabriel, R. C., Ferreira, A. J., Geuna, S., and Winter, D. A. (2003a). Ankle kinematics to evaluate functional recovery in crushed rat sciatic nerve. *Muscle Nerve* **27,** 706–714.

Varejão, A. S. P., Cabrita, A. M., Geuna, S., Melo-Pinto, P., Filipe, M., Gramsbergen, A., and Meek, M. F. (2003b). Toe out angle: A functional index for the evaluation of sciatic nerve recovery in the rat model. *Exp. Neurol.* **183,** 695–699.

Varejão, A. S. P., Cabrita, A. M., Meek, M. F., Bulas-Cruz, J., Melo-Pinto, P., Raimondo, S., Geuna, S., and Giacobini-Robecchi, M. G. (2004). Functional and morphological assessment of a standardized rat sciatic nerve crush injury with a non-serrated clamp. *J. Neurotrauma* **21,** 1652–1670.

Walker, J. L., Resig, P., Guarnieri, S., Sisken, B. F., and Evans, J. M. (1994). Improved footprint analysis using video recording to assess functional recovery following injury to the rat sciatic nerve. *Restor. Neurol. Neurosci.* **6,** 189–193.

Wang, H., Spinner, R. J., Sorenson, E. J., and Windebank, A. J. (2008). Measurement of forelimb function by digital video motion analysis in rat nerve transection models. *J. Peripher. Nerv. Syst.* **13,** 92–102.

Westerga, J., and Gramsbergen, A. (1990). Development of locomotion in the rat. *Dev. Brain Res.* **57,** 163–174.

Zellem, R. T., Miller, D. W., Kenning, J. A., Hoening, E. M., and Buchheit, W. A. (1989). Experimental peripheral nerve repair: Environmental control directed at the cellular level. *Microsurgery* **10,** 290–301.

CURRENT TECHNIQUES AND CONCEPTS IN PERIPHERAL NERVE REPAIR

Maria Siemionow and Grzegorz Brzezicki

Cleveland Clinic, Department of Plastic Surgery, 9500 Euclid Avenue,
Cleveland, Ohio 44195, USA

Despite the progress in understanding the pathophysiology of peripheral nervous system injury and regeneration, as well as advancements in microsurgical techniques, peripheral nerve injuries are still a major challenge for reconstructive surgeons. Thorough knowledge of anatomy, pathophysiology, and surgical reconstruction is a prerequisite of proper peripheral nerve injury management. This chapter reviews the currently available surgical treatment options for different types of nerve injuries in clinical conditions. In overview of direct nerve repair, various end-to-end coaptation techniques and the role of end-to-side repair for proximal nerve injuries is described. When primary repair cannot be performed without undue tension, nerve grafting or tubulization techniques are required. Current gold standard for bridging nerve gaps is nerve autografting. However, disadvantages of this approach, such as donor site morbidity and limited length of available graft material encouraged the search for alternative means of nerve gap reconstruction. Nerve allografting was introduced for repair of extensive nerve injuries. Tubulization techniques with natural or artificial conduits are applicable as an alternative for bridging short nerve defects without the morbidities associated with harvesting of autologous nerve grafts. Achieving better outcomes

INTERNATIONAL REVIEW OF
NEUROBIOLOGY, VOL. 87
DOI: 10.1016/S0074-7742(09)87008-6

depends both on the advancements in microsurgical techniques and introduction of molecular biology discoveries into clinical practice. The field of peripheral nerve research is dynamically developing and concentrates on more sophisticated approaches tested at the basic science level. Future directions in peripheral nerve reconstruction including, tolerance induction and minimal immunosuppression for nerve allografting, cell based supportive therapies and bioengineering of nerve conduits are also reviewed in this chapter.

I. Introduction

Peripheral nerve injury is a common condition in both civil and military circumstances. Approximately 100,000 patients undergo peripheral nerve surgery in the United States and Europe annually (Kelsey *et al.*, 1997). Etiologies of the injury include penetrating injuries, crush, ischemia, and traction as well as less common electric shock and vibration (Robinson, 2000, 2004). Stretch-related injuries sustained in motor vehicle accidents account for the most common civilian nerve trauma. Thirty percent of peripheral nerve injuries derive from lacerations by sharp objects and long bone fractures (Stanec *et al.*, 1997). Up to 73.5% of nerve injuries occur in the upper extremities, with the ulnar nerve, alone or in combination, being the most often affected (Kouyoumdjian, 2006). Blast injury from bombs and other explosive devices is a common cause of nerve trauma in military conditions (Maricevic and Erceg, 1997). High-energy trauma involves massive soft tissue and bony loss accompanied by vascular injuries requiring an emergency repair. As much as one-third of nerve injuries can be associated with arterial damage. Loss of long segments of nerves in the injured limb frequently requires nerve grafting for function restoration.

Severe nerve injury has a devastating impact on patients' quality of life. Typical symptoms are sensory and motor function defects that could result in complete paralysis of an affected limb or development of intractable neuropathic pain. Nerve fibers of the transected nerve regenerate spontaneously to the extent limited by the size of the nerve gap, neuroma formation, and scar tissue formation. In the cases of failed recovery, surgical repair of the affected nerve becomes necessary (Matsuyama *et al.*, 2000). The primary goal of nerve repair is to allow reinnervation of the target organs by guiding regenerating sensory, motor, and autonomic axons into the environment of the distal nerve with minimal loss of fibers at the suture line (Brushart, 1991). Many factors have to be taken into consideration when predicting outcome of peripheral nerve repair including: type, location, and extent of nerve injury; timing of surgery; type of repair; proper alignment of fascicles; surgical technique; and patient comorbidities. The first successful nerve regeneration after surgical repair was reported by Cruikshank (1795). Introduction of microsurgical nerve repair to clinical practice was preceded by extensive experimental and clinical studies on peripheral nerve anatomy and physiology (Levi-Montalcini and Hamburger, 1951; Lundborg *et al.*, 1982) (Fig. 1). However, despite meticulous surgical techniques and different repair methods, full functional outcome, especially of motor function, is rarely achievable. Nerve autografting is still a golden standard for nerve gap repair. Shorter defects, up to 3 cm, can be repaired with natural or artificial nerve conduits that eliminate the donor site morbidity. Large nerve defects repair, exceeding the length of available autograft, remains the biggest surgical challenge in peripheral nerve surgery. Currently, nerve allografting is the only available option for massive nerve tissue loss. Disadvantages of nerve allografting include the risk of rejection and complications related to immunosuppression (Mackinnon *et al.*, 2001). Current experimental studies concentrate on development of low-dose immunosuppression, and tolerance induction strategies (Brenner *et al.*, 2005; Jensen *et al.*, 2005; Siemionow *et al.*, 2003) as well as tissue engineering of natural and artificial conduits capable of supporting nerve regeneration over longer distances (Aebischer *et al.*, 1989; Calder and Green, 1995; Chen *et al.*, 1994; Chiu *et al.*, 1982; Heijke *et al.*, 1993; Mackinnon and Dellon, 1990; Madison *et al.*, 1985, 1987). Recent literature also emphasizes the emerging role of cell-based supportive therapies in nerve repair. Schwann cells and bone marrow stromal cells are of particular interest because of their abilities to express neurotrophic cytokines and support axonal growth (Funakoshi *et al.*, 1993; Rodriguez *et al.*, 2000; Sorensen *et al.*, 2001; Strauch *et al.*, 2001). This chapter reviews techniques

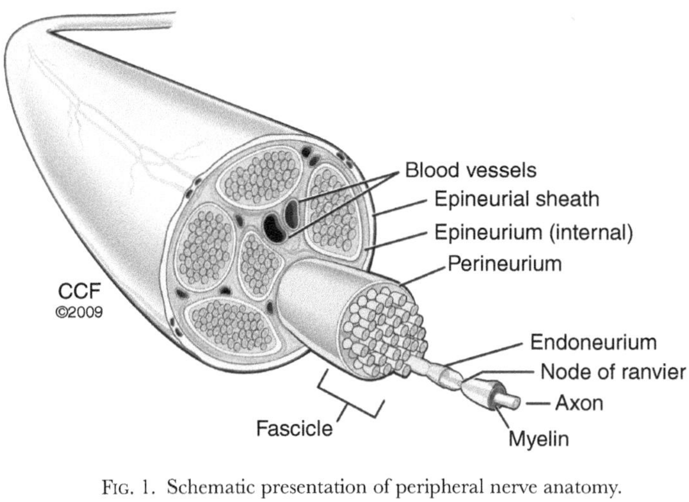

FIG. 1. Schematic presentation of peripheral nerve anatomy.

of nerve repair applied in clinical practice and presents future directions, which are currently developed in experimental conditions.

II. Timing of Nerve Repair

Timing of nerve repair depends on the type of nerve injury sustained, condition of the wound, and vascular supply to nerve bed (Fig. 2) (Siemionow and Sari, 2004a; Spinner and Kline, 2000). In cases of sharp nerve transections with none or minimal crush component, good blood supply and clean wound, primary nerve repair is the best option for restoring the function. Historically,

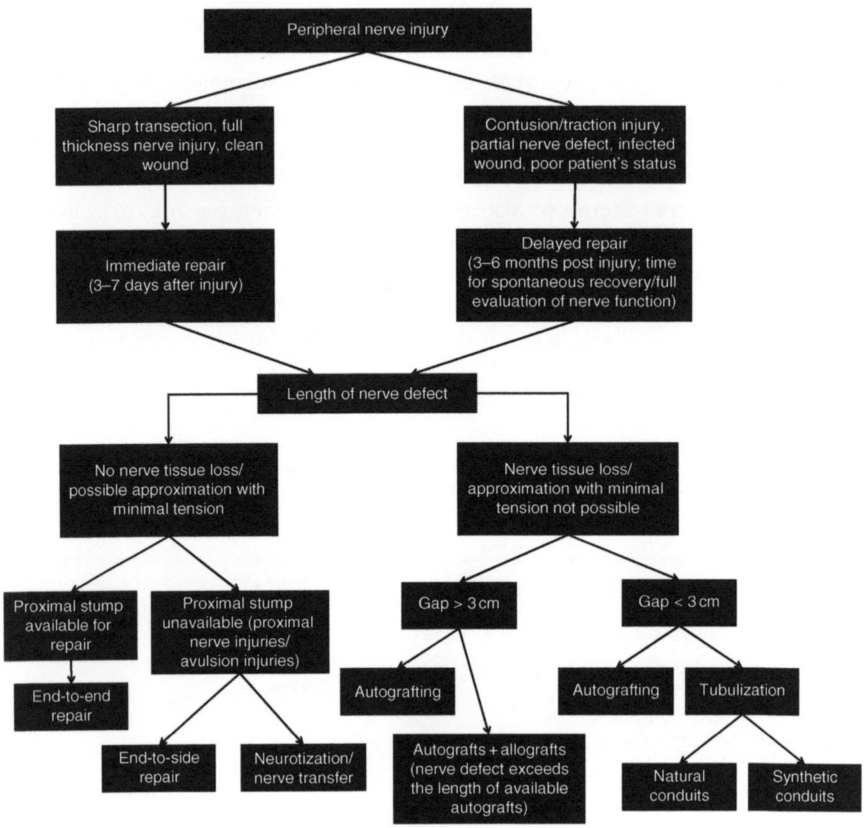

FIG. 2. Algorithm of peripheral nerve repair.

primary nerve repair was performed 3 weeks after initial injury to provide time for full Wallerian degeneration. In studies by Mackinnon *et al.*, better outcomes were associated with immediate reconnection of the transected nerve (Mackinnon, 1989). Currently, primary surgery is performed within 72 h up to 7 days after nerve injury (Dvali and Mackinnon, 2003). However, when immediate repair criteria are not met, delayed repair is required. Secondary reconstruction is also preferred in incomplete nerve injuries and when natural recovery could yield better outcome than achieved by surgical repair. Lack of clinical and electrophysiological signs of spontaneous recovery after 3–6 months imposes surgical intervention (Campbell, 2008). The degree of damage to the nerve stumps after blunt nerve transection or avulsion can also be more precisely evaluated several weeks after trauma. The time of delay is, however, limited by viability of Schwann cells and endoneurial tubes. It is essential that endoneurial tubes will be in contact with regenerating axons within 18–24 months after injury, otherwise degeneration will occur. Target muscle atrophy becomes irreversible after 12–18 months of denervation, which limits the functional outcome of repair (Campbell, 2008). Sensory receptors survive for a much longer time though sensory function can be restored after longer time intervals, even years after initial trauma. The retraction of the nerve stumps which may reach as much as 8% of nerve length within first three weeks, can make the secondary repair more technically demanding than primary repair or may require nerve grafting (Trumble, 1999).

III. Direct Repair

Direct nerve repair is preferred when the gap is small and two ends can be approximated with minimal tension (Diao and Vannuyen, 2000). Better results are obtained when the nerves are purely motor or purely sensory and when the amount of intraneural connective tissue, varying from 22 to 80%, is relatively small (Townsend, 1994; Trumble, 1999). For optimal nerve regeneration after repair nerve stumps must be precisely aligned without tension, and repaired atraumatically with minimal tissue damage and minimal number of sutures (Dahlin and Lundborg, 2001; Dvali and Mackinnon, 2003; Harris and Tindall, 1991; Mackinnon, 1989; Maggi *et al.*, 2003; Siemionow, 2002; Sunderland, 1990; Tetik *et al.*, 2002; Waller, 1850).

A. End-To-End Repair

End-to-end nerve repair techniques include epineural repair, group-fascicular repair, and fascicular repair. Epineural repair is a commonly used technique following sharp nerve injury of proximal portion of nerves without nerve tissue loss and for partial injuries with good fascicle alignment. Usually, 4–8 single sutures are passed through the epineural sheath (Matsuyama *et al.*, 2000) (Fig. 3). The main goals of epineural repair are to obtain continuity of the nerve stumps without tension and with proper fascicular alignment (Wilgis, 1991). Correct fascicle positioning can be confirmed by the continuity of the nerve's surface structures such as blood vessels (vasa nervorum) within the epineurium (Ogata and Naito, 1986; Wilgis, 1991). Alternatively, fibrin glue can be used instead of standard nylon sutures (Ornelas *et al.*, 2006a,b). Grouped fascicle repair is best applied in a crush nerve injury or delayed nerve repair requiring trimming of the nerve ends, especially in mixed nerves where matching groups of fascicles can be easily identified. Epineurium is retracted prior to debridement of damaged

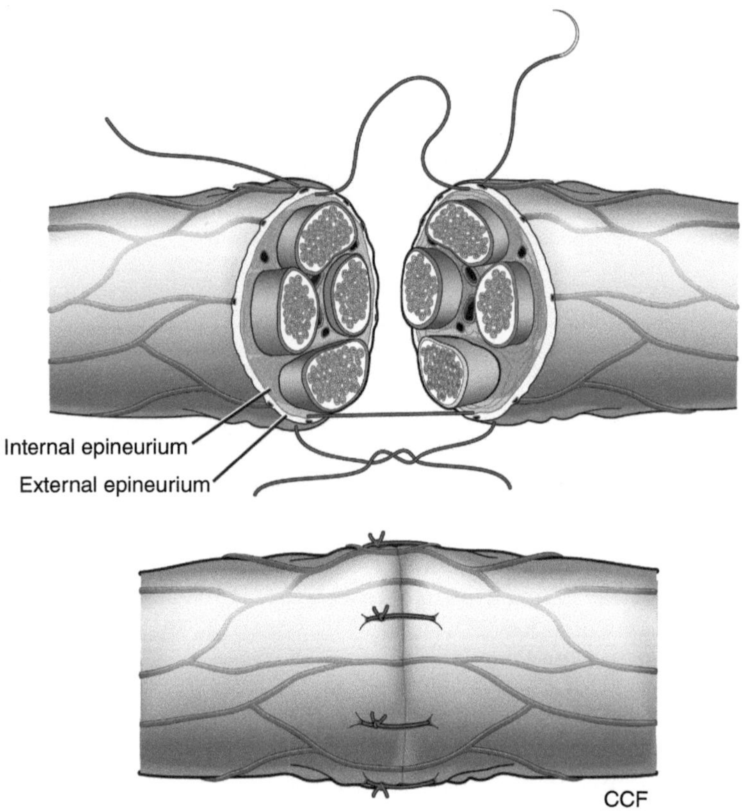

FIG. 3. Diagram of standard epineural repair. Stumps are aligned based on position of blood vessels within epineurium. Sutures are passed through the internal and the external epineurium of proximal and distal stump.

nerve ends and fascicle dissection. Corresponding group of fascicles are approximated with two to three sutures passing through the interfascicular epineurium (Fig. 4). The number of sutures and tension should be minimized to avoid scar tissue formation. In fascicular repair, the sutures are placed within perineurium. This technique requires dissection of the interfascicular epineurium and separation of fascicles. Fascicular repair is not widely used anymore because of the number of sutures required (2–3 per fascicle), adding to scar tissue formation, and is also technically demanding (Trumble, 1999). Currently, the most frequent indication to apply fascicular technique is repair of partially damaged nerves. Both the group fascicular and the fascicular repair provide better fascicular alignment decreasing misdirection of regenerating axons; however, additional dissection and higher number of sutures required compared to epineural repair can lead to increased intraneural scarring and disruption of intraneural blood flow (Brushart *et al.*, 1983; Ogata and Naito, 1986; Sunderland, 1978). Experimental studies did not show advantage of more precise fascicle alignment achieved in group fascicular repair over standard epineural technique when comparing functional outcomes (Bratton *et al.*, 1979; Hudson *et al.*, 1979; Levinthal *et al.*, 1977; Lundborg *et al.*, 1997).

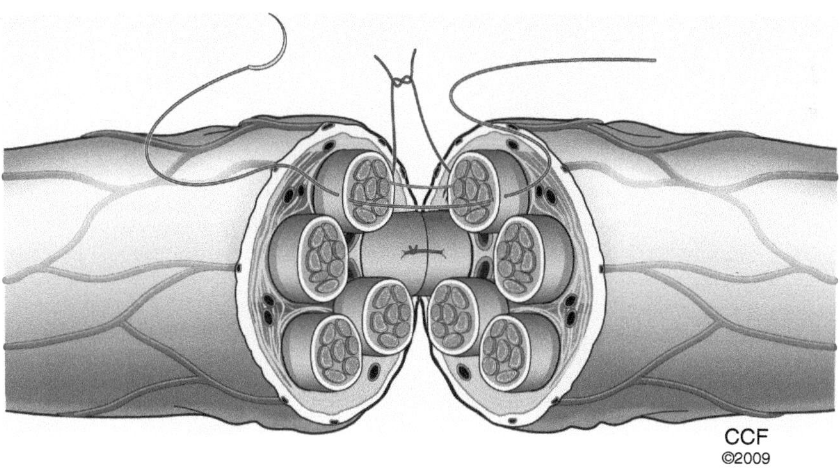

FIG. 4. Schematic presentation of grouped fascicular repair. Epineurium is retracted prior to coaptation. Corresponding group of fascicles are sutured with 2–3 sutures passing through the interfascicular epineurium.

B. Epineural Sleeve Repair

Another possibility for direct nerve repair is epineural sleeve neurorrhaphy. Primarily described as epineural cuff technique by Snyder *et al.* (1968), it was modified and reevaluated by Siemionow group (Lubiatowski *et al.*, 2008; Siemionow, 2002; Tetik *et al.*, 2002). In this technique, epineurium covering the distal stump is rolled back and a 2-mm nerve segment is resected. The created epineural sleeve is pulled-over the proximal nerve end and is sutured to the epineurium 2 mm proximal to the coaptation site with two sutures (Fig. 5). This study showed faster functional recovery compared to standard epineural end-to-end repair. Various factors can be attributed to the superior results of this technique. The compression and tension is transferred from the repair site to the proximally located epineurium. The epineural sleeve provides a biological chamber for the axoplasmic fluid leakage from transected nerve ends providing a neuropermissive environment for growing axons. Additionally, this technique provides guidance for regenerating nerve fibers enabling higher number of

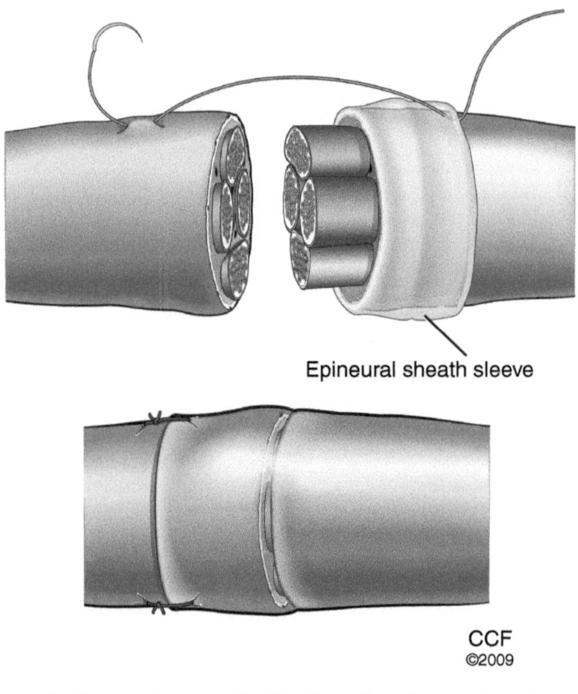

Fig. 5. Diagram of epineural sleeve repair. The free edge of epineurium from the distal stump is rolled back distally and a 2-mm epieneural sleeve is created. A 2-mm fragment of distal nerve is resected. Epineural sleeve is pulled over the proximal nerve and is anchored to the epineurium 2 mm proximal to the coaptation site with two 10-0 sutures placed 180° apart.

axons to reach target organs and prevents neuroma formation (Martini and Fromm, 1989). Better functional results presented in this study have to be confirmed in clinical practice.

C. End-To-Side Repair

End-to-side repair offers a promising technique for repair of peripheral nerve injuries, where the proximal nerve stump is unavailable or a significant nerve gap exists. It is particularly useful in brachial plexus injuries and facial nerve reanimation. Recent studies evaluated application of end-to-side repair in distal nerve defects of the upper limb (Luo *et al.*, 1997; Mennen, 1999, 2003). Since its reintroduction by Viterbo *et al.* (1992) many studies were performed to assess advantages of this approach in nerve repair. One of the major advantages of the end-to-side repair is the recovery of function of the injured nerve without compromising the function of the donor nerve. The origin of the reinnervating axons is currently widely discussed. Some authors assume that the nerve fibers invade from the donor nerve axons damaged during nerve preparation for coaptation (Rovak *et al.*, 2001). Others provide evidence based on double-labeling studies for collateral (nodal) sprouting from the undamaged axons of the donor nerve at the coaptation side (Zhang *et al.*, 1999). It was also shown that uninjured sensory axons spontaneously sprout *de novo* from the donor nerve, but motor axons may need to be injured in order to sprout and provide proper target muscle reinnervation (Tarasidis *et al.*, 1997, 1998). Largest cohort of patients treated with this method was reported by Mennen (1999, 2003) and consisted of 50 cases with various peripheral nerve lesions including ulnar, median, musculocutaneous, and digital nerves. Good sensory and motor recovery was reported in some cases; however, no control groups and statistical analysis was provided in these studies. Other authors reported partial nerve recovery, mostly of the sensory component, in experimental studies (Liu *et al.*, 1999; Lutz *et al.*, 2000; McCallister *et al.*, 1999; Tarasidis *et al.*, 1998; Yuksel *et al.*, 1999; Zhang *et al.*, 1999). Felici *et al.* (2003) reported 39 cases of brachial plexus injury treated with end-to-side neurorrhaphy and 80% were considered fair or poor results. Authors assume that end-to-side technique might be useful, but the outcomes are unpredictable and highly depend on the surgical technique itself. On the other hand, excellent results were noted in 80% of patients undergoing facial nerve reanimation by means of end-to-side coaptation of facial nerve stumps to partially transected hypoglossal nerve with an interpositional jump-graft (Hammerschlag, 1999; Manni *et al.*, 2001; May *et al.*, 1991). In this technique the donor nerve is purposely partially neurectomized allowing motor neurons to sprout more effectively.

IV. Nerve Grafting

The first review on management of peripheral nerve gaps was published by Huber (1895) followed by Sanders (1942) who presented two options for nerve gap management: (1) manipulative operations and (2) bridging. Manipulative operations included techniques, which decreased tension at the primary repair site to achieve end-to-end stump apposition. Nerve stretching, stump mobilization, extremity positioning, and bone shortening procedures were employed, but are only of historical importance now, because it was clearly demonstrated that tension had deleterious effects on nerve regeneration (Millesi, 1991; Miyamoto, 1979; Moneim, 1982; Nunley, 1991).

In case of nerve gaps, which cannot be approximated and coaptated without tension, current gold standard of repair is autologous nerve grafting (Fig. 6). Nerve grafts revealed superior results when compared with direct repairs performed under undue tension that produced nerve ischemia (Birch and Raji, 1991; Kline *et al.*, 1972; Lundborg and Rydevik, 1973; Millesi *et al.*, 1972). Experimental study on rat sciatic nerve ischemia model showed that blood flow was inversely proportional to nerve tension. Following nerve repair acutely causes only 4% nerve retraction, making primary end-to-end repair possible. In contrast, 28% retraction reported in delayed nerve repairs requiring long gap management would produce a profound irreversible decrease in blood flow; hence, nerve grafting is required (Trumble and Shon, 2000). There is still discussion about best treatment options for small to moderate size nerve gaps. Extensive mobilization and coaptation with mild tension carries risk of decreased blood flow and proliferation of

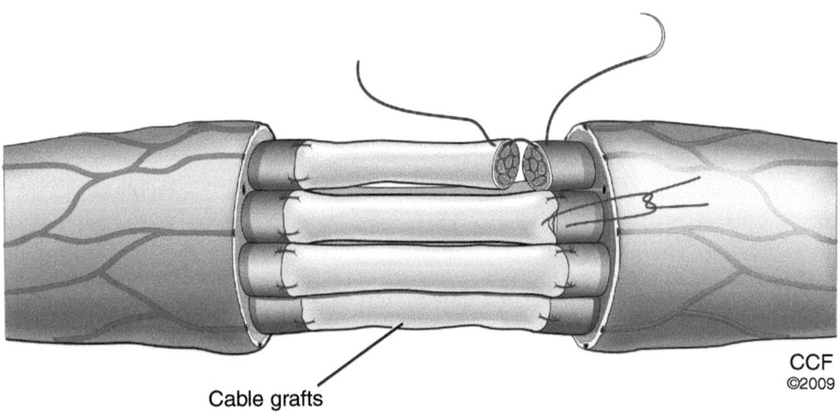

Fig. 6. Schematic presentation of cable graft repair. Graft segments are coaptated to the corresponding groups of fascicles of proximal and distal stumps.

connective tissue blocking axonal regeneration (Rodkey *et al.*, 1980). Acute stretching may result in intraneural hemorrhage leading to scar formation. Scar maturation may compress nerve fibers and ultimately result in neuroma formation. On the other hand, nerve grafting imposes the regenerating fibers to cross two coaptation sites increasing the risk of axonal loss due to scar formation and misdirection into the perifascicular and epineurial connective tissue at each suture line (Bratton *et al.*, 1979; Hudson *et al.*, 1979). Outcomes of many research studies comparing direct epineurial repair under moderate tension with interfascicular nerve grafting (Bratton *et al.*, 1979; Hudson *et al.*, 1979; Rodkey *et al.*, 1980; Wilgis, 1991; Wong and Scott, 1991) revealed that functional results are comparable and the recovery attained was similar. When the gap is relatively small, the tension would be less deleterious for axonal regeneration than crossing two coaptation sites, although the nerve blood flow may be impaired (Wong and Scott, 1991). Hence, if the nerve stumps can be coaptated under mild tension, a direct repair technique is preferred. It is less technically demanding, less time consuming, avoids donor site morbidity, and yields results comparable to nerve grafting recovery rates. Millesi (1967a) was the first to demonstrate the value of autologous nerve grafting in animal studies. It is widely accepted that nerve gaps, which would require relatively large tension in order to perform direct coaptation should be repaired by means of nerve grafting or application of tubulization techniques. Currently, tubulization techniques are feasible only in short nerve gaps, which are not exceeding 3 cm. For larger nerve defects, nerve grafting is required.

A. Nerve Autografts

First experimental work on nerve grafting was performed by Philipeaux and Vulpian (1870). Albert (1885) introduced nerve grafts to clinical practice for the first time in 1885. Unfortunately, high infection rate, poor surgical technique, and inadequate knowledge of nerve anatomy yielded poor outcomes. Seddon (1947) studies on nerve grafting compiled with introduction of microsurgical techniques by Millesi *et al.* (1967b) and detailed knowledge of intraneural anatomy of the peripheral nerves reported by Sunderland (1945) established the free autograft repair technique as a gold standard in nerve gap management. The harvested fascicular graft undergoes Wallerian degeneration (Millesi, 1990) and provides mechanical guidance by creating supportive structure for the ingrowing axons. Autologous nerve grafts fulfill the criteria for an ideal nerve conduit because they provide a permissive and stimulating scaffold including Schwann cell basal laminae, neurotrophic factors, and adhesion molecules (Almgren, 1975). Grafted Schwann cells increase their expression of surface cell adhesion molecules and provide a permissive scaffold—basement membrane containing extracellular matrix proteins such as laminin and fibronectin (Almgren, 1975; Fu and Gordon, 1997). The Schwann cells' ability to promote axonal growth was presented by Maeda *et al.* (1993) who using "stepping stones" procedure of placement of small fragments of the graft into the synthetic conduit obtained improved nerve trunk regeneration. It was reported that loss of Schwann cells in preconditioned conduits resulted in suppression of nerve regeneration (Hall, 1986). It was also stressed that other important source of neurotrophic factors is the axoplasmic fluid leaking from transected axons. Siemionow *et al.* (2002) collected axoplasmic fluid leaking from transected nerve ends and injected the fluid under perineurium of repaired nerve which improved and accelerated nerve regeneration. The most important factors influencing axonal outgrowth through the grafted nerve are: the diameter of the grafted nerve fragment and the vascularity of the surrounding tissue bed. Nonvascularized autografts survive by the diffusion from the surrounding tissues, and as a result of the direct capillary in-growth from the periphery and nerve ends, leading to nerve revascularization (Almgren, 1975; Millesi, 2000). Application of thicker nerve grafts carries the risk of central necrosis and scar formation because of the poor diffusion and delayed revascularization (Bielschowsky and Unger, 1916; Nunley, 1991). Clinical observations show that smaller caliber grafts result in better outcomes (Kline, 1990). Similarly, poorly vascularized tissue bed increases scar formation, delays nerve regeneration, and worsens functional outcome (Matsuyama *et al.*, 2000). Sensory cutaneous nerves are commonly used as donor nerves for autografting because their harvest results in acceptable morbidity, mainly sensory loss in the area supplied by the harvested sensory branch. Currently used donor nerves include: sural nerve, lateral antebrachial cutaneus nerve (LCAN), anterior division of the medial antebrachial cutaneous nerve (MACN), dorsal cutaneous branch of the ulnar nerve (DCBUN), and superficial sensory branch of the radial nerve (SSR) (Matsuyama *et al.*, 2000) (Table I). In order to choose the best autologous nerve graft, a surgeon has to take into consideration the caliber of the nerve to be repaired, length of the defect, and donor site

TABLE I
EXAMPLES OF DONOR NERVES AVAILABLE FOR AUTOGRAFTING

Donor nerve		Available length (cm)	Hypesthesia
Medial antebrachial cutaneous nerve	MACN	10–12 (above elbow) 8–10 (below elbow)	Medial forearm
Lateral antebrachial cutaneous nerve	LACN	10–12	Lateral forearm
Superficial sensory branch of the radial nerve	SSR	20–30	Radial dorsal hand
Dorsal cutaneous branch of the ulnar nerve	DCBUN	4–6	Ulnar dorsal hand
Sural nerve		30–50	Lateral foot

morbidity. The standard cable grafting technique emphasizes the need to obtain maximal coverage of the cross-sectional area of the severed nerve. In small nerves, such as digital nerves, only one cable is usually required for the repair, while in more proximal lesions of major nerves such as the median nerve more cables are needed to cover the cross-sectional area of the nerve stumps. In bridging long nerve defects in large nerve trunks, multiple-site harvest may be required to obtain the proper amount of material for repair. To decrease the donor site morbidity and to expand the availability of the autograft material, Siemionow *et al.* (2004b) investigated the single-fascicle method of nerve repair. Experimental study on rat sciatic nerves revealed faster regeneration and better functional results in group treated with single fascicle, covering 25–59% of the cross-sectional area when compared to the conventional graft. The study also introduced techniques of single-fascicle microcoaptation using single pullout suture, which minimizes the foreign body reaction and fibrosis at the repair site (Fig. 7). The length of the graft required to bridge the defect also depends on the mobility of the severed nerve (Wilgis, 1991). For example, extension

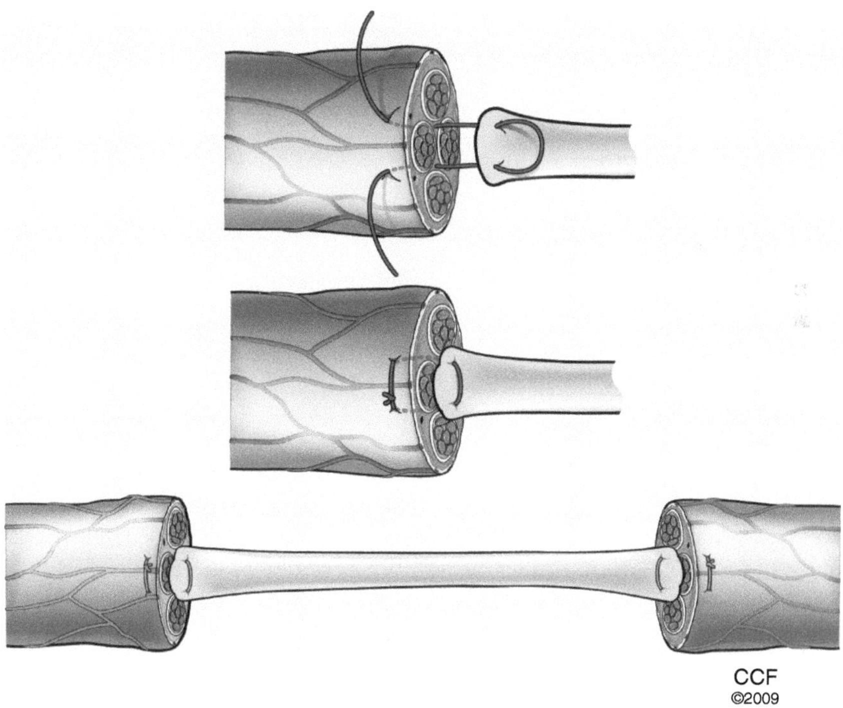

FIG. 7. Single fascicle nerve repair. The single fascicle dissected from the donor nerve is coaptated by the single horizontal mattress suture. The central alignment of the graft is achieved, bridging proximal and distal sites of repair.

of the wrist and fingers causes 7 mm excursion of the median nerve whereas a mean of 4 mm excursion results from flexion of the elbow (McLellan and Swash, 1976). Finally, during the nerve autografting procedure, the donor site morbidity cannot be overcome. In ideal circumstances the graft should be obtained from the same skin incision as the repair site facilitating regional anesthesia and reduce chance for wound healing complications (Nunley, 1991). However, this situation is rarely possible and usually graft is harvested from different anatomical region of the body. When harvesting the autograft the area of permanent anesthesia should be minimized. Critical area for sensation and area adjacent to sensory loss from the primary nerve injury should be avoided. Sural nerve is the most commonly chosen donor nerve. It supplies sensation to the posterior and lateral lower third of the leg and lateral foot. Over 20 cm of the graft material can be obtained when sural nerve proper is harvested alone, and up to 50 cm when harvested in the complex with medial sural cutaneous nerve (Ortiguela *et al.*, 1987). Small diameters of the sural nerve (2–4 mm) and nutrient vessels within the nerve enable fast revascularization of the grafts. Adverse effects of harvesting the sural nerve range from sensory deficit around the lateral foot (reported as a main complaint in 9.1–44% patients), to neuroma formation and unbearable pain noted in 6.1–8.1% of cases (Ortiguela *et al.*, 1987; Staniforth and Fisher, 1978). Mostly the symptoms are longstanding but quite well tolerated.

B. Nerve Allografts

Nerve autografts has remained the gold standard for peripheral nerve gap repair for over 50 years. However, donor site morbidity and limited amount of available graft material remain as the limitations of this technique. In gaps where the reconstruction requires length of the graft exceeding available nerve autografts the application of allograft material from cadaver donors is the only clinical option currently available. Allograft nerve provides guidance and viable donor Schwann cells enabling growing host axons to pass from the proximal to distal stump and reinnervate target organs. The first reconstructions with allograft material were performed on short gaps and yielded promising results. Introduction of freeze-drying and irradiation techniques resulted in some successful outcomes reported in late 1960s (Campbell *et al.*, 1963; Marmor, 1964). Recent extensive experimental studies on nerve allograft immunogenicity, immunosuppressive regimens and allograft revascularization gave the basis for reintroduction of this technique to clinical practice. In 2001, Mackinnon reported the outcome of 10 clinical cases treated with nerve allografts (Mackinnon *et al.*, 2001). The main advantage of nerve allografts compared to autografts is the lack of donor site morbidity and virtually unlimited length of nerve tissue available for transplantation. Moreover, the injured nerve in recipient can be replaced with the same nerve type from the donor; for example, mixed sensory-motor ulnar nerve can be replaced with the same mixed type ulnar nerve harvested from the donor providing better motor recovery compared to sensory only sural nerve graft (Chu *et al.*, 2008; Moradzadeh *et al.*, 2008). Allogenic nerve tissue in spite of having low immunogenic potential compared to skin, muscle, or bone, requires immunosuppressive treatment to prevent rejection of the graft. Immunogenicity of the nerve allograft tends to decrease over time as the process of exchanging Schwann cells from donor origin to host ones proceeds (Midha *et al.*, 1993, 1994). Without immunosuppressive treatment, after transplantation, donor nerves blood–nerve barrier (BNB) is broken down, graft is revascularized and infiltration of immune cells begins. Macrophages actively phagocytose both the intact and damaged myelin. Activated in periphery CD4 and CD8 T-cells infiltrate the graft and target the donor origin Schwann cells which increase their major histocompatibility (MHC) class II alloantigens acting as antigen presenting cells (APC). Cytokines released from activated T-helper cells and APCs, including tumor necrosis factor-α (TNF-α), interleukin-1 and -2 (IL-1, -2), and interferon-γ (IFN-γ) start the cascade of different mechanisms leading to allograft rejection. As a result the graft is rejected, becomes fibrotic, and nonfunctional (Bain, 2000; Gulati, 1998; Hettiaratchy *et al.*, 2004).

Strategies for allograft rejection prevention include MHC matching, allograft pretreatment, host immunosuppression, and tolerance-inducing strategies.

1. *MHC Matching*

Similar to solid organ transplantation MHC matching between donor and recipient results in better recovery as presented by Mackinnon *et al.* (1985) in studies on rats. Clinically, patients receiving nerve allografts from ABO matched donors presented good sensory and motor outcomes (Mackinnon *et al.*, 2001).

2. *Graft Pretreatment*

Graft pretreatment methods have been extensively studied from the late 1960s when irradiation and freeze-drying techniques were reported (Campbell *et al.*, 1963; Marmor, 1964). Other methods include cryopreservation, lyophilization, freeze-thawing, predegeneration, irradiation, and cold-storage in the University of Wisconsin Storage Solution. Most promising results were obtained with cold storage of nerve allografts in University of Wisconsin Storage Solution (Evans *et al.*, 1999). One-week storage at 5 °C did not decrease the number of viable donor Schwann cells, but reduced the expression of MHC class II molecules triggering activation of T-cells (Atchabahian *et al.*, 1999; Evans *et al.*, 1998; Levi *et al.*, 1994; Strasberg *et al.*, 1996b). Increasing the duration of cold storage resulted in reduction of antigenic load and subsequently decreased lymphocytic migration to nerve graft. Recipients of cold-stored allografts required lower doses of immunosuppression compared to fresh allografts (Evans *et al.*, 1995, 1999; Strasberg *et al.*, 1996a). Cryopreservation of nerve allografts treated with cryoprotectant 10% dimethyl sulfoxide at −196 °C in liquid nitrogen did not alter the cells' viability and may be an option for nerve tissue banking. Other methods did not prove to be clinically relevant because either good results were limited to short grafts only (Trumble and Shon, 2000) or theoretical advantages of proposed method were not proven experimentally (Mackinnon *et al.*, 1984; Pollard and McLeod, 1981; Trumble, 1992) or the used method severely damaged donor Schwann cells in an attempt to decrease antigenicity, altering nerve regeneration (Mackinnon *et al.*, 1984).

3. *Immunosuppression*

Current immunosuppressive protocols in nerve allografting procedures are based on experimental data and experience from solid organ and hand transplantations. Until recently, cyclosporine A (CsA) was the mainstay of immunosuppressive protocol (Bain *et al.*, 1988; Mackinnon *et al.*, 1992). The CsA application demonstrated inhibition of host rejection response in rat model of nerve allotransplantation (Zalewski and Gulati, 1980, 1982). Some experimental studies report that outcomes following nerve allografting in CsA-immunosuppressed recipients are comparable with autograft repair (Ishida *et al.*, 1993; Zalewski and Gulati, 1980). Cold-preservation of allografts had synergistic effect with CsA administration expressed by decreased required dose of CsA compared to fresh allografts (Strasberg *et al.*, 1996a). Recently, tacrolimus (FK506) immunosuppression was reported to result in improved functional recovery and axonal regeneration compared to CsA protocol. Moreover, FK506-based immunosuppressive protocol combined with cold-preserved allografts yielded superior and accelerated regeneration outcomes to that observed in isografts (Gold *et al.*, 1995; Jensen *et al.*, 2005). It was also reported that FK506 can rescue short rat nerve allografts within 10 days of occurrence of rejection (Feng *et al.*, 2001). CsA, however, is not capable of reversing the rejection process in rat model (Chen *et al.*, 2002). Atchabahian *et al.* (1998) demonstrated that immunosuppressive therapy in peripheral nerve allografts is not lifelong, unlike in solid organ transplantation, and can be discontinued after axons bridged the defect and reestablished end organ connections. Therefore, immunosuppressive regimen based on FK506 can be shortened because of the accelerated axon growth and stopped when the regenerating axons reach the target.

4. *Tolerance Induction*

Elimination of immunocompetent T-cells promotes nerve allograft tolerance and may be used as an induction protocol. This goal can be achieved by both selective and nonselective T-cell depletion. Supportive pretreatment with donor-origin cells can also be considered. Nonselective depletive polyclonal or monoclonal antibodies target all T-cells and not only alloreactive T-cells. Antibodies used in clinical practice include polyclonal antibodies—polyantilymphocyte globulin (ALG) and antithymocyte globulin (ATG)—and monoclonal antibodies—anti-CD3 (muromonab) and anti-CD52 Campath-1H Campath® (alemtuzumab) (Genzyme Corporation, Cambridge, MA). In rat nerve allograft models 5 week selective T-cell depletion regimen (alfabeta T-cell receptor monoclonal antibody supported with CsA) investigated by Scharpf *et al.* (2006) yielded favorable outcome. Costimulatory blockade achieved by inhibition of second signal of T-cell activation and proliferation is another therapeutical option for prevention of acute and chronic nerve allograft rejection. Administration of anti-CD40 ligand antibody prevents rejection and allows regeneration of peripheral nerve allografts in nonhuman primates (Brenner *et al.*, 2004a). However, the immunosuppressive effect is transient and does not establish long-lasting tolerance but allows fast recovery of immunocompetence (Brenner *et al.*, 2004b). New approaches in tolerance induction include host's pretreatment with donor antigens. Single intraportal injection of ultraviolet-B irradiated splenocytes induced tolerance in rat nerve allograft model as presented by Genden *et al.* (1998, 2001).

The first thorough report on successful nerve allotransplantation clinical series with clinical outcomes, inclusion criteria, and details on immunosuppressive therapy was published by Mackinnon *et al.* (2001). Seven patients aged 3–24 years (four male, three female) who sustained severe trauma resulting in long peripheral nerve defects untreatable with autografts were included in this series. Four patients presented with upper extremity nerve lesions whereas in three cases lower limb was affected. Repair was performed with nerve allografts harvested from fresh cadavers and cold-stored at 5 °C in University of Wisconsin Storage Solution for 7 days prior to transplantation. In 5 cases the allograft repair was augmented with autografts to minimize the antigenic load. Total length of nerve grafts (allo- and autografts) ranged from 96 to 413 cm (including the 350 cm allograft). First group of patients ($n = 3$) were immunosuppressed with CsA, azathioprine, and prednisone protocol. Promising outcomes in experimental studies with FK506-based regimen resulted in exchange of CsA to FK506 in the second group of patients ($n = 4$). Immunosuppressive treatment was started few days before surgery and lasted for about 18 months (range, 12–26 weeks). There were no serious complications of this treatment. The immunosuppression was continued for 6 weeks after signs of nerve regeneration distal to the nerve allograft. No deterioration of nerve function after immunosuppression withdrawal was observed. However, one patient rejected nerve allograft after 4 months due to sub-therapeutical levels of cyclosporin. Significant sensory recovery including light touch, temperature, and pain sensation was observed in all patients. However, measurable 2-point discrimination was not present in any case. Motor recovery occurred only in patients with upper extremity nerve lesions. The outcomes presented by this study cannot be compared to other clinical reports because of the severity of nerve lesions (gap length, multiple nerve injuries) sustained by enrolled patients. Mackinnon series enrolled relatively young population of patients with high nerve regeneration capabilities, but still the recovery especially of the motor function remained scant. On the other hand, patients with severe trauma with limited or lack of function in the affected limb will appreciate any improvement in his/her status, such as protective sensation. It is worth mentioning that in series of clinical cases of hand transplantation good functional recovery of the nerves was achieved, including return of 2-point discrimination and intrinsic muscles function, which was contributed to regenerative properties of FK506 (Lanzetta *et al.*, 2005; Schneeberger *et al.*, 2006; Schuind *et al.*, 2007; Tobin *et al.*, 2001).

V. Conduit Repair

The disadvantages of nerve gap repair with autografts such as donor site morbidity, need of second incision side for autograft harvest, and prolonged surgery durations encouraged the development of conduit/nerve guides based strategies for nerve gap repair. The entubulation model provides an environment for outgrowing axons, growth of Schwann cells, and neurotrophic stimulation by distal stump that are necessary elements for optimal return of nerve function (Meek and Coert, 2008). Currently, tubulization technique is reserved only for small nerve defects up to few centimeters (Strauch, 2000). Various natural and synthetic materials have been tested in experimental and clinical conditions.

A. Biological Conduits

Biological conduits include arteries (Itoh *et al.*, 1996), veins (Chiu *et al.*, 1982; Tang *et al.*, 2008; Walton *et al.*, 1989), mesothelial chambers (Lundborg and Hansson, 1980), predegenerated or fresh skeletal muscle (Battiston *et al.*, 2007; Itoh *et al.*, 1996), and epineural sheath (Karacaoglu, 2001). Application of vein grafts was studied most extensively. In clinical applications, they showed good functional outcomes particularly in pure sensory digital nerves repair (Chiu and Strauch, 1990; Risitano *et al.*, 2002; Walton *et al.*, 1989). Turn-over vein graft was proposed to promote axonal growth by providing an environment rich in collagen, laminin, and Schwann cells encountered in adventitia of vessels (Wang *et al.*, 1995). The experimental results did not show significant functional improvement obtained by this technique over standard vein graft, but confirmed slightly faster regeneration (Tang *et al.*, 2008). Brunelli *et al.* (1993) proposed a combined conduit where vein graft was filled with muscle. This approach prevents vein graft collapse and provides guidance, which muscle alone lacks. Moreover, the muscle provides adequate adhesion for the growing nerve fibers provided by basal lamina (rich in lamimin and fibronectin) resembling the Schwann cells adhesion role (Battiston *et al.*, 2007). Migratory Schwann cells invade muscle-in-vein conduit early and proliferate within mimicking the processes undergoing in nerve graft (Fornaro *et al.* 2001; Geuna *et al.*, 2003). This technique was applied in repair of 40 cases of sensory and mixed nerve defects (0.5–6 cm) with good results achieved in 85% of patients (Battiston *et al.*, 2000). Another material that was experimentally tested was an empty epineural tube. The neural origin is the potential advantage of this conduit over other biological tubes. High laminin B2 and VEGF expression provides a highly neuropermissive environment for Schwann cell attachment and axonal ingrowth (Siemionow *et al.*, unpublished data). Karacaoglu *et al.* (2001) compared epineural sheath repair with vein and nerve grafts in a rat sciatic nerve 1-cm gap model. Functional outcomes measured by walking tract analysis in epineural repair were comparable to autologous graft and significantly better compared to the vein graft group. However, morphometric analysis revealed best results in nerve graft augmentation compared to epineural sheath while the worst results were seen in vein repair group. The epineural conduit, augmented with different supportive cell therapies is currently thoroughly tested in our laboratory on rat models of 2 and 4 cm sciatic nerve gaps (Fig. 8). Preliminary results show favorable outcomes, both functional and histological, which encourages us to test this technique in a larger animal model as an intermediate step towards clinical

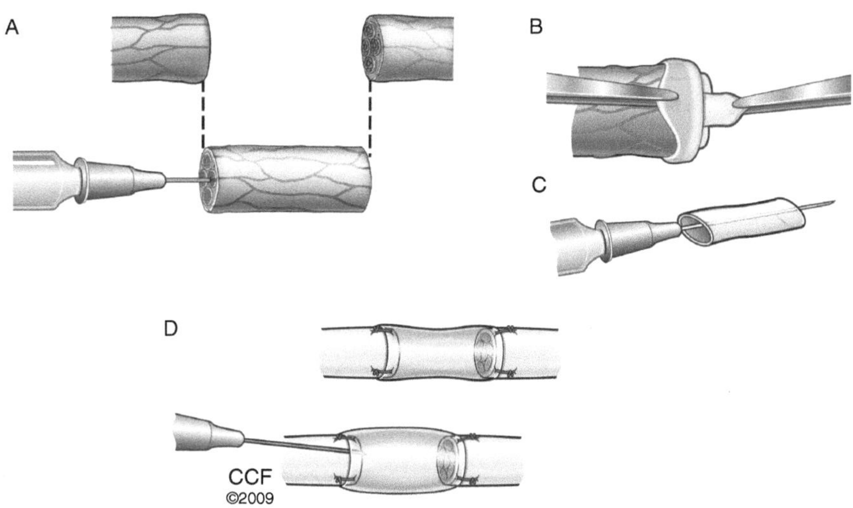

FIG. 8. Diagram of epineural tube repair. (A) Harvested fragment of the donor nerve is suspended on straight irrigator. (B) Fascicles are pulled out and the empty epineural tube (C) is created. Epineural conduit is coaptated using epineural technique to bridge proximal and distal nerve stumps and is filled with suspension of supportive cells (D).

application. Low antigenicity of epineural tissue will enable transplantation of allogenic epineural conduit harvested from cadavers possibly without a need of immunosuppression in the recipient of the graft.

B. Artificial Conduits

Progress in tissue engineering enabled introduction of artificial nerve conduits as a new method of peripheral nerve defect repair (Fig. 9). Various materials including synthetic polymers, extracellular matrix components, as well as different designs (mono- and multichannel conduits) and supportive therapies (cytokines, cells) were used in animal studies (Lietz *et al.*, 2006; Raimondo *et al.*, 2005; Schlosshauer *et al.*, 2003). However, application in humans was reported for 6 types of materials (Battiston *et al.*, 2005; Ichihara *et al.*, 2008; Meek and Coert, 2002): 2 of them are nonresorbable polymers, silicone and expanded polytetrafluoroethylene (PTFE), and 4 others include resorbable conduits: polyglycolic acid polymer (PGA), polylactide-caprolactone polymer (PLCL), polyglycolic acid polymer coated with cross-linked collagen (PGA-c), and type I collagen (Schlosshauer *et al.*, 2006). First clinical application, augmentation of ulnar nerve defect with silicone tube, was reported by Lundborg *et al.* (1991). Following studies on upper extremity nerve lesions showed primarily good functional recovery; however, in few cases longer follow-ups revealed intolerance to cold

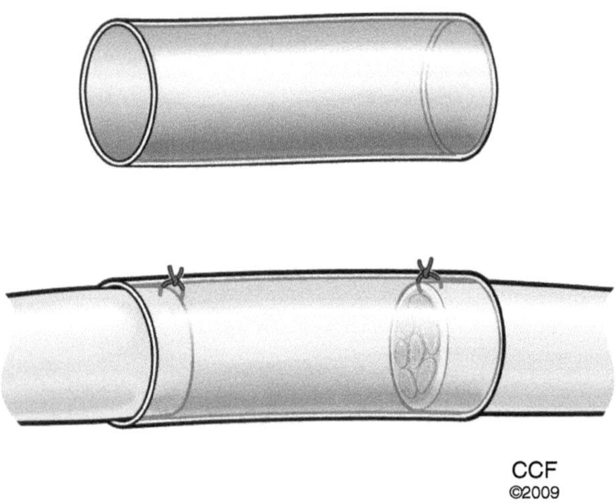

Fig. 9. Schematic presentation of synthetic conduit repair. The proximal and distal stumps are pulled inside synthetic tube and sutured to the tube ends.

temperature, hypersensitivity, and local discomfort. Consequently, implants were removed in these patients without disrupting initial recovery (Dahlin *et al.*, 2001; Lundborg *et al.*, 1997). Short gap repair of median and ulnar nerves (<4 mm) with silicone tubes was found equally efficient in sensory function restoration to direct end-to-end coaptation in 5-year follow-up. Less clinical data is available for PTFE implants. Two studies presented similar outcomes using silicone tubes (Sinis *et al.*, 2005; Stanec and Stanec, 1998), but in two other studies, outcomes were considered poor (Pitta *et al.*, 2001; Pogrel *et al.*, 1998). Neither silicone nor PTFE showed good recovery over defects exceeding 4 cm (Stanec and Stanec, 1998). Majority of data suggests that nonresorbable implants support initial regeneration but can interfere with maturating nerve fibers causing compression of pressure-sensitive axons. The disadvantages of nonresorbable conduits led to development of resorbable nerve tubes. First clinical study with application of PGA implants was performed by Mackinnon and Dellon*et al.* (1990). In this study, 15 patients with 5–30 mm nerve lesions within hands were enrolled. At 11–32 months follow-up, 13 patients achieved good and very good level of sensory function recovery measured by 2-point discrimination, whereas only 2 patients displayed poor regeneration. In conclusion, PGA synthetic implant was found to be equal to repair with autologous nerve grafts. Largest multicenter, prospective, randomized study for PGA implants enrolled 98 patients with 136 nerve lesions in the hand (Weber *et al.*, 2000). Synthetic implants were compared to autologous nerve grafting or direct coaptation. In both groups, the recovery was graded good and very good by a blind observer in approximately 75% of patients without significant differences in groups. Interestingly, nerve defects longer than 4 mm obtained better recovery with PGA tube repair. Navissano *et al.* (2005) reported good motor recovery in 5 of 7 patients with lesions of terminal branches of facial nerve repaired with PGA tube. The efficacy of PLCL tubes repair was reported in randomized trial, in which 30 patients were enrolled with up two 20-mm upper limb nerve defects (Bertleff *et al.*, 2005). No significant difference in functional outcomes was noted between PLCL conduits and autologous nerve grafts. Also, collagen tubes were proved to successfully aid in nerve gap repair (Ashley *et al.*, 2006; Lohmeyer *et al.*, 2007). Only single patients with nerve defects repaired with PGA-c tube were reported inlcuding two cases with successful motor recovery after facial nerve defect repair (Inada *et al.*, 2004, 2005, 2007). In summary, application of artificial resorbable nerve guides to bridge nerve defects up to 3 cm has the same success rate as nerve autograft repair, which results in recovery in up to 69% of cases (Evans, 2000; Lee and Wolfe, 2000; Mackinnon and Dellon, 1988). The first FDA approval for synthetic nerve guide was issued in 1999. Currently, five products, including four resorbable (made of PGA, PLCL, and collagen I) presented in Table II and one nonresorbable, polyvinyl alcohol hydrogel are FDA approved for clinical use. All commercial synthetic tubes are restricted for use in short nerve defects only. A number of studies indicated that

TABLE II

FDA Approved Resorbable Synthetic Guides

Product name	Neurotube	Neurolac	NeuraGen	NeuroMatrix NeuroFlex
Material	Polyglycolic acid	Poly-DL-lactide-ε-caprolactone	Type I Collagen	Type I Collagen
Company	Synovis	Polyganics BV	Integra NeuroSciences	Collagen Matrix Inc
Maximum length (cm)	4	3	3	2.5
Diameter (mm)	2.3–8	1.5–10	1.5–7	2–6
Degradation time	3 months	16 months	4 years	7 months
Clinical data	Yes	Yes	Yes	No

repair of longer nerve defects requires further progress in implant design, both in the field of biomaterial development as well as biology of nerve regeneration. Much experimental work is being done on improving mechanical properties of conduits—tensile strength, flexibility (Yannas *et al.*, 2007), biocompability—decrease inflammation and foreign body reaction, and absorbability—completely dissolve after regeneration is warranted (Chang *et al.*, 2007b; Jansen *et al.*, 2004). Ideal synthetic conduit should be permeable enough to provide sufficient diffusion of oxygen and metabolite for supporting Schwann cells proliferation, but should also prevent fibroblast infiltration (Chang *et al.*, 2007a). Multichannel tubes were developed to avoid misdirection of regenerating axons and polyinnervation of different targets by the same axon. However, until now, experimental data from animal studies did not show any benefit of these devices over standard single lumen tubes (de Ruiter *et al.*, 2008). Another approach is to support artificial tubes with cells or to add neurotrophic factors or cytokines directly into the tube. Animal models showed considerably better nerve regeneration when synthetic nerve guides were seeded with Schwann cells (Lietz *et al.*, 2006; Sinis *et al.*, 2005), which act by guiding growing axons and expressing neurotrophic factors. Nerve growth factor (NGF), vascular endothelial factor (VEGF), and erythropoietin were shown to support growth of regenerating nerve fibers and improve neuron survival in experimental conditions (Kemp *et al.*, 2007; Kim *et al.*, 2004; Lykissas *et al.*, 2007). Despite significant progress in tissue engineering repair of nerve gaps exceeding 3 cm remains a challenge. Therefore, more experimental and preclinical studies are needed to fully evaluate neuroregenerative potential of these new approaches.

VI. Conclusions

Last century expanded our knowledge of peripheral nerve pathophysiology and refined microsurgical skills improved the outcomes of peripheral nerve repair. Introduction of nerve allografting technique made possible bridging large nerve defects, which was previously unachievable with application of standard autografting methods. Tubulization techniques can eliminate morbidities of autograft harvesting and provide comparable outcomes in short nerve gap repair. However, while in directly repaired nerves observed sensory and motor nerve recovery ranges from good to very good, the results of large nerve gap reconstruction are mostly fair especially for motor recovery. Future approaches will introduce new technologies based on basic science discoveries, immunology, neurobiology, molecular biology, and tissue engineering tested in laboratory models and translated into clinical practice.

References

Aebischer, P., Salessiotis, A. N., and Winn, S. R. (1989). Basic fibroblast growth factor released from synthetic guidance channels facilitates peripheral nerve regeneration across long nerve gaps. *J. Neurosci. Res.* **23,** 282–289.

Albert, E. (1885). Einige Operationem an Nerven. *Wien. Med. Wochenschr.* **26,** 1285.

Almgren, K. G. (1975). Revascularization of free peripheral nerve grafts. An experimental study in the rabbit. *Acta. Orthop. Scand. Suppl.* **154,** 1–104.

Ashley, W. W., Jr., Weatherly, T., and Park, T. S. (2006). Collagen nerve guides for surgical repair of brachial plexus birth injury. *J. Neurosurg.* **105,** 452–456.

Atchabahian, A., Doolabh, V. B., Mackinnon, S. E., Yu, S., Hunter, D. A., and Flye, M. W. (1998). Indefinite survival of peripheral nerve allografts after temporary Cyclosporine A immunosuppression. *Restor. Neurol. Neurosci.* **13,** 129–139.

Atchabahian, A., Mackinnon, S. E., and Hunter, D. A. (1999). Cold preservation of nerve grafts decreases expression of ICAM-1 and class II MHC antigens. *J. Reconstr. Microsurg.* **15,** 307–311.

Bain, J. R. (2000). Peripheral nerve and neuromuscular allotransplantation: Current status. *Microsurgery* **20,** 384–388.

Bain, J. R., Mackinnon, S. E., Hudson, A. R., Falk, R. E., Falk, J. A., and Hunter, D. A. (1988). The peripheral nerve allograft: A dose–response curve in the rat immunosuppressed with cyclosporin A. *Plast. Reconstr. Surg.* **82,** 447–457.

Battiston, B., Geuna, S., Ferrero, M., and Tos, P. (2005). Nerve repair by means of tubulization: Literature review and personal clinical experience comparing biological and synthetic conduits for sensory nerve repair. *Microsurgery* **25,** 258–267.

Battiston, B., Tos, P., Conforti, L. G., and Geuna, S. (2007). Alternative techniques for peripheral nerve repair: Conduits and end-to-side neurorrhaphy. *Acta. Neurochir. Suppl.* **100,** 43–50.

Battiston, B., Tos, P., Cushway, T. R., and Geuna, S. (2000). Nerve repair by means of vein filled with muscle grafts I. Clinical results. *Microsurgery* **20,** 32–36.

Bertleff, M. J., Meek, M. F., and Nicolai, J. P. (2005). A prospective clinical evaluation of biodegradable Neurolac nerve guides for sensory nerve repair in the hand. *J. Hand. Surg. [Am.]* **30,** 513–518.

Bielschowsky, M., and Unger, E. (1916). Die Ueberbrueckung grosser Nervenluecken. Beitraege zur Kennetnis der Degeneration and Regeneration peripher Nerven. *J. Physiol. Neurol.* **22,** 267.

Birch, R., and Raji, A. R. (1991). Repair of median and ulnar nerves. Primary suture is best. *J. Bone Joint Surg. Br.* **73,** 154–157.

Bratton, B. R., Kline, D. G., Coleman, W., and Hudson, A. R. (1979). Experimental interfascicular nerve grafting. *J. Neurosurg.* **51,** 323–332.

Brenner, M. J., Jensen, J. N., Lowe, J. B., 3rd, Myckatyn, T. M., Fox, I. K., Hunter, D. A., Mohanakumar, T., and Mackinnon, S. E. (2004a). Anti-CD40 ligand antibody permits regeneration through peripheral nerve allografts in a nonhuman primate model. *Plast. Reconstr. Surg.* **114,** 1802–1814 (discussion, pp. 1815–1817).

Brenner, M. J., Mackinnon, S. E., Rickman, S. R., Jaramillo, A., Tung, T. H., Hunter, D. A., and Mohanakumar, T. (2005). FK506 and anti-CD40 ligand in peripheral nerve allotransplantation. *Restor. Neurol. Neurosci.* **23,** 237–249.

Brenner, M. J., Tung, T. H., Mackinnon, S. E., Myckatyn, T. M., Hunter, D. A., and Mohanakumar, T. (2004b). Anti-CD40 ligand monoclonal antibody induces a permissive state, but not tolerance, for murine peripheral nerve allografts. *Exp. Neurol.* **186,** 59–69.

Brunelli, G. A., Battiston, B., Vigasio, A., Brunelli, G., and Marocolo, D. (1993). Bridging nerve defects with combined skeletal muscle and vein conduits. *Microsurgery* **14,** 247–251.

Brushart, T. M. E. (1991). The mechanical and humoral control of specificity in nerve repair. *In* "Operative Nerve Repair and Reconstruction" (R. H. Gelberman, Ed.), pp. 215–230. JB Lippincott, Philadelphia.

Brushart, T. M., Tarlov, E. C., and Mesulam, M. M. (1983). Specificity of muscle reinnervation after epineurial and individual fascicular suture of the rat sciatic nerve. *J. Hand Surg. [Am.]* **8,** 248–253.

Calder, J. S., and Green, C. J. (1995). Nerve–muscle sandwich grafts: The importance of Schwann cells in peripheral nerve regeneration through muscle basal lamina conduits. *J. Hand. Surg. [Br.]* **20,** 423–428.

Campbell, J. B., Bassett, A. L., and Boehler, J. (1963). Frozen-irradiated homografts shielded with microfilter sheaths in peripheral nerve surgery. *J. Trauma* **3,** 303–311.

Campbell, W. W. (2008). Evaluation and management of peripheral nerve injury. *Clin. Neurophysiol.* **119,** 1951–1965.

Chang, C. J., Hsu, S. H., Yen, H. J., Chang, H., and Hsu, S. K. (2007a). Effects of unidirectional permeability in asymmetric poly(DL-lactic acid-*co*-glycolic acid) conduits on peripheral nerve regeneration: An *in vitro* and *in vivo* study. *J. Biomed. Mater. Res. B. Appl. Biomater.* **83,** 206–215.

Chang, J. Y., Lin, J. H., Yao, C. H., Chen, J. H., Lai, T. Y., and Chen, Y. S. (2007b). *In vivo* evaluation of a biodegradable EDC/NHS-cross-linked gelatin peripheral nerve guide conduit material. *Macromol. Biosci.* **7,** 500–507.

Chen, D. L., Mackinnon, S. E., Jensen, J. N., Hunter, D. A., and Grand, A. G. (2002). Failure of cyclosporin A to rescue peripheral nerve allografts in acute rejection. *Ann. Plast. Surg.* **49,** 660–667.

Chen, L. E., Seaber, A. V., Urbaniak, J. R., and Murrell, G. A. (1994). Denatured muscle as a nerve conduit: A functional, morphologic, and electrophysiologic evaluation. *J. Reconstr. Microsurgery* **10,** 137–144.

Chiu, D. T., Janecka, I., Krizek, T. J., Wolff, M., and Lovelace, R. E. (1982). Autogenous vein graft as a conduit for nerve regeneration. *Surgery* **91,** 226–233.

Chiu, D. T., and Strauch, B. (1990). A prospective clinical evaluation of autogenous vein grafts used as a nerve conduit for distal sensory nerve defects of 3 cm or less. *Plast. Reconstr. Surg.* **86,** 928–934.

Chu, T. H., Du, Y., and Wu, W. (2008). Motor nerve graft is better than sensory nerve graft for survival and regeneration of motoneurons after spinal root avulsion in adult rats. *Exp. Neurol.* **212,** 562–565.

Cruikshank, H. (1795). Experiments on the nerves, particulary on their reproduction; and on the spinal marrow of living animals. *Phil. Trans. R. Soc. London.* 177–189.

Dahlin, L. B., Anagnostaki, L., and Lundborg, G. (2001). Tissue response to silicone tubes used to repair human median and ulnar nerves. *Scand. J. Plast. Reconstr. Surg. Hand Surg.* **35,** 29–34.

Dahlin, L. B., and Lundborg, G. (2001). Use of tubes in peripheral nerve repair. *Neurosurg. Clin. N. Am.* **12,** 341–352.

de Ruiter, G. C., Spinner, R. J., Malessy, M. J., Moore, M. J., Sorenson, E. J., Currier, B. L., Yaszemski, M. J., and Windebank, A. J. (2008). Accuracy of motor axon regeneration across autograft, single-lumen, and multichannel poly(lactic-*co*-glycolic acid) nerve tubes. *Neurosurgery* **63,** 144–153 (discussion, pp. 153–155).

Diao, E., and Vannuyen, T. (2000). Techniques for primary nerve repair. *Hand Clin.* **16,** 53–66 (see also p. viii).

Dvali, L., and Mackinnon, S. (2003). Nerve repair, grafting, and nerve transfers. *Clin. Plast. Surg.* **30,** 203–221.

Evans, G. R. (2000). Challenges to nerve regeneration. *Semin. Surg. Oncol.* **19,** 312–318.

Evans, P. J., Mackinnon, S. E., Best, T. J., Wade, J. A., Awerbuck, D. C., Makino, A. P., Hunter, D. A., and Midha, R. (1995). Regeneration across preserved peripheral nerve grafts. *Muscle Nerve.* **18,** 1128–1138.

Evans, P. J., Mackinnon, S. E., Levi, A. D., Wade, J. A., Hunter, D. A., Nakao, Y., and Midha, R. (1998). Cold preserved nerve allografts: Changes in basement membrane, viability, immunogenicity, and regeneration. *Muscle Nerve.* **21,** 1507–1522.

Evans, P. J., MacKinnon, S. E., Midha, R., Wade, J. A., Hunter, D. A., Nakao, Y., and Hare, G. M. (1999). Regeneration across cold preserved peripheral nerve allografts. *Microsurgery* **19,** 115–127.

Felici, N., Del Bene, M., Battiston, B., and Amadei, F. (2003). Functional results of end-to-side nerve anastomosis in 39 consecutive patients, Abstract Volume, Second Congress of the World Society for Reconstructive Microsurgery, Heidelberg. p. 42.

Feng, F. Y., Ogden, M. A., Myckatyn, T. M., Grand, A. G., Jensen, J. N., Hunter, D. A., and Mackinnon, S. E. (2001). FK506 rescues peripheral nerve allografts in acute rejection. *J. Neurotrauma* **18,** 217–229.

Fornaro, M., Tos, P., Geuna, S., Giacobini-Robecchi, M. G., and Battiston, B. (2001). Confocal imaging of Schwann-cell migration along muscle–vein combined grafts used to bridge nerve defects in the rat. *Microsurgery* **21,** 153–155.

Funakoshi, H., Frisen, J., Barbany, G., Timmusk, T., Zachrisson, O., Verge, V. M., and Persson, H. (1993). Differential expression of mRNAs for neurotrophins and their receptors after axotomy of the sciatic nerve. *J. Cell. Biol.* **123,** 455–465.

Fu, S. Y., and Gordon, T. (1997). The cellular and molecular basis of peripheral nerve regeneration. *Mol. Neurobiol.* **14,** 67–116.

Genden, E. M., Mackinnon, S. E., Yu, S., and Flye, M. W. (1998). Induction of donor-specific tolerance to rat nerve allografts with portal venous donor alloantigen and anti-ICAM-1/LFA-1 monoclonal antibodies. *Surgery* **124,** 448–456.

Genden, E. M., Mackinnon, S. E., Yu, S., Hunter, D. A., and Flye, M. W. (2001). Portal venous ultraviolet B-irradiated donor alloantigen prevents rejection in circumferential rat tracheal allografts. *Otolaryngol. Head Neck Surg.* **124,** 481–488.

Geuna, S., Raimondo, S., Nicolino, S., Boux, E., Fornaro, M., Tos, P., Battiston, B., and Perroteau, I. (2003). Schwann-cell proliferation in muscle–vein combined conduits for bridging rat sciatic nerve defects. *J. Reconstr. Microsurg.* **19,** 119–123 (discussion, p. 124).

Gold, B. G., Katoh, K., and Storm-Dickerson, T. (1995). The immunosuppressant FK506 increases the rate of axonal regeneration in rat sciatic nerve. *J. Neurosci.* **15,** 7509–7516.

Gulati, A. K. (1998). Immune response and neurotrophic factor interactions in peripheral nerve transplants. *Acta Haematol.* **99,** 171–174.

Hall, S. M. (1986). The effect of inhibiting Schwann cell mitosis on the re-innervation of acellular autografts in the peripheral nervous system of the mouse. *Neuropathol. Appl. Neurobiol.* **12,** 401–414.

Hammerschlag, P. E. (1999). Facial reanimation with jump interpositional graft hypoglossal facial anastomosis and hypoglossal facial anastomosis: Evolution in management of facial paralysis. *Laryngoscope* **109,** 1–23.

Harris, M. E., and Tindall, S. C. (1991). Techniques of peripheral nerve repair. *Neurosurg. Clin. N. Am.* **2,** 93–104.

Heijke, G. C., Klopper, P. J., and Dutrieux, R. P. (1993). Vein graft conduits versus conventional suturing in peripheral nerve reconstructions. *Microsurgery* **14,** 584–588.

Hettiaratchy, S., Melendy, E., Randolph, M. A., Coburn, R. C., Neville, D. M., Jr., Sachs, D. H., Huang, C. A., and Lee, W. P. (2004). Tolerance to composite tissue allografts across a major histocompatibility barrier in miniature swine. *Transplantation* **77,** 514–521.

Huber, G. C. (1895). A study of the operative treatment for loss of nerve substance in peripheral nerves. *J. Morphol.* **11,** 629–735.

Hudson, A. R., Hunter, D., Kline, D. G., and Bratton, B. R. (1979). Histological studies of experimental interfascicular graft repairs. *J. Neurosurg.* **51,** 333–340.

Ichihara, S., Inada, Y., and Nakamura, T. (2008). Artificial nerve tubes and their application for repair of peripheral nerve injury: An update of current concepts. *Injury* **39**(Suppl. 4), 29–39.

Inada, Y., Hosoi, H., Yamashita, A., Morimoto, S., Tatsumi, H., Notazawa, S., Kanemaru, S., and Nakamura, T. (2007). Regeneration of peripheral motor nerve gaps with a polyglycolic acid-collagen tube: Technical case report. *Neurosurgery* **61,** E1105–E1107 (discussion, p. E1107).

Inada, Y., Morimoto, S., Moroi, K., Endo, K., and Nakamura, T. (2005). Surgical relief of causalgia with an artificial nerve guide tube: Successful surgical treatment of causalgia (Complex Regional Pain Syndrome Type II) by *in situ* tissue engineering with a polyglycolic acid-collagen tube. *Pain* **117,** 251–258.

Inada, Y., Morimoto, S., Takakura, Y., and Nakamura, T. (2004). Regeneration of peripheral nerve gaps with a polyglycolic acid-collagen tube. *Neurosurgery* **55,** 640–646 (discussion, pp. 646–648).

Ishida, O., Daves, J., Tsai, T. M., Breidenbach, W. C., and Firrell, J. (1993). Regeneration following rejection of peripheral nerve allografts of rats on withdrawal of cyclosporine. *Plast. Reconstr. Surg.* **92,** 916–926.

Itoh, S., Shinomiya, K., Samejima, H., Ohta, T., Ishizuki, M., and Ichinose, S. (1996). Experimental study on nerve regeneration through the basement membrane tubes of the nerve, muscle, and artery. *Microsurgery* **17,** 525–534.

Jansen, K., van der Werff, J. F., van Wachem, P. B., Nicolai, J. P., de Leij, L. F., and van Luyn, M. J. (2004). A hyaluronan-based nerve guide: *In vitro* cytotoxicity, subcutaneous tissue reactions, and degradation in the rat. *Biomaterials* **25,** 483–489.

Jensen, J. N., Brenner, M. J., Tung, T. H., Hunter, D. A., and Mackinnon, S. E. (2005). Effect of FK506 on peripheral nerve regeneration through long grafts in inbred swine. *Ann. Plast. Surg.* **54,** 420–427.

Karacaoglu, E., Yuksel, F., Peker, F., and Guler, M. M. (2001). Nerve regeneration through an epineurial sheath: Its functional aspect compared with nerve and vein grafts. *Microsurgery* **21,** 196–201.

Kelsey, J., Praemer, A., Nelson, L., Felberg, A., and Rice, D. (1997). "Upper Extremity Disorders: Frequency, Impact, and Cost." Churchhill-Livingstone, London.

Kemp, S. W., Walsh, S. K., Zochodne, D. W., and Midha, R. (2007). A novel method for establishing daily *in vivo* concentration gradients of soluble nerve growth factor (NGF). *J. Neurosci. Methods* **165,** 83–88.

Kim, B. S., Yoo, J. J., and Atala, A. (2004). Peripheral nerve regeneration using acellular nerve grafts. *J. Biomed. Mater. Res. A.* **68,** 201–209.

Kline, D. G. (1990). Surgical repair of peripheral nerve injury. *Muscle Nerve.* **13,** 843–852.

Kline, D. G., Hackett, E. R., Davis, G. D., and Myers, M. B. (1972). Effect of mobilization on the blood supply and regeneration of injured nerves. *J. Surg. Res.* **12,** 254–266.

Kouyoumdjian, J. A. (2006). Peripheral nerve injuries: A retrospective survey of 456 cases. *Muscle Nerve.* **34,** 785–788.

Lanzetta, M., Petruzzo, P., Margreiter, R., Dubernard, J. M., Schuind, F., Breidenbach, W., Lucchina, S., Schneeberger, S., van Holder, C., Granger, D., Pei, G., Zhao, J., and Zhang, X. (2005). The International Registry on hand and composite tissue transplantation. *Transplantation* **79,** 1210–1214.

Lee, S. K., and Wolfe, S. W. (2000). Peripheral nerve injury and repair. *J. Am. Acad. Orthop. Surg.* **8,** 243–252.

Levi, A. D., Evans, P. J., Mackinnon, S. E., and Bunge, R. P. (1994). Cold storage of peripheral nerves: An *in vitro* assay of cell viability and function. *Glia* **10,** 121–131.

Levi-Montalcini, R., and Hamburger, V. (1951). Selective growth stimulating effects of mouse sarcoma on the sensory and sympathetic nervous system of the chick embryo. *J. Exp. Zool.* **116,** 321–361.

Levinthal, R., Brown, W. J., and Rand, R. W. (1977). Comparison of fascicular, interfascicular and epineural suture techniques in the repair of simple nerve lacerations. *J. Neurosurg.* **47,** 744–750.

Lietz, M., Ullrich, A., Schulte-Eversum, C., Oberhoffner, S., Fricke, C., Muller, H. W., and Schlosshauer, B. (2006). Physical and biological performance of a novel block copolymer nerve guide. *Biotechnol. Bioeng.* **93,** 99–109.

Liu, K., Chen, L. E., Seaber, A. V., Goldner, R. V., and Urbaniak, J. R. (1999). Motor functional and morphological findings following end-to-side neurorrhaphy in the rat model. *J. Orthop. Res.* **17,** 293–300.

Lohmeyer, J., Zimmermann, S., Sommer, B., Machens, H. G., Lange, T., and Mailander, P. (2007). Bridging peripheral nerve defects by means of nerve conduits. *Chirurg* **78,** 142–147.

Lubiatowski, P., Unsal, F. M., Nair, D., Ozer, K., and Siemionow, M. (2008). The epineural sleeve technique for nerve graft reconstruction enhances nerve recovery. *Microsurgery* **28,** 160–167.

Lundborg, G., Dahlin, L. B., and Danielsen, N. (1991). Ulnar nerve repair by the silicone chamber technique. Case report. *Scand. J. Plast. Reconstr. Surg. Hand Surg.* **25,** 79–82.

Lundborg, G., and Hansson, H. A. (1980). Nerve regeneration through preformed pseudosynovial tubes. A preliminary report of a new experimental model for studying the regeneration and reorganization capacity of peripheral nerve tissue. *J. Hand. Surg. [Am.]* **5,** 35–38.

Lundborg, G., Longo, F. M., and Varon, S. (1982). Nerve regeneration model and trophic factors *in vivo*. *Brain. Res.* **232,** 157–161.

Lundborg, G., Rosen, B., Dahlin, L., Danielsen, N., and Holmberg, J. (1997). Tubular versus conventional repair of median and ulnar nerves in the human forearm: Early results from a prospective, randomized, clinical study. *J. Hand Surg. [Am.]* **22,** 99–106.

Lundborg, G., and Rydevik, B. (1973). Effects of stretching the tibial nerve of the rabbit. A preliminary study of the intraneural circulation and the barrier function of the perineurium. *J. Bone Joint Surg. Br.* **55,** 390–401.

Luo, Y., Wang, T., and Fang, H. (1997). [Preliminary investigation of treatment of ulnar nerve defect by end-to-side neurorrhaphy]. *Zhongguo. Xiu. Fu. Chong. Jian. Wai. Ke. Za. Zhi.* **11,** 338–339.

Lutz, B. S., Chuang, D. C., Hsu, J. C., Ma, S. F., and Wei, F. C. (2000). Selection of donor nerves—An important factor in end-to-side neurorrhaphy. *Br. J. Plast. Surg.* **53,** 149–154.

Lykissas, M. G., Sakellariou, E., Vekris, M. D., Kontogeorgakos, A., Batistatou, A. K., Mitsionis, G. I., and Beris, A. E. (2007). Axonal regeneration stimulated by erythropoietin: An experimental study in rats. *J. Neurosci. Methods* **164,** 107–115.

Mackinnon, S. E. (1989). New directions in peripheral nerve surgery. *Ann. Plast. Surg.* **22,** 257–273.

Mackinnon, S. E., and Dellon, A. L. (1988). "Surgery of the Peripheral Nerve." Thieme Medical Publishers, New York.

Mackinnon, S. E., and Dellon, A. L. (1990). Clinical nerve reconstruction with a bioabsorbable polyglycolic acid tube. *Plast. Reconstr. Surg.* **85,** 419–424.

Mackinnon, S. E., Doolabh, V. B., Novak, C. B., and Trulock, E. P. (2001). Clinical outcome following nerve allograft transplantation. *Plast. Reconstr. Surg.* **107,** 1419–1429.

Mackinnon, S. E., Hudson, A. R., Falk, R. E., and Hunter, D. A. (1985). The nerve allograft response—An experimental model in the rat. *Ann. Plast. Surg.* **14,** 334–339.

Mackinnon, S. E., Hudson, A. R., Falk, R. E., Kline, D., and Hunter, D. (1984). Peripheral nerve allograft: an immunological assessment of pretreatment methods. *Neurosurgery* **14,** 167–171.

Mackinnon, S. E., Midha, R., Bain, J., Hunter, D., and Wade, J. (1992). An assessment of regeneration across peripheral nerve allografts in rats receiving short courses of cyclosporin A immunosuppression. *Neuroscience* **46,** 585–593.

Madison, R., da Silva, C. F., Dikkes, P., Chiu, T. H., and Sidman, R. L. (1985). Increased rate of peripheral nerve regeneration using bioresorbable nerve guides and a laminin-containing gel. *Exp. Neurol.* **88,** 767–772.

Madison, R. D., da Silva, C., Dikkes, P., Sidman, R. L., and Chiu, T. H. (1987). Peripheral nerve regeneration with entubulation repair: Comparison of biodegradeable nerve guides versus polyethylene tubes and the effects of a laminin-containing gel. *Exp. Neurol.* **95,** 378–390.

Maeda, T., Mackinnon, S. E., Best, T. J., Evans, P. J., Hunter, D. A., and Midha, R. T. (1993). Regeneration across 'stepping-stone' nerve grafts. *Brain Res.* **618,** 196–202.

Maggi, S. P., Lowe, J. B., 3rd, and Mackinnon, S. E. (2003). Pathophysiology of nerve injury. *Clin. Plast. Surg.* **30,** 109–126.

Manni, J. J., Beurskens, C. H., van de Velde, C., and Stokroos, R. J. (2001). Reanimation of the paralyzed face by indirect hypoglossal–facial nerve anastomosis. *Am. J. Surg.* **182,** 268–273.

Maricevic, A., and Erceg, M. (1997). War injuries to the extremities. *Mil. Med.* **162,** 808–811.

Marmor, L. (1964). The repair of peripheral nerves by irradiated homografts. *Clin. Orthop. Relat. Res.* **34,** 161–169.

Martini, A., and Fromm, B. (1989). A new operation for the prevention and treatment of amputation neuromas. *J. Bone Joint Surg. Br.* **71,** 379–382.

Matsuyama, T., Mackay, M., and Midha, R. (2000). Peripheral nerve repair and grafting techniques: A review. *Neurol. Med. Chir. (Tokyo)* **40,** 187–199.

May, M., Sobol, S. M., and Mester, S. J. (1991). Hypoglossal–facial nerve interpositional-jump graft for facial reanimation without tongue atrophy. *Otolaryngol. Head Neck Surg.* **104,** 818–825.

McCallister, W. V., Tang, P., and Trumble, T. E. (1999). Is end-to-side neurorrhaphy effective? A study of axonal sprouting stimulated from intact nerves. *J. Reconstr. Microsurg.* **15,** 597–603 (discussion, pp. 603–604).

McLellan, D. L., and Swash, M. (1976). Longitudinal sliding of the median nerve during movements of the upper limb. *J. Neurol. Neurosurg. Psychiatry* **39,** 566–570.

Meek, M. F., and Coert, J. H. (2002). Clinical use of nerve conduits in peripheral-nerve repair: Review of the literature. *J. Reconstr. Microsurg.* **18,** 97–109.

Meek, M. F., and Coert, J. H. (2008). US Food and Drug Administration/Conformit Europe—Approved absorbable nerve conduits for clinical repair of peripheral and cranial nerves. *Ann. Plast. Surg.* **60,** 466–472.

Mennen, U. (1999). End-to-side nerve suture—A technique to repair peripheral nerve injury. *S. Afr. Med. J.* **89,** 1188–1194.

Mennen, U. (2003). End-to-side nerve suture in clinical practice. *Hand Surg.* **8,** 33–42.

Midha, R., Mackinnon, S. E., and Becker, L. E. (1994). The fate of Schwann cells in peripheral nerve allografts. *J. Neuropathol. Exp. Neurol.* **53,** 316–322.

Midha, R., Mackinnon, S. E., Evans, P. J., Best, T. J., Hare, G. M., Hunter, D. A., and Falk-Wade, J. A. (1993). Comparison of regeneration across nerve allografts with temporary or continuous cyclosporin A immunosuppression. *J. Neurosurg.* **78,** 90–100.

Millesi, H. (1967a). Nerve transplantation for reconstruction of peripheral nerves injured by the use of the microsurgical technic. *Minerva. Chir.* **22,** 950–951.

Millesi, H. (1990). Progress in peripheral nerve reconstruction. *World. J. Surg.* **14,** 733–747.

Millesi, H. (1991). Indications and techniques of nerve grafting. *In* "Operative Nerve Repair and Reconstruction" (R. H. Gelberman, Ed.), vol. 1, pp. 525–543. JB Lippincott, Philadelphia.

Millesi, H. (2000). Techniques for nerve grafting. *Hand Clin.* **16,** 73–91 (see also p. viii).

Millesi, H., Ganglberger, J., and Berger, A. (1967b). Erfahrungen mit der Microchirurgie peripher Nerven. *Chir. Plast.* **3,** 47.

Millesi, H., Meissl, G., and Berger, A. (1972). The interfascicular nerve-grafting of the median and ulnar nerves. *J. Bone Joint Surg. Am.* **54,** 727–750.

Miyamoto, Y. (1979). Experimental study of results of nerve suture under tension vs. nerve grafting. *Plast. Reconstr. Surg.* **64,** 540–549.

Moneim, M. S. (1982). Interfascicular nerve grafting. *Clin. Orthop. Relat. Res.* **65–74.**

Moradzadeh, A., Borschel, G. H., Luciano, J. P., Whitlock, E. L., Hayashi, A., Hunter, D. A., and Mackinnon, S. E. (2008). The impact of motor and sensory nerve architecture on nerve regeneration. *Exp. Neurol.* **212,** 370–376.

Navissano, M., Malan, F., Carnino, R., and Battiston, B. (2005). Neurotube for facial nerve repair. *Microsurgery* **25,** 268–271.

Nunley, J. A. (1991). Donor nerves for grafting. *In* "Operative Nerve Repair nad Reconstruction" (R. H. Gelberman, Ed.), vol. 1, pp. 545–552. JB Lippincott, Philadelphia.

Ogata, K., and Naito, M. (1986). Blood flow of peripheral nerve effects of dissection, stretching and compression. *J. Hand Surg. [Br.]* **11,** 10–14.

Ornelas, L., Padilla, L., Di Silvio, M., Schalch, P., Esperante, S., Infante, P. L., Bustamante, J. C., Avalos, P., Varela, D., and Lopez, M. (2006a). Fibrin glue: An alternative technique for nerve coaptation—Part I. Wave amplitude, conduction velocity, and plantar-length factors. *J. Reconstr. Microsurg.* **22,** 119–122.

Ornelas, L., Padilla, L., Di Silvio, M., Schalch, P., Esperante, S., Infante, R. L., Bustamante, J. C., Avalos, P., Varela, D., and Lopez, M. (2006b). Fibrin glue: An alternative technique for nerve coaptation—Part II. Nerve regeneration and histomorphometric assessment. *J. Reconstr. Microsurg.* **22,** 123–128.

Ortiguela, M. E., Wood, M. B., and Cahill, D. R. (1987). Anatomy of the sural nerve complex. *J. Hand Surg. [Am.]* **12,** 1119–1123.

Philipeaux, J. M., and Vulpian, A. (1870). Note sur des esseys de greffe d'un troncon du nerf lingual entre les deux bouts du nerf hypoglosse, apres excision d'un segment de ce derniernerf. *Arch. Physiol. Norm. Path. Paris* 618–620.

Pitta, M. C., Wolford, L. M., Mehra, P., and Hopkin, J. (2001). Use of Gore-Tex tubing as a conduit for inferior alveolar and lingual nerve repair: Experience with 6 cases. *J. Oral. Maxillofac. Surg.* **59,** 493–496 (discussion, p. 497).

Pogrel, M. A., McDonald, A. R., and Kaban, L. B. (1998). Gore-Tex tubing as a conduit for repair of lingual and inferior alveolar nerve continuity defects: A preliminary report. *J. Oral. Maxillofac. Surg.* **56,** 319–321 (discussion, pp. 321–322).

Pollard, J. D., and McLeod, J. G. (1981). Fresh and predegenerate nerve allografts and isografts in trembler mice. *Muscle Nerve.* **4,** 274–281.

Raimondo, S., Nicolino, S., Tos, P., Battiston, B., Giacobini-Robecchi, M. G., Perroteau, I., and Geuna, S. (2005). Schwann cell behavior after nerve repair by means of tissue-engineered muscle–vein combined guides. *J. Comp. Neurol.* **489,** 249–259.

Risitano, G., Cavallaro, G., Merrino, T., Coppolino, S., and Ruggeri, F. (2002). Clinical results and thoughts on sensory nerve repair by autologous vein graft in emergency hand reconstruction. *Chir. Main.* **21,** 194–197.

Robinson, L. R. (2000). Traumatic injury to peripheral nerves. *Muscle Nerve* **23,** 863–873.

Robinson, L. R. (2004). Traumatic injury to peripheral nerves. *Suppl. Clin. Neurophysiol.* **57,** 173–186.

Rodkey, W. G., Cabaud, H. E., and McCarroll, H. R., Jr. (1980). Neurorrhaphy after loss of a nerve segment: Comparison of epineurial suture under tension versus multiple nerve grafts. *J. Hand Surg. [Am.]* **5,** 366–371.

Rodriguez, F. J., Verdu, E., Ceballos, D., and Navarro, X. (2000). Nerve guides seeded with autologous Schwann cells improve nerve regeneration. *Exp. Neurol.* **161,** 571–584.

Rovak, J. M., Cederna, P. S., and Kuzon, W. M., Jr. (2001). Terminolateral neurorrhaphy: A review of the literature. *J. Reconstr. Microsurg.* **17,** 615–624.

Sanders, F. K. (1942). The repair of large gaps in peripheral nerves. *Brain* **65,** 281–337.

Scharpf, J., Strome, M., and Siemionow, M. (2006). Immunomodulation with anti-alphabeta T-cell receptor monoclonal antibodies in combination with cyclosporine A improves regeneration in nerve allografts. *Microsurgery* **26,** 599–607.

Schlosshauer, B., Dreesmann, L., Schaller, H. E., and Sinis, N. (2006). Synthetic nerve guide implants in humans: A comprehensive survey. *Neurosurgery* **59,** 740–747 (discussion, pp. 747–748).

Schlosshauer, B., Muller, E., Schroder, B., Planck, H., and Muller, H. W. (2003). Rat Schwann cells in bioresorbable nerve guides to promote and accelerate axonal regeneration. *Brain Res.* **963,** 321–326.

Schneeberger, S., Ninkovic, M., Piza-Katzer, H., Gabl, M., Hussl, H., Rieger, M., Loescher, W., Zelger, B., Brandacher, G., Bonatti, H., Boesmueller, C., Mark, W., *et al.* (2006). Status 5 years after bilateral hand transplantation. *Am. J. Transplant.* **6,** 834–841.

Schuind, F., Abramowicz, D., and Schneeberger, S. (2007). Hand transplantation: The state-of-the-art. *J. Hand. Surg. Eur.* **32,** 2–17.

Seddon, H. J. (1947). The use of autogenous grafts for repair of large gaps in peripheral nerves. *Br. J. Surg.* **3535,** 151.

Siemionow, M., and Sari, A. (2004a). A contemporary overview of peripheral nerve research from the Cleveland Clinic microsurgery laboratory. *Neurol. Res.* **26,** 218–225.

Siemionow, M., Tetik, C., Ozer, K., Ayhan, S., Siemionow, K., and Browne, E. (2002). Epineural sleeve neurorrhaphy: Surgical technique and functional results—A preliminary report. *Ann. Plast. Surg.* **48,** 281–285.

Siemionow, M., Zielinski, M., and Meirer, R. (2004b). The single-fascicle method of nerve grafting. *Ann. Plast. Surg.* **52,** 72–79.

Siemionow, M. Z., Izycki, D. M., and Zielinski, M. (2003). Donor-specific tolerance in fully major histocompatibility complex-mismatched limb allograft transplants under an anti-alphabeta T-cell receptor monoclonal antibody and cyclosporine A protocol. *Transplantation* **76,** 1662–1668.

Sinis, N., Schaller, H. E., Schulte-Eversum, C., Schlosshauer, B., Doser, M., Dietz, K., Rosner, H., Muller, H. W., and Haerle, M. (2005). Nerve regeneration across a 2-cm gap in the rat median nerve using a resorbable nerve conduit filled with Schwann cells. *J. Neurosurg.* **103,** 1067–1076.

Snyder, C. C., Webster, H. D., Pickens, J. E., Hines, W. A., and Warden, G. (1968). Intraneural neurorrhaphy: A preliminary clinical and histological evaluation. *Ann. Surg.* **167,** 691–696.

Sorensen, J., Fugleholm, K., Moldovan, M., Schmalbruch, H., and Krarup, C. (2001). Axonal elongation through long acellular nerve segments depends on recruitment of phagocytic cells from the near-nerve environment. Electrophysiological and morphological studies in the cat. *Brain Res.* **903,** 185–197.

Spinner, R. J., and Kline, D. G. (2000). Surgery for peripheral nerve and brachial plexus injuries or other nerve lesions. *Muscle Nerve* **23,** 680–695.

Stanec, S., and Stanec, Z. (1998). Reconstruction of upper-extremity peripheral-nerve injuries with ePTFE conduits. *J. Reconstr. Microsurg.* **14,** 227–232.

Stanec, S., Tonkovic, I., Stanec, Z., Tonkovic, D., and Dzepina, I. (1997). Treatment of upper limb nerve war injuries associated with vascular trauma. *Injury* **28,** 463–468.

Staniforth, P., and Fisher, T. R. (1978). The effects of sural nerve excision in autogenous nerve grafting. *Hand* **10,** 187–190.

Strasberg, S. R., Hertl, M. C., Mackinnon, S. E., Lee, C. K., Watanabe, O., Tarasidis, G., Hunter, D. A., and Wong, P. Y. (1996a). Peripheral nerve allograft preservation improves regeneration and decreases systemic cyclosporin A requirements. *Exp. Neurol.* **139,** 306–316.

Strasberg, S. R., Mackinnon, S. E., Hare, G. M., Narini, P. P., Hertl, C., and Hay, J. B. (1996b). Reduction in peripheral nerve allograft antigenicity with warm and cold temperature preservation. *Plast. Reconstr. Surg.* **97,** 152–160.

Strauch, B. (2000). Use of nerve conduits in peripheral nerve repair. *Hand Clin.* **16,** 123–130.

Strauch, B., Rodriguez, D. M., Diaz, J., Yu, H. L., Kaplan, G., and Weinstein, D. E. (2001). Autologous Schwann cells drive regeneration through a 6-cm autogenous venous nerve conduit. *J. Reconstr. Microsurg.* **17,** 589–595 (discussion, pp. 596–597).

Sunderland, S. (1945). The intraneural topography of the radial, median, and ulnar nerves. *Brain* **68,** 243.

Sunderland, S. (1978). "Nerve and Nerve Injuries." 2nd ed. Churchill-Livingstone, Edinburgh.

Sunderland, S. (1990). The anatomy and physiology of nerve injury. *Muscle Nerve* **13,** 771–784.

Tang, J., Wang, X. M., Hu, J., Luo, E., and Qi, M. C. (2008). Autogenous standard versus inside-out vein graft to repair facial nerve in rabbits. *Chin. J. Traumatol.* **11,** 104–109.

Tarasidis, G., Watanabe, O., Mackinnon, S. E., Strasberg, S. R., Haughey, B. H., and Hunter, D. A. (1997). End-to-side neurorrhaphy resulting in limited sensory axonal regeneration in a rat model. *Ann. Otol. Rhinol. Laryngol.* **106,** 506–512.

Tarasidis, G., Watanabe, O., Mackinnon, S. E., Strasberg, S. R., Haughey, B. H., and Hunter, D. A. (1998). End-to-side neurorraphy: A long-term study of neural regeneration in a rat model. *Otolaryngol. Head Neck Surg.* **119,** 337–341.

Tetik, C., Ozer, K., Ayhan, S., Siemionow, K., Browne, E., and Siemionow, M. (2002). Conventional versus epineural sleeve neurorrhaphy technique: Functional and histomorphometric analysis. *Ann. Plast. Surg.* **49,** 397–403.

Tobin, G. R., Granger, D. K., and Breidenbach, W. C. (2001). Hand transplantation follow-up. *Lancet* **358,** 1018.

Townsend, P. L. (1994). Microsurgical techniques in reconstructive surgery. *In* "Operative Surgery and Management" (G. Keen and J. R. Farndon, Eds.), 3rd ed., pp. 434–435. Butterworth-Heinemann, Oxford, UK.

Trumble, T. E. (1992). Peripheral nerve transplantation: The effects of predegenerated grafts and immunosuppression. *J. Neural. Transplant. Plast.* **3,** 39–49.

Trumble, T. E. (1999). Peripheral nerve injury: Pathophysiology and repair. *In* "Trauma" (D. V. Feliciano, E. E. Moore, and K. L. Mattox, Eds.), pp. 2048–2053. McGraw-Hill, New York, NY.

Trumble, T. E., and Shon, F. G. (2000). The physiology of nerve transplantation. *Hand Clin.* **16,** 105–122.

Viterbo, F., Trindade, J. C., Hoshino, K., and Mazzoni Neto, A. (1992). Latero-terminal neurorrhaphy without removal of the epineural sheath. Experimental study in rats. *Rev. Paul. Med.* **110,** 267–275.

Waller, A. (1850). Experiments on the section of the glossopharyngeal and hypoglossal nerves of the frog, and observations of the alterations produced thereby in the structure of their primitive fibres. *Phil. Trans. R. Soc. London.* **140,** 423–429.

Walton, R. L., Brown, R. E., Matory, W. E. Jr., Borah, G. L., and Dolph, J. L. (1989). Autogenous vein graft repair of digital nerve defects in the finger: A retrospective clinical study. *Pleast. Reconstr. Surg.* **84,** 944–949 (discussion, pp. 950–952).

Wang, K. K., Costas, P. D., Bryan, D. J., Eby, P. L., and Seckel, B. R. (1995). Inside-out vein graft repair compared with nerve grafting for nerve regeneration in rats. *Microsurgery* **16,** 65–70.

Weber, R. A., Breidenbach, W. C., Brown, R. E., Jabaley, M. E., and Mass, D. P. (2000). A randomized prospective study of polyglycolic acid conduits for digital nerve reconstruction in humans. *Plast. Reconstr. Surg.* **106,** 1036–1045 (discussion, pp. 1046–1048).

Wilgis, E. F. S. (1991). Techniques of epineural and group fascicular repair. *In* "Operative Nerve Repair and Reconstruction" (R. H. Gelberman, Ed.), Vol. 1, pp. 287–293. JB Lippincott, Philadelphia.

Wong, A. Y., and Scott, J. J. (1991). Functional recovery following direct or graft repair of nerve gaps in the rat. *Exp. Neurol.* **114,** 364–366.

Yannas, I. V., Zhang, M., and Spilker, M. H. (2007). Standardized criterion to analyze and directly compare various materials and models for peripheral nerve regeneration. *J. Biomater. Sci. Polym. Ed.* **18,** 943–966.

Yuksel, F., Karacaoglu, E., and Guler, M. M. (1999). Nerve regeneration through side-to-side neurorrhaphy sites in a rat model: A new concept in peripheral nerve surgery. *Plast. Reconstr. Surg.* **104,** 2092–2099.

Zalewski, A. A., and Gulati, A. K. (1980). Survival of nerve and Schwann cells in allografts after cyclosporin A treatment. *Exp. Neurol.* **70,** 219–225.

Zalewski, A. A., and Gulati, A. K. (1982). Evaluation of histocompatibility as a factor in the repair of nerve with a frozen nerve allograft. *J. Neurosurg.* **56,** 550–554.

Zhang, Z., Soucacos, P. N., Bo, J., and Beris, A. E. (1999). Evaluation of collateral sprouting after end-to-side nerve coaptation using a fluorescent double-labeling technique. *Microsurgery* **19,** 281–286.

ARTIFICIAL SCAFFOLDS FOR PERIPHERAL NERVE RECONSTRUCTION

Valeria Chiono, Chiara Tonda-Turo, and Gianluca Ciardelli

Department of Mechanics, Politecnico di Torino, Corso Duca Degli Abruzzi 24, 10129 Torino, Italy

Posttraumatic peripheral nerve repair is one of the major challenges in restorative medicine and microsurgery. Despite the recent progresses in the field of tissue engineering, functional recovery after severe nerve lesions is generally partial and unsatisfactory. Autograft is still the best method to treat peripheral nerve lesions, although it has several drawbacks and does not allow complete functional recovery. Full recovery of nerve functionality could ideally be achieved by proper guiding axon regeneration toward the original target tissues, through the use of purposely engineered artificial nerve guidance channels (NGCs). In the last decade, artificial NGCs have been produced using a variety of both natural and synthetic, biodegradable and nonbiodegradable polymers. Several techniques have been developed to obtain porous and nonporous NGCs and to realize and incorporate bioactive fillers for NGCs. Some of the developed products have been approved for clinical applications. Many other NGC typologies have been object of interest and are currently under investigation. The current trend of nerve tissue engineering is the realization of biomimetic NGCs, providing chemotactic, topological, and haptotactic signalling to cells, respectively by surface functionalization with cell binding domains, the use of internal-oriented matrices/fibres and the sustained release of neurotrophic factors. The present

INTERNATIONAL REVIEW OF NEUROBIOLOGY, VOL. 87
DOI: 10.1016/S0074-7742(09)87009-8

contribution provides a balanced integration of the most recent achievements of tissue engineering in the field of peripheral nerve repair. By an accurate evaluation of the status of research, the review delineates the most promising directions to which research should address for consistent progress in the field of peripheral nerve repair.

I. Introduction

Peripheral nerve regeneration is a challenging scientific field (Battiston *et al.*, 2005; Brunelli, 2005; Geuna *et al.*, 2006; Lundborg *et al.*, 2004; Pfister *et al.*, 2007a) with relevant clinical implications since nerve injuries are much more frequent than spinal cord injuries (Ciardelli and Chiono, 2006; Evans *et al.*, 2002). Peripheral nerve lesions are common and serious injuries affecting 2.8% of trauma patients annually, and may result in loss of motor function and sensory function that generally lead to lifelong disability (Wilberg and Terenghi, 2003). Peripheral nerve lesions are caused primarily by traumatic accidents, tumour resection, or iatrogenic side effects of various types of surgery, including orthopaedic intervention, intravenous aspiration, and cosmetic facial surgery (Kretschmer *et al.*, 2001). In the United States, 360,000 people suffer from upper extremity paralytic syndromes on an annual basis, whereas in Europe more than 300,000 cases of peripheral nerve injuries occur annually (Belkas *et al.*, 2004). The repair of peripheral nerve lesions has been attempted in many different ways, which have in common the goal of directing the regenerating nerve fibers into the proper distal endoneurial tubes. The strategies developed for nerve repair can be roughly classified into two categories: (1) bridging, which includes grafting and tubulization techniques; (2) end-to-end suturing of the nerve stumps. The former technique has been shown to be more effective, as it avoids tension across the repair site. Autologous nerve graft is still the gold standard in repairing injured peripheral nerve as it provides a scaffold containing both Schwann cells (SCs) and their basal lamina. These cells can produce growth factors that promote the regeneration (NGF, BDNF, NT4/5, GDNF, and IGFs). The basal lamina contains many adhesion molecules, such as laminin, fibronectin, proteoglicans, that can promote neurite elongation. However, autograft has several disadvantages such as limited sources of donor nerve, the need for a second surgery to obtain the donor nerve, loss of nerve function in transplantation, and mismatch between the injured nerve and donor nerve (Battiston *et al.*, 2005). Allograft may be an alternative but it requires an adequate immunosuppressive therapy which may interfere with nerve regeneration (Weinzweig *et al.*, 1996).

A particular model involving the use of vein and muscle has also been developed (Geuna *et al.*, 2004). This muscle-in-vein conduit allows to fill gaps between 2 and 6 cm. Moreover, it has been proved that the clinical results are comparable with nerve autografting (Battiston *et al.*, 2000; Tos *et al.*, 2000). The muscle offers strut to the vein and its basal lamina attract the regenerating axons. When the muscle degenerates, the gap is filled by the proliferation of SCs. In this way, it is possible repairing longer gaps than those repaired with empty tubes.

Thanks to researches in the field of tissue engineering, it appears increasingly possible to use artificial conduits for reconstruction of nerve gaps. Artificial nerve guidance channels (NGCs) offer a promising alternative to conventional treatments, which obviates the sacrifice of a healthy nerve and supports and guides the axons during their growth, while avoiding scar tissue infiltration in the gap. Artificial NGCs are applied to connect the trunked nerve stumps with the aim to facilitate neurotrophic and neurotrophic communication between the nerve stumps and to provide physical guidance for the regenerating axons. The closely fitting tube facilitates axonal regeneration by inducing rapid development of a highly organized capsule that isolates the repair site and guides endoneurial components allowing trophic factors emitted from the distal stump to reach the proximal segment, which enhances physiological conditions for nerve regeneration. The presence of internal matrices inside the conduits acts as physical guidance and regulate axonal growth at the site of injury by providing a structured environment for cell regeneration and organization. Moreover, the conduits can also be functionalized with contact-mediated cues, such as proteins and peptides, chemotactic cues, such as neurotrophic factors (NTFs) and biological cues, such as SCs and astrocytes (satellite cells of the peripheral and central nervous system) (Meinel *et al.*, 2002).

Nerve tissue regeneration requires a tubular scaffold that should ideally be biocompatible, have sufficient mechanical stability during nerve regeneration, be flexible (with a Young's modulus values close to that of nerve tissues to prevent compression of the regenerating nerve), be porous to ensure supply of nutrients, and degrade into nontoxic products to prevent long-term irritation. Artificial NGCs should also provide a biological microenvironment allowing cell proliferation and migration and the production of extracellular matrix (ECM) to form functional tissues (Lietz *et al.*, 2006; Teixeira *et al.*, 2007).

Various materials have been studied to prepare nerve conduits of either synthetic (Ciardelli and Chiono, 2006; Li and Shi, 2007) or natural origin (Chen *et al.*, 2005; Lauto *et al.*, 2008; Muzzarelli, 2009; Pfister *et al.*, 2007b). Moreover, both nondegradable (Wang-Bennett and Coker, 1990) and (bio)degradable materials (Johnson and Soucacos, 2008) have been used for NGC fabrication.

The main objection for using nondegradable conduits is that they remain *in situ* as foreign bodies after the nerve has regenerated. Permanent tubes show a higher risk of infection, can provoke a chronic inflammatory response, and have the potential to compress the nerve after regeneration. A second surgery might then be necessary to remove the conduits, causing possible damage to the nerve. Therefore, biodegradable NGCs appear as a more promising apparatus to reconstruct nerve gaps in clinical (Taras *et al.*, 2005). Biodegradable materials offer several advantages, such as the possibility of attaching SCs or bioactive molecules on the polymer surface through physicochemical modifications and to deliver them during biodegradation. The chemical agents or cells have to be released over a long time, remaining available during the critical stages of nerve growth. Biodegradable synthetic polymers are advantageous because of their flexibility, as variations in their chemical or engineering properties may change biocompatibility, degradation behavior, porosity, and mechanical strength.

Natural-derived materials possess good cell compatibility, but they often need extensive purification and characterization. Furthermore, most materials of natural origin lack adequate mechanical strength and water stability and thus need cross linking (Harley *et al.*, 2007).

In this work, the state-of-the art of materials for peripheral nerve repair will be provided, including both nonbiodegradable and biodegradable synthetic and natural polymers. This review aims at evidencing the main advantages and disadvantages of the more commonly used polymers for NGCs, to find out the most promising research lines to be followed to improve artificial NGCs. The review will also illustrate the fabrication techniques which have been developed for the production of NGCs. More techniques and materials have been often combined to obtain tubular scaffolds mimicking the microenvironment of the natural nerves, and provided with haptotactic and chemotactic cues promoting axon regrowth. The analysis of the up-to-date experimented NGCs will allow a selection of the most promising solutions, as a base for future research in the field.

II. Materials for Peripheral Nerve Repair

A. Nondegradable Materials for Nerve Guides

Silicone NGCs have been frequently used for peripheral nerve regeneration (Wang *et al.*, 1990). As they are nonbiodegradable and not permeable to large molecules thus creating an isolated regeneration environment, they have been widely employed to study the effect of different types of guide fillers, particularly ECM analogs—on axonal elongation. For such studies, autografts, empty silicone tubes, or silicone tubes filled with saline have been used as typical controls. As an example, Chen *et al.* (2000) have shown that silicone nerve conduits filled with a collagen, laminin, and fibronectin-based gels led to a more mature level of axon regeneration as compared to controls. Other reports have been focused on the regulation of neuronal outgrowth by filling silicone tubes with longitudinal oriented fibers based on nonabsorbable materials (e.g., polyamide (Yannas *et al.*, 2004)) or absorbable materials (e.g., collagen or poly-(lactide) (Itoh *et al.*, 2001)) or with oriented gels (e.g., composed of collagen or laminin (Verdu *et al.*, 2002)). Disadvantages of the use of nondegradable artificial nerve guides are chronic foreign body reaction due to excessive scar tissue formation, inflexibility, and lack of stability. Particularly, the inflammatory response of nondegradable NGCs may lead to fibrotic capsule formation around the guide, and consequent chronic nerve compression.

B. Biodegradable Synthetic Materials for Nerve Guides

Recent research has been focused on the production of biodegradable artificial nerve guides, which degrade within a reasonable period and only show mild foreign body reaction. Biodegradable materials offer several advantages, such as the possibility of incorporating SCs or bioactive molecules through physicochemical modifications of the polymers and to deliver them during biodegradation. Another interesting property is their flexibility, as variations in their chemical or engineering properties may change biocompatibility, degradation behavior, porosity, and mechanical strength.

Materials selected for the production of NGCs should be slowly degradable into biocompatible products and have a low degree of swelling during degradation. NGCs should also possess suitable mechanical properties for the bearing of stresses during the surgical procedure (due to handling and suturing) and the implantation time (due to the patient's movements). Precisely, NGCs should be flexible, tough, and capable of bending without kinking. NGC wall permeability is also necessary to allow the exchange of fluids between the regeneration environment and the surrounding tissue, avoiding the build up of pressure due to fluid retention (Rodriguez *et al.*, 1999). Regarding the thermal properties of NGC materials, crystallinity degree as well as morphology of crystalline domains affect permeability and biodegradation rate of NGCs. As the crystalline phase is inaccessible to water and other permeable molecules, biodegradation and permeation decrease with increasing crystallinity degree. Moreover, crystalline debris formed during degradation may cause an inflammatory response which may jeopardize the regeneration process and the recovery of nerve function. On the other hand, some degree of crystallinity is required, as the existence of physical cross links between the polymer chains stabilizes the structure of nerve guides.

Among the bioresorbable materials, aliphatic polyesters and copolyesters have been frequently used for nerve regeneration. Examples include poly(L-lactic acid) (PLLA) (Evans *et al.*, 1999), poly(glycolic acid) (PGA) (Donoghoe *et al.*, 2007; Weber *et al.*, 2000), poly(lactic acid-ε-caprolactone) (Bertleff *et al.*, 2005;

Luis *et al.*, 2007a,b; Rodriguez *et al.*, 1999), poly(DL-lactide-co-glycolide) (Chang *et al.*, 2006, 2007; Li *et al.*, 2007; Yao *et al.*, 2009), poly(1,3-trimethylenecarbonate-ε-caprolactone) (Pego *et al.*, 2001, 2003), and poly(caprolactone) (PCL) (Bender *et al.*, 2004; Chiono *et al.*, 2008; Ciardelli and Chiono, 2006; Clavijo-Alvarez *et al.*, 2007; Ghasemi-Mobarakeh *et al.*, 2008; Waddell *et al.*, 2003).

Copolymers have gained attention due to the possibility of finely tuning their properties by varying the block structure. Several investigations have been focused on copolymers, such as poly(DL-lactic-co-ε-coprolactone) (poly(DLLA-CL), synthesized poly(ester-urethane)s (PU) with polyester macrodiols, poly(trimethylenecarbonate-co-ε-caprolactone) (poly(TMC-CL)), and microbial polyhydroxyalkanoates (e.g., poly(hydroxybutyrate-co-hydroxyhexanoate), PHBHHx).

The PLLA NGCs have been found to have a high *in vivo* structural stability and to be degradable. However, their degradation products, particularly crystalline debris and lactic acid, may adversely affect axonal growth and nerve function (Evans *et al.*, 1999).

Several studies on PCL have been performed to evaluate the ability of this polymer to be shaped into suitable tubular guides (Bender *et al.*, 2004) for peripheral nerve repair or its *in vitro* affinity toward the attachment and proliferation of PC12 cells (Waddell *et al.*, 2003), SCs and rat cortical neurons (Bender *et al.*, 2004). In a recent study (Ciardelli and Chiono, 2006), the authors have shown that PCL supports the *in vitro* adhesion and proliferation of S5Y5 neuroblastoma cells and can be shaped into tubular guides by melt extrusion. The authors have also evaluated the ability of PCL NGCs to support the repair of small nerve defects in a rat model.

On the other hand, poly(DLLA-CL) NGCs have been found to be advantageous for peripheral nerve repair, as they are transparent and possess suitable mechanical properties allowing to accommodate movements over surrounding tissues. Poly(DLLA-CL) NGCs degrade completely within 1 year from implantation; however, their mechanical performance becomes negligible after only 2 months (Johnson *et al.*, 2008). Another drawback of Poly(DLLA-CL) NGCs is their high swelling during degradation, especially during the first 3 months (Johnson *et al.*, 2008). Therefore, poly(DLLA-CL) conduits are suitable for the repair of short nerve gaps or small nerves (e.g., the digital nerve) (Johnson *et al.*, 2008).

A preliminary investigation has been carried out by our research group concerning NGCs produced from a synthesized elastomeric PU, having poly-(ε-caprolactone) as macrodiol and two commercial molecules as chain extender and chain linker (cyclohexane dimethanol and hexamethylene diisocyanate, respectively) (Ciardelli and Chiono, 2006) The advantage of segmented PUs derives from their versatility, as their physicochemical, mechanical, and biodegradation properties can be easily varied by changing the chemistry of constituting blocks. PU NGCs have been found to support the *in vitro* adhesion and proliferation of S5Y5 neuroblastoma cells. *In vivo* tests for the repair of a small defect (5 mm) in the rat peroneal nerve have shown that PU guides are suitable for applications in the field of peripheral nerve regeneration. Further *in vivo* tests are in progress for the repair of more severe defects in a rat model, which will be the object of a future publication. Innovative biocompatible PUs can be synthesized by a proper selection of block composition which allows a fine tuning of their mechanical properties, degradation rate, biocompatibility, and biomimickry (Rechichi *et al.*, 2008). Recently, bioactive PUs have been produced by our research group by different techniques (Sartori *et al.*, 2007, 2008).

Porous conduits for peripheral nerve repair have been recently produced from PHBHHx (Bian *et al.*, 2009), a microbial polyhydroxyalkanoate, which is very promising as a tissue engineering material due to its adjustable mechanical properties, biocompatibility, and biodegradability (Chen *et al.*, 2005). Porous PHBHHx conduits with 1.5 mm diameter have been produced by a dipping-leaching technique, using NaCl sized particles as porogen. Both conduits with an uniform and a nonuniform porosity have been evaluated, the former having pores of around 30–50 μm, the latter having an external layer with similar porosity (30–50 μm) and an inner layer with pores of around 10 μm. Pore size of 10 μm (or lower) prevents the connective and scar tissues from growing into the internal wall, hindering axon regeneration. *In vivo* tests were performed for the repair of 10 mm defects in the rat sciatic nerve showing good results for both conduit types. Due to the high mechanical properties, nerve regeneration ability, and nontoxicity of degradation products, PHBHHx nerve conduits have been shown to be useful to repair peripheral nerve damages.

The synthetic nerves guides marketed up-to-date for nerve repair are: Neurolac® [Poliganics B.V.—poly(DL-lactic-co-ε-coprolactone)] (Bertleff *et al.*, 2005; Luis *et al.*, 2007a,b), CultiGuide® [Pittsburgh Tissue Engineering Initiative, Inc.—composite made of poly-caprolactone and porous collagen-based beads] (Bender *et al.*, 2004; Clavijo-Alvarez *et al.*, 2007), SaluBridge® [Salumedica L.L.C.—polyvinyl alcohol] (Lundborg *et al.*, 1997), and Neurotube® [Synovis Life Technologies Inc.—polyglycolic acid] (Donoghoe *et al.*, 2007; Weber *et al.*, 2000). Except for the CultiGuide®, all tubes are approved by the Food and Drug Administration (FDA). Neurolac®, Salubridge®, and Neurotube® have been found to support axonal regeneration and have granted an EU authorization. Although approved for human use, the efficacy for all the tubes marketed up to date is limited to the repair of short defects (<3 cm) of the small caliber nerves (Meek *et al.*, 2002).

C. Natural Polymers for Nerve Guides

Natural polymers are advantageous materials for tissue engineering of nerves as they are biocompatible, favor the migration of supporting cells, and avoid the occurrence of toxic effects. However, the poor mechanical properties, the high swelling behavior, and the relatively fast *in vivo* biodegradation rate of natural polymers generally limit their applications as constituent materials of the external tubular structure of NGCs. In some cases, blends between natural and synthetic polymers have been proposed for NGCs, to combine the biocompatibility of the natural component with the advantageous processing properties and mechanical performance of the synthetic material (Chiono *et al.*, 2008). More commonly, natural polymers have been used as inner fillers for NGCs, in the form of fibers, channels, porous sponges, or hydrogel matrices; or as delivery vehicles for cells, growth factors, or drugs.

Collagens constitute a family of 28 proteins which share a triple helical structure in the form of an extended rod. Collagen generally displays a very weak antigenic activity, due to its phylogenetically conserved primary sequence and a helical structure. However, in a few cases, anticollagen antibodies have been produced by heterogenic collagens (Ellingsworth *et al.*, 1986; Siegle *et al.*, 1984). Collagen immunogenity has been found to be reduced by enzymatic treatment (e.g., by pepsin), which removes collagen telopeptide fraction, and by physical and chemical cross linking (Winn *et al.*, 1998). Disadvantages of collagen also include its relatively high cost and mechanical weakness.

Interestingly, it has been shown that repair of peripheral nerves with a collagen-based nerve guide conduit over a short nerve gap (4 mm) is as effective as a standard nerve autograft (Li *et al.*, 1992). Moreover, collagen filaments have been found to guide axon repair (Millesi, 2000; Yoshii *et al.*, 2002). Among the realized NGCs based on natural polymers, collagen conduits, such a NeuroMatrix®, Neuroflex® [Collagen Matrix Inc.], and NeuraGen® [Integra Lifesciences Corp.] have been approved by FDA. Neuromas are formed as a result of random axonal regeneration at the site of injury, which creates a painful mass of neural tissue. When transected nerves lacking a sufficient distal stimulus attempt to regenerate, axons have a propensity to sprout aberrantly, forming dense nerve tangles called neuromas. A NeuraGen nerve guide has recently been applied for the repair of rat sciatic nerve to examine the effects of the NeuraGen nerve guide on the regrowth of the terminally transected nerve and subsequent development of neuropathic pain in the denervated limb (Tyner *et al.*, 2007). The enhanced axonal outgrowth promoted by the NeuraGen tubes reduced observed neuropathic behavior. NeuraGen tubes temporarily provided a physical barrier to physical external stimuli, thus preventing neuropathic sensations and reducing autotomy. A new EU-approved semipermeable and bioresorbable porcine collagen type I + III nerve conduit (Revolnerv®) has been recently investigated to repair a 10 mm gap in the rat peroneal nerve with successful motor axonal regeneration and locomotor recovery (Alluin *et al.*, 2009).

A collagen-based microstructured three-dimensional (3D) NGC containing numerous longitudinal guidance channels with dimensions resembling natural endoneurial tubes have been produced (Bozkurt *et al.*, 2009). The NGCs have been functionalized by SC seeding. Viable SCs within the guidance channels formed cellular columns reminiscent of "Bands of Büngner," which are crucial structures in the natural process of peripheral nerve regeneration during the Wallerian degeneration. The orientated 3D nerve guides (decorated with SCs) with their physical and molecular properties showed great promise in the repair of peripheral nerve lesion. Gelatin is a protein derived from collagen by thermal denaturation or physical and chemical degradation. Gelatin does not to express antigenicity in physiological conditions and is cheaper and easier to obtain in concentrated solutions than collagen. Other interesting properties of gelatin are its biocompatibility, biodegradability, and adhesiveness. The mechanical and chemical properties of gelatin can be modulated by proper cross linking. Gelatin has been the first biodegradable material to be tested as a nerve guide (Ijkema-Paassen *et al.*, 2004). In their previous feature article, the authors have illustrated several examples of gelatin scaffolds produced for the purpose of peripheral nerve repair (Ciardelli and Chiono, 2006). One of the main advantages of gelatin-based scaffolds is the possibility to covalently incorporate bioactive molecules (mainly neurotrophic factors) which are then gradually released during polymer biodegradation (Chen *et al.*, 2004; Hanthamrongwit *et al.*, 1996; Laemmel *et al.*, 1998; Pieper *et al.*, 1999). Gelatin conduits have been produced using cross linked gelatin by photocuring (Gamez *et al.*, 2004) or genipin treatment (Chen *et al.*, 2005). In both cases, NGCs have been implanted for the repair of a 10 mm nerve gap in the rat sciatic nerve. Neural regeneration was assessed in terms of functional recovery, electrophysiological responses, and tissue morphological regeneration. Photocured gelatin tubes packed with bioactive substances coimmobilized in multifilament fibers (Gamez *et al.*, 2004) have shown promising results after 1 year from implantation due to a contact guidance effect and a high surface area with immobilized bioactive substances.

Gelatin has also been investigated as an ECM for nerve regeneration. Random and aligned PCL/gelatin biocomposite matrices have been fabricated by varying the ratios of PCL and gelatin concentrations (Ghasemi-Mobarakeh *et al.*, 2008). The synthetic component provided mechanical properties while the natural polymer promoted cellular attachment and growth. Although randomly oriented nanofibrous scaffolds are useful in tissue engineering, the results have shown that aligned nanofibers highly support the nerve cells and improve the neurite outgrowth and cell differentiation process.

Recently, PLGA–gelatin compound scaffolds consisting of different weight ratios between PLGA and gelatin, have been produced and investigated for their mechanical properties, *in vitro* degradation behavior, morphology, and ability toward SC adhesion (Li *et al.*, 2007).

Chitosan (CS) is a polysaccharide obtained from *N*-deacetylation of chitin and it is a copolymer of D-glucosamine and *N*-acetyl-D-glucosamine, which interest in tissue engineering is due to its antitumor and antibacterial activity, biodegradability, and biocompatibility (Rinaudo *et al.*, 2006). CS has a similar molecular structure to the glycosaminoglycans in the basal membrane and ECM, which allows interactions between CS and extracellular adhesive molecules such as laminin, fibronectin, and collagen. In a recent work, glial cell line-derived nerve growth factor (GDNF) and laminin have been added to CS for the production of multifunctional NGCs. *In vivo* tests have shown that CS guides functionalized with growth factors and laminin enhanced both functional and sensory recovery (Patel *et al.*, 2007). Gastrocnemius muscle weight measurements and sensitivity testing have been performed as they are correlated to functional nerve recovery. Results indicated an increase in the functional and sensory recovery of the multifunctional guides when compared to the unblended CS nerve guides.

A variety of layered CS NGCs have been developed. An example is provided by bilayered tubes comprising an outer layer of CS film and an inner layer of electrospun nonwoven CS (Wang *et al.*, 2008). Gycine spacers were introduced into the CYIGSR sequence domain of laminin that enhances SCs migration and attachment, as well as neural outgrowth, resulting in the aminoacid sequences CGGYIGSR and CGGGGGGYIGSR. These peptides were covalently bound to the nano/microfiber mesh surface of the CS tube. The efficacy of nerve regeneration into CS tubes with immobilized CGGGGGGYIGSR peptide was similar to that of the isograft (Wang *et al.*, 2008).

Recently, NGCs have been obtained consisting of CS conduits filled with porous collagen sponges simulating the 3D structure of ECM (Guo *et al.*, 2009). The inner sponge was imbibed with nerve growth factor (NGF). Conduits have been tested for the repair of a 10 mm defect in the rabbit facial nerve: in one case, a suspension of neural stem cells (NSCs) has been injected into the tube during *in vivo* implantation, whereas, in the control test, a saline solution has been used. Autograft has been employed as a positive control. After 12 weeks from implantation, nerves treated with autografts have shown similar recovery as compared to the ones treated with NGCs containing NSCs.

Silk fibroin (SF) has been recently investigated as a promising biomaterial for the production of protein delivery systems integrated into NGCs, due to its biocompatibility, high resilience, and slow biodegradation (Meinel *et al.*, 2005). SF is water soluble and becomes water insoluble by physical induction of β-sheet formation. The use of SF as a substrate for neuronal cells (Uebersax *et al.*, 2006) as well as a biomaterial for the delivery of proteins (Hofmann *et al.*, 2006) has been documented. In a recent study, SF has been found to support the proliferation of PC12 cells with subsequent cell differentiation by neurite outgrowth (Uebersax *et al.*, 2007). In the same work, SF has been found to be adequate for coprocessing with NGF to produce sustained delivery matrices. However, NGF delivery from these substrates has been incomplete, due to both the strong ionic interactions between NGF and SF and the high molecular weight of NGF.

Alginate is a linear polysaccharide copolymer of (1-4)-linked D-mannuronic acid and L-guluronic acid; it is derived primarily from brown seaweed and bacteria (Remminghorst *et al.*, 2006). The physical properties of alginate gels vary widely depending on the proportion of guluronic to mannuronic acid residues and the overall molecular weight of the polymer. Sodium alginate conduits for the immobilization of Sertoli cells have been produced by extrusion and further gelling of a polymer solution containing cells (Mazzitelli *et al.*, 2008). The viability of immobilized cells as well as *in vivo* biocompatibility of the system have been assessed. Finally, alginate NGCs have been employed for the successful repair of a 7.0 mm defect in the rat sciatic nerve.

Agarose is a polysaccharide derived from red algae. Agarose hydrogel has been found to support neurite extension *in vitro* (Labrador *et al.*, 1995). The advantage of agarose derives from the possibility to easily couple proteins and glycosaminoglycans to the polymer, resulting in improved cell response. Anisotropic agarose hydrogel scaffolds with gradients of laminin have been shown to promote enhanced neurite extension from chick dorsal root ganglia (DRG) *in vitro* as compared to isotropic scaffolds (Dodla and Bellamkonda, 2006). In a recent work, polysulfone NGCs have been filled with anisotropic agarose hydrogels containing gradients of laminin and NGF and used for the repair of 2 cm defects in the rat sciatic nerve (Dodla and Bellamkonda, 2008). As control, syngenic grafts, NGCs filled with isotropic agarose hydrogels with uniform concentration of NGF and laminin, and NGCs filled with anisotropic agarose with gradient concentration of either NGF or laminin have been used. After 4 months from surgery, NGCs containing anisotropic agarose hydrogels with gradients of both laminin and NGF have shown much better results than for isotropic fillers, also suggesting a synergism between the two bioactive molecules.

III. Techniques for the Production of Scaffolds for Peripheral Nerve Repair from Synthetic Polymers

Many fabrication techniques have been developed to produce scaffolds for peripheral nerve regeneration from synthetic and natural polymers. Material composition, macro and microstructure, and mechanical properties have been shown to significantly affect scaffold cellular response. The structural and chemical characteristics of substrate has the ability to actively induce or influence specific cell behaviors (Christopherson *et al.*, 2009; Ren *et al.*, 2008). After injury, peripheral nerves can be induced to regenerate if the transected nerve ends are inserted into a tubular implant (Johnson and Soucacos, 2008). There are several key properties that all scaffolds for peripheral nerve repair should possess: (1) they must be readily formed into a conduit having a defined and constant diameter and wall thickness; (2) they must be fabricated using biocompatible and preferentially biodegradable materials; (3) they should have good structural stability to avoid scaffold collapse; (4) they should be simple to implant using microsurgical techniques; and (5) they should not show changes in mechanical, chemical, and biological characteristics upon sterilization. One of the more common way to produce NGCs from polymers is by melt extrusion (Chiono *et al.*, 2008; Ciardelli and Chiono, 2006). Melt extrusion is the process of converting a thermoplastic raw material into a product of uniform shape and density by forcing it through a die of desired shape. Many biocompatible synthetic polymers are not easily processable by melt extrusion therefore alternative fabrication techniques have been described in literature to prepare hollow guides for nerve regeneration. The low thermal stability of some synthetic polymers, such as PLA and poly(hydroxybutyrate) (PHB), do not allow their melt processing by extrusion. For the natural-based materials, proteins may undergo denaturation at high temperatures, while polysaccharides may degrade during melt processing. The extrusion process has been successfully applied to fabricate PCL/gelatin melt extruded guides without the occurrence of polymer degradation (Chiono *et al.*, 2008).

As an alternative technique, steel or Teflon spinning-mandrel of adjustable diameter can be dipped into the polymer solution and then left air-drying to obtain a conduit with a defined diameter and a wall thickness depending on the polymer solution density and on the number of dipping cycles (Bian *et al.*, 2009; Chen and Wu, 2005; Wang *et al.*, 2001). Both synthetic- and natural-based nerve guides have been obtained by spinning-mandrel techniques. The main requirement of these methods is the use of highly viscous solutions, to avoid dropping during the fabrication process and to reduce the evaporation rate.

Moulds of different shapes and materials have been shown to be suitable for tube preparation, although appropriate solution viscosity and homogeneity are required to obtain well-defined NGCs (Freiera *et al.*, 2005; Wang *et al.*, 2009).

Other combined techniques have been developed to fabricate NGCs for peripheral nerve repair: for instance, Li *et al.* have described a multiwalled PLLA conduit prepared by means of solvent casting, physical imprinting, and rolling-fusing methods having a microstructured tube wall surface to guide the healing process (Li and Shi, 2007).

Scaffold permeability is a fundamental parameter to obtain a complete regeneration after injury (Chang *et al.*, 2006, 2007; Harley *et al.*, 2006). A uniform, repeatable, and adjustable porosity is desirable for NGCs. The nerve guide wall should be semipermeable to provide sufficient diffusion of oxygen and metabolites between the external environment and tube lumen. Moreover, the pore dimension has to be small enough to prevent infiltration of scar-forming fibroblasts that could alter the regeneration process. As demonstrated by Rodriguez *et al.*, permeable poly(L-lactide-*co*-ε-caprolactone) (PLC) guides allow faster and higher levels of reinnervation than impermeable or low-permeable PLC guides

(Rodriguez *et al.*, 1999). Oh *et al.* have fabricated a bilayer NGCs having two different layer porosity: the inner tube surface had nanosized pores (50 nm) preventing fibrous tissue infiltration and allowing permeation of nutrients and retention of neurotrophic factors; the outer tube surface had microsized pores (50 m) allowing vascular ingrowth for effective supply of nutrients into the tube (Oh *et al.*, 2007). A number of processing techniques based on textile technologies, particulate leaching, phase separation processes, and gas foaming methods have been developed with different success in the production of porous biodegradable polymeric scaffolds for peripheral nerve regeneration. As described by Pego *et al.* (2003), 1,3-trimethylene carbonate and ε-caprolactone copolymers (poly(TMC-CL)) have been processed into porous two-ply tubes by means of dip-coating and fiber winding techniques. To prepare the inner layer, glucose particles (diameter <20 μm) have been added to the polymer solution and, then, a glass mandrel has been dipped into the sugar–polymer solution. For the outer layer, polymer fibers have been spun and wound on the rotating dip-coated tube. The mean pore diameter obtained has been 20 and 60 μm, respectively, for the inner and outer layer (Pego *et al.*, 2003).

Collagen/CS blend conduits have been prepared by filling a concentric tubular mould with the blend solution, further gelling, freezing, and overnight lyophilization to obtain a porous scaffold. The porous collagen-blended CS NGCs were successful in enhancing motor and sensory nerve recovery after 12 weeks in a rat animal model having a nerve gap between the proximal and distal nerve of 1 cm (Patel *et al.*, 2008).

In an attempt to mimic the guidance structure of peripheral nerve autografts, proper fillers have been frequently inserted into the guide lumen, as topological cues favoring glial cell attachment, proliferation, and migration, as well as axon regrowth. For instance, natural-based gels, such as collagen, gelatin, and laminin gels, filling artificial NGCs improved the recovery of neural gaps in rat or mouse sciatic nerves (Ciardelli and Chiono, 2006; Verdú *et al.*, 2002). Macroporous 3D sponges obtained by freeze-drying from CS solutions represent an effective approach for the improvement of the peripheral nerve regeneration after injury (Amado *et al.*, 2008). Many studies have demonstrated that nanofibrous scaffolds are a suitable environment for cell attachment and proliferation due to their mimicking the natural fibrous structure of the ECM (Agarwal *et al.*, 2008; Sill and von Recum, 2008). In peripheral nerve regeneration, nanofiber internal matrices to NGCs can be applied to support and direct the axon regenerations. Many authors have reported the influence of nanofiber orientation on cell growth and function (Christopherson *et al.*,2009; Ghasemi-Mobarakeh *et al.*, 2008; Kim *et al.*, 2007; Mahoney *et al.*, 2005; Matsumoto *et al.*, 2000). Cells have a higher probability of migrating in direction which are associated with structural or chemical properties of the substrate: in this contest, aligned nanofibers provide a maximum contact guidance effect during neurite ingrowth. Fibrous matrices mimic the complex biological structures and provide the mechanical support to allow the cells of the damaged tissue to remodel and repair forming 3D tissue structures that resemble the original tissue. In recent years, electrospinning has been extensively used to construct tissue-engineered scaffolds because it is a simple fabrication process that can easily produce nano- and microsized synthetic polymeric fibers. In brief, this technique utilizes an electric field generated by an applied voltage that subsequently introduces surface charges on the polymer solution. This induces the formation of a Taylor cone polymeric droplet at the tip of the spinneret. As the electric potential that is created at the droplet surface exceeds a critical value, the electrostatic forces will overcome the solution surface tension to initiate polymer jet stream. Nanofibers can then be drawn from the polymer jet stream, and collected on the grounded collector as the solvent evaporates (Koh *et al.*, 2008). Fibers can be collected to have a randomly or aligned orientation; for collecting aligned nanofibers, a rotating disk can be used whereas a flat aluminium plate is generally used for preparing randomly oriented membranes.

In recent years, microfabrication process has been applied to study the cellular response on 2D or 3D structures with a well-defined geometry in the range of a few microns prepared by rapid prototyping system (Goubko and Cao, 2009; Vozzi *et al.*, 2007; Yamada *et al.*, 2008). Yao *et al.* (2009) showed that microgrooves on PLGA films exerted a guidance effect on both early stage neurite outgrowth and neurite elongation. Neurites showed significantly more parallel growth on small groove sizes (5 μm) than on larger groove sizes (10 μm) (Yao *et al.*, 2009). The use of rapid prototyping microfabrication techniques is of fundamental importance to fabricate internal matrices with well-defined structures, able to induce a specific cellular response enhancing the regeneration process.

IV. Functionalized Bioactive Materials for Axon Regeneration

During the embryonic development of the nervous system, the developing axons are stimulated by a variety of *haptotactic* (or contact-mediated) and *chemotactic* (or diffusible) cues to find their target organs. Some of these signals are naturally present in a gradient concentration. Analogously, a number of cellular and molecular cues stimulating regeneration have been identified during the spontaneously occurring axon regeneration response after axonotmesis injuries. Following nerve injury, regenerating axon sprouts show an orientated framework of proliferating SCs in the distal nerve stump. SCs of both myelinated and unmyelinated axons multiply, the former resulting in formation of "Bands of Büngner," arrays of SCs and their interdigitating processes within a space circumscribed by the basement membrane ("Schwann tube") (Chaudhry *et al.*, 1992). SCs express diffusible growth factors and cell adhesion molecules which together with the basal lamina are key factors for the regeneration process (Ide *et al.*, 1983; Morris *et al.*, 1972).

Bioinspired strategies have been implemented to improve the regenerative effectiveness of artificial conduits, by an integration of contact-mediated cues and the incorporation and delivery of exogenous growth factors into the tube lumen.

A. Haptotactic Cues

The filler structures for NGCs—in the form of fibres, guidance channels, sponges, or hydrogels—ideally require a surface functionalization with haptotactic cues to promote axonal regeneration. Contact-mediated cues, such as ECM proteins, mainly collagen, laminin, and fibronectin, are haptotactic cues guiding the axon growth cones during regeneration (Archibald *et al.*, 1991; Bunge, 1993; Giannini *et al.*, 1990; Gundersen *et al.*, 1987; Sorenson *et al.*, 1993).

As the major organic component of the natural ECM of most vertebrate tissues, collagen is arguably the most versatile substrate for supporting cell proliferation and differentiation. Additionally, the mechanical properties and degradation rate of collagen can be tailored by altering the degree to which it is cross linked. Furthermore, the abundance of functional groups along its polypeptide backbone makes it highly receptive to the binding of genes, growth factors and other biological molecules.

Laminin is one of the most investigated proteins of the ECM. Its interest derives from its being an abundant component of the basement membrane during the development of the embryonic nervous system, and from its presence in the mature nervous system with apparently important functions not restricted just to guidance or adhesion. Laminin is a multidomain protein (Beck *et al.*, 1990) with many binding sites for different cell receptors (Castronovo, 1993). For instance, the Ile-Lys-Val-Ala-Val (IKVAV) sequence, which is located on the C-terminal end of the long arm of the laminin-1 chain, is a bioactive peptide promoting neurite outgrowth.

Laminin plays a crucial role in the developing and maturing central nervous system, for example, in cell migration, differentiation, and axonal growth (Martin *et al.*, 1988). Laminin 5 is expressed in the developing central nervous system (Libby *et al.*, 2000) and regulates neurite outgrowth and extension (Culley *et al.*, 2001; Smith *et al.*, 1996). Some studies have shown the guidance effect of laminin-coated micropatterns on neurite growth (Mahoney *et al.*, 2005; Miller *et al.*, 2002) and others have studied the synergistic effect of the bioactive peptides of laminin 5 with physical guidance cues on neurite growth (Yao *et al.*, 2009).

Fibronectin is an ECM protein which is dispersed in interstitial matrices. Fibronectin is composed of several rod-like domains, one of which (the Type III repeat) contains the RGDS sequence (L-arginine, L-glycine, L-aspartic acid, and L-serine) which regulates cell adhesion. This repeating sequence is characteristic of other ECM proteins, including collagen, laminin, and thrombospondin. Fibronectin has been found to play a role in axonal growth and cell migration (Zhang *et al.*, 2003, 2005). Fibronectin and fibronectin-derived peptides containing the RGD (arginine-glycine-aspartic acid) binding motif were immobilized on glass substrates, the effect being that to enhance neurite outgrowth (Zhang *et al.*, 2005). However, RGD-containing peptides showed reduced effectiveness in promoting neurite outgrowth when compared to fibronectin, due to their short sequence and limited conformational flexibility. Moreover, fibronectin/comb polymer pattered surfaces were found to regulate neurite attachment and outgrowth: cells attached and outgrew only on fibronectin stripes with a similar or larger size than the cell body of neurons (30–40 mm). Several investigators have shown that the incorporation of specific biomimetic peptides on scaffolds may enhance nerve regeneration (Pierschbacher *et al.*, 1984). The YIGSR sequence on the $\beta 1$ chain of laminin has been found to favor cell adhesion of neural cells, the IKVAV sequence on the α chain of laminin has been found to promote neurite outgrowth of PC12 cells, whereas the RGD sequence has a recognized role in generic cell binding. YIGSR and IKVAV sequences have also been found to interact synergistically directing neuron adhesion and outgrowth (Tong *et al.*, 2001). The GRGDS fibronectin peptide fragment has also been found to guide axonal outgrowth *in vitro* (Luo *et al.*, 2004).

B. Chemotactic Cues

Regulation of cellular activities through signaling molecules such as neurotransmitters, hormones, and growth factors are complex events. They often involve multiple signaling cascades and cross talk between pathways. Analyzing the sequence of events in signaling cascades is important not only for understanding of disease regulation, but in developing and improving strategies for regenerative medicine. Nerve regeneration has been found to be enhanced by filling nerve guides with neurotrophic factors, such as NGF, GDNF, and neurotrophin NT-3 (Rich *et al.*, 1989; Whitworth *et al.*, 1996). Neurotrophin family members are small homolog proteins which bind to their specific receptors, activating important intracellular signaling. Neurotrophic factors promote neuronal regrowth, sprouting, and ultimately new connections between the transected axons. NGF is a neurotrophic factor produced by the target organs of sympathetic and sensory nerves (Barde, 1989). NGF has been found to stimulate neurite outgrowth and promote the survival of sensory ganglia (Levi Montalcini, 1987). The spatial and temporal delivery of regulatory molecules is important for successful neural tissue engineering. When a concentration gradient of NGF is present, growth cones guide growing axons to their proper innervating target (Tessier-Lavigne *et al.*, 1991). Photoreactive moieties activated by a pulsed infrared or ultraviolet (UV) laser within the confocal microscope have been found to be particularly useful to obtain protein patterns on substrate surfaces. The desired patterns on a molecular scale can be produced with great precision when combined with photochemistry (Wosnick and Shoichet, 2008). This method of spatially controlled pattern deposition on substrate surfaces can be used to create various immobilized protein patterns in 2D, thereby mimicking those found *in vivo*, such as concentration gradient patterns. This system is a useful tool to investigate and identify the signaling molecules involved in cell behavior and function. For example, surface patterns of cell adhesive peptides or concentration gradients of signaling molecules have been found to influence cell elongation, formation of neuron aggregates, and controlled axon guidance in 2D cultures (Cao *et al.*, 2001; Dodla and Bellamkonda, 2006, 2008). The stage movement of the confocal microscope may be controlled through computer programs, obtaining 3D patterning (Wosnick and Shoichet, 2008). The creation of complex 3D patterns of biological molecules is a powerful tool to control and guide cell growth and proliferation. For instance, NGF has been chemically immobilized on CS films using such a method (Yu *et al.*, 2008). Using rhodamine as a model for NGF, a series of immobilized concentration gradients have been obtained, by varying the number of rastering scans within a defined area and the distance between each area. The same method has been applied to create NGF patterns on CS films, which have been found to remain bioactive. Neuron survival was 73.2 $\pm$ 1.3% after 3 days of culture on CS films with 30 ng/cm^2 of homogenously immobilized NGF, which was comparable to 74.8 $\pm$ 3.4% neuron survival on CS with 50 ng/ml of soluble NGF present. Neurons cultured on a CS film with distinct immobilized NGF-patterned areas have been found to predominantly remain as single cells on the NGF-patterned regions, forming aggregates on the plain CS areas. Thus, the immobilized NGF patterns have been found to influence neuron behavior. Importantly, the versatility of the confocal laser patterning technique allows functionalization with other factors with the aim to elucidate fundamental cell functions, and hence to design strategies in regenerative medicine. Growth factor therapy is a difficult task due to the high biological activity (in pico- to nanomolar range), pleiotrophic effects (acting on various targets), and short biological half-life (few minutes to hours) of growth factors. A detailed review on NGCs with integrated delivery systems for NGF delivery has been recently written by Pfister *et al.* (2007a,b). Briefly, localized growth factor delivery has been obtained in three different ways: (i) direct delivery from the NGC wall or lumen; (ii) seeding of cells able to produce growth factors within NGC lumen; and (iii) use of gene therapy to transfect resident cells to express certain proteins.

V. Conclusion

The recent trend of tissue engineering of the peripheral nerves is focused on the use of artificial guides based on biodegradable materials with tailored properties to the clinical case, specifically:

- The choice of the materials for NGCs is currently addressed to biodegradable polymers, to avoid the need for an additional surgery to remove the conduits after regeneration has occurred.
- Multifunctional NGCs may be obtained by the combination of different materials, each imparting specific properties: for instance, a material with advantageous mechanical properties assuring the *in vivo* structure stability of the tube can be enriched with another material for the release of bioactive agents (e.g., growth factors).
- Both natural and synthetic polymers are currently under investigation for peripheral nerve repair; however, the more promising choice is likely to be the use of biodegradable synthetic polymers as main components of NGC wall and natural polymers as inner fillers or minor components of the NGC wall.
- For short gaps (<2.0 cm), mechanical properties of NGCs have not to respond to rigorous specifications; however, for longer gaps (≥ 2.0 cm), NGCs should be resistant to compressive, flexional, and torsion stresses, and easily fit the body movements to avoid NGC detachment from the implant site.
- The thickness of NGCs should primarily be selected allowing an easy suturing.
- For short gaps (<2.0 cm), guide fillers are not necessary and indeed may hinder fast regeneration; on the other hand, for more severe gaps (≥ 2.0 cm), an internal structure able to support glial cell adhesion, proliferation, and migration is necessary.
- Porous NGCs are necessary for the treatment of long defects: pore size lower than 10 μm is recommended to avoid scar invasion.

From the literature data and our experience, optimal materials for NGCs are synthesized biodegradable and biocompatible block copolymers, which physicochemical, mechanical, and biological properties may be easily tailored to the application. In this context, segmented polyurethanes (PUs) are promising as they are elastomeric biomaterials with tunable mechanical properties, processability, and high biocompatibility in a variety of applications (Sartori *et al.*, 2007, 2008). The materials for NGCs are believed not to have a profound effect on nerve repair outcome; on the other hand, the molecular biological stimulation of the internal conduit milieu may result in a plethora of new materials added as bioactive substrates within conduits (Strauch, 2000). As topographic and chemical micro- or nanoscaled patterns have shown a strong influence on cell shape and migration, protein synthesis and gene expressions (Goddard *et al.*, 2007; Shin, 2007), biomaterials to be used as NGC fillers should allow a geometrical control of regulatory directional guidance signals for a spatial organization of neural cells into a linear structure. Biological cues are thus necessary to direct cells to assemble into desirable tissue structures imitating the native microenvironment. Longitudinally aligned biodegradable fibers based on natural polymers and functionalized with concentration gradients of specific peptides for nerve regeneration may combine chemical, mechanical, and geometrical signals leading to an artificial environment closely mimicking the native tissue architecture (Khademhosseini *et al.*, 2006). Although to date, engineered NGCs has not been found more effective than autografts, bioengineered grafts are a promising alternative, as they can incorporate all the new developing strategies for nerve regeneration which continually develop side by side with the knowledge of the mechanism of regeneration. Nanotechnology or the use of nanomaterials (defined as those materials with constituent dimensions less than 100 nm) could finally solve the numerous problems associated with traditional implants, by mimicking the properties of natural tissues. Since cells directly interact with (and create) nanostructured ECMs, the biomimetic features and excellent physicochemical properties of nanomaterials play a key role in stimulating cell growth as well as guide tissue regeneration. Progress in nanotechnology provides a platform to develop novel and improved neural tissue engineering synthetic materials, including designing nanofiber/nanotube scaffolds with excellent cytocompatibility, conductivity properties to boost neuron activities, and in some cases able to encapsulate neural stem cells and SCs into biomimetic scaffolds to enhance nerve repair.

References

Agarwal, S., Wendorff, J. H., and Greiner, A. (2008). Use of electrospinning technique for biomedical applications. *Polymer* **49,** 5603–5621.

Alluin, O., Wittmann, C., Marqueste, T., Chabas, J. F., Garcia, S., Lavaut, M. N., Guinard, D., Feron, F., and Decherchi, P. (2009). Functional recovery after peripheral nerve injury and implantation of a collagen guide. *Biomaterials* **30,** 363–373.

Amado, S., Simoes, M. J., Armada da Silva, P. A. S., Luıs, A. L., Shirosaki, Y., Lopes, M. A., Santos, J. D., Fregnan, F., Gambarotta, G., Raimondo, S., Fornaro, M., Veloso, A. P., *et al.* (2008). Use of hybrid chitosan membranes and N1E-115 cells for promoting nerve regeneration in an axonotmesis rat model. *Biomaterials* **29,** 4409–4419.

Archibald, S. J., Krarup, C., and Shefner, J. (1991). A collagen-based nerve guide conduit for peripheral nerve repair: An electrophysiological study of nerve regeneration in rodents and non-human primates. *J. Comp. Neurol.* **306,** 685–696.

Barde, Y. A. (1989). Trophic factors and neuronal survival. *Neuron* **2,** 1525–1534.

Battiston, B., Tos, P., Geuna, S., Giacobini-Robecchi, M. G., and Guglielmone, R. (2000). Nerve repair by means of vein filled with muscle grafts. II. Morphological analysis of regeneration. *Microsurgery* **20,** 37–41.

Battiston, B., Geuna, S., Ferrero, M., and Tos, P. (2005). Nerve repair by means of tubulization: Literature review and personal clinical experience comparing biological and synthetic conduits for sensory nerve repair. *Microsurgery* **25**(4), 258–267.

Beck, K., Hunter, I., and Engel, J. (1990). Structure and function of laminin: Anatomy of a multidomain glycoprotein. *FASEB J.* **4,** 148–160.

Belkas, J. S., Shoichet, M. S., and Midha, R. (2004). Axonal guidance channels in peripheral nerve regeneration. *Oper. Tech. Orthop.* **14,** 190–198.

Bender, M. D., Bennett, J. M., Waddell, R. L., Doctor, J. S., and Marra, K. G. (2004). Multi-channeled biodegradable polymer/cultispher composite nerve guides. *Biomaterials* **25,** 1269–1278.

Bertleff, M. J., Meek, M. F., and Nicolai, J. P. (2005). A prospective clinical evaluation of biodegradable neurolac nerve guides for sensory nerve repair in the hand. *J. Hand Surg.* **30,** 513–518.

Bian, Y. Z., Wang, Y., Aibaidoula, G., Chen, G. Q., and Wu, Q. (2009). Evaluation of poly(3-hydroxybutyrate-co-3-hydroxyhexanoate) conduits for peripheral nerve regeneration. *Biomaterials* **30,** 217–225.

Bozkurt, A., Deumens, R., Beckmannm, C., Damink, L. O., Schügner, F., Heschel, I., Sellhaus, B., Weis, J., Jahnen-Dechent, W., Brook, G. A., and Pallua, N. (2009). *In vitro* cell alignment obtained with a Schwann cell enriched microstructured nerve guide with longitudinal guidance channels. *Biomaterials* **30,** 169–179.

Brunelli, G. A. (2005). Direct muscular neurotization. *J. Am. Soc. Surg. Hand* **5**(4), 193–200.

Bunge, R. P. (1993). Expanding roles for the Schwann cell: Ensheathment, myelination, trophism and regeneration. *Curr. Opin. Neurobiol.* **3,** 805–809.

Cao, X., and Shoichet, M. S. (2001). Defining the concentration gradient of nerve growth factor for guided neurite outgrowth. *Neuroscience* **103,** 831–840.

Castronovo, V. (1993). Laminin receptors and laminin-binding proteins during tumor invasion and metastasis. *Invasion Metastasis* **13,** 1–30.

Chang, C. J., and Hsu, S. H. (2006). The effect of high outflow permeability in asymmetric poly(DL-lactic acid-co-glycolic acid) conduits for peripheral nerve regeneration. *Biomaterials* **27,** 1035–1042.

Chang, C. J., Hsu, S. H., Yen, H. J., Chang, H., and Hsu, S. K. (2007). Effects of unidirectional permeability in asymmetric poly(DL-lactic acid-co-glycolic acid) conduits on peripheral nerve regeneration: An *in vitro* and *in vivo* study. *J. Biomed. Mater. Res. B* **83,** 206–215.

Chaudhry, V., Glass, J. D., and Griffin, J. W. (1992). Wallerian degeneration in peripheral nerve disease. *Neurol. Clin.* **10,** 613–627.

Chen, G. Q., and Wu, Q. (2005). The application of polyhydroxyalkanoates as tissue engineering materials. *Biomaterials* **26,** 6565–6578.

Chen, Y. S., Hsieh, C. L., Tsai, C. C., Chen, T. H., Cheng, W. C., Hu, C. L., and Yao, C. H. (2000). Peripheral nerve regeneration using silicone rubber chambers filled with collagen, laminin and fibronectin. *Biomaterials* **21,** 1541–1547.

Chen, P. R., Chen, M. H., Sun, J. S., Chen, M. H., Tsai, C. C., and Lin, F. H. (2004). Biocompatibility of NGF-grafted GTG membranes for peripheral nerve repair using cultured Schwann cells. *Biomaterials* **25,** 5667–5673.

Chen, Y. S., Chang, J. Y., Cheng, C. Y., Tsai, F. J., Yao, C. H., and Liu, B. S. (2005). An *in vivo* evaluation of a biodegradable genipin-cross-linked gelatin peripheral nerve guide conduit material. *Biomaterials* **26,** 3911–3918.

Chiono, V., Ciardelli, G., Vozzi, G., Cortez, J., Barbani, N., Gentile, P., and Giusti, P. (2008). Enzymatically- modified melt-extruded guides for peripheral nerve repair. *Eng. Life Sci.* **8,** 1–12.

Christopherson, G. T., Song, H., and Mao, H. Q. (2009). The influence of fiber diameter of electrospun substrates on neural stem cell differentiation and proliferation. *Biomaterials* **30,** 556–564.

Ciardelli, G., and Chiono, V. (2006). Materials for peripheral nerve regeneration. *Macromol. Biosci.* **6,** 13–26.

Clavijo-Alvarez, J. A., Nguyen, V. T., Santiago, L. Y., Doctor, J. S., Lee, W. P., and Marra, K. G. (2007). Comparison of biodegradable conduits within aged rat sciatic nerve defects. *Plast. Reconstr. Surg.* **119,** 1839–1851.

Culley, B., Murphy, J., Babaie, J., Nguyen, D., Pagel, A., Rousselle, P., and Clegg, D. O. (2001). Laminin-5 promotes neurite outgrowth from central and peripheral chick embryonic neurons. *Neurosci. Lett.* **301,** 83–86.

Dodla, M. C., and Bellamkonda, R. V. (2006). Anisotropic scaffolds facilitate enhanced neurite extension *in vitro*. *J. Biomed. Mater. Res. A* **78,** 213–221.

Dodla, M. C., and Bellamkonda, R. V. (2008). Differences between the effect of anisotropic and isotropic laminin and nerve growth factor presenting scaffolds on nerve regeneration across long peripheral nerve gaps. *Biomaterials* **29,** 33–46.

Donoghoe, N., Rosson, G. D., and Dellon, A. L. (2007). Reconstruction of the human median nerve in the forearm with the neurotube. *Microsurgery* **27,** 595–600.

Ellingsworth, L. R., De Lustro, F., Brennan, J. E., Sawamura, S., and McPherson, J. (1986). The human immune response to reconstituted bovine collagen. *J. Immunol.* **136,** 877–882.

Evans, G. R., Brandt, K., Widmer, M. S., Lu, L., Meszlenyi, R. K., Gupta, P. K., Mikos, A. G., Hodges, J., Williams, J., Gürlek, A., Nabawi, A., Lohman, R., *et al.* (1999). *In vivo* evaluation of poly(L-lactic acid) porous conduits for peripheral nerve regeneration. *Biomaterials* **20,** 1109–1115.

Evans, G. R. D., Brandt, K., Katz, S., Chauvin, P., Otto, L., Bogle, M., Wang, B., Meszlenyi, R. K., Lu, L., Mikos, A. G., and Patrick, C. W. Jr. (2002). Bioactive poly(L-lactic acid) conduits seeded with Schwann cells for peripheral nerve regeneration. *Biomaterials* **23,** 841–848.

Freiera, T., Montenegro, R., Koh, H. S., and Shoichet, M. S. (2005). Chitin-based tubes for tissue engineering in the nervous system. *Biomaterials* **26,** 4624–4632.

Gámez, E., Yoshinobu, G., Nagata, K., Iwaki, T., Sasaki, T., and Matsuda, T. (2004). Photofabricated gelatin-based nerve conduits: Nerve tissue regeneration potentials. *Cell. Tranplant.* **13,** 549–564.

Geuna, S., Tos, P., Battiston, B., and Giacobini-Robecchi, M. G. (2004). Bridging peripheral nerve defects with muscle–vein combined guides. *Neurol. Res.* **26,** 139–144.

Geuna, S., Papalia, I., and Tos, P. (2006). End-to-side (terminolateral) nerve regeneration: A challenge for neuroscientists coming from an intriguing nerve repair concept. *Brain Res.* **52,** 381–388.

Ghasemi-Mobarakeh, L., Prabhakaran, M. P., Morshed, M., Nasr-Esfahani, M. H., and Ramakrishna, S. (2008). Electrospun poly(ε-caprolactone)/gelatin nanofibrous scaffolds for nerve tissue engineering. *Biomaterials* **29,** 4532–4539.

Giannini, C., and Dyck, P. J. (1990). The fate of Schwann cell basement membranes in permanently transected nerves. *J. Neuropathol. Exp. Neurol.* **49,** 550–563.

Goddard, J. M., and Hotchkiss, J. H. (2007). Polymer surface modification for the attachment of bioactive compounds. *Prog. Polym. Sci.* **32,** 698–725.

Goubko, C. A., and Cao, X. (2009). Patterning multiple cell types in co-cultures: A Review. *Mater. Sci. Eng. C.* Doi: 10.1016/j.msec.2009.02.016.

Gundersen, R. W. (1987). Response of sensory neurites and growth cones to patterned substrata of laminin and fibronectin *in vitro*. *Dev. Biol.* **121,** 423–431.

Guo, B. F., and Dong, M. M. (2009). Application of neural stem cells in tissue-engineered artificial nerve. *Otolaryngol. Head Neck Surg.* **140,** 159–164.

Hanthamrongwit, M., Reid, W. H., and Grant, M. H. (1996). Chondroitin-6-sulphate incorporated into collagen gels for the growth of human keratinocytes: The effect of cross-linking agents and diamines. *Biomaterials* **17,** 775–780.

Harley, B. A., Hastings, A. Z., Yannas, I. V., and Sannino, A. (2006). Fabricating tubular scaffolds with a radial pore size gradient by a spinning technique. *Biomaterials* **27,** 866–874.

Harley, B. A., Leung, J. H., Silva, E. C. C. M., and Gibson, L. J. (2007). Mechanical characterization of collagen–glycosaminoglycan scaffolds. *Acta Biomater.* **3,** 463–474.

Hofmann, S., Wong Po Foo, C. T., Rossetti, F., Textor, M., Vunjak-Novakovic, G., Kaplan, D. L., Merkle, H. P., and Meinel, L. (2006). Silk fibroin as an organic polymer for controlled drug delivery. *J. Control. Release* **111,** 219–227.

Ide, C., Tohyama, K., Yokota, R., Nitatori, T., and Onodera, S. (1983). Schwann cell basal lamina and nerve regeneration. *Brain Res.* **288,** 61–75.

Ijkema-Paassen, J., Jansen, K., Gramsbergen, A., and Meek, M. F. (2004). Transection of peripheral nerves, bridging strategies and effect evaluation. *Biomaterials* **25,** 1583–1592.

Itoh, S., Takakuda, K., Ichinose, S., Kikuchi, M., and Schinomiya, K. A. (2001). A study of induction of nerve regeneration using bioabsorbable tubes. *J. Reconstr. Microsurg.* **17,** 115–123.

Johnson, E. O., and Soucacos, P. N. (2008). Nerve repair: Experimental and clinical evaluation of biodegradable artificial nerve guides. *Injury Int. J. Care Injured* **39S,** S30–S36.

Khademhosseini, A., Langer, R., Borenstein, J., and Vacanti, J. P. (2006). Microscale technologies for tissue engineering and biology. *Proc. Natl. Acad. Sci. USA* **103,** 2480–2487.

Kim, T. G., Lee, D. S., and Park, T. G. (2007). Controlled protein release from electrospun biodegradable fiber mesh composed of poly(ε-caprolactone) and poly(ethylene oxide). *Int. J. Pharm.* **338,** 276–283.

Koh, H. S., Yong, T., Chan, C. K., and Ramakrishna, S. (2008). Enhancement of neurite outgrowth using nano-structured scaffolds coupled with laminin. *Biomaterials* **29**(26), 3574–3582.

Kretschmer, T., Antoniadis, G., and Braun, V. (2001). Evaluation of iatrogenic lesions in 722 surgically treated cases of peripheral nerve trauma. *J. Neurosurg.* **94,** 905–912.

Labrador, R. O., Buti, M., and Navarro, X. (1995). Peripheral nerve repair: Role of agarose matrix density on functional recovery. *Neuroreport* **6,** 2022–2026.

Laemmel, E., Penhoat, J., Warocquier-Clerout, R., and Sigot-Luizard, M. F. (1998). Heparin immobilized on proteins usable for arterial prosthesis coating: Growth inhibition of smooth-muscle cells. *J. Biomed. Mater. Res.* **39,** 446–452.

Lauto, A., Foster, L. J., Avolio, A., Sampson, D., Raston, C., Sarris, M., McKenzie, G., and Stoodley, M. (2008). Sutureless nerve repair with laser-activated chitosan adhesive: A pilot *in vivo* study. *Photomed. Laser Surg.* **26,** 227–234.

Levi Montalcini, R. (1987). The nerve growth factor 35 years later. *Science* **237,** 1154–1162.

Li, J., and Shi, R. (2007). Fabrication of patterned multi-walled poly-L-lactic acid conduits for nerve regeneration. *J. Neurosci. Methods* **165,** 257–264.

Li, S. T., Archibald, S. J., Krarup, C., and Madison, R. D. (1992). Peripheral nerve repair with collagen conduits. *Clin. Mater.* **9,** 195–200.

Li, X. K., Cai, S. X., Liu, B., Xu, Z. L., Dai, X. Z., Ma, K. W., Li, S. Q., Yang, L., Sunge, K. L. P., and Fu, X. B. (2007). Characteristics of PLGA–gelatin complex as potential artificial nerve scaffold. *Colloids Surf. B Biointerfaces* **57,** 198–203.

Libby, R. T., Champliaud, M. F., Claudepierre, T., Xu, Y., Gibbons, E. P., Koch, M., Burgeson, R. E., Hunter, D. D., and Brunken, W. J. (2000). Laminin expression in adult and developing retinae: Evidence of two novel CNS laminins. *J. Neurosci.* **20,** 6517–6528.

Lietz, M., Dreesmann, L., Hoss, M., Oberhoffner, S., and Schlosshauer, B. (2006). Neuro tissue engineering of glial nerve guides and the impact of different cell types. *Biomaterials* **27,** 1425–1436.

Luis, A. L., Rodrigues, J. M., Amado, S., Veloso, A. P., Armada-Da-Silva, P. A. S., Raimondo, S., Fregnan, F., Ferreira, A. J., Lopes, M. A., Santos, J. D., Geuna, S., Varejão, A. S. P., *et al.* (2007a). PLGA 90/10 and caprolactone biodegradable nerve guides for the reconstruction of the rat sciatic nerve. *Microsurgery* **27,** 125–137.

Luis, A. L., Rodrigues, J. M., Lobato, J. V., Lopes, M. A., Amado, S., Veloso, A. P., Armada-da-Silva, P. A., Raimondo, S., Geuna, S., Ferreira, A. J., Varejão, A. S., Santos, J. D., *et al.* (2007b). Evaluation of two biodegradable nerve guides for the reconstruction of the rat sciatic nerve. *Biomed. Mater. Eng.* **17,** 39–52.

Lundborg, G., Dahlin, L., Dohi, D., Kanje, M., and Terada, N. (1997). A new type of "bioartificial" nerve graft for bridging extended defects in nerves. *J. Hand Surg.* **22,** 299–303.

Lundborg, G., Rosén, B., Dahlin, L., Holmberg, J., and Rosén, I. (2004). Tubular repair of the median or ulnar nerve in the human forearm: A 5-year follow-up. *J. Hand Surg.* **29,** 100–107.

Luo, Y., and Shoichet, M. S. (2004). A photolabile hydrogel for guided three-dimensional cell growth and migration. *Nat. Mater.* **3,** 249–253.

Mahoney, M. J., Chen, R. R., Tan, J., and Saltzman, W. M. (2005). The influence of microchannels on neurite growth and architecture. *Biomaterials* **26,** 771–778.

Martin, G. R., Timpl, R., and Kühn, K. (1988). Basement membrane proteins: Molecular structure and function. *Adv. Protein Chem.* **39,** 1–50.

Matsumoto, K., Ohnishi, K., Sekine, T., Ueda, H., Yamamoto, Y., Kiyotani, T., Nakamura, T., Endo, K., and Shimizu, Y. (2000). Use of a newly developed artificial nerve conduit to assist peripheral nerve regeneration across a long gap in dogs. *ASAIO J.* **46,** 415–420.

Mazzitelli, S., Capretto, L., Tosi, A., Bilancetti, L., Luca, G., and Nastruzzi, C. (2008). Living matrixes for peripheral nerve regeneration: Effect of process parameters on conduit characteristics. *J. Control Release* **132,** 58–60.

Meek, M. F., and Coert, J. H. (2002). Clinical use of nerve conduits in peripheral-nerve repair: Review of the literature. *J. Reconstr. Microsurg.* **18,** 97–109.

Meinel, L., Hofmann, S., Karageorgiou, V., Kirker-Head, C., McCool, J., Gronowicz, G., Miller, C., Jeftinija, S., and Mallapragada, S. (2002). Synergistic eVects of physical and chemical guidance cues on neurite alignment and outgrowth on biodegradable polymer substrates. *Tissue Eng.* **8,** 367–378.

Meinel, L., Hofmann, S., Karageorgiou, V., Kirker-Head, C., McCool, J., Gronowicz, G., Zichner, L., Langer, R., Vunjak-Novakovic, G., and Kaplan, D. L. (2005). The inflammatory responses to silk films *in vitro* and *in vivo*. *Biomaterials* **26,** 147–155.

Miller, C. A., Jeftinija, S., and Mallapragada, S. K. (2002). Synergistic effects of physical and chemical guidance cues on neurite alignment and outgrowth on biodegradable polymer substrates. *Tissue Eng.* **8,** 367–378.

Millesi, H. (2000). Techniques for nerve grafting. *Hand Clin.* **16,** 73–91.

Morris, J. H., Hudson, A. R., and Weddell, G. (1972). A study of degeneration and regeneration in the divided rat sciatic nerve based on electron microscopy. I. The traumatic degeneration of myelin in the proximal stump of the divided nerve. *Z. Zellforsch. Mikrosk. Anat.* **124,** 76–102.

Muzzarelli, R. A. A. (2009). Chitins and chitosans for the repair of wounded skin, nerve, cartilage and bone. *Carbohydr. Polym.* **76,** 167–182.

Oh, S. H., Park, I. K., Kim, J. M., and Lee, J. H. (2007). *In vitro* and *in vivo* characteristics of PCL scaffolds with pore size gradient fabricated by a centrifugation method. *Biomaterials* **28,** 1664–1671.

Patel, M., Mao, L., Wu, B., and Vandevord, P. J. (2007). GDNF-chitosan blended nerve guides: A functional study, 1. *J. Tissue Eng. Regen. Med.* **1,** 360–367.

Patel, M., VandeVord, P. J., Matthew, H. W., DeSilva, S., Wu, B., and Wooley, P. H. (2008). Functional gait evaluation of collagen chitosan nerve guides for sciatic nerve repair. *Tissue Eng. Part C* **14,** 365–370.

Pego, P., Poot, A. A., Grijpma, D. W., and Feijen, J. (2001). Copolymers of trimethylene carbonate and epsilon-caprolactone for porous nerve guides: Synthesis and properties. *J. Biomater. Sci. Polym. Ed.* **12,** 35–53.

Pego, A. P., Poot, A. A., Grijpma, D. W., and Feijen, J. (2003). Biodegradable elastomeric scaffolds for soft tissue engineering. *J. Control Release* **87,** 69–79.

Pfister, L. A., Papaloïzos, M., Merkle, H. P., and Gander, B. (2007a). Nerve conduits and growth factor delivery in peripheral nerve repair. *J. Peripher. Nerv. Syst.* **12,** 65–82.

Pfister, L. A., Papaloızos, M., Merkle, H. P., and Gander, B. (2007b). Hydrogel nerve conduits produced from alginate/chitosan complexes. *J. Biomed. Mater. Res.* **80A,** 932–937.

Pieper, J. S., Oosterhof, A., Dijkstra, P. J., Veerkamp, J. H., and van Kuppevelt, T. H. (1999). Preparation and characterization of porous crosslinked collagenous matrices containing bioavailable chondroitin sulphate. *Biomaterials* **20,** 847–858.

Pierschbacher, M. D., and Ruoslahti, E. (1984). Cell attachment activity of fibronectin can be duplicated by small synthetic fragments of the molecule. *Nature* **309,** 30–33.

Rechichi, A., Ciardelli, G., D'Acunto, M., Vozzi, G., and Giusti, P. (2008). Degradable block polyurethanes from nontoxic building blocks as scaffold materials to support cell growth and proliferation. *J. Biomed. Mater. Res.* **84A,** 847–855.

Remminghorst, U., and Rehm, B. H. A. (2006). Bacterial alginates: From biosynthesis to applications. *Biotechnol. Lett.* **28,** 1701–1712.

Ren, Y. J., Zhang, H., Huang, H., Wang, X. M., Zhou, Z. Y., Cui, F. Z., and An, Y. H. (2008). *In vitro* behavior of neural stem cells in response to different chemical functional groups. *Biomaterials* **30,** 1036–1044.

Rich, K. M., Alexander, T. D., Pryor, J. C., and Hollowell, J. P. (1989). Nerve growth factor enhances regeneration through silicone chambers. *Exp. Neurol.* **105,** 162–170.

Rinaudo, M. (2006). Chitin and chitosan: Properties and applications. *Prog. Polym. Sci.* **31,** 603–632.

Rodriguez, F. J., Gomez, N., Perego, G., and Navarro, X. (1999). Highly permeable polylactide-caprolactone nerve guides enhance peripheral nerve regeneration through long gaps. *Biomaterials* **16,** 1489–1500.

Sartori, S., Rechichi, A., Ciardelli, G., Caporale, A., Vozzi, G., and Mazzucco, L. (2007). New strategies in polymeric biomaterials functionalisation. *Eur. Cells Mater.* **14**(18), 8.

Sartori, S., Rechichi, A., Vozzi, G., D'Acunto, M., Heine, E., Giusti, P., and Ciardelli, G. (2008). Surface modification of a synthetic polyurethane by plasma glow discharge: Preparation and characterization of bioactive monolayers. *React. Funct. Polym.* **68,** 809–821.

Shin, H. (2007). Fabrication methods of an engineered microenvironment for analysis of cell-biomaterial interactions. *Biomaterials* **28,** 126–133.

Siegle, R. J., McCoy, I. P. Jr., Schade, W., and Swansan, N. A. (1984). Intradermal implantation of bovine collagen. Humoral immune responses associated with clinical reactions. *Arch. Dermatol.* **120,** 183–187.

Sill, T. J., and von Recum, H. A. (2008). Electrospinning: Applications in drug delivery and tissue engineering. *Biomaterials* **29,** 1989–2006.

Smith, B. E., Bradshaw, A. D., Choi, E. S., Rouselle, P., Wayner, E. A., and Clegg, D. O. (1996). Human SY5Y neuroblastoma cell interactions with laminin isoforms: Neurite outgrowth on laminin-5 is mediated by integrin alpha 3 beta 1. *Cell Adhes. Commun.* **3,** 451–462.

Sorenson, E. J., and Windebank, A. J. (1993). Relative importance of basement membrane and soluble growth factors in delayed and immediate regeneration of rat sciatic nerve. *J. Neuropathol. Exp. Neurol.* **52,** 216–222.

Strauch, B. (2000). Use of nerve conduits in peripheral nerve repair. *Hand Clin.* **16,** 123–130.

Taras, J. S., Nanavati, V., and Steelman, P. (2005). Nerve conduits. *J. Hand Ther.* **18**(2), 191–197.

Teixeira, A. I., Duckworth, J. K., and Hermanson, O. (2007). Getting the right stuff: Controlling neural stem cell state and fate *in vivo* and *in vitro* with biomaterials. *Cell Res.* **17,** 56–61.

Tessier-Lavigne, M., and Placzek, M. (1991). Target attraction: Are developing axons guided by chemotropism? *Trends Neurosci.* **14,** 303–310.

Tong, Y. W., and Shoichet, M. S. (2001). Enhancing the neuronal interaction on fluoropolymer surfaces with mixed peptides or spacer group linkers. *Biomaterials* **22,** 1029–1034.

Tos, P., Battiston, B., Geuna, S., Giacobini-Robecchi, M. G., Hill, M. A., Lanzetta, M., and Owen, E. R. (2000). Tissue specificity in rat peripheral nerve regeneration through combined skeletal muscle and vein conduit grafts. *Microsurgery* **20,** 65–71.

Tyner, T. R., Parks, N., Faria, S., Simons, M., Stapp, B., Curtis, B., Sian, K., and Yamaguchi, K. T. (2007). Effects of collagen nerve guide on neuroma formation and neuropathic pain in a rat model. *Am. J. Surg.* **193,** 1–6.

Uebersax, L., Fedele, D. E., Schumacher, C., Kaplan, D. L., Merkle, H. P., Boison, D., and Meinel, L. (2006). The support of adenosine release from adenosine kinase deficient ES cells by silk substrates. *Biomaterials* **27,** 4599–4607.

Uebersax, L., Mattotti, M., Papaloïzos, M., Merkle, H. P., Gander, B., and Meinel, L. (2007). Silk fibroin matrices for the controlled release of nerve growth factor (NGF). *Biomaterials* **28,** 4449–4460.

Verdú, E., Labrador, R. O., Rodríguez, F. J., Ceballos, D., Forés, J., and Navarro, X. (2002). Alignment of collagen and laminin-containing gels improve nerve regeneration within silicone tubes. *Restor. Neurol. Neurosci.* **20,** 169–179.

Vozzi, G., Rechichi, A., Dini, F., Salvatori, C., Vozzi, F., Burchielli, S., Carlucci, F., Aruspici, M., Ahluwalia, A., Ciardelli, G., and Giusti, P. (2007). PAM-microfabricated polyurethane scaffolds: *In vivo* and *in vitro* preliminary studies. *Macromol. Biosci.* **8,** 60–68.

Waddell, R. L., Marra, K. G., Collins, K. L., Leung, J. T., and Doctor, J. S. (2003). Using PC12 cells to evaluate poly(caprolactone) and collagenous microcarriers for applications in nerve guide fabrication. *Biotechnol. Prog.* **19,** 1767–1774.

Wang, S., Wan, A. C. A., Xu, X., Gao, S., Mao, H. Q., Leong, K. W., and Yu, H. (2001). A new nerve guide conduit material composed of a biodegradable poly(phosphoester). *Biomaterials* **22,** 1157–1169.

Wang, W., Itoh, S., Matsuda, A., Aizawa, T., Demura, M., Ichinose, S., Shinomiya, K., and Tanaka, J. (2008). Enhanced nerve regeneration through a bilayered chitosan tube: The effect of introduction of glycine spacer into the CYIGSK sequence. *J. Biomed. Mat. Res.* **85A,** 919–928.

Wang, S., Yaszemski, M. J., Knight, A. M., Gruetzmacher, J. A., Windebank, A. J., and Lu, L. (2009). Photo-crosslinked poly(e-caprolactone fumarate) networks for guided peripheral nerve regeneration: Material properties and preliminary biological evaluations. *Acta Biomater.* Doi 10.1016/j. actbio.2008.12.015.

Wang-Bennett, L. T., and Coker, N. J. (1990). Analysis of axonal regeneration through the silicone regeneration chamber: A retrograde tracing study in the rabbit facial nerve. *Exp. Neurol.* **107,** 209–220.

Weber, R. A., Breidenbach, W. C., Brown, R. E., Jabaley, M. E., and Mass, D. P. (2000). A randomized prospective study of polyglycolic acid conduits for digital nerve reconstruction in humans. *Plast. Reconstr. Surg.* **106,** 1036–1045.

Weinzweig, N., Grindel, S., Gonzalez, M., Kuy, D., Fang, J., Shahani, B., and Van Beek, A. L. (1996). Peripheral-nerve allotransplantation in rats immunosuppressed with transient or long-term FK-506. *J. Reconstr. Microsurg.* **12,** 451–459.

Whitworth, I. H., Brown, R. A., Dore, C. J., Anand, P., Green, C. J., and Terenghi, G. (1996). Nerve growth factor enhances nerve regeneration through fibronectin grafts. *J. Hand Surg. (Br.)* **21,** 514–522.

Wilberg, M., and Terenghi, G. (2003). Will it be possible to produce peripheral nerve? *Surg. Technol. Int.* **11,** 303–310.

Winn, S. R., Uludag, H., and Hollingner, J. O. (1998). Sustained release emphasizing recombinant human bone morphogenetic protein-2. *Adv. Drug Del. Rev.* **31,** 303–318.

Wosnick, J. H., and Shoichet, M. S. (2008). Three-dimensional chemical patterning of transparent hydrogels. *Chem. Mater.* **20,** 55–60.

Yamada, A., Niikura, F., and Ikuta, K. (2008). A three-dimensional microfabrication system for biodegradable polymers with high resolution and biocompatibility. *J. Micromech. Microeng.* **18,** 1–9.

Yannas, I. V., and Hill, B. J. (2004). Selection of biomaterials for peripheral nerve regeneration using data from the nerve chamber model. *Biomaterials* **25,** 1593–1600.

Yao, L., Wang, S., Cui, W., Sherlock, R., O'Connell, C., Damodaran, G., Gorman, A., Windebank, A., and Pandit, A. (2009). Effect of functionalized micropatterned PLGA on guided neurite growth. *Acta Biomater.* **5,** 580–588.

Yoshii, S., Oka, M., Shima, M., Taniguchi, A., and Akagi, M. (2002). 30 mm regeneration of rat sciatic nerve along collagen filaments. *Brain Res.* **949,** 202–208.

Yu, L. M., Wosnick, J. H., and Shoichet, M. S. (2008). Miniaturized system of neurotrophin patterning for guided regeneration. *J. Neurosci. Methods* **171,** 253–263.

Zhang, J., Oswald, T. M., Lineaweaver, W. C., Chen, Z. R., Zhang, G. J., Chen, Z. W., and Zhang, F. (2003). Enhancement of rat sciatic nerve regeneration by fibronectin and laminin through a silicone chamber. *J. Reconstr. Microsurg.* **19,** 467–472.

Zhang, Z., Yoo, R. J., Wells, M., Biran, R., Tresco, P., and Beebe, T. P., Jr. (2005). Neurite outgrowth on well characterized surfaces: Preparation and characterization of chemically and spatially controlled fibronectin and RGD substrates with good bioactivity, neuron pathfinding on peptide- and fibronectin-covered surfaces. *Biomaterials* **26,** 47–61.

CONDUIT LUMINAL ADDITIVES FOR PERIPHERAL NERVE REPAIR

Hede Yan,* Feng Zhang,* Michael B. Chen,* and William C. Lineaweaver†

*Division of Plastic Surgery, University of Mississippi Medical Center, Jackson, Mississippi, USA

†Rankin Plastic Surgery Center, Brandon, Mississippi, USA

The use of nerve conduits as an alternative for nerve grafting has a long experimental and clinical history. Luminal additives, factors introduced into these nerve conduits, were later developed to enhance the nerve regeneration through conduits. This chapter generalizes the types of additives used, and the reported performance of luminal additives in conduits to present a preference list for the most effective additives to use over specific distances of nerve defect.

INTERNATIONAL REVIEW OF
NEUROBIOLOGY, VOL. 87
DOI: 10.1016/S0074-7742(09)87010-4

0074-7742/09 $35.00

I. Introduction

Injuries to peripheral nerves may occur due to trauma or surgical procedures, which can result in the loss of muscle function, impaired sensation, and/or painful neuropathies. Sacrifice of peripheral nerves can also complicate tumor resection (Young *et al.*, 1991). Peripheral nerve injury is an important cause of patient morbidity today, affecting 2.8% of trauma patients, many of whom experience long-term disability and decreased function (Noble *et al.*, 1998). The standard technique for repair is to transplant autologous nerve grafts from uninjured sites to the injured site (Birch and Raji, 1991). Primary repair is the most desirable approach for peripheral nerve injuries, as it gives the best long-term preservation of function (Birch and Raji, 1991). However, primary repair is limited to injuries in which there is little or no defect and in which the nerve ends may be approximated without tension, while for those defects with significant gaps precluding primary repair. Nerve autografting is limited by graft availability, secondary deformities, and potential differences in tissue structure and size (Danielsen *et al.*, 1988; Ducker and Hayes, 1970).

Nerve conduits arose out of the limitations of autografting. Although many of synthetic conduits (Archibald *et al.*, 1995; Barcelos *et al.*, 2003; Hall, 1997; Lundborg *et al.*, 1994) have been inferior to autografts in supporting nerve regeneration (Evans *et al.*, 1999; Mersa *et al.*, 2004), there are a successful few which return regeneration results comparable to nerve autografts over shorter distances in various animal models (Berger *et al.*, 1994; Chiu and Strauch, 1990; Dellon and Mackinnon, 1988; Mackinnon and Dellon, 1990; Matsumoto *et al.*, 2000; Strauch *et al.*, 1996, 2001; Tang, 1995; Toba *et al.*, 2002).

In attempt to not only enhance conduit efficacy at shorter distances, but also to extend the operational distance of conduits in nerve reconstruction for wider clinical application, the concept of introducing luminal additives into conduits has thus been explored. Based on the facts that within hours of nerve transection, the damaged nerve ends secrete a fluid containing various cells, structural components, and neurotrophic factors which are important for nerve regeneration (Danielsen and Varon, 1995; Seckel, 1990), researchers have experimented with the introduction of similar substances as luminal additives into nerve conduits. The purpose of this chapter is to evaluate the effectiveness of these factors and thus make recommendations to meet clinical needs regarding specific distances of nerve defects.

II. Cellular Components

The observations of injury response manifest that the natural process of nerve regeneration comprises a largely acellular fibrin matrix acting as a scaffold for the cells (Schwann cells, fibroblasts, and macrophages) from the damaged nerve ends to aggregate on within the first week of peripheral nerve injury (Williams *et al.*, 1983). The cellular components are essential, as structural and neurotrophic components exist primarily to support and guide their journey.

A. Schwann Cells

Schwann cells (SCs), as the main support cells of the peripheral nervous system, are responsible for supporting axons through roles such as debris removal and forming the myelin sheath to provide axonal insulation and to improve conduction speed (Matsuoka *et al.*, 1997), and are also critical for successful axonal regeneration over nerve defects. They perform these functions not only through producing and secreting neurotrophic factors (Ide, 1996; Strauch *et al.*, 2001; Terenghi, 1995), but also through organizing themselves into a series of cylinders called bands of Bungner (Abernethy *et al.*, 1994; Lietz *et al.*, 2006) which help guide regenerating axons (Abernethy *et al.*, 1994; Fornaro *et al.*, 2001).

In 1992, Guenard *et al.* introduced an elaborate study on synthetic conduits seeded with different concentrations of cultured syngeneic and heterologous SCs to spin an 8 mm rat sciatic nerve gap (Guenard *et al.*, 1992). They observed that there was a positive correlation between the concentration of syngeneic SCs used and the number of myelinated axons: $40 \times 10^6\ ml^{-1}$ SCs returned 1.2×10^4 myelinated axons at graft midpoint, $80 \times 10^6\ ml^{-1}$ SCs returned 2.5×10^4 axons, and $120 \times 10^6\ ml^{-1}$ SCs returned 3.2×10^4 axons, which was the highest count among all the groups excepting the autograft group. In contrast, heterologous SCs elicited a strong immune reaction which impeded nerve regeneration. Matrigel alone or empty conduits also produced poor results.

Further studies utilizing SCs (syngeneic (Ansselin *et al.*, 1997), allogeneic (Evans *et al.*, 2002; Mosahebi *et al.*, 2002), autologous (Sinis *et al.*, 2005; Strauch *et al.*, 2001), or transduced (Mosahebi *et al.*, 2001, 2003), in bioartifical conduits bridging rat sciatic nerve gaps confirmed this significant increase in nerve regeneration (Koshimune *et al.*, 2003). The most successful were the allogeneic and autologous SCs. Allogeneic SCs combined with a collagen matrix and injected into poly L-lactic acid (PLA) conduits produced similar SFI (sciatic functional index) and gastrocnemius weight to isograft at 4 months, but lower nerve fiber density (Evans *et al.*, 2002). Autologous SCs produced slightly better recovery and higher numbers of regenerating axons than either their syngeneic or isogeneic counterparts, but still fall short of autograft results (Rodriguez *et al.*, 2000). Rutkowski further extended the understanding of the effect of SC concentration. While higher SC seeding densities enhanced nerve regeneration, he added the observation that the increased branching of the axons may require additional resources to maintain directed regeneration and can lead to painful neuroma formation if not in an optimal environment (Rutkowski and Heath, 2002).

In an effort to boost SCs' effectiveness, simulation of the macro- and microarchitecture of native peripheral nerves was attempted in other studies. Copolymer foam conduits with multiple continuous channels resulted in higher axonal cross sectional areas during nerve regeneration (Hadlock *et al.*, 2000). Micropatterned biodegradable conduits with seeded SCs improved recovery time and SFI (Rutkowski *et al.*, 2004). Self-organizing gel-contracting collagen conduits (Phillips *et al.*, 2005), and synthetic conduits resembling polyfascicular peripheral nerves (Hadlock *et al.*, 1998), also returned improvements in neural regeneration. Biological conduit of small intestinal submucosa (SIS) showed a good biocompatibility with SCs (Su *et al.*, 2007). It appears an ideal artificial nerve could be composed of a biodegradable nerve conduit and bioactive SCs distributed in the Bungner's band organization (Zhu *et al.*, 2000), and indeed experiments using microcell chips and conduits with microstructured internal polymer filaments to achieve this organization resulted in more rapid, highly oriented axonal growth without meandering (Fansa *et al.*, 1999).

SCs proved that they could also be used in biological conduits with success. In 1999, Fansa introduced SCs into an acellular gracilis muscle bridging a rat sciatic nerve defect, increasing the number of axons compared to control (Fansa *et al.*, 1999a,b). In 2001, Zhang used a rabbit tibial 4 cm gap nerve repair model with autogenous vein grafting filled with SCs (Zhang *et al.*, 2002). The addition of SCs resulted in new nerve fascicles and evoked muscle action potentials as in the nerve autograft, whereas the vein graft alone showed neither. Other materials such as muscle-vein tissue for conduits were also successfully used in conjunction with SCs (Raimondo *et al.*, 2005). The basal lamina of the grafted muscle fibers act as an early migration pathway for them to colonize the conduit and thus aid nerve regeneration. Further advancements such as low-intensity ultrasound stimulation (Chang and Hsu, 2004) and gingko biloba application to SCs in conduits also return promising results (Hsu *et al.*, 2004).

B. Bone Stromal Cells

SCs are relatively difficult to obtain to use as luminal additives, despite recent advances (Komiyama *et al.*, 2003), as there is only a small resource of them in the body and the process of extracting them is relatively difficult. An interesting alternative are bone stromal cells, which have the ability to differentiate into cells phenotypically similar to SCs (Keilhoff *et al.*, 2005; Zhao *et al.*, 2005), but are easier to obtain via aspiration and are found in larger amounts.

In 2005, Zhang introduced both SCs and bone stromal cells into biodegradable conduits using the rat sciatic nerve model (Zhang *et al.*, 2005). At 6 weeks both groups with cultured cells showed a significantly higher number of axons, more well-shaped remyelinated axons, and an advance in clinical functional recovery than the empty conduit-bridging group. A further study confirmed that bone stromal cells have the potential to differentiate into SC-like cells, and thus improve nerve regeneration (Keilhoff *et al.*, 2005). Single bone stromal cells could even myelinate more than one axon. Although research into this area is nascent, results are promising and future studies will help elucidate the role of bone stromal cells.

C. Fibroblasts

Fibroblasts are connective tissue cells important in the formation of various tissues in the human body. Within the first week of nerve transection, fibroblasts and macrophages are the primary cellular component of the fluid secreted from the nerve ends (Li *et al.*, 2004). While fibroblasts have been introduced into conduits bridging 5 mm rat sciatic nerve defects with an increase in neural regeneration (Phillips *et al.*, 2005), they may also form scar tissue, hampering the regeneration process (Lietz *et al.*, 2006). Further research to clarify their role in nerve regeneration is warranted.

D. Other Cellular Components

In 1995, Tang experimented with vein conduits filled with interposed nerve segments and produced good results (Tang, 1995), and in 2008 Zhang noted the similar results with better outcomes regarding the nerve conduction velocity and the remyelinated axons counting of the deacetyl chitin conduit filled in nerve fibers spinning 10 mm sciatic nerve defect in rats than that of simple conduit-bridging group (Zhang *et al.*, 2008), albeit uncompared with nerve autografts in these two reports. A further study performed by Lloyd demonstrated that no significant differences in nerve recovery were observed between the motor, sensory, and mixed nerve fragments used as the conduit additives to repair nerve defect (Lloyd *et al.*, 2007). While in 1997, Francel showed that introducing short 2-mm interposed nerve segments into a silicon conduit bridging a 13–15 mm sciatic nerve gap in rats was just as effective as nerve autografts on neural regeneration, assessed histologically, electrophysiologically, and functionally (Francel *et al.*, 1997). However, the practicality of acquiring multiple nerve segments for such a conduit is questionable given that other cellular luminal additives have been proven as effective (Bunting *et al.*, 2005; Evans *et al.*, 2002; Ide *et al.*, 1998; Rosen *et al.*, 1989, 1992), and without donor site issues.

In 2007, Nie utilized ectomesenchymal stem cells (EMSCs) as the differentiated cells implanted into tissue-engineered nerve to repair 10 mm nerve defect in rats and achieved similar results compared with autografts (Nie *et al.*, 2007), showing that when EMSCs are transplanted to a peripheral nerve defect, they differentiate into supportive cells that contribute to the promotion of axonal regeneration.

Muscle-vein grafts involve introducing skeletal muscle in veins to provide a scaffold for the nerve fibers to correctly grow and orientate, allowing more myelination and a thicker perineurium (Di Benedetto *et al.*, 1998; Tos *et al.*, 2000). Similarly, it appears that enrichment of synthetic tubes with fresh skeletal muscle promotes axonal regeneration and SC migration in early nerve repair stages (Nicolino *et al.*, 2003; Varejao *et al.*, 2003). Although the availability of skeletal muscle compared to nerve makes this option more attractive, interest in muscle as a luminal additive has dwindled, probably as there are more convenient options available.

Although no cellular components by themselves have proven to produce nerve regeneration on par with autografting, they remain a critical part of the mechanism by which some structural and neurotrophic factors do produce autograft-level results.

III. Structural Components

After a nerve is transected, fluid from the damaged nerve ends accumulates from which a matrix, consisting mainly of fibrin polymers, coalesces (Williams and Varon, 1985; Williams *et al.*, 1983). This scaffold can be used by the cells from the damaged nerve ends on which to aggregate and start rebuilding neural tissue. While the natural scaffold is composed mainly of fibrin matrix, other materials have also been trialed with success.

A. Fibrin

Fibrin is a fibrillar protein that when formed into matrices acts as structural support for growth and regeneration of tissues in the body, including neural tissue (Williams and Varon, 1985).

In 1987, Williams and associates, in an attempt to mimic this natural process of nerve regeneration following injury, prefilled conduits spanning a 10 mm rat sciatic nerve gap with a solution of dialyzed plasma, which gradually formed a fibrin matrix (Williams, 1987). The rate of regeneration, SC migration, and axonal elongation was increased. Williams soon applied this method to longer nerve defects, achieving an increase in functional nerve recovery over 15 and even 20 mm gaps (Williams *et al.*, 1987). Also, following earlier studies using collagen (Ceballos *et al.*, 1999; Dubey *et al.*, 1999; Verdu *et al.*, 2002), it was thought that if the structure of the fibrin matrix could be magnetically aligned longitudinally in the conduit, it would act as a guide for regenerating neurons and thus improve nerve regeneration, proven in 2001 by Dubey (Dubey *et al.*, 2001). Recently in 2007, Nakayama and coworkers (Nakayama *et al.*, 2007), in an effort to determine the effects of the fibrin gel filling and nerve conduit shape (straight or bellows-shaped tubes), a 10 mm sciatic nerve gap in rats was created and prefilled conduit with fibrin gel was utilized to bridge the gap. They found that the fibrin gel filling induced bridge formation between the transected nerve endings as early as 4 and 8 weeks after implantation, suggesting that the fibrin filling and the bellows shape enhance the nerve regenerative performance of nerve guide tubes constructed from bioabsorbable materials.

Kalbermatten *et al.* (2008) developed a matrix based on fibrin to fill optimally the nerve conduits with cells for spinning large nerve gap in 2008. *In vitro* analysis showed that both SCs and differentiated mesenchymal stem cells (MSC) adhered significantly better to poly-3-hydroxybutyrate (PHB) in the presence of fibrin and cells continued to maintain their differentiated state. Cells were more optimally distributed throughout the conduit when seeded in fibrin than by delivery in growth medium alone. Transplantation of the nerve conduits *in vivo* to bridge a 10 mm gap in the sciatic nerve of adult rats showed that cells in combination with fibrin matrix significantly increased nerve regeneration distance when compared with empty PHB conduits. They concluded that the beneficial combinatory effect of an optimized fibrin matrix, cells, and conduit material as a step towards bridging nerve gaps which should ultimately lead to improved functional recovery following nerve injury.

Seckel discovered that hyaluronic acid in conduits produced better conduction velocity, higher axon counts, and a trend towards earlier myelination compared with saline (Seckel *et al.*, 1995). His team postulated that it may achieve these effects through improving fibrin matrix formation and decreasing the scarring which hinders the speed of nerve regeneration (Cajal, 1928; Hudson *et al.*, 1972; Williams *et al.*, 1983). A study in a rabbit model by Mohammad, using hyaluronic acid in conjunction with nerve growth factor, reported a 45% increase in myelinated axon count compared to control (Mohammad *et al.*, 2000). Fibronectin mats and cables have also been used successfully, as they act as contact guides for SCs and fibroblasts and thus the regenerating nerve (Ahmed *et al.*, 2003).

B. Laminin

Laminin, found in basal lamina, is an important adhesion molecule for growth and regeneration of tissues, including neural tissue, in the body. In 1985, Madison found that introducing a gel containing 80% laminin into a nerve conduit spanning murine transected sciatic nerve produced, at 2 weeks, regenerating neuronal cells in the conduit compared to none in control (Madison *et al.*, 1985).

In 1987, in a quantitative extension of this experiment, Madison confirmed that laminin gel significantly increased the initial rate at which axons from primary sensory and motor neurons cross a transection site (Madison *et al.*, 1987). However, the presence of laminin actually inhibited the same regenerative process at 6 weeks. A subsequent study by his team a year later proved that laminin gel could also be used to extend nerve regeneration across defects thought previously too wide for this process (20–25 mm) (Madison *et al.*, 1988).

Laminin's usefulness in other animal models was also demonstrated (Matsumoto *et al.*, 2000). A laminin-soaked collagen sponge introduced into synthetic conduits across an 80 mm gap in canine peroneal nerve produced near-normal walking patterns after 12 months (Toba *et al.*, 2002).

C. Collagen

Collagen is the fibrous protein constituent of nearly all the tissues in the human body. The use of collagen as a structural matrix for regenerating axons was investigated in the 1988 study performed by Madison which had produced promising results with laminin on nerve regeneration (Madison *et al.*, 1988). All of the tubes with collagen and laminin, but only one-sixth of the initially empty tubes, displayed a regenerated nerve when assessed histologically. Further studies cemented collagen's role as a factor for improving nerve regeneration (Rosen *et al.*, 1989, 1990, 1992; Satou *et al.*, 1986).

It appears that the mere presence of these matrices is insufficient to guarantee a good result. The concentration of collagen or laminin is a crucial factor in determining their effectiveness on nerve regeneration. Labrador studied murine sciatic nerve defects of 4 and 6 mm bridged by silicon tubes filled with saline solution, and different concentrations of collagen gels or laminin gels (Labrador *et al.*, 1998). Reinnervation started earlier and achieved slightly higher levels when more dilute collagen (1.28 mg/ml) or laminin (4 mg/ml) gels were used, compared with higher concentrations of these agents, or saline solution alone. Labrador postulated that the higher concentrations interfere structurally with the regeneration process. Collagen was more effective over the shorter 4 mm gap while laminin more effective over the longer 6 mm gap. The significance of these observations is yet to be determined through more rigorous investigation.

In 1990, Rosen revealed the true potential of collagen when he bridged a 5 mm rat peroneal nerve defect with a polyglycolic acid (PGA) conduit filled with collagen extracellular matrix (Rosen *et al.*, 1990). The axonal regeneration of the conduit was equal to sutured autografts at 11–12 months, as measured by axonal counts, physiological, and functional methods, although the sutured autografts demonstrated superior axonal diameters (Rosen *et al.*, 1990). He supported these findings in 1992, when he repeated the experiment with similar results using the same model with collagen in glycolide trimethylene carbonate (GTMC) conduits (Rosen *et al.*, 1992). More recent research has focused on magnetically aligning the collagen fibrils to enhance regeneration. Dubey in 1999 showed that there was a positive correlation between depth and axial bias of neurite elongation and the intensity of the magnetic field used to align the collagen in conduits entubing rat dorsal root ganglia (Dubey *et al.*, 1999). Ceballos and Verdu quickly confirmed these findings in a peripheral nerve model, using murine sciatic nerves (Ceballos *et al.*, 1999; Verdu *et al.*, 2002).

Interestingly, Allmeling and colleagues (Allmeling *et al.*, 2008) introduced the use of spider silk fibers as a new material in nerve tissue engineering, in a 20-mm sciatic nerve defect in rats in 2008. Natural materials are favored with regard to their reduced cytotoxicity and enhanced biocompatibility, moreover, in contrast to silkworm silk, which evoked fibroblastic response when used as nerve suture material in a rabbit model (DeLee *et al.*, 1977), spider silk is devoid of sericin glue-like proteins, which have been shown to exert no influence on biocompatibility and immunological response (Altman *et al.*, 2003). They compared isogenic nerve grafts to vein grafts with spider silk fibers, either alone or supplemented with SCs, or SCs and matrigel. Their results demonstrated that bridging an extensive gap by cell-free constructs based on vein and spider silk was highly effective in nerve regeneration and suggested that spider silk is a viable guiding material for SC migration and proliferation as well as for axonal regrowth in a long-distance model for peripheral nerve regeneration.

Similarly, Yang *et al.* (2007) introduced another investigation utilizing the oriented SF filaments silk fibroin (SF) as luminal additive in a SF nerve guidance conduit (NGC). The SF graft was used for bridge implantation across a 10-mm long sciatic nerve defect in rats, and the examined functional and morphological parameters show that SF grafts could promote peripheral nerve regeneration with effects approaching those elicited by nerve autografts.

In a mouse tibial nerve model, Apel and colleagues (Apel *et al.*, 2008) revealed that keratin-based hydrogel additive significantly improved electrophysiological recovery, compared with empty conduits and sensory nerve autografts, at an early time point of regeneration, and also produce long-term electrical and histological results superior to empty conduits and equivalent to sensory nerve autografts. Sierpinski (Sierpinski *et al.*, 2008) discovered that the keratin gel additive improved functional recovery compared to sensory nerve autograft, indicating that keratin biomaterials appear to be neuroconductive and contain regulatory molecules capable of enhancing nerve tissue regeneration by inductive mechanisms as well.

D. Synthetic Longitudinal Matrices

Alternatives to biological meshworks are synthetic filaments, which offer the advantage of being able to be manufactured to the dimensions and characteristics desired and regulated more closely.

Empty silicon tubes are effective as regeneration aids as long as the rat nerve defect is not too wide, for example, shorter than 10 mm (Cheng *et al.*, 2001; Heijke *et al.*, 2001). Lundborg showed this operational distance could be extended to 15 mm in rat sciatic nerve if synthetic filaments were inserted into the tubes (Lundborg *et al.*, 1997). Eight polyamide filaments with a diameter of 250 μm, placed inside silicone tubes with an inner diameter of 1.8 mm, helped sensory fibers bridge the gap and extend into the distal nerve segment, supported by a positive pinch reflex and positive staining for neurofilaments in the distal segment. In contrast, there were minimal structures regenerated in the control silicon tube, and a negative pinch reflex. Arai, in 2000, showed that filling silicone tubes across a 15 mm rat sciatic nerve gap with various synthetic filaments (polyamide, catgut, polydioxanone, and polyglactin) all induced nerve tissue to regenerate and resulted in functional recovery not achieved in empty tubes (Arai *et al.*, 2000). Wu, in 2008, found that a silicon tube implanted with Valproic acid (VPA) to spin a 10 mm nerve defect achieved better results than that in empty tubes, suggesting the potential clinical application of VPA for the treatment of peripheral nerve injury in humans (Wu *et al.*, 2008).

As our knowledge of synthetic materials grows and the technology of manufacturing processes advance, we will no doubt see more novel materials, such as glass (Bunting *et al.*, 2005), being used as luminal additives to enhance nerve regeneration.

IV. Neurotrophic Components

Certain factors are important for supporting survival, differentiation, and regeneration of nervous tissue of humans and animals (Lundborg *et al.*, 1994), and are released from peripheral nerve ends after transection for this purpose (Politis *et al.*, 1982; Zheng and Kuffler, 2000). This knowledge has stimulated research into the effect of artificially introducing these factors into conduits to improve nerve regeneration.

A. Fibroblast Growth Factor (FGF)

FGFs are a family of at least 23 cytokines that are involved in cell growth and regeneration, and are naturally secreted by damaged nerve ends after injury (Richardson, 1991). The earliest published account of FGF use in nerve repair appeared in 1992, when it was used for rat cavernous nerve defects to improve erectile function recovery (Ball *et al.*, 1992). In 1993, Walter investigated the introduction of recombinant acidic FGF into a synthetic conduit bridging a 15 mm peripheral nerve gap in the rat (Walter *et al.*, 1993). The FGF-treated group produced functional motor recovery as assessed through the amplitude and latency of compound muscle action potentials, in contrast to no recovery in control. Subsequent studies supported this initial report of FGF efficacy (Midha *et al.*, 2003).

Basic FGF had an advantageous effect in other animal models as well. Ide's work in 1998 with bFGF in canine sciatic defects showed increased numbers of myelinated axons, though of a quality still inferior to autografts (Ide *et al.*, 1998). Li's 2000 study using 0.2 ml bFGF solution (4000 U/ml) intravenously injected into vein graft conduits produced an increased rate of axonal regeneration with more regularity than controls (Li and Cao, 2000). More recent studies have combined FGF with structural components. Midha showed that 10 μg/ml FGF-1 and collagen matrix added to synthetic tube bridging a 10 mm nerve defect demonstrated regeneration comparable to nerve autografts and significantly better than collagen matrix alone (Midha *et al.*, 2003).

B. Nerve Growth Factor (NGF)

NGFs are a family of neurotrophins which play an important role in the natural process of nerve growth and regeneration (Richardson, 1991). In 1989, Rich studied NGF's effects on rat sciatic nerve injury bridged by silicone conduits. When NGF 1 mg/ml saline was injected into silicon tubes the number of myelinated axons in the chamber approximately doubled, the myelin sheath increased 58% in thickness, and the nerve appeared more mature and organized when compared with normal saline (Rich *et al.*, 1989). These findings have been supported by subsequent studies (He *et al.*, 1992; Mohammad *et al.*, 2000; Pu *et al.*, 1999).

Further studies utilized more sophisticated NGF delivery systems to more closely mimic the natural process of secretion and return very promising results. Preliminary reports in 1999 by Hadlock concluded that a nerve regeneration conduit could be successfully created to deliver growth promoting substances (in this case inosine, a neurotrophic purine analog) over a protracted time course to improve nerve regeneration (Hadlock *et al.*, 1999). In 2003, Lee used heparin to immobilize NGF and slow its diffusion from a fibrin matrix inside a conduit bridging a 13-mm rat sciatic nerve defect to produce similar numbers of nerve fibers to isograft (Lee *et al.*, 2003). Other controlled release systems involving microspheres (Yang *et al.*, 2005) and adenoviruses expressing NGF have also reported increased recovery (Tang *et al.*, 2004). The introduction of SCs into conduits as NGF producers has also been suggested and described by simple models, but definitive results on nerve regeneration are yet to come (Rutkowski and Heath, 2002). Recently, Unezaki *et al.* also noted that NGF injected directly into the silicone tube bridging a 5 mm nerve gap accelerated the regeneration of the sciatic nerve (Unezaki *et al.*, 2009).

C. Glial Growth Factor (GGF)

GGF is produced by neurons and stimulates proliferation of SCs, underlining the close interaction between neuronal and glial cells during peripheral nerve regeneration (Terenghi, 1999). GGF introduced into conduits spanning 2–4 cm rabbit common peroneal nerve gaps increased the quantity of SCs and the rate of axonal regeneration and significantly reduced the muscle mass percentage loss compared to controls (Mohanna *et al.*, 2005).

D. Ciliary Neurotrophic Factor (CNTF)

CNTF is an important trophic factor that is known to improve nerve regeneration, proven by directly introducing CNTF around the neurons (Newman *et al.*, 1996). CNTF was used as a luminal additive in a silicon conduit bridging 10-mm rat sciatic nerve gap (Zhang *et al.*, 2004). The results show an increase in the axon diameter and number of axons, improved myelination, resulting in higher motor nerve conduction velocity of the sciatic nerve and higher muscle action potential amplitude of the anterior tibial muscle, compared to the controls (Zhang *et al.*, 2004).

V. VEGF

Vascular endothelial growth factor (VEGF), although acting primarily on vascular tissue, also helps nerve regeneration as there is a close relationship between nerve fibers and blood vessels during this process (Terenghi, 1995). In 2000, Hobson introduced VEGF (500–700 ng/ml) in a laminin-based gel (Matrigel) into silicon conduit bridging 1-cm rat sciatic nerve defects (Hobson *et al.*, 2000). The addition of VEGF significantly increased blood vessel penetration within the chamber in a dose-dependent manner, and was correlated with an increase of axonal regeneration and SC migration.

VI. GDNF

Glial-derived neurotrophic factor (GDNF) promotes both sensory and motor neuron survival. The delivery of GDNF to the peripheral nervous system has been shown to enhance regeneration following injury. In a latest study, Matthew *et al.*

(Wood *et al.*, 2009) evaluated the effect of affinity-based delivery of GDNF from a fibrin matrix in a NGC on nerve regeneration in a 13-mm rat sciatic nerve defect. They concluded that the delivery of GDNF via the affinity-based delivery system can enhance peripheral nerve regeneration through a silicone conduit across a critical nerve gap and offers insight into potential future alternatives to the treatment of peripheral nerve injuries. However, Unezaki observed that GDNF showed no effects on the acceleration the axonal growth during sciatic nerve regeneration or myelination administrated by osmotic pump into the silicone tube spinning 5-mm nerve gap (Unezaki *et al.*, 2009).

A. Other Factors

Neurotropin-3 (NT-3) and brain-derived neurotrophic factor (BDNF) have both been shown to increase rat sciatic nerve regeneration in a synthetic hydrogel tube, but are inferior to FGF-1 in this respect (Midha *et al.*, 2003). BDNF at high doses, delivered through osmotic pump directly to the injured rat sciatic nerve even appears to inhibit axonal regeneration (Gordon *et al.*, 2003).

VII. Combined Additives

In later studies, the addition of bioactive substances (laminin, fibronectin, and NGF), coimmobilized in multifilament fibers in the tube produced the higher regenerative potentials than the bioactive substances alone (Rosen *et al.*, 1989).

In 2003, Yu and Bellamkonda (Yu, 2003) evaluated the effectiveness of combined additives of laminin 1 (LN-1) and NGF inserted into a semipermeable polysulfone tube to bridge a 10-mm sciatic nerve gap in adult rats. They demonstrated that tissue-engineered scaffolds with combined luminal additives match the performance of autografts *in vivo*. Also, in 2007, Dodla (Dodla and Bellamkonda, 2008) modified the former investigation with anisotropic scaffolds of agarose hydrogels containing gradients of LN-1 and NGF molecules to promote sciatic nerve regeneration across a challenging 20-mm nerve gap in rats. Their data demonstrates that nerve regeneration occurs only when gradients of both LN-1 and NGF are present, but not when these proteins are presented in an isotropically distributed manner, and concludes that anisotropic distribution of trophic and extracellular matrix (ECM) proteins in 3D hydrogels represents a significant step forward in the design of tissue-engineered nerve bridges.

VIII. Recommendations

There are a myriad of effective luminal additives available, and those reviewed can be found in Table I along with the conduits they were tested in and their reported effects on nerve regeneration. The additives recommended by the authors over specific distances in different animal models can be found below. Please see the relevant section in the main text for more detail.

The rat sciatic nerve model is the most popular model for studying nerve repair. At distances >2 cm (i.e., 2.5 cm) collagen or laminin fibrils are the only viable option (Madison *et al.*, 1987). At 2.0 cm SCs are an alternative. Both autologous SCs (Sinis *et al.*, 2005) or bone stromal cells (Keilhoff *et al.*, 2005), which differentiate into cells phenotypically similar to SCs result in regeneration and clinical improvement compared to none in controls, but bone stromal cells are preferred due to their relative availability and ease of procurement (Keilhoff *et al.*, 2005). Collagen or laminin fibrils would be more practical though, as, unlike cellular components, they are limited only by the manufacturing process, they have been proven to return results similar to autografts (albeit only over 0.5 cm defects) (Rosen *et al.*, 1990), and it is easier to enhance their potential through magnetic alignment (Ceballos *et al.*, 1999; Dubey *et al.*, 1999; Verdu *et al.*, 2002). Syngeneic SCs increase regeneration over 1.8 cm gaps, but have not been compared to autografts (Ansselin *et al.*, 1997). At 1.5 cm interposed nerve segments are the only luminal additive which return regeneration that is histologically, electrophysiologically, and functionally similar to nerve autografts (Francel *et al.*, 1997). At 1.3 cm, use of heparin to slow the diffusion of 20 and 50 ng/ml NGF from a fibrin matrix in a conduit also produced similar numbers of nerve fibers to isografts at 6 weeks (Lee *et al.*, 2003). At 1.0 cm, while collagen matrix impregnated with 10 μg/ml FGF-1 returned results histomorphometrically similar to that of autografts (Midha *et al.*, 2003), the 0.33 g/ml NGF results are more relevant, having been better characterized with similar electrophysiological (conduction velocity and latency) and clinical (improvement in walking pattern) results to autografting (Pu *et al.*, 1999). The interposed nerve segments would be a third choice due to the difficulty in obtaining neural tissue. Conduits filled with allogeneic SCs, although returning a similar SFI and gastrocnemius weight to isograft, suffer lower nerve fiber density (Evans *et al.*, 2002). At smaller distances of <1 cm, the inherent ease of introducing soluble neurotrophic factors into conduits would still place NGF and FGF-1 at the top of the list, followed by the collagen fibers which at 0.5 cm return autograft-like results (Rosen *et al.*, 1990), and then the interposed nerve segments.

The most impressive luminal additive used in larger animal models are laminin-coated collagen and laminin-soaked sponge in conduits over 8-cm canine

TABLE I
LUMINAL ADDITIVES IN NERVE CONDUITS FOR NERVE REPAIR

Category	Additives	Conduits	Animal model	Defect (cm)	Outcomes
Cellular	SC (Fansa *et al.*, 1999; Zhang *et al.*, 2002)	Acellular muscle	Rat sciatic	1.0	↑ number of axons, improved organization, and orientation of fibers
	SC (Raimondo *et al.*, 2005)	AVNC	Rabbit tibial	4.0	Growth of new nerve fascicles and presence of evoked MAPs cf to none in control, although still less than nerve autografts
	SC (autologous) (Strauch *et al.*, 2001)	AVNC	Rabbit peroneal	6.0	Neuronal regeneration over 6 cm, compared to no regeneration in control
	SC (autologous) (Sinis *et al.*, 2005)	TMC/CL	Rat median	2.0	↑ muscle potentials, ↑ FDS weight, and +ve grasping test cf −ve in control
	SC (autologous) (Rodriguez *et al.*, 2000)	L-Lactide, epsilon-caprolactone	Mouse sciatic	0.6	↑ recovery and number of regenerated fibers reaching the distal nerve than isologous and syngeneic, but slightly inferior to nerve autografts
	SC (syngeneic) (Guenard *et al.*, 1992)	Permselective synthetic	Rat sciatic	0.8	Larger, more organotypic fascicles with higher axonal population, although still lower than nerve autografts
	SC (syngeneic) (Ansselin *et al.*, 1997)	Reinforced collagen	Rat sciatic	1.8	↑ number of myelinated fibers and ↑ functional recovery with $>0.5 \times 10^6$ SCs/ml, cf to none in control
	SC (syngeneic) (Mosahebi *et al.*, 2002)	PHB	Rat sciatic	1.0	↑ axonal regeneration distance and ↑ quantity of axons cf to allogeneic SCs

(*continued*)

TABLE I (*continued*)

Category	Additives	Conduits	Animal model	Defect (cm)	Outcomes
	SC (allogeneic) (Evans *et al.*, 2002)	PLLA	Rat sciatic	1.2	Similar SFI and gastrocnemius weight to isograft, but lower nerve fiber density
	SC (lacZ-transduced) (Mosahebi *et al.*, 2001, 2003) autol	PHB	Rat sciatic	1.0	2× rate of axonal regeneration
	SC (heterologous) (Guenard *et al.*, 1992)	Permselective synthetic	Rat sciatic	0.8	↓ Nerve regeneration due to strong immune reaction
	Bone stromal cells (Keilhoff *et al.*, 2005)	Autologous muscle	Rat sciatic	2.0	↑ Rate of regeneration and myelination
	Bone stromal cells (Zhang *et al.*, 2005)	Chitin (biodegradable)	Rat sciatic		↑ Axon count, more well-shaped remyelinated axons, and ↑ SFI
	Fibroblasts (Phillips *et al.*, 2005)	Silicone	Rat sciatic	0.5	↑ Neural regeneration
	Interposed nerve segments (Francel *et al.*, 1997)	Silicone	Rat sciatic	1.3–1.5	Histological, electrophysiological, and functional assessment of regeneration similar to nerve autografts
	Skeletal muscle(Varejao *et al.*, 2003)	Poly(DLLA-epsilon-CL)	Rat sciatic	1.0	↑ Axon regeneration and SC migration
	Interposed nerve segments (Zhang *et al.*, 2008)	Deacetyl chitin	Rat sciatic	1.0	↑ The nerve conduction velocity and the remyelinated axons counting
	EMSCs (Nie *et al.*, 2007)	PLGA	Rat sciatic	1.0	↑Axonal regeneration ↑ SFI
Structural	Fibrin (Williams, 1987; Williams *et al.*, 1987)	Silicone	Rat sciatic	1.0	↑ Rate of regeneration, SC migration, and axonal elongation
	Fibrin (from HA) (Seckel *et al.*, 1995)	Silicone	Rat sciatic		Better conduction velocity, higher axon counts, earlier remyelination
	Fibrin gel (Nakayama *et al.*, 2007)	Bioabsorbable polymer	Rat sciatic	1.0	↑ myelinated axons in bellow tubes
	Fibrin, SC, dMSC (Kalbermatten *et al.*, 2008)	PHB	Rat sciatic	1.0	↑ nerve regeneration distance

Laminin (Madison *et al.*, 1985)	Bioresorbable	Mouse sciatic		↑ rate of axonal regeneration
Laminin (Madison *et al.*, 1987)	Biodegradable and polyethylene	Mouse peripheral motor and sensory		↑ rate of axonal regeneration
Laminin (Madison *et al.*, 1988)	Silicone	Rat sciatic	2.0–2.5	Presence of regenerated nerve cable cf to none in controls
Laminin-coated collagen (Matsumoto *et al.*, 2000) and laminin-soaked sponge (Toba *et al.*, 2002)	PGA-collagen	Canine peroneal	8.0	↑ myelinated axon number, ↑ evoked nerve potentials, ↑ functional recovery (near-normal walking)
Collagen (Madison *et al.*, 1988)	Silicone	Rat sciatic	2.0–2.5	Presence of regenerated nerve cable cf to none in controls
Collagen (Satou *et al.*, 1986)	Silicone	Rat sciatic	0.5	More rapid growth of sprouting axons
Collagen (Rosen *et al.*, 1990)	PGA	Rat peroneal	0.5	Axonal regeneration similar to sutured autografts
Collagen (Rosen *et al.*, 1992)	GTMC	Rat peroneal	0.5	Axonal regeneration similar to sutured autografts
Collagen (magnetically aligned) (Ceballos *et al.*, 1999)	Bioresorbable	Mouse sciatic	0.6	Presence of regenerating myelinated fibers cf to none in control
Collagen (magnetically aligned) (Verdu *et al.*, 2002)	Silicone	Mouse sciatic	0.6	↑ Reinnervation, and recovery of motor, sensory, and sudomotor functions
Spider silk fibers (Allmeling *et al.*, 2008)	Vein	Rat sciatic	2.0	Degeneration of the gastrocnemius muscle ↑ alignment of regenerated axons
Silk fibroin filaments (Yang *et al.*, 2007)	Silk fibroin-nerve guidance conduit	Rat sciatic	1.0	↑ Peripheral nerve regeneration (functionally and morphologically)
Polyamide filaments (Lundborg *et al.*, 1997)	Silicone	Rat sciatic	1.5	Presence of regenerating neurofilaments and +ve pinch test cf no regeneration and −ve pinch test in control

(*continued*)

TABLE I (*continued*)

Category	Additives	Conduits	Animal model	Defect (cm)	Outcomes
	keratin-based hydrogel (Apel *et al.*, 2008)	Silicone	Mouse tibial	0.4	↑ electrophysiological recovery
	Keratin gel (Sierpinski *et al.*, 2008)	Silastic	Mouse tibial	0.4	↑ functional recovery
	Polyamide, catgut, polydioxanone, polyglactin (Arai *et al.*, 2000)	Silicone	Rat sciatic	1.5	Presence of neural regeneration and functional recovery cf to none in controls
	VPA (Wu *et al.*, 2008)	Silicone	Rat sciatic	0.8	↑ functional recovery ↑ SFI
	Bioglass 45S5 (Bunting *et al.*, 2005)	Silastic	Rat sciatic	0.5	Axonal regeneration similar to nerve autografts
Neurotrophic	FGF (acidic) (Walter *et al.*, 1993)	Synthetic	Rat sciatic	1.5	↑ Neuronal regeneration and MAPs (cf to none in controls)
	FGF (acidic) (Midha *et al.*, 2003)	PHEMA-MMA	Rat sciatic	1.0	Neuronal regeneration similar to nerve autografts
	FGF (basic) (Li and Cao, 2000)	Vein graft	Rabbit sciatic		↑ Axonal regeneration and nerve conduction velocity
	NGF (Rich *et al.*, 1989)	Silicone	Rat sciatic		2× Myelinated axon count, + 58% myelin thickness, and improved organization
	NGF (Pu *et al.*, 1999)	Vein	Rat sciatic	1.0	↑ Conduction velocity, ↓ latency, improvement in walk pattern, all comparable to nerve autografts
	NGF (He *et al.*, 1992)	Silicone	Rat sciatic	0.5	↑ conduction velocity
	NGF/HA (Mohammad *et al.*, 2000)	Human amnionic membrane	Rabbit peripheral	2.5	+45% myelinated axon count
	NGF/GDNF (Unezaki *et al.*, 2009)	Silicone	Mouse sciatic	0.5	Earlier mechanical response ↑density of unmyelinated axons in NGF group

	GGF (Mohanna *et al.*, 2005)	PHB	Rabbit peroneal n	4.0	↑ SC count, ↑ neuronal fascicle count, ↑ muscle mass
	NT-3 (Midha *et al.*, 2003)	PHEMA-MMA	Rat sciatic	1.0	↑ Nerve regeneration, but less than autografts or FGF-1
	BDNF (Midha *et al.*, 2003)	PHEMA-MMA	Rat sciatic	1.0	↑ Nerve regeneration, but less than autografts or FGF-1
	CNTF (Zhang *et al.*, 2004)	Silicone	Rat sciatic	1.0	↑ Axon diameter, number of axons, improved myelination, ↑ MNCV, ↑ MAP
	VEGF (Hobson *et al.*, 2000)	Silicone	Rat sciatic	1.0	↑ Axonal regeneration and SC migration
	GDNF (Wood *et al.*, 2009)	Silicone	Rat sciatic	1.3	↑Nerve fiber density, percent neural tissue, and myelinated area
Combined	Laminin and NGF (Yu, 2003)	Semipermeable polysulfone	Rat sciatic	1.0	Match the performance of autografts
	Laminin and NGF (Dodla and Bellamkonda, 2008)	Tubular Polysulfone	Rat sciatic	2.0	↑ Nerve regeneration with gradients of both LN-1 and NGF

AVNC, Autogenous venous nerve conduit; BDNF, brain-derived neurotrophic factor; FDS, flexor digitorum superficialis muscle; GTMC, glycolide trimethylene carbonate; EMSCs, ectomesenchymal stem cells; PLGA, porous biodegradable dl-lactic-co-glycolic acid polymers; HA, hyaluronic acid; dMSC, differentiated mesenchymal stem cells; MAP, muscle action potential; VPA, valproic acid; MNCV, motor nerve conduction velocity; PGA, polyglycolic acid; PHB, poly-3-hydroxybutyrate; PHEMA-MMA, poly(2-hydroxyethyl methacrylate-co-methyl methacrylate); PLLA, poly(L-lactic acid); SC, Schwann cell; SFI, sciatic functional index; GDNF, glial-derived neurotrophic factor; TMC/CL, trimethylenecarbonate-co-epsilon-caprolactone.

peripheral nerve gaps, resulting in increased myelinated axon number, evoked nerve potentials, and most impressively functional recovery almost to normal (Ceballos *et al.*, 1999; Verdu *et al.*, 2002). The most cost-effective and convenient luminal additive would be NGF, which can be directly injected as a solution into the nerve conduit to produce autograft-like results over shorter distances (Pu *et al.*, 1999). Longer distances might require more additional components and more complicated delivery systems such as magnetically aligned, laminin-coated collagen fibrils impregnated with controlled-release NGF, which based on the results of the individual components, in combination should return superior results to autograft. However, this combination has never been tested in any model, and may be too cost ineffective for clinical application.

IX. Conclusion

Many luminal additives in nerve conduits have been demonstrated to have a beneficial effect on nerve regeneration across nerve defects. Most impressive are neurotrophic factors alone such as NGF (Lee *et al.*, 2003; Pu *et al.*, 1999) or FGF-1 impregnated into collagen/laminin fibrils (Midha *et al.*, 2003), which are readily manufactured, easily introduced, and return autograft-level results. Several other additives are alternatives, but are less practical or only return autograft-level results at short distances (Bunting *et al.*, 2005; Evans *et al.*, 2002; Ide *et al.*, 1998; Rosen *et al.*, 1989, 1992). Because these results are achieved without the drawbacks of autografting, namely no requirement for a donor site and quantities limited only by the manufacturing process, would ensure their widespread adoption in various clinical applications of nerve repair.

There are several further directions for research that would seem beneficial to pursue. Firstly, the combination of preexisting additives has already returned promising results (Gamez *et al.*, 2004; Muller *et al.*, 1987), as the natural process involves cellular, structural, and hormonal components. Secondly, the enhancement of preexisting additives such as magnetically aligning collagen/laminin fibrils (Ceballos *et al.*, 1999; Dubey *et al.*, 1999; Verdu *et al.*, 2002) which already produce autograft-like results (Rosen *et al.*, 1990), 106 or using controlled released systems of NGF (Lee *et al.*, 2003) would also be fruitful. Thirdly, as we discover more about the natural process of nerve regeneration and the factors involved, new additives will be manufactured. Our arsenal of luminal additives will grow, ensuring that increasingly serious peripheral nerve injury will be able to be successfully repaired, possibly even with results superior to autografts (Tang, 1995).

References

Abernethy, D. A., Thomas, P. K., Rud, A., and King, R. H. (1994). Mutual attraction between emigrant cells from transected denervated nerve. *J. Anat.* **184,** 239–249.

Ahmed, Z., Underwood, S., and Brown, R. A. (2003). Nerve guide material made from fibronectin: Assessment of *in vitro* properties. *Tissue. Eng.* **9,** 219–231.

Allmeling, C., Jokuszies, A., Reimers, K., Kall, S., Choi, C. Y., Brandes, G., Kasper, C., Scheper, T., Guggenheim, M., and Vogt, P. M. (2008). Spider silk fibres in artificial nerve constructs promote peripheral nerve regeneration. *Cell. Prolif.* **41**(3), 408–420.

Altman, G. H., Diaz, F., Jakuba, C., Calabro, T., Horan, R. L., Chen, J., Lu, H., Richmond, J., and Kaplan, D. L. (2003). Silk-based biomaterials. *Biomaterials* **24**(3), 401–416.

Ansselin, A. D., Fink, T., and Davey, D. F. (1997). Peripheral nerve regeneration through nerve guides seeded with adult Schwann cells. *Neuropathol. Appl. Neurobiol.* **23,** 387–398.

Apel, P. J., Garrett, J. P., Sierpinski, P., Ma, J., Atala, A., Smith, T. L., Koman, L. A., and Van Dyke, M. E. (2008). Peripheral nerve regeneration using a keratin-based scaffold: Long-term functional and histological outcomes in a mouse model. *J. Hand Surg. (Am.)* **33**(9), 1541–1547.

Arai, T., Lundborg, G., and Dahlin, L. B. (2000). Bioartificial nerve graft for bridging extended nerve defects in rat sciatic nerve based on resorbable guiding filaments. *Scand. J. Plast. Reconstr. Surg. Hand Surg.* **34,** 101–108.

Archibald, S. J., Shefner, J., Krarup, C., and Madison, R. D. (1995). Monkey median nerve repaired by nerve graft or collagen nerve guide tube. *J. Neurosci.* **15,** 4109–4123.

Ball, R. A., Lipton, S. A., Dreyer, E. B., Richie, J. P., and Vickers, M. A. (1992). Entubulization repair of severed cavernous nerves in the rat resulting in return of erectile function. *J. Urol.* **148,** 211–215.

Barcelos, A. S., Rodrigues, A. C., Silva, M. D., and Padovani, C. R. (2003). Inside-out vein graft and inside-out artery graft in rat sciatic nerve repair. *Microsurgery* **23,** 66–71.

Berger, A., Lassner, F., and Schaller, E. (1994). The Dellon tube in injuries of peripheral nerves. *Handchir. Mikrochir. Plast. Chir.* **26**(1), 44–47.

Birch, R., and Raji, A. R. (1991). Repair of median and ulnar nerves. Primary suture is best. *J. Bone Joint Surg. Br.* **73,** 154–157.

Bunting, S., di Silvio, L., Deb, S., and Hall, S. (2005). Bioresorbable glass fibres facilitate peripheral nerve regeneration. *J. Hand Surg. (Br.)* **30,** 242–247.

Cajal, R. y. S. (1928). "Degeneration and Regeneration of the Nervous System," vol. 1. Oxford University Press, London.

Ceballos, D., Navarro, X., Dubey, N., Wendelschafer-Crabb, G., Kennedy, W. R., and Tranquillo, R. T. (1999). Magnetically aligned collagen gel filling a collagen nerve guide improves peripheral nerve regeneration. *Exp. Neurol.* **158,** 290–300.

Chang, C. J., and Hsu, S. H. (2004). The effects of low-intensity ultrasound on peripheral nerve regeneration in poly(Dl-lactic acid-co-glycolic acid) conduits seeded with Schwann cells. *Ultrasound Med. Biol.* **30,** 1079–1084.

Cheng, Y. S., Cheng, W. C., Yao, C. H., Hsieh, C. L., Lin, J. G., Lai, T. Y., Lin, C. C., and Tsai, C. C. (2001). Effects of buyang huanwu decoction on peripheral nerve regeneration using silicone rubber chambers. *Am. J. Chin. Med.* **29,** 423–432.

Chiu, D. T., and Strauch, B. (1990). A prospective clinical evaluation of autogenous vein grafts used as a nerve conduit for distal sensory nerve defects of 3 cm or less. *Plast. Reconstr. Surg.* **86**(5), 928–934.

Danielsen, N., and Varon, S. (1995). Characterization of neurotrophic activity in the silicone-chamber model for nerve regeneration. *J. Reconstr. Microsurg.* **11,** 231–235.

Danielsen, N., Pettmann, B., Vahlsing, H. L., Manthorpe, M., and Varon, S. (1988). Fibroblast growth factor effects on peripheral nerve regeneration in a silicone chamber model. *J. Neurosci. Res.* **20,** 320–330.

DeLee, J. C., Smith, M. T., and Green, D. P. (1977). The reaction of nerve tissue to various suture materials: A study in rabbits. *J. Hand Surg. (Am.)* **2**(1), 38–43.

Dellon, A. L., and Mackinnon, S. E. (1988). An alternative to the classical nerve graft for the management of the short nerve gap. *Plast. Reconstr. Surg.* **82**(5), 849–856.

Di Benedetto, G., Zura, G., Mazzucchelli, R., Santinelli, A., Scarpelli, M., and Bertani, A. (1998). Nerve regeneration through a combined autologous conduit (vein plus acellular muscle grafts). *Biomaterials* **19,** 173–181.

Dodla, M. C., and Bellamkonda, R. V. (2008). Differences between the effect of anisotropic and isotropic laminin and nerve growth factor presenting scaffolds on nerve regeneration across long peripheral nerve gaps. *Biomaterials* **29**(1), 33–46.

Dubey, N., Letourneau, P., and Tranquillo, R. T. (1999). Guided neurite elongation and schwann cell invasion into magnetically aligned collagen in simulated peripheral nerve regeneration. *Exp. Neurol.* **158,** 338–350.

Dubey, N., Letourneau, P., and Tranquillo, R. T. (2001). Neuronal contact guidance in magnetically aligned fibrin gels: Effect of variation in gel mechano-structural properties. *Biomaterials* **22,** 1065–1075.

Ducker, T. B., and Hayes, G. J. (1970). Peripheral nerve grafts: Experimental studies in the dog and chimpanzee to define homograft limitations. *J. Neurosurg.* **32,** 236–243.

Evans, G. R., Brandt, K., Widmer, M. S., Lu, L., Meszlenyi, R. K., Gupta, P. K., Mikos, A. G., Hodges, J., Williams, J., Gurlek, A., Nabawi, A., Lohman, R., *et al.* (1999). *In vivo* evaluation of poly(l-lactic acid) porous conduits for peripheral nerve regeneration. *Biomaterials* **20**(12), 1109–1115.

Evans, G. R., Brandt, K., Katz, S., Chauvin, P., Otto, L., Bogle, M., Wang, B., Meszlenyi, R. K., Lu, L., Mikos, A. G., and Patrick, C. W., Jr. (2002). Bioactive poly(l-lactic acid) conduits seeded with Schwann cells for peripheral nerve regeneration. *Biomaterials* **23,** 841–848.

Fansa, H., Keilhoff, G., Forster, G., Seidel, B., Wolf, G., and Schneider, W. (1999a). Acellular muscle with Schwann-cell implantation: An alternative biologic nerve conduit. *J. Reconstr. Microsurg.* **15,** 531–537.

Fansa, H., Keilhoff, G., Plogmeier, K., Frerichs, O., Wolf, G., and Schneider, W. (1999b). Successful implantation of Schwann cells in acellular muscles. *J. Reconstr. Microsurg.* **15,** 61–65.

Fornaro, M., Tos, P., Geuna, S., Giacobini-Robecchi, M. G., and Battiston, B. (2001). Confocal imaging of Schwann-cell migration along muscle-vein combined grafts used to bridge nerve defects in the rat. *Microsurgery* **21,** 153–155.

Francel, P. C., Francel, T. J., Mackinnon, S. E., and Hertl, C. (1997). Enhancing nerve regeneration across a silicone tube conduit by using interposed short-segment nerve grafts. *J. Neurosurg.* **87,** 887–892.

Gamez, E., Goto, Y., Nagata, K., Iwaki, T., Sasaki, T., and Matsuda, T. (2004). Photofabricated gelatin-based nerve conduits: Nerve tissue regeneration potentials. *Cell Transplant.* **13,** 549–564.

Gordon, T., Sulaiman, O., and Boyd, J. G. (2003). Experimental strategies to promote functional recovery after peripheral nerve injuries. *J. Peripher. Nerv. Syst.* **8,** 236–250.

Guenard, V., Kleitman, N., Morrissey, T. K., Bunge, R. P., and Aebischer, P. (1992). Syngeneic Schwann cells derived from adult nerves seeded in semipermeable guidance channels enhance peripheral nerve regeneration. *J. Neurosci.* **12,** 3310–3320.

Hadlock, T., Elisseeff, J., Langer, R., Vacanti, J., and Cheney, M. (1998). A tissue-engineered conduit for peripheral nerve repair. *Arch. Otolaryngol. Head Neck Surg.* **124,** 1081–1086.

Hadlock, T., Sundback, C., Koka, R., Hunter, D., Cheney, M., and Vacanti, J. (1999). A novel, biodegradable polymer conduit delivers neurotrophins and promotes nerve regeneration. *Laryngoscope* **109,** 1412–1416.

Hadlock, T., Sundback, C., Hunter, D., Cheney, M., and Vacanti, J. P. (2000). A polymer foam conduit seeded with Schwann cells promotes guided peripheral nerve regeneration. *Tissue Eng.* **6,** 119–127.

Hall, S. (1997). Axonal regeneration through acellular muscle grafts. *J. Anat.* **190,** 57–71.

He, C., Chen, Z., and Chen, Z. (1992). Enhancement of motor nerve regeneration by nerve growth factor. *Microsurgery* **13,** 151–154.

Heijke, G. C., Klopper, P. J., Baljet, B., and van Doorn, I. B. (2001). Silicone rubber tubulization in peripheral sensory nerve reconstruction: An experimental study in rabbits. *Microsurgery* **21,** 306–316.

Hobson, M. I., Green, C. J., and Terenghi, G. (2000). VEGF enhances intraneural angiogenesis and improves nerve regeneration after axotomy. *J. Anat.* **197,** 591–605.

Hsu, S. H., Chang, C. J., Tang, C. M., and Lin, F. T. (2004). *In vitro* and *in vivo* effects of Ginkgo biloba extract egb 761 on seeded Schwann cells within poly(dl-lactic acid-co-glycolic acid) conduits for peripheral nerve regeneration. *J. Biomater. Appl.* **19,** 163–182.

Hudson, A. R., Morris, J., Weddell, G., and Drury, A. (1972). Peripheral nerve autografts. *J. Surg. Res.* **12,** 267–274.

Ide, C. (1996). Peripheral nerve regeneration. *Neurosci. Res.* **25,** 101–121.

Ide, C., Tohyama, K., Tajima, K., Endoh, K., Sano, K., Tamura, M., Mizoguchi, A., Kitada, M., Morihara, T., and Shirasu, M. (1998). Long acellular nerve transplants for allogeneic grafting and the effects of basic fibroblast growth factor on the growth of regenerating axons in dogs: A preliminary report. *Exp. Neurol.* **154,** 99–112.

Kalbermatten, D. F., Kingham, P. J., Mahay, D., Mantovani, C., Pettersson, J., Raffoul, W., Balcin, H., Pierer, G., and Terenghi, G. (2008). Fibrin matrix for suspension of regenerative cells in an artificial nerve conduit. *J. Plast. Reconstr. Aesthet. Surg.* **61**(6), 669–675.

Keilhoff, G., Goihl, A., Langnase, K., Fansa, H., and Wolf, G. (2005). Transdifferentiation of mesenchymal stem cells into Schwann cell-like myelinating cells. *J. Reconstr. Microsurg.* **21,** 565–571.

Komiyama, T., Nakao, Y., Toyama, Y., Asou, H., Vacanti, C. A., and Vacanti, M. P. (2003). A novel technique to isolate adult Schwann cells for an artificial nerve conduit. *J. Neurosci. Methods* **122,** 195–200.

Koshimune, M., Takamatsu, K., Nakatsuka, H., Inui, K., Yamano, Y., and Ikada, Y. (2003). Creating bioabsorbable Schwann cell coated conduits through tissue engineering. *Biomed. Mater. Eng.* **13,** 223–229.

Labrador, R. O., Buti, M., and Navarro, X. (1998). Influence of collagen and laminin gels concentration on nerve regeneration after resection and tube repair. *Exp. Neurol.* **149,** 243–252.

Lee, A. C., Yu, V. M., Lowe, J. B. III, Brenner, M. J., Hunter, D. A., Mackinnon, S. E., and Sakiyama-Elbert, S. E. (2003). Controlled release of nerve growth factor enhances sciatic nerve regeneration. *Exp. Neurol.* **184,** 295–303.

Li, C. Y., and Cao, D. C. (2000). Experimental study on repair of peripheral nerve defect by basic fibroblast growth factor combined with autogenous vein graft conduit. *Zhongguo Xiu Fu Chong Jian Wai Ke Za Zhi* **14,** 14–16.

Li, J., Yan, J. G., Ai, X., Hu, S., Gu, Y. D., Matloub, H. S., and Sanger, J. R. (2004). Ultrastructural analysis of peripheral-nerve regeneration within a nerve conduit. *J. Reconstr. Microsurg.* **20,** 565–569.

Lietz, M., Dreesmann, L., Hoss, M., Oberhoffner, S., and Schlosshauer, B. (2006). Neuro tissue engineering of glial nerve guides and the impact of different cell types. *Biomaterials* **27,** 1425–1436.

Lloyd, B. M., Luginbuhl, R. D., Brenner, M. J., Rocque, B. G., Tung, T. H., Myckatyn, T. M., Hunter, D. A., Mackinnon, S. E., and Borschel, G. H. (2007). Use of motor nerve material in peripheral nerve repair with conduits. *Microsurgery* **27**(2), 138–145.

Lundborg, G., Dahlin, L., Danielsen, N., and Zhao, Q. (1994a). Trophism, tropism, and specificity in nerve regeneration. *J. Reconstr. Microsurg.* **10,** 345–354.

Lundborg, G., Rosen, B., Abrahamson, S. O., Dahlin, L., and Danielsen, N. (1994b). Tubular repair of the median nerve in the human forearm. Preliminary findings. *J. Hand Surg. (Br.)* **19,** 273–276.

Lundborg, G., Dahlin, L., Dohi, D., Kanje, M., and Terada, N. (1997). A new type of "bioartificial" nerve graft for bridging extended defects in nerves. *J. Hand Surg. (Br.)* **22,** 299–303.

Mackinnon, S. E., and Dellon, A. L. (1990). A study of nerve regeneration across synthetic (Maxon) and biologic (collagen) nerve conduits for nerve gaps up to 5 cm in the primate. *J. Reconstr. Microsurg.* **6**(2), 117–121.

Madison, R., Dsc, S., Dikkes, P., Chiu, T. H., and Sidman, R. L. (1985). Increased rate of peripheral nerve regeneration using bioresorbable nerve guides and a laminin-containing gel. *Exp. Neurol.* **88,** 767–772.

Madison, R. D., da Silva, C. F., Dikkes, P., Sidman, R. L., and Chiu, T. H. (1987). Peripheral nerve regeneration with entubulation repair: Comparison of biodegradeable nerve guides versus polyethylene tubes and the effects of a laminin-containing gel. *Exp. Neurol.* **95,** 378–390.

Madison, R. D., da Silva, C. F., and Dikkes, P. (1988). Entubulation repair with protein additives increases the maximum nerve gap distance successfully bridged with tubular prostheses. *Brain Res.* **447,** 325–334.

Matsumoto, K., Ohnishi, K., Kiyotani, T., Sekine, T., Ueda, H., Nakamura, T., Endo, K., and Shimizu, Y. (2000). Peripheral nerve regeneration across an 80-mm gap bridged by a polyglycolic acid (PGA)-collagen tube filled with laminin-coated collagen fibers: A histological and electrophysiological evaluation of regenerated nerves. *Brain Res.* **868,** 315–328.

Matsuoka, I., Nakane, A., and Kurihara, K. (1997). Induction of LIF-mrna by TGF-beta 1 in Schwann cells. *Brain Res.* **776,** 170–180.

Mersa, B., Agir, H., Aydin, A., and Sen, C. (2004). Comparison of expanded polytetrafluoroethylene (eptfe) with autogenous vein as a nerve conduit in rat sciatic nerve defects. *Kulak Burun Bogaz Ihtis. Derg.* **13**(5–6), 103–111.

Midha, R., Munro, C. A., Dalton, P. D., Tator, C. H., and Shoichet, M. S. (2003). Growth factor enhancement of peripheral nerve regeneration through a novel synthetic hydrogel tube. *J. Neurosurg.* **99,** 555–565.

Mohammad, J. A., Warnke, P. H., Pan, Y. C., and Shenaq, S. (2000). Increased axonal regeneration through a biodegradable amnionic tube nerve conduit: Effect of local delivery and incorporation of nerve growth factor/hyaluronic acid media. *Ann. Plast. Surg.* **44,** 59–64.

Mohanna, P. N., Terenghi, G., and Wiberg, M. (2005). Composite PHB-GGF conduit for long nerve gap repair: A long-term evaluation. *Scand. J. Plast. Reconstr. Surg. Hand Surg.* **39,** 129–137.

Mosahebi, A., Woodward, B., Wiberg, M., Martin, R., and Terenghi, G. (2001). Retroviral labeling of Schwann cells: *In vitro* characterization and *in vivo* transplantation to improve peripheral nerve regeneration. *Glia* **34,** 8–17.

Mosahebi, A., Fuller, P., Wiberg, M., and Terenghi, G. (2002). Effect of allogeneic Schwann cell transplantation on peripheral nerve regeneration. *Exp. Neurol.* **173,** 213–223.

Mosahebi, A., Wiberg, M., and Terenghi, G. (2003). Addition of fibronectin to alginate matrix improves peripheral nerve regeneration in tissue-engineered conduits. *Tissue Eng.* **9,** 209–218.

Muller, H., Williams, L. R., and Varon, S. (1987). Nerve regeneration chamber: Evaluation of exogenous agents applied by multiple injections. *Brain Res.* **413,** 320–326.

Nakayama, K., Takakuda, K., Koyama, Y., Itoh, S., Wang, W., Mukai, T., and Shirahama, N. (2007). Enhancement of peripheral nerve regeneration using bioabsorbable polymer tubes packed with fibrin gel. *Artif. Organs* **31**(7), 500–508.

Newman, J. P., Verity, A. N., Hawatmeh, S., Fee, W. E., Jr., and Terris, D. J. (1996). Ciliary neurotrophic factors enhances peripheral nerve regeneration. *Arch. Otolaryngol. Head Neck Surg.* **122,** 399–403.

Nicolino, S., Raimondo, S., Tos, P., Battiston, B., Fornaro, M., Geuna, S., and Perroteau, I. (2003). Expression of alpha2a-2b neuregulin-1 is associated with early peripheral nerve repair along muscle-enriched tubes. *Neuroreport* **14,** 1541–1545.

Nie, X., Zhang, Y. J., Tian, W. D., Jiang, M., Dong, R., Chen, J. W., and Jin, Y. (2007). Improvement of peripheral nerve regeneration by a tissue-engineered nerve filled with ectomesenchymal stem cells. *Int. J. Oral Maxillofac. Surg.* **36**(1), 32–38.

Noble, J., Munro, C. A., Prasad, V. S., and Midha, R. (1998). Analysis of upper and lower extremity peripheral nerve injuries in a population of patients with multiple injuries. *J. Trauma* **45,** 116–122.

Phillips, J. B., Bunting, S. C., Hall, S. M., and Brown, R. A. (2005). Neural tissue engineering: A self-organizing collagen guidance conduit. *Tissue Eng.* **11,** 1611–1617.

Politis, M. J., Ederle, K., and Spencer, P. S. (1982). Tropism in nerve regeneration *in vivo*. Attraction of regenerating axons by diffusible factors derived from cells in distal nerve stumps of transected peripheral nerves. *Brain Res.* **253,** 1–12.

Pu, L. L., Syed, S. A., Reid, M., Patwa, H., Goldstein, J. M., Forman, D. L., and Thomson, J. G. (1999). Effects of nerve growth factor on nerve regeneration through a vein graft across a gap. *Plast. Reconstr. Surg.* **104,** 1379–1385.

Raimondo, S., Nicolino, S., Tos, P., Battiston, B., Giacobini-Robecchi, M. G., Perroteau, I., and Geuna, S. (2005). Schwann cell behavior after nerve repair by means of tissue-engineered muscle-vein combined guides. *J. Comp. Neurol.* **489,** 249–259.

Rich, K. M., Alexander, T. D., Pryor, J. C., and Hollowell, J. P. (1989). Nerve growth factor enhances regeneration through silicone chambers. *Exp. Neurol.* **105,** 162–170.

Richardson, P. M. (1991). Neurotrophic factors in regeneration. *Curr. Opin. Neurobiol.* **1,** 401–406.

Rodriguez, F. J., Verdu, E., Ceballos, D., and Navarro, X. (2000). Nerve guides seeded with autologous schwann cells improve nerve regeneration. *Exp. Neurol.* **161,** 571–584.

Rosen, J. M., Pham, H. N., Abraham, G., Harold, L., and Hentz, V. R. (1989). Artificial nerve graft compared to autograft in a rat model. *J. Rehabil. Res. Dev.* **26,** 1–14.

Rosen, J. M., Padilla, J. A., Nguyen, K. D., Padilla, M. A., Sabelman, E. E., and Pham, H. N. (1990). Artificial nerve graft using collagen as an extracellular matrix for nerve repair compared with sutured autograft in a rat model. *Ann. Plast. Surg.* **25,** 375–387.

Rosen, J. M., Padilla, J. A., Nguyen, K. D., Siedman, J., and Pham, H. N. (1992). Artificial nerve graft using glycolide trimethylene carbonate as a nerve conduit filled with collagen compared to sutured autograft in a rat model. *J. Rehabil. Res. Dev.* **29,** 1–12.

Rutkowski, G. E., and Heath, C. A. (2002). Development of a bioartificial nerve graft. I. Design based on a reaction-diffusion model. *Biotechnol. Prog.* **18,** 362–372.

Rutkowski, G. E., Miller, C. A., Jeftinija, S., and Mallapragada, S. K. (2004). Synergistic effects of micropatterned biodegradable conduits and Schwann cells on sciatic nerve regeneration. *J. Neural. Eng.* **1,** 151–157.

Satou, T., Nishida, S., Hiruma, S., Tanji, K., Takahashi, M., Fujita, S., Mizuhara, Y., Akai, F., and Hashimoto, S. (1986). A morphological study on the effects of collagen gel matrix on regeneration of severed rat sciatic nerve in silicone tubes. *Acta Pathol. Jpn.* **36,** 199–208.

Seckel, B. R. (1990). Enhancement of peripheral nerve regeneration. *Muscle Nerve* **13,** 785–800.

Seckel, B. R., Jones, D., Hekimian, K. J., Wang, K. K., Chakalis, D. P., and Costas, P. D. (1995). Hyaluronic acid through a new injectable nerve guide delivery system enhances peripheral nerve regeneration in the rat. *J. Neurosci. Res.* **40,** 318–324.

Sierpinski, P., Garrett, J., Ma, J., Apel, P., Klorig, D., Smith, T., Koman, L. A., Atala, A., and Van Dyke, M. (2008). The use of keratin biomaterials derived from human hair for the promotion of rapid regeneration of peripheral nerves. *Biomaterials* **29**(1), 118–128.

Sinis, N., Schaller, H. E., Schulte-Eversum, C., Schlosshauer, B., Doser, M., Dietz, K., Rosner, H., Muller, H. W., and Haerle, M. (2005). Nerve regeneration across a 2-cm gap in the rat median nerve using a resorbable nerve conduit filled with Schwann cells. *J. Neurosurg.* **103,** 1067–1076.

Strauch, B., Ferder, M., Lovelle-Allen, S., Moore, K., Kim, D. J., and Llena, J. (1996). Determining the maximal length of a vein conduit used as an interposition graft for nerve regeneration. *J. Reconstr. Microsurg.* **12**(8), 521–527.

Strauch, B., Rodriguez, D. M., Diaz, J., Yu, H. L., Kaplan, G., and Weinstein, D. E. (2001). Autologous Schwann cells drive regeneration through a 6-cm autogenous venous nerve conduit. *J. Reconstr. Microsurg.* **17**(8), 589–595; discussion 596–597.

Su, Y., Zeng, B. F., Zhang, C. Q., Zhang, K. G., and Xie, X. T. (2007). Study of biocompatibility of small intestinal submucosa (SIS) with Schwann cells *in vitro*. *Brain Res.* **1145**(11), 41–47.

Tang, J. B. (1995). Vein conduits with interposition of nerve tissue for peripheral nerve defects. *J. Reconstr. Microsurg.* **11**(1), 21–26.

Tang, X. Q., Cai, J., Nelson, K. D., Peng, X. J., and Smith, G. M. (2004). Functional repair after dorsal root rhizotomy using nerve conduits and neurotrophic molecules. *Eur. J. Neurosci.* **20,** 1211–1218.

Terenghi, G. (1995). Peripheral nerve injury and regeneration. *Histol. Histopathol.* **10,** 709–718.

Terenghi, G. (1999). Peripheral nerve regeneration and neurotrophic factors. *J. Anat.* **194,** 1–14.

Toba, T., Nakamura, T., Lynn, A. K., Matsumoto, K., Fukuda, S., Yoshitani, M., Hori, Y., and Shimizu, Y. (2002). Evaluation of peripheral nerve regeneration across an 80-mm gap using a polyglycolic acid (PGA)—collagen nerve conduit filled with laminin-soaked collagen sponge in dogs. *Int. J. Artif. Organs* **25,** 230–237.

Tos, P., Battiston, B., Geuna, S., Giacobini-Robecchi, M. G., Hill, M. A., Lanzetta, M., and Owen, E. R. (2000). Tissue specificity in rat peripheral nerve regeneration through combined skeletal muscle and vein conduit grafts. *Microsurgery* **20,** 65–71.

Unezaki, S., Yoshii, S., Mabuchi, T., Saito, A., and Ito, S. (2009). Effects of neurotrophic factors on nerve regeneration monitored by *in vivo* imaging in thy1-YFP transgenic mice. *J. Neurosci. Methods* **178**(2), 308–315.

Varejao, A. S., Cabrita, A. M., Meek, M. F., Fornaro, M., and Geuna, S. (2003). Nerve regeneration inside fresh skeletal muscle-enriched synthetic tubes: A laser confocal microscope study in the rat sciatic nerve model. *Ital. J. Anat. Embryol.* **108,** 77–82.

Verdu, E., Labrador, R. O., Rodriguez, F. J., Ceballos, D., Fores, J., and Navarro, X. (2002). Alignment of collagen and laminin-containing gels improve nerve regeneration within silicone tubes. *Restor. Neurol. Neurosci.* **20,** 169–179.

Walter, M. A., Kurouglu, R., Caulfield, J. B., Vasconez, L. O., and Thompson, J. A. (1993). Enhanced peripheral nerve regeneration by acidic fibroblast growth factor. *Lymphokine Cytokine Res.* **12,** 135–141.

Williams, L. R. (1987). Exogenous fibrin matrix precursors stimulate the temporal progress of nerve regeneration within a silicone chamber. *Neurochem. Res.* **12,** 851–860.

Williams, L. R., and Varon, S. (1985). Modification of fibrin matrix formation in situ enhances nerve regeneration in silicone chambers. *J. Comp. Neurol.* **231,** 209–220.

Williams, L. R., Longo, F. M., Powell, H. C., Lundborg, G., and Varon, S. (1983). Spatial-temporal progress of peripheral nerve regeneration within a silicone chamber: Parameters for a bioassay. *J. Comp. Neurol.* **218,** 460–470.

Williams, L. R., Danielsen, N., Muller, H., and Varon, S. (1987). Exogenous matrix precursors promote functional nerve regeneration across a 15-mm gap within a silicone chamber in the rat. *J. Comp. Neurol.* **264,** 284–290.

Wood, M. D., Moore, A. M., Hunter, D. A., TuVaha, S., Borschel, G. H., Mackinnon, S. E., and Sakiyama-Elbert, S. E. (2009). Affinity-based release of glial-derived neurotrophic factor from fibrin matrices enhances sciatic nerve regeneration. *Acta. Biomater.* **5**(4), 959–968.

Wu, F., Xing, D., Peng, Z., and Rao, T. (2008). Enhanced rat sciatic nerve regeneration through silicon tubes implanted with valproic acid. *J. Reconstr. Microsurg.* **24**(4), 267–276.

Yang, Y., De Laporte, L., Rives, C. B., Jang, J. H., Lin, W. C., Shull, K. R., and Shea, L. D. (2005). Neurotrophin releasing single and multiple lumen nerve conduits. *J. Control Release* **104,** 433–446.

Yang, Y., Ding, F., Wu, J., Hu, W., Liu, W., Liu, J., and Gu, X. (2007). Development and evaluation of silk fibroin-based nerve grafts used for peripheral nerve regeneration. *Biomaterials* **28**(36), 5526–5535. Epub 2007 Sep 19.

Young, J. N., Friedman, A. H., Harrelson, J. M., Rossitch, E. Jr., Alston, S., and Rozear, M. (1991). Hemangiopericytoma of the sciatic nerve. Case report. *J. Neurosurg.* **74**(3), 512–515.

Yu, X. (2003). Tissue-engineered scaffolds are effective alternatives to autografts for bridging peripheral nerve gaps. *Tissue Eng.* **9**(3), 421–430.

Zhang, F., Blain, B., Beck, J., Zhang, J., Chen, Z., Chen, Z. W., and Lineaweaver, W. C. (2002). Autogenous venous graft with one-stage prepared Schwann cells as a conduit for repair of long segmental nerve defects. *J. Reconstr. Microsurg.* **18,** 295–300.

Zhang, J., Lineaweaver, W. C., Oswald, T., Chen, Z., Chen, Z., and Zhang, F. (2004). Ciliary neurotrophic factor for acceleration of peripheral nerve regeneration: An experimental study. *J. Reconstr. Microsurg.* **20,** 323–327.

Zhang, P., He, X., Zhao, F., Zhang, D., Fu, Z., and Jiang, B. (2005). Bridging small-gap peripheral nerve defects using biodegradable chitin conduits with cultured schwann and bone marrow stromal cells in rats. *J. Reconstr. Microsurg.* **21,** 565–571.

Zhang, P., Xue, F., Kou, Y., Fu, Z., Zhang, D., Zhang, H., and Jiang, B. (2008). The experimental study of absorbable chitin conduit for bridging peripheral nerve defect with nerve fasciculu in rats. *Artif. Cells Blood Substit. Immobil. Biotechnol.* **36**(4), 360–371.

Zhao, F. Q., Zhang, P. X., He, X. J., Du, C., Fu, Z. G., Zhang, D. Y., and Jiang, B. G. (2005). Study on the adoption of Schwann cell phenotype by bone marrow stromal cells *in vitro* and *in vivo*. *Biomed. Environ. Sci.* **18,** 326–333.

Zheng, M., and Kuffler, D. P. (2000). Guidance of regenerating motor axons *in vivo* by gradients of diffusible peripheral nerve-derived factors. *J. Neurobiol.* **42,** 212–219.

Zhu, Q. T., Zhu, J. K., and Cheng, G. (2000). Advance in study of artificial nerve. *Zhongguo Xiu Fu Chong Jian Wai Ke Za Zhi* **14,** 369–371.

TISSUE ENGINEERING OF PERIPHERAL NERVES

Bruno Battiston,*[,1] Stefania Raimondo,[†,1] Pierluigi Tos,* Valentina Gaidano,[†] Chiara Audisio,[‡] Anna Scevola,* Isabelle Perroteau,[‡] and Stefano Geuna[†]

*Reconstructive Microsurgery Unit, Department of Orthopedics, C.T.O. Hospital, Turin 10126, Italy

[†]Department of Clinical and Biological Sciences, San Luigi Gonzaga School of Medicine, University of Turin, 10043 Turin, Italy

[‡]Department of Animal and Human Biology, University of Turin, Turin, Italy

[1]These two authors equally contributed to this work

Tissue engineering of peripheral nerves has seen an increasing interest over the last years and, similarly to many other fields of regenerative medicine, great expectations have risen within the general public to its potential clinical application in the treatment of damaged nerves. However, in spite of the scientific advancements, applications to the patients is still very limited and it appears that to optimize the strategy for the tissue engineering of the peripheral nerves in the clinical view, researchers have to strive for a new level of innovation which will bring together (in a multitranslational approach) the main pillars of tissue engineering: namely (1) microsurgery, (2) cell and tissue transplantation, (3) material science, and (4) gene transfer. This review paper provides an overview of these four key approaches to peripheral nerve tissue engineering. While some of these issues will also be specifically addressed in other papers in this special issue on peripheral nerve regeneration of the *International Review of Neurobiology*, in this paper we will focus on an example of successful translational research in tissue engineering, namely nerve reconstruction by muscle–vein-combined nerve scaffolds.

INTERNATIONAL REVIEW OF NEUROBIOLOGY, VOL. 87
DOI: 10.1016/S0074-7742(09)87011-6

I. Introduction

Nerve reconstruction by tissue engineering has seen an increasing interest over the last years (de Ruiter *et al.*, 2009; Leach and Schmidt, 2005; Li and Hoffman-Kim, 2008; Pfister *et al.*, 2007). However, peripheral nerve do spontaneously regenerate without any treatment (Geuna *et al.*, 2009, this issue; Hall, 2005) and it could thus be wondered why do we have to spend so much efforts and money too for finding out new tissue engineering strategies for nerve regeneration? The answer is that, in spite of the spontaneous regeneration potential of peripheral nerves, almost never complete recovery of nerve function occurs and clinical results are still unsatisfactory (Casha *et al.*, 2008; Ferreira *et al.*, 1994; Gordon *et al.*, 2003; Höke, 2006; Lundborg, 2002; Midha, 2006; Millesi, 2006; Ruijs *et al.*, 2005; Samii *et al.*, 1997; Shieh *et al.*, 2007). Figure 1 illustrates how the questionable "cultural" approach that nerve fiber "in the peripheral nervous system (PNS) can regenerate" while "in the central nervous system (CNS) cannot regenerate" (spread also among scientists and clinicians) can prevent scientific research and innovation.

On the other hand, a more rational and constructive approach should be that nerve fiber "in the PNS regenerate more" while "in the CNS regenerate less," a view that provides a significant impulse to innovative research in this fields. Yet, the consideration that, in spite of the scientific advancements, applications to the

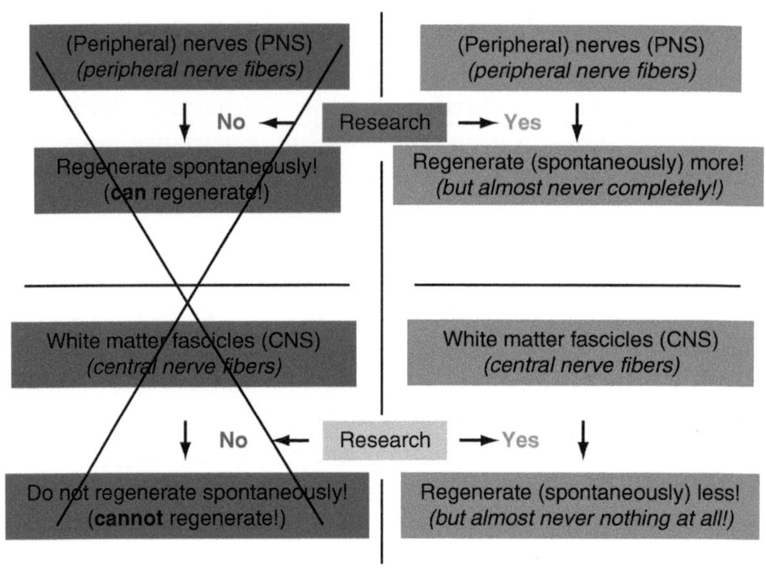

FIG. 1. How shall we think about central and peripheral nerve fiber regeneration potential?

patients are still limited suggests that researchers should optimize the strategy for tissue engineering of peripheral nerves striving for a new level of innovation which brings together, in a multitranslational approach, the main pillars of tissue engineering, namely (1) microsurgery, (2) cell and tissue transplantation, (3) material science, and (4) gene transfer (Fig. 2).

The aim of this review is to provide the reader with an integrated and translational overview on these four key approaches to peripheral nerve tissue engineering. In addition, we will focus on a particular tissue engineering approach for nerve reconstruction, muscle–vein-combined nerve guides (Battiston *et al.*, 2000a,b), that we have extensively investigated over the last years and that represents an example of successful translational research in biomedicine.

II. Microsurgery

Microsurgery is the key scientific discipline in nerve tissue engineering (Fig. 2), not only because it represents the last step for most clinical applications, but also because it should give directions to the other disciplines in order to avoid the production of "sterile" basic science results.

The surgical approach to the peripheral nerves has a lengthy history (Battiston *et al.*, 2009, this issue; Naff and Ecklund, 2001; Papalia *et al.*, 2007; Terzis *et al.*, 1997). For many years, the common surgical approach for the repair of a transected nerve was end-to-end neurorrhaphy of the two stumps. When nerve substance loss occurs, tension microsuturing precludes nerve regeneration (Millesi, 1970; Millesi *et al.*, 1972, 1976; Naff and Ecklund, 2001) and thus bridging strategies have to be sought in order to provide a scaffold for the regenerating nerve (Kine, 2008; Siemionow and Brzezicki, 2009, this issue). The demonstration, in the early 1970s, that grafting an autologous nerve segment

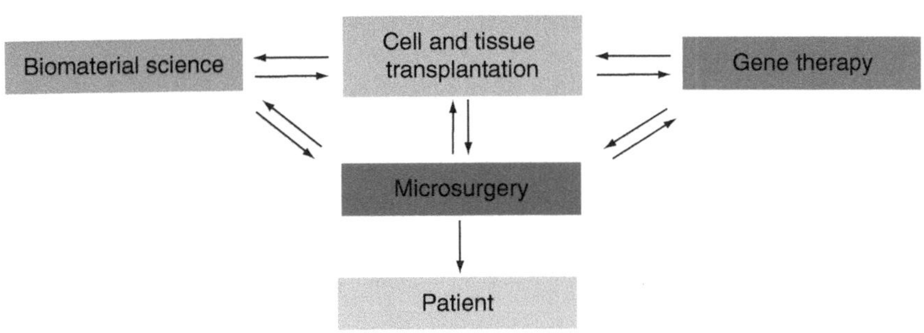

FIG. 2. The pillars of peripheral nerve tissue engineering.

to bridge a nerve defect led to better clinical results than suturing the two stumps under tension (Millesi, 1970; Millesi *et al.*, 1972, 1976) opened a new era in peripheral nerve surgery.

Unfortunately, nerve grafting has some disadvantages since it does require an extra surgical incision for the withdrawal of a healthy sensory nerve and the removal of the healthy nerve might result in sensory residual deficits. Yet, graft material is limited (in terms of length). Thus, possibility of repairing nerve defects by bridging the gap using nonnervous tubes has been widely studied, both experimentally and in clinical practice (Battiston *et al.*, 2005; Chiono *et al.*, 2009, this issue; Geuna *et al.*, 2007; Tang, 1995). This surgical approach is called tubulization (or entubulation) and it represents a great challenge for tissue engineering as it offers the possibility of premanipulating in the laboratory different tissues and biomaterials in order to fashion conduits that mimics the important elements of the nerve environment that are essential for promoting axonal regeneration and that are missing in nonnervous scaffolds (Amado *et al.*, 2008; de Ruiter *et al.*, 2009; Luis *et al.*, 2007; Pfister *et al.*, 2007; Terzis and Kostas, 2007; Tos *et al.*, 2007).

Today, microsurgery also offers alternative concepts for nerve regeneration in case of nerve defects, such as end-to-side neurorrhaphy, that also represent challenging fields for tissue engineering and that are addressed by other reviews of this special Issue of the *International Review of Neurobiology* (Beris and Lykissas, 2009; Bontioti and Dahlin, 2009; Tos *et al.*, 2009, this issue).

III. Cell and Tissue Transplantation

The sought for alternatives to nerve autografts for bridging nerve defects, together with the improvements in stem cell manipulation and microsurgical technologies, have provided the basis for the employment of cell and tissue transplantation for nerve reconstruction. Unlike other organs, in peripheral nerves transplantation strategies have mainly focused on autologous rather than heterologous transplants (Siemionow and Brzezicki, 2009, this issue) and efforts of the researchers have been mainly aimed at enriching the nerve conduits with various cell elements that can promote axonal regeneration. Actually, much focus has been dedicated to the employment of Schwann cells since these cells play a pivotal role in peripheral nerve regeneration forming the bands of Büngner to direct regenerating axons across the lesion site and releasing neurotrophic factors (Frostick *et al.*, 1998; Geuna *et al.*, 2009, this issue; Hall, 2005; Ide, 1996).

Many studies showed that, when seeded in artificial nerve conduits, Schwann cells enhance nerve regeneration. Noteworthy are the studies by Zhang *et al.* (2002) and Strauch *et al.* (2001) who showed that a vein conduit filled with

Schwann cells allowed successful bridging of rabbit nerve defects up to 40 and 60 mm, respectively. In addition, advances in stem cell biology and manipulation (Geuna *et al.*, 2001; Tohill and Terenghi, 2004) open new perspectives since it has been shown that Schwann cells can be derived from various stem cell niches, such as adipocyte-derived stem cells (Kingham *et al.*, 2007). This opportunity also opens interesting perspectives in the clinical view since it may permit to obtain an adequate quantity of autologous Schwann cells from the same patient.

As regards tissue transplantation for nerve regeneration, blood vessels and skeletal muscles are those that receive the most attention by researchers. Conduits made by small segments of an artery were first successfully employed by Bungner (1891). However, interest shifted then to veins for their larger availability and reduced side effects related to their withdrawal (Wrede, 1909). A milestone for the use of this surgical procedure was represented by the demonstration in the 1980s that vein grafting was capable of fostering nerve repairs comparable to autogenous nerve grafting (Chiu and Strauch, 1990; Chiu *et al.*, 1982; Risitano *et al.*, 1989; Suematsu, 1989; Walton *et al.*, 1989). Similarly to veins, also the use of skeletal muscle autografts for nerve repair was already reported many years ago (Kraus and Reisner, 1940) and was comprehensively explored starting in the 1980s (Fawcett and Keynes, 1986; Glasby *et al.*, 1986; Keynes *et al.*, 1984; Kong *et al.*, 1986). The idea for employing muscle fibers for axonal regeneration is on the similarities between the muscle basal lamina and the endoneurial tubes (Fawcett and Keynes, 1986; Glasby *et al.*, 1986). Experimental and clinical studies showed that both fresh and denatured (to avoid the presence of impeding material) muscle conduits can lead to successful nerve repair (Glasby *et al.*, 1986; Lundborg, 2003; Meek and Coert, 2002; Mligiliche *et al.*, 2001; Norris *et al.*, 1988; Pereira *et al.*, 1991a,b, 1996; Rath, 2002).

A. A Combined Tissue Autotransplantation Approach: The Muscle–Vein-Combined Technique

Although the efficacy monotissue vein and muscle conduits has been proven both experimentally and with patients (Lundborg, 2003; Meek and Coert, 2002; Pereira *et al.*, 1991a,b, 1996; Pogrel and Maghen, 2001; Stahl and Goldberg, 1999; Terzis and Kostas, 2007), its effectiveness is usually limited to reconstruction of short nerve gaps (Chiu, 1999). In fact, the vein tends to collapse and axon dispersion occurs when muscle autografts alone are used to bridge long nerve gaps (Brunelli *et al.*, 1993). For this reason, Brunelli, Battiston, and coworkers decided to investigate the possibility of engineering a combined conduit by enriching vein segments with fresh skeletal muscle fibers to improve effectiveness of tubulization nerve repair (Fig. 3) (Battiston *et al.*, 2000a,b; Brunelli *et al.*, 1993).

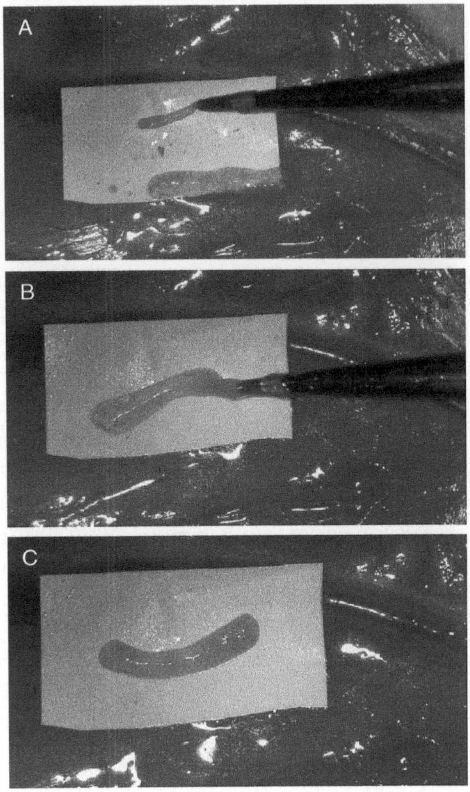

FIG. 3. Preparation of the muscle–vein-combined nerve scaffold in the rat. (A) The vein is gently dilated with the forceps. (B) The vein is filled up with a longitudinal piece of biceps muscle. (C) The fashioned tissue-engineered conduit, ready to be grafted for bridging the nerve defect.

The original rationale of the muscle–vein-combined approach was that muscle enrichment prevents vein collapse while the vein wall prevents axon dispersion (Brunelli *et al.*, 1993). Yet, the choice of fresh instead of predegenerated muscle was aimed at reducing surgical times by avoiding the need of a predegeneration procedure. Basic morphological investigation by confocal and electron microscopy (Fig. 4) showed that basal lamina scaffolds of fresh muscle fibers are available to migratory Schwann cells without the need of any preliminary degeneration of the skeletal muscle (Battiston *et al.*, 2000a,b; Geuna *et al.*, 2000; Raimondo *et al.*, 2005; Tos *et al.*, 2007), as claimed by other authors in order to reduce the presence of impeding material inside the graft and maintain only basal lamina scaffolds that drive axon regeneration (Meek *et al.*, 1999) (Fig. 4A and B). Yet it was shown that most muscle fibers degenerate during the first postoperative

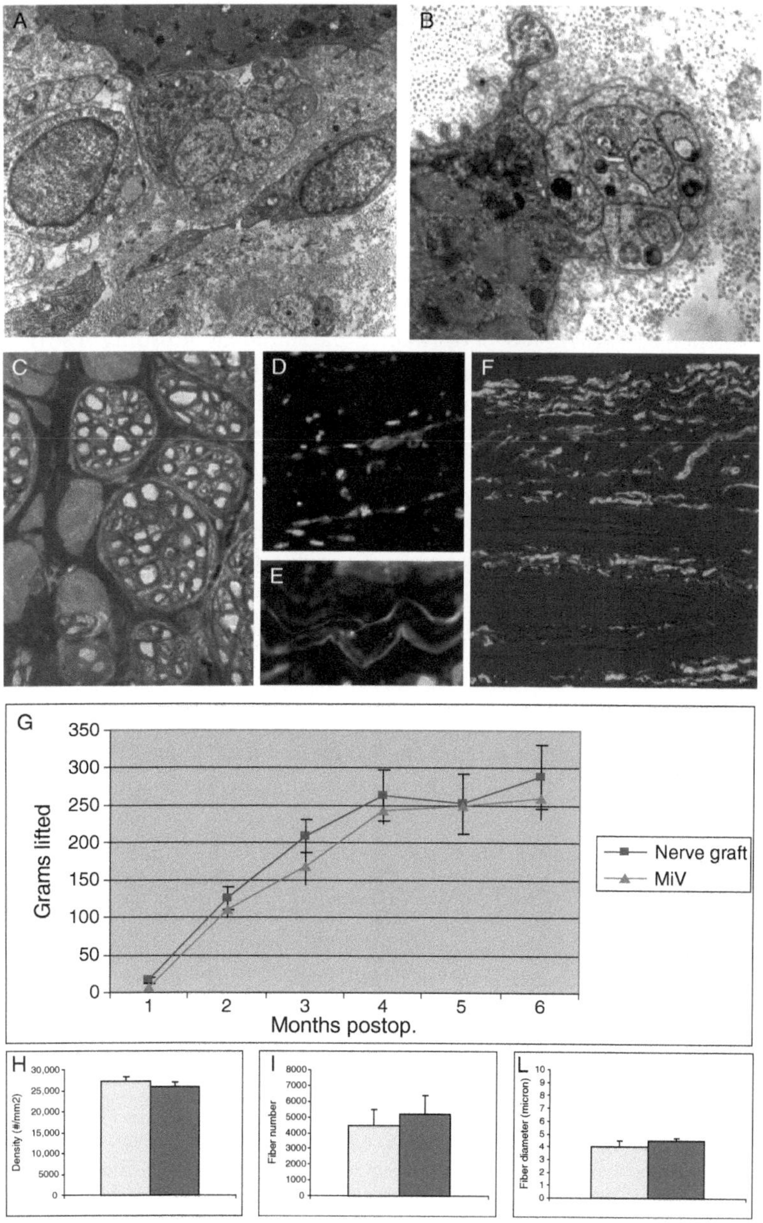

FIG. 4. Results of muscle–vein-combined nerve reconstruction in the rat. (A, B) Electron microscopy showing Schwann cells, regenerating axons, perineurial cells and muscle fibers (A). Migrating Schwann cells and regrowing axons located under the muscle basal lamina (B). (C) High resolution light microscopy showing muscle fibers among nerve fascicles completely separated by clearly

weeks and only some of them remain alive over time (Geuna *et al.*, 2000), when they do not play any further role in relation to nerve fibers because, at late postoperative times, nerve fascicles are always completely separated from muscle by clearly delineated perineurial tubes (Raimondo *et al.*, 2005) (Fig. 4C).

Results of confocal imaging of Schwann cells and regenerating axons showed that muscle–vein combined grafts were massively colonized by a number of actively proliferating Schwann cells (Fig. 4D) coming from the two nerve stumps starting from the first postoperative days (Fornaro *et al.*, 2001; Geuna *et al.*, 2003; Raimondo *et al.*, 2005) while axon regeneration can be clearly detected inside the graft only at week 2 postoperatively (Fig. 4E and F).

Assessment of functional performance measured by grasping test (Fig. 4G) showed a statistical trend towards a slightly lower that recovery of median nerve function was in muscle–vein-combined guides in comparison to autografts. The same was true for the stereological estimates of regenerated myelinated nerve fibers (Fig. 4H–L) that showed that nerve grafts, in this experimental model, performed slightly better than muscle–vein-combined conduits.

Interestingly, this type of autologous tissue-engineered scaffolds can be easily adapted to perform simultaneous repair of two distal nerve stumps using only one single proximal stump by means Y-shaped muscle–vein-combined conduits (Tos *et al.*, 2000) that, in the rat, permit to obtain a partial recovery of voluntary control of the motor function of both nerves (Lee *et al.*, 2007; Tos *et al.*, 2004).

Moreover, we have also explored the molecular mechanisms at the basis of the effectiveness of this tissue engineering approach for peripheral nerve reconstruction focusing in particular on the gliotrophic system based on NRG1/ErbB signaling (Hoke, 2006). NRG1 is the consensus name for a group of molecules encoded by NRG1 gene. From the single NRG1 gene, various isoforms of the protein are produced, mRNA that encode the isoforms are transcribed by several promoters and alternative splicing contributes to their hererogeneity (Britsch, 2007). Some of these isoforms are known to be strongly involved in the regulation of myelination in the PNS (Chen *et al.*, 2006; Nave and Salzer, 2006); other isoforms promote the gliogenic fate of neural crest cells, the migration of Schwann cells precursors along axons and their subsequent proliferation and survival

delineated perineurial tubes. (D) Confocal microscopy showing muscle–vein combined scaffold massively colonized by a number of actively proliferating Schwann cells (green: PCNA, red: GFAP). (E) Confocal microscopy showing regenerating axons inside the scaffold (green: Neurofilament) guided by Schwann cells (red: S100). (F) Confocal microscopy of a longitudinal section of the scaffold showing regenerating axons (green: neurofilament) along muscle fibers. (G) Comparison of functional recovery (by means grasping test) after autograft and muscle–vein-combined nerve gap reconstruction. (H, I, L) Stereological analysis of regenerated myelinated nerve fibers after 6 months of surgery comparing autograft and muscle–vein-combined scaffold. Magnifications: A = 8000×; B = 10,000×; C = 600×; D = 200×; E, F = 400×. (See Color Insert.)

induced by axons (Birchmeier, 2009; Birchmeier and Nave, 2008). Due to the pivotal role exerted by NRG1, through the interaction with the ErbB tyrosine receptor family (Britsch, 2007), in Schwann cells development and functions, we have studied how it can affect Schwann cells behavior during nerve regeneration. Results showed that, during early postoperative times, only the nonmitogenic isoform α_{2a-2b} of NRG1 (Oka *et al.*, 2000; Raabe *et al.*, 1996; Wen *et al.*, 1994) is overexpressed inside muscle–vein combined tubes (Nicolino *et al.*, 2003). Similar results were obtained in denervated skeletal muscles where expression of the NRG1/ErbB system is detectable at low levels in normal skeletal muscle (Suárez *et al.*, 2001) and increases after muscle denervation (Nicolino *et al.*, 2009), suggesting that skeletal muscle fibers and Schwann cells share a common autocrine trophic loop that is overactivated in the case of loss of contact with axons. Interference between the two (muscle and glial) autotrophic loops can be one of the mechanisms for explaining the effectiveness of the muscle–vein-combined technique for nerve tissue engineering (Geuna *et al.*, 2007; Nicolino *et al.*, 2009).

It should be noted that fresh muscle enrichment is particularly effective when vein conduits are used while in combination with of synthetic conduits its effectiveness is reduced (Varejão *et al.*, 2003a,b). Finally, some evidence was obtained about the involvement of NRG1/ErbB system also in different types of peripheral nerve injury. In fact, ErbB receptor mRNA expression is modulated in the early phases of nerve regeneration after end-to-end and end-to-side coaptation (Audisio *et al.*, 2008). These results rise the perspective that the manipulation of this gliotrophic system by means of gene transfer in the site of injury, can be useful for improving nerve regeneration and functional recovery after nerve lesion both with and without substance loss.

IV. Material Science—Biomaterials for Nerve Reconstruction

Concurrently with the attempts to use autologous tissues for engineering the damaged peripheral nerves, many efforts have also been spend to explore the use of biomaterials as substitutes of tissues. The use of materials for nerve reconstruction has a lengthy history started at the beginning of the twentieth century (Payr, 1900) and many attempts to use various nonbiological materials, such as metals, permeable cellulose esters, gelatine tubes, rubber, plastics, etc., were carried out (reviewed in Fields *et al.*, 1989).

Similar to what has happened for biological transplants, the last 30 years saw an impressive increase of experimental studies aimed at testing new biomaterials for nerve regeneration, especially for tubulization reconstruction of nerve gaps (Battiston *et al.*, 2005; Chiono *et al.*, 2009, this issue; de Ruiter *et al.*, 2009; Pfister *et al.*, 2007; Schmidt and Leach, 2003). The results have been in general very successful and some authors even proposed the use of tubes as an alternative to direct nerve sutures for lesions without substance loss (Dahlin and Lundborg, 2001). The underlying concept is that intentionally leaving a short gap between the two nerve stumps enhances the accumulation of cells and extracellular matrix, which can stimulate correct axonal regrowth, and long-term follow-up clinical examination of patients suggested that silicone tubular repair of median and ulnar nerves was at least as good as end-to-end neurorrhaphy (Dahlin and Lundborg, 2001; Dahlin *et al.*, 2001; Lundborg *et al.*, 2004).

While both absorbable and nonabsorbable synthetic materials have been investigated for nerve regeneration, the concerns about occurrence of complications due to local fibrosis and nerve compression in case of the latter approach (Dahlin *et al.*, 2001; Merle *et al.*, 1989), focused much interest on bioabsorbable tubes. Many experimental studies have shown that their effectiveness is similar to traditional nerve autografts and sometimes even superior to them (Dellon *et al.*, 1988; Den Dunnen *et al.*, 1995, 1996; Mackinnon and Dellon, 1990a,b; Meek, 2000; Meek *et al.*, 1999; Navarro *et al.*, 1996; Nicoli-Aldini *et al.*, 1996; Robinson *et al.*, 1991; Valero-Cabre *et al.*, 2001; Yannas and Hill, 2004; Young *et al.*, 2002). In particular, tubes made of polyglycolic acid (PGA) have been successfully used both experimentally and clinically for bridging nerve defects (Dellon and Mackinnon, 1988; Mackinnon and Dellon, 1990a; Molander *et al.*, 1983; Reid *et al.*, 1978). PGA tubes were shown to support successfully regeneration across 3 cm defects in the ulnar nerves of monkeys (Dellon and Mackinnon, 1988). In other experiments in monkeys, also glycolide trimethylene carbonate (Maxon) conduits were used to bridge 2 cm nerve gaps with good results (Mackinnon and Dellon, 1990b). These conduits supported some regeneration even across defects as long as 5 cm. Navarro and coworkers have performed extensive studies on the use of nerve guides in mice with special reference to tube material and tube contents (Butí *et al.*, 1996; Navarro *et al.*, 1994, 1996; Rodríguez *et al.*, 1999, 2000; Valero-Cabre *et al.*, 2001). They used a mouse sciatic nerve model in which various types of tubular conduit were used to bridge a nerve gap of 6 mm. They found resorbable guides made of collagen or polylactate caprolactone (PLC) superior to nonresorbable guides such as silicone, Teflon, or Polysylfone. Cultured Schwann cells suspended in Matrigel, introduced in the tubes, were important factors contributing to successful nerve regeneration (Rodriguez *et al.*, 2000).

Collagen tubes have also been tested for bridging nerve defects in mice (Navarro *et al.*, 1996), rabbits (Kim *et al.*, 1993), and primates (Archibald *et al.*, 1995; Li *et al.*, 1992; Mackinnon and Dellon, 1990b). Krarup *et al.* performed extensive studies on factors that influence peripheral nerve regeneration through collagen tubes in monkeys (Krarup *et al.*, 2002). Nerve gap distances of various lengths were repaired with collagen-based nerve guides, and extensive neurophysiological investigations were performed postoperatively over a period of 3–4 years.

It was found that nerve gap distance and the type of repair strongly influenced the time to the earliest muscle reinnervation. Nerve gaps up to 5 cm were successfully bridged by the collagen-based nerve guide tube (Krarup *et al.*, 2002).

Description and critical evaluation of the various types of biomaterials currently proposed for nerve regeneration is also addressed in several recent reviews (Chiono *et al.*, 2009, this issue; Li and Hoffman-Kim, 2008; Pfister *et al.*, 2007), and it is beyond the aims of this article to review that body of literature. Within this wide and polymorphic offer, we would like to focus on chitosan, a partially deacetylated polymer of acetyl glucosamine obtained after the alkaline deacetylation of chitin (Senel and McClure, 2004), that has recently attracted much attention because of its biocompatibility, biodegradability, low toxicity, low cost, enhancement of wound-healing, and antibacterial effects (Amado *et al.*, 2008). The effectiveness of chitosan for promoting in nerve regeneration has been demonstrated both *in vitro* and *in vivo* (Freier, 2005). Chitosan and chitosan-based materials have been proven to promote adhesion, survival, and neurite outgrowth of neurons (Shirosaki *et al.*, 2005). The potential usefulness of chitosan in nerve regeneration has been demonstrated both *in vitro* and *in vivo* (Chandy and Sharma, 1990; Freier *et al.*, 2005; Ishikawa *et al.*, 2007; Yamaguchi *et al.*, 2003; Yuan *et al.*, 2004) and we have recently published the results of a study in which we have shown that enwrapping with type-III chitosan membranes, characterized by a highly porous microstructure, improve nerve regeneration after contusion sciatic nerve injury, representing also a good substrate for the local delivery of stem cell therapy (Amado *et al.*, 2008).

V. Gene Transfer

Biotechnological progress that today makes it possible to induce therapeutic changes through gene transfer represents one of the pillars of tissue engineering and has engendered much excitement and opens great perspective in many disciplines of biomedicine (Giacca, 2007) including improvement nerve regeneration (Haastert and Grothe, 2007; Zacchigna and Giacca, 2009, this issue). Gene therapy may contribute to stimulate regeneration of the peripheral nerve by locally supplying several neurotrophic factors the efficacy of which in case of exogenous application, is limited because of their fast degradation. Yet, systemic application of trophic factors can have side effects that are reduced if they are produced locally. The use of viral vectors provides a high rate of transduction and expression and the recent development of nontoxic, nonimmunogenic viral vectors that drive local, long-term transgene expression, makes their use much more safe.

We have recently focused our attention on viral vectors based on the adeno-associated virus (AAV), a nonpathogenic and widespread parvovirus, incapable of autonomous replication and able to transduce both dividing and nondividing cells and show a specific tropism for postmitotic cells, including skeletal and cardiac muscle (Su *et al.*, 2000), neurons (Kaplitt *et al.*, 1994), and liver (Xiao *et al.*, 1998). Because these vectors do not contain any viral genes—which are transiently transfected in trans for the packaging process—they elicit virtually no inflammatory or immune response (Kay *et al.*, 2000). As a consequence, transgene expression from these vectors persists for several months in a variety of animal tissues *in vivo* (Monahan and Samulski, 2000). Vectors based on AAV have recently been used in phase I clinical trials for the treatment of neurological disorders, such as Parkinson's and Canavan's diseases. Indeed, AAV-mediated gene transfer is a promising tool for the delivery of therapeutic gene into the central and PNSs.

In depth description of the perspectives of application of gene therapy to nerve tissue engineering can be found in another paper of this special issue on peripheral nerve regeneration of the *International Review of Neurobiology* (Zacchigna and Giacca, 2009, this issue). It should just be noted that the high effectiveness of skeletal muscle infection by AAVs makes it possible to foresee that this tissue can be the vehicle for transferring genes that can improve nerve regeneration either by infecting the muscles that surround the nerve lesion site, or even by creating muscle–vein-combined scaffolds (see Section III.1) previously potentiated by AAV gene transfer.

VI. Clinical Experience

Today, the use of artificial nerve conduits is limited to nerve gaps up to 30 mm. Results deteriorate with extended gap length. For certain indications, nerve conduits have become a useful tool to avoid donor site morbidity associated with autologous nerve grafting. However, published data on this technique are still limited, and further studies are needed to explore all assets and drawbacks. Artificial nerve conduits can be constructed to maintain the appropriate degree of mechanical strength to optimize the time needed for resorption and to strictly avoid any donor site morbidity. Biological tubes are less expensive and seem to give similar results.

A. Polymeric Scaffolds

Tubes consisting of various biodegradable materials, such as PGA have been successfully used for bridging nerve defects in patients (Mackinnon and Dellon, 1990b). PGA tubes were successfully used to bridge digital nerve gaps of 0.5–3.0 cm (mean 1.7 cm). A randomized prospective multicenter study on the use of

PGA conduits for human digital nerve construction was performed by Weber *et al.* (1999). PGA conduits were found to lead to similar functional outcome, in comparison to nerve grafts and end-to-end repair, for nerve gaps of 4 mm or less. One year later, Casanas reported on 17 secondary reconstructed nerves with subjective sensory improvement in every case.

Nonresorbable conduits made of silicone can lead to secondary nerve compression and usually prevent nutrient diffusion into the lumen. Experimental studies displayed worse results compared with degradable materials, and the clinical application may lead to nerve compression that sometimes required secondary conduit removal (Braga-Silva, 1999; Lundborg *et al.*, 2004; Merle *et al.*, 1989).

In Lundborg and coworkers' clinical studies, silicone tubes were used to give a chance, in primary nerve repair, to self orientation of regenerating axons (chemiotropism) for repair of mixed nerves in the human forearm. In a randomized prospective clinical study, comparison of the clinical outcome was carried out between routine microsurgical repair and a silicone conduit repair where the nerve ends were intentionally left 3–4 mm apart inside the tube. At 12-month follow-up, there were no differences in the outcome of both sensory and motor function. At 5-year follow-up, the functional outcome was not significantly different in the two groups although the conduit repair group showed less cold intolerance (Lundborg, 2003).

B. Biomimetic and Biological Scaffolds

Lohmeyer and coworkers employed Collagen I conduits (NeuraGen; Integra LifeSciences, Plainsboro, NJ) in 12 patients with good results: 4 out of 12 patients, assessed 12 months postoperatively, showed excellent sensibility with s2PD of 7 mm (S4). Five patients achieved good sensibility (s2PD 15 mm, S3þ), one poor (s2PD > 15 mm, S2), and two no sensibility (Lohmeyer *et al.*, 2009).

Basal membranes can also represent a good matrix for guiding nerve fiber regeneration. In clinical trials, coaxial autografts of skeletal muscle which had been frozen then thawed has been used to repair injured digital nerves in eight patients. Assessment from 3 to 11 months after surgery showed an excellent clinical outcome, namely recovery to MRC (British Medical Research Council) sensory category S3 + in all but one patient (Norris *et al.*, 1988). When used to bridge mixed nerves, results of this technique were not so encouraging: on a total of 13 nerves (six median and seven ulnar) five patients achieved an S3 + level of sensory recovery. Two underwent revision of their muscle grafts to nerve grafts and motor recovery was in general poor (Calder and Norris, 1993).

Autologous organs such as veins, alone and in combination with muscle inlays, have also been assessed clinically. Unfortunately, the tendency to collapse is high. Yet, scarring of the surrounding tissue might subsequently prevent the vein to expand later when the nerve growth cone is passing the nerve gap (Tang *et al.*, 1995). Veins have been used successfully in patients for bridging nerve gaps less than 3 cm long (Chiu and Strauch, 1990). Risitano reported a retrospective study on 22 sensory nerves repaired using vein grafts in cases in which primary suture was not feasible, in emergency hand reconstruction. Results were classified according to Sakellarides and 20/22 were classified as very good or good. In addition, also in the cases classified as poor, patients were satisfied and no secondary nerve grafting has been carried out (Risitano *et al.*, 2002). An interesting strategy that has been proposed for avoiding vein collapse is filling the vein lumen with small pieces of nerve tissue (Terzis and Kostas, 2007).

C. Muscle–Vein-Combined Scaffolds

On the basis of previously reported experimental evidence, muscle–vein-combined conduits have been used in the clinical practice to fill gap up to 6 cm (mean 2.5) both for sensory and mixed nerves (Battiston *et al.*, 2000a, 2005). This technique was applied for bridging both sensory and mixed nerve defects (21 cases). We reported good results in 85% of our cases with a minimum follow-up of 14 months. These results seem to be superior to those reported with other kinds of artificial or biological conduits. The conduits are cost free and prepared according to reconstructive needs after consideration of nerve size and length defect. In a second review of our case series, we reported 47 patients treated: 13 patients had sensitive nerve lesions, 9 had motor nerve lesions, and 25 had mixed nerve lesions. Indications were very restricted (as shown by the little series of patients who undergone this type of surgery over 12-year lapse of time): (i) treatment in emergency (especially in case of crush lesions), (ii) not enough nerve graft to cover large gaps (brachial plexus surgery), (iii) no will of the patient in harvesting a healthy sensory nerve. For mixed nerves, good or very good recovery was obtained in 52% of patient, while 13% had unsatisfactory recovery, according to the grading system proposed by Sakellarides. A partial recovery was obtained in 35% of patients with mixed nerves lesions: in these patients only the motor or the sensory nerve fibers had a good recovery. For pure motor or sensory nerves, only 10% of patients had an unsatisfactory outcome (Battiston *et al.*, 2008).

As general rule, an open nerve injury should be early treated and repaired directly when optimal conditions such as (a) a clean uncontaminated wound and (b) a sharp cut injury are present (Dvali and Mackinnon, 2003). Primary nerve repair is not recommended in crush injuries and when a soft tissue damage is present because the extension of nerve damage beyond the gap is not directly evaluable. In these circumstances, the ends of the damaged nerve should be identified and sutured to avoid retraction, waiting for a secondary reconstruction when definitive repair or graft can be performed. Though the sacrifice of a donor nerve is not justified for primary repair, an attempt of emergency reconstruction with conduits can be performed in selected cases.

In a last study, we assessed results of nerve reconstruction in a group of 13 patients which underwent in emergency a primary repair for crush injuries of sensory and mixed nerves of the upper limb with muscle–vein combined graft. The segments involved were sensory digital nerves in five cases, and mixed nerves in eight cases (four median nerve and four ulnar nerve). The nerve defect ranged from 0.8 to 4 cm (sensory nerve mean defect 1.2 cm and mixed nerve mean defect 2.5 cm). All the patients which underwent digital nerve repair had favorable results graded as S4 in one case and S3 + in four cases. As for mixed nerve repair, we observed two S4, two S3 +, two S3, one S2, and one S0 sensory recovery. Less favorable results were observed for motor function with one M5, two M4, one M3, two M2, and two M0 recovery.

Tubulization in emergency may restore the continuity of the nerve avoiding secondary nerve graft when a short gap is present. We conclude that an attempt of nerve reconstruction in emergency with muscle–vein combined graft or alternative conduits is justified considering the favorable results that can be expected with these techniques. Furthermore, in case of unsuccessful primary nerve repair, the possibility for secondary reconstruction was not precluded and the anatomical continuity restored (Tos *et al.*, 2009). On the other hand, nerve graft remains the gold standard in neat lesion or secondary procedures.

VII. Conclusions

Tissue engineering of damaged peripheral nerve is a challenging branch of regenerative medicine with a very high potential clinical impact. Although the use of tissue engineering for nerve repair must proceed with caution because of possible interference with the normal mechanism of axonal regeneration, most experimental studies proved that this approach can lead to very good results. At least two considerations arise from the large body of literature. First, various reports demonstrated that the newly devised tissue-engineered conduits can lead to similar, and sometimes better, histomorphometrical and functional results in comparison to autogenous nerve grafts (Archibald *et al.*, 1995; Den Dunnen *et al.*, 1996; Meek *et al.*, 1999; Tountas *et al.*, 1993; Weber *et al.*, 2000). Second, evidence exists that the critical maximum length limit (3 cm), as defined in the recent past, can be overcome by some new types of tissue-engineered conduit (Battiston, 2000; Geuna *et al.*, 2004; Matsumoto *et al.*, 2000; Strauch *et al.*, 2001; Zhang *et al.*, 2002).

From a more general point of view, it appears that the success of tissue engineering approaches based on single strategies, such as monotissue biological transplants with veins or skeletal muscles, is usually limited (Chiu, 1999) and an interdisciplinary and multitranslational approach which brings together different disciplines needs to be sought. A successful example of interdisciplinary and translational research is represented by the collaboration which brought together experts from various disciplines and permitted to disclose the potentiality, as well as the limits of employment, of the muscle–vein-combined nerve grafts (see Sections III.1 and VI.2). Today, this technique is well established in the clinical practice as well (Battiston *et al.*, 2000a, 2008). The indications are represented, at the moment, by crush injuries with no neat lesions of sensory and mixed nerve, gaps up to 3–4 cm, primary repair when local conditions does not permit an emergency nerve graft (Tos *et al.*, 2009). If regeneration doesn't occur and the anatomy has been restored, a secondary nerve grafting is always possible.

We wish to conclude by recalling a paper by Brunelli *et al.* published in 1994 where these authors outlined the requirements for an "ideal" for a tissue-engineered peripheral nerve which should (1) be compatible with the surrounding tissues; (2) be of adequate size and length; (3) contain substances that enable and support axonal regeneration; and (4) protect regeneration of nerve fibers from scar invasion. The analysis of what have been published since then tells us that those ideal requirements are only partially met by the large variety of tissue-engineered constructs that have been devised so far and, thus, more basic and applied research is definitely needed in this field.

Acknowledgments

This work was supported by grants from the MUR (Italian Ministry of University and Research), ex-60% fund, FIRB fund (code: RBAU01BJ95), PRIN2005 fund (code: 2005057088), the Compagnia di San Paolo (Bando Programma Neuroscienze), and the Regione Piemonte (Progetto Ricerca Sanitaria Finalizzata). Stefania Raimondo is recipient of a PostDoc grant partially supported by the Regione Piemonte (Azione Contenimento del Brain Drain).

References

Amado, S., Simões, M. J., Armada da Silva, P. A., Luís, A. L., Shirosaki, Y., Lopes, M. A., Santos, J. D., Fregnan, F., Gambarotta, G., Raimondo, S., Fornaro, M., Veloso, A. P., *et al.* (2008). Use of hybrid chitosan membranes and N1E-115 cells for promoting nerve regeneration in an axonotmesis rat model. *Biomaterials* **29,** 4409–4419.

Archibald, S. J., Shefner, J., Krarup, C., and Madison, R. D. (1995). Monkey median nerve repaired by nerve graft or collagen nerve guide tube. *J. Neurosci.* **15,** 4109–4123.

Audisio, C., Nicolino, S., Sccvola, A., Tos, P., Geuna, S., Battiston, B., and Perroteau, I. (2008). ErbB receptors modulation in different types of peripheral nerve regeneration. *Neuroreport* **19,** 1605–1609.

Battiston, B., Tos, P., Cushway, T., and Geuna, S. (2000a). Nerve repair by means of vein filled with muscle grafts. I. Clinical results. *Microsurgery* **20,** 32–36.

Battiston, B., Tos, P., Geuna, S., Giacobini-Robecchi, M. G., and Guglielmone, R. (2000b). Nerve repair by means of vein filled with muscle grafts. II. Morphological analysis of regeneration. *Microsurgery* **20,** 37–41.

Battiston, B., Geuna, S., Ferrero, M., and Tos, P. (2005). Nerve repair by means of tubulization: Literature review and personal clinical experience comparing biological and synthetic conduits for sensory nerve repair. *Microsurgery* **25,** 258–267.

Battiston, B., Tos, P., Conforti, L. G., Ciclamini, D., Artiaco, S., and Geuna, S. (2008). Riparazione delle perdite di sostanza nervosa attraverso tubulizzazione biologica con muscolo in vena: Tecnica chirurgica, risultati clinici ed indicazioni. *G.I.O.T. Giornale Italiano di Ortopedia e Traumatologia* **34,** 111–120.

Battiston, B., Papalia, I., Tos, P., and Geuna, S. (2009). Peripheral nerve repair and regeneration research: A historical note. *Int. Rev. Neurobiol.* **87,** 1–8.

Beris, A. E., and Lykissa, M. G. (2009). Experimental results in end-to-side neurorrhaphy. *Int. Rev. Neurobiol.* **87,** 269–280.

Birchmeier, C. (2009). ErbB receptors and the development of the nervous system. *Exp. Cell Res.* **315,** 611–618.

Birchmeier, C., and Nave, K. A. (2008). Neuregulin-1, a key axonal signal that drives Schwann Cell growth and differentiation. *Glia* **56,** 1491–1497.

Bontioti, E., and Dahlin, L. B. (2009). Mechanisms underlying the end-to-side nerve regeneration. *Int. Rev. Neurobiol.* **87,** 251–268.

Braga-Silva, J. (1999). The use of silicone tubing in the late repair of the median and ulnar nerves in the forearm. *J. Hand Surg. (Br.)* **24,** 703–706.

Britsch, S. (2007). The Neuregulin-1/ErbB signaling system in development and disease. *Adv. Anat. Embryol. Cell. Biol.* **190,** 1–65.

Brunelli, G. A., Battiston, B., Vigasio, A., Brunelli, G., and Marocolo, D. (1993). Bridging nerve defects with combined skeletal muscle and vein conduits. *Microsurgery* **14,** 247–251.

Bungner, O. V. (1891). Ueber die degenerations und regenerations—Vorgange am Nerven nach Verletzungen. *Beitr. Pathol. Anal.* **10,** 321.

Butí, M., Verdú, E., Labrador, R. O., Vilches, J. J., Forés, J., and Navarro, X. (1996). Influence of physical parameters of nerve chambers on peripheral nerve regeneration and reinnervation. *Exp. Neurol.* **137,** 26–33.

Calder, J. S., and Norris, R. W. (1993). Repair of mixed peripheral nerves using muscle autografts: A preliminary communication. *Br. J. Plast. Surg.* **46,** 557–564.

Casha, S., Yong, V. W., and Midha, R. (2008). Minocycline for axonal regeneration after nerve injury: A double-edged sword. *Exp. Neurol.* **213,** 245–248.

Chandy, T., and Sharma, C. P. (1990). Chitosan—As a biomaterial. *Biomater. Artif. Cells Artif. Organs* **18,** 1–24.

Chen, S., Velardez, M. O., Warot, X., Yu, Z. Y., Miller, S. J., Cros, D., and Corfas, G. (2006). Neuregulin1-erbB signaling is necessary for normal myelination and sensory function. *J. Neurosci.* **26,** 3079–3086.

Chiono, V., Tonda-Turo, C., and Ciardelli, G. (2009). Artificial scaffolds for peripheral nerve reconstruction. *Int. Rev. Neurobiol.* **87,** 173–198.

Chiu, D. T. (1999). Autogenous venous nerve conduits. A review. *Hand Clin.* **15,** 667–671.

Chiu, D. T., and Strauch, B. (1990). A prospective clinical evaluation of autogenous vein grafts used as a nerve conduit for distal sensory nerve defects of 3 cm or less. *Plast. Reconstr. Surg.* **86,** 928–934.

Chiu, D. T., Janecka, I., Krizek, T. J., Wolff, M., and Lovelace, R. E. (1982). Autogenous vein graft as a conduit for nerve regeneration. *Surgery* **91,** 226–233.

Dahlin, L. B., and Lundborg, G. (2001). Use of tubes in peripheral nerve repair. *Neurosurg. Clin. N. Am.* **12,** 341–352.

Dahlin, L. B., Anagnostaki, L., and Lundborg, G. (2001). Tissue response to silicone tubes used to repair human median and ulnar nerves. *Scand. J. Plast. Reconstr. Surg. Hand Surg.* **35,** 29–34.

de Ruiter, G. C., Malessy, M. J., Yaszemski, M. J., Windebank, A. J., and Spinner, R. J. (2009). Designing ideal conduits for peripheral nerve repair. *Neurosurg. Clin. N. Am.* **20,** 91–105.

Dellon, A. L., and Mackinnon, S. E. (1988). An alternative to the classical nerve graft for the management of the short nerve gap. *Plast. Reconstr. Surg.* **82,** 849–856.

den Dunnen, W. F., van der Lei, B., Robinson, P. H., Holwerda, A., Pennings, A. J., and Schakenraad, J. M. (1995). Biological performance of a degradable poly(lactic acid-epsilon-caprolactone) nerve guide: Influence of tube dimensions. *J. Biomed. Mater. Res.* **29,** 757–766.

den Dunnen, W. F., van der Lei, J. M., Schakenraad, J. M., Stokroos, I., Blaaw, E., Bartels, H., Pennings, A. J., and Robinson, P. H. (1996). Poly (DL-lactidee-caprolactone) nerve guides perform better than autologous nerve grafts. *Microsurgery* **17,** 348–357.

Dvali, L., and Mackinnon, S. (2003). Nerve repair, grafting and nerve transfers. *Clin. Plast. Surg.* **30,** 203–221.

Fawcett, J. W., and Keynes, R. J. (1986). Muscle basal lamina: A new graft material for peripheral nerve repair. *J. Neurosurg.* **65,** 354–363.

Ferreira, M. C., Besteiro, J. M., and Tuma Júnior, P. (1994). Results of reconstruction of the facial nerve. *Microsurgery* **15,** 5–8.

Fields, R. D., Le Beau, J. M., Longo, F. M., and Ellisman, M. H. (1989). Nerve regeneration through artificial tubular implants. *Prog. Neurobiol.* **33,** 87–134.

Fornaro, M., Tos, P., Geuna, S., Giacobini-Robecchi, M. G., and Battiston, B. (2001). Confocal imaging of Schwann-cell migration along muscle-vein combined grafts used to bridge nerve defects in the rat. *Microsurgery* **21,** 153–155.

Freier, T., Montenegro, R., Shan Koh, H., and Shoichet, M. S. (2005). Chitin-based tubes for tissue engineering in the nervous system. *Biomaterials* **26,** 4624–4632.

Frostick, S. P., Yin, Q., and Kemp, G. J. (1998). Schwann cells, neurotrophic factors, and peripheral nerve regeneration. *Microsurgery* **18,** 397–405.

Geuna, S., Tos, P., Battiston, B., Guglielmone, R., and Giacobini-Robecchi, M. G. (2000). Morphological analysis of peripheral nerve regenerated by means of vein grafts filled with fresh skeletal muscle. *Anat. Embryol.* **201,** 475–482.

Geuna, S., Borrione, P., Fornaro, M., and Giacobini-Robecchi, M. G. (2001). Adult stem cells and neurogenesis: Historical roots and state of the art. *Anat. Rec.* **265,** 132–141.

Geuna, S., Raimondo, S., Nicolino, S., Boux, E., Fornaro, M., Tos, P., Battiston, B., and Perroteau, I. (2003). Schwann-cell proliferation in muscle–vein combined conduits for bridging rat sciatic nerve defects. *J. Reconstr. Microsurg.* **19,** 119–123.

Geuna, S., Tos, P., Battiston, B., and Giacobini-Robecchi, M. G. (2004). Bridging peripheral nerve defects with muscle-vein combined guides. *Neurol. Res.* **26,** 139–144.

Geuna, S., Nicolino, S., Raimondo, S., Gambarotta, G., Battiston, B., Tos, P., and Perroteau, I. (2007). Nerve regeneration along bioengineered scaffolds. *Microsurgery* **27,** 429–438.

Geuna, S., Raimondo, S., Ronchi, G., Di Scipio, F., Tos, P., and Fornaro, M. (2009). Histology of the peripheral nerve and changes occurring during nerve regeneration. *Int. Rev. Neurobiol.* **87,** 27–46.

Giacca, M. (2007). Virus-mediated gene transfer to induce therapeutic angiogenesis: Where do we stand? *Int. J. Nanomed.* **2,** 527–540.

Glasby, M. A., Gschmeissner, S. G., Hitchcock, R. J. I., Huang, C. L. H., and de Souza, B. A. (1986). A comparison of nerve regeneration through nerve and muscle grafts in rat sciatic nerve. *Neuroorthopaedics* **2,** 21–28.

Gordon, T., Sulaiman, O., and Boyd, J. G. (2003). Experimental strategies to promote functional recovery after peripheral nerve injuries. *J. Peripher. Nerv. Syst.* **8,** 236–250.

Haastert, K., and Grothe, C. (2007). Gene therapy in peripheral nerve reconstruction approaches. *Curr. Gene Ther.* **7,** 221–228.

Hall, S. (2005). The response to injury in the peripheral nervous system. *J. Bone Joint Surg. Br.* **87,** 1309–1319.

Höke, A. (2006). Neuroprotection in the peripheral nervous system: Rationale for more effective therapies. *Arch. Neurol.* **63,** 1681–1685.

Ide, C. (1996). Peripheral nerve regeneration. *Neurosci. Res.* **25,** 101–121.

Ishikawa, N., Suzuki, Y., Ohta, M., Cho, H., Suzuki, S., Dezawa, M., and Ide, C. (2007). Peripheral nerve regeneration through the space formed by a chitosan gel sponge. *J. Biomed. Mater. Res. A* **83,** 33–40.

Kaplitt, M. G., Leone, P., Samulski, R. J., Xiao, X., Pfaff, D. W., O'Malley, K. L., and During, M. J. (1994). Long-term gene expression and phenotypic correction using adeno-associated virus vectors in the mammalian brain. *Nat. Genet.* **8,** 148–154.

Kay, M. A., Manno, C. S., Ragni, M. V., Larson, P. J., Couto, L. B., McClelland, A., Glader, B., Chew, A. J., Tai, S. J., Herzog, R. W., Arruda, V., Johnson, F., *et al.* (2000). Evidence for gene transfer and expression of factor IX in haemophilia B patients treated with an AAV vector. *Nat. Genet.* **24,** 257–261.

Keynes, R. J., Hopkins, W. G., and Huang, L. H. (1984). Regeneration of mouse peripheral nerves in degenerating skeletal muscle: Guidance by residual muscle fibre basement membrane. *Brain Res.* **295,** 275–281.

Kim, D. H., Connolly, S. E., Zhao, S., Beuerman, R. W., Voorhies, R. M., and Kline, D. G. (1993). Comparison of macropore, semipermeable, and nonpermeable collagen conduits in nerve repair. *J. Reconstr. Microsurg.* **9,** 415–420.

Kingham, P. J., Kalbermatten, D. F., Mahay, D., Armstrong, S. J., Wiberg, M., and Terenghi, G. (2007). Adipose-derived stem cells differentiate into a Schwann cell phenotype and promote neurite outgrowth *in vitro*. *Exp. Neurol.* **207,** 267–274.

Kong, J. M., Zhong, S. Z., Bo, S., and Zhu, S. X. (1986). Experimental study of bridging the peripheral nerve gap with skeletal muscle. *Microsurgery* **7,** 183–189.

Krarup, C., Archibald, S. J., and Madison, R. D. (2002). Factors that influence peripheral nerve regeneration: An electrophysiological study of the monkey median nerve. *Ann. Neurol.* **51,** 69–81.

Kraus, H., and Reisner, H. (1940). Behandlungsergenisse von Verletungen pripherer Nerven mit besonderer Beru¨ cksichtigung der Schussverletzungen delJahre 1919, 1927, und 1934. *Arch. Klin. Chir.* **199,** 318–336.

Leach, J. B., and Schmidt, C. E. (2005). Characterization of protein release from photocrosslinkable hyaluronic acid-polyethylene glycol hydrogel tissue engineering scaffolds. *Biomaterials* **26,** 125–135.

Lee, J. M., Tos, P., Raimondo, S., Fornaro, M., Papalia, I., Geuna, S., and Giacobini-Robecchi, M. G. (2007). Lack of topographic specificity in nerve fiber regeneration of rat forelimb mixed nerves. *Neuroscience* **144,** 985–990.

Li, G. N., and Hoffman-Kim, D. (2008). Tissue-engineered platforms of axon guidance. *Tissue Eng. Part B Rev.* **14,** 33–51.

Li, S. T., Archibald, S. J., Krarup, C., and Madison, R. D. (1992). Peripheral nerve repair with collagen conduits. *Clin. Mater.* **9,** 195–200.

Lohmeyer, J. A., Siemers, F., Machens, H. G., and Mailänder, P. (2009). The clinical use of artificial nerve conduits for digital nerve repair: A prospective cohort study and literature review. *J. Reconstr. Microsurg.* **25,** 55–61.

Luis, A. L., Rodrigues, J. M., Lobato, J. V., Lopes, M. A., Amado, S., Veloso, A. P., Armada-da-Silva, P. A., Raimondo, S., Geuna, S., Ferreira, A. J., Varejão, A. S., Santos, J. D., *et al.* (2007).

Evaluation of two biodegradable nerve guides for the reconstruction of the rat sciatic nerve. *Biomed. Mater. Eng.* **17,** 39–52.

Lundborg, G. (2002). Enhancing posttraumatic nerve regeneration. *J. Peripher. Nerv. Syst.* **7,** 139–140.

Lundborg, G. (2003). Bridging nerve defects: The role of tissue interpositioning. *In* "Severe Traumatic Defects of the Upper Limb" (A. C. Masquelet, and A. C. Ferreira, Eds.), Martin Dunitz, London.

Lundborg, G., Rosen, B., Dahlin, L., Holmberg, J., and Rosen, I. (2004). Tubular repair of the median or ulnar nerve in the human forearm: A 5-year follow-up. *J. Hand Surg. (Br.)* **29,** 100–107.

Mackinnon, S., and Dellon, A. L. (1990a). Clinical nerve reconstruction with a bioabsorbable polyglycolic acid tube. *Plast. Reconstr. Surg.* **85,** 419–422.

Mackinnon, S., and Dellon, A. L. (1990b). A study of nerve regeneration across synthetic (maxon) and biologic (collagen) nerve conduits for nerve gaps up to 5 cm in the primate. *J. Reconstr. Microsurg.* **6,** 117–121.

Matsumoto, K., Ohnishi, K., Kiyotani, T., Sekine, T., Ueda, H., Nakamura, T., Endo, K., and Shimizu, Y. (2000). Peripheral nerve regeneration across an 80-mm gap bridged by a polyglycolic acid (PGA)-collagen tube filled with laminin-coated collagen fibers: A histological and electrophysiological evaluation of regenerated nerves. *Brain. Res.* **868,** 315–328.

Meek, M. F. (2000). Artificial nerve guides-assessment of nerve function. Dissertation. University of Groningen, Groningen, The Netherlands. ISBN: 90–3671303-x.

Meek, M. F., and Coert, J. H. (2002). Clinical use of nerve conduits in peripheralnerve repair: Review of the literature. *J. Reconstr. Microsurg.* **18,** 97–109.

Meek, M. F., Dijkstra, J. R., Den Dunnen, W. F. A., Ijkema-Paassen, J., Schakenraad, J. M., Gramsbergen, A., and Robinson, P. H. (1999). Functional assessment of sciatic nerve reconstruction: Biodegradable poly(DLLA-e-CL) nerve guides versus autologous nerve grafts. *Microsurgery* **19,** 381–388.

Merle, M., Dellon, A. L., Campbell, J. N., and Chang, P. S. (1989). Complications from silicone polymer intubulation of nerves. *Microsurgery* **10,** 130–133.

Midha, R. (2006). Emerging techniques for nerve repair: Nerve transfers and nerve guidance tubes. *Clin. Neurosurg.* **53,** 185–190.

Millesi, H. (1970). Interfascicular nerve grafting. *Orthop. Clin. N. Am.* **2,** 419–435.

Millesi, H. (2006). Factors affecting the outcome of peripheral nerve surgery. *Microsurgery* **26,** 295–302.

Millesi, H., Meissl, G., and Berger, A. (1972). The interfascicular nerve-grafting of the median and ulnar nerves. *J. Bone Joint Surg.* **54,** 727–750.

Millesi, H., Meissl, G., and Berger, A. (1976). Further experience with interfascicular grafting of the median, ulnar, and radial nerves. *J. Bone Joint Surg.* **58,** 209–218.

Mligiliche, N., Tabata, Y., Endoh, K., and Ide, C. (2001). Peripheral nerve regeneration through a long detergent-denatured muscle autografts in rabbits. *Neuroreport* **12,** 1719–1722.

Molander, H., Engkvist, O., Hägglund, J., Olsson, Y., and Torebjörk, E. (1983). Nerve repair using a polyglactin tube and nerve graft: An experimental study in the rabbit. *Biomaterials* **4,** 276–280.

Monahan, P. E., and Samulski, R. J. (2000). AAV vectors: Is clinical success on the horizon? *Gene Ther* **7,** 24–30.

Naff, N. J., and Ecklund, J. M. (2001). History of peripheral nerve surgery techniques. *Neurosurg. Clin. N. Am.* **12,** 197–209.

Navarro, X., Verdú, E., and Butí, M. (1994). Comparison of regenerative and reinnervating capabilities of different functional types of nerve fibers. *Exp. Neurol.* **129,** 217–224.

Navarro, X., Rodriguez, F. J., Labrador, R. O., Buti, M., Ceballos, D., Gomez, N., Cuadras, J., and Perego, G. (1996). Peripheral nerve regeneration through bioresorbable and durable nerve guides. *J. Peripher. Nerv. Syst.* **1,** 53–64.

Nave, K. A., and Salzer, J. L. (2006). Axonal regulation of myelination by neuregulin 1. *Curr. Opin. Neurobiol.* **16,** 492–500.

Nicoli-Aldini, N., Perego, G., Cella, G. D., Maltarello, M. C., Rocca, M., and Giardino, R. (1996). Effectiveness of a bioabsorbable conduit in the repair of peripheral nerves. *Biomaterials* **17,** 959–962.

Nicolino, S., Raimondo, S., Tos, P., Battiston, B., Fornaro, M., Geuna, S., and Perroteau, I. (2003). Expression of alpha2a-2b neuregulin-1 is associated with early peripheral nerve repair along muscle-enriched tubes. *Neuroreport* **14,** 1541–1545.

Nicolino, S., Panetto, A., Raimondo, S., Gambarotta, G., Guzzini, M., Fornaro, M., Battiston, B., Tos, P., Geuna, S., and Perroteau, I. (2009). Denervation and reinnervation of adult skeletal muscle modulate mRNA expression of neuregulin-1 and ErbB receptors. *Microsurgery* [Epub ahead of print].

Norris, R. W., Glasby, M. A., and Gattuso, J. M. (1988). Peripheral nerve repair in humans using muscle autografts: A new technique. *J. Bone Joint Surg.* **70,** 530–533.

Oka, N., Kawasaki, T., Matsui, M., Tachibana, H., Sugita, M., and Akiguchi, I. (2000). Neuregulin is associated with nerve regeneration in axonal neuropathies. *Neuroreport* **11,** 3673–3676.

Papalia, I., Geuna, S., D'Alcontres, F. S., and Tos, P. (2007). Origin and history of end-to-side neurorrhaphy. *Microsurgery* **27,** 56–61.

Payr, E. (1900). Beitrage zur Technik der Blutgefass und Nervennaht nebst Mitteilungen uber die Verwendung eines resorbibaren Metalles in der Chirurgie. *Arch. Klin. Chir.* **62,** 67–111.

Pereira, J. H., Bowden, R. E. M., Gattuso, J. M., and Norris, R. W. (1991a). Comparison of results of repair of digital nerves by denatured muscle grafts and end-to-end sutures. *J. Hand Surg.* **16,** 519–523.

Pereira, J. H., Palande, D. D., Subramanian, A., Narayanakumar, T. S., Curtis, J., and Turk, J. L. (1991b). Denatured autologous muscle graft in leprosy. *Lancet* **338,** 1239–1240.

Pereira, J. H., Bowden, R. E., Narayanakumar, T. S., and Gschmeissner, S. E. (1996). Peripheral nerve reconstruction using denatured muscle autografts for restoring protective sensation in hands and feet of leprosy patients. *Indian J. Lepr.* **68,** 83–91.

Pfister, B. J., Huang, J. H., Kameswaran, N., Zager, E. L., and Smith, D. H. (2007). Neural engineering to produce *in vitro* nerve constructs and neurointerface. *Neurosurgery* **60,** 137–141.

Pogrel, M. A., and Maghen, A. (2001). The use of autogenous vein grafts for inferior alveolar and lingual nerve reconstruction. *J. Oral Maxillofac. Surg.* **59,** 985–988.

Raabe, T. D., Clive, D. R., Neuberger, T. J., Wen, D., and DeVries, G. H. (1996). Cultured neonatal Schwann cells contain and secrete neuregulins. *J. Neurosci. Res.* **46,** 263–270.

Raimondo, S., Nicolino, S., Tos, P., Battiston, B., Giacobini-Robecchi, M. G., Perroteau, I., and Geuna, S. (2005). Schwann cell behavior after nerve repair by means of tissue-engineered muscle–vein combined guides. *J. Comp. Neurol.* **489,** 249–255.

Rath, E. M. (2002). Skeletal muscle autograft for repair of the human inferior alveolar nerve: A case report. *J. Oral Maxillofac. Surg.* **60,** 330–334.

Reid, R. L., Cutright, D. E., and Garrison, J. S. (1978). Biodegradable cuff an adjunct to peripheral nerve repair: A study in dogs. *Hand* **10,** 259–266.

Risitano, G., Cavallaro, G., and Lentini, M. (1989). Autogenous vein and nerve grafts: A comparative study of nerve regeneration in the rat. *J. Hand Surg.* **14,** 102–104.

Risitano, G., Cavallaro, G., Merrino, T., Coppolino, S., and Ruggeri, F. (2002). Clinical results and thoughts on sensory nerve repair by autologous vein graft in emergency hand reconstruction. *Chir. Main.* **21,** 194–197.

Robinson, P. H., van der Lei, B., Hoppen, H. J., Leenslag, J. W., Pennings, A. J., and Nieuwenhuis, P. (1991). Nerve regeneration through a two-ply biodegradable nerve guide in the rat and the influence of ACTH4–9 nerve growth factor. *Microsurgery* **12,** 412–419.

Rodríguez, F. J., Gómez, N., Perego, G., and Navarro, X. (1999). Highly permeable polylactide-caprolactone nerve guides enhance peripheral nerve regeneration through long gaps. *Biomaterials* **20,** 1489–1500.

Rodríguez, F. J., Verdú, E., Ceballos, D., and Navarro, X. (2000). Nerve guides seeded with autologous schwann cells improve nerve regeneration. *Exp. Neurol.* **161,** 571–584.

Ruijs, A. C., Jaquet, J. B., Kalmijn, S., Giele, H., and Hovius, S. E. (2005). Median and ulnar nerve injuries: A meta-analysis of predictors of motor and sensory recovery after modern microsurgical nerve repair. *Plast. Reconstr. Surg.* **116,** 484–494.

Samii, M., Carvalho, G. A., Nikkhah, G., and Penkert, G. (1997). Surgical reconstruction of the musculocutaneous nerve in traumatic brachial plexus injuries. *J. Neurosurg.* **87,** 881–886.

Schmidt, C. E., and Leach J. B. (2003). Neural tissue engineering: Strategies for repair and regeneration. *Annu. Rev. Biomed. Eng.* **5,** 293–347.

Senel, S., and McClure, S. J. (2004). Potential applications of chitosan in veterinary medicine. *Adv. Drug Deliv. Rev.* **56,** 1467–1480.

Shieh, S. J., Lee, J. W., and Chiu, H. Y. (2007). Long-term functional results of primary reconstruction of severe forearm injuries. *J. Plast. Reconstr. Aesthet. Surg.* **60,** 339–348.

Shirosaki, Y., Tsuru, K., Hayakawa, S., Osaka, A., Lopes, M. A., Santos, J. D., and Fernandes, M. H. (2005). *In vitro* cytocompatibility of MG63 cells on chitosan-organosiloxane hybrid membranes. *Biomaterials* **26,** 485–493.

Siemionow, M., and Brzezicki, G. (2009). Current techniques and concepts in peripheral nerve repair. *Int. Rev. Neurobiol* **87,** 141–172.

Stahl, S., and Goldberg, J. A. (1999). The use of vein grafts in upper extremity surgery. *Eur. J. Plast. Surg.* **22,** 255–259.

Strauch, B., Rodriguez, D. M., Diaz, J., Yu, H. L., Kaplan, G., and Weinstein, D. E. (2001). Schwann cells drive regeneration through a 6-cm autogenous venous nerve conduit. *J. Reconstr. Microsurg.* **17,** 589–595.

Su, H., Lu, R., and Kan, Y. W. (2000). Adeno-associated viral vector-mediated vascular endothelial growth factor gene transfer induces neovascular formation in ischemic heart. *Proc. Natl. Acad. Sci. USA* **97,** 13801–13806.

Suárez, E., Bach, D., Cadefau, J., Palacin, M., Zorzano, A., and Gumá, A. (2001). A novel role of neuregulin in skeletal muscle. Neuregulin stimulates glucose uptake, glucose transporter translocation, and transporter expression in muscle cells. *J. Biol. Chem.* **276,** 18257–18264.

Suematsu, N. (1989). Tubulation for peripheral nerve gap: Its history and possibility. *Microsurgery* **10,** 71–74.

Tang, J. B. (1995). Vein conduits with interposition of nerve tissue for peripheral nerve defects. *J. Reconstr. Microsurg.* **11,** 21–26.

Tang, J. B., Shi, D., and Zhou, H. (1995). Vein conduits for repair of nerves with a prolonged gap or in unfavourable conditions: An analysis of three failed cases. *Microsurgery* **16,** 133–137.

Terzis, J. K., and Kostas, I. (2007). Vein grafts used as nerve conduits for obstetrical brachial plexus palsy reconstruction. *Plast. Reconstr. Surg.* **120,** 1930–1941.

Terzis, J. K., Sun, D. D., and Thanos, P. K. (1997). Historical and basic science review: Past, present, and future of nerve repair. *J. Reconstr. Microsurg.* **13,** 215–225.

Tohill, M. P., and Terenghi, G. (2004). Stem cell plasticity and therapy for injuries of the peripheral nervous system. *Biotechnol. Appl. Biochem.* **40,** 17–24.

Tos, P., Battiston, B., Geuna, S., Giacobini-Robecchi, M. G., Hill, M. A., Lanzetta, M., and Owen, E. R. (2000). Tissue specificity in rat peripheral nerve regeneration through combined skeletal muscle and vein conduit grafts. *Microsurgery* **20,** 65–71.

Tos, P., Calcagni, M., Gigo-Benato, D., Boux, E., Geuna, S., and Battiston, B. (2004). Use of muscle–vein-combined Y-chambers for repair of multiple nerve lesions: Experimental results. *Microsurgery* **24,** 459–464.

Tos, P., Battiston, B., Nicolino, S., Raimondo, S., Fornaro, M., Lee, J. M., Chirila, L., Geuna, S., and Perroteau, I. (2007). Comparison of fresh and predegenerated muscle-vein-combined guides for the repair of rat median nerve. *Microsurgery* **27,** 48–55.

Tos, P., Artiaco, S., Titolo, P., Geuna, S., and Battiston, B. (2009). Biological conduits for primary repair of crush nerve injuries. *Injury* submitted for publication.

Tountas, C. P., Bergman, R. A., Lewis, T. W., Stone, H. E., Pyrek, J. D., and Mendenhall, H. V. (1993). A comparison of peripheral nerve repair using an absorbable tubulization device and conventional suture in primates. *J. Appl. Biomater.* **4,** 261–268.

Valero-Cabre, A., Tsironis, K., Skouras, E., Perego, G., Navarro, X., and Neiss, W. F. (2001). Superior muscle reinnervation after autologous nerve graft or poly-L-lactide-epsilon-caprolactone (PLC) tube implantation in comparison to silicone tube repair. *J. Neurosci. Res.* **63,** 214–223.

Varejão, A. S., Cabrita, A. M., Geuna, S., Patrício, J. A., Azevedo, H. R., Ferreira, A. J., and Meek, M. F. (2003a). Functional assessment of sciatic nerve recovery: Biodegradable poly (DLLA-epsilon-CL) nerve guide filled with fresh skeletal muscle. *Microsurgery* **23,** 346–353.

Varejão, A. S., Cabrita, A. M., Meek, M. F., Fornaro, M., and Geuna, S. (2003b). Nerve regeneration inside fresh skeletal muscle-enriched synthetic tubes: A laser confocal microscope study in the rat sciatic nerve model. *Ital. J. Anat. Embryol.* **108,** 77–82.

Walton, R. L., Brown, R. E., Matory, W. E., Jr., Borah, G. L., and Dolph, J. L. (1989). Autogenous vein graft repair of digital nerve defects in the finger: A retrospective clinical study. *Plast. Reconstr. Surg.* **84,** 944–949.

Weber, C. J., Safley, S., Hagler, M., and Kapp, J. (1999). Evaluation of graft-host response for various tissue sources and animal models. *Ann. N. Y. Acad. Sci.* **875,** 233–254.

Weber, R. A., Breidenbach, W. C., Brown, R. E., Jabaley, M. E., and Mass, D. P. (2000). A randomized prospective study of polyglycolic acid conduits for digital nerve reconstruction in humans. *Plast. Reconstr. Surg.* **106,** 1036–1045.

Wen, D., Suggs, S. V., Karunagaran, D., Liu, N., Cupples, R. L., Luo, Y., Janssen, A. M., Ben-Baruch, N., Trollinger, D. B., Jacobsen, V. L., Meng, S. Y., Lu, H. S., *et al.* (1994). Structural and functional aspects of the multiplicity of Neu differentiation factors. *Mol. Cell. Biol.* **14,** 1909–1919.

Wrede, L. (1909). Uberbrueckung eines Nervendefektes mittels Seidennahtund leben Venenstueckes. *Dtsch. Med. Wochenschr.* **35,** 1125–1160.

Xiao, W., Berta, S. C., Lu, M. M., Moscioni, A. D., Tazelaar, J., and Wilson, J. M. (1998). Adeno-associated virus as a vector for liver-directed gene therapy. *J. Virol.* **72,** 10222–10226.

Yamaguchi, I., Itoh, S., Suzuki, M., Osaka, A., and Tanaka, J. (2003). The chitosan prepared from crab tendons: II. The chitosan/apatite composites and their application to nerve regeneration. *Biomaterials* **24,** 3285–3292.

Yannas, V., and Hill, B. J. (2004). Selection of biomaterials for peripheral nerve regeneration using data from the nerve chamber model. *Biomaterials* **25,** 1593–1600.

Young, R. C., Wiberg, M., and Terenghi, G. (2002). Poly-3-hydroxybutyrate (PHB): A resorbable conduit for long-gap repair in peripheral nerves. *Br. J. Plast. Surg.* **55,** 235–240.

Yuan, Y., Zhang, P., Yang, Y., Wang, X., and Gu, X. (2004). The interaction of Schwann cells with chitosan membranes and fibers *in vitro*. *Biomaterials* **25,** 4273–4278.

Zacchigna, S., and Giacca, M. (2009). Gene therapy perspectives for nerve repair. *Int. Rev. Neurobiol* **87,** 381–392.

Zhang, F., Blain, B., Beck, J., Zhang, J., Chen, Z., Chen, Z. W., and Lineaweaver, W. C. (2002). Autogenous venous graft with one-stage prepared Schwann cells as a conduit for repair of long segmental nerve defects. *J. Reconstr. Microsurg.* **18,** 295–300.

MECHANISMS UNDERLYING THE END-TO-SIDE NERVE REGENERATION

Eleana Bontioti* and Lars B. Dahlin†

*Department of Hand Surgery, Evgenidio Hospital, Athens, Greece
†Department of Hand Surgery, Malmö University Hospital, SE-205 02 Malmö, Sweden

End-to-side (ETS) nerve repair is used in selected clinical cases. The mechanisms, by which regeneration into the attached nerve segment is initiated and occur, are still not fully understood. Based on numerous experimental studies, different mechanisms have been suggested by which regenerating axons are recruited, such as contamination from the proximal nerve segment, collateral sprouting, and terminal regenerating sprouting from the donor nerve. A variety of experimental models, most commonly in the lower and upper extremity of rats, and techniques have been used to shed light on the mechanisms. Retrograde labeling techniques have revealed that collateral sprouting do occur, but is probably, at least as observed in long-term experiments, less important over time. Pruning of branching nerve fibers, induced by the collateral sprouting, is an additional mechanism in this context. Experiments have also focused on the stimuli, including the question of epineurial or perineurial windows, that trigger the sprouting of axons form the donor nerve, which can detected by the use of markers of cellular injury. In the present article, we review studies contributing to clarifications of mechanisms of end-to-side nerve repair, including used experimental techniques. We also stress the importance of the plastic brain.

INTERNATIONAL REVIEW OF
NEUROBIOLOGY, VOL. 87
DOI: 10.1016/S0074-7742(09)87012-8

I. Introduction

The severity of upper extremity nerve injuries varies from a simple cut to a digital nerve up to extensive lacerations and avulsions of the roots of the brachial plexus. Conventional surgical methods, which are proved to be effective, are sometimes not applicable to situations where the proximal nerve stump is not available. These situations and the poor functional outcome, which is the major concern for any surgical repair, have prompted the development of alternative nerve repair methods.

One of them—end-to-side nerve repair—was described as early as 1876 by Despres and 1899 by Kennedy. The concept is that the distal stump of an injured nerve is attached in an end-to-side fashion to an intact peripheral nerve trunk—end-to-side nerve repair (ETS nerve repair; also referred as ETS coaptation, ETS neurorrhaphy, and terminolateral neurorrhaphy). For nearly a century, no encouraging results were obtained with the ETS nerve repair. The method was abandoned until 1992 when Viterbo and coworkers demonstrated axons in an ETS attached nerve in a rat model (Viterbo *et al.*, 1992). After "reintroducing" the concept of ETS nerve repair, several researchers have turned their attention to this alternative nerve repair technique.

Still some controversies exist among them about the efficacy or not of ETS nerve repair and the usefulness of this repair in clinical practice (Dvali and Myckatyn, 2008; Pannucci *et al.*, 2007). Authors have shed light to different technical aspects of the method, to morphological issues, and mostly to the functional outcome and its enhancement (Bontioti, 2005; Bontioti *et al.*, 2005, 2006a,b; Hayashi *et al.*, 2008; Kovacic *et al.*, 2007; Lundborg *et al.*, 1994; Rovak *et al.*, 2000; Sanapanich *et al.*, 2002; Tarasidis *et al.*, 1998; Tham and Morrison, 1998; Yamauchi *et al.*, 2000; Yan *et al.*, 2002; Yuksel *et al.*, 1999). The opponents of this technique have mostly focused on functional recovery studies and have, most of them, reabandonded the ETS nerve repair even as an alternative surgical method (Bertelli *et al.*, 1996; Jaberi *et al.*, 2003; Pondaag and Gilbert, 2008). Thus, there are still some controversies and questions that need to be answered. In this paper, we will review experimental papers of ETS nerve repair concerning the biological questions of how axons are recruited into the attached nerve segment and the stimuli that are trigger outgrowth.

II. Proposed Mechanisms and Experimental Techniques

Three neuronal mechanisms have, so far, been proposed to be involved with ETS nerve repair (Fig. 1) leading to reinnervation. (1) Reinnervation by means of terminal sprouting from the proximal stump of the injured nerve, thus reinnervation by *contamination*. (2) Reinnervation by *collateral sprouting*, where a healthy uninjured axon could give an extra branch to arborize the abandoned neuronal

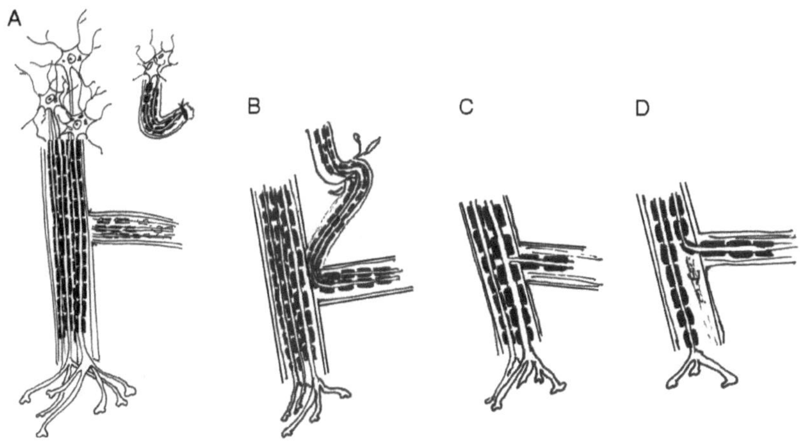

FIG. 1. Schematic drawing of the end-to-side nerve repair (A) and three possible mechanisms by which axons are recruited to the attached distal nerve end (B: contamination; C: collateral sprouting; D: terminal regenerating sprouting).

tube of an injured attached nerve. (3) Reinnervation by elongation of regenerating injured axons, also referred as *terminal/regenerating sprouting*. In addition to the above mechanisms, *pruning* is a phenomenon that should be given attention, as well as the role of the neuronal and brain *plasticity*.

To support one of the above-mentioned mechanisms in different animal models different investigating methods have been used, such as electrophysiological techniques (action potentials, conduction velocity, and/or muscle electrical stimulation results (contractile force)), histological methods (morphometry: measurement of total nerve area, number of axons, thickness of myelin, and signs of degeneration; immunocytochemistry: neurofilaments, S100, Growth Associated Protein 43 (GAP43) staining, etc.), "advanced morphological" techniques (e.g., visualization of growth cone, microtease technique), retrograde labeling of sensory and motor neurons (tracers as fast blue (FB), diamidino yellow (DY), etc.), and expression of different cell substances and signals (e.g., neuropeptides, cytokines, transcription factors, genes, etc.).

Since the donor nerve damage is considered to be of a great importance (Bontioti, 2005; Bontioti *et al.*, 2006a; Dvali and Myckatyn, 2008) and key to the explanation of the mechanisms of ETS nerve repair, different technical aspects have been introduced to the donor nerve, which vary from a traumatic repairs to minimal or major nerve trauma (i.e., silicon tubes, epineurial and perineurial window, and partial donor nerve neurectomy). In addition, the difference between sensory and motor neurons to regenerate into one ETS attached distal nerve segment is also an important issue (Bontioti *et al.*, 2005, 2006a,b). The role of degeneration and of signals elicited from the recipient nerve has also been investigated in the ETS nerve repair.

III. Proximal Stump Contamination

Some authors have concluded that a donor intact nerve serves as a "bridge" for neurons to travel from the proximal to the distal nerve stump (McCallister *et al.*, 2001a,b). They attached both the proximal and the distal ends, with a distance of 10 mm between them, of an injured peroneal nerve to an intact tibial nerve of the rat. Thus, the intact tibial nerve acted as a bridge for any axons growing from the proximal peroneal to the distal peroneal nerve ends. Besides showing functional recovery equivalent to a nerve autograft, the authors showed a larger amount of myelinated axons traveling in the outer epineurium of the tibial nerve. In the same direction Ozbek *et al.* (2005), using the same experimental settings with the only important difference being the creation of an epineurial window to the donor nerve, conclude that regeneration in the distal peroneal nerve was gained through contamination from the proximal stump, while the intact nerve serves as a perfect conduit. Function was observed with gait analysis and electrophysiology. Morphology revealed axons traveling in the inner surface of the epineurium of the bridging intact nerve. Other authors have walked on the same path (Eren *et al.*, 2005; Lykissas *et al.*, 2007).

Efforts have been made to rule out contamination from the proximal nerve stump in a number of studies by ligating, curving back, and burying it to the adjacent musculature (Bontioti *et al.*, 2005; Tarasidis *et al.*, 1998; Viterbo *et al.*, 1994; Yan *et al.*, 2002). Others have used silicon tubes to envelope the coaptation site and thus insulating it from "parasitic" outgrowth (Akeda *et al.*, 2006; Gavazzi, 1995; Yamauchi *et al.*, 2000) or applied a silicon tube or glue at the tip of the proximal stump thereby eluding elongation of growth cones from the cut end of the injured nerve (Goheen-Robillard *et al.*, 2002). Most of these studies have shown no evidence of nerve contamination. The contralateral intact nerve has also been used as a donor with the same results, even though the long cross-leg grafts used alter the results on functional outcome due to more complex regenerating pathways (Lundborg *et al.*, 1994; Rovak *et al.*, 2000). To summarize, most studies show that contamination from the injured proximal end of a recipient nerve is not the main source of regenerating axons in the ETS nerve repair.

IV. Collateral Sprouting

Collateral sprouting, a mechanism well-known in the CNS and the synaptic terminals on end-organs, was thought to be the main biological explanation of the efficacy of ETS nerve repair. As early as 1928, Ramon y Cajal stated that sensory axons develop collateral branches even in the absence of any detectable traumatic stimulus. Early evidence of collateral sprouting by means of ElectroMyoGraphic (EMG) findings in one patient was published by Esslen (1960), where the author concluded that "electrophysiographic analysis of associated movements after peripheral paresis reveals a rarely occurring type of misdirection, where two or more sprouts of one axon are connected to different muscle." Hayashi *et al.* (2004, 2008) showed collateral sprouting by means of electrophysiological measurements and histological evaluation; the latter technique showed regenerating axons directly at the coaptation site. Jiang *et al.* (2007) have provided evidence that one axon can regenerate and maintain up to 3–4 collaterals.

Retrograde labeling techniques have particularly been used to show signs of collateral sprouting in other models. Tarasidis *et al.* (1998), also ruled out the possibility of proximal stump contamination, concluded that induction of collateral sprouting occurred even if only poor sensory reinnervation can be expected. Visualization of the growth cone, penetrating the neural sheath from inside to outside, using an anti-GAP43 antibody, was made by Yamauchi *et al.* (2000), indicating collateral sprouting phenomenon.

Following elegant studies with noninjured ETS nerve repair coaptation devices (Y shaped silicon tubes), Matsumoto *et al.* (1999) as well as Hayashi *et al.* (2004) demonstrated regeneration into the recipient nerve without any type of epineurial damage to the donor nerve, thus providing support to the collateral sprouting theory.

Using a microtease technique, Zhu *et al.* (2008) successfully removed the connective tissue from a nerve trunk and separated the nerve fibers. They were able to follow a single sprout growing from the donor nerve at the node of Ranvier and entering the ETS attached nerve segment. Other researchers base their assumption of collateral sprouting only by histomorphometric nerve analyses (Bajrovic *et al.*, 2002; De Sa *et al.*, 2004; Giovanoli *et al.*, 2000; Goheen-Robillard *et al.*, 2002; Jaberi *et al.*, 2003).

Most authors though, agree that the best evidence of collateral sprouting could be given only by addressing the issue with double retrograde labeling techniques. Several studies have used different neuronal tracers for this purpose (Bontioti *et al.*, 2005; Chung and Chung, 2001; Hayashi *et al.*, 2004; Kanje *et al.*, 2000; Kovacic *et al.*, 2007; Kubek *et al.*, 2004; Lutz *et al.*, 2000; Matsuda *et al.*, 2005; Samal *et al.*, 2006; Sanapanich *et al.*, 2002; Zhang *et al.*, 1999). Researchers have applied different tracers to the donor and the recipient(s) nerves either by soluble solution with the use of a Hamilton syringe or by crystals applied at the distal cut ends of the nerves, care been taken to avoid perfusion of the tracers. The most commonly used tracers were FB, DY, FG (fluorogold), and FE (fluoroemerald).

Kanje *et al.* (2000), using crystals of FB and DY at the cut ends of the sciatic nerve in a cross-leg graft model, found 4–14 double labeled sensory neuronal cells and only up to 11% motor neurons. Lutz, using the same tracers, found a ratio of distal recipient/donor labeled cells that equals 10:3.3 for the motor neurons in the ventral horn and 10:7 for the sensory neurons in the Dorsal Root Ganglia (DRGs), which, besides providing evidence for collateral sprouting, also showed a better capacity of sensory neurons to sprouting (Lutz *et al.*, 2000). This view is supported by other authors (Dvali and Myckatyn, 2008). Dvali and Myckatyn (2008) suggested several explanations for the observed better sprouting of sensory neurons than motor neurons. These arguments include: (1) a more efficient tracer transport or administration in sensory neurons; (2) more abundant sensory neurons in nerve trunks, and (3) collateral spontaneous sprouting of nonmyelinated axons (some sensory afferents), while myelinated axons only can sprout at nodes of Ranvier.

Using two other tracers (FR (fluororuby) and FE), Haninec *et al.* (2007) showed very few double labeled neurons at the C6–C8 spinal cord levels after ETS nerve attachment of the musculocutaneous to ulnar nerves. The observation of double labeled neurons (either motor or sensory) with FR and FE in the experimental settings of Kubek (Kubek *et al.*, 2004) provided evidence that both afferent and motor axons are able to send out collateral sprouts. Samal *et al.* (2006) correlated the amount of double labeled neuronal cells with the results of the functional grooming test in the rat and also commented on a profound decrease of staining with time.

After introducing and using the upper extremity of the rat as a more suitable experimental model (Bontioti *et al.*, 2003), retrograde labeling has been used to determine the normal neuronal pool of the terminal branches of the brachial plexus in the rat, together which, with the measurements in any experiments, could give us a more accurate estimate of our results. In experiments at our laboratory and more recently, also by Sananpanich and coworkers (Bontioti *et al.*, 2005, 2006a,b; Sananpanich *et al.*, 2007), we used FB, DY, TMRD (tetramethylrhodamin dextran), and FG to shed light to the tracing uptake. In the work of Sananpanich, up to 10% double labeled axons showed motor axons projecting into both nerves and the corresponding value for sensory neurons were 12%.

All the retrograde labeling studies have found a number of double labeled cells indicating the existence of collateral sprouting as one of the mechanism underlying efficacy of ETS nerve repair. In addition, several studies have demonstrated differences between motor and sensory neuronal cell sprouting capacity as pointed above. Nevertheless, the amount of double labeled neurons was not as high as it could be expected in order to "have a winner" for the explanation of the origin of the regenerating axons in the ETS nerve repair. Most researchers agree that other mechanisms accompanying collateral sprouting exist.

In our studies of retrograde labeling only few double labeled cells were counted either from the motor or the sensory neuronal pool, when the radial nerve was attached to the musculocutaneous, after 180 days evaluation time (Fig. 2). In another experiment, where the observation period was even longer (260 days),

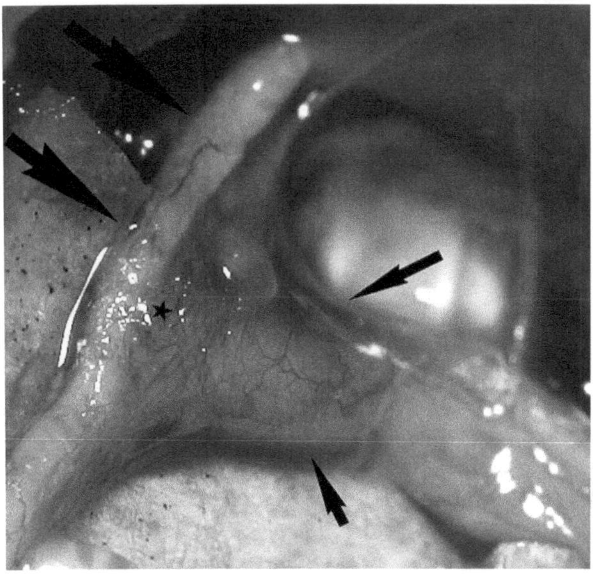

FIG. 2. Photo from the upper extremity of a rat, where a transected distal radial nerve (thin arrows) are attached end-to-side (*) to an intact musculocutaneous nerve (thick arrows) 6 months previously. Reproduced by kind permission from *Journal of Peripheral Nervous System* (Bontioti *et al.*, 2005).

the count of double labeled cells was even less (Bontioti *et al.*, 2005, 2006a,b). From the later study, where we also used the proximal nerve stump of the recipient attached to the musculocutaneous along with the distal hoping that it could serve as an extra source of regenerating axons, an interesting result from retrograde labeling was a shift of the distribution of DY (recipient) labeled neurons towards higher levels. These were findings for which we could give no reasonable explanation, unless we also shift our attention to more higher levels (Bontioti *et al.*, 2006a,b).

Intraneuronal signaling is a mechanism that should gain some of our attention. Bajrovic *et al.* (2002) suggested that cells (probably proliferating Schwann cells) in the degenerated neural pathways are necessary, but not sufficient to induce collateral sprouting of nociceptive axons. The interaction between the injured and noninjured neurons within the dorsal root ganglia (i.e., direct or indirect interneuronal signaling) are important as well as transneuronal transfer of the signal for growth across the spinal cord from the cell bodies of axotomized motor neurons to the somata of intact motor neurons.

The different sprouting capacity between sensory and motor neurons has also been a topic of interest among the researchers (Bontioti *et al.*, 2005; Dvali and Myckatyn, 2008). It has been an attempt to clarify it with the use of double retrograde labeling, as it has been shown in most of the above described experimental works. In addition, Chung and Chung, using double immunostaining techniques with antibodies to Tyrosine Hydroxilase (TH) and GAP-43, could not rule out the possibility of regenerating sprouting that contributed to the increase in sympathetic fibers in the DRGs after L5 spinal nerve ligation (Chung and Chung, 2001). Their morphological observations strongly suggested that collateral sprouting from the perivascular plexus contributes to this increase together with regenerating collateral sprouting.

The only pitfall in retrograde tracing experiments is a difficulty in interpretation of the obtained data in quantitative terms. This could represent the inability of labeling all the neurons that sent sprouts into an ETS attached nerve segment (see above) due to diffusion of the tracers that are usually water soluble, or, again due to apoptosis or pruning. Collateral sprouting does occur. However, as a mechanism alone it does not support efficient functional recovery as it is stated by many researchers (Jancso and Kiraly, 1983; Kovacic *et al.*, 2007; Tarasidis *et al.*, 1998). Maybe, the conclusion provided by Tarasidis saying that normal, uninjured sensory axons spontaneously sprout *de novo*, but motor axons may need to be injured to sprout, is an explanation between different properties among motor and sensory neurons as well as an indication on which is the major mechanism underlying ETS nerve repair (Dvali and Myckatyn, 2008; Pannucci *et al.*, 2007; Tarasidis *et al.*, 1997, 1998).

V. Terminal/Regenerating Sprouting

How about the role of an amount of damage to the donor nerve? Although most agree that no or limited damage is created in the ETS nerve repair where an epineurial window is used, the functional recovery after non- or limited injurious coaptation techniques is not as high as it would be desirable. Terminolateral regeneration can occur also without even a superficial lesion to the donor nerve (Hayashi *et al.*, 2004; Matsumoto *et al.*, 1999). Several experimental studies have shown that regeneration is superior when a window is opened in the donor nerve (Liu *et al.*, 1999; Noah *et al.*, 1997; Okajima and Terzis, 2000; Zhang *et al.*, 2006).

Comparing functional and morphometrical results of ETS nerve repair, with or without an epineurial window, Liu (Liu *et al.*, 1999) concluded that direct contact between the severed stump and the intact nerve appears to be a prerequisite for successful reinnervation to occur, since this nerve contact provides a continuous pathway to allow communication and interaction of Schwann cells between the two nerves and facilitate the effect of multiple growth factors. Furthermore, the holes created by needle penetration during suture could play an important role.

Giovanoli *et al.* (2000) recognized, by means of histomorphometry, regenerating axons that represented terminal sprouts from axons damaged during surgical procedures. Rovak and coworkers revealed, by means of histology and electrophysiology, a deficit at the part of the donor nerve (Rovak *et al.*, 2000). There exists a hypothesis that functional neuromuscular connections are sacrificed from the donor nerve and that terminal sprouting from damaged axons occurs. Still, the evidence is not strong enough for this mechanism to stay alone.

Introducing a new helicoid ETS nerve repair technique, Yan *et al.* (2002) showed a superiority in means of morphometry and functional recovery when this greatly increased area of coaptation allowed a greater amount of axons to sprout from the intact donor nerve.

From another group of experiments, Hayashi *et al.* (2004) compared atraumatic techniques (the recipient was grasped around the donor or a muscle aponeurosis was used to wrap around the recipient) with a perineurial window and partial neurectomy performed at the donor nerve. Results from electrophysiological and histological evaluation as well as fluorescent dye staining showed that nerves were regenerated by collateral sprouting. However, the same research group has visualized axonal sprouting and, expanding their evaluation armamentarium with stereology of nerve regeneration, motor end-plate evaluation, and transcutaneous imaging of sural nerve. They concluded that donor nerve injury is a prerequisite for clinically relevant nerve regeneration through ETS nerve repair. The regeneration occurred coincidently with the local upregulation of Cyclic Adenosine MonoPhosphate (cAMP) Response Element Binding (CREB) phosphorylation and downregulation of Myelin Associated Glycoprotein (MAG) expression issues that will be discussed further in this chapter.

In an interesting study using an experimental model in bigger animals, such as baboons, Kelly *et al.* (2007) concluded that in order to achieve good axonal regeneration, donor nerve is expected to sustain damage, so that the connective tissue barriers to be excluded from the battlefield that regenerating axons have to cross. A side result from their studies was that rats have a superior sprouting and regenerating capacity than any other species and that an interindividual variability exists between experimental animals (and humans) of the same species.

Going back to the standard animal models allowed more easily to be used (i.e., rats), the role of the connective tissues of the nerve trunk as a form of barriers is widely investigated. Experimental groups representing a continuum of donor nerve injury, with least (perineurial window) and maximal injury (partial neurectomy), were used by Brenner *et al.* (2007). Evaluation with retrograde labeling, histology, and wet muscle weight revealed that partial donor nerve neurectomy proved to be a significantly more potent stimulus for regeneration. As they comment, injured axons strongly favor reinnervating their motor targets by traversing preexisting pathways rather than establishing new ones. Furthermore, none of the ETS repair models achieved regeneration comparable to a transection followed by end-to-end repair. Considered together, these data suggested a dose-response relationship between donor nerve axotomy and motor neuron regeneration. A conclusion that we should already have in our minds by the knowledge that end-to-end nerve repair (where the model is equivalent to total axotomy) is the gold standard method of nerve repair, concerning the functional outcome.

Further evidence of the need of some amount of donor nerve damage is given by Akeda *et al.* (2006), who state that donor nerve injury is required and that structural and constitutional changes are a prerequisite. In their study, regenerating axons traverse the perineurium circumferentially around the coaptation site and about 3 mm proximally, travel within the epineurium and enter the distal nerve through the edge of ETS repair. Their concluding remark is that the mode of axonal regeneration in ETS without perineurial window is not a collateral sprouting. It seems to rely on the physiologic reaction of the sensory nervous system to nerve injury. The latest study found in literature promoting terminal sprouting as the main mechanism of ETS efficacy comes from Hilliard (2009) with experiments performed in different animal species. Their concluding result is that in some cases, as in vertebrates, a certain level of degeneration appears to be necessary for axonal regeneration to occur.

Both collateral and terminal/regenerating sprouting seem to contribute, so far, to the efficiency of ETS nerve repair with terminal sprouting being a better or a more prominent candidate (Pannucci *et al.*, 2007). So far, the evidence we get concerning the sprouting (collateral or terminal) capacity of the nerves comes from evaluation mostly of the changes happen at the site of the coaptation itself or some centimeters proximally or distally from this "war zone." Brief or more detailed description of the events occurring at the microenvironment of nerve and the shift mode from maintenance to regeneration has long been the main focus in nerve research. We have managed to reach a somehow detailed information level on this issue and also have improved our knowledge on how we could enhance nerve regeneration (Edds, 1953; Hall, 2005; McDonald *et al.*, 2006; Okajima and Terzis, 2000). A well-documented article explaining some of the alterations occurring in the regenerating axonal pathways comes from Witzel and coauthors (Witzel *et al.*, 2005) and intriguing information is hinted. Using mice expressing a yellow variant of green fluorescent protein (YFP) in a subset of their axons and backlabeling experimental nerves with FG they concluded first that neurons differ in their ability to reinnervate the distal stump. There exist an issue of regeneration stagger and, last and most importantly, they proved that the repair and distal stump environments are rapidly changing. The latter confirmation leads to the knowledge of the complex and plastic ability neurons have, since, "axons crossing the repair site at different times are confronted with an evolving menu of environmental cues upon which they have to base pathfinding decisions."

VI. Stimuli Needed for Triggering Nerve Sprouting

Researchers' attention has turned as well to the factors that trigger sprouting from nerve fibers (Pannucci *et al.*, 2007). What type of ETS-related stimuli is required to initiate cell activation? In an attempt to shed light on this question, we investigated three different types of manipulations: (1) piece of muscle or nerve applied close to an intact nerve trunk, (2) creation of an epineurial window and application of epineurial sutures to the intact nerve, and (3) attachment of a nerve segment in an epineurial window to a donor nerve without the existence of a target-organ (Fig. 3). The transcription factor, activating transcription factor3 (ATF3), was used as a marker of neuronal and Schwann cell injury (see Chapter 28 by Dahlin *et al.*, this volume). Following the activation of this factor to the spinal cord levels, to the DRGs and to the nerve trunks involved, we conclude that an injury to a peripheral nerve trunk by the surgical procedure invoked in ETS may be a main prerequisite for activation of neurons and nonneuronal cells leading to sprouting of axons into the ETS attached nerve (Bontioti *et al.*, 2006a). This

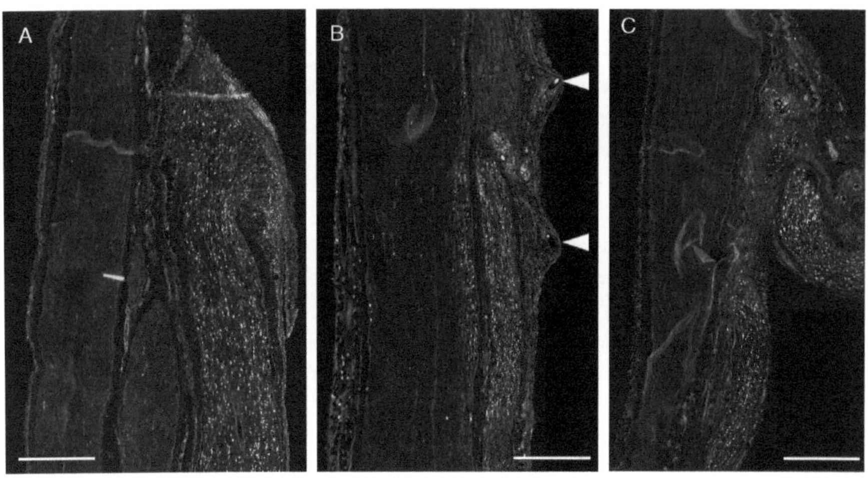

FIG. 3. Longitudinal sections of a sciatic nerve from a rat. A peroneal nerve was applied close to the sciatic nerve without sutures (A). In another model two sutures (arrow heads) were applied in the epineurium of the sciatic nerve (B). A peroneal nerve was attached end-to-side with an epineurial window and applied sutures (C). Note that Schwann cells express ATF3 in their nuclei in the two models where the suture was applied and a piece of a peroneal nerve (to the right) is applied (B, C). The staining in A represent activation of Schwann cells by Wallerian degeneration in the peroneal nerve. Reproduced by kind permission from *Scandinavian Journal of Plastic and Reconstructive Surgery and Hand Surgery* (Bontioti *et al.*, 2006).

conclusion is also later stressed by Pannucci *et al.* (2007). The number of motor and sensory neurons that express ATF3 correlates with axonal outgrowth (Saito and Dahlin, 2008) with downregulation of activation in accordance with time or after target-organ reinnervation being successful. Schwann cells are also rapidly activated at the site of the lesion and in the distal nerve segment with proliferation and remodeling of the extracellular matrix (Martensson *et al.*, 2007). The induction of ATF3 is more robust after transection and has a more slow decrease than after a crush lesion. Different mechanisms with respect to ATF3 regulation appear to exist between motor and sensory neurons where ATF3 is differentially regulated (Kataoka *et al.*, 2007). Concerning the specific neuronal marker and its induction in sympathetic neurons after axotomy, the results given by Sachs are summarized at the following statement: "ATF3 has different dimerization partners in different cell types within the ganglion and it might alter the transcription of different genes depending on its dimerization partners" (Hyatt Sachs *et al.*, 2007). For more detailed description on nerve injury signaling, we recommend the article by Abe and Cavalli (2008).

VII. Pruning

One more well-known mechanism involved in the ETS mechanism is pruning. It represents a neurostructural reassembly. It is a neurological regulatory process, which facilitates a productive change in neural structure by favoring more efficient synaptic configurations and thus playing a role into neuronal competition.

When collateral sprouting occurs in an ETS nerve repair, the result is that one parent cell, one nerve cell body innervates more than one nerve trunk and in the sequence more than one target-organs. Theoretically, it is not possible that one motor or sensory neuron can regenerate into two different target-organs. Thus, it is expected from the synapses to adapt to that type of new innervation, trigger the mechanism of synaptic competition, and eventually result in elimination of the nonproductive ones. This pruning mechanism should exist after ETS collateral sprouting in order for the function to be efficient.

Lutz and Papalia stressed the importance of donor nerve selection, since when antagonistic nerves are chosen the functional outcome is disappointing due to cocontractions of muscles innervated by the donor and the recipient nerves (Lutz *et al.*, 2000; Papalia *et al.*, 2007). The nervous system has an unfathomable capacity to adapt in complex situations with different plasticity mechanisms over time. Cocontraction of the flexors and extensors of the digits after ETS nerve repair of the median to radial nerve of the rats can be eliminated, when the investigation time is prolonged (Papalia *et al.*, 2007).

The pruning mechanism has also been assumed in ETS experimental studies, especially ones that use retrograde labeling tracing, from the decrease number of counted cells. In some of our studies the number of double labeled neuronal cells representing collateral sprouts was reduced with different time points (Bontioti *et al.*, 2005); also shown in a quantitative assessment of collateral sprouting (Samal *et al.*, 2006). The distal segment of an injured musculocutaneous nerve was attached in the latter study to a perineurial window on the ulnar nerve of the rat. The ratio of sensory/motor neurons in the nerves suggested a pruning process (Samal *et al.*, 2006). As explained by many authors, misdirection of axons after any type of repair has a consequence of abnormal or poor function. Accurate regeneration of axons to their original targets is required, based on matching of the regenerating axons back to their corresponding endoneurial tubes. On this basis, Brushart (1988, 1993) has stated that sprouts of motor axons initially grow in both cutaneous and muscular branches of a damaged femoral nerve in the rat but over time they are pruned from afferent endoneurial tubes while maintained in motor ones. The arborizing morphology of many YFP (+) axons is a useful tool for proving collateral pruning (Witzel *et al.*, 2005). The stimulus that induces pruning is yet to be determined, but possible explanations like selective trafficking of structural precursors and differential trophic support of collaterals have already been proposed (see Witzel *et al.*, 2005, for discussion).

VIII. Brain Plasticity

After addressing this correcting mechanism that work either at the most distal level of nerve regeneration, that is, the target-organs, or along the whole nerve pathway and up to the nerve cell bodies; the next level of neuronal plasticity evidence should be the maestro of our nervous system, the brain. Research has uncovered that injuries to a peripheral nerve trunk induces numerous cascade events at cellular and molecular levels that progress throughout plastic changes at the spinal cord levels, the DRG, the brainstem nuclei, the thalamus, and the brain cortex. Even though these types of events have not yet been investigated in the ETS nerve repair model, their importance in nerve injuries and the positive implications that they could have in therapeutic decisions should gain some of our attention. After profound functional reorganization changes in the somatosensory cortex, the focus of the physicians and of the brain as such, is the learning of "a new language spoken by the hand" (see e.g., Lundborg and Rosen (Lundborg *et al.*, 2007)). What really happens in the brain after a nerve injury in the hand? Several authors have concluded that, within minutes after injury, deafferentiation leads to immediate and long standing changes in cortical hand representation. Adjacent cortical areas expand and occupy the former injured nerve territory (Silva *et al.*, 1996). The originally well-organized hand representation is changed to a mosaic-like pattern. This phase 1 deorganization is followed after several months by the phase 2 reorganization period after nerve repair and regeneration (Lundborg and Rosen, 2007).

Brain plasticity in adults may overcome neural adversity, but one must try hard and for a very long period (Merzenich *et al.*, 1983a,b; Sereno, 2005). Exquisite work is presented by Ramachandran in numerous papers that is mostly, but not only, focused on reorganization of the brain and the phantom-limb phenomenon (Ramachandran *et al.*, 1992, 1995). In review articles, detailed description of the cellular and molecular events involved in neuronal plasticity after nerve injury and regeneration are described (Borrelli *et al.*, 2008). If the brain has to learn the new language spoken by the injured hand then we should also try to learn the language spoken by the brain. However, a detailed description of the delicate mechanisms behind brain plasticity is beyond the scope of this review.

Taken together, standard evaluation methods, such as histology, electrophysiology, retrograde labeling, gait analysis accompanied by more sophisticated ones, like visualization of regenerating axons and stereology, as well as the use of transgenic animals, have been used to clarify the mechanism underlying the ETS nerve repair. Mechanisms have been discussed behind axonal outgrowth into the attached recipient nerve segment in ETS nerve repair. Collateral sprouting does occur, but is not the only mechanism, and also pruning and particularly terminal sprouting are relevant for the understanding of the mechanisms. In addition, there may be a difference between motor and sensory neurons in efficiency of collateral sprouting. However, a suitable stimulus is needed to trigger the outgrowth of axons in the distal nerve segment (Bontioti, 2005).

Acknowledgments

The authors are supported by grants from the Swedish Research Council (Medicine and Natural Sciences), Crafoord's Fund for Medical Research, Konsul Thure Carlsson Fund for Medical Research, Region Skåne and Funds from the University Hospital Malmö, Sweden. Due to limited space for references, we do apologize for not including all relevant articles.

References

Abe, N., and Cavalli, V. (2008). Nerve injury signaling. *Curr. Opin. Neurobiol.* **18,** 276–283.

Akeda, K., Hirata, H., Matsumoto, M., Fukuda, A., Tsujii, M., Nagakura, T., Ogawa, S., Yoshida, T., and Uchida, A. (2006). Regenerating axons emerge far proximal to the coaptation site in end-to-side nerve coaptation without a perineurial window using a T-shaped chamber. *Plast. Reconstr. Surg.* **117,** 1194–1203; discussion 1204–1205.

Bajrovic, F., Kovacic, U., Pavcnik, M., and Sketelj, J. (2002). Interneuronal signalling is involved in induction of collateral sprouting of nociceptive axons. *Neuroscience* **111,** 587–596.

Bertelli, J. A., Dos Santos, A. R., and Calixto, J. B. (1996). Is axonal sprouting able to traverse the conjunctival layers of the peripheral nerve? A behavioural motor, motor and sensory study of end-to-side nerve anastomosis. *J. Reconstr. Microsurg.* **12,** 559–563.

Bontioti, E. (2005). End-to-side nerve repair. A study in the forelimb of the rat. *In* "Hand Surgery/ Dept of Clinical sciences in Malmö." Lund University, Malmö.

Bontioti, E. N., Kanje, M., and Dahlin, L. B. (2003). Regeneration and functional recovery in the upper extremity of rats after various types of nerve injuries. *J. Peripher. Nerv. Syst.* **8,** 159–168.

Bontioti, E., Kanje, M., Lundborg, G., and Dahlin, L. B. (2005). End-to-side nerve repair in the upper extremity of rat. *J. Peripher. Nerv. Syst.* **10,** 58–68.

Bontioti, E., Dahlin, L. B., Kataoka, K., and Kanje, M. (2006a). End-to-side nerve repair induces nuclear translocation of activating transcription factor 3. *Scand. J. Plast. Reconstr. Surg. Hand Surg.* **40,** 321–328.

Bontioti, E., Kanje, M., and Dahlin, L. B. (2006b). End-to-side nerve repair: Attachment of a distal, compared with a proximal and distal, nerve segment. *Scand. J. Plast. Reconstr. Surg. Hand Surg.* **40,** 129–135.

Borrelli, E., Nestler, E. J., Allis, C. D., and Sassone-Corsi, P. (2008). Decoding the epigenetic language of neuronal plasticity. *Neuron* **60,** 961–974.

Brenner, M. J., Dvali, L., Hunter, D. A., Myckatyn, T. M., and Mackinnon, S. E. (2007). Motor neuron regeneration through end-to-side repairs is a function of donor nerve axotomy. *Plast. Reconstr. Surg.* **120,** 215–223.

Brushart, T. M. E. (1988). Preferential reinnervation of motor nerves by regenerating motor axons. *J. Neurosci.* **8,** 1026–1031.

Brushart, T. M. E. (1993). Motor axons preferentially reinnervate motor pathways. *J. Neurosci.* **13,** 2730–2738.

Chung, K., and Chung, J. M. (2001). Sympathetic sprouting in the dorsal root ganglion after spinal nerve ligation: Evidence of regenerative collateral sprouting. *Brain Res.* **895,** 204–212.

De Sa, J. M., Mazzer, N., Barbieri, C. H., and Barreira, A. A. (2004). The end-to-side peripheral nerve repair. Functional and morphometric study using the peroneal nerve of rats. *J. Neurosci. Methods* **136,** 45–53.

Dvali, L. T., and Myckatyn, T. M. (2008). End-to-side nerve repair: Review of the literature and clinical indications. *Hand Clin.* **24,** 455–460, vii.

Edds, M. V., Jr. (1953). Collateral nerve regeneration. *Q. Rev. Biol.* **28,** 260–276.

Eren, F., Yuksel, F., Ulkur, E., Cavdar, S., Ercan, F., and Celikoz, B. (2005). Nerve regeneration through a healthy nerve trunk: A new and hopeful conduit for bridging nerve defects. *Plast. Reconstr. Surg.* **116,** 1697–1705.

Esslen, E. (1960). Electromyographic findings on two types of misdirection of regenerating axons. *Electroencephalogr. Clin. Neurophysiol.* **12,** 738–741.

Gavazzi, I. (1995). Collateral sprouting and responsiveness to nerve growth factor of ageing neurons. *Neurosci. Lett.* **189,** 47–50.

Giovanoli, P., Koller, R., Meuli-Simmen, C., Rab, M., Haslik, W., Mittlbock, M., Meyer, E., and Frey, M. (2000). Functional and morphometric evaluation of end-to-side neurorrhaphy for muscle reinnervation. *Plast. Reconstr. Surg.* **106,** 383–392.

Goheen-Robillard, B., Myckatyn, T. M., Mackinnon, S. E., and Hunter, D. A. (2002). End-to-side neurorrhaphy and lateral axonal sprouting in a long graft rat model. *Laryngoscope* **112,** 899–905.

Hall, S. (2005). The response to injury in the peripheral nervous system. *J. Bone Joint Surg. Br.* **87,** 1309–1319.

Haninec, P., Samal, F., Tomas, R., Houstava, L., and Dubovwy, P. (2007). Direct repair (nerve grafting), neurotization, and end-to-side neurorrhaphy in the treatment of brachial plexus injury. *J. Neurosurg.* **106,** 391–399.

Hayashi, A., Yanai, A., Komuro, Y., Nishida, M., Inoue, M., and Seki, T. (2004). Collateral sprouting occurs following end-to-side neurorrhaphy. *Plast. Reconstr. Surg.* **114,** 129–137.

Hayashi, A., Pannucci, C., Moradzadeh, A., Kawamura, D., Magill, C., Hunter, D. A., Tong, A. Y., Parsadanian, A., Mackinnon, S. E., and Myckatyn, T. M. (2008). Axotomy or compression is required for axonal sprouting following end-to-side neurorrhaphy. *Exp. Neurol.* **211,** 539–550.

Hilliard, M. A. (2009). Axonal degeneration and regeneration: a mechanistic tug-of-war. *J. Neurochem.* **108,** 23–32.

Hyatt Sachs, H., Schreiber, R. C., Shoemaker, S. E., Sabe, A., Reed, E., and Zigmond, R. E. (2007). Activating transcription factor 3 induction in sympathetic neurons after axotomy: Response to decreased neurotrophin availability. *Neuroscience* **150,** 887–897.

Jaberi, F. M., Abbas, B. P., Nezhad, S. T., and Tanideh, N. (2003). End-to-side neurorrhaphy: An experimental study in rabbits. *Microsurgery* **23,** 359–362.

Jancso, G., and Kiraly, E. (1983). Cutaneous nerve regeneration in the rat: Reinnervation of the denervated skin by regenerative but not by collateral sprouting. *Neurosci. Lett.* **36,** 133–137.

Jiang, B. G., Yin, X. F., Zhang, D. Y., Fu, Z. G., and Zhang, H. B. (2007). Maximum number of collaterals developed by one axon during peripheral nerve regeneration and the influence of that number on reinnervation effects. *Eur. Neurol.* **58,** 12–20.

Kanje, M., Arai, T., and Lundborg, G. (2000). Collateral sprouting from sensory and motor axons into an end-to-side attached nerve segment. *Neuroreport* **11,** 2455–2459.

Kataoka, K., Kanje, M., and Dahlin, L. B. (2007). Induction of activating transcription factor 3 after different sciatic nerve injuries in adult rats. *Scand. J. Plast. Reconstr. Surg. Hand Surg.* **41,** 158–166.

Kelly, E. J., Jacoby, C., Terenghi, G., Mennen, U., Ljungberg, C., and Wiberg, M. (2007). End-to-side nerve coaptation: A qualitative and quantitative assessment in the primate. *J. Plast. Reconstr. Aesthet. Surg.* **60,** 1–12.

Kovacic, U., Tomsic, M., Sketelj, J., and Bajrovic, F. F. (2007). Collateral sprouting of sensory axons after end-to-side nerve coaptation—A longitudinal study in the rat. *Exp. Neurol.* **203,** 358–369.

Kubek, T., Kyr, M., Haninec, P., Samal, F., and Dubovy, P. (2004). Morphological evidence of collateral sprouting of intact afferent and motor axons of the rat ulnar nerve demonstrated by one type of tracer molecule. *Ann. Anat.* **186,** 231–234.

Liu, K., Chen, L. E., Seaber, A. V., Goldner, R. V., and Urbaniak, J. R. (1999). Motor functional and morphological findings following end-to-side neurorrhaphy in the rat model. *J. Orthop. Res.* **17,** 293–300.

Lundborg, G., and Rosen, B. (2007). Hand function after nerve repair. *Acta. Physiol. (Oxf.)* **189,** 207–217.

Lundborg, G., Zhao, Q., Danielsson, N., and Kerns, J. M. (1994). Can sensory and motor collateral sprouting be induced from intact peripheral nerve by end-to-side anastomosi? *J. Hand Surg.* **19B,** 277–282.

Lundborg, G., Bjorkman, A., and Rosen, B. (2007). Enhanced sensory relearning after nerve repair by using repeated forearm anaesthesia: Aspects on time dynamics of treatment. *Acta Neurochir. Suppl.* **100,** 121–126.

Lutz, B. S., Chuang, D. C., Hsu, J. C., Ma, S. F., and Wei, F. C. (2000). Selection of donor nerves—An important facor in end-to-side neurorrhaphy. *Br. J. Plast. Surg.* **53,** 149–153.

Lykissas, M. G., Korompilias, A. V., Batistatou, A. K., Mitsionis, G. I., and Beris, A. E. (2007). Can end-to-side neurorrhaphy bridge large defects? An experimental study in rats. *Muscle Nerve* **36,** 664–671.

Martensson, L., Gustavsson, P., Dahlin, L. B., and Kanje, M. (2007). Activation of extracellular-signal-regulated kinase-1/2 precedes and is required for injury-induced Schwann cell proliferation. *Neuroreport* **18,** 957–961.

Matsuda, K., Kakibuchi, M., Fukuda, K., Kubo, T., Madura, T., Kawai, K., Yano, K., and Hosokawa, K. (2005). End-to-side nerve grafts: Experimental study in rats. *J. Reconstr. Microsurg.* **21,** 581–591.

Matsumoto, M., Hirata, H., Nishiyama, M., Morita, A., Sasaki, H., and Uchida, A. (1999). Schwann cells can induce collateral sprouting from intact axons: Experimental study of end-to-side neurorrhaphy using a y-chamber model. *J. Reconstr. Microsurg.* **15,** 281–286.

McCallister, W. V., Cober, S. R., Norman, A., and Trumble, T. E. (2001a). Using intact nerve to bridge peripheral nerve defects: An alternative to the use of nerve grafts. *J. Hand Surg. (Am.)* **26,** 315–325.

McCallister, W. V., Tang, P., Smith, J., and Trumble, T. E. (2001b). Axonal regeneration stimulated by the combination of nerve growth factor and ciliary neurotrophic factor in an end-to-side model. *J. Hand Surg. (Am.)* **26,** 478–488.

McDonald, D., Cheng, C., Chen, Y., and Zochodne, D. (2006). Early events of peripheral nerve regeneration. *Neuron Glia Biol.* **2,** 139–147.

Merzenich, M. M., Kaas, J. H., Wall, J. T., Sur, M., Nelson, R. J., and Felleman, D. J. (1983a). Progression of change following median nerve section in the cortical representation of the hand in areas 3b and 1 in adult owl and squirrel monkeys. *Neuroscience* **10,** 639–665.

Merzenich, M. M., Kaas, J. H., Wall, R. J., Nelson, M., Sur, D., and Felleman, D. (1983b). Topographic reorganization of somatosensory cortical areas 3B and 1 in adult monkeys following restricted desfferentiation. *Neuroscience* **8,** 33–55.

Noah, E. M., Williams, A., Jorgenson, C., Skoulis, T. G., and Terzis, J. K. (1997). End-to-side neurorraphy: A histologic and morphometric study of axonal sprouting into an end-to-side nerve graft. *J. Reconstr. Microsurg.* **12,** 99–106.

Okajima, S., and Terzis, J. K. (2000). Ultrastructure of early axonal regeneration in an end-to-side neurorrhaphy model. *J. Reconstr. Microsurg.* **16,** 313–23; discussion 323–326.

Ozbek, S., Ozcan, M., Noyan, B., Kurt, M. A., Tirelioglu, S., Bozkurt, C., Karaca, K., and Filiz, G. (2005). End-to-side nerve coaptation: Is an additional proximal coaptation useful when available? *Ann. Plast. Surg.* **55,** 281–288.

Pannucci, C., Myckatyn, T. M., Mackinnon, S. E., and Hayashi, A. (2007). End-to-side nerve repair: Review of the literature. *Restor. Neurol. Neurosci.* **25,** 45–63.

Papalia, I., Cardaci, A., d'Alcontres, F. S., Lee, J. M., Tos, P., and Geuna, S. (2007). Selection of the donor nerve for end-to-side neurorrhaphy. *J. Neurosurg.* **107,** 378–382.

Pondaag, W., and Gilbert, A. (2008). Results of end-to-side nerve coaptation in severe obstetric brachial plexus lesions. *Neurosurgery* **62,** 656–63; discussion 656–663.

Ramachandran, V. S., Stewart, M., and Rogers-Ramachandran, D. C. (1992). Perceptual correlates of massive cortical reorganization. *Neuroreport* **3,** 583–586.

Ramachandran, V., Rogers-Ramachandran, D., and Cobb, S. (1995). Touching the phantom limb. *Nature* **377,** 489–490.

Rovak, J. M., Cederna, P. S., Macionis, V., Urbanchek, M. S., Van Der Meulen, J. H., and Kuzon, W. M., Jr. (2000). Termino-lateral neurorrhaphy: The functional axonal anatomy. *Microsurgery* **20,** 6–14.

Saito, H., and Dahlin, L. B. (2008). Expression of ATF3 and axonal outgrowth are impaired after delayed nerve repair. *BMC Neurosci.* **9**(88).

Samal, F., Haninec, P., Raska, O., and Dubovy, P. (2006). Quantitative assessment of the ability of collateral sprouting of the motor and primary sensory neurons after the end-to-side neurorrhaphy of the rat musculocutaneous nerve with the ulnar nerve. *Ann. Anat.* **188,** 337–344.

Sanapanich, K., Morrison, W. A., and Messina, A. (2002). Physiologic and morphologic aspects of nerve regeneration after end-to-end or end-to-side coaptation in a rat model of brachial plexus injury. *J. Hand Surg. (Am.)* **27,** 133–142.

Sananpanich, K., Galea, M. P., Morrison, W. A., and Messina, A. (2007). Quantitative characterization of regenerating axons after end-to-side and end-to-end coaptation in a rat brachial plexus model: A retrograde tracer study. *J. Neurotrauma* **24,** 864–875.

Sereno, M. I. (2005). Neuroscience: Plasticity and its limits. *Nature* **435,** 288–289.

Silva, A. C., Rasey, S. K., Wu, X., and Wall, J. T. (1996). Initial cortical reactions to injury of the median and radial nerves to the hands of adult primates. *J. Comp. Neurol.* **366,** 700–716.

Tarasidis, G., Watanabe, O., Mackinnon, S. E., Strasberg, S. R., Haughey, B. H., and Hunter, D. A. (1997). End-to-side neurorrhaphy resulting in limited sensory axonal regeneration in a rat model. *J. Otol. Rhinol. Laryngol.* **106,** 506–512.

Tarasidis, G., Watanabe, O., Mackinnon, S. E., Strasberg, S. R., Haughey, B. H., and Hunter, D. A. (1998). End-to-side neurorraphy: A long-term study of neural regeneration in a rat model. *Otolaryngol. Head Neck Surg.* **119,** 337–341.

Tham, S. K., and Morrison, W. A. (1998). Motor collateral sprouting through an end-to-side nerve repair. *J. Hand Surg.* **23A,** 844–851.

Viterbo, F., Trindade, J. C., Hoshini, K., and Mazzonineto, A. (1992). Latero-terminal neurorrhaphy without removal of the epineural sheath. Experimental study in rats. *Rev. Paul. Med.* **110,** 267.

Viterbo, F., Trindade, J. C., Hoshino, K., and Mazzoni Neto, A. (1994). End-to-side neurorrhaphy with removal of the epineurial sheath: An experimental study in rats. *Plast. Reconstr. Surg.* **94,** 1038–1047.

Witzel, C., Rohde, C., and Brushart, T. M. (2005). Pathway sampling by regenerating peripheral axons. *J. Comp. Neurol.* **485,** 183–190.

Yamauchi, T., Maeda, M., Tamai, S., Tamai, M., Yajima, H., Takakura, Y., Haga, S., and Yamamoto, H. (2000). Collateral sprouting mechanism after end-to-side nerve repair in the rat. *Med. Electron Microsc.* **33,** 151–156.

Yan, J. G., Matloub, H. S., Sanger, J. R., Zhang, L. L., Riley, D. A., and Jaradeh, S. S. (2002). A modified end-to-side method for peripheral nerve repair: Large epineurial window helicoid technique versus small epineurial window standard end-to-side technique. *J. Hand Surg. (Am.)* **27,** 484–492.

Yuksel, F., Karacaoglu, E., and Guler, M. M. (1999). Nerve regeneration through side-to-side neurorrhaphy sites in a rat model: A new concept in peripheral nerve surgery. *Plast. Reconstr. Surg.* **104,** 2092–2099.

Zhang, Z., Soucacos, P. N., Bo, J., and Beris, A. E. (1999). Evaluation of collateral sprouting after end-to-side nerve coaptation using a fluorescent double-labeling technique. *Microsurgery* **19,** 281–286.

Zhang, Z., Johnson, E. O., Vekris, M. D., Zoubos, A. B., Bo, J., Beris, A. E., and Soucacos, P. N. (2006). Repair of the main nerve trunk of the upper limb with end-to-side neurorrhaphy: An experimental study in rabbits. *Microsurgery* **26,** 245–252.

Zhu, Q. T., Zhu, J. K., and Chen, G. Y. (2008). Location of collateral sprouting of donor nerve following end-to-side neurorrhaphy. *Muscle Nerve* **38,** 1506–1509.

EXPERIMENTAL RESULTS IN END-TO-SIDE NEURORRHAPHY

Alexandros E. Beris and Marios G. Lykissas

Department of Orthopaedic Surgery, University of Ioannina, School of Medicine, Ioannina, Greece

Experimental evidence suggests that reinnervation of the distal stump of a transected nerve may occur if the former is coapted end-to-side to the trunk of an adjacent nerve. Axonal regeneration occurs by collateral sprouting of healthy donor nerve axons, induced by neurotrophic factors. End-to-side neurorrhaphy can provide satisfactory functional recovery for the recipient nerve, without any deterioration of the donor nerve function. Various experimental models have been proposed in order to increase regeneration efficiency after end-to-side neurorrhaphy. End-to-side neurorrhaphy has already been used in the clinical practise, but there are still some issues that have not been completely clarified yet: (i) the origin of regenerating axons, (ii) collateral sprouting molecular mechanisms, and (iii) the degree of donor nerve axotomy needed for motor functional recovery. The results of experimental studies trying to investigate these parameters are briefly discussed in this review article.

I. Introduction

End-to-side neurorrhaphy is defined as coaptation of the distal stump of a transected nerve to the trunk of a donor nerve. It has been used in a large number of both experimental and clinical studies as a surgical alternative when the proximal stump of an injured nerve in unavailable or obliterated or the nerve gap is too long to be bridged by a nerve graft. In such cases, the use of conventional nerve repair or grafting is not applicable or likely to result in poor

INTERNATIONAL REVIEW OF NEUROBIOLOGY, VOL. 87
DOI: 10.1016/S0074-7742(09)87013-X

functional outcome. End-to-side neurorrhaphy was first described by Létiévant (1873) as a reconstructive strategy of peripheral nerves in cases of large substance loss. Huber (1895) was the first author to publish experimental data on end-to-side neurorrhaphy in 1895, with disappointing results. Ever since, this modality was employed by a limited number of investigators, until the early 1990s when Viterbo and colleagues (1992) supported successful collateral sprouting after end-to-side neurorrhaphy in a rat model. Since then, a numerous experimental studies and a large number of different animal models have been described in the literature (Battiston *et al.*, 2000; Bertelli *et al.*, 1996; Hayashi *et al.*, 2004; Lundborg *et al.*, 1994; Yamauchi *et al.*, 2000; Zhang *et al.*, 1999).

In end-to-side neurorrhaphy, investigation in animals is now focusing in four main elements: (i) the source of regenerating axons, (ii) molecular mechanism of collateral sprouting, (iii) degree of motor versus sensory regeneration, and (iv) different surgical models in end-to-side neurorrhaphy.

II. Source of Regenerating Axons

There is still a great controversy regarding nerve regeneration following end-to-side neurorrhaphy. There are three main hypothesis to explain the source of regenerating axons found into the distal nerve segment: (i) "terminal sprouting" from inadvertent axonal injury of the donor nerve during preparation or suturing, (ii) "collateral sprouting" from the nodes of Ranvier just proximal to the area of coaptation, and (iii) contamination from the proximal stump of the transected nerve.

Several investigators utilizing various methods have tried to demonstrate the origin of regenerating axons. Histological studies have shown axonal regeneration from the donor nerve and successful reinnervation of the recipient nerve after penetration of the conjuctival layers of the nerve, by utilizing toluidine blue, silver impregnation, and immunocytochemical staining of neurofilament protein globin (Matsumoto *et al.*, 1999; Noah *et al.*, 1997a,b; Zhang *et al.*, 2000). However, none of them is effective to distinguish if regenerating axons within the recipient nerve represent collateral or terminal sprouts. Using the fluorescent double-labeling technique, Zhang and colleagues (1999) showed that one parent nerve fiber can emanate additional action by collateral sprouting following end-to-side nerve coaptation. According to the authors, the mechanism causing collateral sprouting may result from switching signals and/or switching factors, presumably neurotrophic (Zhang *et al.*, 1999). These findings were in accordance with other studies in which similar staining methods were used (Matsuda *et al.*, 2005; Saémal *et al.*, 2006; Xiong *et al.*, 2003). Double-labeling technique is an effective method to eliminate contamination from the proximal nerve stump as a possible source of regenerating axons, but it is of limited value when regenerating axons are evaluated directly at the coaptation site (collateral vs. terminal sprouting).

In another experimental model, Hayashi *et al.* (2004) used a free sciatic nerve graft with noninjurious end-to-side coaptation between a free sciatic nerve graft and the right and left median nerves, without additional artificial objects. The regenerated axons were observed directly at the coaptation site using fluorescent carbocyanine dye in conjunction with confocal microscopy. Fluorescent dye staining determined that nerve regeneration had occurred by collateral sprouting from the donor nerve. This observation was confirmed in a preliminary study of Zhu *et al.* (2008) by utilizing a microtease technique. In a rat model, the distal stump of the common peroneal nerve was attached in an end-to-side fashion to the tibial nerve after creation of an epineurial window. Three months postoperatively, epineurium and perineurium were stripped off and nerve fibers were teased. Examination under light microscopy revealed axons sprouted collaterally from myelinated nerve fibers at the node of Ranvier.

III. Molecular Mechanism of Collateral Sprouting

Al-Qattan and Al-Thunyan (1998) evaluated nerve regeneration following end-to-side neurorrhaphy by utilizing electron microscopy. They found that the repair process following end-to-side neurorrhaphy is structurally similar to that observed after end-to-end neurorrhaphy, but the molecular mechanism of collateral sprouting has not been completely clarified yet. Factors that may induce collateral sprouting include axotomy of the donor nerve, insertion of sutures into the donor nerve, and biological response of the donor neuron to factors emanating from the transected nerve and the denervated end organs.

Bontioti *et al.* (2006a), in an end-to-side neurorrhaphy model, suggested increased expression of activating transcription factor 3 (ATF3), a marker of cell activation that is induced in sensory and motor neurons after peripheral nerve injury. They revealed robust ATF3 expression after creation of an epineurial window and/or suturing, but limited ATF3 expression after placement of a piece of nerve or muscle close to the donor nerve. Based on their findings, the authors supported that an operative injury to the donor nerve during end-to-side neurorrhaphy is the main prerequisite for axonal sprouting. In another study, Akeda *et al.* (2006) used immunofluorescence with antibody against NG2, a marker of axonal regeneration. Because there was no NG2 accumulating at the coaptation site, the authors suggested that regenerating axons have to be emerged far more proximal to the coaptation site. In the same study, the authors by using an end-to-side neurorrhaphy model with an isogenic peroneal nerve demonstrated that nerve damage is required for axonal regeneration.

The role of Schwann cells in collateral sprouting has been extensively investigated (Audisio *et al.*, 2008; Bajrović *et al.*, 2002; Matsumoto *et al.*, 1999). After nerve injury, Schwann cells proliferate rapidly and secrete a great number of chemotactic substances, like nerve growth factor (NGF). According to Bajrović *et al.* (2002), Schwann cells are necessary for induction and/or maintenance of collateral sprouting of certain axons. However, Schwann cells are probably not sufficient, or they need more time, to induce collateral sprouting of sensory axons in the absence of other induction mechanisms (Bajrović *et al.*, 2002). Matsumoto and colleagues (1999) proposed three steps in axonal sprouting stimulation by Schwann cells: (i) column formation by Schwann cells migrated at the coaptation site, (ii) invasion of the Schwann cells into the epineurium of the donor nerve, and (iii) perforation of the perineurium by regenerating axons which interfere with Schwann cells within the epineurial layer. In a recent study, Audisio *et al.* (2008) investigated variations in expression of ErbBs, a family of tyrosine kinases receptor involved in the trophic maintenance of Schwann cells, after end-to-end and end-to-side neurorrhaphy. They concluded that there are no differences in ErbBs mRNA variation in relation with the type of nerve reconstruction.

Trying to elucidate collateral sprouting mechanism in a noninjurious rat model, Yamauchi *et al.* (2000) transferred the medial gastrocnemius nerve to the side of an intact lateral gastrocnemius nerve using a Y-shaped silicone tube. By utilizing an antibody against growth-associated protein-43 (GAP-43), a component of growth cones, the authors demonstrated growth cone direction from the intact donor nerve to the peripheral nerve segment of the transected nerve. In the same study, association of neurotrophin-3 (NT-3) in the process of collateral sprouting of motor nerves was also suggested by increased expression of NT-3 and its receptor Trk C at the coaptation site. McCallister *et al.* (2001) supported that reinnervation of the recipient nerve following end-to-side neurorrhaphy is the result of contamination from the proximal stump and can be enhanced by the combination of NGF and ciliary neurotrophic factor (CNTF). Likewise, Fortes *et al.* (1999) demonstrated association between axonal sprouting and growth and increased insulin-like growth factor (IGF I–II) levels at the coaptation site, while recently Chen *et al.* (2008) stimulated nerve regeneration after end-to-side neurorrhaphy by short-course FK506. It has also been shown that systemic administration of erythropoietin stimulates axonal regeneration after end-to-side neurorrhaphy (Lykissas *et al.*, 2007b). Much of the neurological benefit associated with erythropoietin administration occurs within the first weeks after nerve injury.

Collateral sprouting of cutaneous nociceptive axons may occur by three different mechanisms: (i) factors derived from denervated target tissues, (ii) factors released by degenerated neural pathways, and (iii) an interneuronal signaling within the dorsal root ganglion (DRG) (Bajrović *et al.*, 2002). According to interneuronal mechanism hypothesis, cell bodies of the injured neurons communicate directly or indirectly with the cell bodies of the noninjured neurons within the same DRG. Using a rat model in which the peroneal nerve was transected and sutured to the side of the intact sural nerve, Bajrović *et al.* supported that this neuron interaction in the DRG may stimulate collateral sprouting (Bajrović *et al.*, 2002).

IV. Degree of Motor Versus Sensory Regeneration

When end-to-side neurorrhaphy first applied in peripheral nerve reconstruction, most authors supported that transection of at least some axons of the donor nerve was a prerequisite for successful functional outcome. There is now enough evidence showing that sensory sprouting is generally easier than motor sprouting and deliberate axonal injury enhances motor axonal regeneration (Brenner *et al.*, 2007). Although many authors reveal no deterioration of the donor nerve function in their models, unpredictable morbidity to the donor nerve may occur as a result of axonal disruption.

Researchers suggest that donor nerve injury from surgical procedures, such as creation of an epineurial or perineurial window and application of sutures is required for functional motor or sensory reinnervation (Bontioti *et al.*, 2006a). Experimental work of Brenner *et al.* (2007), using three end-to-side repair models with progressively greater donor nerve injury (conventional end-to-side, end-to-side plus crush injury, and end-to-side plus neurotomy), demonstrated a dose-response relationship between axotomy of the donor nerve and motor axons regeneration. According to Bontioti *et al.* (2005), these motor fibers from the donor nerve may enter the recipient nerve segment to supply muscles which were normally innervated by motor fibers from the recipient nerve, resulting in a functional recovery of 60–72% compared with the contralateral side. On the other hand, other authors (Battal *et al.*,1996; Matsuda *et al.*, 1995) showed significant motor functional recovery without axotomy after end-to-side neurorrhaphy.

In order to shed light on the effect of axotomy to motor reinnervation of the recipient nerve, several noninjurious models have been proposed. Bertelli *et al.* (1996) failed to demonstrate functional recovery by using fibrin glue at the coaptation sites. In another experimental model, Akeda *et al.* (2006) used a T-shaped silicone chamber to minimize the donor nerve damage during end-to-side coaptation of the peroneal nerve to the ipsilateral tibial nerve. They similarly noticed delay of axonal regeneration in end-to-side neurorrhaphy without a perineurial window. Based on their findings, they suggested that in their non-injurious model the mode of axonal regeneration is not a collateral sprouting, but physiologic reaction of the sensory nervous system to nerve injury. In another study, the same authors utilized retrograde labeling technique in a Y-shaped chamber model (Matsumoto *et al.*, 1999). Staining of the dorsal root ganglia alone indicated that only sensory axons were regenerated.

On the contrary, Hayashi *et al.* (2004) showed functional motor and sensory reinnervation following end-to-side coaptation with two different noninjurious techniques in a rat model. In the first one, the sciatic nerve was separated in the midline and the median nerve was enveloped by the separated sciatic nerve, while in the second technique the aponeurosis of the spinal muscles was used to wrap the sciatic and median nerves in an end-to-side fashion. However, when these techniques were repeated in transgenic mice expressing fluorescent protein in motor axons, no axonal regeneration was noticed for at least 30 days after surgery. Motor nerves functional recovery in a noninjurious model of end-to-side neurorrhaphy was also obtained by measuring choline-acetyltransferase activity (ChAT) (Yamauchi *et al.*, 2001). By using a Y-shaped silicone tube, the authors calculated ChAT activity in their end-to-side model of approximately two-thirds of the value in end-to-end neurorrhaphy, and 55% of the value in normal nerves, three months postoperatively.

V. Results in Various End-to-Side Surgical Models

When the concept of end-to-side neurorrhaphy was revived by Viterbo *et al.* (1992), a "double" end-to-side technique was proposed. It is considered that coaptation of both proximal and distal nerve segment to the adjacent donor nerve stimulates axonal growth by a supercharged effect compared with traditional end-to-side repair. According to Bontioti *et al.* (2006b), two sources of axons may contribute to the increased number of regenerating nerve fibers; axons sprouted collaterally from myelinated nerve fibers at the node of Ranvier of the donor nerve and axons that arise from the proximally coapted nerve segment. In our double end-to-side model, the reinnervation of the distal segment of the recipient nerve seems possible to occur through enhanced contamination from the proximal segment rather than collateral sprouting (Lykissas *et al.*, 2007a). In such case, the epineurium of the donor nerve serves as a bridge for the regenerated axons (Fig. 1). Superior functional results after "double" end-to-side neurorrhaphy have been obtained in both upper and lower extremity in rat models (Bontioti *et al.*, 2006b; Lykissas *et al.*, 2007a; Viterbo *et al.*, 1992). However, when the transected radial nerve was sutured in an end-to-side fashion to the musculocutaneous nerve, the additional coaptation of the proximal nerve stump was not of any advantage over the single attachment of the distal nerve segment (Bontioti *et al.*, 2006b).

Reverse end-to-side neurorrhaphy has been proposed in case of an injury in continuity. By switching the roles of the donor and recipient nerves,

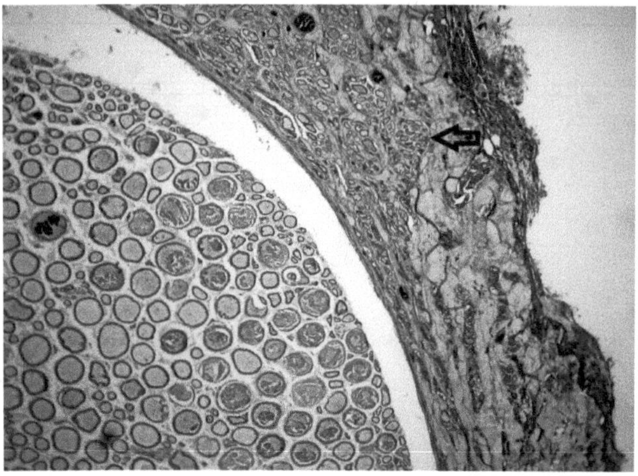

FIG. 1. Transverse section of the bridging part of the donor nerve following "double" end-to-side neurorrhaphy (toluidine blue, original magnification ×400). Note the small myelinated fibers travelling in the outer epineurium (arrow). Reinnervation of the recipient nerve may occur through contamination from the proximally coapted nerve segment, collateral sprouting, or both.

Isaacs *et al.* (2005) transected the donor nerve and coapted its proximal end to the side of the damaged recipient nerve just distal to the injury site after creating an epineurial window. According to the authors, successful reinnervation of the recipient nerve end organ can be established by axons arising from the donor nerve. In order to shorten the denervation time and prevent irreversible muscle atrophy, Fujiwara and colleagues (2005) evaluated augmentation of nerve regeneration after end-to-end neurorrhaphy by reverse end-to-side coaptation. In a rat model, the right sciatic nerve was cut and repaired using end-to-end neurorrhaphy. The contralateral sciatic nerve was also transected, passed through a subcutaneous tunnel to the right side and sutured to the side of the right sciatic nerve 20 mm distal to the end-to-end coaptation. Augmentation of the regenerating axons derived from the donor nerve and double innervation of the denervated muscles by both donor and recipient nerves was noted.

End-to-side nerve repair has been successfully used for reinnervation of more than one nerve by a single donor nerve (Bontioti *et al.*, 2005; Ozbek and Kurt, 2006; Zhang *et al.*, 1998). In a rat model, when the radial and median/ulnar nerves were sutured to the musculocutaneous nerve in an end-to-side fashion, no statistically significant differences in functional recovery were noted between the two groups (Bontioti *et al.*, 2005). Nevertheless, a greater number of less mature nerve fibers were found in the median/ulnar nerve group than in the radial nerve group. End-to-side nerve graft has also been used experimentally for multiple branch reconstruction. Matsuda *et al.* (2008) in order to perform facial nerve reconstruction with less graft requirement as compared with end-to-end cable graft method, coapted the four branches of the bilateral facial nerves to a single sciatic nerve graft in an end-to-side fashion. End-to-end neurorrhaphy of the sciatic nerve graft to the right facial nerve and ligation of the proximal stump of the left facial nerve was also performed at the same time. The authors supported that this method is technically easy, minimize graft requirements, and thus donor side morbidity. In end-to-side nerve graft technique, reinnervation of the multiple recipient nerves occurs not by collateral sprouting but by terminal sprouting from the donor nerve via end-to-end neurorrhaphy (Matsuda *et al.*, 2008).

Although end-to-side neurorrhaphy was reported as an alternative in cases that the proximal nerve stump is not accessible, the donor nerve might not be close enough. In these cases, an end-to-side neurorrhaphy may be performed with the use of a nerve conduit that will bridge the gap between the distal end of the injured nerve and the donor nerve. Ülkür *et al.* (2003) showed that the use of a nerve graft or a vein graft in end-to-side neurorrhaphy is possible. Although the muscle-filled vein graft yielded adequate results in end-to-end neurorrhaphy, its use in end-to-side neurorrhaphy had disappointing functional and morphological results (Battiston *et al.*, 2000).

When a transected peripheral nerve is coapted in an end-to-side fashion to a healthy donor nerve, simultaneous cocontractions of muscles, innervated by the donor nerve and muscles, reinnervated by collateral sprouting, is most likely to be induced. (Beris *et al.*, 2007; Bertelli *et al.*, 1996; Lutz *et al.*, 2000; Özbek *et al.*, 2005; Tarasidis *et al.*, 1997). Therefore, it has been suggested that the use of an agonistic source of regenerated axons is required for voluntary motor function to recover (Lutz *et al.*, 2000). This is in conflict with recent data, supporting recovery of voluntary motor function when an antagonistic muscle is used as a donor nerve. More specifically, Papalia *et al.* (2007), after end-to-side neurorrhaphy between median and radial nerves in a rat model, demonstrated motor function recovery by grasping test. This finding was attributed to adaptation capacities of nervous system to complex surgical reconstructions.

VI. Conclusions

The results of the majority of experimental studies suggest that reinnervation of the distal stump of a transected nerve may occur if the former is coapted end-to-side to the trunk of an adjacent nerve. The source of the regenerating axons traveling in the epineurium of the donor nerve is thought to be the proximal Ranvier's nodes at the site of end-to-side neurorrhaphy, but histological evidence is still lucking. Although a great number of humoral factors have been identified, molecular mechanisms of nerve regeneration after end-to-side neurorrhaphy have not been completely clarified yet. Partial neurotomy of the donor nerve may enhance regeneration of motor neurons through end-to-side neurorrhaphy and reinnervation of motor targets. Further experimental studies are needed to determine the degree of focal axotomy and improve motor recovery after end-to-side nerve repair without deterioration of the donor nerve function.

References

Akeda, K., Hirata, H., Matsumoto, M., Fukuda, A., Tsujii, M., Nagakura, T., Ogawa, S., Yoshida, T., and Uchida, A. (2006). Regenerating axons emerge far proximal to the coaptation site in end-to-side nerve coaptation without a perineurial window using a T-shaped chamber. *Plast. Reconstr. Surg.* **117,** 1194–1203.

Al-Qattan, M. M., and Al-Thunyan, A. (1998). Variables affecting axonal regeneration following end-to-side neurorraphy. *Br. J. Plast. Surg.* **51,** 238–242.

Audisio, C., Nicolino, S., Scevola, A., Tos, P., Geuna, S., Battiston, B., and Perroteau, I. (2008). ErbB receptors modulation in different types of peripheral nerve regeneration. *Neuroreport* **19,** 1605–1609.

Bajrović, F., Kovačič, U., Pavčnik, M., and Sketelj, J. (2002). Interneuronal signaling is involved in induction of collateral sprouting of nociceptive axons. *Neuroscience* **111,** 587–596.

Battal, M. N., Hata, Y., Matsuka, K., Ito, O., and Matsuda, H. (1996). Cross-facial nerve grafting by end-to-side neurorrhaphy with or without removal of the epineurium in rats. *J. Jpn. Plast. Reconstr. Surg.* **16,** 641–647.

Battiston, B., Tos, P., Geuna, S., Giacobini-Robecchi, M. G., and Guglielmone, R. (2000). Nerve repair by means of vein filled with muscle grafts. II. Morphological analysis of regeneration. *Microsurgery* **20,** 37–41.

Beris, A., Lykissas, M., Korompilias, A., and Mitsionis, G. (2007). End-to-side nerve repair in peripheral nerve injury. *J. Neurotrauma* **24,** 909–916.

Bertelli, J. A., Dos Santos, A. R., and Calixto, J. B. (1996). Is axonal sprouting able to traverse the conjunctival layers of the peripheral nerve? A behavioral, motor, and sensory study of end-to-side nerve anastomosis. *J. Reconstr. Microsurg.* **12,** 559–563.

Bontioti, E., Kanje, M., Lundborg, G., and Dahlin, L. B. (2005). End-to-side nerve repair in the upper extremity of rat. *J. Peripher. Nerv. Syst.* **10,** 58–68.

Bontioti, E., Dahlin, L. B., Kataoka, K., and Kanje, M. (2006a). End-to-side nerve repair induces nuclear translocation of activating transcription factor 3. *Scand. J. Plast. Reconstr. Surg. Hand Surg.* **40,** 321–328.

Bontioti, E., Kanje, M., and Dahlin, L. B. (2006b). End-to-side nerve repair: Attachment of a distal, compared with a proximal and distal, nerve segment. *Scand. J. Plast. Reconstr. Surg. Hand Surg.* **40,** 129–135.

Brenner, M. J., Dvali, L., Hunter, D. A., Myckatyn, T. M., and Mackinnon, S. E. (2007). Motor neuron regeneration through end-to-side repairs is a function of donor nerve axotomy. *Plast. Reconstr. Surg.* **120,** 215–223.

Chen, B., Song, Y., and Liu, Z. (2008). Promotion of nerve regeneration in peripheral nerve by short-course FK506 after end-to-side neurorrhaphy. *J. Surg. Res.* (Epub ahead of print).

Fortes, W. M., Noah, E. M., Liuzzi, F. J., and Terzis, J. K. (1999). End-to-side neurorrhaphy: Evaluation of axonal response and upregulation of IGF-I and IGF-II in a non-injury model. *J. Reconstr. Microsurg.* **15,** 449–457.

Fujiwara, T., Matsuda, K., Kubo, T., Tomita, K., Hattori, R., Masuoka, T., Yano, K., and Hosokawa, K. (2005). Axonal supercharging technique using reverse end-to-side neurorrhaphy in peripheral nerve repair: An experimental study in the rat model. *J. Neurosurg.* **107,** 821–829.

Hayashi, A., Yanai, A., Komuro, Y., Nishida, M., Inoue, M., and Seki, T. (2004). Collateral sprouting occurs following end-to-side neurorrhaphy. *Plast. Reconstr. Surg.* **114,** 129–137.

Huber, G. C. (1895). A study on the operative treatment for loss of nerve substance in peripheral nerves. *J. Morphol.* **11**(Pt. 2), 629–740.

Isaacs, J., Allen, D., Chen, L. A., and Nunley, L. (2005). Reverse end-to-side neurotization. *J. Reconstr. Microsurg.* **21,** 43–50.

Létiévant, E. (1873). "Traité des Sections Nerveuses." J. B. Baillière et Fils, Paris.

Lundborg, G., Zhao, Q., Kanje, M., Danielsen, N., and Kerns, J. M. (1994). Can sensory and motor collateral sprouting be induced from intact peripheral nerve by end-to-side anastomosis? *J. Hand Surg.* **19B,** 227–282.

Lutz, B. S., Chuang, C. C., Hsu, J. C., Ma, S. F., and Wei, F. C. (2000). Selection of donor nerves-an important factor in end-to-side neurorrhaphy. *Br. J. Plast. Surg.* **53,** 149–154.

Lykissas, M. G., Korompilias, A. V., Batistatou, A. K., Mitsionis, G. I., and Beris, A. E. (2007a). Can end-to-side neurorrhaphy bridge large defects? An experimental study in rats. *Muscle Nerve* **36,** 664–671.

Lykissas, M. G., Sakellariou, E., Vekris, M. D., Kontogeorgakos, V. A., Batistatou, A. K., Mitsionis, G. I., and Beris, A. E. (2007b). Axonal regeneration stimulated by erythropoietin: An experimental study in rats. *J. Neurosci. Methods* **164,** 107–115.

Matsuda, H., Hata, Y., Matsuka, K., Ito, O., and Battal, N. M. (1995). Experimental study of nerve regeneration in end-to-side nerve coaptation. *J. Jpn. Plast. Reconstr. Surg.* **15,** 910–918.

Matsuda, K., Kakibuchi, M., Fukuda, K., Kubo, T., Madura, T., Kawai, K., Yano, K., and Hosokawa, K. (2005). End-to-side nerve grafts: Experimental study in rats. *J. Reconstr. Microsurg.* **21,** 581–591.

Matsuda, K., Kakibuchi, M., Kubo, T., Tomita, K., Fujiwara, T., Hattori, R., Yano, K., and Hosokawa, K. (2008). A new model of end-to-side nerve graft for multiple branch reconstruction: End-to-side cross-face nerve graft in rats. *J. Plast. Reconstr. Aesthet. Surg.* **61,** 1357–1367.

Matsumoto, M., Hirata, H., Nishiyama, M., Morita, A., Sasaki, H., and Uchida, A. (1999). Schwann cells can induce collateral sprouting intact axons: experimental study of end-to-side neurorrhaphy using a Y-chamber model. *J. Reconstr. Microsurg.* **15,** 281–286.

McCallister, W. V., Tang, P., Smith, J., and Trumble, T. E. (2001). Axonal regeneration stimulated by combination of nerve growth factor and ciliary neurotrophic factor in an end-to-side model. *J. Hand Surg.* **26A,** 478–488.

Noah, E. M., Williams, A., Jorgenson, C., Skoulis, T. G., and Terzis, J. K. (1997a). End-to-side neurorraphy: A histologic and morphometric study of axonal sprouting into an end-to-side nerve graft. *J. Reconstr. Microsurg.* **12,** 99–106.

Noah, E., Williams, A., Fortes, W., and Terzis, J. (1997b). A new animal model to investigate axonal sprouting after end-to-side neurorrhaphy. *J. Reconstr. Microsurg.* **13,** 317–325.

Ozbek, S., and Kurt, M. A. (2006). Simultaneous end-to-side coaptations of two severed nerves to a single healthy nerve in rats. *J. Neurosurg. Spine* **4,** 43–50.

Ozbek, S., Ozcan, M., Noyan, B., Kurt, M. A., Tirelioğlu, S., Bozkurt, C., Karaca, K., and Filiz, G. (2005). End-to-side nerve coaptation. Is an additional proximal coaptation useful when available. *Ann. Plast. Surg.* **55,** 281–288.

Papalia, I., Cardaci, A., d'Alcontres, F. S., Lee, J. M., Tos, P., and Geuna, S. (2007). Selection of the donor nerve for end-to-side neurorrhaphy. *J. Neurosurg.* **107,** 378–382.

Sámal, F., Haninec, P., Raska, O., and Dubový, P. (2006). Quantitative assessment of the ability of collateral sprouting of the motor and primary sensory neurons after the end-to-side neurorrhaphy of the rat musculocutaneous nerve with the ulnar nerve. *Ann. Anat.* **188,** 337–344.

Tarasidis, G., Watanabe, O., Mackinnon, S. E., Strasberg, S. R., Haughey, B. H., and Hunter, D. A. (1997). End-to-side neurorrhaphy resulting in limited sensory axonal regeneration in a rat model. *Ann. Otol. Rhinol. Laryngol.* **106,** 506–512.

Ülkür, E., Yüksel, F., Acikel, C., Okar, I., and Celiköz, B. (2003). Comparison of functional results of nerve graft, vein graft, and vein filled with muscle graft in end-to-side neurorrhaphy. *Microsurgery* **23,** 40–48.

Viterbo, F., Trindade, J. C., Hoshino, K., and Mazzoni Neto, A. (1992). Latero terminal neurorrhaphy without removal of the epineural sheath: Experimental study in rats. *Rev. Paul. Med.* **110,** 267–275.

Xiong, G., Ling, L., Nakamura, R., and Sugiura, Y. (2003). Retrograde tracing and electrophysiological findings of collateral sprouting after end-to-side neurorrhaphy. *Hand Surg.* **8,** 145–150.

Yamauchi, T., Maeda, M., Tamai, S., Tamai, M., Yajima, H., Takakura, Y., Haga, S., and Yamamoto, H. (2000). Collateral sprouting mechanism after end-to-side nerve repair in the rat. *Med. Electron Microsc.* **33,** 151–156.

Yamauchi, T., Yajima, H., Tamai, S., Ohgushi, H., Tamai, M., Maeda, M., Kizaki, K., Kobata, Y., Fukui, A., and Takakura, Y. (2001). Neurohistochemical analysis of regeneration in rat peripheral nerve after end-to-side neurorrhaphy. *J. Orthop. Sci.* **6,** 82–87.

Zhang, F., Cheng, C., Chin, B. T., Ho, P. R., Weibel, T. J., Lineaweaver, W. C., and Buncke, H. J. (1998). Results of termino-lateral neurorrhaphy to original and adjacent nerves. *Microsurgery* **18,** 276–281.

Zhang, Z., Soucacos, P. N., Bo, J., and Beris, A. E. (1999). Evaluation of collateral sprouting after end-to-side nerve coaptation using a fluorescent double-labeling technique. *Microsurgery* **19,** 281–286.

Zhang, Z., Soucacos, P., Beris, A., Bo, J., Ioachim, E., and Johnson, E. O. (2000). Long-term evaluation of rat peripheral nerve repair with end-to-side neurorrhaphy. *J. Reconstr. Microsurg.* **16,** 303–311.

Zhu, Q. T., Zhu, J. K., and Chen, G. Y. (2008). Location of collateral sprouting of donor nerve following end-to-side neurorrhaphy. *Muscle Nerve* **38,** 1506–1509.

END-TO-SIDE NERVE REGENERATION: FROM THE LABORATORY BENCH TO CLINICAL APPLICATIONS

Pierluigi Tos,* Stefano Artiaco,† Igor Papalia,‡ Ignazio Marcoccio,§ Stefano Geuna,¶ and Bruno Battiston*

*Reconstructive Microsurgery Unit, Department of Orthopedics, C.T.O. Hospital, Turin 10126, Italy

†Department of Orthopaedics, Traumatology, Rehabilitation, Plastic and Reconstructive Sciences, Second University of Naples, Naples, Italy

‡Department of Surgical Disciplines, University of Messina, Messina, Italy

§Hand and Microsurgery Unit, Istituto Clinico Città di Brescia, Brescia, Italy

¶Department of Clinical and Biological Sciences, San Luigi Gonzaga School of Medicine, University of Turin, Turin, Brescia, Italy

Translation of laboratory results to the patient is a critical step in biomedical research and sometimes promising basic science and preclinical results fail to meet the expectations when translated to the clinics. End-to-side (ETS) nerve regeneration is an example of an innovative neurobiological concept, which, after having generated great expectations in experimental and preclinical studies, provided very conflicting results when applied to clinical case series.

A number of basic science studies have shown that ETS neurorrhaphy, in fact, is able to induce collateral sprouting from donor nerve's axons, allowing for massive repopulation of the distal nerve stump. Experimental studies have also shown that ETS neurorrhaphy can recover voluntary control of skeletal muscles and that voluntary motor function recovery can be achieved both with agonistic and antagonistic donor nerves, thus widening the potential clinical indications. However, clinical case series reported so far, did not meet these promises and results have been rather conflicting, especially regarding repair of proximally located mixed nerves. In contrast, ETS reconstruction of distal sensory nerve

INTERNATIONAL REVIEW OF NEUROBIOLOGY, VOL. 87
DOI: 10.1016/S0074-7742(09)87014-1

lesions led to a more positive outcome and, most importantly, consistent results among international centers carrying out clinical trials.

Concluding, ETS is a promising microsurgical approach for nerve coaptation, based on a convincing and innovative neurobiological concept. However, conflicting clinical results and disagreement among surgeons regarding its employment suggest that this technique should still be considered an *ultima ratio*, reserved for cases where no other repair technique can be attempted. New data coming from neurobiological research will help further enlarge the clinical indications of ETS nerve reconstruction, explain the different results found in laboratory animals and humans, and contribute to new treatments and rehabilitation strategies aimed at improving the efficacy of nerve regeneration after ETS neurorrhaphy.

I. Introduction

End-to-side (ETS) coaptation for repairing a severed peripheral nerves has received a growing attention beginning with Viterbo's studies published 15 years ago (Viterbo *et al.*, 1992, 1994). Early studies describing ETS coaptation as a treatment option in nerve lesions were already published in 1873 by Letievant in the "Traitè des Sections Nerveuses" (Letievant, 1873; Papalia *et al.*, 2007a). Along the last years of the nineteenth century and beginning of the twentieth century, several interesting clinical and experimental studies were carried out on this nerve repair technique (Battiston *et al.*, 2009a; Papalia *et al.*, 2007a). Surprisingly, interest in ETS neurorrhaphy disappeared for most of the twentieth century, until its rediscovery in the early 1990s.

The basic neurobiological concept, underlying ETS nerve repair, is disarmingly simple: to obtain nerve fiber regeneration along the distal stump of a transected nerve, of which the proximal stump is no longer available, by inducing collateral axonal sprouting from a healthy neighbor donor nerve (Geuna *et al.*, 2006).

This is the third review in a series of papers included with this special issue of the *International Review of Neurobiology*, that focus on the basic science research in ETS nerve regeneration (Beris and Lykissas, 2009, this issue; Bontioti and Dahlin, 2009, this issue). A brief overview of key issues from the last 15 years of preclinical and clinical literature is presented in this chapter outlining some possible explications for the discrepancy between basic and clinical data. Finally, some potential investigation lines are described toward which future basic science research should be directed, in order to delineate effective treatments and rehabilitation strategies for improving the outcome of ETS nerve regeneration.

II. Basic Science Studies

The nervous system can undergo adaptive changes in relation to environmental perturbations, a phenomenon that is usually referred to as neuroplasticity. The occurrence of ETS-induced collateral axonal sprouting represents a long-known paradigm exemplifying the plastic properties of the nervous system (Edds, 1953). However, axonal sprouting in adults is usually limited to the terminal portion of an axon and occurs when a neighboring axon degenerates, leaving its motor unit denervated. Until recently, the possibility that a healthy adult axon could generate collateral sprouting from a proximal portion of its length, that is, far from its peripheral ends, was considered unsuitable (Geuna *et al.*, 2006). Yet, the possibility that collateral sprouted axons, having reinnervated the target organs of another nerve, could eventually lead to a recovery of the lost function, was considered highly unlikely. But looking at the experimental data accumulated over the last 15 years, these beliefs have been disproved, as it was clearly demonstrated that ETS neurorrhaphy leads to the above-mentioned phenomena (Geuna *et al.*, 2006).

One of the main questions that gathered the researchers' attention is the origin of new axons found in ETS regeneration. Some authors still support the idea of ETS axonal regeneration originates from terminal sprouting of the donor nerve's axons which have been injured during surgery (Akeda *et al.*, 2006; Fernandez *et al.*, 2007; Hayashi *et al.*, 2008), this being claimed especially in case of motor axons (Pannucci *et al.*, 2007). In response to this, various experimental studies have shown that terminolateral axon regeneration may occur also without opening an epineurial window (Lundborg *et al.*, 1994; Viterbo *et al.*, 1992). Studies which employed noninjurious ETS cooptation devices, Y-shaped silicone chamber and wrapped muscle aponeurosis, respectively, (Hayashi *et al.*, 2004; Matsumoto *et al.*, 1999), demonstrated the occurrence of collateral sprouting and terminolateral regeneration in absence of any type of epineurial damage. Further evidence supporting the collateral sprouting model came from double retrograde labeling studies, which showed the presence of some double-labeled neurons after ETS nerve repair (Bontioti *et al.*, 2005; Furukawa *et al.*, 2008; Kanje *et al.*, 2000; Kubek *et al.*, 2004; Matsuda *et al.*, 2005; Samal *et al.*, 2006; Sananpanich *et al.*, 2007; Xiong *et al.*, 2003; Zhang *et al.*, 1999). An elegant recent study by Zhu *et al.* (2008) using a microtease technique showed that axons sprouted collaterally from myelinated nerve fibers at the node of Ranvier.

Despite the evidence of terminolateral nerve regeneration occurring also without any surface lesion to the donor nerve, several experimental studies have shown superior results with nerve regeneration when a window is opened in the donor nerve's side (Kokkalis *et al.*, 2009; Lunborg *et al.*, 1994; Tham and Morrison, 1998; Zhang *et al.*, 2000; Zhao *et al.*, 1997). Therefore, it is now widely accepted that resection of a small part of the epineurium improves the effectiveness of ETS nerve repair (Dvali and Myckatyn, 2008; Lundborg, 2005). Several experimental studies have also shown that a resection of the deeper connective layers of the nerve (the perineurium) further increases the effectiveness of terminolateral regeneration (Goheen-Robillard *et al.*, 2002; Liu *et al.*, 1999; Noah *et al.*, 1997; Okajima and Terzis, 2000; Yan *et al.*, 2002; Zhang *et al.*, 2001, 2006). Despite experimental studies showing that even large perineurial lesions induce only a limited injury and no functional impairment to the donor nerve (Walker *et al.*, 2004), the potential risk of damaging the donor nerve makes the clinical employment of the perineurial window opening still controversial (Lundborg, 2005).

Factors triggering collateral sprouting from nerve fibers in a large nerve trunk still need to be identified, as well as the basic biochemical mechanisms which regulate the early phases of this phenomenon. A class of molecules potentially implicated in this process are metalloproteinases (MMPs) (Dzwonek *et al.*, 2004) that, together with their endogenous tissue inhibitors (TIMPs), participate in remodeling of the extracellular matrix (Chang and Werb, 2001). An increasing interest in the role of the TIMP/MMP system in axonal pathfinding (McFarlane, 2003) can be seen in the scientific community, possibly leading to a better understanding of the basic mechanisms underlying collateral axonal sprouting after ETS neurorrhaphy.

Interestingly, interposition of a nonnervous conduit in ETS nerve repair leads to successful axon regeneration (Manasseri *et al.*, 2007; Ulkur *et al.*, 2003), supporting the idea that the triggers of axonal collateral sprouting can exert their action also at distance. In the other hand, two very effective techniques in end-to-end nerve repair, interposition of a fresh-muscle-enriched vein segment (Raimondo *et al.*, 2005), and fibrin glue coaptation (Ornelas *et al.*, 2006), inhibit collateral axonal regeneration. This suggests that important differences exist between the mechanisms which promote and regulate end-to-end and end-to-side nerve fiber regeneration.

The interaction between injured and noninjured neurons within the dorsal root ganglion, through direct or indirect interneuronal signaling, has also been suggested as a possible trigger of collateral sprouting in case of sensory axons (Bajrovic *et al.*, 2002). Finally, activation of specific transcription factors, in particular activating transcription factor 3 (ATF3), has also been proposed as a possible trigger of both neuronal and nonneuronal cells after ETS neurorrhaphy (Bontioti *et al.*, 2006, 2009, this issue).

Whereas most authors now accept that repopulation of a severely damaged distal nerve stump through ETS neurorrhaphy is possible, the question as to how can axons function when they are connected simultaneously to two different peripheries is still unanswered. Experimental studies have shown a variable degree of functional recovery after newly generated axons had reinnervated the periphery of the severed nerve (Bontioti *et al.*, 2005; Giovanoli *et al.*, 2000; Papalia *et al.*, 2003). Yet, a partial recovery of lost function has been demonstrated by simultaneous ETS suturing of two severed nerve trunks to a single healthy nerve in rats (Bontioti *et al.*, 2005; Ozbek and Kurt, 2006). Contrary to present belief (Lutz *et al.*, 2000), recovery of voluntary motor function in the rat has been demonstrated using an antagonistic donor nerve as well (Geuna *et al.*, 2007; Papalia *et al.*, 2007b).

These observations raise several questions: How can voluntary functional recovery occur after ETS nerve repair? What happens to the central neuronal circuitries? Peripherally, axonal pruning appears to be the basic mechanism for the functional adaptation to the new connections (Bontioti *et al.*, 2005; Hayashi *et al.*, 2004) but it is clear that central (brain) adaptations are crucial to ETS nerve repair. Brain reorganization related to major changes in the peripheral connections is an important emerging neuroscientific issue (Dahlin *et al.*, 2009, this issue; Lou *et al.*, 2006; Lundborg, 2000; Navarro, 2009, this issue) and its study in the case of ETS nerve repair might provide important information on brain plasticity mechanisms.

III. Clinical Studies

Literature from the last two decades has reported an increasing number of studies on clinical application of ETS neurorraphy. Results of ETS nerve coaptation were addressed in case reports and selected clinical series, used to treat a wide pattern of peripheral nerve conditions, including (a) brachial plexus injuries (Amr and Moharram, 2005; Ferraresi *et al.*, 2002; Haninec *et al.*, 2007; Malessy *et al.*, 1999; Mennen, 2003; Pienaar *et al.*, 2004; Viterbo *et al.*, 1995), (b) mixed nerves lesions (Kayikcioglu *et al.*, 2000; Kostakoglu, 1999; Luo *et al.*, 1997; Mennen, 2003; Mouilhade *et al.*, 2008; Ogun *et al.*, 2003; Yuksel *et al.*, 2004), (c) digital nerves injuries (Frey and Giovanoli, 2003; Landwehrs *et al.*, 2008; Mennen, 2003; Pelissier *et al.*, 2001; Voche, 2005), (d) painful neuroma of sensory branches of the radial nerve (Al-Qattan, 2001; Marcoccio and Adani, 2008), and (e) facial nerve injuries (Hammerschlag, 1999; Manni *et al.*, 2001; May *et al.*, 1991).

Despite growing knowledge, emerging from experimental studies, a clear assessment on the clinical effectiveness of ETS nerve suture is far from being reached. Several papers base their findings on small clinical case series which are often difficult to compare. Randomized controlled trials have never been performed in order to evaluate the results of nerve repair using ETS coaptation versus autologous nerve grafting or alternative reconstructive techniques, such as biological and synthetic nerve conduits. Until now, the largest clinical series of

ETS nerve sutures, reported by Mennen in 2003, included 56 patients with a variety of peripheral nerve injuries, affecting almost always brachial plexus or upper limb nerve trunks. The authors concluded that, although the outcome may still be unpredictable to some degree, more than 85% of compliant patients have partial or near normal sensory and/or motor function. Unfortunately, successful results have not been consistently reported in the literature. Compared to some successful methods, as those performed in sensory nerves repair (Frey and Giovanoli, 2003; Landwehrs and Bruser, 2008; Pelissier *et al.*, 2001; Voche, 2005) or in cases of mixed nerves with long nerve defect (Kostakoglu, 1999; Luo *et al.*, 1997; Mennen, 2003; Mouilhade *et al.*, 2009; Ogun *et al.*, 2003; Yuksel *et al.*, 2004), ETS nerve sutures showed poor or at best unclear results in other studies (Amr and Moharram, 2005; Ferraresi *et al.*, 2002; Kayikcioglu, 2000; Malessy *et al.*, 1999; Pienaar *et al.*, 2004).

The opinions reported, so far, in literature obviously reflect personal experiences and are, for this reason, somewhat contrasting. As for the treatment of painful neuromas and isolated radial nerve injuries, scarce data is available. A systematic analysis of published clinical studies indicates as most documented fields of application for ETS neurorraphy, the repair of digital nerve lesions and the treatment of mixed nerve and brachial plexus injuries.

A. Sensory Nerves

Regarding digital nerves injuries, our clinical experience and analysis of the literature supports the view that ETS nerve suture can be a reliable technique for recovery of distal sensory reinnervation. During 6 years, we performed ETS nerve coaptation in eight patients with traumatic or postsurgical digital nerve injuries (Artiaco *et al.*, 2009). All patients had sensory recovery, graded as S3+ in seven cases and S3 in one case, according to the classification of British Medical Research Council (MRC) modified by Mackinnon and Dellon (1985). The average two-point discrimination distance was 13 mm. From five retrospective clinical studies reported in the literature a comprehensive number of 26 further patients have been described and in all but one case a successful result has been documented (Frey and Giovanoli, 2003; Landwehrs and Bruser, 2008; Mennen, 2003; Pelissier *et al.*, 2001; Voche, 2005). Most patients showed a S3 or S3+ sensory recovery and in four cases the result was particularly good, with a two-point discrimination distance equal or inferior to 6 mm (S4).

Therefore, successful outcomes may be expected in the treatment of digital nerve lesions by employing the terminolateral nerve suture technique. In our opinion, this reconstructive method is indicated in late digital nerve lesions and acute lesions, when direct suturing of the nerve is difficult or not possible at all due to the gap's length. In case of nerve gaps of less than 3 cm, ETS coaptation may prove as an alternative to biological and synthetic tubulization, especially when digital nerve reconstruction by means of nerve autograft is refused by the patient. In cases of digital sensory nerve defects greater than 3 cm, use of ETS suture technique may avoid donor site morbidity resulting from autograft repair. In addition, this technique can be considered a good therapeutic option when another previous repair attempt failed.

B. Mixed Nerves and Brachial Plexus

Long defects of mixed nerves have been often treated with ETS coaptation. For anatomical and functional reasons, the median and ulnar nerves are particularly suitable for this kind of repair in forearm lesions. Mennen (2003) reported a clinical series with 33 cases of ulnar to median and 7 cases of median to ulnar ETS coaptation. Although less than half of the patients were available for clinical follow-up, a satisfactory sensory recovery was demonstrated in most cases. Instead, motor recovery (M3) was recorded only in 3 out of 14 patients with ulnar to median ETS coaptation, as previously reported by Luo (1997) in a clinical case description. Later reports confirmed the effectiveness of median to ulnar ETS coaptation in recovering protective sensibility (Kostakoglu, 1999; Mouilhade *et al.*, 2009; Ogun *et al.*, 2003; Yuksel *et al.*, 2004).

As for brachial plexus injury, the first ETS neurotization was reported in the recent medical literature by Viterbo who sutured C5 and C6 roots to the phrenic nerve (Viterbo *et al.*, 1995). Since then, further studies have been published describing heterogeneous results (Amr and Moharram, 2005; Ferraresi *et al.*, 2002; Haninec *et al.*, 2007; Mennen, 2003; Pienaar *et al.*, 2004).

Mennen (2003) performed ETS nerve sutures in eight patients who underwent brachial plexus repair. Seven out of eight patients were followed up between 18 and 36 months postoperatively, reporting contrasting outcomes. In fact, according to MRC scale, motor (deltoid and biceps) and sensory recovery were successful in four cases (M4), partially successful in one case (M3), and unsuccessful in two (M0). These findings were not confirmed in a following study performed by Pienaar *et al.* (2004), who observed, in nine patients with brachial plexus lesion, two cases of partial sensory recovery, without useful motor recovery due to ETS neurotization. Recently, Haninec *et al.* (2007) reported their experience with ETS nerve suture in a homogeneous series of incomplete brachial plexus injuries. They employed intraplexual donor nerves (ulnar–median–radial) and the axillary nerve as a recipient, reporting a motor recovery of the deltoid muscle in 64% of patients (9 out of 14).

We used ETS nerve sutures, associated with standard neurotizations, in 11 patients with brachial plexus injuries (Battiston *et al.*, 2009b). In this case series, the ETS coaptations, performed to support shoulder function recovery (abduction–external rotation), showed successful results (M4) in one case, partially successful (M3) in two cases, and poor or absent result (M0-M1-M2) in five cases. We did not observe a significant benefit from ETS neurotization in primary reconstructive surgery for patients affected by traumatic closed brachial plexus injuries. Interestingly, in one patient with a complete brachial plexus injury, the ETS coaptation of hypoglossal nerve to suprascapularis nerve resulted in the recovery of some active shoulder abduction (M3), with a typical lingual synkinesia. This observation contradicted the results of previous studies, in which the hypoglossal nerve seemed to be an unsuitable donor for terminolateral nerve suture in brachial plexus repair (Ferraresi *et al.*, 2002; Malessy *et al.*, 1999).

Currently, we believe that this technique should not substitute standard neurotizations which are more reliable in adult brachial plexus surgery. Occasionally, ETS coaptation may support standard reconstructive procedures in case of severe brachial plexus injuries when few undamaged donor nerves are available.

IV. Future Perspectives

A large review of the comprehensive literature on ETS nerve regeneration shows that, in laboratory animals, transected nerves with missing proximal end can be repaired by ETS microsurgical suture, both on agonistic and antagonistic donor nerves, leading to axonal regeneration and the partial restoration of voluntary control of the motor function lost after nerve damage.

In spite of good experimental results, reports published, so far, on clinical application of ETS coaptation did not show the same positive outcome observed in laboratory animals. Noteworthy, application of ETS nerve repair in a primate model also gave rise to unpredictable results (Kelly *et al.*, 2007).

Despite a limited number of treated patients, which makes it difficult to draw conclusions regarding the clinical efficacy of ETS neurorrhaphy, literature analysis supports the view that this technique can be efficient for distal sensory reinnervation and that, for the repair of digital sensory nerve defects superior to 3–4 cm, this approach may avoid the donor site morbidity required for autograft repair. Moreover, ETS reconstruction may be a good therapeutic option when previous attempts of nerve repair, using another method, have failed. By contrast, results in brachial plexus surgery are more difficult to interpret, due to frequent use of associated neurotization procedures that hinders the discrimination of the real effectiveness of ETS neurorrhaphy in functional recovery.

The recent history of ETS nerve regeneration research represents a typical example of the complexity in translating results obtained on the laboratory bench to the clinical application. In fact, while experimental results have opened exciting perspectives for the employment of ETS nerve repair with human patients, the clinical results obtained, so far, have been often disappointing. It is clear that more basic research is needed to explain the discrepancy between basic and clinical data and especially, to find out how to ameliorate the outcome of ETS nerve repair by adequate treatment and rehabilitation. In particular, one of the possible reasons for poor outcome, observed in patients, is that ETS axon regeneration is slower than end-to-end regeneration (De Sa *et al.*, 2004; Sanapanich *et al.*, 2002). Therefore, the quest for chemical and physical agents that can stimulate nerve regeneration after ETS repair appears to be one of the next key goals for basic science research since it could improve the effectiveness of this technique and thus extend the indications for its clinical employment. Experimental studies have shown that ETS nerve regeneration can be significantly improved by phototherapy (Gigo-Benato *et al.*, 2004) and FK506 (Chen *et al.*, 2009) and acetyl-L-carnitine (Kostopoulos *et al.*, 2009) administration and several papers in this special issue highlight various strategies that can indicate possible future directions for improving the effectiveness of ETS nerve repair, such as drugs (Magnaghi *et al.*, 2009, this issue) and gene therapy (Zacchigna and Giacca, 2009, this issue). Interdisciplinary and multitranslational research must be sought, since discovering new ways for improving ETS axonal regeneration is not just a matter of testing many different agents, but it should also be directed towards understanding the basic mechanisms through which the agents exert their action.

Acknowledgments

The authors wish to thank Josette Legagnaux and Jean Luc Vignes and the Laboratoire de Microchirurgie of the Ecole de Chirurgie de Paris for the valuable expert and technical assistance. This work was supported by grants from the MUR (Italian Ministry of University and Research), ex-60% fund, FIRB fund (code: RBAU01BJ95), PRIN2005 fund (code: 2005057088), the Compagnia di San Paolo (Bando Programma Neuroscienze), and the Regione Piemonte (Progetto Ricerca Sanitaria Finalizzata).

References

Akeda, K., Hirata, H., Matsumoto, M., Fukuda, A., Tsujii, M., Nagakura, T., Ogawa, S., Yoshida, T., and Uchida, A. (2006). Regenerating axons emerge far proximal to the coaptation site in end-to-side nerve coaptation without a perineurial window using a T-shaped chamber. *Plast. Reconstr. Surg.* **117,** 1194–1203.

Al-Qattan, M. M. (2001). Terminolateral neurorrhaphy: Review of experimental and clinical studies. *J. Reconstr. Microsurg.* **17,** 99–108.

Amr, S. M., and Moharram, A. N. (2005). Repair of brachial plexus lesions by end-to-side side-to-side grafting neurorrhaphy: Experience based on 11 cases. *Microsurgery* **25,** 126–146.

Artiaco, S., Tos, P., Conforti, L. G., Geuna, S., and Battiston, B. (2009). Termino-lateral nerve suture in lesions of the digital nerves: Clinical experience and literature review. *J. Hand Surg.* (in press).

Bajrovic, F., Kovacic, U., Pavcnik, M., and Sketelj, J. (2002). Interneuronal signalling is involved in induction of collateral sprouting of nociceptive axons. *Neuroscience* **111,** 587–596.

Battiston, B., Raimondo, S., Tos, P., Gaidano, V., Audisio, C., Scevola, A., Perroteau, I., and Geuna, S. (2009a). Tissue engineering of peripheral nerves. *Int. Rev. Neurobiol.* **87,** 227–250.

Battiston, B., Artiaco, S., Conforti, L. G., and Tos, P. (2009b). End-to-side nerve suture in traumatic injuries of brachial plexus. Review of the literature and personal case series. *J. Hand Surg.* (in press).

Beris, A. E., and Lykissas, M. G. (2009). Experimental results in end-to-side neurorrhaphy. *Int. Rev. Neurobiol.* **87,** 269–279.

Bontioti, E., Kanje, M., Lundborg, G., and Dahlin, L. B. (2005). End-to-side nerve repair in the upper extremity of rat. *J. Peripher. Nerv. Syst.* **10,** 58–68.

Bontioti, E., Dahlin, L. B., Kataoka, K., and Kanje, M. (2006). End-to-side nerve repair induces nuclear translocation of activating transcription factor 3. *Scand. J. Plast. Reconstr. Surg. Hand Surg.* **40,** 321–328.

Bontioti, E., and Dahlin, L. B. (2009). Mechanisms underlying the end-to-side nerve regeneration. *Int. Rev. Neurobiol.* **87,** 251–268.

Chang, C., and Werb, Z. (2001). The many faces of metalloproteases: Cell growth, invasion, angiogenesis and metastasis. *Trends Cell Biol.* **11,** 37–43.

Chen, B., Song, Y., and Liu, Z. (2009). Promotion of nerve regeneration in peripheral nerve by short-course FK506 after end-to-side neurorrhaphy. *J. Surg. Res.* **152,** 303–310.

Dahlin, L., Johansson, F., Lindwall, C., and Kanje, M. (2009). Future perspective in peripheral nerve reconstruction. *Int. Rev. Neurobiol.* **87,** 507–530.

De Sa, J. M., Mazzer, N., Barbieri, C. H., and Barriera, A. A. (2004). The end-to-side peripheral nerve repair. Functional and morphometric study using the peroneal nerve of rats. *J. Neurosci. Methods* **136,** 45–53.

Dvali, L. T., and Myckatyn, T. M. (2008). End-to-side nerve repair: Review of the literature and clinical indications. *Hand Clin.* **24,** 455–460.

Dzwonek, J., Rylski, M., and Kaczmarek, L. (2004). Matrix metalloproteinases and their endogenous inhibitors in neuronal physiology of the adult brain. *FEBS Lett.* **567,** 129–135.

Edds, M. V. (1953). Collateral nerve regeneration. *Q. Res. Biol.* **28,** 260–276.

Fernandez, E., Lauretti, L., Tufo, T., D'Ercole, M., Ciampini, A., and Foglietto, F. (2007). End-to-side nerve neurorrhaphy: Critical appraisal of experimental and clinical data. *Acta Neurochir. Suppl.* **100,** 77–84.

Ferraresi, S., Garozzo, D., Ravenni, R., Dainese, R., De Grandis, D., and Buffatti, P. (2002). Hemihypoglossal nerve transfer in brachial plexus repair: Technique and results. *Neurosurgery* **50,** 332–335.

Frey, M., and Giovanoli, P. (2003). End-to-side neurorrhaphy of sensory nerves. *Eur. J. Plast. Surg.* **26,** 85–88.

Furukawa, H., Saito, A., Mol, W., Sekido, M., Sasaki, S., and Yamamoto, Y. (2008). Double innervation occurs in the facial mimetic muscles after facial-hypoglossal end-to-side neural repair: Rat model for neural supercharge concept. *J. Plast. Reconstr. Aesthet. Surg.* **61,** 257–264.

Geuna, S., Papalia, I., and Tos, P. (2006). End-to-side (terminolateral) nerve regeneration: A challenge for neuroscientists coming from an intriguing nerve repair concept. *Brain Res. Rev.* **52,** 381–388.

Geuna, S., Tos, P., Raimondo, S., Lee, J. M., Gambarotta, G., Nicolino, S., Fornaro, M., Papalia, I., Perroteau, I., and Battiston, B. (2007). Functional, morphological and biomolecular assessment of posttraumatic neuro-muscular recovery in the rat forelimb model. *Acta Neurochir. Suppl.* **100,** 173–177.

Gigo-Benato, D., Geuna, S., de Castro Rodrigues, A., Tos, P., Fornaro, M., Boux, E., Battiston, B., and Giacobini-Robecchi, M. G. (2004). Low-power laser biostimulation enhances nerve repair after end-to-side neurorrhaphy: A double-blind randomized study in the rat median nerve model. *Lasers Med. Sci.* **19,** 57–65.

Giovanoli, P., Koller, R., Meuli-Simmen, C., Rab, M., Haslik, W., Mittlbock, M., Meyer, E., and Frey, M. (2000). Functional and morphometric evaluation of end-to-side neurorrhaphy for muscle reinnervation. *Plast. Reconstr. Surg.* **106,** 383–392.

Goheen-Robillard, B., Myckatyn, T. M., Mackinnon, S. E., and Hunter, D. A. (2002). End-to-side neurorrhaphy and lateral axonal sprouting in a long graft rat model. *Laryngoscope* **112,** 899–905.

Hammerschlag, P. E. (1999). Facial reanimation with jump interpositional graft hypoglossal facial anastomosis and hypoglossal facial anastomosis: Evolution in management of facial paralysis. *Laryngoscope* **109,** 1–23.

Haninec, P., Samal, F., Tomas, R., Houstava, L., and Dubovy, P. (2007). Direct repair (nerve grafting), neurotization and end-to-side neurorrhaphy in the treatment of brachial plexus injury. *J. Neurosurg.* **106,** 391–399.

Hayashi, A., Yanai, A., Komuro, Y., Nishida, M., Inoue, M., and Seki, T. (2004). Collateral sprouting occurs following end-to-side neurorrhaphy. *Plast. Reconstr. Surg.* **114,** 129–137.

Hayashi, A., Pannucci, C., Moradzadeh, A., Kawamura, D., Magill, C., Hunter, D. A., Tong, A. Y., Parsadanian, A., Mackinnon, S. E., and Myckatyn, T. M. (2008). Axotomy or compression is required for axonal sprouting following end-to-side neurorrhaphy. *Exp. Neurol.* **211,** 539–550.

Kanje, M., Arai, T., and Lundborg, G. (2000). Collateral sprouting from sensory and motor axons into an end to side attached nerve segment. *Neuroreport* **11,** 2455–2459.

Kayikcioglu, A., Karamursel, S., Agaoglu, G., Kecik, A., Celiker, R., and Cetin, A. (2000). End-to-side neurorrhaphies of the ulnar and median nerves at the wrist: Report of two cases without sensory or motor improvement. *Ann. Plast. Surg.* **45,** 641–643.

Kelly, E. J., Jacoby, C., Terenghi, G., Mennen, U., Ljungberg, C., and Wiberg, M. (2007). End-to-side nerve coaptation: A qualitative and quantitative assessment in the primate. *J. Plast. Reconstr. Aesthet. Surg.* **60,** 1–12.

Kokkalis, Z. T., Soucacos, P. N., and Terzis, J. K. (2009). Effect of donor nerve injury distal to an end-to-side neurorrhaphy model. *J. Reconstr. Microsurg.* (in press).

Kostakoglu, N. (1999). Motor and sensory reinnervation after end-to-side median to ulnar nerve coaptation in the forearm. *Br. J. Plast. Surg.* **52,** 402–407.

Kostopoulos, K., Davis, C. L., and Terzis, J. K. (2009). Effects of acetylo-L-carnitine in end-to-side neurorrhaphy: A pilot study. *Microsurgery* (in press).

Kubek, T., Kyr, M., Haninec, P., Samal, F., and Dubovy, P. (2004). Morphological evidence of collateral sprouting of intact afferent and motor axons of the rat ulnar nerve demonstrated by one type of tracer molecule. *Ann. Anat.* **186,** 231–234.

Landwehrs, G. M., and Bruser, P. (2008). Clinical results of terminolateral neurorraphy in digital nerves. *Handchir. Mikrochir. Plast. Chir.* **40,** 318–321.

Létiévant, E. (1873). "Traité des Sections Nerveuses," pp. 421–429. J.B. Baillière et Fils, Paris.

Liu, K., Chen, L. E., Seaber, A. V., Goldner, R. V., and Urbaniak, J. R. (1999). Motor functional and morphological findings following end-to-side neurorrhaphy in the rat model. *J. Orthop. Res.* **17,** 293–300.

Lou, L., Shou, T., Li, Z., Li, W., and Gu, Y. (2006). Transhemispheric functional reorganization of the motor cortex induced by the peripheral contralateral nerve transfer to the injured arm. *Neuroscience* **138,** 1225–1231.

Lundborg, G. (2000). Brain plasticity and hand surgery: An overview. *J. Hand Surg.* **25B,** 242–252.

Lundborg, G. (2005). "Nerve Injury and Repair," pp. 151–155. Churchill Livingstone, Edinburgh.

Lundborg, G., Zhao, Q., Kanje, M., Danielsen, N., and Kerns, J. M. (1994). Can sensory and motor collateral sprouting be induced from intact peripheral nerve by end-to-side anastomosis? *J. Hand Surg.* **19B,** 277–282.

Luo, Y., Wang, T., and Fang, H. (1997). Preliminary investigation of treatment of ulnar nerve defect by end-to-side neurorrhaphy. *Zhongguo Xiu Fu Chong Jian Wai Ke Za Zhi* **11,** 338–339.

Lutz, B. S., Chuang, D. C. C., Hsu, J. C., Ma, S. F., and Wei, F. C. (2000). Selection of donor nerves—An important factor in end-to-side neurorrhaphy. *Br. J. Plast. Surg.* **53,** 149–154.

Mackinnon, S. E., and Dellon, A. L. (1985). Two-point discrimination tester. *J. Hand Surg.* **10A,** 906–907.

Magnaghi, V., Procacci, P., and Tata, A. M. (2009). Novel pharmacological approaches to Schwann cells as neuroprotective agents for peripheral nerve regeneration. *Int. Rev. Neurobiol.* **87,** 295–315.

Malessy, M. J., Hoffmann, C. F., and Thomeer, R. T. (1999). Initial report on the limited value of hypoglossal nerve transfer to treat brachial plexus root avulsions. *J. Neurosurg.* **91,** 601–604.

Manasseri, B., Raimondo, S., Geuna, S., Risitano, G., and Stagno D'alcontres, F. (2007). Ulnar nerve repair by end-to-side neurorrhaphy on the median nerve with interposition of a vein: An experimental study. *Microsurgery* **27,** 27–31.

Manni, J. J., Beurskens, C. H., Van de Velde, C., and Stokroos, R. J. (2001). Reanimation of the paralyzed face by indirect hypoglossal-facial nerve anastomosis. *Am. J. Surg.* **182,** 268–273.

Marcoccio, I., and Adani, R. (2008). Digital neuromas: Sensation recovery through termino-lateral nerve repair. *Riv. Chir. Mano* **1,** 64–71.

Matsuda, K., Kakibuchi, M., Fukuda, K., Kubo, T., Madura, T., Kawai, K., Yano, K., and Hosokawa, K. (2005). End-to-side nerve grafts: Experimental study in rats. *J. Reconstr. Microsurg.* **21,** 581–591.

Matsumoto, M., Hirata, H., Nishiyama, M., Morita, A., Sasaki, H., and Uccida, A. (1999). Schwann cells can induce collateral sprouting from intact axons: Experimental study of end-to-side neurorrhaphy using a Y-chamber model. *J. Reconstr. Microsurg.* **15,** 281–286.

May, M., Sobol, S. M., and Mester, S. J. (1991). Hypoglossal-facial nerve interpositional-jump graft for facial reanimation without tongue atrophy. *Otolaryngol. Head Neck Surg.* **104,** 818–825.

McFarlane, S. (2003). Metalloproteases: Carving out a role in axon guidance. *Neuron* **37,** 559–562.

Mennen, U. (2003). End-to-side nerve suture in clinical practice. *J. Hand Surg.* **8,** 33–42.

Mouilhade, F., Barbary, S., Apard, T., and Dautel, G. (2009). End-to-side neurorrhaphy for median nerve repair after elbow tumor resection: Case report. *J. Hand Surg.* **34A,** 83–86.

Navarro, X. (2009). Neural plasticity after nerve injury and regeneration. *Int. Rev. Neurobiol.* **87,** 483–505.

Noah, E. M., Williams, A., Jorgenson, C., Skoulis, T. G., and Terzis, J. K. (1997). End-to-side neurorrhaphy: A histologic and morphometric study of axonal sprouting into an end-to-side nerve graft. *J. Reconstr. Microsurg.* **13,** 99–106.

Ogun, T. C., Ozdemir, M., Senaran, H., and Ustun, M. E. (2003). End-to-side neurorrhaphy as a salvage procedure for irreparable nerve injuries. *J. Neurosurg.* **99,** 180–185.

Okajima, S., and Terzis, J. K. (2000). Ultrastructure of early axonal regeneration in an end-to-side neurorrhaphy model. *J. Reconstr. Microsurg.* **16,** 313–323.

Ornelas, L., Padilla, L., Di Silvio, M., Schalch, P., Esperante, S., Infante, R. L., Bustamante, J. C., Avalos, P., Varela, D., and Lopez, M. (2006). Fibrin glue: An alternative technique for nerve coaptation—Part II. Nerve regeneration and histomorphometric assessment. *J. Reconstr. Microsurg.* **22,** 123–128.

Ozbek, S., and Kurt, M. A. (2006). Simultaneous end-to-side coaptations of two severed nerves to a single healthy nerve in rats. *J. Neurosurg. Spine* **4,** 43–50.

Pannucci, C., Myckatyn, T. M., Mackinnon, S. E., and Hayashi, A. (2007). End-to-side nerve repair: Review of the literature. *Restor. Neurol. Neurosci.* **25,** 45–63.

Papalia, I., Geuna, S., Tos, P., Boux, E., Battiston, B., and Stagno d'Alcontres, F. (2003). Morphologic and functional study of rat median nerve repair by terminolateral neurorrhaphy of the ulnar nerve. *J. Reconstr. Microsurg.* **19,** 257–264.

Papalia, I., Geuna, S., D'Alcontres, F. S., and Tos, P. (2007a). Origin and history of end-to-side neurorrhaphy. *Microsurgery.* **27,** 56–61.

Papalia, I., Cardaci, A., Stagno d'Alcontres, F., Lee, J. M., Tos, P., and Geuna, S. (2007b). Selection of the donor nerve for end-to-side neurorrhaphy. *J. Neurosurg.* **107,** 378–382.

Pelissier, P., Rihai, R., Casoli, V., Martin, D., and Baudet, J. (2001). Les anastomoses nerve uses terminolaterales. Rapport clinique a propos de dix cases. *Ann. Chir. Plast. Esthet.* **46,** 129–133.

Pienaar, C., Swan, M. C., De Jager, W., and Solomons, M. (2004). Clinical experience with end-to-side nerve transfer. *J. Hand Surg. Br.* **29B**(5), 438–443.

Raimondo, S., Nicolino, S., Tos, P., Battiston, B., Giacobini-Robecchi, M. G., Perroteau, I., and Geuna, S. (2005). Schwann cell behavior after nerve repair by means of tissue-engineered muscle-vein combined guides. *J. Comp. Neurol.* **489,** 249–259.

Sámal, F., Haninec, P., Raska, O., and Dubovy, P. (2006). Quantitative assessment of the ability of collateral sprouting of the motor and primary sensory neurons after the end-to-side neurorrhaphy of the rat musculocutaneous nerve with the ulnar nerve. *Ann. Anat.* **188,** 337–344.

Sananpanich, K., Galea, M. P., Morrison, W. A., and Messina, A. (2007). Quantitative characterization of regenerating axons after end-to-side and end-to-end coaptation in a rat brachial plexus model: A retrograde tracer study. *J. Neurotrauma* **24,** 864–875.

Sanapanich, K., Morrison, W. A., and Messina, A. (2002). Physiologic and morphologic aspects of nerve regeneration after end-to-end or end-to-side coaptation in a rat model of brachial plexus injury. *J. Hand Surg.* **27A,** 133–142.

Tham, S. K., and Morrison, W. A. (1998). Motor collateral sprouting through an end-to-side nerve repair. *J. Hand Surg.* **23A,** 844–851.

Ulkur, E., Yuksel, F., Acikel, C., Okar, I., and Celikoz, B. (2003). Comparison of functional results of nerve graft, vein graft, and vein filled with muscle graft in end-to-side neurorrhaphy. *Microsurgery* **23,** 40–48.

Viterbo, F., Trinidade, J. C., Hoshino, K., and Mazzoni, A. (1992). Latero-terminal neurorrhaphy without removal of the epineural sheat: Experimental study in rats. *Sao Paulo Med. J.* **110,** 267–275.

Viterbo, F., Franciosi, L. F., and Phalares, A. (1995). Nerve grafting and end-to-side neurorraphies connecting the phrenic nerve to the brachial plexus. *Plast. Reconstr. Surg.* **96,** 94–95.

Voche, P. (2005). Our experience about ten end-to-side sensitive nerve sutures at the hand. *Ann. Chir. Plast. Esthet.* **50,** 264–269.

Walker, J. C., Brenner, M. J., Mackinnon, S. E., Winograd, J. M., and Hunter, D. A. (2004). Effect of perineurial window size on nerve regeneration, blood-nerve barrier integrity, and functional recovery. *J. Neurotrauma* **21,** 217–227.

Xiong, G., Ling, L., Nakamura, R., and Sugiura, Y. (2003). Retrograde tracing and electrophysiological findings of collateral sprouting after end-to-side neurorrhaphy. *J. Hand Surg.* **8,** 145–150.

Yan, J. G., Matloub, H. S., Sanger, J. R., Zhang, L. L., Riley, D. A., and Jaradeh, S. S. (2002). A modified end-to-side method for peripheral nerve repair: Large epineurial window helicoid technique versus small epineurial window standard end-to-side technique. *J. Hand Surg.* **27A,** 484–492.

Yuksel, F., Peker, F., and Celikoz, B. (2004). Two applications of end-to-side nerve neurorrhaphy in severe uppe-extremity nerve injuries. *Microsurgery* **24,** 363–368.

Zacchigna, S., and Giacca, M. (2009). Gene therapy perspectives for nerve repair. *Int. Rev. Neurobiol.* **87,** 381–392.

Zhang, Z., Soucacos, P. N., Bo, J., and Beris, A. E. (1999). Evaluation of collateral sprouting after end-to-side nerve coaptation using a fluorescent double-labeling technique. *Microsurgery* **19,** 281–286.

Zhang, Z., Soucacos, P. N., Beris, A. E., Bo, J., Ioachim, E., and Johnson, E. O. (2000). Long-term evaluation of rat peripheral nerve repair with end-to-side neurorrhaphy. *J. Reconstr. Microsurg.* **16,** 303–311.

Zhang, Z., Soucacos, P. N., Bo, J., Beris, A. E., Malizos, K. N., Ioachim, E., and Agnantis, N. J. (2001). Reinnervation after end-to-side nerve coaptation in a rat model. *Am. J. Orthop.* **30,** 400–406.

Zhang, Z., Johnson, E. O., Vekris, M. D., Zoubos, A. B., Bo, J., Beris, A. E., and Soucacos, P. N. (2006). Long-term evaluation of rabbit peripheral nerve repair with end-to-side neurorrhaphy in rabbits. *Microsurgery* **26,** 262–267.

Zhao, J. Z., Chen, Z. W., and Chen, T. Y. (1997). Nerve regeneration after terminolateral neurorrhaphy: Experimental study in rats. *J. Reconstr. Microsurg.* **13,** 31–37.

Zhu, Q. T., Zhu, J. K., and Chen, G. Y. (2008). Location of collateral sprouting of donor nerve following end-to-side neurorrhaphy. *Muscle Nerve* **38,** 1506–1509.

NOVEL PHARMACOLOGICAL APPROACHES TO SCHWANN CELLS AS NEUROPROTECTIVE AGENTS FOR PERIPHERAL NERVE REGENERATION

Valerio Magnaghi,*,† Patrizia Procacci,‡ and Ada Maria Tata§

*Department of Endocrinology, Physiopathology, Applied Biology, Università degli Studi di Milano, 20133 Milan, Italy

†C.I.Ma.I.Na., Interdisciplinary Centre for Nanostructured Materials and Interfaces, University of Milan, 20133 Milan, Italy

‡Department of Human Morphology and Biomedical Sciences-Città Studi, Università degli Studi di Milano, 20133 Milan, Italy

§Department of Cell and Developmental Biology, Neurobiology Center "Daniel Bovet", "La Sapienza" University, Rome, Italy

Peripheral neuropathies are common neurologic disorders, but current treatments are limited. Among the different approaches to treat the acquired neuropathies due to traumatic injuries, the pharmacological interventions directed to Schwann cell may represent a useful and challenging opportunity. Following nerve damage the distal axon and the ensheathing Schwann cells degenerate, ensuing a process known as "Wallerian degeneration". Schwann cells then dedifferentiate and proliferate to support neurite outgrowth.

In the recent years, several pharmacological agents that may promote the Schwann cell in its role of supporting nerve regeneration have been proposed. However, in view of increased understanding of the cellular mechanisms controlling neuron–glial interactions, a great attention has focused on neurotransmitters, neuroactive steroids, and neurohormones. In this review, we survey the latest findings on these factors and assess their potential as novel promising treatments for peripheral neuropathies caused by injury.

DOI: 10.1016/S0074-7742(09)87015-3

0074-7742/09 $35.00

I. Introduction

The acquired neuropathies include those due to mechanical and traumatic injuries to peripheral nerves (Head, 2006). These pathologies mainly result from motor vehicle, sport, or job accidents and are the cause of considerable disability across the world (Robinson, 2000). Although microsurgical procedures are the first clinical aid to nerve repair and regeneration, functional recovery is difficult and rarely complete, thus much research has been directed to improving functional outcomes of traumatic nerve damage.

The Schwann cells in the peripheral nervous system (PNS) are cells deputed to ensheath axons, forming the myelin, and allowing the faster propagation of action potentials (Jessen and Mirsky, 1997). The myelin sheath consists of multiple layers of Schwann cell membrane containing some typical characteristic proteins, including glycoprotein P0 (P0) and peripheral myelin protein 22 (PMP22), which collectively account for over 55% of total myelin proteins and exert important roles for the maintenance of the multilamellar structure of myelin. Several mutations or changes in gene expression of P0 and PMP22 have been associated with a group of hereditary peripheral neuropathies in humans (e.g., Charcot-Marie-Tooth type 1A, CMT1A, and 1B, CMT1B) (Scherer and Wrabetz, 2008). The Schwann cells that do not activate the program of myelin gene expression become nonmyelinating Schwann cells (Jessen and Mirsky, 1997).

The Schwann cells are fundamental for the development, maintenance, and regeneration of peripheral nerves and these processes are controlled by similar molecular mechanisms. After a peripheral nerve injury the distal nerve segment degenerates, following a process known as "Wallerian degeneration," so that the proximal Schwann cells dedifferentiate, reentering into the proliferative state. As the axon regrow, new Schwann cells promote neurite outgrowth from the proximal stump by secreting substances such as laminin, neuron-glia cell adhesion molecules (NgCAM), integrins, and growth factors (for review see Fu and Gordon, 1997). Unlike their counterparts in the central nervous system (CNS), the axons of motoneurons and primary sensory neurons of the PNS, however, have marked capacity of regeneration (Donnerer, 2003).

The distal nerve degeneration, however, is an early event that may be amenable to treatment. Over the last decade, based on the increasing knowledge of the regulatory mechanisms during nerve damage, several substances including laminin, neurotrophins (NTs), or cAMP modulators have been investigated for their ability to promote nerve regeneration (reviewed in Chen *et al.*, 2007). Unfortunately, none of these substances have been introduced in therapy, and currently available therapies are mainly addressed to control painful symptoms rather than to treat nerve degeneration and/or regeneration. Recent studies focused on the importance of bidirectional crosstalk between neuron and glial cells also in the PNS. For instance, the stimulation of peripheral motor axons induces Ca^{2+} waves in the terminal presynaptic Schwann cells (Jahromi *et al.*, 1992), suggesting that molecules released upon synaptic transmission bind to the receptors on the surrounding glial cells, which in turn determine the intracellular Ca^{2+} increase. On the other hand, the neural impulse activity during fetal and perinatal life may influence the development of myelinating glia, mainly through the release of nonsynaptic neurotransmitters (Stevens and Fields, 2000). Therefore, classic neurotransmitters might be synthesized and/or interact with uncommon target, such as glial Schwann cells in the PNS.

The study of pharmacological modulation of Schwann cells, therefore, represents an interesting and potential tool for studies addressed to identify new therapeutic strategies. Among all factors controlling Schwann cell biology, the neurotransmitters and neurohormones are presently considered to be the most interesting for nerve regeneration. In this paper, we review current understanding of these promising factors, which might be a useful treatment for peripheral neuropathies caused by injury.

II. GABAergic System

GABA is generally considered the main inhibitory neurotransmitter in the CNS, although emerging findings revealed that the GABAergic system is also active in the PNS, particularly in Schwann cells. In the middle '70s date the first studies on the presence of GABA and its receptors outside the CNS. The accumulation of ^{3}H-GABA (and glutamate) was shown in Schwann cells of the taste buds of the amphibian *Necturus maculosus* (Nagai *et al.*, 1998). At about the same time it was also shown that myelinated and unmyelinated fibers possess GABA receptors and GABA carriers (Brown and Marsh, 1978; Brown *et al.*, 1979; Morris *et al.*, 1983; Olsen *et al.*, 1984).

GABA interacts with specific type A (GABA-A) ionotropic, and type B (GABA-B) metabotropic receptors; GABA-A is a ligand-gated ion channel, giving a chloride ion influx, while GABA-B is a seven-transmembrane domains (TMS) receptor coupled to G protein and linked to the adenylate cyclase system (Bettler *et al.*, 2004). GABA-A receptors are on normal mammalian sensory axons and are reestablished after regeneration (Bhisitkul *et al.*, 1987). Additionally, Schwann cells express different subunits of the GABA-A receptor, such as alpha2, alpha3, and beta1–3 (Magnaghi *et al.*, 2001, 2006; Melcangi *et al.*, 1999a). The presence of GABA-B receptors has been initially demonstrated in the rat dorsal root ganglion (DRG), in peripheral axons, in autonomic nerve terminals, and in pig nodose ganglion cells (Bowery *et al.*, 1981; Desarmenien *et al.*, 1984; Liske and Morris, 1994; Sun and Chiu, 1999; Towers *et al.*, 2000; Zagorodnyuk *et al.*, 2002).

Very recently, Schwann cells were also shown to express different isoforms of GABA-B receptor, such as -1a, -1b, -1c, and -2 (Magnaghi *et al.*, 2004a, 2006). Interestingly, GABA-A and GABA-B receptors in Schwann cells revealed active, because of its capacity to respond to specific ligands. For instance, exposure of Schwann cells to the specific GABA-A agonist muscimol stimulates PMP22 levels, suggesting that this protein might be under GABA-A control (Magnaghi *et al.*, 2001; Melcangi *et al.*, 2005). The specific GABA-B agonist baclofen, instead, decreases the cell proliferation and the percentage of Schwann-BrdUrd immunopositive cells. Baclofen also decreases the levels of some myelin proteins, particularly of PMP22 and P0 (Magnaghi *et al.*, 2004a). The GABA-B-mediated control of Schwann cells likely goes across the activation of the intracellular cAMP signaling system (Magnaghi *et al.*, 2004a). However, possible mechanisms tempting to explain how ionotropic GABA-A receptor activation may control transcriptional activity includes increase of Ca^{2+} influx, mitogen-activated protein kinase (MAPK) activation, and cAMP response element binding protein (CREB) phosphorylation (Obrietan *et al.*, 2002). Some new findings demonstrate the GABA synthesis in Schwann cells (Fig. 1), that autocrinally activate the GABA-A receptor and the downstream MAPK–CREB signaling system (Magnaghi *et al.*, unpublished observations).

Altogether, these data emphasizes how the PNS is a potential target for GABA's action, suggesting that GABA-A and GABA-B receptors likely cross-interact in a complex mechanism influencing the Schwann cell biology (Fig. 1). The activation of GABA-A exerts a stimulatory effect on PMP22, whereas the GABA-B activation decreases the proliferation and PMP22 expression. This mechanistic explanation is fully consistent with the literature and underpins that, depending on the GABA-A or GABA-B receptor involved, GABA may control the PMP22 levels.

A recent analysis of GABA-B1 knockout mice further evidences the role of this receptor in PNS. GABA-B1-deficient mice exhibit morphological and molecular changes in peripheral myelin, including an increase in the number of irregular fibers and in the expression levels of PMP22 and P0 (Magnaghi *et al.*, 2008). This finding appears to be contradictory. However, since GABA-B receptor activation decreases intracellular cAMP levels (Ulrich and Bettler, 2007), which in turn reduces protein kinase A (PKA)-dependent phosphorylation of transcription factors such as CREB or ATF-4 (Ren and Mody, 2003; Steiger *et al.*, 2004) it has been proposed that the loss of tonic or phasic GABA-B receptor activity might increase the cAMP levels and consequently the PMP22 expression in Schwann cells. These mice were also hyperalgesic to thermal and mechanical stimuli and showed gait impairment, which were correlated to the dysmyelinating process observed (Magnaghi *et al.*, 2008).

Putative GABAergic modulation during nerve regeneration was poorly investigated, providing only some controversial findings that deserve further

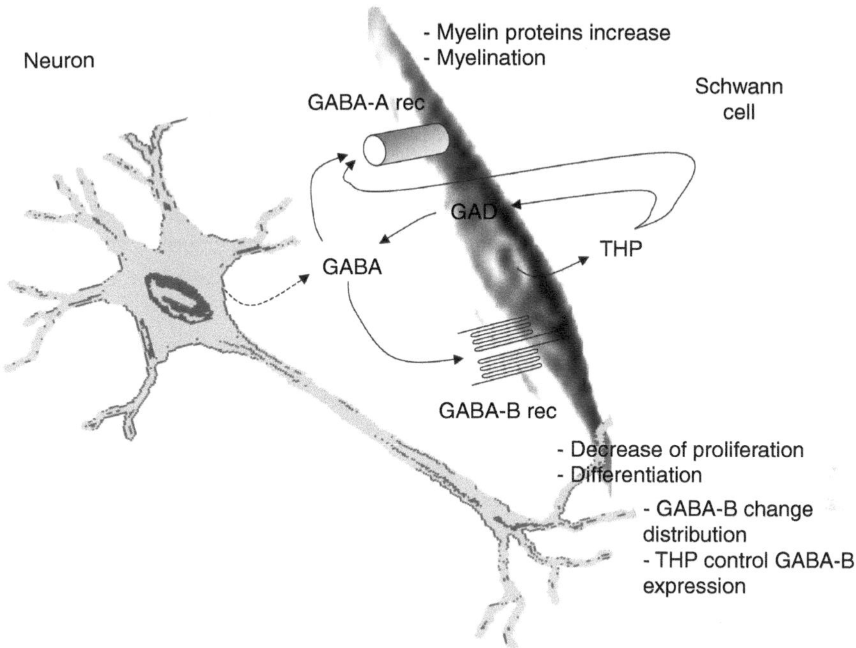

FIG. 1. GABA-A, GABA-B receptors and the neuroactive steroid THP cross-interact in the PNS (for review see Magnaghi, 2007). GABA coming from neuron or produced by Schwann cells, which express GAD enzymes, may affect the paracrine cross talk between these cells. In particular, extracellular GABA may interact with the GABA-B receptor on the Schwann cells, decreasing cell proliferation and inducing to start differentiation. The neuroactive steroid THP (produced by Schwann cells) modulate the expression and the responsivity of GABA-B receptor, and in turn its desensitization. THP further decreases Schwann cell proliferation and simultaneously stimulates their differentiation. By a direct interaction with the GABA-A receptor, in fact, THP increases some myelin protein expressions, inducing myelination. Therefore, the GABA-mediated control of the Schwann cell proliferation/differentiation might be particularly relevant to explain the mechanisms affecting the Schwann cell biology under normal or pathologic conditions.

investigations. For instance, valproic acid (a GABA-increasing drug) enhances the sciatic nerve regeneration in rats (Wu *et al.*, 2008), although pregabalin (a GABA analog used to attenuate neuropathic pain) did not ameliorate nerve regeneration (Whitlock *et al.*, 2007).

III. Neuroactive Steroids

The term "neurosteroids" was introduced early in the 1980s by Baulieu (1997), referring to steroids that are synthesized *de novo* in the CNS. Now, the term "neuroactive steroids" is generally preferred to indicate classes of steroids which are converted into neuroactive metabolites able to act directly in the nervous system (Melcangi *et al.*, 1999b).

Neuroactive steroids, like progesterone (P) and its metabolites (i.e., dihydroprogesterone (DHP) and tetrahydroprogesterone (THP), also named allopreganolone), dehydroepiandrosterone, estrogens, androgens, etc., exert important neuroprotective effects in the neuronal and nonneuronal compartments of the nervous system. These properties candidate neuroactive steroids as new therapeutic tools to treat neurodegenerative events (Ibanez *et al.*, 2003; Roglio *et al.*, 2008b). For instance, estrogens in the PNS exert different effects on sensory and autonomic neurons, influencing development, plasticity, and repair of DRG neurons, but also controlling the neuritogenesis of sympathetic neurons and the proliferation of Schwann cells (Koszykowska *et al.*, 2008). Furthermore, neuroactive steroids such as P, DHP, and THP modulate mitogenic activity, proliferation, and synthesis of peripheral myelin proteins (P0 and PMP22) by Schwann cells (Desarnaud *et al.*, 1998; Magnaghi *et al.*, 2001; Melcangi *et al.*, 1998, 1999a). In this context, while P and DHP act via the classic progesterone receptor (PR), THP is a potent allosteric modulator of GABA-A (Lambert *et al.*, 2003). However, Schwann cells are contemporarily a target but also a source of neuroactive steroids (Celotti *et al.*, 1992; Melcangi *et al.*, 1999b).

The effect of progestagens in PNS has been extensively studied also *in vivo* (Magnaghi *et al.*, 2001; Melcangi *et al.*, 2000b). Progestagens stimulate the expression of P0 and PMP22 in the sciatic nerve of young and old male rats (Magnaghi *et al.*, 2001; Melcangi *et al.*, 1998, 1999a, 2000b). Moreover, P and its derivatives reduce myelin abnormalities and fiber loss of aged sciatic nerve (Azcoitia *et al.*, 2003). Concerning nerve regeneration, progestagens proved to exert beneficial effects on sciatic nerve remyelination after cryolesion (Koenig *et al.*, 1995), transection (Melcangi *et al.*, 2000a), or crush injury (Roglio *et al.*, 2008a). In particular, DHP ameliorates myelin protein expression and fiber density in crushed animals (Roglio *et al.*, 2008a). Chitosan prosthesis, used as scaffold to promote rabbit facial nerve regeneration, yields good results when filled with P or its precursor pregnenolone, thus providing a sort of *in situ* drug delivery (Chavez-Delgado *et al.*, 2005). However, these results suggest that also axons may be an important progestagens target (Melcangi *et al.*, 2003). Treatment with mifepristone (selective PR antagonist) *in vivo*, during development, induces a reduction of axon diameter and an increase in neurofilament density (Melcangi *et al.*, 2003).

Mifepristone was recently considered because of its ability in decreasing the PMP22 overexpression in a model of peripheral inherited neuropathy (i.e., CMT1A) (Sereda *et al.*, 2003). This result led to the conclusion that PR likely participated in the control of PMP22 synthesis. However, in CMT1A pathology the levels of neuroactive steroids 3beta-THP and 3alpha-diol, respectively, an antagonist and an agonist of GABA-A receptor (Backstrom *et al.*, 2005; Lundgren *et al.*, 2003), are also strongly changed (Caruso *et al.*, 2008), suggesting their putative neuroprotective role in such a pathology.

Therefore, progestagens modulate the expression of myelin proteins through different mechanisms. Although the effects of P and DHP on P0 seem to be mediated by the classic PR in Schwann cells (Magnaghi *et al.*, 2001; Melcangi *et al.*, 2000b), the effect of THP on PMP22 appears to be mediated via GABA-A receptor (Magnaghi *et al.*, 2001, 2006). Bicuculline, a specific GABA-A receptor antagonist, abolishes the effect of THP on PMP22, while muscimol (specific GABA-A agonist) mimics the THP action (Magnaghi *et al.*, 2001, 2006). Also the androgenic metabolite 3alpha-diol, another GABA-A ligand, is able to stimulate the PMP22 expression (Magnaghi *et al.*, 2001, 2004b), further indicating that PMP22 expression might be under the control of GABA-A receptor. Collectively these data underpin the role of neuroactive steroids (i.e., mainly THP) in the control of Schwann cell biology via a GABAergic-mediated mechanism.

In Schwann cells the neuroactive steroids THP also controls the expression of different GABA-B receptor subunits (Magnaghi *et al.*, 2006). However, to make the issue even more puzzling, it has been hypothesized that THP regulates GABA-B receptor via a GABA-A-mediated mechanism, since THP effects on GABA-B receptor expression were mimicked by muscimol and GABA (Magnaghi, 2007). The downstream intracellular pathways regulating this mechanism are complicated and presently not completely identified.

Therefore, in the PNS the interactions among GABA-A, GABA-B receptors and neuroactive steroids are relevant for the bidirectional cross talk between neurons and Schwann cells (Fig. 1). GABA, coming from the neuronal compartment or produced by Schwann cells, may affect the paracrine interplay between these cells. Data obtained with baclofen, in fact, suggest that extracellular GABA might interact with GABA-B receptors on Schwann cells, decreasing their proliferation. This effect promotes the Schwann cell differentiation. THP produced by Schwann cells (Melcangi *et al.*, 1999b) modulates the GABA-B receptor expression, distribution, and responsivity, and in turn its desensitization. Successively, THP increases some myelin protein expressions by a direct GABA-A receptor modulation. Muscimol mimicked this effect on Schwann cells, further supporting the hypothesis that GABA might also participate in the control of the Schwann cell proliferation/differentiation.

THP is an interesting neuroactive steroid also participating in nociception. Painful peripheral nerve injury induces changes in gene expression and activity of the P450scc enzyme, a key regulator of neurosteroidogenesis (reviewed in Patte-Mensah and Mensah-Nyagan, 2008), and in THP synthesis in the spinal dorsal horn (Patte-Mensah *et al.*, 2006). The neuroactive steroids then change the neurochemical architecture of circuits involved in analgesia and nociception. Given that THP is a GABA modulator, it is hypothesizable that its peripheral effects in nociception are partially mediated via interaction with the GABAergic system on Schwann cells, as recently demonstrated studying the GABA-B1 knockout mice (Magnaghi *et al.*, 2008).

IV. Glutamate

In CNS, the excitatory glutamatergic system is generally correlated but functionally opposed to the GABAergic system. The presence of glutamate transporters, however, can provide a source of precursor for GABA synthesis.

The role of glutamatergic system has been recently considered also in the PNS. Pharmacological, electrophysiological, and immunohistochemical analysis evidenced that glutamate is a neurotransmitter in many DRG and spinal cord neurons (Carlton, 2001; De Biasi and Rustioni, 1988; Huettner *et al.*, 2002). Moreover, glutamate may act together with acetylcholine (ACh) at the vertebrate neuromuscular junction (NMJ) (De Biasi and Rustioni, 1988). The normal mammalian NMJs, in fact, contain the quiescent machinery required for synaptic release and action of glutamate, which may be activated following damage. This evidences the ability of denervated muscle to be reinnervated by glutamatergic axons from the spinal cord neurons of CNS. This hypothesis is in accordance with previous observations, showing that connecting skeletal muscle to the lateral white matter of the spinal cord produced functional muscle reinnervation (Brunelli *et al.*, 2005). The restored neuromuscular activity is resistant to common curare blockers but sensitive to a glutamate receptor antagonist, such as alpha-amino-3-hydroxy-5-methyl-4-isoxazolepropionic acid, confirming that under glutamatergic transmission the NMJ switches from cholinergic to glutamatergic synapse (Brunelli *et al.*, 2005).

Some observations indicate that Schwann cells *in vivo* or *in vitro* express a variety of ionotropic glutamate receptors (i.e., NMDA, AMPA, or KA receptor), suggesting that glutamate may play a role also in the Schwann cell biology (Fink *et al.*, 1999; Kinkelin *et al.*, 2000). Schwann cells surrounding giant axons of the small squid hyperpolarize in response to glutamate, revealing a dose-dependent effect (Lieberman *et al.*, 1989). Nevertheless, Schwann cells release relevant concentrations of excitotoxins, such as glutamate and D-serine, which can affect neuronal morphology and viability (Wu *et al.*, 2005). These findings, however, are opposed to previous observations showing that the Schwann and satellite cells of the rat nodose ganglion were immunonegative for glutamate (Schaffar *et al.*, 1997). Additionally, the identification of the membrane-bound glutamate transporters system in the PNS has been recently performed *in vivo*. In particular the GLT1 transporter is expressed in the cytoplasm of Schwann cells of the rat sciatic nerve, the glutamate-aspartate transporter (GLAST) localized in satellite cells, DRG neurons, and in peripheral myelin, while EAAC1 transporter was shown only in the myelin layer (Carozzi *et al.*, 2008). Conversely, preliminary data addressed to specifically characterize the glutamate transporters in Schwann cell *in vitro*, evidenced that these cells express GLAST and EAAC1 but not GLT1 (Magnaghi *et al.*, unpublished observation). However, GLAST is present on perisynaptic Schwann cells, where it may regulate the buffering of extracellular glutamate (Pinard *et al.*, 2003).

Glutamate excess is implicated in cases of peripheral neuropathies (Watkins, 2000), and metabotropic glutamate receptors (mGluRs) were raised in the adult rat motoneurons after sciatic nerve transection (Anneser *et al.*, 2000; Popratiloff *et al.*, 1996). However, some therapeutic approaches for nerve injuries, aimed to block glutamate receptors are disadvantageous and revealed contradictory, given the induction of severe side effects (Cvrcek, 2008; Yashpal *et al.*, 2001).

V. Cholinergic System

ACh was the first molecule identified as neurotransmitter and phylogenetically it seems to be one of older signaling molecules. ACh, in fact, has been detected in bacteria, protozoa, fungi, algae, and primitive plants, indicating that the cholinergic system was widely distributed in living organisms before its appearance in the nervous system (Wessler *et al.*, 2001). During embryogenesis (e.g., sea urchin, chick) ACh acts as a morphogen, controlling the cell migration (Buznikov *et al.*, 1996; Oettling *et al.*, 1992), whereas in adult mammals, its ability to regulate basic cell functions including growth, survival, differentiation, and apoptosis has been described (Eglen, 2006; Tata, 2008). However, before acquiring its neurotransmitter function ACh also modulates neurogenesis by controlling neural stem and progenitor cell proliferation, as well as neuron and glial cell survival and differentiation (Loreti *et al.*, 2007a; Ma *et al.*, 2004). Several *in vitro* studies have shown that ACh can regulate different aspect of nervous system morphogenesis. For instance, ACh via alpha-7 nicotinic receptor activation induces neurite outgrowth in neonatal rat primary olfactory bulb cultures (Coronas *et al.*, 2000), and it stimulates ChAT-transfected N18TG2 neuroblastoma cells through muscarinic receptors (Salani *et al.*, 2009). ACh also shows chemoattractant property, regulating nerve growth cone guidance (Zheng *et al.*, 1994). Unlike the inhibition of neurite extension in embryonic mouse spinal cord neurons (Owen and Bird, 1995), ACh induces neurite elongation in sensory neurons of chick embryos (Tata *et al.*, 2003). In fact, the fiber number, length, and neurite fasciculation resulted in increased DRG neurons cultured in absence of nerve growth factor (NGF) but in presence of muscarinic agonist (e.g., carbachol or muscarine). These effects, however, appear to be modulated by cooperation between muscarinic and nicotinic receptors.

Several studies suggested that after nerve injury, ACh may have a relevant role in motor nerve terminal outgrowth and muscle repair. In fact, the block of the nicotinic receptor expressed in the muscles by alpha-bungarotoxin, significantly inhibits nerve outgrowth (Pestronk and Drachman, 1985). This suggests that local release of ACh is necessary, either to directly trigger nerve terminal outgrowth or to modulate the release of other trophic/chemical factors. Furthermore, the relevance of ACh in nerve regrowth has recently demonstrated by studying biodegradable polymers containing different concentrations of ACh. These scaffolds promoted neurite extension, suggesting that they could be used in the treatment of nervous system injury, and to recover sensorimotor and cognitive functions (Gumera and Wang, 2007). Altogether, these data clearly indicate the ability of ACh to modulate growth cone guidance, fiber elongation, and fasciculation, whereas a relevant question remains still opened: who and where ACh is synthesized? (i.e., growth cone, tissue target, or Schwann cells).

Cultured Schwann cells express several ACh receptors of muscarinic types (Loreti *et al.*, 2006) and that selective activation of M2 subtype with arecaidine causes an arrest of Schwann cell proliferation in G1 phase. This block appears reversible, since removing M2 agonist from the culture medium, the Schwann cells recover their ability to proliferate (Table I) (Loreti *et al.*, 2007a). M2 activation also increases the expression of several myelin proteins such as P0 (Loreti *et al.*, 2007b) and PMP22 (unpublished observations). This indicates that ACh via the activation of M2 receptor subtype may be one of the molecules involved in Schwann cell switching from the proliferative to differentiated phenotype. Further support to this hypothesis comes from the finding that M2 receptors are also expressed by Schwann cells in neonatal rat sciatic nerves (Loreti *et al.*, 2007a), remaining expressed towards adult life (Bernardini *et al.*, 1999). However, Schwann cells also express M1/M3 receptor subtypes and their selective activation appears to stimulate cell proliferation (Loreti *et al.*, 2007b). Collectively, these data suggest that periaxonal Schwann cells are responsive to ACh and the muscarinic receptor type expressed could differently contribute to address

TABLE I

Distribution of Schwann Cells in Cell Cycle Phases G1, S, G2 After Treatment with M2 Agonist Arecaidine

Treatment	% G1	% S	% G2
Control (+Fsk)	82.85 ± 6.33	11.99 ± 4.88	5.16 ± 1.60
Fsk + arecaidine 24 h	94.35 ± 0.95	1.08 ± 0.38	4.57 ± 1.20
Fsk + arecaidine 48 h	95.91 ± 2.66	0.54 ± 0.50	3.57 ± 2.21
Fsk + arecaidine 72 h	97.77 ± 1.88	0.40 ± 0.26	1.83 ± 1.64
Fsk + arecaidine 48 h + 48 h recovery	78.22 ± 3.84	16.58 ± 3.22	4.49 ± 0.40

Data are expressed as percentage (mean ± S.E.M.); Fsk, forskolin.

Schwann cells towards proliferation (M1/M3 prevalence) or differentiation (M2 as main receptor).

After nerve injury, perisynaptic Schwann cells synthesize ACh, and spontaneously released it in a quantal secretion (Dennis and Miledi, 1974). This phenomenon may be relevant to maintain the nicotinic receptor clustering on the muscle, driving the nerve regeneration. Recently, the ACh synthesis by muscle fibers has been also demonstrated, supporting the mechanical muscle activity in absence of nerve, during myogenesis, and perhaps during muscle denervation (Bandi *et al.*, 2005).

Collectively, it may be hypothesized that ACh released in extrasynaptic or in synaptic regions (alone or likely combined to other molecules) may control Schwann cell proliferation and/or differentiation, and simultaneously, it supports growth cone guide, fiber elongation, and the maintenance of the postsynaptic element organization on the muscle fibers. All these aspects are relevant to promote fast nerve regeneration, emphasizing that ACh may play an important role not only during nervous system development but also following nerve injury.

For long time, the lack of selective ligands capable of binding to specific ACh receptors limited the development of therapies based on cholinergic receptor activation. The use of the muscarinic agonists, for instance, causes important side effects, such as decrease of body temperature, alteration of the cardiac rate, and ptyalism. In the last decade, however, the progress in molecular biochemistry allowed the production of new selective muscarinic ligands, bearing minimal side effects and opening new therapeutic perspectives for the use of cholinergic receptor ligands in the treatment of several diseases (Table II) (Tata, 2008). In this context, the use of selective muscarinic or nicotinic agonist/antagonists, alone or nanostructurated in biodegradable polymers, could represent a profitable tool to accelerate nerve regeneration, reducing muscle atrophy.

VI. Purinergic System

Early in the '70s date the first identification of adenosine 5′-triphosphate (ATP) as neurotransmitter in nerves. However, ATP is now recognized as a cotransmitter in all nerves of PNS and CNS (Abbracchio *et al.*, 2009; Burnstock, 2009). Purinergic cotransmission in fact is present, not only in sympathetic nerves, but also in parasympathetic, sensory-motor, and enteric nerves. ATP is usually coreleased with GABA, glutamate, ACh, dopamine, NA, and 5-hydroxytryptamine in different populations of nerve fibers (Burnstock, 2009). Indeed, ATP released from rat sciatic nerve trunks during electrical stimulation, is blocked by the sodium channel inhibitor tetrodotoxin and the non-NMDA glutamate receptor antagonist 6-cyano-7-nitroquinoxaline-2,3-dione (CNQX) (Liu and Bennett, 2003).

TABLE II
MUSCARINIC RECEPTOR AGONISTS AND ANTAGONISTS APPROVED FOR THERAPEUTIC PROTOCOLS OR USED IN CLINICAL TRIALS

Ligands	Receptor selectivity	Functions	Status	Therapeutic application
Solifenacin	M3	Antagonist	Phase III	Overactive Bladder
Darifenacin	M3	Antagonist	Phase III	Overactive Bladder
Iprotropium	M1/M3	Antagonist	Approved	Chronic Obstructive Pulmonary disease
Oxitropium	M1/M3	Antagonist	Approved	Chronic Obstructive Pulmonary disease
Tiotropium	Nonselective	Antagonist	Approved	Chronic Obstructive Pulmonary disease
Tolterodine	Nonselective	Antagonist	Approved	Overactive Bladder
Alvameline	M1	Agonist	Phase III discontinued	Alzheimer's disease
Xanomeline	M1>M4	Agonist	Phase III	Alzheimer's disease
Cevimeline	M1	Agonist	Approved	Sjogren's disease
Vadaclidine (LY297802)	M4	Agonist	Phase II	Antinociceptive effects

However, Schwann cells isolated from the nerve trunks did not release ATP when electrically stimulated, but did it in response to glutamate, following a concentration-dependent manner. This effect was inhibited by a specific noncompetitive AMPA receptor antagonist (GYKI 52466) and by CNQX (Liu and Bennett, 2003), suggesting that is a glutamate-mediated indirect effect.

Among purinergic receptors, Schwann cells express P2Y receptors, which respond to ATP eliciting a Ca^{2+} wave, either in Schwann cell cultures or in myelinated/unmyelinated fibers (Ansselin *et al.*, 1997; Green *et al.*, 1997). In addition, adenosine receptors A_{2A} have been recently identified in mouse Schwann cells, where they inhibit cell proliferation through a different mechanism from P2Y (Stevens *et al.*, 2004).

Altogether, these findings evidenced an activity-dependent corelease of ATP with glutamate in peripheral nerves that in turn propagate on neighbor Schwann cells, determining functional consequences. ATP proved able to arrest the Schwann cells maturation in a morphological immature state, before differentiation into the myelinating or nonmyelinating phenotypes (Stevens and Fields, 2000). Namely, the axonal electrical activity, through the release of ATP can strongly influence the Schwann cell proliferation, differentiation, and gene expression.

VII. Mitogen-Activated Protein Kinases (MAPKs)

MAPKs are major components of the pathways controlling cell proliferation, differentiation, and death (Chen *et al.*, 2001; Pearson *et al.*, 2001). There are three main subfamilies of MAPKs (Chen *et al.*, 2001): (a) extracellular signal regulated kinases (ERK), comprising ERK1 and ERK2, that are activated by serum, growth factors, cytokines, and stress (Pearson *et al.*, 2001); (b) c-Jun N terminal kinases also known as stress-activated protein kinases (JNK/SAPKs), a class of ubiquitously expressed protein which are increased in response to cytokines or growth factor deprivation; (c) p38 and its isoforms, which are activated in response to extracellular stress or inflammation, often mediated by a variety of signaling events such as proinflammatory cytokines and growth factors.

In the PNS, modulation of MAPKs occurs under normal and pathological conditions. In particular, during cell damage or nerve regeneration MAPKs are expressed in sensory neurons and in Schwann cells. Axotomy or nerve crush determines a distal Wallerian degeneration, inducing an increased expression of neurotrophic factors such as NGF, brain-derived neurotrophic factor (BDNF) or glial-derived neurotrophic factor (GDNF), neuregulins, and proinflammatory cytokines, which in turn participate in nerve fiber regeneration and injury responses (reviewed in Chen *et al.*, 2007). Experiments by Wiklund *et al.* (2002) on cultured L4–L5 DRG demonstrated that GDNF and NT-3 exert their growth effects by activating MAPK pathways.

Also BDNF can enhance axonal regeneration after nerve injury (Lindsay, 1988) and its upregulation in rat DRG seems to be correlated to the increased levels of phosphorylated ERK after axotomy (Obata *et al.*, 2003). Moreover, also in the proximal and distal stumps adjacent to the sciatic nerve transection site the ERK phosphorylation is increased (Sheu *et al.*, 2000). Also after crushing the nerve, ERK increases rapidly, persisting for at least 16 days (Sheu *et al.*, 2000). ERK activation is also predominant in Schwann cells (Agthong *et al.*, 2006), particularly in those proliferating to form the tubular structure (Bungner's band) for nerve regeneration.

Similarly, also JNK/SAPKs are the key regulators of several cellular processes. After sciatic nerve injury, JNK is locally activated, then conveyed via retrograde transport from the injury site to the nerve cell body (Lindwall and Kanje, 2005). In the injured sensory neurons JNK mediates *c-jun* nuclear activation (Lindwall *et al.*, 2004), contributing to the initiation of the cell body response to survival and to the regenerative processes (Lindwall and Kanje, 2005). In primary Schwann cell cultures from rat sciatic nerves, JNK is activated by NT-3 stimulation, supporting the hypothesis that the enhancement of Schwann cell migration induced by NT-3 is mediated by JNK (Yamauchi *et al.*, 2003).

The p38 subfamily of MAPKs is also activated following nerve injury. p38 activation, in fact, occurs in regenerating spinal motoneurons as well as in the distal segment of the regenerating sciatic nerve, and is presumably transported to the growth cone (Murashov *et al.*, 2001). By contrast, other authors suggested a proapoptotic role for p38 after peripheral nerve injury (Myers *et al.*, 2003, 2006).

Preclinical studies on the available MAPK inhibitors elucidated the differences among MAPK-activated signaling pathways (Wiklund *et al.*, 2002). However, in recent years, the therapeutic use of some of these inhibitors has been investigated. In particular, the p38 blocker SD-169 may enhance axonal regeneration by interfering with proinflammatory cytokine expression and inhibiting neuronal and Schwann cell apoptosis (Myers *et al.*, 2003).

VIII. Other Approaches

Erythropoietin (Epo) is an acidic glycoprotein, firstly identified as a cytokine stimulating erythropoiesis. Unlike it is not a classic neurohormone, Epo and its receptor (Epo-R) are synthesized in several cell populations and in non-hematopoietic tissues, particularly in CNS and PNS, where they exert neuroprotective roles. In the PNS, Epo and Epo-R are expressed in about 50% of the nerve cell bodies of both small and large diameter axons, as well as in axons of rat's DRG. In normal peripheral nerves Epo and Epo-R are also localized in Schwann cells and endothelial cells (Campana and Myers, 2001).

Epo counteracts toxic molecules, enhancing antioxidative enzyme production. Epo prevents *in vivo* and *in vitro* neuronal death caused by glutamate cytotoxicity (Brines *et al.*, 2000; Digicaylioglu *et al.*, 2004), metabolizes free radicals, stimulates neoangiogenesis, and affects neurotransmitter release (Buemi *et al.*, 2003). Several observations have also shown that Epo exerts neuroprotective effects by reducing inflammatory response, emphasizing its direct contribution in peripheral nerve regeneration.

In a model of painful chronic constriction injury (CCI), it has been shown that Epo expression is upregulated in Schwann cells, whereas expression of its receptor did not change (Campana and Myers, 2001). This increase in Epo expression determines, via an autocrine/paracrine mechanism, the Schwann cell proliferation and migration at the injury site (Campana and Myers, 2001). By studying the CCI model, it has been also identified the Epo's mechanism of action. Epo binding to its receptor provokes Epo-R homodimerization, phosphorylation of Janus cytoplasmic tyrosine kinase-2 (JAK-2) (Campana and Myers, 2001), which in turn mediates ERK/MAPK phosphorylation pathway (Li *et al.*, 2005). As described above, ERK/MAPK activation is involved in Schwann cell proliferation.

Therefore, pharmacological treatments with Epo enhance recovery of sciatic functions in injured mice, and these protective effects are evident up to 1 week after injury. However, Epo's neuroregenerative properties have also been demonstrated in the nerve regeneration after neurorraphy. In particular, in an end-to-side nerve repair model, Epo systemic administrations enhance peripheral nerve regeneration, bearing significant functional recovery of muscle during the early phase of treatment (Lykissas *et al.*, 2007).

The use of Epo or recombinant human Epo (rHuEpo, produced in the middle '80s) became relevant for therapeutic intervention. Erbayraktar *et al.* (2003) evaluated the AsialoEpo (obtained by total enzymatic desialylation of rHuEPO) in a rat model of spinal cord compression and injury of sciatic nerve. These authors showed that AsialoEpo maintains neuroprotective properties, without the erythropoietic side effects on bone marrow. More recently, Brines *et al.* (2008) evaluated the therapeutic effects of the 58–82 terminal portion of Epo helix B, that facing away from the ligand binding site of the Epo-R homodimer, demonstrating that this fragment exhibits neuroprotective properties *in vitro* and *in vivo*, lacking erythropoietic side effects.

Additionally, some classic vitamins are interesting for PNS regenerative medicine, mainly based on their ability to modulate Schwann biology. As a consequence of nerve trauma, occurs a series of events, generally referred as oxidative stress, which play a central role in neurodegeneration. The aberrant production of ROS (reactive oxygen species) and RNS (reactive nitrogen species) alters the permeability and functionality of Schwann cells, leading to severe myelin impairment. It is known that vitamin E and C possess antioxidant characteristics and are neuroprotective (Rock *et al.*, 1996). Therefore a possible use of these compounds for nerve regeneration has been hypothesized. Moreover, also vitamin D2 (ergocalciferol) modulates Schwann cells activity, regulating vitamin D receptors, and NGF expression. In a model of nerve injury, vitamin D2 proved to potentiate axon regeneration and to increase the axonal diameter (Chabas *et al.*, 2008).

IX. Conclusions

Despite surgical interventions are fundamental to promote peripheral nerve regeneration, the possibility to modulate Schwann cell that physiologically support nerve regrowth is an interesting therapeutic challenge. In the last decades, some pharmacological approaches yielded unsatisfactory results in regeneration; therefore, the identification of alternative mechanisms was necessary. Giving the novel role of neurotransmitters and neurohormones in glial cells, like Schwann cells of the PNS, the use of ligands targeting these systems represents a promising opportunity. In this review, we described some of these systems such as the GABAergic, the glutamatergic, or the cholinergic system. We also described the cross interaction among them, hoping that their study in PNS might provide the molecular basis for future human pharmacological or adjuvant therapies for nerve regeneration.

Acknowledgment

The authors are grateful to Dr. Vincenzo Conte and to Mrs. Marinella Ballabio for technical assistance.

References

Abbracchio, M. P., *et al.* (2009). Purinergic signalling in the nervous system: An overview. *Trends Neurosci.* **32,** 19–29.

Agthong, S., *et al.* (2006). Activation of MAPK ERK in peripheral nerve after injury. *BMC Neurosci.* **7,** 45.

Anneser, J. M., *et al.* (2000). Axotomy of the sciatic nerve differentially affects expression of metabotropic glutamate receptor mRNA in adult rat motoneurons. *Brain Res.* **868,** 215–221.

Ansselin, A. D., *et al.* (1997). Extracellular ATP increases intracellular calcium in cultured adult Schwann cells. *Neuroscience* **76,** 947–955.

Azcoitia, I., *et al.* (2003). Progesterone and its derivatives dihydroprogesterone and tetrahydroprogesterone reduce myelin fiber morphological abnormalities and myelin fiber loss in the sciatic nerve of aged rats. *Neurobiol. Aging* **24,** 853–860.

Backstrom, T., *et al.* (2005). Isoallopregnanolone; an antagonist to the anaesthetic effect of allopregnanolone in male rats. *Eur. J. Pharmacol.* **512,** 15–21.

Bandi, E., *et al.* (2005). Autocrine activation of nicotinic acetylcholine receptors contributes to Ca2+ spikes in mouse myotubes during myogenesis. *J. Physiol.* **568,** 171–180.

Baulieu, E. E. (1997). Neurosteroids: A role in aging? New functions in the central and peripheral nervous systems. *Aging (Milano)* **9,** 12.

Bernardini, N., *et al.* (1999). Rat dorsal root ganglia express m1-m4 muscarinic receptor proteins. *J. Peripher. Nerv. Syst.* **4,** 222–232.

Bettler, B., *et al.* (2004). Molecular structure and physiological functions of GABA(B) receptors. *Physiol. Rev.* **84,** 835–867.

Bhisitkul, R. B., *et al.* (1987). Axonal GABA receptors are selectively present on normal and regenerated sensory fibers in rat peripheral nerve. *Exp. Brain Res.* **66,** 659–663.

Bowery, N. G., *et al.* (1981). Bicuculline-insensitive GABA receptors on peripheral autonomic nerve terminals. *Eur. J. Pharmacol.* **71,** 53–70.

Brines, M. L., *et al.* (2000). Erythropoietin crosses the blood–brain barrier to protect against experimental brain injury. *Proc. Natl. Acad. Sci. USA* **97,** 10526–10531.

Brines, M., *et al.* (2008). Nonerythropoietic, tissue-protective peptides derived from the tertiary structure of erythropoietin. *Proc. Natl. Acad. Sci. USA* **105,** 10925–10930.

Brown, D. A., and Marsh, S. (1978). Axonal GABA-receptors in mammalian peripheral nerve trunks. *Brain Res.* **156,** 187–191.

Brown, D. A., *et al.* (1979). Distribution of gaba-receptors and gaba-carriers in the mammalian nervous system. *J. Physiol. (Paris)* **75,** 667–671.

Brunelli, G., *et al.* (2005). Glutamatergic reinnervation through peripheral nerve graft dictates assembly of glutamatergic synapses at rat skeletal muscle. *Proc. Natl. Acad. Sci. USA* **102,** 8752–8757.

Buemi, M., *et al.* (2003). The pleiotropic effects of erythropoietin in the central nervous system. *J. Neuropathol. Exp. Neurol.* **62,** 228–236.

Burnstock, G. (2009). Purinergic signalling: Past, present and future. *Braz. J. Med. Biol. Res.* **42,** 3–8.
Buznikov, G. A., *et al.* (1996). From oocyte to neuron: Do neurotransmitters function in the same way throughout development? *Cell. Mol. Neurobiol.* **16,** 537–559.
Campana, W. M., and Myers, R. R. (2001). Erythropoietin and erythropoietin receptors in the peripheral nervous system: Changes after nerve injury. *FASEB J.* **15,** 1804–1806.
Carlton, S. M. (2001). Peripheral excitatory amino acids. *Curr. Opin. Pharmacol.* **1,** 52–56.
Carozzi, A., *et al.* (2008). Expression and distribution of 'high affinity' glutamate transporters GLT1, GLAST, EAAC1 and of GCPII in the rat peripheral nervous system. *J. Anat.* **213,** 539–546.
Caruso, D., *et al.* (2008). Neuroactive steroid levels in a transgenic rat model of CMT1A neuropathy. *J. Mol. Neurosci.* **34,** 249–253.
Celotti, F., *et al.* (1992). The 5 alpha-reductase in the brain: Molecular aspects and relation to brain function. *Front. Neuroendocrinol.* **13,** 163–215.
Chabas, J. F., *et al.* (2008). Vitamin D2 potentiates axon regeneration. *J. Neurotrauma* **25,** 1247–1256.
Chavez-Delgado, M. E., *et al.* (2005). Ultrastructural analysis of guided nerve regeneration using progesterone- and pregnenolone-loaded chitosan prostheses. *J. Biomed. Mater. Res. B Appl. Biomater.* **74,** 589–600.
Chen, Z., *et al.* (2001). MAP kinases. *Chem. Rev.* **101,** 2449–2476.
Chen, Z. L., *et al.* (2007). Peripheral regeneration. *Annu. Rev. Neurosci.* **30,** 209–233.
Coronas, V., *et al.* (2000). Acetylcholine induces neuritic outgrowth in rat primary olfactory bulb cultures. *Neuroscience* **98,** 213–219.
Cvrcek, P. (2008). Side effects of ketamine in the long-term treatment of neuropathic pain. *Pain Med.* **9,** 253–257.
De Biasi, S., and Rustioni, A. (1988). Glutamate and substance P coexist in primary afferent terminals in the superficial laminae of spinal cord. *Proc. Natl. Acad. Sci. USA* **85,** 7820–7824.
Dennis, M. J., and Miledi, R. (1974). Electrically induced release of acetylcholine from denervated Schwann cells. *J. Physiol.* **237,** 431–452.
Desarmenien, M., *et al.* (1984). Coexistence of GABAA and GABAB receptors on A delta and C primary afferents. *Br. J. Pharmacol.* **81,** 327–333.
Desarnaud, F., *et al.* (1998). Progesterone stimulates the activity of the promoters of peripheral myelin protein-22 and protein zero genes in Schwann cells. *J. Neurochem.* **71,** 1765–1768.
Digicaylioglu, M., *et al.* (2004). Acute neuroprotective synergy of erythropoietin and insulin-like growth factor I. *Proc. Natl. Acad. Sci. USA* **101,** 9855–9860.
Donnerer, J. (2003). Regeneration of primary sensory neurons. *Pharmacology* **67,** 169–181.
Eglen, R. M. (2006). Muscarinic receptor subtypes in neuronal and non-neuronal cholinergic function. *Auton. Autacoid Pharmacol.* **26,** 219–233.
Erbayraktar, S., *et al.* (2003). Asialoerythropoietin is a nonerythropoietic cytokine with broad neuroprotective activity *in vivo*. *Proc. Natl. Acad. Sci. USA* **100,** 6741–6746[erratum appears in *Proc. Natl. Acad. Sci. USA*. 2003 Jul 22; 100(15), 9102].
Fink, T., *et al.* (1999). Glutaminergic and adrenergic receptors expressed on adult guinea pig Schwann cells *in vitro*. *Can. J. Physiol. Pharmacol.* **77,** 204–210.
Fu, S. Y., and Gordon, T. (1997). The cellular and molecular basis of peripheral nerve regeneration. *Mol. Neurobiol.* **14,** 67–116.
Green, A. C., *et al.* (1997). ATP acting on P2Y receptors triggers calcium mobilization in Schwann cells at the neuroelectrocyte junction in skate. *Neuroscience* **80,** 635–651.
Gumera, C. B., and Wang, Y. (2007). Modulating neuronal responses by controlled integration of acetylcholine-like functionalities in biomimetic polymers. *Adv. Mater.* **19,** 4404–4409.
Head, K. A. (2006). Peripheral neuropathy: Pathogenic mechanisms and alternative therapies. *Altern. Med. Rev.* **11,** 294–329.
Huettner, J. E., *et al.* (2002). Glutamate and the presynaptic control of spinal sensory transmission. *Neuroscientist* **8,** 89–92.

Ibanez, C., *et al.* (2003). Steroids and the reversal of age-associated changes in myelination and remyelination. *Prog. Neurobiol.* **71,** 49–56.

Jahromi, B. S., *et al.* (1992). Transmitter release increases intracellular calcium in perisynaptic Schwann cells *in situ*. *Neuron* **8,** 1069–1077.

Jessen, K. R., and Mirsky, R. (1997). Embryonic Schwann cell development: The biology of Schwann cell precursors and early Schwann cells. *J. Anat.* **191**(Pt. 4), 501–505.

Kinkelin, I., *et al.* (2000). Localization of ionotropic glutamate receptors in peripheral axons of human skin. *Neurosci. Lett.* **283,** 149–152.

Koenig, H. L., *et al.* (1995). Progesterone synthesis and myelin formation by Schwann cells. *Science* **268,** 1500–1503.

Koszykowska, M., *et al.* (2008). Effect of steroid hormones on the peripheral nervous system. *J. Anim. Feed Sci.* **17,** 3–18.

Lambert, J. J., *et al.* (2003). Neurosteroid modulation of GABAA receptors. *Prog. Neurobiol.* **71,** 67–80.

Li, X., *et al.* (2005). Schwann cells express erythropoietin receptor and represent a major target for Epo in peripheral nerve injury. *Glia* **51,** 254–265.

Lieberman, E. M., *et al.* (1989). Evidence that glutamate mediates axon-to-Schwann cell signaling in the squid. *Glia* **2,** 94–102.

Lindsay, R. M. (1988). Nerve growth factors (NGF, BDNF) enhance axonal regeneration but are not required for survival of adult sensory neurons. *J. Neurosci.* **8,** 2394–2405.

Lindwall, C., *et al.* (2004). Inhibition of c-Jun phosphorylation reduces axonal outgrowth of adult rat nodose ganglia and dorsal root ganglia sensory neurons. *Mol. Cell. Neurosci.* **27,** 267–279.

Lindwall, C., and Kanje, M. (2005). Retrograde axonal transport of JNK signaling molecules influence injury induced nuclear changes in p-c-Jun and ATF3 in adult rat sensory neurons. *Mol. Cell. Neurosci.* **29,** 269–282.

Liske, S., and Morris, M. E. (1994). Extrasynaptic effects of GABA (gamma-aminobutyric acid) agonists on myelinated axons of peripheral nerve. *Can. J. Physiol. Pharmacol.* **72,** 368–374.

Liu, G. J., and Bennett, M. R. (2003). ATP secretion from nerve trunks and Schwann cells mediated by glutamate. *Neuroreport* **14,** 2079–2083.

Loreti, S., *et al.* (2006). Rat Schwann cells express M1-M4 muscarinic receptor subtypes. *J. Neurosci. Res.* **84,** 97–105.

Loreti, S., *et al.* (2007a). Acetylcholine inhibits cell cycle progression in rat Schwann cells by activation of the M2 receptor subtype. *Neuron Glia Biol.* **3,** 269–279.

Loreti, S., *et al.* (2007b). *In* "Acetylcholine Controls the Schwann Cell Transition from Proliferating to Mature Phenotype, Counteracting the NRG1 Action", (Medimond, Ed.), pp. 47–53. VIII European Meeting of the Glial Cells in Health and Disease, London.

Lundgren, P., *et al.* (2003). Allopregnanolone-stimulated GABA-mediated chloride ion flux is inhibited by 3beta-hydroxy-5alpha-pregnan-20-one (isoallopregnanolone). *Brain Res.* **982,** 45–53.

Lykissas, M. G., *et al.* (2007). Axonal regeneration stimulated by erythropoietin: An experimental study in rats. *J. Neurosci. Methods* **164,** 107–115.

Ma, W., *et al.* (2004). Signaling cascades implicated in muscarinic regulation of proliferation of neural stem and progenitor cells. *Drug News Perspect.* **17,** 258–266.

Magnaghi, V. (2007). GABA and neuroactive steroid interactions in glia: New roles for old players? *Curr. Neuropharmacol.* **5,** 47–64.

Melcangi, R. C., *et al.* (1998). Age-induced decrease of glycoprotein Po and myelin basic protein gene expression in the rat sciatic nerve. Repair by steroid derivatives. *Neuroscience* **85,** 569–578.

Melcangi, R. C., *et al.* (1999a). Progesterone derivatives are able to influence peripheral myelin protein 22 and P0 gene expression: Possible mechanisms of action. *J. Neurosci. Res.* **56,** 349–357.

Melcangi, R. C., *et al.* (1999b). Steroid metabolism and effects in central and peripheral glial cells. *J. Neurobiol.* **40,** 471–483.

Melcangi, R. C., *et al.* (2000a). The action of steroid hormones on peripheral myelin proteins: A possible new tool for the rebuilding of myelin? *J. Neurocytol.* **29,** 327–339.

Melcangi, R. C., *et al.* (2000b). Aging in peripheral nerves: Regulation of myelin protein genes by steroid hormones. *Prog. Neurobiol.* **60,** 291–308.

Magnaghi, V., *et al.* (2001). Neuroactive steroids and peripheral myelin proteins. *Brain Res. Brain Res. Rev.* **37,** 360–371.

Melcangi, R. C., *et al.* (2003). Mifepristone (RU 38486) influences expression of glycoprotein Po and morphological parameters at the level of rat sciatic nerve: In vivo observations. *Exp. Neurol.* **184,** 930–938.

Magnaghi, V., *et al.* (2004a). GABAB receptors in Schwann cells influence proliferation and myelin protein expression. *Eur. J. Neurosci.* **19,** 2641–2649.

Magnaghi, V., *et al.* (2004b). The synthesis of glycoprotein Po and peripheral myelin protein 22 in sciatic nerve of male rats is modulated by testosterone metabolites. *Brain Res. Mol. Brain Res.* **126,** 67–73.

Melcangi, R. C., *et al.* (2005). Peripheral nerves: A target for the action of neuroactive steroids. *Brain Res. Brain Res. Rev.* **48,** 328–338.

Magnaghi, V., *et al.* (2006). GABA receptor-mediated effects in the peripheral nervous system: A cross-interaction with neuroactive steroids. *J. Mol. Neurosci.* **28,** 89–102.

Magnaghi, V., *et al.* (2008). Altered peripheral myelination in mice lacking GABAB receptors. *Mol. Cell. Neurosci.* **37,** 599–609.

Morris, M. E., *et al.* (1983). Depolarizing action of GABA (gamma-aminobutyric acid) on myelinated fibers of peripheral nerves. *Brain Res.* **278,** 117–126.

Murashov, A. K., *et al.* (2001). Crosstalk between p38, Hsp25 and Akt in spinal motor neurons after sciatic nerve injury. *Brain Res. Mol. Brain Res.* **93,** 199–208.

Myers, R. R., *et al.* (2003). Inhibition of p38 MAP kinase activity enhances axonal regeneration. *Exp. Neurol.* **184,** 606–614.

Myers, R. R., *et al.* (2006). The role of neuroinflammation in neuropathic pain: Mechanisms and therapeutic targets. *Drug Discov. Today* **11,** 8–20.

Nagai, T., *et al.* (1998). Uptake and release of neurotransmitter candidates, [3H]serotonin, [3H] glutamate, and [3H]gamma-aminobutyric acid, in taste buds of the mudpuppy, *Necturus maculosus. J. Comp. Neurol.* **392,** 199–208.

Obata, K., *et al.* (2003). Differential activation of extracellular signal-regulated protein kinase in primary afferent neurons regulates brain-derived neurotrophic factor expression after peripheral inflammation and nerve injury. *J. Neurosci.* **23,** 4117–4126.

Obrietan, K., *et al.* (2002). Excitatory actions of GABA increase BDNF expression via a MAPK–CREB-dependent mechanism—A positive feedback circuit in developing neurons. *J. Neurophysiol.* **88,** 1005–1015.

Oettling, G., *et al.* (1992). Characterization of the Ca2+ influx into embryonic cells after stimulation of the embryonic muscarinic receptor. *J. Dev. Physiol.* **17,** 147–155.

Olsen, R. W., *et al.* (1984). Autoradiographic localization of low affinity GABA receptors with [3H] bicuculline methochloride. *Eur. J. Pharmacol.* **99,** 247–248.

Owen, A., and Bird, M. (1995). Acetylcholine as a regulator of neurite outgrowth and motility in cultured embryonic mouse spinal cord. *Neuroreport* **6,** 2269–2272.

Patte-Mensah, C., and Mensah-Nyagan, A. G. (2008). Peripheral neuropathy and neurosteroid formation in the central nervous system. *Brain Res. Rev.* **57,** 454–459.

Patte-Mensah, C., *et al.* (2006). Neurogenic pain and steroid synthesis in the spinal cord. *J. Mol. Neurosci.* **28,** 17–31.

Pearson, G., *et al.* (2001). Mitogen-activated protein (MAP) kinase pathways: Regulation and physiological functions. *Endocr. Rev.* **22,** 153–183.

Pestronk, A., and Drachman, D. B. (1985). Motor nerve terminal outgrowth and acetylcholine receptors: Inhibition of terminal outgrowth by alpha-bungarotoxin and anti-acetylcholine receptor antibody. *J. Neurosci.* **5,** 751–758.

Pinard, A., *et al.* (2003). Glutamatergic modulation of synaptic plasticity at a PNS vertebrate cholinergic synapse. *Eur. J. Neurosci.* **18,** 3241–3250.

Popratiloff, A., *et al.* (1996). Glutamate receptors in spinal motoneurons after sciatic nerve transection. *Neuroscience* **74,** 953–958.

Ren, X., and Mody, I. (2003). Gamma-hydroxybutyrate reduces mitogen-activated protein kinase phosphorylation via GABA B receptor activation in mouse frontal cortex and hippocampus. *J. Biol. Chem.* **278,** 42006–42011.

Robinson, L. R. (2000). Traumatic injury to peripheral nerves. *Muscle Nerve* **23,** 863–873.

Rock, C. L., *et al.* (1996). Update on the biological characteristics of the antioxidant micronutrients: Vitamin C, vitamin E, and the carotenoids. *J. Am. Diet Assoc.* **96,** 693–702; quiz 703–704.

Roglio, I., *et al.* (2008a). Neuroprotective effects of dihydroprogesterone and progesterone in an experimental model of nerve crush injury. *Neuroscience* **155,** 673–685.

Roglio, I., *et al.* (2008b). Neuroactive steroids and peripheral neuropathy. *Brain Res. Rev.* **57,** 460–469.

Salani, M., *et al.* (2009). Acetylcholine-induced neuronal differentiation: Muscarinic receptor activation regulates EGR-1 and REST expression in neuroblastoma cells. *J. Neurochem.* **108,** 821–834.

Schaffar, N., *et al.* (1997). Immunohistochemical detection of glutamate in rat vagal sensory neurons. *Brain Res.* **778,** 302–308.

Scherer, S. S., and Wrabetz, L. (2008). Molecular mechanisms of inherited demyelinating neuropathies. *Glia* **56,** 1578–1589.

Sereda, M. W., *et al.* (2003). Therapeutic administration of progesterone antagonist in a model of Charcot-Marie-Tooth disease (CMT-1A). *Nat. Med.* **9,** 1533–1537.

Sheu, J. Y., *et al.* (2000). Differential patterns of ERK and STAT3 phosphorylation after sciatic nerve transection in the rat. *Exp. Neurol.* **166,** 392–402.

Steiger, J. L., *et al.* (2004). Camp response element-binding protein, activating transcription factor-4, and upstream stimulatory factor differentially control hippocampal GABABR1a and GABABR1b subunit gene expression through alternative promoters. *J. Neurosci.* **24,** 6115–6126.

Stevens, B., and Fields, R. D. (2000). Response of Schwann cells to action potentials in development. *Science* **287,** 2267–2271.

Stevens, B., *et al.* (2004). Adenosine: An activity-dependent axonal signal regulating MAP kinase and proliferation in developing Schwann cells. *Neuron Glia Biol.* **1,** 23–34.

Sun, B. B., and Chiu, S. Y. (1999). N-type calcium channels and their regulation by GABAB receptors in axons of neonatal rat optic nerve. *J. Neurosci.* **19,** 5185–5194.

Tata, A. M. (2008). Muscarinic acetylcholine receptors: New potential therapeutic targets in antinociception and in cancer therapy. *Recent Pat. CNS Drug Discov.* **3,** 94–103.

Tata, A. M., *et al.* (2003). Cholinergic modulation of neurofilament expression and neurite outgrowth in chick sensory neurons. *J. Neurosci. Res.* **73,** 227–234.

Towers, S., *et al.* (2000). GABAB receptor protein and mRNA distribution in rat spinal cord and dorsal root ganglia. *Eur. J. Neurosci.* **12,** 3201–3210.

Ulrich, D., and Bettler, B. (2007). GABA(B) receptors: Synaptic functions and mechanisms of diversity. *Curr. Opin. Neurobiol.* **17,** 298–303.

Watkins, J. C. (2000). L-Glutamate as a central neurotransmitter: Looking back. *Biochem. Soc. Trans.* **28,** 297–309.

Wessler, I., *et al.* (2001). The biological role of non-neuronal acetylcholine in plants and humans. *Jpn. J. Pharmacol.* **85,** 2–10.

Whitlock, E. L., *et al.* (2007). Pregabalin does not impact peripheral nerve regeneration after crush injury. *J. Reconstr. Microsurg.* **23,** 263–268.

Wiklund, P., *et al.* (2002). Mitogen-activated protein kinase inhibition reveals differences in signalling pathways activated by neurotrophin-3 and other growth-stimulating conditions of adult mouse dorsal root ganglia neurons. *J. Neurosci. Res.* **67,** 62–68.

Wu, S. Z., *et al.* (2005). Schwann cells exhibit excitotoxicity consistent with release of NMDA receptor agonists. *J. Neurosci. Res.* **79,** 638–643.

Wu, F., *et al.* (2008). Enhanced rat sciatic nerve regeneration through silicon tubes implanted with valproic acid. *J. Reconstr. Microsurg.* **24,** 267–276.

Yamauchi, J., *et al.* (2003). Neurotrophin 3 activation of trkc induces Schwann cell migration through the c-Jun N-terminal kinase pathway. *Proc. Natl. Acad. Sci. USA* **100,** 14421–14426.

Yashpal, K., *et al.* (2001). Differential effects of NMDA and group I mglur antagonists on both nociception and spinal cord protein kinase C translocation in the formalin test and a model of neuropathic pain in rats. *Pain* **94,** 17–29.

Zagorodnyuk, V. P., *et al.* (2002). Functional GABAB receptors are present in guinea pig nodose ganglion cell bodies but not in peripheral mechanosensitive endings. *Auton Neurosci.* **102,** 20–29.

Zheng, J. Q., *et al.* (1994). Turning of nerve growth cones induced by neurotransmitters. *Nature* **368,** 140–144.

MELATONIN AND NERVE REGENERATION

Ersan Odaci* and Suleyman Kaplan[†]

*Department of Histology and Embryology, Karadeniz Technical University School of Medicine, 61080 Trabzon, Turkey

[†]Department of Histology and Embryology, Ondokuz Mayis University School of Medicine, 55139 Samsun, Turkey

Melatonin is a widely distributed and important signal molecule that occurs in unicellular organisms, plants, and fungi in addition to animals and humans. It is the main hormone of the pineal gland and its synthesis occurs mainly in this gland. It has free radical scavenging and antioxidative properties and shows clinical antibacterial and analgesic effects. By means of these properties, it is able to protect cells, tissues, and organs against oxidative damage from free radicals. Recently, widespread interest has grown among researchers regarding the apparent protective effects of melatonin following traumatic events to peripheral nerves, especially the sciatic nerve and its pathological conditions, as melatonin administration could be beneficial following surgery. Although there are great numbers of studies that have mentioned protective effects of melatonin on peripheral nerve pathologies, there are also some studies that report toxic effects of melatonin on peripheral nerves. This paper reviews the available literature in terms of both the beneficial and the toxic effects of melatonin on peripheral nerves. Short descriptions of the structure of pineal gland and synthesis and secretion of melatonin are also given.

INTERNATIONAL REVIEW OF NEUROBIOLOGY, VOL. 87
DOI: 10.1016/S0074-7742(09)87016-5

I. Introduction

The pineal gland has been recognized as an anatomical structure for more than a thousand years, being variously described over the centuries as a mysterious unpaired organ of the brain, the "third eye," "the seat of the soul," or a "calcified vestigial organ with no function" (Arendt, 1995). However, in the latter half of the twentieth century, the physiological importance of the pineal gland, and its primary hormone, melatonin, were recognized and the gland was widely investigated with new research techniques such as stereological, light and electron microscopy (Arendt, 1995; Turgut *et al.*, 2005a,b). Today, there are many clinical or experimental studies in the literature on structure of pineal gland; as well as hundreds of papers on melatonin chemistry and its synthesis, secretion, molecular architecture, its effects on other organs, and its therapeutic effects on chronic diseases or other pathologic conditions (Lerner *et al.*, 1958, 1959; Møller and Baeres, 2002; Shokouhi *et al.*, 2008; Stavisky *et al.*, 2005; Turgut *et al.*, 2005a,b; Vollrath, 1981, 1984).

There has also been a tremendous increase in studies on peripheral nerves and their regeneration. This is largely due to the fact that nerve reconstruction is becoming a more frequent and widespread type of surgery following major nerve trauma. Towards this aim, the effects of experimental administration of some chemicals, drugs, or exogenous hormones on peripheral nerves have been widely investigated. Melatonin, the neurohormone from the pineal gland, is one such substance that has apparent beneficial effects on peripheral nerves that have been damaged by pathologic conditions or traumatic events (Sayan *et al.*, 2004; Shokouhi *et al.*, 2008; Stavisky *et al.*, 2005; Turgut *et al.*, 2005a,b). However, some researchers have reported that melatonin can have a toxic effect on peripheral nerve regeneration (Piezzi and Cavicchia, 1981; Prevedello *et al.*, 1979). Therefore, in this chapter, we review the literature reporting the effects of melatonin on peripheral nerves in response to different pathological conditions, such as injury, transaction, or traumatic events. Additionally, very brief descriptions of human pineal gland structure at the macroscopic and microscopic levels and of melatonin synthesis, secretion, and antioxidant action are also presented.

Pineal gland is also known as epiphysis cerebri, cervical body, pineal body, or pineal organ and is about 5–10 mm in length, 1–5 mm wide, and 3–5 mm in thickness in adults. It has an approximate weight of about 100–180 mg, with little apparent variation related to either age or gender in humans (Macchi and Bruce, 2004). It has a pyramidal shape and is attached by a short pineal stalk to the posterior extremity of the third ventricle of the forebrain, above the roof of the diencephalons. The gland receives afferent fibers from postganglionic sympathetic fibers that arise from the paired superior cervical ganglia in the neck. It is highly vascularized although it lacks a blood–brain barrier (Macchi and Bruce, 2004;

Touitou, 2001; Vollrath, 1981). The parenchyma of the pineal gland is similar to that of other endocrine glands, consisting of cords of cells separated by capillaries with fairly wide perivascular spaces. In humans, the parenchyma is divided into lobules separated by loose connective tissue septa that are derived from the pial tissue surrounding the gland (Møller and Baeres, 2002). The main cell types of the human pineal gland are the neuroglia and the hormone-producing pinealocytes (Korf, 2000; Møller and Baeres, 2002). Neuroglia are predominantly fibrous astrocytes, although some may be microglia. They are scattered in an irregular manner throughout the pineal gland and are generally seen either surrounding the pinealocytes or in peripheral patches (Macchi and Bruce, 2004; Vollrath, 1984).

Melatonin (*N*-acetyl-5-methoxytryptamine) is an important signal molecule that is widely distributed in nature. It is found in vertebrate animals and humans, but is also a component of unicellular organisms, plants, and fungi (Bob and Fedor-Freybergh, 2008; Reiter *et al.*, 2007). It is a small lipid and water-soluble indoleamine that can easily diffuse through cell membranes (Bob and Fedor-Freybergh, 2008) and is the main hormone secreted by the pineal gland, which is the major site of its synthesis in vertebrates. It is named based on its effect on lightening the melanin pigmentation in frog skin (Touitou, 2001).

Melatonin synthesis in the pineal gland begins when pinealocytes take up the precursor, tryptophan, from the blood. Tryptophan is first hydroxylated to 5-hydroxytryptophan (5HTP) by the enzyme tryptophan-5-hydroxylase and then decarboxylated to serotonin by the enzyme 5HTP-decarboxylase. Serotonin is then converted to *N*-acetyl-serotonin by the rate-limiting enzyme *N*-acetyl transferase (NAT), and finally *N*-acetyl-serotonin is converted to melatonin by the enzyme hydroxyindole-*O*-methyl transferase (Axelrod, 1974; Klein *et al.*, 1997; Macchi and Bruce, 2004; Møller and Baeres, 2002; Reiter, 1981, 1991; Sugden and Klein, 1987; Touitou, 2001) (Fig. 1).

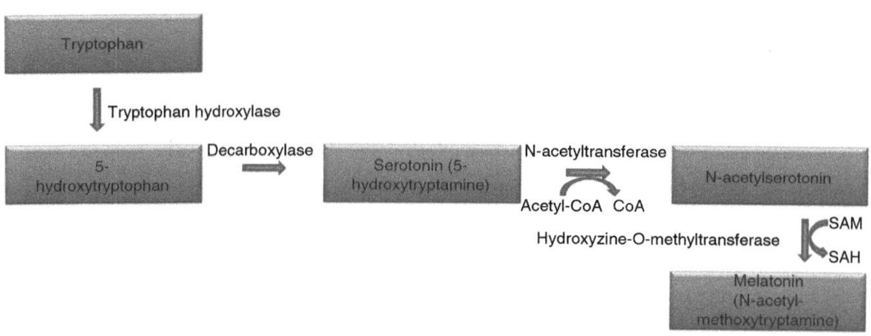

FIG. 1. A schematic diagram of melatonin synthesis (Modified from Touitou, 2001).

Pineal melatonin secretion occurs predominantly at night and its synthesis is closely related to sleep regulation and other cyclic metabolic activities (Bob and Fedor-Freybergh, 2008). It shows a circadian rhythmicity that is reflected in plasma levels that are low during the day and high at night. In the dark phase of the diurnal period, the plasma concentration of melatonin is three to ten times higher than it is during the light period. This circadian rhythmicity continues into old age and can still be seen in elderly humans (Touitou, 2001; Touitou *et al.*, 1981). The release of norepinephrine (NE) into the synaptic clefts between the sympathetic nerve endings and the pinealocyte is the initial stimulation for melatonin synthesis in the pineal gland (Touitou, 2001). During the dark part of the diurnal period, NE is released and activates adenylate cyclase, which in turn induces cAMP production. cAMP then activates serotonin-*N*-acetyltransferase, the key enzyme in melatonin biosynthesis (Touitou, 2001).

Free radicals are continuously produced in cells as byproducts of oxidative phosphorylation in mitochondria and during fatty acid oxidation in peroxisomes (Touitou, 2001). The normal production of free radicals is disrupted in several human pathological conditions, leading to detrimental effects on normal cellular functions, and in some cases leading to death (Harman, 1984). During the abnormal production of free radicals, aggressive oxygen species can damage all of the biochemical components of the organism (Touitou, 2001). For example, some free radical reactions may irreversibly change the structure of DNA, RNA, proteins, carbohydrates, and unsaturated lipids (Harman, 1984, 1995; Touitou, 2001). Melatonin has been reported to be a free radical scavenger and to possess antioxidant properties. It protects cells, tissues, and organs against oxidative damage induced by a variety of free radical generating agents and processes (Reiter *et al.*, 1997). Melatonin is effective as an antioxidant in protecting nuclear DNA, membrane lipids, and possibly also cytosolic proteins from oxidative damage. It is also able to cross morphophysiological barriers such as the blood–brain or blood–testis barrier and can readily enter cells and subcellular compartments. This feature of melatonin may simplify the scavenging of free radicals (Reiter *et al.*, 1997; Touitou, 2001).

In addition to the effectiveness of melatonin as an electron donor and free radical scavenger (Hardeland *et al.*, 1995; Reiter *et al.*, 2001), it is suggested that melatonin may provide antioxidant protection without the benefit of receptors (Ceraulo *et al.*, 1999) and can act locally as a free radical scavenger (Touitou, 2001). Some results indicate that melatonin stabilizes microsomal membranes and resists the rigidity induced by free radical attack. Therefore, the ability of melatonin to stabilize cellular membranes appeared to be a result of its free radical scavenging activity and was related to its ability to reduce lipid peroxidation (Garcia *et al.*, 1997). However, it has been noted that there is no evidence that physiological concentrations of melatonin can affect the human antioxidative defense system *in vivo* (Touitou, 2001).

II. The Effects of Melatonin on Peripheral Nerves

A. The Effects of Melatonin on Ischemia–Reperfusion Injury in the Peripheral Nerves

Several studies have investigated ischemia–reperfusion (I/R)-induced alterations in peripheral nerves, especially the sciatic nerve. I/R is a common and serious problem that occurs in a variety of clinical conditions and leads to endoneurial edema, segmental demyelization, axonal degeneration, and multifocal-diffuse loss of nerve fibers (Bagdatoglu *et al.*, 2002; Li *et al.*, 1996; Mitsui *et al.*, 1999a,b; Nagamatsu *et al.*, 1996; Nukada *et al.*, 1993). Morphological alterations of peripheral nerves after ischemic injury secondary to occlusion of major arteries were investigated by Nukada *et al.* (1993). In this study, the nerve blood flow (NBF) was measured, the nerves were examined histopathologically and areas of injury were identified at various levels distal to the ligature site. The most severe ischemia, induced by focally reducing NBF by 80% in the tibial nerve at a level just below the knee after femoral artery ligation, caused degenerative changes in nerve fibers in the ischemic nerve segments, especially in the subperineurial region. Following ligation of the internal iliac artery, approximately 60% reductions in NBF at the upper and mid-thigh levels of the sciatic nerve occurred and endoneurial edema was seen only at this level of the sciatic nerve. NBF was reduced about 20% at the pelvic level of the sciatic nerve and there were no any pathological findings following superior gluteal artery ligation. Therefore, it has been proposed that mild levels of ischemia cause endoneurial edema, while moderate levels of ischemia produce demyelization and severe ischemia produces Wallerian degeneration (Nukada *et al.*, 1993).

Several studies have investigated the effect of melatonin or melatonin precursors on I/R-induced alterations in the tissues, organs, or biochemical markers in the blood, etc. (Cervantes *et al.*, 2008; Genade *et al.*, 2008; Kim and Lee, 2008; Nagai *et al.*, 2008; Wang *et al.*, 2006). However, only one study reports on the effects of melatonin on I/R injury in the peripheral nerves (Sayan *et al.*, 2004). This study was conducted to investigate the protective effect of melatonin on sciatic nerves of rats subjected to 2 h ischemia followed by 3 h of reperfusion. The major finding of this study was that pretreatment with melatonin ameliorated I/R injury of the sciatic nerve. In I/R groups, axonal damage occurred in the most of the myelinated fibers, which showed common axonal shrinkage and swollen axons. The most striking morphologic changes occurring in the myelin sheath were vacuolization and lamellar separation. Total destruction of the axons and a honeycomb appearance of the fibers were found in some nerves. Degenerative changes were also seen in Schwann cells, which showed vacuolization in the cytoplasm (Sayan *et al.*, 2004). Vacuolization and degeneration of some of the unmyelinated fibers were also common, and this fiber degeneration was attributed to endoneurial edema. In animals pretreated with melatonin, fewer morphologic changes were observed. After treatment with melatonin, not only was myelin breakdown decreased but also a remarkable improvement in the axons was also seen, since vacuolization and lamellar separation of the axonal myelin sheath was less evident. Schwann cells were seemingly normal after I/R of the melatonin-treated rats and there were no significant changes in the ultrastructure of unmyelinated fibers. Additionally, the area of vacuolization of the myelin sheath in the sciatic nerve of the rats in the melatonin group was significantly decreased (Sayan *et al.*, 2004).

B. The Effects of Melatonin on Painful Situations During Peripheral Nerve Injury

It has been reported that melatonin has both anti-inflammatory and analgesic (Yu *et al.*, 2000) effects in animal models of local and systemic inflammation (Cuzzocrea *et al.*, 1997, 1999; El-Shenawy *et al.*, 2002; Yu *et al.*, 2000). Cuzzocrea *et al.* (1999) investigated the effect of melatonin on the production of prostaglandins, which are inflammatory mediators, in carrageenan-induced pleurisy, a model of acute inflammation where prostaglandins are known to play a crucial role (Cuzzocrea *et al.*, 1999). They showed that melatonin inhibits the inflammatory response and that pretreatment of carrageenan-treated rats with a higher dose of melatonin prevented COX-2 expression. Melatonin, therefore, exerted potent anti-inflammatory effects and part of these anti-inflammatory effects may have been related to a reduction of prostaglandin production during the inflammatory process (Cuzzocrea *et al.*, 1999).

In another study, the effects of melatonin treatment in carrageenan-induced paw edema and pleurisy models of acute inflammation were investigated (Cuzzocrea *et al.*, 1997). The oxyradical peroxynitrite played a crucial role in the inflammatory process, while melatonin was seen to exert an inhibitory effect on the expression of the inducible isoform of nitric oxide synthase (NOS) and also prevented the formation of nitrotyrosine (Cuzzocrea *et al.*, 1997). Therefore, melatonin showed a potent anti-inflammatory effect, and this effect may have been related to inhibition of the expression of the inducible NOS, or was directly related to oxyradical and peroxynitrite scavenging (Cuzzocrea *et al.*, 1997).

Analgesic effects of melatonin in animal models have also been studied. In a study designed to explore the site and mechanism of the analgesic action of melatonin (Yu *et al.*, 2000), melatonin was administered intraperitoneally (IP) and intracerebroventricularly (ICV) and produced an antinociceptive effect in a dose-dependent manner. Therefore, it is suggested that melatonin has an analgesic effect in rats and that the central nervous system (CNS) may be the primary site for elicitation of the response (Yu *et al.*, 2000). Since melatonin has both anti-inflammatory and analgesic properties, as described above, studies have also been conducted on pain condition of peripheral nerves following injury. Injury to the peripheral nerves leads to abnormal pain conditions known as neuropathic pain. Two clinical symptoms, mechanical allodynia and thermal hyperalgesia are frequently seen associated with neuropathic pain (Ulugol *et al.*, 2006). Melatonin can exert sedative/hypnotic, anticonvulsant, and antianxiety effects (Geoffriau *et al.*, 1998; Golombek *et al.*, 1996; Sugden, 1983) and is commonly used clinically for this reason to treat neuropathic pain. The effects of ICV and IP melatonin on mechanical allodynia and thermal hyperalgesia have been studied in mice following partial tight ligation of the sciatic nerve (Ulugol *et al.*, 2006). For this purpose, a marked mechanical allodynia and thermal hyperalgesia were created by surgery in sciatic nerve-ligated neuropathic mice. Melatonin by ICV and IP, at higher doses, produced a blockade of thermal hyperalgesia, but not of mechanical allodynia (Ulugol *et al.*, 2006). Therefore, melatonin appeared to be useful for the clinical management of chronic pain.

In a separate study, both mechanical allodynia and depression-like behavior were exacerbated after peripheral nerve injury (PNI) in Wistar-Kyoto (WKY) rats, a genetic variation of Wistar rats with demonstrable depression-like behavior (Zeng *et al.*, 2008). After melatonin administration into the anterior cingular cortex contralateral, PNI was investigated. Melatonin prevented the exacerbation of mechanical allodynia and also concurrently improved the depression-like behavior in WKY rats. This suggested the existence of a reciprocal relationship between mechanical allodynia and depression-like behavior, and implicated the melatoninergic system in the anterior cingular cortex in the interaction between pain and depression (Zeng *et al.*, 2008).

C. The Effects of Melatonin After Peripheral Nerve Transection and PNI

The effects of melatonin on peripheral nerves after the peripheral nerve transection have been widely studied. In one study, the effect of melatonin on the suppression of collagen production and neuroma formation after peripheral nerve transection was investigated (Turgut *et al.*, 2005a). An ameliorative effect of melatonin on the development of neuroma formation at the cut end of the proximal stump and on its collagen content was seen (Turgut *et al.*, 2005a). In another study, Turgut and coworkers (2005b) investigated the effects of melatonin on nerve repair and neuronal regeneration in rat sciatic nerve suture repair. This study demonstrates that exogenous melatonin administration significantly inhibits collagen accumulation in the neuroma of a suture repair site and thereby improves nerve regeneration. Clinically, the positive effect of melatonin administration on neuroma formation and nerve regeneration seems a particularly attractive treatment option. Therefore, it is proposed that peripheral nerve repair with addition of melatonin may be a worthwhile option, in addition to other treatment modalities, in cases of melatonin deficiency (Turgut *et al.*, 2005b).

Transforming growth factor (TGF)-β and basic fibroblast growth factor (bFGF) play an important role in collagen production by fibroblasts and in Schwann cell activity (Bunge and Bunge, 1978; Einheber *et al.*, 1995; Rufer *et al.*, 1994; Tatagiba *et al.*, 2002). The immunohistochemical profile of TGF-β1 and bFGF in the anastomotic region of the sciatic nerve, after pinealectomy and exogenous melatonin administration, were also investigated by Turgut *et al.* (2006). In this study, strong TGF-β1 and/or bFGF expression was observed in the epineurium of nerves from animals that had undergone pinealectomy, but no or weak staining was observed in animals in the control and melatonin treatment groups. Therefore, it was suggested that both TGF-β1 and bFGF play important roles in control of collagen accumulation and neuroma formation at the anastomotic site, and that melatonin has a beneficial effect on nerve regeneration (Turgut *et al.*, 2006).

Another study was designed to examine the effects of melatonin on polyethylene glycol (PEG)-induced repair *in vitro* and/or *in vivo* by plasmalemmal fusion of rat sciatic axons severed by crushing (Stavisky *et al.*, 2005). In this study, the conduction of compound action potentials through the lesion site was significantly higher and of greater amplitude in melatonin-treated animals following PEG-fusion. Melatonin also significantly increased the ability of sciatic axons to PEG-fuse *in vivo*, compared to the other treatment groups (Stavisky *et al.*, 2005). Therefore, PEG combined with melatonin might be highly effective for repair of severed axons following crush-type injuries to sciatic nerves in humans (Stavisky *et al.*, 2005).

The potent antioxidant properties of melatonin in biological systems are well known (Hardeland *et al.*, 1993). One of the beneficial effects is the prevention of trauma-induced PNIs. For example, the first line of antioxidant defense by melatonin is the prevention of peroxidative injuries during oxidizing conditions in the nervous system (Hsu *et al.*, 2002; Pieri and Marra, 1994). Melatonin can reduce the harmful effects of reactive oxygen species via free radical scavenging or by decreasing NOS activity and radical generation in the CNS (Baydas *et al.*, 2003; Nam *et al.*, 2005; Reiter, 1996; Reiter *et al.*, 2000; Rogério *et al.*, 2002). Besides inhibiting posttraumatic polymorphonuclear infiltration (El-Abhar *et al.*, 2002), melatonin also stimulates superoxide dismutase (SOD), glutathione peroxidase, and glutathione reductase (Sayan *et al.*, 2004). In addition, it protects the sciatic nerve from I/R injury by attenuating neural lipid peroxidation (Sayan *et al.*, 2004). Shokouhi *et al.* (2008) investigated the neuroprotective effects of melatonin on neural fiber damage and lipid peroxidation after a blunt sciatic nerve trauma, comparing the effects of low dose (10 mg/kg) and high dose (50 mg/kg). According to their results, low-dose melatonin reduced trauma-induced myelin breakdown and axonal changes in the sciatic nerve. However, high-dose melatonin almost entirely neutralized any ultrastructural changes. Therefore, it may be proposed that melatonin has a potent neuroprotective effect at a dose of 50 mg/kg and it can preserve peripheral neural fibers from lipid peroxidative damage after blunt trauma (Shokouhi *et al.*, 2008).

The effects of melatonin supplementation on peripheral nerve function in the ovariectomized (OVX)-aged rats have also been examined (Ek *et al.*, 2007). In this study, OVX-aged rats received melatonin (5 or 20 mg/kg) daily for either 2 or 6 weeks. Nerve conduction velocities and distal latencies were determined from the propagation of action potential recorded by an extracellular electrophysiological technique. The mean distal latencies of the melatonin-treated groups were shorter than were those of the control group and nerve conduction velocity was significantly greater in both melatonin-treated groups. These results suggested that melatonin was able to alleviate the electrophysiological degeneration of the sciatic nerve in OVX-aged rats. Based on these results, it was postulated that melatonin might have a potential clinical application for the treatment of postmenopausal peripheral nerve degeneration (Ek *et al.*, 2007).

Neonatal rat motoneurons are vulnerable to sciatic nerve transection (Greensmith and Vrobová, 1996). Since different neurons in the both central and peripheral mammalian nervous systems express NOS, it is possible that NOS activity can be induced in related neurons normally devoid of it following PNI (Chang *et al.*, 2000). Some researchers have suggested that motoneuron death in the lumbar enlargement of newborn rats after sciatic nerve transection could be mediated via a neuronal isoform of NOS (Ando *et al.*, 1996; Clowry, 1993; Estévez *et al.*, 1998; Yu, 1997), a nicotinamide adenine dinucleotide phosphate dependent diaphorase (NADPH-d) present in the cytoplasm (Clowry, 1993; Hope *et al.*, 1991). The injury-induced expression of NOS seems to signal the impending death of lesioned cells, and this enzyme may function as a killer protein that produces neurotoxic levels of NO (Clowry, 1993).

Recently, a possible protective effect of melatonin as an antioxidant agent and an inhibitor of neuronal NOS (nNOS) were investigated on spinal motoneurons after axonal injury (Rogério *et al.*, 2002). In this experiment, sciatic nerve transections were carried out on rat pups at postnatal day (PND) 2 and examined at PND7. In this study, melatonin decreased neuronal death at doses of 1–50 mg/kg, which indicated that lower doses of melatonin had a neuroprotective effect. Higher doses of melatonin (50 and 100 mg/kg) were toxic. Additionally, there were no differences in nNOS expression between the treated and the control rats, and surviving motoneurons in the sciatic pool did not express the enzyme. These results suggest that nNOS may not be involved in neuronal death or survival in these experimental conditions (Rogério *et al.*, 2002).

The expression of SOD1 and 2 were also studied following similar treatment of newborn rats to determine whether daily doses of melatonin (Rogério *et al.*, 2005) induced a protective free radical scavenger effect (Reiter, 1998; Reiter *et al.*, 2000). Melatonin protects motoneuron loss by about 75% following axotomy at PND3 and PND7. However, neither sciatic transection nor melatonin induced any detectable changes in the immunoreactivity patterns of the enzymes (Rogério *et al.*, 2005).

Melatonin is able to reduce not only apoptotic cell death in the CNS (Reiter, 1998) but also the neuronal death induced by peripheral axotomy in neonatal rats. For this reason, apoptotic events after sciatic axotomy and after administration of melatonin were investigated by Rogério *et al.* (2006a), focusing on the antiapoptotic protein Bcl-2 and the cell death promoter Bax in the spinal cord of neonatal rats. This study results can be summarized as follows: first, sciatic transection at PND2 increased Bax mRNA in the lumbar enlargement; second, Bax immunoreactivity in immature motoneuron was not altered by axotomy; and final, melatonin protected motoneuron and dorsal horn cells through a mechanism independent of Bax and Bcl-2. Therefore, it has been suggested that both physiological and axotomy-induced cell death in the dorsal horn of neonatal rat lumbar enlargement are associated with Bax expression. However, this expression does not seem to be related to motoneuron death. Melatonin not only protected axotomized motoneurons but also reduced the loss of dorsal horn cells 1 day after lesion. In both cases, the mechanism of action of the neurohormone was not related to changes in Bax or Bcl-2 expression (Rogério *et al.*, 2006a).

In earlier studies, nNOS was determined in the spinal motoneurons after sciatic transection in newborn rats. nNOS was observed in lesioned motoneurons of adult animals and was associated with cell death (Chang *et al.*, 2000; Wu *et al.*, 1994; Yu, 2002). Although neuronal loss was confirmed in these studies, nNOS was not detected immunohistochemically in the injured cells 1 or 5 days following axotomy (Rogério et al., 2002, 2005) until the study of Rogério *et al.* (2006b). In this latter study, enzymatic activity measurement was used to study NOS isoforms during the first week after lesion, the time range in which peripheral sciatic section in neonatal rats causes the death of most of the axotomized motoneurons (Rogério *et al.*, 2005). Sciatic axotomy in PND2 rats causes lumbar motoneuron loss, which may be associated with NO production. Therefore, NOS expression and NO synthesis in the lumbar enlargement of rats after sciatic nerve transection at PND2 and treatment with the antioxidant melatonin were examined (Rogério *et al.*, 2006b). According to this study results, melatonin did not alter NOS expression and did not depend on calcium to change NOS activity (Rogério *et al.*, 2006b).

The effects of melatonin treatment have also been examined on cranial motor neurons whose axons had been lesioned peripherally (Chang *et al.*, 2000). Oxidative stress and massive production of NO occur in the neuropathogenesis following PNI.

Therefore, Chang and coworkers (2000) conducted a study to determine whether melatonin would exert its neuroprotective effect on the lesioned hypoglossal neurons after peripheral axotomy, because it is known that melatonin reduces the oxidative damage in a variety of experimental neuropathologies in which NO is involved. This study morphologically demonstrated that injury-induced neuronal NADPH-d/NOS expression in the hypoglossal motoneurons can be effectively reduced by different doses of melatonin. As a result, it has been suggested that melatonin may be efficacious in reducing oxidative stress after PNI (Chang *et al.*, 2000). This finding has not only shed light on better understanding the functional roles of NO involved in the processes of neuronal damage but also offers the possibilities of potential therapeutic use of melatonin for the prevention of oxidative damage following traumatic nerve injury (Chang *et al.*, 2000). A further study investigated whether melatonin could exert beneficial effects by preserving SOD activity following PNI (Chang *et al.*, 2008). In this study, adult rats exposed to hypoglossal nerve transection were IP injected with melatonin. Their results showed that melatonin appeared to serve as an effective therapeutic agent with a significant capacity for preserving SOD activity, thus reducing PNI-relevant oxidative damage, in agreement with previous studies (Chang *et al.*, 2000). However, the neuroprotective role of melatonin went beyond conserving the antioxidant levels, and also promoted the regeneration process after PNI. As exogenous melatonin could markedly improve the function of the antioxidative defense system, it is reasonable to suggest that melatonin may prove to be a useful candidate in clinical trials as a promising therapy for PNI-relevant oxidative injury (Chang *et al.*, 2008).

D. Optic Nerve Studies with Melatonin

It is known that transection of the optic nerve (ON) close to the posterior pole of the eye leads to neuronal degeneration without concomitant pathophysiologic conditions, such as brain edema, secondary ischemia, or local bleeding (Kilic *et al.*, 2002). Transection of the ON axons near the cell body leads to death of retinal ganglion cells (RGCs) in adult animals (Garcia-Valenzuela *et al.*, 1994; Isenmann *et al.*, 1997). Recently, it was shown that retrograde degeneration of RGCs after ON transection is a result of an apoptotic process (Garcia-Valenzuela *et al.*, 1994; Isenmann *et al.*, 1997) involving the activation of Bax (Isenmann *et al.*, 1997) and caspase-3 (Kermer *et al.*, 1998).

RGC axotomy has been widely used to explain degeneration mechanisms of CNS neurons in experimental studies (Isenmann *et al.*, 1997, 1998; Kermer *et al.*, 1998; Klöcker *et al.*, 1998; Levkovitch-Verbin e*t al.*, 2003). In monkeys, when the superior one-third of the orbital ON on one side was transected, secondary RGC injury was noted. The number of RGC bodies in the superior and inferior halves of the retina and the number and diameter of axons in the ON were compared by means of histomorphometric methods. Primary RGC death due to ON injury was clearly associated with secondary death of surrounding RGCs that had not been directly injured by the surgery (Levkovitch-Verbin *et al.*, 2001). Transection of the ON is a useful experimental design for searching for neuroprotective strategies and for evaluating endogenous or exogenous neuroprotective compounds in animal models (Isenmann *et al.*, 1998; Kermer *et al.*, 1998; Kilic *et al.*, 2002; Klöcker *et al.*, 1997, 1998). There are several studies in the literature that have mentioned protective or adverse effects of endogenous or exogenous agents on ONs pathologies (Aktaş *et al.*, 2007; Bernstein *et al.*, 2007; Dimitriu *et al.*, 2008; Munemasa *et al.*, 2008; Weishaupt *et al.*, 2004; Ying *et al.*, 2008). Melatonin administration has been shown to have protective effects on eye pathologies such as retinal edema during the experimental uveitis or I/R in animal models (Kükner *et al.*, 2004; Yilmaz *et al.*, 2002). Regarding ON pathologies (i.e., ON injury), one study (Kilic *et al.*, 2002) investigated the effects of pinealectomy and of IP administration of melatonin on retrograde degeneration of RGCs by means of ON transection in C57BL/6J mice. RGCs were prelabeled with the fluorescent tracer 1,1′-dioctadecyl-3,3,3′,3′-tetramethyl indocarbocyanine perchlorate (Di-I), and the ON was cut inside the orbital cavity 7 days later. The degree of RGC injury was assessed by counting viable Di-I labeled RGCs in various locations in the retina. After axotomy, cell density markedly declined at 14 days. Sham-pinealectomy did not influence the density of RGCs at 14 days after ON transection (Kilic *et al.*, 2002). However, the RGC number was significantly reduced in pinealectomized animals compared with untreated and sham-pinealectomized animals. This effect of pinealectomy was reversed after IP administration of melatonin. On the other hand, IP melatonin did not influence the RGC density in nonpinealectomized animals. Therefore, the results of this study showed that endogenous melatonin prevents the delayed degeneration of neurons in CNS.

Additionally, it has been shown that exogenous melatonin may be useful for protection of injured neurons against cell death under conditions of melatonin deficiency, when melatonin synthesis and secretion have decreased (Kilic *et al.*, 2002). Further study investigating the possible effects of pinealectomy and exogenous melatonin, and the exact mechanism of melatonin effects on the RGCs following ON transaction, is still needed to confirm this study.

III. Melatonin Toxicity on Peripheral Nerves

Although there are great number of studies that mention positive effects of melatonin on peripheral nerve pathologies, there are some experimental and clinical studies that report a toxic effect of melatonin on peripheral nerves.

For example, melatonin, when combined with other drugs, can lead to optic neuropathy, ON pathology (Lehman and Johnson, 1999). It is known that melatonin is naturally found in the retina (Liu *et al.*, 2004) in addition to the pineal gland. It is involved in light–dark circadian rhythms and mediates retinal processes in a manner antagonistic to that of dopamine. The reuptake of serotonin, a precursor of melatonin, at the neural synapse is blocked by sertraline, an antidepressant drug. Recently, a case was reported of a 42-year-old woman suffering from visual acuity loss, dyschromatopsia, and altered light adaptation (Lehman and Johnson, 1999). The patient had been treated with sertraline for 4 years and began a high-protein diet with melatonin supplementation 2 weeks before onset of visual symptoms. Her visual acuity and color vision improved within 2 months after melatonin and the high-protein diet were discontinued. Therefore, it was postulated that combined use of sertraline, melatonin, and a high-protein diet may have resulted in a melatonin/dopamine imbalance in the retina, manifesting as a toxic optic neuropathy (Lehman and Johnson, 1999).

The effects of melatonin and cooling on the disintegration and reassembly of microtubules were investigated in the sciatic nerve of the toad by Piezzi and Cavicchia (1981). In this study, the majority of microtubules of both control and melatonin-treated sciatic nerves showed disintegration after 2 h of cooling. Thereafter, while microtubules were reformed in the control group nerves, the disorganization of microtubules persisted in the sciatic nerves treated with melatonin when room temperature was restored (Piezzi and Cavicchia, 1981). Therefore, these results demonstrate a toxic effect of melatonin that interfered with microtubular reassembly after cooling of the axons of the sciatic nerve of the toad.

In 1979, the effect of pineal indoles and melatonin on fast axoplasmic transport of proteins was investigated in the sciatic nerves of rats injected with [3H] leucine in the sixth lumbar dorsal root ganglion (Prevedello *et al.*, 1979). In this study, local application of melatonin to the sciatic nerve significantly impaired axonal transport. Other indoles (5-hydroxyindoleacetic acid, 5-methoxyindoleacetic acid, serotonin, *N*-acetylserotonin, tryptamine, 5-methoxytryptamine) were less potent than melatonin itself in the impairment fast axonal flow (Prevedello *et al.*, 1979). Therefore, it may be suggested that melatonin can impair axonal transport; however, no further study in support of this idea has yet been carried out.

IV. Conclusion

Melatonin is a small lipid and water-soluble indoleamine that can easily diffuse through cell membranes (Bob and Fedor-Freybergh, 2008). It easily crosses morphophysiological barriers such as the blood–brain or blood–testis barrier, and enters the cells and subcellular divisions. This feature of melatonin may be simplified to scavenge of free radicals. The free radical scavenging (Hardeland *et al.*, 1995; Reiter *et al.*, 2001) and antioxidant actions of melatonin were demonstrated in hundreds of experimental studies (Pappolla *et al.*, 2000; Reiter, 1998; Reiter *et al.*, 2000). Therefore, it protects cells, tissues, and organs against oxidative damage induced by a variety of free radical generating agents and processes (Reiter *et al.*, 1997). Melatonin is also effective in protecting nuclear DNA, membrane lipids, and possibly cytosolic proteins from oxidative damage as an antioxidant. In the literature, there are great numbers of studies that have mentioned protective effects of melatonin on peripheral nerves pathologies as given above (Cuzzocrea *et al.*, 1997, 1999; El-Shenawy *et al.*, 2002; Sayan *et al.*, 2004; Turgut *et al.*, 2005a,b; Yu *et al.*, 2000). These studies suggested that melatonin has anti-inflammatory, analgesic, and protective effect on peripheral nerve pathologies. However, there are some studies that report toxic effects of melatonin on peripheral nerves (Lehman and Johnson, 1999; Piezzi and Cavicchia, 1981; Prevedello *et al.*, 1979). Therefore, further studies are required to clearly refine protective or toxic effects of melatonin administration in studying the peripheral nerves.

References

Aktaş, Z., Gürelik, G., Akyürek, N., Onol, M., and Hasanreisoğlu, B. (2007). Neuroprotective effect of topically applied brimonidine tartrate 0.2% in endothelin-1-induced optic nerve ischaemia model. *Clin. Exp. Ophthalmol.* **35,** 527–534.

Ando, A., Domodo, T., Tsumori, T., and Yasui, Y. (1996). Changes of NADPH-diaphorase activity in the lumbosacral intermediolateral neurons of the rat after pelvic axotomy. *Brain Res. Bull.* **40,** 37–42.

Arendt, J. (1995). "Melatonin and the Mammalian Pineal Gland." Springer, Berlin.

Axelrod, J. (1974). The pineal gland: A neurochemical transducer. *Science* **1184,** 1341–1348.

Bagdatoglu, C., Saray, A., Surucu, H. S., Ozturk, H., and Tamer, L. (2002). Effect of trapidil in ischemia–reperfusion injury of peripheral nerves. *Neurosurgery* **51,** 212–220.

Baydas, G., Reiter, R. J., Nedzvetskii, V. S., Yasar, A., Tuzcu, M., Ozveren, F., and Canatan, H. (2003). Melatonin protects the central nervous system of rats against toluene-containing thinner intoxication by reducing reactive gliosis. *Toxicol. Lett.* **137,** 169–174.

Bernstein, S. L., Mehrabyan, Z., Guo, Y., and Moianie, N. (2007). Estrogen is not neuroprotective in a rodent model of optic nerve stroke. *Mol. Vis.* **13,** 1920–1925.

Bob, P., and Fedor-Freybergh, P. (2008). Melatonin, consciousness, and traumatic stress. *J. Pineal Res.* **44,** 341–347.

Bunge, R. P., and Bunge, M. B. (1978). Evidence that contact with connective tissue matrix is required for normal interaction between Schwann cell and nerve fibers. *J. Cell Biol.* **78,** 943–950.

Ceraulo, L., Ferrugia, M., Tesoriere, L., Segreto, S., Livrea, M. A., and Turco Liveri, V. (1999). Interactions of melatonin with membrane models: Portioning of melatonin in AOT and lecithin reversed micelles. *J. Pineal Res.* **26,** 108–112.

Cervantes, M., Moralí, G., and Letechipía-Vallejo, G. (2008). Melatonin and ischemia–reperfusion injury of the brain. *J. Pineal Res.* **45,** 1–7.

Chang, H. M., Ling, E. A., Lue, J. H., Wen, C. Y., and Shieh, J. Y. (2000). Melatonin attenuates neuronal NADPH-d/NOS expression in the hypoglossal nucleus of adult rats following peripheral nerve injury. *Brain Res.* **873,** 243–251.

Chang, H. M., Huang, Y. L., Lan, C. T., Wu, U. I., Hu, M. E., and Youn, S. C. (2008). Melatonin preserves superoxide dismutase activity in hypoglossal motoneurons of adult rats following peripheral nerve injury. *J. Pineal Res.* **44,** 172–180.

Clowry, G. J. (1993). Axotomy induces NADPH diaphorase activity in neonatal but not adult motoneurons. *Neuroreport* **5,** 361–364.

Cuzzocrea, S., Zingarelli, B., Gilad, E., Hake, P., Salzman, A. L., and Szabo, C. (1997). Protective effect of melatonin in carrageenan-induced models of local inflammation: Relationship to its inhibitory effect on nitric oxide production and its peroxynitrite scavenging activity. *J. Pineal Res.* **23,** 106–116.

Cuzzocrea, S., Costantino, G., Mazzon, E., and Caputi, A. P. (1999). Regulation of prostaglandin production in carrageenan-induced pleurisy by melatonin. *J. Pineal Res.* **27,** 9–14.

Dimitriu, C., Bach, M., Lagrèze, W. A., and Jehle, T. (2008). Methylprednisolone fails to preserve retinal ganglion cells and visual function after ocular ischemia in rats. *Invest. Ophthalmol. Vis. Sci.* **49,** 5003–5007.

Einheber, S., Hannocks, M. J., Metz, C. N., Rifkin, D. B., and Salzer, J. L. (1995). Transforming growth factor-beta 1 regulates axon/Schwann cell interactions. *J. Cell Biol.* **129,** 443–458.

Ek, R. O., Zencirci, S. G., Dost, T., Birincioglu, M., and Bilgin, M. D. (2007). Effects of melatonin supplementary on the sciatic nerve conduction velocity in the ovariectomized-aged rat. *Neuro Endocrinol. Lett.* **28,** 666–670.

El-Abhar, H. S., Shaalan, M., Barakat, M., and El-Denshary, E. S. (2002). Effect of melatonin and nifedipine on some antioxidant enzymes and different energy fuels in the blood and brain of global ischemic rats. *J. Pineal Res.* **33,** 87–94.

El-Shenawy, S. M., Abdel-Salam, O. M., Baiuomy, A. R., El-Batran, S., and Arbid, M. S. (2002). Studies on the anti-inflammatory and anti-nociceptive effects of melatonin in the rat. *Pharmacol. Res.* **46,** 235–243.

Estévez, A. G., Spear, N., Manuel, S. M., Radi, R., Henderson, C. E., Barbeito, L., and Beckman, J. S. (1998). Nitric oxide and superoxide contribute to motor neuron apoptosis induced by trophic factor deprivation. *J. Neurosci.* **18,** 923–931.

Garcia, J. J., Reiter, R. J., Guerrero, J. M., Escames, G., Yu, B. P., Oh, C. S., and Muñoz-Hoyos, A. (1997). Melatonin presents changes in microsomal membrane fluidity during induced lipid peroxidation. *FEBS Lett.* **408,** 297–308.

Garcia-Valenzuela, E., Gorczyca, W., Darzynkiewicz, Z., and Sharma, S. C. (1994). Apoptosis in adult retinal ganglion cells after axotomy. *J. Neurobiol.* **25,** 431–438.

Genade, S., Genis, A., Ytrehus, K., Huisamen, B., and Lochner, A. (2008). Melatonin receptor-mediated protection against myocardial ischaemia/reperfusion injury: Role of its anti-adrenergic actions. *J. Pineal Res.* **45,** 449–458.

Geoffriau, M., Brun, J., Chazot, G., and Claustrat, B. (1998). The physiology and pharmacology of melatonin in humans. *Horm. Res.* **49,** 136–141.

Golombek, D. A., Pevet, P., and Cardinali, D. P. (1996). Melatonin effects on behavior: Possible mediation by the central GABAergic system. *Neurosci. Biobehav. Rev.* **20,** 403–412.

Greensmith, L., and Vrobová, G. (1996). Motoneurone survival: A functional approach. *Trends Neurosci.* **19,** 450–455.

Hardeland, R., Reiter, R. J., Poeggeler, B., and Tan, D. X. (1993). The significance of the metabolism of the neurohormone melatonin: Antioxidative protection and formation of bioactive substances. *Neurosci. Biobehav. Rev.* **17,** 347–357.

Hardeland, R., Balzer, I., Poeggeler, B., Fuhrberg, B., Uría, H., Behrmann, G., Wolf, R., Meyer, T. J., and Reiter, R. J. (1995). On the primary functions of melatonin in evolution: Mediation of photoperiodic signals in a unicell, photooxidation, and scavenging of free radicals. *J. Pineal Res.* **18,** 104–111.

Harman, D. (1984). Free radical theory of aging: The free radical diseases. *Age* **7,** 111–131.

Harman, D. (1995). Role of antioxidant nutrients in aging: Overview. *Age* **18,** 51–62.

Hope, B. T., Michael, G. J., Knigge, K. M., and Vincent, S. R. (1991). Neuronal NADPH diaphorase is a nitric oxide synthase. *Proc. Natl. Acad. Sci. USA* **86,** 2811–2814.

Hsu, C. H., Chi, B. C., and Casida, J. E. (2002). Melatonin reduces phosphineinduced lipid and DNA oxidation *in vitro* and *in vivo* in rat brain. *J. Pineal Res.* **32,** 53–58.

Isenmann, S., Wahl, C., Krajewski, S., Reed, J. C., and Bähr, M. (1997). Up-regulation of Bax protein in degenerating retinal ganglion cells precedes apoptotic cell death after optic nerve lesion in the rat. *Eur. J. Neurosci.* **9,** 1763–1772.

Isenmann, S., Klöcker, N., Gravel, C., and Bähr, M. (1998). Protection of axotomized retinal ganglion cells by adenovirally delivered BDNF *in vivo*. *Eur. J. Neurosci.* **10,** 2751–2756.

Kermer, P., Klöcker, N., Labes, M., and Bähr, M. (1998). Inhibiton of CPP-32-like proteases rescues axotomized retinal ganglion cells from secondary death *in vivo*. *J. Neurosci.* **18,** 4656–4662.

Kilic, E., Hermann, D. M., Isenmann, S., and Bähr, M. (2002). Effects of pinealectomy and melatonin on the retrograde degeneration of retinal ganglion cells in a novel model of intraorbital optic nerve transection in mice. *J. Pineal Res.* **32,** 106–111.

Kim, S. H., and Lee, S. M. (2008). Cytoprotective effects of melatonin against necrosis and apoptosis induced by ischemia/reperfusion injury in rat liver. *J. Pineal Res.* **44,** 165–171.

Klein, D. C., Coon, S. L., Roseboom, P. H., Weller, J. L., Bernard, M., Gastel, J. A., Zatz, M., Iubone, M., Rodriguez, I. R., Begay, V., Falcon, J., Cahill, G., *et al.* (1997). The melatonin rhythm generating enzyme: Molecular regulation of serotonin *N*-acetyltransferase in the pineal gland. *Recent Prog. Horm. Res.* **52,** 307–358.

Klöcker, N., Bräunling, F., Isenmann, S., and Bähr, M. (1997). *In vivo* neurotrophic effects of GDNF on axotomized retinal ganglion cells. *Neuroreport* **8,** 3439–3442.

Klöcker, N., Cellerino, A., and Bähr, M. (1998). Free radical scavenging and inhibition of nitric oxide synthase potentiates the neurotrophic effects of brain-derived neurotrophic factor on axotomized retinal ganglion cells *in vivo*. *J. Neurosci.* **18,** 1038–1046.

Korf, H. W. (2000). Evolution of melatonin-producing pinealocytes. *In* "Melatonin after Four Decades" (J. Olcese, Ed.), pp. 17–29. Kluwer Academic/Plenum Press, New York.

Kükner, A. S., Kükner, A., Naziroðlu, M., Colakoðlu, N., Celebi, S., Yilmaz, T., and Aydemir, O. (2004). Protective effects of intraperitoneal vitamin C, aprotinin and melatonin administration on retinal edema during experimental uveitis in the guinea pig. *Cell Biochem. Funct.* **22,** 299–305.

Lehman, N. L., and Johnson, L. N. (1999). Toxic optic neuropathy after concomitant use of melatonin, zoloft, and a high-protein diet. *J. Neuroophthalmol.* **19,** 232–234.

Lerner, A. B., Case, J. D., Takahashi, Y., Lee, T. H., and Mori, W. (1958). Isolation of melatonin, the pineal gland factor that lightens melanocytes. *J. Am. Chem. Soc.* **80,** 2587.

Lerner, A. B., Case, J. D., Mori, W., and Wright, M. R. (1959). Melatonin in peripheral nerve. *Nature* **183,** 1821.

Levkovitch-Verbin, H., Quigley, H. A., Kerrigan-Baumrind, L. A., D'Anna, S. A., Kerrigan, D., and Pease, M. E. (2001). Optic nerve transection in monkeys may result in secondary degeneration of retinal ganglion cells. *Invest. Ophthalmol. Vis. Sci.* **42,** 975–982.

Levkovitch-Verbin, H., Quigley, H. A., Martin, K. R., Zack, D. J., Pease, M. E., and Valenta, D. F. (2003). A model to study differences between primary and secondary degeneration of retinal ganglion cells in rats by partial optic nerve transection. *Invest. Ophthalmol. Vis. Sci.* **44,** 3388–3393.

Li, Y., Bickel, K. D., Im, M. J., Hu, L., Dellon, A. L., Vander Kolk, C. A., and Manson, P. N. (1996). Effects of deferoxamine on ischemia/reperfusion injury after peripheral nerve compression. *Ann. Plast. Surg.* **36,** 365–369.

Liu, C., Fukuhara, C., Wessel, J. H., III, Iuvone, P. M., and Tosini, G. (2004). Localization of Aa-nat mRNA in the rat retina by fluorescence *in situ* hybridization and laser capture microdissection. *Cell Tissue Res.* **315,** 197–201.

Macchi, M. M., and Bruce, J. N. (2004). Human pineal physiology and functional significance of melatonin. *Front. Neuroendocrinol.* **25,** 177–195.

Mitsui, Y., Schmelzer, J. D., Zollman, P. J., Mitsui, M., Tritschler, H. J., and Low, P. A. (1999a). Alpha-lipoic acid provides neuroprotection from ischemia–reperfusion injury of peripheral nerve. *J. Neurol. Sci.* **163,** 11–16.

Mitsui, Y., Schmelzer, J. D., Zollman, P. J., Kihara, M., and Low, P. A. (1999b). Hypothermic neuroprotection of peripheral nerve of rats from ischaemia-reperfusion injury. *Brain* **122,** 161–169.

Møller, M., and Baeres, F. M. (2002). The anatomy and innervation of the mammalian pineal gland. *Cell Tissue Res.* **309,** 139–150.

Munemasa, Y., Kim, S. H., Ahn, J. H., Kwong, J. M., Caprioli, J., and Piri, N. (2008). Protective effect of thioredoxins 1 and 2 in retinal ganglion cells after optic nerve transection and oxidative stress. *Invest. Ophthalmol. Vis. Sci.* **49,** 3535–3543.

Nagai, R., Watanabe, K., Wakatsuki, A., Hamada, F., Shinohara, K., Hayashi, Y., Imamura, R., and Fukaya, T. (2008). Melatonin preserves fetal growth in rats by protecting against ischemia/reperfusion-induced oxidative/nitrosative mitochondrial damage in the placenta. *J. Pineal Res.* **45,** 271–276.

Nagamatsu, M., Schmelzer, J. D., Zollman, P. J., Smithson, I. L., Nickander, K. K., and Low, P. A. (1996). Ischemic reperfusion causes lipid peroxidation and fiber degeneration. *Muscle Nerve* **19,** 37–47.

Nam, E., Lee, S. M., Koh, S. E., Joo, W. S., Maeng, S., Im, H. I., and Kim, Y. S. (2005). Melatonin protects against neuronal damage induced by 3-nitropropionic acid in rat striatum. *Brain Res.* **1046,** 90–96.

Nukada, H., Powell, H. C., and Myers, R. R. (1993). Spatial distribution of nerve injury after occlusion of individual major vessels in rat sciatic nerve. *J. Neuropathol. Exp. Neurol.* **52,** 452–459.

Pappolla, M. A., Chyan, Y. J., Poeggeler, B., Frangione, B., Wilson, G., Ghiso, J., and Reiter, R. J. (2000). An assessment of the antioxidant and antiamyloidogenic properties of melatonin: Implications for Alzheimer's disease. *J. Neural Transm.* **107,** 203–231.

Pieri, C., and Marra, M. (1994). Melatonin: A peroxyl radical scavenger more effective than vitamin E. *Life Sci.* **55,** PL271–PL276.

Piezzi, R. S., and Cavicchia, J. C. (1981). Effects of cold and melatonin on the microtubules of the toad sciatic nerve. *Anat. Rec.* **200,** 115–120.

Prevedello, M. R., Ritta, M. N., and Cardinali, D. P. (1979). Fast axonal transport in rat sciatic nerve. Inhibition by pineal indoles. *Neurosci. Lett.* **13,** 29–34.

Reiter, R. J. (1981). The mammalian pineal gland: Structure and function. *Am. J. Anat.* **162,** 287–313.

Reiter, R. J. (1991). Pineal melatonin: Cell biology of its synthesis and of its physiological interactions. *Endocr. Rev.* **12,** 151–180.

Reiter, R. J. (1996). Functional diversity of the pineal hormone melatonin: Its role as an antioxidant. *Exp. Clin. Endocrinol. Diabetes* **104,** 10–16.

Reiter, R. J. (1998). Oxidative damage in the central nervous system: Protection by melatonin. *Prog. Neurobiol.* **56,** 359–384.

Reiter, R., Tang, L., Garcia, J. J., and Muñoz-Hoyos, A. (1997). Pharmacological actions of melatonin in oxygen radical pathophysiology. *Life Sci.* **60,** 2255–2271.

Reiter, R. J., Tan, D. X., Osuna, C., and Gitto, E. (2000). Actions of melatonin in the reduction of oxidative stress. *J. Biomed. Sci.* **7,** 444–458.

Reiter, R. J., Tan, D. X., Manchester, L. C., and Qi, W. (2001). Biochemical reactivity of melatonin with reactive oxygen and nitrogen species: A review of the evidence. *Cell Biochem. Biophys.* **34,** 237–256.

Reiter, R. J., Tan, D. X., Manchester, L. C., Simopoulos, A. P., Maldonado, M. D., Flores, L. J., and Terron, M. P. (2007). Melatonin in edible plants (Phytomelatonin): Identification, concentrations, bioavailability and proposed functions. *World Rev. Nutr. Diet.* **97,** 211–230.

Rogério, F., de Souza, L., Queiroz, L., Teixeira, S. A., Oliveira, A. L., de Nucci, G., and Langone, F. (2002). Neuroprotective action of melatonin on neonatal rat motoneurons after sciatic nerve transection. *Brain Res.* **926,** 33–41.

Rogério, F., Teixeira, S. A., de Rezende, A. C., de Sá, R. C., de Souza, L., Queiroz, L., De Nucci, G., Muscará, M. N., and Langone, F. (2005). Superoxide dismutase isoforms 1 and 2 in lumbar spinal cord of neonatal rats after sciatic nerve transection and melatonin treatment. *Brain Res. Dev. Brain Res.* **154,** 217–225.

Rogério, F., Jordão, H., Jr., Vieira, A. S., Maria, C. C., Santos de Rezende, A. C., Pereira, G. A., and Langone, F. (2006a). Bax and Bcl-2 expression and TUNEL labeling in lumbar enlargement of neonatal rats after sciatic axotomy and melatonin treatment. *Brain Res.* **1112,** 80–90.

Rogério, F., Teixeira, S. A., Júnior, H. J., Maria, C. C., Vieira, A. S., de Rezende, A. C., Pereira, G. A., Muscará, M. N., and Langone, F. (2006b). mRNA and protein expression and activities of nitric oxide synthases in the lumbar spinal cord of neonatal rats after sciatic nerve transection and melatonin administration. *Neurosci. Lett.* **407,** 182–187.

Rufer, M., Flanders, K., and Unsicker, K. (1994). Presence and regulation of transforming growth factor beta mRNA and protein in the normal and lesioned rat sciatic nerve. *J. Neurosci. Res.* **39,** 412–423.

Sayan, H., Ozacmak, V. H., Ozen, O. A., Coskun, O., Arslan, S. O., Sezen, S. C., and Aktas, R. G. (2004). Beneficial effects of melatonin on reperfusion injury in rat sciatic nerve. *J. Pineal Res.* **37,** 143–148.

Shokouhi, G., Tubbs, R. S., Shoja, M. M., Hadidchi, S., Ghorbanihaghjo, A., Roshangar, L., Farahani, R. M., Mesgari, M., and Oakes, W. J. (2008). Neuroprotective effects of high-dose vs low-dose melatonin after blunt sciatic nerve injury. *Childs Nerv. Syst.* **24,** 111–117.

Stavisky, R. C., Britt, J. M., Zuzek, A., Truong, E., and Bittner, G. D. (2005). Melatonin enhances the *in vitro* and *in vivo* repair of severed rat sciatic axons. *Neurosci. Lett.* **376,** 98–101.

Sugden, D. (1983). Psychopharmacological effects of melatonin in mouse and rat. *J. Pharmacol. Exp. Ther* **227,** 587–591.

Sugden, D., and Klein, D. C. (1987). A cholera toxin substrate regulates cyclic GMP content of rat pinealocytes. *J. Biol. Chem.* **262,** 7447–7450.

Tatagiba, M., Rosahl, S., Gharabaghi, A., Blömer, U., Brandis, A., Skerra, A., Sami, M., and Schwab, M. E. (2002). Regeneration of auditory nerve following complete sectioning and intrathecal application of the IN-1 antibody. *Acta Neurochir. (Wien.)* **144,** 181–187.

Touitou, Y. (2001). Human aging and melatonin. Clinical relevance. *Exp. Gerontol.* **36,** 1083–1100.

Touitou, Y., Fèvre, M., Lagoguey, M., Carayon, A., Bogdan, A., Reinberg, A., Cesselin, F., and Touitou, C. (1981). Age- and mental health-related circadian rhythms of plasma levels of melatonin, prolactin, luteinizing hormone and follicle stimulating hormone in man. *J. Endocrinol.* **91,** 467–475.

Turgut, M., Uyanikgil, Y., Baka, M., Tunç, A. T., Yavaþoðlu, A., Yurtseven, M. E., and Kaplan, S. (2005a). Pinealectomy exaggerates and melatonin treatment suppresses neuroma formation of transected sciatic nerve in rats: Gross morphological, histological and stereological analysis. *J. Pineal Res.* **38,** 284–291.

Turgut, M., Uysal, A., Pehlivan, M., Oktem, G., and Yurtseven, M. E. (2005b). Assessment of effects of pinealectomy and exogenous melatonin administration on rat sciatic nerve suture repair: An electrophysiological, electron microscopic, and immunohistochemical study. *Acta Neurochir. (Wien.)* **147,** 67–77.

Turgut, M., Oktem, G., Uysal, A., and Yurtseven, M. E. (2006). Immunohistochemical profile of transforming growth factor-beta1 and basic fibroblast growth factor in sciatic nerve anastomosis following pinealectomy and exogenous melatonin administration in rats. *J. Clin. Neurosci.* **13,** 753–758.

Ulugol, A., Dokmeci, D., Guray, G., Sapolyo, N., Ozyigit, F., and Tamer, M. (2006). Antihyperalgesic, but not antiallodynic, effect of melatonin in nerve-injured neuropathic mice: Possible involvements of the L-arginine-NO pathway and opioid system. *Life Sci.* **78,** 1592–1597.

Vollrath, I. (1981). "The Pineal Organ." Springer, Berlin.

Vollrath, L. (1984). Functional anatomy of the human pineal gland. *In* "The Pineal Gland" (R. J. Reiter, Ed.), pp. 285–322. Raven Press, New York.

Wang, W. Z., Fang, X. H., Stephenson, L. L., Khiabani, K. T., and Zamboni, W. A. (2006). Melatonin reduces ischemia/reperfusion-induced superoxide generation in arterial wall and cell death in skeletal muscle. *J. Pineal Res.* **41,** 255–260.

Weishaupt, J. H., Rohde, G., Pölking, E., Siren, A. L., Ehrenreich, H., and Bähr, M. (2004). Effect of erythropoietin axotomy-induced apoptosis in rat retinal ganglion cells. *Invest. Ophthalmol. Vis. Sci.* **45,** 1514–1522.

Wu, W., Liuzzi, F. J., Schinco, F. P., Depto, A. S., Li, Y., Mong, J. A., Dawson, T. M., and Snyder, S. H. (1994). Neuronal nitric oxide synthase is induced in spinal neurons by traumatic injury. *Neuroscience* **6,** 719–726.

Yilmaz, T., Celebi, S., and Kükner, A. S. (2002). The protective effects of melatonin, vitamin E and octreotide on retinal edema during ischemia–reperfusion in the guinea pig retina. *Eur. J. Ophthalmol.* **12,** 443–449.

Ying, X., Zhang, J., Wang, Y., Wu, N., Wang, Y., and Yew, D. T. (2008). Alpha-crystallin protected axons from optic nerve degeneration after crushing in rats. *J. Mol. Neurosci.* **35,** 253–258.

Yu, W. H. A. (1997). Regulation of nitric oxide synthase expression in motoneurons following nerve injury. *Dev. Neurosci.* **19,** 247–254.

Yu, W. H. (2002). Spatial and temporal correlation of nitric oxide synthase expression with cuzn-superoxide dismutase reduction in motor neurons following axotomy. *Ann. N. Y. Acad. Sci.* **962,** 111–121.

Yu, C. X., Zhu, B., Xu, S. F., Cao, X. D., and Wu, G. C. (2000). The analgesic effects of peripheral and central administration of melatonin in rats. *Eur. J. Pharmacol.* **403,** 49–53.

Zeng, Q., Wang, S., Lim, G., Yang, L., Mao, J., Sung, B., Chang, Y., Lim, J. A., Guo, G., and Mao, J. (2008). Exacerbated mechanical allodynia in rats with depression-like behavior. *Brain Res.* **1200,** 27–38.

TRANSTHYRETIN: AN ENHANCER OF NERVE REGENERATION

Carolina E. Fleming, Fernando Milhazes Mar, Filipa Franquinho, and Mónica M. Sousa

Nerve Regeneration Group, Instituto de Biologia Molecular e Celular—IBMC, R. Campo Alegre 823, 4150-180 Porto, Portugal

Transthyretin (TTR), a plasma and cerebrospinal fluid protein secreted by the liver and choroid plexus, is mainly known as the physiological carrier of thyroxine (T_4) and retinol. Under pathological conditions, various TTR mutations are related to familial amyloid polyneuropathy (FAP), a neurodegenerative disorder characterized by deposition of TTR amyloid fibrils, particularly in the peripheral nervous system (PNS), leading to axonal loss and neuronal death. Recently, a number of TTR functions in neurobiology have been described; these may explain the preferential TTR deposition, when mutated, in the PNS of FAP patients. In this respect, and with a particular relevance in the PNS, TTR has been shown to have the ability to enhance neurite outgrowth *in vitro* and nerve regeneration following injury, *in vivo*. In the following pages, this novel TTR function, as well as its importance in nerve biology and repair will be discussed.

I. Introduction

Transthyretin (TTR) is mainly synthesized by the liver and choroid plexus of the brain (Aleshire *et al.*, 1983; Gitlin and Gitlin, 1975), which are the sources of the protein found in the plasma and cerebrospinal fluid (CSF), respectively. When mutated in particular residues, TTR leads to the development of familial amyloid

INTERNATIONAL REVIEW OF NEUROBIOLOGY, VOL. 87
DOI: 10.1016/S0074-7742(09)87017-7

polyneuropathy (FAP), an autosomal dominant lethal disorder characterized by extracellular deposition of TTR amyloid fibrils, principally in the peripheral nervous system (PNS). In the PNS of FAP patients, TTR accumulation leads to axonal loss and ultimately neuronal death (Andrade, 1952), although the detailed mechanism ending up in neurodegeneration is largely unknown. Similarly, the reason for the deposition of TTR particularly in the PNS of FAP patients is unclear but suggests that TTR might play a role in nerve physiology. Physiologically, TTR has access to the nerve both through the blood–nerve barrier (which is effective in slowing but not in preventing the entry of plasma proteins into the endoneurium) and through the CSF, as peripheral nerve roots contact with this fluid where TTR is present in high levels. Similarly, in the central nervous system (CNS) TTR may access the brain both through the blood–brain barrier and the CSF barrier.

In addition to FAP, several lines of evidence have recently drawn the attention to the importance of TTR in the biology of the nervous system. In this review, we will concentrate on the recently described TTR function as a regeneration enhancer.

II. Transthyretin

TTR, a soluble tetrameric protein composed of four identical subunits, was first termed prealbumin as a consequence of its electrophoretic migration prior to albumin. TTR was first described in the CSF (Kabat *et al.*, 1942) and shortly after in the plasma (Seibert and Nelson, 1942). As already mentioned, the name TransThyRetin discloses its dual physiological role as a transporter of thyroxine (Woeber and Ingbar, 1968) and retinol, in the latter case through the formation of a complex with retinol-binding protein-RBP (Kanai *et al.*, 1968).

The two major sites of TTR synthesis are the liver (Gitlin and Gitlin, 1975) and the choroid plexus (Aleshire *et al.*, 1983) that are respectively the sources of TTR in the plasma and CSF. It is interesting to note that the major site of TTR expression, as a ratio of tissue/mass, is the choroid plexus (Duan *et al.*, 1991). As a consequence of its high expression levels by the choroid plexus, TTR is an abundant protein in the CSF as it constitutes up to 20% of total ventricular protein (Weisner and Roethig, 1983).

During embryonic development, TTR synthesis occurs around the eighth week of gestation in humans (Jacobsson, 1989), being first expressed in the tela choroidea, the precursor of the choroid plexus, followed by expression in the liver. In evolutionary terms, TTR synthesis occurs in fish (Santos and Power, 1999), reptiles, birds, and mammalian ancestors (Richardson *et al.*, 1994). In fish, TTR is produced mainly by the liver whereas in reptiles TTR is produced by the choroid plexus but not by the liver (Achen *et al.*, 1993), while birds and mammals produce TTR in both tissues (Harms *et al.*, 1991). The pattern of TTR expression in the choroid plexus conserved throughout evolution and starting early in embryonic development, supports a pivotal role for TTR in neurobiology.

III. TTR KO Mice

In 1993, TTR KO mice were generated (Episkopou *et al.*, 1993) to investigate the physiological role of TTR in embryonic development and in the adult, specially the role of its ligands, T_4 and retinol. In general terms, this strain revealed to be fertile, to have a normal life span, displaying no obvious phenotypic abnormalities postnatally (Episkopou *et al.*, 1993). As expected, plasma from TTR KO mice had decreased levels of both T_4 and T_3 (Episkopou *et al.*, 1993). In terms of tissue content in thyroid hormones, no significant differences in T_4 levels were found in TTR KO mice when compared to WT mice (Palha *et al.*, 1997). Probably underlying their euthyroid status, free T_4 was 50% increased in the serum of TTR KO mice (Palha, 2002). In conclusion, the absence of TTR does not seem to affect thyroid hormone function. In the case of retinol, TTR KO mice were expected to have a vitamin A deficiency. However, although their retinol plasma levels were below the level of detection, mice lacking TTR did not show any symptoms of vitamin A deficiency (Episkopou *et al.*, 1993). In agreement with the lack of symptoms of vitamin A deficiency, the total retinol levels in tissues were not significantly different from WT mice (Wei *et al.*, 1995). In conclusion, similarly to thyroid hormones, in the case of vitamin A, the above findings suggest that TTR KO mice present no major defects related to retinol deficiency. Since the initial characterization, TTR KO mice revealed to be one of the most valuable tools to link the biology of TTR with that of the CNS and PNS.

IV. TTR Mutations as the Cause of FAP

FAP was first described in Portuguese patients who had a lethal hereditary amyloidosis characterized by sensorimotor peripheral polyneuropathy accompanied by autonomic dysfunction (Andrade, 1952). At the biochemical level, TTR was identified as the major protein component in the amyloid deposits of FAP patients (Costa *et al.*, 1978) and later, a substitution of methionine for valine at position 30 of the protein was found to be the most common molecular defect causing the disease (Saraiva *et al.*, 1984). To date, besides the V30M mutation, a large number of other amyloidogenic TTR variants have been described, most of them associated with polyneuropathy (Saraiva, 2001).

In FAP, widespread systemic extracellular deposition of TTR aggregates and amyloid fibrils occurs throughout the connective tissue, particularly in the PNS (Sousa and Saraiva, 2003). In the PNS, TTR amyloid deposits are found within nerve trunks, plexuses, sensory, and autonomic ganglia. Following TTR deposition, axonal degeneration occurs, starting in the unmyelinated and myelinated fibers of low diameter, and ultimately leading to neuronal loss at ganglionic sites (Sousa and Saraiva, 2003).

In the recent years, significant progresses as to the understanding of the molecular mechanisms of TTR-mediated cellular toxicity were made. The hypothesis that nonfibrillar TTR aggregates might be the pathogenic agent in FAP was investigated and unexpectedly, it was shown that mature TTR fibrils are essentially harmless, whereas TTR aggregates are toxic to cells, therefore being able to induce neurodegeneration (Andersson *et al.*, 2002; Sousa *et al.*, 2001). From the interaction of TTR aggregates with membranes, to the resulting intracellular events leading to neuronal loss, advances have been made that allowed to enlarge the understanding of FAP pathophysiology (Hou *et al.*, 2007; Sousa and Saraiva 2003). However, the full characterization of the molecular cues leading to neuronal loss in FAP haven't still been elucidated.

V. TTR Enhances Nerve Regeneration

To address the possibility that TTR might have a role in peripheral nerve biology that would explain its preferential deposition in the nerve of FAP patients, the PNS of TTR KO mice was assessed at the functional and morphological levels. These animals were initially shown to present a sensorimotor impairment (Fleming *et al.*, 2007) which was primarily characterized through SHIRPA analysis (Rogers *et al.*, 1997). The sensorimotor deficits of TTR KO mice progressed with age and included limb clasping (Fig. 1), poor performance in the vertical pole, and decreased locomotor activity in aged animals. In relation to locomotor activity, it is interesting to note that young TTR KO mice were reported to be more active than WT littermates (Sousa *et al.*, 2004); however, in aged animals, this tendency was inverted, probably as a consequence of the motor discoordination. In addition to the SHIRPA parameters, in the hot plate test, a standard procedure to measure the nociceptive response to a thermal stimulus, TTR KO mice had an increased latency to react to heat (Fleming *et al.*, 2007). The above observations suggested a specific function for TTR in nerve physiology. To identify the reasons underlying this phenotype, morphometric and electrophysiological analysis of sciatic nerves were performed but no differences were found

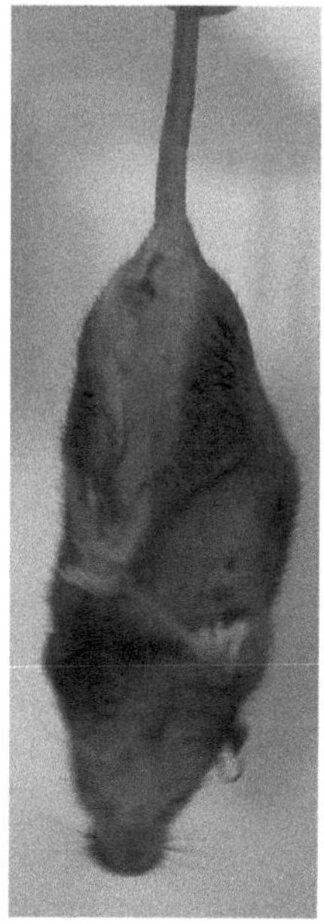

FIG. 1. Limb clasping in a 12-month-old TTR KO mouse.

between WT and TTR KO littermates (Fleming *et al.*, 2007). Moreover, since the cerebellum is related to the coordination of body movement, it was compared in both strains; similarly to the sciatic nerve, no differences between the two strains were detected.

As such, despite that the above findings clearly show that TTR participates in nerve biology, the reason underlying the sensorimotor impairment of TTR KO mice, awaits identification. To further unravel the putative role of TTR in nerve physiology, and given the possibility that upon injury the consequences arising from the absence of TTR might be exacerbated, nerve crush was performed in WT and TTR KO mice. As discussed below, the data obtained led to the significant finding that TTR enhances nerve regeneration. The assignment of such a TTR function in nerve biology and repair, may explain its preferential deposition in the PNS of FAP patients.

A. Lack of TTR Leads to Delayed Nerve Regeneration

As a paradigm for nerve injury, sciatic nerve crush was performed in WT and TTR KO mice and regeneration was assessed 15 and 30 days postcrush, at the functional level and by morphometric analysis (Fleming *et al.*, 2007). Functionally, nerve conduction velocity was determined and locomotor activity was assessed. Following nerve crush, TTR KO mice presented a lower nerve conduction velocity throughout the course of regeneration, which was the first evidence of a possible decreased regenerative capacity in the absence of TTR. Moreover, irrespective of their higher locomotor activity prior crush, following injury, TTR KO mice presented a decreased locomotor activity, reinforcing their slower regeneration.

To determine if the decreased functional performance of TTR KO mice was related to neuropathological findings, nerve regeneration was evaluated by morphometry. Fifteen days after crush, the number of myelinated fibers was approximately 20% lower in TTR KO mice when compared to WT littermates. In relation to unmyelinated fibers, their density was approximately 40% decreased in TTR KO mice, following 30 days of regeneration.

To assess whether TTR expression in the nerve would lead to a gain of regenerative capacity, supporting the finding that TTR enhances nerve regeneration, Thy1.2-TTR mice, that is, human TTR transgenic mice with TTR expression both in sensory and motor neurons (Sousa *et al.*, 2004), were backcrossed to the TTR KO background. In these animals (Thy1.2-TTRxTTR KO mice) endogenous mouse TTR is not expressed whereas human TTR is expressed in neurons (Fleming *et al.*, 2007). As expected, neuronal TTR expression in a TTR KO background lead to an accelerated regenerative capacity: after 15 days of regeneration, Thy1.2-TTRxTTR KO mice had an approximately 50% increase both in the number of myelinated fibers and density of unmyelinated fibers, when compared to TTR KO mice.

This functional and morphometric data demonstrates that TTR enhances nerve regeneration; following injury, this effect may be crucial for timely target innervation and regain of functional capacity.

B. Cellular Mechanism Underlying the Effect of TTR on Nerve Regeneration

Following the assignment of a regeneration enhancer capacity to TTR, efforts concentrated at understanding this function at the cellular and molecular levels. The results obtained, so far, will be discussed in the following paragraphs.

To evaluate at the cellular level the cause for the delayed regeneration of TTR KO mice, neuronal survival was assessed after nerve crush (Fleming *et al.*, 2007). For that, the number of fibers in the proximal nerve stump (which reflects the number of DRG neurons and motoneurons surviving to injury), was evaluated 30 days after crush. No differences were found in the density of myelinated or unmyelinated fibers between WT and TTR KO littermates, showing that TTR does not affect neuronal survival/death. To validate this observation, the density of L4–6 DRG neurons was determined at the same time point; again, no differences between strains were found, reinforcing that the effect of TTR in nerve regeneration is unrelated to neuronal survival/death (Fleming *et al.*, 2007).

The hypothesis that TTR enhances neurite outgrowth and extension was next assessed (Fleming *et al.*, 2007). In cell culture, neuronal-like cell lines exposed to TTR KO serum displayed an approximate 25% decrease in the neurite number per cell and in the length of the longest neurite, when compared to cells grown with WT serum. The fact that TTR is a carrier of T_4 and retinol raised the possibility that its effect on nerve regeneration might be indirect. However, as referred to above, TTR KO mice are euthyroid and lack symptoms of vitamin A deficiency (Episkopou *et al.*, 1993), suggesting that the effect of this protein on nerve regeneration is unlikely to result from impaired thyroid hormone homeostasis or retinoic acid metabolism. To establish whether the absence of TTR was directly responsible for this decreased neurite number and size, TTR KO serum was supplemented with WT TTR; addition of TTR was able to totally rescue the phenotype observed in the absence of the protein. Moreover, a TTR mutant with very low affinity for both T_4 and RBP, behaved similarly to WT TTR and was able to rescue the phenotype of cells grown in TTR KO serum (Fleming *et al.*, 2007).

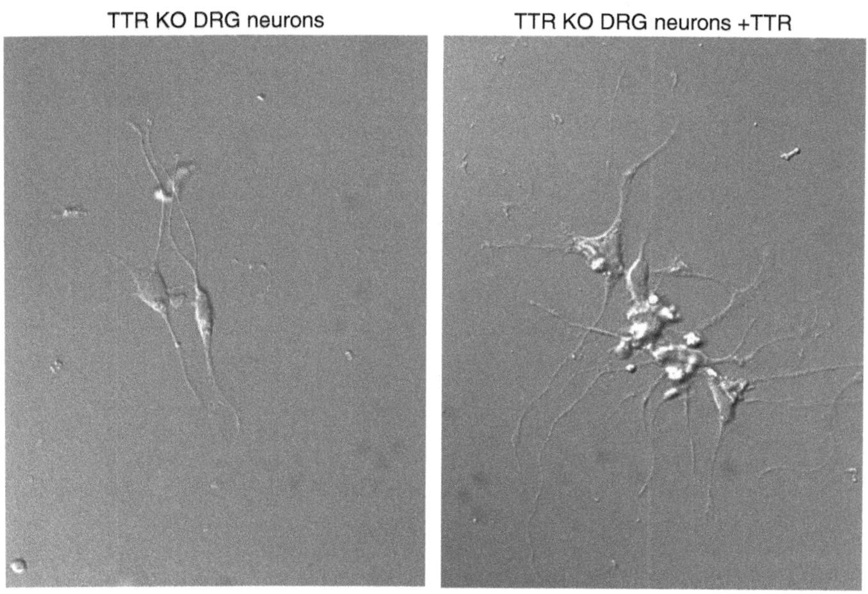

FIG. 2. Neurite outgrowth of TTR KO DRG neurons grown in the absence (left) and in the presence (right) of TTR.

To determine whether TTR KO sensory neurons display an intrinsic decreased neurite outgrowth, resulting from their physiological TTR-free environment, DRG neurons from WT and TTR KO mice were grown in the absence of TTR, and their neurite number and size were compared (Fleming *et al.*, 2007). Similarly to neuronal-like cell lines, DRG neurons from TTR KO mice presented an approximately 20% decreased neurite number and length of the longest neurite, when compared to WT DRG neurons. Further supporting that TTR is directly responsible for this phenotype, addition of the protein rescued the decreased intrinsic ability to grow neurites of TTR KO DRG neurons cultivated in a T_4- and retinol-free medium (Fig. 2).

VI. Conclusion

Summing up, the data discussed above demonstrate that TTR enhances nerve regeneration, affecting primarily neurons. Future work should now disclose at the molecular level the mechanism underlying this finding.

Acknowledgments

The work from the author's laboratory was supported by grants from Association Française contre les Myopathies (AFM), France, and Fundação para a Ciência e a Tecnologia (FCT), Portugal (PTDC/BIA-PRO/64437/2006). Fleming CE was the recipient of POCI fellowship (POCI/V.5/A0107), Portugal. Mar F is the recipient of a FCT fellowship (SFRH/BD/43484/2008).

References

Achen, M. G., Duan, W., Pettersson, T., Harms, P. J., Richardson, S. J., Lawrence, M. C., Wettenhall, R. E. H., Aldred, A. R., and Schreiber, G. (1993). Transthyretin gene expression in choroid plexus first evolved in reptiles. *Am. J. Physiol.* **265,** R982–R989.

Aleshire, S. L., Bradley, C. A., Richardson, L. D., and Parl, F. F. (1983). Localization of human prealbumin in choroid plexus epithelium. *J. Histochem. Cytochem.* **31,** 608–612.

Andersson, K., Olofsson, A., Nielsen, E. H., Svehag, S. E., and Lundgren, E. (2002). Only amyloidogenic intermediates of transthyretin induce apoptosis. *Biochem. Biophys. Res. Commun.* **294,** 309–314.

Andrade, C. (1952). A peculiar form of peripheral neuropathy; familiar atypical generalized amyloidosis with special involvement of the peripheral nerves. *Brain* **75,** 408–427.

Costa, P. P., Figueira, A. S., and Bravo, F. R. (1978). Amyloid fibril protein related to prealbumin in familial amyloidotic polyneuropathy. *Proc. Natl. Acad. Sci. USA* **75,** 4499–4503.

Duan, W., Achen, M. G., Richardson, S. J., Lawrence, M. C., Wettenhall, R. E. H., Jaworowski, A., and Schreiber, G. (1991). Isolation, characterization, cdna cloning and gene expression of an avian transthyretin. *Eur. J. Biochem.* **200,** 679–687.

Episkopou, V., Maeda, S., Nishiguchi, S., Shimada, K., Gaitanaris, G. A., Gottesman, M. E., and Robertson, E. J. (1993). Disruption of the transthyretin gene results in mice with depressed levels of plasma retinol and thyroid hormone. *Proc. Natl. Acad. Sci. USA* **90,** 2375–2379.

Fleming, C. E., Saraiva, M. J., and Sousa, M. M. (2007). Transthyretin enhances nerve regeneration. *J. Neurochem.* **103,** 831–839.

Gitlin, D., and Gitlin, J. D. (1975). Fetal and neonatal development of human plasma proteins. *In* "The plasma proteins" (F. W. Putnam, Ed.), vol. 2. Academic Press, New York.

Harms, P. J., Tu, G. F., Richardson, S. J., Aldred, A. R., Jaworowski, A., and Schreiber, G. (1991). Transthyretin (prealbumin) gene expression in choroid plexus is strongly conserved during evolution of vertebrates. *Comp. Biochem. Physiol.* **99B,** 239–249.

Hou, X., Aguilar, M. I., and Small, D. H. (2007). Transthyretin and familial amyloidotic polyneuropathy. Recent progress in understanding the molecular mechanism of neurodegeneration. *FEBS J.* **274,** 1637–1650.

Jacobsson, B. (1989). Localization of transthyretin mrna and immunoreactive transthyretin in the human fetus. *Virchows Arch. Pathol. Anat.* **415,** 259–263.

Kabat, E. A., Moore, D. H., and Landow, H. (1942). An electrophoretic study of the protein components in cerebrospinal fluid and their relationship to the serum proteins. *J. Clin. Invest.* **21,** 571–577.

Kanai, M., Raz, A., and Goodman, D. S. (1968). Retinol-binding protein: The transport protein for vitamin A in human plasma. *J. Clin. Invest.* **47,** 2025–2044.

Palha, J. A. (2002). Transthyretin as a thyroid hormone carrier: Function revisited. *Clin. Chem. Lab. Med.* **40,** 1292–1300.

Palha, J. A., Hays, M. T., Morreale de Escobar, G., Episkopou, V., Gottesman, M. E., and Saraiva, M. J. (1997). Transthyretin is not essential for thyroxine to reach the brain and other tissues in transthyretin-null mice. *Am. J. Physiol.* **272,** E485–E493.

Richardson, S. J., Bradley, A. J., Duan, W., Wettenhall, R. E., Harms, P. J., Babon, J. J., Southwell, B. R., Nicol, S., Donnellan, S., and Schreiber, G. (1994). Evolution of marsupial and other vertebrate thyroxine-binding plasma proteins. *Am. J. Physiol.* **266,** R1359–R1370.

Rogers, D. C., Fisher, E. M., Brown, S. D., Peters, J., Hunter, A. J., and Martin, J. E. (1997). Behavioral and functional analysis of mouse phenotype: SHIRPA, a proposed protocol for comprehensive phenotype assessment. *Mamm. Genome* **8,** 711–713.

Santos, C. R., and Power, D. M. (1999). Identification of transthyretin in fish (*Sparus aurata*): cDNA cloning and characterisation. *Endocrinology* **140,** 2430–2433.

Saraiva, M. J. (2001). Transthyretin mutations in hyperthyroxinemia and amyloid diseases. *Hum. Mutat.* **17,** 493–503.

Saraiva, M. J., Birken, S., Costa, P. P., and Goodman, D. S. (1984). Amyloid fibril protein in familial amyloidotic polyneuropathy, Portuguese type. Definition of molecular abnormality in transthyretin (prealbumin). *J. Clin. Invest.* **74,** 104–119.

Seibert, F. B., and Nelson, J. W. (1942). Electrophoretic study of the blood response in tuberculosis. *J. Biol. Chem.* **143,** 29–38.

Sousa, M. M., and Saraiva, M. J. (2003). Neurodegeneration in familial amyloid polyneuropathy: From pathology to molecular signaling. *Prog. Neurobiol.* **71,** 385–400.

Sousa, M. M., Cardoso, I., Fernandes, R., Guimarães, A., and Saraiva, M. J. (2001). Deposition of transthyretin in early stages of familial amyloidotic polyneuropathy: Evidence for toxicity of nonfibrillar aggregates. *Am. J. Pathol.* **159,** 1993–2000.

Sousa, J. C., Grandela, C., Fernandez-Ruiz, J., de Miguel, R., de Sousa, L., Magalhães, A. I., Saraiva, M. J., Sousa, N., and Palha, J. A. (2004). Transthyretin is involved in depression-like behaviour and exploratory activity. *J. Neurochem.* **88,** 1052–1058.

Wei, S., Episkopou, V., Piantedosi, R., Maeda, S., Shimada, K., Gottesman, M. E., and Blaner, W. S. (1995). Studies on the metabolism of retinol and retinol-binding protein in transthyretin-deficient mice produced by homologous recombination. *J. Biol. Chem.* **270,** 866–870.

Weisner, B., and Roethig, H. J. (1983). The concentration of prealbumin in cerebrospinal fluid (CSF), indicator of CSF circulation disorders. *Eur. Neurol.* **22,** 96–105.

Woeber, K. A., and Ingbar, S. H. (1968). The contribution of thyroxine binding prealbumin to the binding of thyroxine in human serum as assessed by immunoprecipitation. *J. Clin. Invest.* **47,** 1710–1721.

ENHANCEMENT OF NERVE REGENERATION AND RECOVERY BY IMMUNOSUPPRESSIVE AGENTS

Damien P. Kuffler

Institute of Neurobiology, University of Puerto Rico Medical Sciences Campus,
San Juan, Puerto Rico 00901, USA

Clinically, little can be done to induce restoration of good to excellent neurological function following nervous system trauma, and time is required before an effective technique is developed and applied clinically. However, there are novel techniques that have not been tested experimentally or clinically that may induce significantly faster, reliable, and extensive neurological recovery following nervous system trauma than is presently possible, even for techniques currently being tested on animal models. To repair peripheral nerves following trauma in which a length of the nerve pathway is destroyed, many clinicians consider autologous sensory nerve grafts to be the "gold standard" for inducing neurological recovery. However, this technique has severe limitations, such as being effective only across gaps less than 2 cm, for repairs performed less than 2 months posttrauma, and in young patients. As a consequence, many patients suffer permanent neurological deficits or recover only limited neurological function, and they frequently develop irreversible neuropathic pain. This review examines the clinical role that immunosuppressants might play, in the presence

INTERNATIONAL REVIEW OF
NEUROBIOLOGY, VOL. 87
DOI: 10.1016/S0074-7742(09)87018-9

or absence of autologous, allografts, or xenografts, in increasing the rate, success, and extent of neurological recovery following nervous system trauma.

I. Introduction

Trauma to the central nervous system (CNS) typically causes neuron death. However, an even larger number of neurons usually die due to sequelae of the initial trauma. Neuron death causes the loss of all synaptic connections, while other neurons that remain viable may have their axons irreversibly damaged, leading to extensive neurological losses. In cases of CNS trauma in which only long nerve tracts are damaged, no regeneration develops because the cellular environment of the axons inhibits their regeneration.

In the peripheral nervous system (PNS), trauma to a nerve pathway can create a gap across which axons cannot regenerate, leading to permanent neurological losses of the function of the damaged axons.

To minimize these neurological losses, it is critical to minimize neuron death, minimize the loss of the synaptic connections, and promote viable neurons to extend axons. In addition, axons must be induced to regenerate across nerve gaps and into their remaining distal pathways, which must be changed from regeneration inhibiting to regeneration permissive or regeneration promoting which will allow the axons to be guided to their appropriate targets.

Many nonclinically tested techniques influence neuron survival following trauma and induce axon regeneration. However, a detailed discussion of these techniques is beyond the scope of this review. Therefore, this paper is restricted to examining what is known about the actions of immunosuppressants as neuroprotectants and promoters of axon regeneration.

When a damaged nerve cannot be repaired by nerve end-to-end anastomosis, autologous (from the animal or person undergoing nerve repair) sensory nerve grafts are the predominant method for repairing peripheral nerves with a gap in their pathway. This success of axon regeneration and neurological recovery in these cases is significantly enhanced by the use of immunosuppressants such as FK506. The immunosuppressant not only prevents graft rejection, in case where allografts (a tissue graft from another animal or person) are used, but also as promoters of axon regeneration (Hayashi *et al.*, 2008). The administration of FK506 following placement of an allograph increases the diameter of the regenerating axons within the graft, improves the responses of sensory neurons to metabolites such as potassium chloride and lactic acid, and induces a fast-to-slow-fiber-type transition of the tibialis anterior muscle. Thus, FK506 potentiates metabosensitive nerve fiber regeneration making it an excellent agent for enhancing the rate and extent of axon regeneration and neurological recovery (Chabas *et al.*, 2009). FK506 also increases the number of regenerating myelinated and unmyelinated axons (Hayashi *et al.*, 2008). Therefore, the expanded use of specific immunosuppressants, simultaneously with performing nerve repairs, can be important for inducing more extensive neurological recoveries by acting as promoters of axon regeneration.

One caveat about the use of immunosuppression is the generally perceived degree of morbidity associated with their nonspecific systemic actions. Therefore, the use of immunosuppressants was previously restricted to patients with particularly severe nerve injuries. However, an increasing understanding of the mechanisms of action of immunnosuppresants in promoting axon regeneration and neurological recovery has brought about a reevaluation of the risk/benefits of their use.

A. Immunophilins

Immunophilins (IP) are a family of proteins that are receptors for immunosuppressive drugs like cyclosporin A (CsA), FK506 (tacrolimus), rapamycin, and their nonimmunosuppressive analogs, which are collectively referred to as "immunophilin ligands" (IPL) (Avramut and Achim, 2003). Cyclosporin A binds to a class of IP called cyclophilins, whereas the receptors for FK506 and rapamycin belong to the family of FK506-binding proteins (FKBP). The latter are designated according to their molecular weight: FKBP12, 25, 52, etc.

Neurotrophic activity of neuroimmunophilin ligands (FK506 and its nonimmunosuppressant derivatives) has been assumed to be mediated by the FK506-binding protein-12 (FKBP-12). However, the activity of FK506 is retained even in hippocampal neurons from FKBP-12 knockout mice, indicating that binding to FKBP-12 is not necessary (Gold *et al.*, 2005).

Three nonimmunosuppressant FK506 derivatives (V-13,450, V-13,629, and V-13,670), which do not bind FKBP-12 ($>$12.5 mM affinity), are equipotent to FKBP-12 ligands (FK506, V-10,367, and V-13,449) for increasing neurite elongation in SH-SY5Y cells (Gold *et al.*, 2005). One non-FKBP-12 ligand (V-13,670) accelerates functional recovery and nerve regeneration in the rat sciatic nerve crush model (Gold *et al.*, 2005).

FK506 is a FDA-approved highly effective immunosuppressive agent preventing organ transplant, and allograft rejection in humans. FK506 and nonimmunosuppressant compounds like, GPI-1046 and L685818, are IPLs that specifically bind to IPs, like FKBP-12 (Galat, 2008; Kang *et al.*, 2008). These ligands also exert neurotrophic properties in neural injury models *in vitro* and *in vivo*.

Therapeutic uses of FK506 include protection against ischemia, neurological diseases, and excitotoxicity (Furuichi *et al.*, 2007; Ghazinezami *et al.*, 2007; Gold and Schneider-Gold, 2008; Katsura *et al.*, 2008). FK506 not only protects neural tissue from focal cerebral ischemia but also significantly ameliorates motor deficits produced by the lesion (Ghazinezami *et al.*, 2007). Thus, FK506 may be useful in the treatment of stroke (Butcher *et al.*, 1997). Because CsA, FK506, and other immunosuppressive derivatives block or activate several intracellular pathways, and exert neuroprotective effects, they may be useful in ameliorating the negative impact of ischemia, Parkinson's disease, and excitotoxic insults (Gold and Nutt, 2002; Udina and Navarro, 2002).

FK506 is a potent survival factor that prevents neuronal cell death following axotomy of CNS neurons even when administered up to 8 days postaxotomy (Iwasaki *et al.*, 2002; Winter *et al.*, 2000). FK506 acts by interfering with the transcriptional program of the axotomy-induced cell body response, such as activating transcription factor-2 suppression and c-Jun expression and phosphorylation (Tashiro *et al.*, 2006; Winter *et al.*, 2000). FK506 and cyclosporin A also provide neuroprotection to axotomized neonatal rat motor neurons (Iwasaki *et al.*, 2002). Therefore, the IPLs may reduce the pathogenesis in the early stages of amyotrophic lateral sclerosis (ALS) (Kihira *et al.*, 2002; Manabe *et al.*, 2002).

As stated, sensory nerve grafts are still generally considered to be the "gold standard" for bridging gaps in the PNS. In addition to their relatively poor promotion of axon regeneration, their small diameter, relative to the nerves being repaired, requires multiple grafts, which in turn require extensive numbers of sutures, which induce inflammation, which inhibits axon regeneration (Fry *et al.*, 2007; Schwab *et al.*, 2004). Sensory nerve allografts and xenografts (a graft from another species that in which the grafts is being used) can also be used for grafts, but without immunosuppression they are rejected by the host body. Therefore, there are strong incentives to identify exogenous agents that may allow the retention of these grafts while also enhancing the rate of axon regeneration and the reestablishment of neurological function.

II. Promoting Axon Regeneration

A. Central Nervous System

Normally axons cannot regenerate into the CNS due to the presence of chondroitin sulfate proteoglycans (CSPG) on the astrocytes (McKeon *et al.*, 1991; Silver, 1994). However, daily subcutaneous injections of CsA or FK506 for 30 days induce axons of dorsal root ganglion neurons to regenerate into the spinal cord where they arborized extensively (Sugawara *et al.*, 1999). In the absence of administering FK506, the only means for inducing axon regeneration into the spinal cord is by digesting the regeneration inhibitory molecule CSPG, by which the regeneration inhibiting environment is changed to one that is regeneration permissive (McKeon *et al.*, 1995). FK506, or its nonimmunosuppressive derivative (FK1706), also exerts a neurotrophic action, by accelerating the rate of nerve regeneration in the PNS and CNS leading to more rapid neurological recovery, including following spinal cord hemisections (Gold *et al.*, 2005; Hayashi *et al.*, 2005; Lagoda *et al.*, 2008; Lopez-Vales *et al.*, 2005, 2006; Udina *et al.*, 2004; Voda *et al.*, 2005). Additional IPLs, including nonimmunosuppressants, which do not inhibit calcineurin, protect axons from degeneration (Valentine *et al.*, 2007).

If rats are administered subcutaneous injections of FK506 (0.5 or 2.0 mg/kg) and GPI-1046 (10 or 40 mg/kg) five times a week for 12 weeks following a partial ascending spinal cord dorsal column lesion, the axons sprout massively at the lesion sites, and regenerate up to 10 mm (Bavetta *et al.*, 1999). Such sprouting and regeneration never takes place in control animals. In addition, the administration induces three 3 times as many nonsectioned axons to survive in the dorsal column, showing that FK506 increases the number of axons that survive death caused by secondary effectors initiated by trauma (Bavetta *et al.*, 1999). Similarly, IPL administration induces extensive axon sprouting and growth of axons into the striatum (Ross *et al.*, 2001).

B. Peripheral Nervous System

Peripheral sensory nerve allotransplantation might be an alternative to autografts. However, allografts can only be successful if they are not rejected, which requires the use of immunophilin-immunosuppressive agents (CsA and FK506).

The actions of FK506 are due to the IP FKBP-52, and not the FKBP-12 (Gold *et al.*, 1999). FK506 also enhances collateral sprouting of peripheral nerve fibers (Udina *et al.*, 2003). FK506 doubles the number of axons that regenerate following a nerve injury, increases the number of myelinated axons by 40%, and significantly increase myelin thickness (Sulaiman *et al.*, 2002). In addition, FK506 reduces, by half, the time to neurological recovery following a nerve lesion (Becker *et al.*, 2000; Chen *et al.*, 2002; Chunasuwankul *et al.*, 2002; Myckatyn *et al.*, 2002). The nonimmunosuppressive IPL GPI-1046 is another potent stimulator of axon outgrowth (Grand *et al.*, 2002). The neurite elongation activity of FK506 involves FKBP-52 (also known as FKBP-59 or Hsp-56), a component of mature steroid receptor complexes. FKBP-52 binds to Hsp-90, which binds to p23 and the steroid receptor protein to form the complex (Gold and Villafranca, 2003).

While the FKBP-12-binding ligand FK506 stimulates nerve regeneration and simultaneously prevents rejection of peripheral nerve allografts, the immunosuppressant rapamycin, another FKBP-12-binding ligand, and CsA, stimulate axon regeneration *in vitro*, but have more limited influences than FK506 on nerve regeneration in isografts (a graft of tissue between two individuals who are genetically identical) and allografts (Wang *et al.*, 1997).

In an animal model, weekly twice injections of FK506 (0.1–0.4 mg/kg) allow immunosuppressive serum FK506 levels to be maintained, along with the allografts (Jensen *et al.*, 2005).

FK506 also enhances axonal regeneration through both the nerve autografts and allografts with defined histocompatibility barriers (Jensen *et al.*, 2005). In one clinical study, seven patients received major nerve reconstructions with nerve allografts under temporary immunosuppression, and six had sensory recovery, three motor recovery, while one had no recovery due to rejection (Bain, 2000).

III. Neuroprotection

CNS trauma leads directly to the death of some neurons. However, during the 48 h following trauma, a far greater number of neurons die due to sequelae of the trauma. Reducing this secondary cycle of neuron death will reduce the loss of neurological function. Thus, the administration of the appropriate immunosuppressant immediately following CNS trauma both maintains neuron viability, while allowing their damaged axons to regenerate and reestablish connections. This results in a significant reduction in trauma-induced neurological losses.

FK506 acts as a powerful neuroprotective agent providing protection *in vivo* against focal cerebral ischemia, even when administered up to 60 min postocclusion (Giordani *et al.*, 2003; Gold *et al.*, 2004; Guzman-Lenis *et al.*, 2008; Lopez-Vales *et al.*, 2005; Pan *et al.*, 2006). The concentration of FK506 providing this neuroprotection is comparable with the dose used in humans to prevent tissue rejections (Brandhorst *et al.*, 2008; Wollenberg *et al.*, 2008).

Beside the widespread therapeutic use of ligands of IPs as immunosuppressants, FK506 and V-10,367 mediate neuroprotection of retinal ganglion neurons against optic nerve crush-induced cell death that occurs when the optic nerve is crushed within 2 mm of the eye (Cui *et al.*, 2007; Rosenstiel *et al.*, 2003). FK506 also increases the regeneration of retinal ganglion cell axons (Cui *et al.*, 2007), further indicating the breadth of beneficial actions of FK506.

Rapamycin is an immunosuppressive IPL with neurotrophic activity. However, modifying rapamycin at its mammalian target, the mTOR binding region, yields IPLs, WYE-592 and ILS-920, with potent neurotrophic activities, efficacy against the consequences of ischemic stroke, and significantly reduced immunosuppressive activity (Ruan *et al.*, 2008). Further, both compounds have higher binding selectivity for FKBP52 versus FKBP12 than other IPLs. These IPLs protect neurons from calcium-induced cell death by modulating calcium channels and promoting neurite outgrowth via FKBP52 binding (Ruan *et al.*, 2008).

Cyclophilin A protects cells from death following expression of mutant Cu/Zn superoxide dismutase, which is associated with familial ALS (Aghdasi *et al.*, 2001; Vinceti *et al.*, 2002). FKBP12 and FKBP52 are expressed in the human nervous system, especially in the substantia nigra-deep gray matter axis (Avramut and Achim, 2003; Nilsson *et al.*, 2007). In neurodegenerative diseases, FKBP12 levels increase in neurons situated in areas of pathology (Nilsson *et al.*, 2007). The IP colocalizes with synaptophysin and α-synuclein, suggesting it may be a marker of pathology. The IPLs also prevent H_2O_2-induced apoptotic cell death by increasing glutathione levels in neuro 2A neuroblastoma cells (Tanaka *et al.*, 2000).

A. Calcium Stabilization

FKBP levels in the rat brain are up to 50 times higher than in the immune system. FKBP12, the 12-kDa FKBP, is a ubiquitously abundant protein that acts as a receptor for FK506, and binds tightly to intracellular calcium release channels. Thus, FK506 appears to play a role in stabilizing intracellular calcium release (Katsura *et al.*, 2008), and calcium stabilization is essential for neuronal survival following trauma which typically leads to the loss of calcium homeostasis.

B. Hypothermia

Even though hypothermia maintains neurons viable during prolonged periods of the ischemic insult (Reyes *et al.*, 2006), many neurons die during the period of reperfusion and warming (Du *et al.*, 2008; Fujita *et al.*, 2006). Administration of FK506 during perfusion provides neuroprotection to neurons that die during reperfusion (Szydlowska *et al.*, 2006). It is also possible that hypothermia, combined with alkalinization and administration of FK506, will provide even greater neuroprotection through the period of reperfusion (Reyes *et al.*, 2006).

IV. Timing of Administration of FK506

Following CNS and PNS trauma, clinicians tend to wait prior to making their diagnosis and plans for treatment. Sometimes these delays are related to the resolution of tissue trauma, but not at other times. Therefore, it is important to know the optimal time window following trauma to the nervous system during which FK506 administration is most effective.

Subcutaneous administration of FK506 (5 mg/kg/day) is equally as effective when given discontinuously, starting immediately following nerve section (days 0–8 and 10–17), or with a delay (days 9–17), with continuous administration starting immediately being most effective (Wang and Gold, 1999). However, the greatest influence of FK506 is when a nerve repair is performed immediately following the lesion and simultaneously with the initiation of administration of FK506 (Brenner *et al.*, 2004). Delaying the repair surgery and the administration of FK506 (even by 7 days), or immediate surgery with delayed administration of FK506 leads to less extensive axon regeneration and fewer motor neurons extending axons (Brenner *et al.*, 2004).

Prolonging FK506 treatment from 2 to 3 months following an acute nerve transection does not increase the extent of the ultimate neurological recovery (Chunasuwankul *et al.*, 2002). On the other hand, when FK506 is administered 2 months after nerve transection it doubles the rate of axon regeneration, which is similar to that seen when it is administered to an acutely damaged peripheral nerve (Sulaiman *et al.*, 2002).

V. Concentration of Neurotrophic Activity

FK506 enhances the rate of axon regeneration by daily subcutaneous injections in doses of from 0.5 to 10 mg/kg/day, and the administration can be continued for weeks to months, depending on the length required to allow for axon regeneration required (Sulaiman*et al.*, 2002; Yang *et al.*, 2003). FK506 acts in a bimodal dose-dependent manner with intermediate doses (0.5 and 1 mg/kg) and a higher dose (10 mg/kg) being the same as controls, and 2 mg/kg/day slightly hindering axon regeneration (Udina *et al.*, 2002, 2003). FK506 is also effective when given orally, with the most effective dose being 15 mg/kg/day. Therefore, orally active FK506 may be useful for the treatment of human peripheral nerve disorders (Gold *et al.*, 1998). However, recently the lower daily dose of 0.05 mg/kg of FK506 was shown to have lower immunosuppression and neurotoxicity, but to induce a faster axonal regeneration rate (Sarikcioglu *et al.*, 2006; Udina *et al.*, 2003). Because sprouting of sympathetic axons is more extensive at a high dose, this suggests that the efficacy of FK506 varies between neuron subtypes (Udina *et al.*, 2003).

VI. Mechanisms of Action of FK506

FK506 induces axon regeneration predominantly by binding to FKBP-12, which activates GAP-43 (growth associated protein) and the TGF beta 1-pathway (Fansa *et al.*, 1999a). FK506 also contributes to enhancing the rate of axon regeneration at very low doses (100 μM) by significantly increasing the extent of Schwann cell proliferation, and inducing Schwann cells to increase their production and release of the neurotrophic factors, which enhances the number of axons and the distance they regenerate (Fansa *et al.*, 1999b). Daily administration of FK506 (0.6 mg/kg) also leads to a decrease in myelin debris in autologous nerve grafts (Fansa *et al.*, 1999b) indicating an acceleration of Wallerian degeneration that would lead to an enhanced rate of axon regeneration.

In the absence of immunosuppression, peripheral nerve allografts contain few axons but many T cells and macrophages. One of the key roles of systemic (intraperitoneal) administration of rapamycin, cyclosporin A, or FK506 is that they significantly reduce cellular infiltration of T cells and macrophages into allografts, with FK506 being more effective than cyclosporin and rapamycin (Grand *et al.*, 2002; Myckatyn *et al.*, 2002; Rigol *et al.*, 2008; Seron *et al.*, 2007), which tends to inhibit axon regeneration (Keilhoff *et al.*, 2007; Zhang *et al.*, 2008). The benefit is that in the absence of these cells axonal regeneration is far more extensive (Gillon *et al.*, 2003; Grand *et al.*, 2002).

VII. Side Effects of FK506

Although FK506 enhances nerve regeneration in various rodent models, it has neurological and other side effects, such as metabolic, cosmetic, and neuropsychiatric, which raises concern about its clinical use to speedup nerve regeneration, and concerns about the possible consequence when FK506 treatment stops (Okajima *et al.*, 2002). One of the more serious complications following the use of FK506 in animal models is the development of posttransplantation diabetes mellitus, assessed by increased glycemia levels. However, even with prolonged use of FK506, up to 240 days, the increased glycemia levels are not chronic, and eventually return to normal, making FK506 treatment effects on glycemia only a time-related side effect (Nassar *et al.*, 2007; Okajima *et al.*, 2002). These results indicate that the fear of long-term damage induced by FK506 is not justified.

When FK506 is administered for up to 12 weeks to prevent rejection of allograft grafts, by 12 weeks after stopping the use of FK506 no histological signs of rejection and cellular infiltration are seen, and normal axon morphology is seen, with action potential conduction recorded for up to 12 weeks (Okajima *et al.*, 2002). Therefore, there appear to be no side effects when FK506 administration is stopped.

One mechanism by which FK506 appears to induce neurological side effects is via the production of reactive oxygen species and a decreased antioxidant status (Jin *et al.*, 2008). It may be possible to inhibit these side effects, by the parallel administration of antioxidants with FK506 administration. However, cyclosporine and FK506-based immunosuppressive protocols induce additional side effects, such as metabolic, cosmetic, and neuropsychiatric, which still raise concern about their use (Srinivas and Meier-Kriesche, 2008).

FK506 without steroid use provides a greater therapeutic benefit in chronic allograft nephropathy compared to the use of cyclosporin (Marcard *et al.*, 2008). Benefits of FK506 over cyclosporin are also related to their use for nephrotic syndrome (Li *et al.*, 2008). In one study of children suffering from steroid-resistant nephrotic syndrome, there was a subgroup that was nonresponsive to the therapeutic benefits of cyclophosphamide and cyclosporine administration, but responded well to FK506 (Gulati *et al.*, 2008). Therefore, subpopulations may exist which respond better to FK506 than cyclosporin. These results suggest that it is safer and potentially more beneficial to use FK506 versus cyclosporin for neuroprotection and inducing axon regeneration.

Due to their side effects, it must be determined whether newer immunosuppressive agents, such as mycophenolate mofetil and sirolimus (Rapa), have similar side effects, and whether they can be used as alternatives for providing neuroprotection and inducing axon regeneration (Srinivas *et al.*, 2008). However, until these alternative compounds have been sufficiently tested to determine that they induce the desired influences, the immunosuppressives cyclosporine and FK506 should continue to be used.

VIII. Clinical Applications

In Institutional Review Board (IRB)-approved studies, this laboratory developed a novel technique that induces neurological recovery following adult human anatomical spinal cord transections, and induces neurological recovery of peripheral nerves with gaps of up to 12 cm in length, and when the nerve repair is performed up to 3.25 years posttrauma. The technique involves the introduction of collagen tube filled with platelet-rich fibrin into the peripheral nerve or spinal cord gap. Although one patient recovered 100% sensory and motor function, and the general data are superior to that of autologous sensory nerve grafts, the extent of axon regeneration must be increased to make the neurological recovery more reliable and extensive. Among techniques, the addition of T effector cells to the bridged region, electrical stimulation of the repaired region, and the administration of the immunosuppressant FK505 are believed to enhance the neurological recovery.

IX. Conclusion

Immunosuppressants act to prevent nerve graft rejection, enhance the rate of peripheral and central axon regeneration, the length axon regeneration, and the degree axon myelination. They also provide neuroprotection of neurons and their axons against ischemia and excitoxicity, leading to improved neurological recovery. The action of immunosuppressants on nerve regeneration is separate from their immunosuppressant action, because FK506-related compounds that bind to FKBP-12, but do not inhibit calcineurin, are also able to increase nerve regeneration. Thus, FK506's ability to increase nerve regeneration is due to a calcineurin-independent mechanism (i.e., one not involving an increase in GAP-43 phosphorylation), and indicates broad functional roles for IPs in the nervous system. Both calcineurin-dependent (e.g., neuroprotection via reduced NO formation) and calcineurin-independent mechanisms (i.e., nerve regeneration) explain the many different neuronal effects of FK506. Thus, multiple IPs mediate FK506's neuronal effects and novel nonimmunosuppressant ligands for FKBPs represent important new drugs that may be important in promoting improved neurological recovery following both acute and long-term PNS and CNS lesions, and for treatment of a variety of neurological disorders, such as ALS. FK506 is used extensively for its immunosuppressive actions because of its limited number of side effects, especially when used in adults. These and newer compounds must be further studied for their potential benefits in enhancing axon regeneration and neurological recovery following both peripheral nerve and spinal cord trauma.

References

Aghdasi, B., Ye, K., Resnick, A., Huang, A., Ha, H. C., Guo, X., Dawson, T. M., Dawson, V. L., and Snyder, S. H. (2001). FKBP12, the 12-kda FK506-binding protein, is a physiologic regulator of the cell cycle. *Proc. Natl. Acad. Sci. USA* **98,** 2425–2430.

Avramut, M., and Achim, C. L. (2003). Immunophilins in nervous system degeneration and regeneration. *Curr. Top. Med. Chem.* **3,** 1376–1382.

Bain, J. R. (2000). Peripheral nerve and neuromuscular allotransplantation: Current status. *Microsurgery.* **20,** 384–388.

Bavetta, S., Hamlyn, P. J., Burnstock, G., Lieberman, A. R., and Anderson, P. N. (1999). The effects of FK506 on dorsal column axons following spinal cord injury in adult rats: Neuroprotection and local regeneration. *Exp. Neurol.* **158,** 382–393.

Becker, D. B., Jensen, J. N., Myckatyn, T. M., Doolabh, V. B., Hunter, D. A., and Mackinnon, S. E. (2000). Effects of FKBP-12 ligands following tibial nerve injury in rats. *J. Reconstr. Microsurg.* **16,** 613–620.

Brandhorst, G., Tenderich, G., Zittermann, A., Oezpeker, C., Koerfer, R., Oellerich, M., and Armstrong, V. W. (2008). Everolimus exposure in cardiac transplant recipients is influenced by concomitant calcineurin inhibitor. *Ther. Drug Monit.* **30,** 113–116.

Brenner, M. J., Fox, I. K., Kawamura, D. H., Yu, V. M., Lowe, J. B. III, Hunter, D. A., and Mackinnon, S. E. (2004). Delayed nerve repair is associated with diminished neuroenhancement by FK506. *Laryngoscope* **114,** 570–576.

Butcher, S. P., Henshall, D. C., Teramura, Y., Iwasaki, K., and Sharkey, J. (1997). Neuroprotective actions of FK506 in experimental stroke: *In vivo* evidence against an antiexcitotoxic mechanism. *J. Neurosci.* **17,** 6939–6946.

Chabas, J. F., Alluin, O., Rao, G., Garcia, S., Lavaut, M. N., Legre, R., Magalon, G., Marqueste, T., Feron, F., and Decherchi, P. (2009). FK506 induces changes in muscle properties and promotes metabosensitive nerve fiber regeneration. *J. Neurotraum.* **26,** 97–108.

Chen, G. F., Gu, L. Q., and Pei, G. X. (2002). [Peripheral nerve regeneration under immunosuppression]. *Zhongguo Xiu Fu Chong Jian Wai Ke Za Zhi* **16,** 139–141.

Chunasuwankul, R., Ayrout, C., Dereli, Z., Gal, A., Lanzetta, M., and Owen, E. (2002). Low dose discontinued FK506 treatment enhances peripheral nerve regeneration. *Int. Surg.* **87,** 274–278.

Cui, Q., Hodgetts, S. I., Hu, Y., Luo, J. M., and Harvey, A. R. (2007). Strain-specific differences in the effects of cyclosporin A and FK506 on the survival and regeneration of axotomized retinal ganglion cells in adult rats. *Neuroscience* **146,** 986–999.

Du, Y., Hirooka, K., Miyamoto, O., Itano, T., Tokuda, M., and Shiraga, F. (2008). Both amacrine and bipolar cells release glutamate in the rat retina after ischemia/reperfusion insult *in vitro*. *Curr. Eye Res.* **33,** 782–788.

Fansa, H., Keilhoff, G., Altmann, S., Plogmeier, K., Wolf, G., and Schneider, W. (1999a). The effect of the immunosuppressant FK 506 on peripheral nerve regeneration following nerve grafting. *J. Hand Surg. (Br.)* **24,** 38–42.

Fansa, H., Keilhoff, G., Horn, T., Altmann, S., Wolf, G., and Schneider, W. (1999b). [Stimulation of Schwann cell growth and axon regeneration of peripheral nerves by the immunosuppressive drug FK 506]. *Handchir. Mikrochir. Plast. Chir.* **31,** 323–329; discussion 330–322.

Fry, E. J., Ho, C., and David, S. (2007). A role for Nogo receptor in macrophage clearance from injured peripheral nerve. *Neuron* **53,** 649–662.

Fujita, K., Kato, T., Shibayama, K., Imada, H., Yamauchi, M., Yoshimoto, N., Miyachi, E., and Nagata, Y. (2006). Protective effect against 17beta-estradiol on neuronal apoptosis in hippocampus tissue following transient ischemia/recirculation in mongolian gerbils via down-regulation of tissue transglutaminase activity. *Neurochem. Res.* **31,** 1059–1068.

Furuichi, Y., Maeda, M., Matsuoka, N., Mutoh, S., and Yanagihara, T. (2007). Therapeutic time window of tacrolimus (FK506) in a nonhuman primate stroke model: Comparison with tissue plasminogen activator. *Exp. Neurol.* **204,** 138–146.

Galat, A. (2008). Functional drift of sequence attributes in the FK506-binding proteins (fkbps). *J. Chem. Inf. Model.* **48,** 1118–1130.

Ghazinezami, B., Rahimpour, S., Gholipour, T., Ghasemi, M., Sadeghipour, H., Mehr, S. E., Emami-Razavi, S. H., and Dehpour, A. R. (2007). Pharmacologic preconditioning of random-pattern skin flap in rats using local cyclosporine and FK-506: Interaction with nitric oxide system. *Ann. Plast. Surg.* **59,** 435–440.

Gillon, R. S., Cui, Q., Dunlop, S. A., and Harvey, A. R. (2003). Effects of immunosuppression on regrowth of adult rat retinal ganglion cell axons into peripheral nerve allografts. *J. Neurosci. Res.* **74,** 524–532.

Giordani, F., Benetolli, A., Favero-Filho, L. A., Lima, K. C., Cestari Junior, L., and Milani, H. (2003). Tacrolimus (FK506) reduces ischemia-induced hippocampal damage in rats: A 7- and 30-day study. *Braz. J. Med. Biol. Res.* **36,** 495–502.

Gold, B. G., and Nutt, J. G. (2002). Neuroimmunophilin ligands in the treatment of Parkinson's disease. *Curr. Opin. Pharmacol.* **2,** 82–86.

Gold, R., and Schneider-Gold, C. (2008). Current and future standards in treatment of myasthenia gravis. *Neurotherapeutics* **5,** 535–541.

Gold, B. G., and Villafranca, J. E. (2003). Neuroimmunophilin ligands: The development of novel neuroregenerative/neuroprotective compounds. *Curr. Top. Med. Chem.* **3,** 1368–1375.

Gold, B. G., Zeleny-Pooley, M., Chaturvedi, P., and Wang, M. S. (1998). Oral administration of a nonimmunosuppressant FKBP-12 ligand speeds nerve regeneration. *Neuroreport* **9,** 553–558.

Gold, B. G., Densmore, V., Shou, W., Matzuk, M. M., and Gordon, H. S. (1999). Immunophilin FK506-binding protein 52 (not FK506-binding protein 12) mediates the neurotrophic action of FK506. *J. Pharmacol. Exp. Ther.* **289,** 1202–1210.

Gold, B. G., Udina, E., Bourdette, D., and Navarro, X. (2004). Neuroregenerative and neuroprotective actions of neuroimmunophilin compounds in traumatic and inflammatory neuropathies. *Neurol. Res.* **26,** 371–380.

Gold, B. G., Armistead, D. M., and Wang, M. S. (2005). Non-FK506-binding protein-12 neuroimmunophilin ligands increase neurite elongation and accelerate nerve regeneration. *J. Neurosci. Res.* **80,** 56–65.

Grand, A. G., Myckatyn, T. M., Mackinnon, S. E., and Hunter, D. A. (2002). Axonal regeneration after cold preservation of nerve allografts and immunosuppression with tacrolimus in mice. *J. Neurosurg.* **96,** 924–932.

Gulati, S., Prasad, N., Sharma, R. K., Kumar, A., Gupta, A., and Baburaj, V. P. (2008). Tacrolimus: A new therapy for steroid-resistant nephrotic syndrome in children. *Nephrol. Dial. Transplant.* **23,** 910–913.

Guzman-Lenis, M. S., Vallejo, C., Navarro, X., and Casas, C. (2008). Analysis of FK506-mediated protection in an organotypic model of spinal cord damage: Heat shock protein 70 levels are modulated in microglial cells. *Neuroscience* **155,** 104–113.

Hayashi, Y., Shumsky, J. S., Connors, T., Otsuka, T., Fischer, I., Tessler, A., and Murray, M. (2005). Immunosuppression with either cyclosporine a or FK506 supports survival of transplanted fibroblasts and promotes growth of host axons into the transplant after spinal cord injury. *J. Neurotrauma* **22,** 1267–1281.

Hayashi, A., Moradzadeh, A., Tong, A., Wei, C., Tuffaha, S. H., Hunter, D. A., Tung, T. H., Parsadanian, A., Mackinnon, S. E., and Myckatyn, T. M. (2008). Treatment modality affects allograft-derived Schwann cell phenotype and myelinating capacity. *Exp. Neurol.* **212,** 324–336.

Iwasaki, Y., Ichikawa, Y., Igarashi, O., Iwamoto, K., Kinoshitata, M., and Ikeda, K. (2002). Neuroprotective actions of FK506 and cyclosporin A on motor neuron survival following neonatal axotomy. *Neurol. Res.* **24,** 573–576.

Jensen, J. N., Brenner, M. J., Tung, T. H., Hunter, D. A., and Mackinnon, S. E. (2005). Effect of FK506 on peripheral nerve regeneration through long grafts in inbred swine. *Ann. Plast. Surg.* **54,** 420–427.

Jin, K. B., Hwang, E. A., Han, S. Y., Park, S. B., Kim, H. C., Ha, E. Y., Suh, S. I., and Mun, K. C. (2008). Effects of tacrolimus on antioxidant status and oxidative stress in glioma cells. *Transplant. Proc.* **40,** 2740–2741.

Kang, C. B., Hong, Y., Dhe-Paganon, S., and Yoon, H. S. (2008). FKBP family proteins: Immunophilins with versatile biological functions. *Neurosignals* **16,** 318–325.

Katsura, K., Takahashi, K., Asoh, S., Watanabe, M., Sakurazawa, M., Ohsawa, I., Mori, T., Igarashi, H., Ohkubo, S., Katayama, Y., and Ohta, S. (2008). Combination therapy with transductive anti-death FNK protein and FK506 ameliorates brain damage with focal transient ischemia in rat. *J. Neurochem.* **106,** 258–270.

Keilhoff, G., Langnaese, K., Wolf, G., and Fansa, H. (2007). Inhibiting effect of minocycline on the regeneration of peripheral nerves. *Dev. Neurobiol.* **67,** 1382–1395.

Kihira, T., Hironishi, M., Utunomiya, H., and Kondo, T. (2002). FKBP12 immunoreactivity in the human spinal cord of motor neuron disease patients. *Neuropathology* **22,** 269–274.

Lagoda, G., Sezen, S. F., Liu, T., Hoke, A., and Burnett, A. L. (2008). FK506-binding protein localizations in human penile innervation. *BJU Int.* **101,** 604–609.

Li, X., Li, H., Chen, J., He, Q., Lv, R., Lin, W., Li, Q., He, X., Qu, L., and Suya, W. (2008). Tacrolimus as a steroid-sparing agent for adults with steroid-dependent minimal change nephrotic syndrome. *Nephrol. Dial. Transplant.* **23,** 1919–1925.

Lopez-Vales, R., Garcia-Alias, G., Fores, J., Udina, E., Gold, B. G., Navarro, X., and Verdu, E. (2005). FK 506 reduces tissue damage and prevents functional deficit after spinal cord injury in the rat. *J. Neurosci. Res.* **81,** 827–836.

Lopez-Vales, R., Fores, J., Navarro, X., and Verdu, E. (2006). Olfactory ensheathing glia graft in combination with FK506 administration promote repair after spinal cord injury. *Neurobiol. Dis.* **24,** 443–454.

Manabe, Y., Warita, H., Murakami, T., Shiote, M., Hayashi, T., Omori, N., Nagano, I., Shoji, M., and Abe, K. (2002). Early decrease of the immunophilin FKBP 52 in the spinal cord of a transgenic model for amyotrophic lateral sclerosis. *Brain Res.* **935,** 124–128.

Marcard, T., Ivens, K., Grabensee, B., Willers, R., Helmchen, U., Rump, L. C., and Blume, C. (2008). Early conversion from cyclosporine to tacrolimus increases renal graft function in chronic allograft nephropathy at BANFF stages I and II. *Transpl. Int.* **21,** 1153–1162.

McKeon, R. J., Schreiber, R. C., Rudge, J. S., and Silver, J. (1991). Reduction of neurite outgrowth in a model of glial scarring following CNS injury is correlated with the expression of inhibitory molecules on reactive astrocytes. *J. Neurosci.* **11,** 3398–3411.

McKeon, R. J., Hoke, A., and Silver, J. (1995). Injury-induced proteoglycans inhibit the potential for laminin-mediated axon growth on astrocytic scars. *Exp. Neurol.* **136,** 32–43.

Myckatyn, T. M., Ellis, R. A., Grand, A. G., Sen, S. K., Lowe, J. B. III, Hunter, D. A., and Mackinnon, S. E. (2002). The effects of rapamycin in murine peripheral nerve isografts and allografts. *Plast. Reconstr. Surg.* **109,** 2405–2417.

Nassar, C. A., Nassar, P. O., Andia, D. C., Guimaraes, M. R., Pepato, M. T., and Spolidorio, L. C. (2007). Biochemical evaluation of glycemic levels of long-term tacrolimus therapy in rats. *Braz. Oral Res.* **21,** 293–297.

Nilsson, A., Skold, K., Sjogren, B., Svensson, M., Pierson, J., Zhang, X., Caprioli, R. M., Buijs, J., Persson, B., Svenningsson, P., and Andren, P. E. (2007). Increased striatal mRNA and protein levels of the immunophilin FKBP-12 in experimental Parkinson's disease and identification of FKBP-12-binding proteins. *J. Proteome Res.* **6,** 3952–3961.

Okajima, S., Hojo, T., Tamai, K., Takai, S., and Hirasawa, Y. (2002). Histological and electrophysiological analysis of the peripheral nerve allografts using an immunosuppressive agent. *Microsc. Res. Tech.* **58,** 52–58.

Pan, F., Chen, A. M., Guo, F. J., and Zhu, C. L. (2006). [Effect of tacrolimus on apoptosis and expression of heat shock protein 70 after acute spinal cord injury in rats]. *Zhonghua Wai Ke Za Zhi* **44,** 1708–1712.

Reyes, O., Sosa, I., and Kuffler, D. P. (2006). Neuroprotection of adult human neurons against ischemia by hypothermia and alkalinization. *P. R. Health Sci. J.* **25,** 43–50.

Rigol, M., Solanes, N., Sionis, A., Galvez, C., Martorell, J., Rojo, I., Brunet, M., Ramirez, J., Roque, M., Roig, E., Perez-Villa, F., Barquin, L., *et al.* (2008). Effects of cyclosporine, tacrolimus and sirolimus on vascular changes related to immune response. *J. Heart Lung Transplant.* **27,** 416–422.

Rosenstiel, P., Schramm, P., Isenmann, S., Brecht, S., Eickmeier, C., Burger, E., Herdegen, T., Sievers, J., and Lucius, R. (2003). Differential effects of immunophilin-ligands (FK506 and V-10,367) on survival and regeneration of rat retinal ganglion cells *in vitro* and after optic nerve crush *in vivo*. *J. Neurotrauma* **20,** 297–307.

Ross, D. T., Guo, H., Howorth, P., Chen, Y., Hamilton, G. S., and Steiner, J. P. (2001). The small molecule FKBP ligand GPI 1046 induces partial striatal re-innervation after intranigral 6-hydroxydopamine lesion in rats. *Neurosci. Lett.* **297,** 113–116.

Ruan, B., Pong, K., Jow, F., Bowlby, M., Crozier, R. A., Liu, D., Liang, S., Chen, Y., Mercado, M. L., Feng, X., Bennett, F., von Schack, D., *et al.* (2008). Binding of rapamycin analogs to calcium channels and FKBP52 contributes to their neuroprotective activities. *Proc. Natl. Acad. Sci. USA* **105,** 33–38.

Sarikcioglu, L., Duygulu, E., Aydin, H., Gurer, E. I., Ozkan, O., and Tuzuner, S. (2006). Effects of intrathecal administration of FK506 after sciatic nerve crush injury. *J. Reconstr. Microsurg.* **22,** 649–654.

Schwab, J. M., Conrad, S., Elbert, T., Trautmann, K., Meyermann, R., and Schluesener, H. J. (2004). Lesional rhoA + cell numbers are suppressed by anti-inflammatory, cyclooxygenase-inhibiting treatment following subacute spinal cord injury. *Glia* **47,** 377–386.

Seron, D., O'Valle, F., Moreso, F., Goma, M., Hueso, M., Grinyo, J. M., and Garcia del Moral, R. (2007). Immunophenotype of infiltrating cells in protocol renal allograft biopsies from tacrolimus-versus cyclosporine-treated patients. *Transplantation* **83,** 649–652.

Silver, J. (1994). Inhibitory molecules in development and regeneration. *J. Neurol.* **242,** S22–S24.

Srinivas, T. R., and Meier-Kriesche, H. U. (2008). Minimizing immunosuppression, an alternative approach to reducing side effects: Objectives and interim result. *Clin. J. Am. Soc. Nephrol.* **3**(Suppl. 2), S101–S116.

Sugawara, T., Itoh, Y., and Mizoi, K. (1999). Immunosuppressants promote adult dorsal root regeneration into the spinal cord. *Neuroreport* **10,** 3949–3953.

Sulaiman, O. A., Voda, J., Gold, B. G., and Gordon, T. (2002). FK506 increases peripheral nerve regeneration after chronic axotomy but not after chronic schwann cell denervation. *Exp. Neurol.* **175,** 127–137.

Szydlowska, K., Zawadzka, M., and Kaminska, B. (2006). Neuroprotectant FK506 inhibits glutamate-induced apoptosis of astrocytes *in vitro* and *in vivo*. *J. Neurochem.* **99,** 965–975.

Tanaka, K., Fujita, N., Asanuma, M., and Ogawa, N. (2000). Immunophilin ligands prevent H2O2-induced apoptotic cell death by increasing glutathione levels in neuro 2A neuroblastoma cells. *Acta Med. Okayama* **54,** 275–280.

Tashiro, M., Dabrowski, A., Guo, L., Sans, M. D., and Williams, J. A. (2006). Calcineurin-dependent and calcineurin-independent signal transduction pathways activated as part of pancreatic growth. *Pancreas* **32,** 314–320.

Udina, E., and Navarro, X. (2002). [Immunophilins: Neuroprotective agents and promoters of neural regeneration]. *Neurologia* **17,** 200–213.

Udina, E., Ceballos, D., Verdu, E., Gold, B. G., and Navarro, X. (2002). Bimodal dose-dependence of FK506 on the rate of axonal regeneration in mouse peripheral nerve. *Muscle Nerve* **26,** 348–355.

Udina, E., Ceballos, D., Gold, B. G., and Navarro, X. (2003a). FK506 enhances reinnervation by regeneration and by collateral sprouting of peripheral nerve fibers. *Exp. Neurol.* **183,** 220–231.

Udina, E., Voda, J., Gold, B. G., and Navarro, X. (2003b). Comparative dose-dependence study of FK506 on transected mouse sciatic nerve repaired by allograft or xenograft. *J. Peripher. Nerv. Syst.* **8,** 145–154.

Udina, E., Verdu, E., and Navarro, X. (2004). Effects of the immunophilin ligand FK506 on nerve regeneration in collagen guides seeded with Schwann cells in rats. *Neurosci. Lett.* **357,** 99–102.

Valentine, H., Chen, Y., Guo, H., McCormick, J., Wu, Y., Sezen, S. F., Hoke, A., Burnett, A. L., and Steiner, J. P. (2007). Neuroimmunophilin ligands protect cavernous nerves after crush injury in the rat: New experimental paradigms. *Eur. Urol.* **51,** 1724–1731.

Vinceti, M., Bergomi, M., Nacci, G., Pietrini, V., Ferrari, A., Fortini, K., Guidetti, D., Sola, P., Rocchi, E., Mancia, D., and Vivoli, G. (2002). Erythrocyte zinc, copper, and copper/zinc superoxide dismutase and risk of sporadic amyotrophic lateral sclerosis: A population-based case-control study. *Amyotroph. Lateral Scler. Other Motor Neuron Disord.* **3,** 208–214.

Voda, J., Yamaji, T., and Gold, B. G. (2005). Neuroimmunophilin ligands improve functional recovery and increase axonal growth after spinal cord hemisection in rats. *J. Neurotrauma* **22,** 1150–1161.

Wang, M. S., and Gold, B. G. (1999). FK506 increases the regeneration of spinal cord axons in a predegenerated peripheral nerve autograft. *J. Spinal Cord Med.* **22,** 287–296.

Wang, M. S., Zeleny-Pooley, M., and Gold, B. G. (1997). Comparative dose-dependence study of FK506 and cyclosporin A on the rate of axonal regeneration in the rat sciatic nerve. *J. Pharmacol. Exp. Ther.* **282,** 1084–1093.

Winter, C., Schenkel, J., Burger, E., Eickmeier, C., Zimmermann, M., and Herdegen, T. (2000). The immunophilin ligand FK506, but not GPI-1046, protects against neuronal death and inhibits c-Jun expression in the substantia nigra pars compacta following transection of the rat medial forebrain bundle. *Neuroscience* **95,** 753–762.

Wollenberg, A., Reitamo, S., Girolomoni, G., Lahfa, M., Ruzicka, T., Healy, E., Giannetti, A., Bieber, T., Vyas, J., and Deleuran, M. (2008). Proactive treatment of atopic dermatitis in adults with 0.1% tacrolimus ointment. *Allergy* **63,** 742–750.

Yang, R. K., Lowe, J. B. III, Sobol, J. B., Sen, S. K., Hunter, D. A., and Mackinnon, S. E. (2003). Dose-dependent effects of FK506 on neuroregeneration in a rat model. *Plast. Reconstr. Surg.* **112,** 1832–1840.

Zhang, Z., Fauser, U., and Schluesener, H. J. (2008). Dexamethasone suppresses infiltration of rhoA + cells into early lesions of rat traumatic brain injury. *Acta Neuropathol.* **115,** 335–343.

THE ROLE OF COLLAGEN IN PERIPHERAL NERVE REPAIR

Guido Koopmans,* Birgit Hasse,* and Nektarios Sinis†

*SCT Spinal Cord Therapeutics GmbH, Max-Planck-Str. 15a, 40699 Erkrath, Germany
†Klinik für Hand-, Plastische-, Rekonstruktive und Verbrennungschirurgie,
Eberhard-Karls-Universität Tübingen, BG-Unfallklinik, Schnarrenbergstr. 95,
72076 Tübingen, Germany

Collagens are extracellular proteins characterized by a triple helical structure and predominantly involved in the formation of fibrillar and microfibrillar networks of the extracellular matrix and basement membranes. There are 29 collagen types which differ in size, structure, and function. In the peripheral nervous system, two classes of collagen molecules are expressed: fibril forming collagens (type-I, III, and V) and basement membrane collagens (type-IV). Collagens are required for normal extracellular matrix assembly and play an important role in the regulation of Schwann cell function. After injury collagen production in the severed nerve often exceeds the ideal response which is suggested to hinder the growth of sprouting axons into the appropriate distal fascicles and therefore delays and limits nerve regeneration. Both surgical techniques and pharmacological agents are developed to reduce injury induced scarring but despite this nerve regeneration is frequently incomplete. The aim of the present review is to provide the reader a clear overview of the current knowledge with respect to collagens in the peripheral nervous system and to emphasize its role after nerve injury.

INTERNATIONAL REVIEW OF
NEUROBIOLOGY, VOL. 87
DOI: 10.1016/S0074-7742(09)87019-0

I. Introduction

A. Outline of the Review

In the present paper, we will review the current knowledge with respect to collagen and its role in the peripheral nervous system (PNS) and after peripheral nerve injury. In the introduction, a historical overview of peripheral nerve repair is presented. The second part of the paper thoroughly describes the peripheral nerve collagens structure, synthesis, and function. In response to peripheral nerve injury, modulated collagen production would ideally enhance tissue strength; however, as in other organs, collagen production in the nerve often exceeds the appropriate response and results in scar formation and incomplete recovery (Millesi, 1977; Siironen *et al.*, 1992a; Sunderland, 1968). The consequences of excessive collagen production in the severed nerve are discussed in the third part of the review. Finally, various pharmacological approaches to reduce injury induced scar formation are discussed.

B. Peripheral Nerve Repair: An Historical Overview

The possibility of regeneration of severed nerves has been a subject of discussion in the earliest medical writings. The first experimental examinations by physiological methods that proved this possibility were conducted by Cruikshank (1795). Cruikshank divided the vago-sympathetic trunk, in the neck of the dog, on both sides and showed that death followed within a short time. In a second experiment the nerve was cut on one side alone, and he observed that the animal survived without noticeable injury. In a final experiment, the nerve was cut on one side, and after an interval of three weeks the other side was cut as well. The animal survived, proving for the first time that a severed nerve has the ability to regenerate and become functional again (Cruikshank, 1795).

About half a century later the first theories with respect to the underlying phenomena have been reported. Nasse (1839) was the first to describe the degenerative changes that take place in the distal nerve stump after transection; however, he believed that these degenerative changes affect the distal nerve end only when functional regeneration with the proximal end does not take place (Howell and Huber, 1892). In contrast, Waller believed and reported in numerous publications (Waller, 1850), that in every case of nerve transection, whether functionally regenerated or not, the entire distal nerve end undergoes degeneration, and, moreover, that the axonal degeneration is complete. It is to the persistence of Waller, that the fundamental fact of Wallerian degeneration is generally accepted within the subject of nerve injury.

The first attempts to repair peripheral nerves have also been performed at this time by Paget (1847). Paget reported primary suturing of a median nerve after laceration in an 11-year-old patient, with complete recovery of nerve function (Wilgis, 1982). In 1870, Philipeaux and Vulpian were the first to successfully apply an autograft of the hypoglossal nerve to bridge a defect in the lingual nerve of a dog (Philipeaux and Vulpian, 1870). A few years later Hueter developed an epineural suturing technique and achieved end-to-end coaptation of nerve stumps, which is, up to date, the standard method of nerve repair (Millesi, 1981). The first clinical experience of nerve grafting was described by Albert (1878), who used a nerve autograft from an amputated foot of one patient to bridge a defect in the median nerve as a consequence of a tumor resection surgery in another patient (Albert, 1878). A breakthrough in peripheral nerve surgery was achieved by Seddon who performed a large and well documented clinical series of autografts on multi traumatic peripheral nerve injuries at all levels (Seddon, 1954). As a consequence of the introduction of the surgical microscope better surgical treatment of the nerve was possible which was followed by continuous refinements of surgical techniques. In 1967, Bora showed that by using a surgical microscope groups of fascicle could be coapted by suturing the perineurium; the perineural suturing technique was born (Bora, 1967).

The role of collagen in peripheral nerve repair became apparent by the time that the methodology of fixation and histological staining improved. As a consequence, Holmes and Young (1942) were able to report that the endoneural connective tissue was rich in collagen, and that during Wallerian degeneration the endoneurial collagen content increases in the nerve stump of a severed peripheral nerve (Holmes and Young, 1942). Holmes and others found evidence that the dense collagen formation at the site of coaptation, impedes the entry of axonal sprouts into appropriate fascicles of the distal nerve stump (Abercrombie and Johnson, 1946, 1947; Holmes and Young, 1942; Sunderland, 1968).

II. Peripheral Nerve Collagens: Structure, Synthesis and Function

Collagens are abundantly present in the extracellular matrix (ECM) of peripheral nerves (Pleasure, 1984; Thomas and Olsson, 1984) and play an important part in the development of the peripheral nervous system as well as in the maintenance of normal peripheral nerve function during adulthood (Hubert *et al.*, 2009). Originally, the ECM, was considered a static mechanical structure that provided support, separation or filter functions in tissues. However, a broader function has been suggested since the identification and detailed study of interactions between the ECM proteins (e.g., laminin, collagens, nidogen, or entactin and proteoglycans) and cellular membrane receptors. These interactions can trigger intracellular signals and control cellular processes such as migration, proliferation, differentiation, and survival (Aszodi *et al.*, 2006; Yurchenco and Cheng, 1994). In the following section, we will focus on the structure, synthesis, and function of peripheral nerve collagens.

A. Collagen Structure and Types

To date there are 29 collagen types, numbered I to XXIX, identified and reported in the literature. Despite the rather high structural diversity among the different collagen types, all members of the collagen family have some common characteristics: (1) all collagens are transmembrane or extracellular molecules, (2) all collagens are formed by a right-handed triple helix composed of three α-chains (Piez, 1984). These can be formed by three identical chains (homotrimers) or by two or three different chains (heterotrimers). Thus, for a defined collagen type, various isoforms with distinct functions can exist, (3) the triple helical structure is determined by a glycine residue in every third position of the polypeptide chains resulting in a $(\text{Gly-X-Y})_n$ repeat structure which characterizes the "collagenous" domains of all collagens. The X and Y position could represent any amino-acid but X is often a proline and Y a 4-hydroxyproline. The latter is essential for the formation of intramolecular hydrogen bonds and contributes to the stability of the triple helical structure. Beside the triple helix which is the key component of all collagens there are also important noncollagenous domains flanking the central helical part (Fig. 1). The C-propeptide is thought to play an important role in the initiation of triple helix formation, whereas the N-propeptide is thought to be involved in the regulation of primary fibril diameters (Bateman *et al.*, 1996). The non-helical telopeptides of the processed collagen monomers are involved in the covalent cross-linking of the collagen molecules and in linking of the collagen molecules to other molecules in the surrounding matrix (Rossert and de Crombrugghe, 2002). Moreover, the collagen

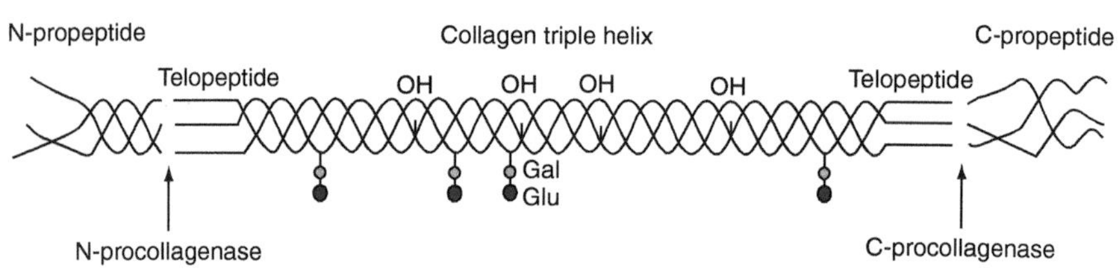

Fig. 1. Molecular structure of fibrillar collagens with the various subdomains, see text for details. Gal = Galactose, Glu = Glucose.

chains also bind glucosyl and galactosyl residues which mediate the interaction with proteoglycans another ECM protein (Fig. 1).

The architectural and functional role of the collagen family in connective tissue has been widely assessed, but although collagens are fundamental components of the ECM, they are rarely present in the mature nervous system. Only recently it was shown that some collagens are expressed by neurons (Hubert *et al.*, 2007; Seppanen *et al.*, 2006; Sund *et al.*, 2001). Predominantly, there are three locations in the mature nervous system where collagens are expressed (1) the connective tissues that surround the central (CNS) and peripheral (PNS) nervous system, (2) the basement membranes (BM) between the nervous system and other tissues (muscular, endothelial), and (3) the sensory end organs.

In the adult PNS collagen fibres are present in the three layers (epineurium, perineurium, and endoneurium) that ensheath the peripheral nerve tissue. In general, these layers consist of fibril forming collagens; type-I, type-III and type-V collagens. The type-I collagen is the most abundant collagen in the human body and the type-I triple helix is usually formed as a hetrotrimer by one $\alpha2$(I)-chain and two identical $\alpha1$(I)-chains. Type-III collagen is a homotrimer of three $\alpha1$(III)-chains and is, in vivo, mostly incorporated in to composite with collagen type-I. The type-V collagen is formed as heterotrimer of three different α-chains ($\alpha1$, $\alpha2$, $\alpha3$) (Chernousov *et al.*, 2000), and typically form heterofibrils with types I and III collagens (Fleischmajer *et al.*, 1990; Niyibizi and Eyre, 1989). Moreover, type-V collagen may function as a core structure of fibrils with type-I and III collagens polymerizing around this central axis (Bateman *et al.*, 1996).

Next to the difference in α-chains composition there are other differences between types I, III, and V collagens. After completion of trimer assembly the C-propeptide and N-propeptide are removed from the type-I and type-III collagen trimers by, respectively, C-procollagenase and N-procollagenase and are not part of the mature collagen molecules. In contrast, the N-propeptide of the type-V collagen trimer is retained in the mature collagen molecule and is suggested to mediate Schwann cell adhesion to type-V collagen and affect Schwann cell function (Erdman *et al.*, 2002). Moreover, types I and III collagens are solely present in small diameter collagen fibrils associated with the external face of the Schwann cell basal lamina (Osawa and Ide, 1986). Collagen type-V on the other hand, co-localizes with types I and III collagen but is also present in the basal lamina enveloping myelinating Schwann cells (Chernousov *et al.*, 2006).

Next to the fibril forming collagens there is a second class of collagens that are associated to the PNS; the basement membrane collagens. The BM collagens are typically type-IV collagens and are a member of the group of network forming collagens because they are the most important structural component of BMs integrating laminins, perlacan, nidogen, and other ECM proteins into a stable supramolecular aggregate (Hudson *et al.*, 1993). There are six type-IV collagen α-chains identified, $\alpha1(IV) - \alpha6(IV)$, associating into three different heterotrimeric molecules. The predominant form consists of two $\alpha1$(IV) chains and one $\alpha2$(IV) chain forming the essential network in most embryonic and adult BMs (Miner and Sanes, 1994). The two other heterotrimers are formed by two $\alpha5$(IV) chains and one $\alpha6$(IV) chain, or by three different chains $\alpha3$(IV) $\alpha4$(IV) $\alpha5$(IV). As in fibril forming collagens, trimer assembly is directed by the noncollagen C-propeptide domains but, in contrast to fibril forming collagens, the noncollagen N-propeptide domains of the type-IV collagen molecules are not removed after trimer assembly (Khoshnoodi *et al.*, 2006). The Collagen domain of type-IV collagen is interrupted by short noncollagen sequences which contributes to the flexibility of the collagen type-IV network. In the PNS, type-IV collagen containing BMs are surrounding Schwann cells and their associated axons and underlie epithelial cells.

B. Collagen Biosynthesis

A characteristic property of collagens is the formation of triple helices composed of three polypeptide chains. Fibril forming collagens consist of uninterrupted triple helices, but other collagens have one or more triple helical domain of various lengths. In peripheral nerve, collagens are expressed by fibroblasts and Schwann cells, and their biosynthesis is characterized by the presence of an extensive number of co- and posttranslational modifications of the polypeptides chains (Prockop and Kivirikko, 1995).

After transcription, mRNA is extensively processed and then translated in the rough endoplasmic reticulum. The first step in intracellular processing of the polypeptide chains is the cleavage of signal peptides by a signal peptidase. Collagen, like most proteins that are destined for transport to the extracellular spaces for their function or activity, is produced initially as a larger precursor molecule called procollagen (Bellamy and Bornstein, 1971). Procollagen contains extension proteins on each end called amino and carboxy procollagen propeptides (N-propeptide and C-propeptide, see Fig. 1). These nonhelical propeptides make it very soluble and therefore easy to move within the cell as it undergoes further post-translational modifications (Prockop and Kivirikko, 1995; Prockop *et al.*, 1979). One of the first modifications to take place is the very critical step of hydroxylation of proline and lysine residues in Y-position to 4-hydroxyproline and hydroxylysine by prolyl-4-hydroxylase and lysyl-hydroxylase. The hydroxylase enzymes require ascorbic acid, 2-oxoglutarate, molecular oxygen and ferrous iron as cofactors, see Fig. 2 (Gelse *et al.*, 2003; Mussini *et al.*, 1967). A few X-position proline residues are hydroxylated to 3-hydroxyproline. However, the function of 3-hydroxyproline is not known (Bateman *et al.*, 1996). The presence of 4-hydroxyproline is essential for intramolecular hydrogenbonds and thus contributes to the thermal stability of the triple helical domain, and therefore also to the integrity of the monomer and collagen fibril (Gelse *et al.*, 2003).

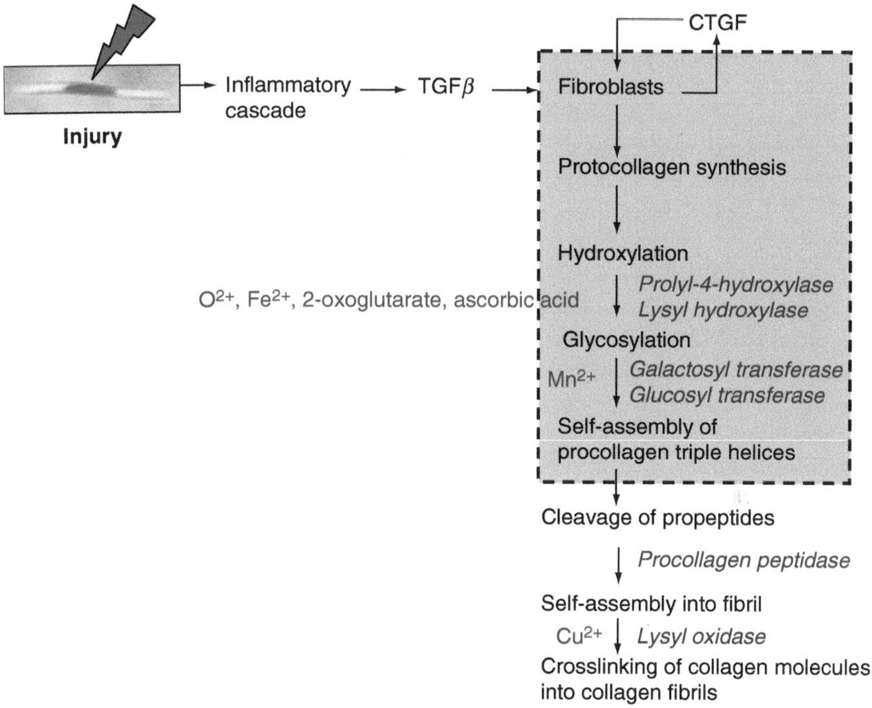

FIG. 2. General process of collagen biosynthesis in fibroblasts, see text for details.

The hydroxylysine residues are able to form stable intermolecular cross-linking of collagen molecules in fibrils and additionally represent sites for the attachment of carbohydrates. Galactose and/or glucose (Fig. 1) are added to some of the hydroxylysine by hydroxylysyl galactosyltransferase and galactosylhydroxylysyl glucosyltransferase (Anttinen *et al.*, 1978) and impart unique chemical and structural characteristics to the newly formed collagen molecule and may influence fibril size (Kivirikko and Myllyla, 1979). The enzymes that catalyze the glycosylation step require the trace metal manganese (see Fig. 2).

The C-propeptides have an essential function in the assembly of the three α-chains into trimeric collagen monomers. The globular structure of the C-propeptides is stabilized by intra-chain disulfide bonds and an N-linked carbohydrate group is added by the oligosaccharyl transferase complex. The formation to triple helices is preceded by the alignment of the C-terminal domains of three α–chains and initiates the formation of the triple helix progressing to the N-terminus. The efficient formation and folding of the procollagen chains depends on the presence of further enzymes like peptidyl-prolyl *cis-trans*-isomerase (PPI; (Lang *et al.*, 1987)). The processing and assembly of fibrillar and nonfibrillar collagens is principally the same, although many nonfibrillar collagens contain N- and/or C-terminal domains that are not removed and therefore not called propeptides. Additionally, different collagens may have special features in their synthesis, e.g. chain association and folding of type I and IV collagens may involve collagen-specific chaperones like heat shock protein 47 (HSP47; (Clarke, 1991)).

While inside the cell and when the procollagen peptides are intact, the molecule is about 1000 times more soluble than it is at a latter stage when the extension propeptides are removed (Prockop *et al.*, 1979). This high degree of solubility allows the procollagen molecule to be transported by microtubules to the cell surface where it is secreted into the extracellular space (Diegelmann and Peterkofsky, 1972). As the procollagen is secreted from the cell, it is acted upon by specialized enzymes called procollagen N- and C-proteinases or procollagenases (see Fig. 1) that remove both of the extension propeptides from the ends of the molecule (Lapiere *et al.*, 1971). Both proteins belong to a family of Zn^{2+}-dependent metalloproteinases (Prockop, 1998). Portions of these digested end pieces are thought to re-enter the cell and regulate the amount of collagen synthesis by a feed-back type of mechanism (Bateman *et al.*, 1996; Lichtenstein *et al.*, 1973; Schlumberger *et al.*, 1988; Wiestner *et al.*, 1979). The processed molecule is referred to as collagen and now begins to be involved in the important process of fiber formation. Collagen then spontaneously self-assembles into fibrils (see Fig. 2). Stabilization of the fibrils and the formation of fibers are provided by intra- and intermolecular covalent cross-links generated by conversion of some of the lysine and hydroxylysine residues to aldehyde derivates. This critical step is catalyzed by the copper-dependent enzyme lysyl oxidase (Bailey *et al.*, 1974; Kadler *et al.*, 1996; Prockop and Kivirikko, 1995) and gives the collagen fibers such tremendous strength (see Fig. 2).

The structure of type IV collagen genes is distinctly different from those of fibril forming collagens. The collagenous domain of type IV collagen is longer and however, is frequently interrupted with noncollagenous sequences (Prockop and Kivirikko, 1995). Type IV collagen molecules form their network with a different process. The N-terminal 7-S domains of four type IV collagens are covalently joined together (Risteli *et al.*, 1980), while the C-terminal noncollagenous globular domains (NC1) of two separate type IV collagen molecules joined together by disulfide bonds (Than *et al.*, 2002). Type IV collagens form a mesh-like structure outside the laminin layer and give stability to the basement membrane (Kuhn, 1995).

C. Collagen Function in Peripheral Nerve Development and Repair

During development of the PNS, bundles of axons innervating their motor or sensory targets become ensheathed by a "family" of immature Schwann cells that initiate the deposition of ECM components which will later ensemble into a surrounding Basal Lamina (BL) (Webster, 1971). This finding was verified by several other studies. Firstly, it was shown *in vitro* that immature Schwann cells are active producers of collagens (Bunge *et al.*, 1980). Secondly, fibril forming collagens in the endoneurium first appear at around E15 in the mouse sciatic nerve (Osawa and Ide, 1986). At that time there are no fibroblasts in the endoneurium so the collagen fibrils must be synthesized by immature Schwann cells that are associated to embryonic axons. Thirdly, in cell culture studies of Schwann cells it was shown that cultures incubated in medium lacking ascorbic acid, an essential co-factor for collagen posttranslational modification, fail to secrete stable collagen trimers that do not assemble ECM (Moya *et al.*, 1980).

The nerve fibres in the PNS are either myelinated or unmyelinated and thus ensheated by either myelinating or unmyelinating Schwann cells. The fate of an immature Schwann cell (e.g. myelin-forming or nonmyelin forming) is predominantly determined by the integration of internal (axon) and external signals (Bunge and Bunge, 1983). The neuregulins expressed by axons and the laminins present in the BL surrounding Schwann cells are key regulators of Schwann cell differentiation and myelination by their interaction with integrins and dystroglycan (reviewed in Bhatheja and Field, 2006; Court *et al.*, 2006; Jessen and Mirsky, 2005). Fibrilar collagen type-V is another BL molecule implicated in the process of myelination. Type-V collagen is often referred to as "minor" fibril forming, although it is relatively abundant in the PNS, where it is present in the BL of myelinated Schwann cells-axon units and in the surrounding ECM (Chernousov *et al.*, 2006; Melendez-Vasquez *et al.*, 2005). The collagen $\alpha3$(V) chain, which forms heterotrimers with $\alpha1$(V) chains, is strongly expressed by Schwann cells during development at the timepoint where myelination occurs (Chernousov *et al.*, 2001). Schwann cell adhesion activity of the type-V collagen is located predominantly in the noncollagenous N-propeptide domain of the collagen $\alpha3$(V) chain (Erdman *et al.*, 2002). Schwann cells have membrane anchored heparin sulphate (HS) proteoglycans on their surface which can mediate cell adhesion by binding to the N-propeptide domain of the collagen $\alpha3$(V) chain (Carey, 1997). There are two main HS-proteoglycans expressed on the cell surface of Schwann cells: syndecan-3 (Carey *et al.*, 1992) and glypican-1 (Carey *et al.*, 1993). The latter has been suggested as the main Schwann cell receptor for $\alpha3$(V) collagen. This was supported by the observation that suppression using small interfering RNA, of glypican-1 expression, but not syndecan-3 expression effectively diminished Schwann cell adhesion to $\alpha3$(V) collagen (Chernousov *et al.*, 2006). Whether collagen-glypican association is involved in BL assembly or triggers an intracellular signalling event remains to be elucidated.

Injury to the peripheral nerve often initiates nerve degeneration, the ensheating Schwann cells assist macrophages in the removal of the debris of both axons and myelin sheaths. However, the original BL that surrounded the previous axon-Schwann cell units, is not degraded and Schwann cells form cordons within these

BL remnants termed "bands of Bungner" (Griffin *et al.*, 1993). When regenerating axons re-enter the peripheral nerve matrix, they grow within the bands of Bungner. During this process of nerve regeneration there is long lasting BM gene expression of collagen types I, III, and IV in the proximal nerve stump which is markedly shorter in the distal nerve stump (Nath *et al.*, 1997; Seyer *et al.*, 1977; Siironen *et al.*, 1992a,b). Remarkably it was shown that axonal reinnervation did not affect gene expression of collagen types I and III after a transection injury (Siironen *et al.*, 1992a). Moreover, next to Schwann cells, endoneurial fibroblasts contribute to the production of collagen type-I. After peripheral nerve injury collagen type I and III are believed to provide mechanical support for axonal growth and regeneration. The gene expression of collagen type-IV on the contrary seemed to be enhanced if axonal reinnervation was allowed (Siironen *et al.*, 1992b). Siironen and colleagues also noticed a relatively high level of type-IV collagen gene expression in the uninjured control nerves which indicated that the maintenance of the Schwann cell BM requires high turnover of collagen type-IV.

III. Excessive Collagen Formation can Act as Mechanical Barrier After PNI

As already mentioned in the introduction, Holmes and Abercrombie were the first to observe excessive collagen formation or scar formation at the severed ends of the nerve after transection (Abercrombie and Johnson, 1947; Holmes and Young, 1942). During regeneration, different types of specific collagenases enable the axonal sprouts to penetrate the scar and proceed via the bands of Bungner until the end organ is reached (Holmes and Young, 1942; Lehman and Hayes, 1967). Nevertheless, nerve regeneration is often incomplete despite technically adequate surgical repair (Brown, 1972; Dellon and Mackinnon, 1988; Sunderland, 1968). It is postulated that scar formation at the coaptation site hinders the growth of sprouting axons into the appropriate distal fascicles and therefore delays and limits nerve regeneration (Abercrombie and Johnson, 1946, 1947; Holmes and Young, 1942; Sunderland, 1968).

In response to severe trauma to the nerve, fibroblasts are recruited to the site of damage and are induced to form collagen. In general, defined growth factors such as transforming growth factor beta (TGFβ) and connective tissue growth factor (CTGF) are believed to be important autocrine mediators of scar deposition in many tissues (Grotendorst, 1997; Pierce *et al.*, 1989; Schwab *et al.*, 2001). TGFβ is the principle growth factor responsible for fibroblast recruitment and collagen production, whereas CTGF is the downstream mediator. In the mammalian system, at least three and up to five isoforms of TGFβ regulate scar formation in virtually every organ system including the PNS (Kiefer *et al.*, 1993;

Roberts and Sporn, 1988). With respect to the different isoforms, TGFβ1 is predominantly expressed after PNI (Rufer *et al.*, 1994; Scherer *et al.*, 1993). In contrast, mRNA expression of TGFβ2 could not be detected after PNI (Rufer *et al.*, 1994; Scherer *et al.*, 1993) although some protein immunoreactivity has been noted. Moreover, TGFβ3 expression was lower than TGFβ1 expression (Rufer *et al.*, 1994) or even depressed according to Scherer and colleagues. Some controversial findings have been reported with respect to the duration of TGFβ1 expression after PNI. TGFβ1 mRNA expression was shown to be increased up to 7 days after injury (Kiefer *et al.*, 1993), whereas Rufer *et al.* (1994) observed by protein immunocytochemistry that the increase in TGFβ1 expression lasted up to 14 days post-transection and moved wave-like from the proximal part of the nerve to the very distal part of the nerve. Notably, TGFβ was shown to be primarily produced by Schwann cells (Rufer *et al.*, 1994; Scherer *et al.*, 1993). Macrophage derived TGFβ only comprised a minor part of the total TGFβ produced. Since PNI increases the proliferation rate of both cell types, the TGFβ-induced recruitment of fibroblasts is extremely high which subsequently results in excessive collagen formation.

Next to the pathological process of epineural scarring there are some disadvantages associated with the currently used surgical techniques in peripheral nerve repair that might increase collagen formation even more. The main disadvantage of the epineural and perineural suturing techniques is the presence of foreign material which leads to additional fibrosis or scarring. Likewise, tension at the site of coaptation has to be avoided because it predisposes to disproportionate collagen production and vascular compromise of the reconstructed structures. These mechanical effects of collagen scar deposition become more pronounced with time (Siironen *et al.*, 1995) and with poor surrounding tissue vascularity (Starkweather *et al.*, 1978).

IV. Inhibition of Collagen Synthesis Affects Peripheral Nerve Regeneration

In the previous section, we elaborately mentioned the collagen scar formation and its underlying mechanisms at the severed end of the nerve. In daily clinical practice, surgeons encounter limited functional recovery after peripheral nerve repair as a result of disproportionate scar formation at the coaptation site. In the past decades researchers attempted to control collagen accumulation in the formation of neuroma by various physical and chemical methods. The improvement of surgical techniques and the introduction of microsurgery reduced the collagen formation and improved the results of neurorrhaphy (Kleinert and Neale, 1974; Millesi, 1977; Seddon, 1963). Despite this, in most cases functional recovery after PNI is still incomplete and as a consequence various pharmacological agents have been developed to overcome this dilemma. One particular branch of these agents aims for the therapeutic collagen reduction but up to now these have yielded only limited functional success.

Several proline analogues like *cis*-hydroxyproline and azetidine-2-carboxylic acid have been shown to be effective inhibitors of collagen accumulation and synthesis *in vitro* and *in vivo* (Bora *et al.*, 1972; Lane *et al.*, 1971a,b, 1972; Pleasure *et al.*, 1974). Proline analogues are incorporated into polypeptides of procollagen and other proteins in place of proline. The analogues prevent formation of normal collagen triple helices, and the unstable, analogue-containing procollagen molecules are degraded within the cell (Uitto *et al.*, 1972). In a very nice study by Pleasure and co-workers it was shown that parenteral administration of *cis*-hydroxyproline for 18 days, beginning 4 days after sciatic nerve transection and reanastomosis, caused a 47% collagen reduction in the distal nerve stump in comparison to controls. Seventy days after surgery an accelerated remyelination of posterior tibial nerves in treated rats was noted and might be an indirect effect of enhanced penetration of axonal sprouts through the anastomotic scar (Pleasure *et al.*, 1974). Despite accelerated nerve regeneration in treated animals, no functional recovery was detectable by clinical examination after follow-up of 10 weeks (Pleasure *et al.*, 1974).

Neutralization of TGFβ by the use of antibodies is another approach to suppress injury induced scar formation after PNI. Nath and colleagues treated rats with a unilateral crush lesion of the sciatic nerve with either a TGFβ antibody or vehicle (Nath *et al.*, 1998). As a consequence of TGFβ antibody treatment the epineural fibroblast numbers were reduced, most likely while the chemotactic effects of TGFβ on fibroblasts decreased (Roberts *et al.*, 1986). Moreover, the overall procollagen signal in the injured tissue was reduced, although the individual fibroblast production of collagen type I was not affected (Nath *et al.*, 1998). Unfortunately, Nath and co-workers did not investigate the effects of TGFβ neutralization on axon regeneration and functional recovery. Davison and colleagues reported improved muscle function of the gastrocnemius-soleus muscle complex after sciatic nerve axotomy and TGFβ neutralization but they did not investigate if reduced scar formation was the mechanism underlying the observed functional improvement (Davison *et al.*, 1999).

Next to proline analogues and TGFβ antibodies there are various pharmacological agents, such as aprotinin, adcon-T/N, hyaluronic acid and citicoline that have been reported to reduce scar formation and facilitate nerve regeneration in experimental studies but have not yet been clinically investigated (Gorgulu *et al.*, 1998; Ozay *et al.*, 2007; Ozgenel, 2003; Petersen *et al.*, 1996). Despite this, we are convinced that a thoroughly designed pharmacological agent for the inhibition of scar formation makes surgical repair of injured nerves in daily clinical practice more effective.

References

Abercrombie, M., and Johnson, M. L. (1946). Collagen content of rabbit sciatic nerve during Wallerian degeneration. *J. Neurol. Neurosurg. Psychiatr.* **9,** 113–118.

Abercrombie, M., and Johnson, M. L. (1947). The effect of reinnervation on collagen formation in degenerating sciatic nerves of rabbits. *J. Neurol. Neurosurg. Psychiatr.* **10,** 89–92.

Albert, E. (1878). Berichte des natuurwissenschaftlich – medizinischen Vereines in Innsbruck. Jahrgang IX, pp. 97–98. Wagner'schen Universitäts-Buchhandlung, Innsbruck.

Anttinen, H., Myllyla, R., and Kivirikko, K. I. (1978). Further characterization of galactosylhydroxylysyl glucosyltransferase from chick embryos. Amino acid composition and acceptor specificity. *Biochem. J.* **175,** 737–742.

Aszodi, A., Legate, K. R., Nakchbandi, I., and Fassler, R. (2006). What mouse mutants teach us about extracellular matrix function. *Annu. Rev. Cell Dev. Biol.* **22,** 591–621.

Bailey, A. J., Robins, S. P., and Balian, G. (1974). Biological significance of the intermolecular crosslinks of collagen. *Nature* **251,** 105–109.

Bateman, J. F., Lamande, S. R., and Ramshaw, J. A. M. (1996). Collagen superfamily. *In* "Extracellular Matrix" (W. D. Comper, Ed.), pp. 22–67. Harwood Academic Press, Melbourne.

Bellamy, G., and Bornstein, P. (1971). Evidence for procollagen, a biosynthetic precursor of collagen. *Proc. Natl. Acad. Sci. USA* **68,** 1138–1142.

Bhatheja, K., and Field, J. (2006). Schwann cells: Origins and role in axonal maintenance and regeneration. *Int. J. Biochem. Cell Biol.* **38,** 1995–1999.

Bora, F. W., Jr. (1967). Peripheral nerve repair in cats. The fascicular stitch. *J. Bone Joint Surg. Am.* **49,** 659–666.

Bora, F. W., Jr., Lane, J. M., and Prockop, D. J. (1972). Inhibitors of collagen biosynthesis as a means of controlling scar formation in tendon injury. *J. Bone Joint Surg. Am.* **54,** 1501–1508.

Brown, P. W. (1972). Factors influencing the success of the surgical repair of peripheral nerves. *Surg. Clin. North Am.* **52,** 1137–1155.

Bunge, M. B., Williams, A. K., Wood, P. M., Uitto, J., and Jeffrey, J. J. (1980). Comparison of nerve cell and nerve cell plus Schwann cell cultures, with particular emphasis on basal lamina and collagen formation. *J. Cell Biol.* **84,** 184–202.

Bunge, R. P., and Bunge, M. B. (1983). Interrelationship between Schwann cell function and extracellular matrix production. *Trends Neurosci.* **6,** 499–505.

Carey, D. J. (1997). Syndecans: Multifunctional cell-surface co-receptors. *Biochem. J.* **327**(Pt. 1), 1–16.

Carey, D. J., Evans, D. M., Stahl, R. C., Asundi, V. K., Conner, K. J., Garbes, P., and Cizmeci-Smith, G. (1992). Molecular cloning and characterization of *N*-syndecan, a novel transmembrane heparan sulfate proteoglycan. *J. Cell Biol.* **117,** 191–201.

Carey, D. J., Stahl, R. C., Asundi, V. K., and Tucker, B. (1993). Processing and subcellular distribution of the Schwann cell lipid-anchored heparan sulfate proteoglycan and identification as glypican. *Exp. Cell Res.* **208,** 10–18.

Chernousov, M. A., Rothblum, K., Stahl, R. C., Evans, A., Prentiss, L., and Carey, D. J. (2006). Glypican-1 and alpha4(V) collagen are required for Schwann cell myelination. *J. Neurosci.* **26,** 508–517.

Chernousov, M. A., Rothblum, K., Tyler, W. A., Stahl, R. C., and Carey, D. J. (2000). Schwann cells synthesize type V collagen that contains a novel alpha 4 chain. Molecular cloning, biochemical characterization, and high affinity heparin binding of alpha 4(V) collagen. *J. Biol. Chem.* **275,** 28208–28215.

Chernousov, M. A., Stahl, R. C., and Carey, D. J. (2001). Schwann cell type V collagen inhibits axonal outgrowth and promotes Schwann cell migration via distinct adhesive activities of the collagen and noncollagen domains. *J. Neurosci.* **21,** 6125–6135.

Clarke, K. A. (1991). Swing time changes contribute to stride time adjustment in the walking rat. *Physiol. Behav.* **50,** 1261–1262.

Court, F. A., Wrabetz, L., and Feltri, M. L. (2006). Basal lamina: Schwann cells wrap to the rhythm of space-time. *Curr. Opin. Neurobiol.* **16,** 501–507.

Cruikshank, W. (1795). Experiments on the nerves, particularly on their reproduction; and on the spinal marrow of living animals. *Philos. Trans. R. Soc. London* **XVII.**

Davison, S. P., McCaffrey, T. V., Porter, M. N., and Manders, E. (1999). Improved nerve regeneration with neutralization of transforming growth factor-beta1. *Laryngoscope* **109,** 631–635.

Dellon, A. L., and Mackinnon, S. E. (1988). Basic scientific and clinical applications of peripheral nerve regeneration. *Surg. Annu.* **20,** 59–100.

Diegelmann, R. F., and Peterkofsky, B. (1972). Inhibition of collagen secretion from bone and cultured fibroblasts by microtubular disruptive drugs. *Proc. Natl. Acad. Sci. USA* **69,** 892–896.

Erdman, R., Stahl, R. C., Rothblum, K., Chernousov, M. A., and Carey, D. J. (2002). Schwann cell adhesion to a novel heparan sulfate binding site in the N-terminal domain of alpha 4 type V collagen is mediated by syndecan-3. *J. Biol. Chem.* **277,** 7619–7625.

Fleischmajer, R., Perlish, J. S., Burgeson, R. E., Shaikh-Bahai, F., and Timpl, R. (1990). Type I and type III collagen interactions during fibrillogenesis. *Ann. N. Y. Acad. Sci.* **580,** 161–175.

Gelse, K., Poschl, E., and Aigner, T. (2003). Collagens—Structure, function, and biosynthesis. *Adv. Drug Deliv. Rev.* **55,** 1531–1546.

Gorgulu, A., Imer, M., Simsek, O., Sencer, A., Kutlu, K., and Cobanoglu, S. (1998). The effect of aprotinin on extraneural scarring in peripheral nerve surgery: An experimental study. *Acta Neurochir. (Wien.)* **140,** 1303–1307.

Griffin, J. W., Kidd, G., and Trapp, B. D. (1993). Interactions between axons and Schwann cells. *In* "Peripheral Neuropathy" (P. J. Dyck, P. K. Thomas, J. W. Griffin, P. A. Low, and J. F. Poduslo, Eds.), pp. 317–330. Saunders, Philadelphia.

Grotendorst, G. R. (1997). Connective tissue growth factor: A mediator of TGF-beta action on fibroblasts. *Cytokine Growth Factor Rev.* **8,** 171–179.

Holmes, W., and Young, J. Z. (1942). Nerve regeneration after immediate and delayed suture. *J. Anat.* **77,** 63–96.

Howell, W. H., and Huber, G. C. (1892). A physiological, histological and clinical study of the degeneration and regeneration in peripheral nerve fibres after severance of their connections with the nerve centres. *J. Physiol.* **13,** 335–406.

Hubert, T., Grimal, S., Carroll, P., and Fichard-Carroll, A. (2009). Collagens in the developing and diseased nervous system. *Cell Mol. Life Sci.* **66,** 1223–1238.

Hubert, T., Grimal, S., Ratzinger, S., Mechaly, I., Grassel, S., and Fichard-Carroll, A. (2007). Collagen XVI is a neural component of the developing and regenerating dorsal root ganglia extracellular matrix. *Matrix Biol.* **26,** 206–210.

Hudson, B. G., Reeders, S. T., and Tryggvason, K. (1993). Type IV collagen: Structure, gene organization, and role in human diseases. Molecular basis of Goodpasture and Alport syndromes and diffuse leiomyomatosis. *J. Biol. Chem.* **268,** 26033–26036.

Jessen, K. R., and Mirsky, R. (2005). The origin and development of glial cells in peripheral nerves. *Nat. Rev. Neurosci.* **6,** 671–682.

Kadler, K. E., Holmes, D. F., Trotter, J. A., and Chapman, J. A. (1996). Collagen fibril formation. *Biochem. J.* **316**(Pt. 1), 1–11.

Khoshnoodi, J., Cartailler, J. P., Alvares, K., Veis, A., and Hudson, B. G. (2006). Molecular recognition in the assembly of collagens: Terminal noncollagenous domains are key recognition modules in the formation of triple helical protomers. *J. Biol. Chem.* **281,** 38117–38121.

Kiefer, R., Lindholm, D., and Kreutzberg, G. W. (1993). Interleukin-6 and transforming growth factor-beta 1 mRNAs are induced in rat facial nucleus following motoneuron axotomy. *Eur. J. Neurosci.* **5,** 775–781.

Kivirikko, K. I., and Myllyla, R. (1979). Collagen glycosyltransferases. *Int. Rev. Connect. Tissue Res.* **8,** 23–72.

Kleinert, H. E., and Neale, H. W. (1974). Microsurgery in hand surgery. *Clin. Orthop. Relat. Res.* 158–161.

Kuhn, K. (1995). Basement membrane (type IV) collagen. *Matrix Biol.* **14,** 439–445.

Lane, J. M., Bora, F. W., Jr., Prockop, D. J., Heppenstall, R. B., and Black, J. (1972). Inhibition of scar formation by the proline analog *cis*-hydroxyproline. *J. Surg. Res.* **13,** 135–137.

Lane, J. M., Dehm, P., and Prockop, D. J. (1971a). Effect of the proline analogue azetidine-2-carboxylic acid on collagen synthesis *in vivo*. I. Arrest of collagen accumulation in growing chick embryos. *Biochim. Biophys. Acta* **236,** 517–527.

Lane, J. M., Parkes, L. J., and Prockop, D. J. (1971b). Effect of the proline analogue azetidine-2-carboxylic acid on collagen synthesis *in vivo*. II. Morphological and physical properties of collagen containing the analogue. *Biochim. Biophys. Acta* **236,** 528–541.

Lang, K., Schmid, F. X., and Fischer, G. (1987). Catalysis of protein folding by prolyl isomerase. *Nature* **329,** 268–270.

Lapiere, C. M., Lenaers, A., and Kohn, L. D. (1971). Procollagen peptidase: An enzyme excising the coordination peptides of procollagen. *Proc. Natl. Acad. Sci. USA* **68,** 3054–3058.

Lehman, R. A., and Hayes, G. J. (1967). Degeneration and regeneration in peripheral nerve. *Brain* **90,** 285–296.

Lichtenstein, J. R., Martin, G. R., Kohn, L. D., Byers, P. H., and McKusick, V. A. (1973). Defect in conversion of procollagen to collagen in a form of Ehlers–Danlos syndrome. *Science* **182,** 298–300.

Melendez-Vasquez, C., Carey, D. J., Zanazzi, G., Reizes, O., Maurel, P., and Salzer, J. L. (2005). Differential expression of proteoglycans at central and peripheral nodes of Ranvier. *Glia* **52,** 301–308.

Millesi, H. (1977). Healing of nerves. *Clin. Plast. Surg.* **4,** 459–473.

Millesi, H. (1981). Reappraisal of nerve repair. *Surg. Clin. North Am.* **61,** 321–340.

Miner, J. H., and Sanes, J. R. (1994). Collagen IV alpha 3, alpha 4, and alpha 5 chains in rodent basal laminae: Sequence, distribution, association with laminins, and developmental switches. *J. Cell Biol.* **127,** 879–891.

Moya, F., Bunge, M. B., and Bunge, R. P. (1980). Schwann cells proliferate but fail to differentiate in defined medium. *Proc. Natl. Acad. Sci. USA* **77,** 6902–6906.

Mussini, E., Hutton, J. J. Jr., and Udenfriend, S. (1967). Collagen proline hydroxylase in wound healing, granuloma formation, scurvy, and growth. *Science* **157,** 927–929.

Nath, R. K., Kwon, B., Mackinnon, S. E., Jensen, J. N., Reznik, S., and Boutros, S. (1998). Antibody to transforming growth factor beta reduces collagen production in injured peripheral nerve. *Plast. Reconstr. Surg.* **102,** 1100–1106. (Discussion, pp. 1107–1108).

Nath, R. K., Mackinnon, S. E., Jensen, J. N., and Parks, W. C. (1997). Spatial pattern of type I collagen expression in injured peripheral nerve. *J. Neurosurg.* **86,** 866–870.

Niyibizi, C., and Eyre, D. R. (1989). Bone type V collagen: Chain composition and location of a trypsin cleavage site. *Connect. Tissue Res.* **20,** 247–250.

Osawa, T., and Ide, C. (1986). Changes in thickness of collagen fibrils in the endo- and epineurium of the mouse sciatic nerve during development. *Acta. Anat. (Basel)* **125,** 245–251.

Ozay, R., Bekar, A., Kocaeli, H., Karli, N., Filiz, G., and Ulus, I. H. (2007). Citicoline improves functional recovery, promotes nerve regeneration, and reduces postoperative scarring after peripheral nerve surgery in rats. *Surg. Neurol.* **68,** 615–622. (Discussion, p. 622).

Ozgenel, G. Y. (2003). Effects of hyaluronic acid on peripheral nerve scarring and regeneration in rats. *Microsurgery* **23,** 575–581.

Petersen, J., Russell, L., Andrus, K., MacKinnon, M., Silver, J., and Kliot, M. (1996). Reduction of extraneural scarring by ADCON-T/N after surgical intervention. *Neurosurgery* **38,** 976–983. (Discussion, pp. 983–974).

Philipeaux, J. M., and Vulpian, A. (1870). Note sur des essais de greffe d'u troncon de nerf lingualentre les deux bouts du nerf hypoglosse. Apres excision d'un segment du dernier nerf. *Arch. Phys. Norm. Pathol.* **3,** 618.

Pierce, G. F., Mustoe, T. A., Lingelbach, J., Masakowski, V. R., Griffin, G. L., Senior, R. M., and Deuel, T. F. (1989). Platelet-derived growth factor and transforming growth factor-beta enhance tissue repair activities by unique mechanisms. *J. Cell Biol.* **109,** 429–440.

Piez, K. A. (1984). Molecular and aggregate structure of the collagens. *In* "Extracellular Matrix Biology" (K. A. Piez and H. Reddi, Eds.), pp. 1–39. Elsevier, Amsterdam.

Pleasure, D. (1984). The structural proteins of peripheral nerve. *In* "Peripheral Neuropathy" (P. J. Dyck, P. K. Thomas, E. H. Lambert, and R. P. Bunge, Eds.), vol. 1, pp. 441–452. Saunders, Philadelphia/London/Toronto.

Pleasure, D., Bora, F. W., Jr., Lane, J., and Prockop, D. (1974). Regeneration after nerve transection: Effect of inhibition of collagen synthesis. *Exp. Neurol.* **45,** 72–78.

Prockop, D. J. (1998). What holds us together? Why do some of us fall apart? What can we do about it? *Matrix Biol.* **16,** 519–528.

Prockop, D. J., and Kivirikko, K. I. (1995). Collagens: Molecular biology, diseases, and potentials for therapy. *Annu. Rev. Biochem.* **64,** 403–434.

Prockop, D. J., Kivirikko, K. I., Tuderman, L., and Guzman, N. A. (1979). The biosynthesis of collagen and its disorders (first of two parts). *N. Engl. J. Med.* **301,** 13–23.

Risteli, J., Bachinger, H. P., Engel, J., Furthmayr, H., and Timpl, R. (1980). 7-S collagen: Characterization of an unusual basement membrane structure. *Eur. J. Biochem.* **108,** 239–250.

Roberts, A. B., and Sporn, M. B. (1988). Transforming growth factor beta. *Adv. Cancer Res.* **51,** 107–145.

Roberts, A. B., Sporn, M. B., Assoian, R. K., Smith, J. M., Roche, N. S., Wakefield, L. M., Heine, U. I., Liotta, L. A., Falanga, V., Kehrl, J. H., and Fauci, A. S. (1986). Transforming growth factor type beta: Rapid induction of fibrosis and angiogenesis *in vivo* and stimulation of collagen formation *in vitro*. *Proc. Natl. Acad. Sci. USA* **83,** 4167–4171.

Rossert, J., and de Crombrugghe, B. (2002). Type I collagen: Structure, synthesis and regulation. *In* "Principles in Bone Biology" (J. P. Bilezkian, L. G. Raisz, and G. A. Rodan, Eds.), pp. 189–210. Academic Press, Orlando.

Rufer, M., Flanders, K., and Unsicker, K. (1994). Presence and regulation of transforming growth factor beta mRNA and protein in the normal and lesioned rat sciatic nerve. *J. Neurosci. Res.* **39,** 412–423.

Scherer, S. S., Kamholz, J., and Jakowlew, S. B. (1993). Axons modulate the expression of transforming growth factor-betas in Schwann cells. *Glia* **8,** 265–276.

Schlumberger, W., Thie, M., Volmer, H., Rauterberg, J., and Robenek, H. (1988). Binding and uptake of Col 1(I), a peptide capable of inhibiting collagen synthesis in fibroblasts. *Eur. J. Cell Biol.* **46,** 244–252.

Schwab, J. M., Beschorner, R., Nguyen, T. D., Meyermann, R., and Schluesener, H. J. (2001). Differential cellular accumulation of connective tissue growth factor defines a subset of reactive astrocytes, invading fibroblasts, and endothelial cells following central nervous system injury in rats and humans. *J. Neurotrauma* **18,** 377–388.

Seddon, H. J. (1954). Peripheral nerve injuries. *Spec. Rep. Ser. Med. Res. Counc. (G. B)* **282.**

Seddon, H. J. (1963). Nerve grafting. *J. Bone Joint Surg. Br.* **45,** 447–461.

Seppanen, A., Autio-Harmainen, H., Alafuzoff, I., Sarkioja, T., Veijola, J., Hurskainen, T., Bruckner-Tuderman, L., Tasanen, K., and Majamaa, K. (2006). Collagen XVII is expressed in human CNS neurons. *Matrix Biol.* **25,** 185–188.

Seyer, J. M., Kang, A. H., and Whitaker, J. N. (1977). The characterization of type I and type III collagens from human peripheral nerve. *Biochim. Biophys. Acta* **492,** 415–425.

Siironen, J., Sandberg, M., Vuorinen, V., and Roytta, M. (1992a). Expression of type I and III collagens and fibronectin after transection of rat sciatic nerve. Reinnervation compared with denervation. *Lab. Invest.* **67,** 80–87.

Siironen, J., Sandberg, M., Vuorinen, V., and Roytta, M. (1992b). Laminin B1 and collagen type IV gene expression in transected peripheral nerve: Reinnervation compared to denervation. *J. Neurochem.* **59,** 2184–2192.

Siironen, J., Vuorinen, V., Taskinen, H. S., and Roytta, M. (1995). Axonal regeneration into chronically denervated distal stump. 2. Active expression of type I collagen mRNA in epineurium. *Acta Neuropathol.* **89,** 219–226.

Starkweather, R. J., Neviaser, R. J., Adams, J. P., and Parsons, D. B. (1978). The effect of devascularization on the regeneration of lacerated peripheral nerves: An experimental study. *J. Hand Surg. [Am]* **3,** 163–167.

Sund, M., Vaisanen, T., Kaukinen, S., Ilves, M., Tu, H., Autio-Harmainen, H., Rauvala, H., and Pihlajaniemi, T. (2001). Distinct expression of type XIII collagen in neuronal structures and other tissues during mouse development. *Matrix Biol.* **20,** 215–231.

Sunderland, S. (1968). "Nerves and Nerve Injuries." Williams & Wlkins, Baltimore.

Than, M. E., Henrich, S., Huber, R., Ries, A., Mann, K., Kuhn, K., Timpl, R., Bourenkov, G. P., Bartunik, H. D., and Bode, W. (2002). The 1.9-Å crystal structure of the noncollagenous (NC1) domain of human placenta collagen IV shows stabilization via a novel type of covalent Met-Lys cross-link. *Proc. Natl. Acad. Sci. USA* **99,** 6607–6612.

Thomas, P. K., and Olsson, Y. (1984). Microscopic anatomy and function of the connective tissue components of peripheral nerve. *In* "Peripheral Neuropathy" (P. J. Dyck, P. K. Thomas, E. H. Lambert, and R. P. Bunge, Eds.), vol. 1, pp. 97–120. Saunders, Philadelphia/London/Toronto.

Uitto, J., Dehm, P., and Prockop, D. J. (1972). Incorporation of *cis*-hydroxyproline into collagen by tendon cells. Failure of the intracellular collagen to assume a triple-helical conformation. *Biochim. Biophys. Acta* **278,** 601–605.

Waller, A. (1850). Experiments on the section of the glossopharyngeal and hypoglossal nerves of the frog, and observations of the alterations produced thereby in the structure of their primitive fibres. *Philos. Trans. R. Soc. London* **140,** 423–429.

Webster, H. D. (1971). The geometry of peripheral myelin sheaths during their formation and growth in rat sciatic nerves. *J. Cell Biol.* **48,** 348–367.

Wiestner, M., Krieg, T., Horlein, D., Glanville, R. W., Fietzek, P., and Muller, P. K. (1979). Inhibiting effect of procollagen peptides on collagen biosynthesis in fibroblast cultures. *J. Biol. Chem.* **254,** 7016–7023.

Wilgis, E. F. S. (1982). Nerve repair and grafting. *In* "Operative Hand Surgery" (D. P. Green, Ed.), Churchill Livingstone, New York.

Yurchenco, P. D., and Cheng, Y. S. (1994). Laminin self-assembly: A three-arm interaction hypothesis for the formation of a network in basement membranes. *Contrib. Nephrol.* **107,** 47–56.

GENE THERAPY PERSPECTIVES FOR NERVE REPAIR

Serena Zacchigna and Mauro Giacca

Molecular Medicine Laboratory, International Centre for Genetic Engineering and Biotechnology (ICGEB), Trieste 34149, Italy

Despite advances in microsurgical techniques and a good rate of structural nerve healing, functional recovery often remains suboptimal, and thus innovative strategies able to provide extra neurotrophic support to the proper re-establishment of functional circuits are warranted. In this review, we will discuss the potential of gene therapy in fostering neuroprotection and neuroregeneration. We will then address a few key emerging concepts in the field, which might help in gaining a better understanding of the cellular and molecular events involved in axonal regeneration, and eventually to the definition of novel targets for intervention. The translation of these new concepts into effective therapies will represent an outstanding challenge for regenerative medicine over the next decades.

I. Introduction

Since peripheral axons have the inherent capacity to regenerate, in theory, complete structural and functional nerve recovery following injury appears possible. However, our poor knowledge of the molecular and cellular mechanisms involved in nerve repair still limits the therapeutic application of this intrinsic property, and the main treatment options currently available aim simply at surgically reconnecting severed nerve ends (Abrams and Widenfalk, 2005). The original idea of a nerve suture belongs to Paul von Aegina, active in the

DOI: 10.1016/S0074-7742(09)87020-7

0074-7742/09 $35.00

seventh century B.C. (Streppel *et al.*, 2000). Subsequently, a great deal of information regarding the evaluation and treatment of traumatic nerve injuries came with the experience of treating war-time injuries. Surprisingly, the theoretical speculations on good functional recovery, which was noted at this time, remain up-to-date. At present, neurosurgical repair typically consists of either direct, end-to-end anastomosis or the insertion of an autologous nerve graft to refill tissue loss (Schmidt and Leach, 2003). Despite advances in microsurgical techniques and a good rate of nerve healing, these therapies often remain suboptimal, providing a poor functional outcome and lead to the development of neuropathic pain (Abrams and Widenfalk, 2005). Therefore, various experimental strategies have been developed aimed at providing extra neurotrophic support for the proper re-establishment of functional circuits (Haastert and Grothe, 2007). In this review, we will first summarize the existing gene-therapy strategies focused on neuroprotection, which may help to optimize axonal regrowth and nerve repair. Second, we will discuss some novel emerging concepts in the field derived from a better understanding of the cellular and molecular events involved in axonal degeneration and regeneration, which have led to the definition of novel targets for intervention. The translation of these new concepts into effective therapies will represent an outstanding challenge for regenerative medicine over the next decades.

II. Gene Transfer Technologies to Target the Peripheral Nervous System

A. Where We Stand Now

To date, gene therapy for peripheral nerve reconstruction has essentially aimed at delivering neurotrophic factors to the healing nerve ends, thus favoring survival and regeneration of both sensory and motor axons and, ultimately, recovery of nervous functions (Haastert and Grothe, 2007). Although it is widely recognized that the recovery of functional nerve circuits is appreciably improved by exogenous neurotrophic support, the delivery of recombinant growth factors to the peripheral nervous system (PNS) has been a major challenge so far (Makwana and Raivich, 2005). The short half-life of most neurotrophic factors would require either multiple administrations or a continuous infusion of the therapeutic molecules in order to achieve an adequate and effective local concentration, thus significantly limiting the acceptance and compliance by patients (Haastert and Grothe, 2007). In contrast, the delivery of the cDNA sequence coding for a neurotrophic gene of interest using available gene transfer technologies permits a sustained expression of the therapeutic factor event after a single administration. Additionally, expression of the gene might be modulated by the introduction of regulatory elements for the controlled or tissue-specific expression of the desired molecule.

Therapeutic factors delivered by gene therapy might improve different steps in the whole process of peripheral nerve regeneration, from sensory and motor neuron survival, to axonal outgrowth, guidance of the regenerating axons along the Schwann cells scaffolds (known as "bands of Büngner"), re-establishment of functional connections between nerve ends, and remyelination (Makwana and Raivich, 2005).

A few approaches have used plasmids as vehicles to deliver therapeutic genes to peripheral nerves. In these cases, the skeletal muscle has been the preferred site for delivery and expression of the transgene. For instance, intramuscular delivery of a plasmid encoding for vascular endothelial growth factor (VEGF) was shown to protect against myelin wasting and axonal loss in different models of ischemic, diabetic, and toxic neuropathies (Kirchmair *et al.*, 2007; Murakami *et al.*, 2006; Schratzberger *et al.*, 2000). These promising results have led to an open-labeled, dose-escalating clinical trial in human patients with chronic ischemic neuropathy, in which both symptoms and functional indexes resulted significantly improved by the administration of a high dose (3–9 mg) of a plasmid encoding human VEGF (Simovic *et al.*, 2001). Similarly, the use of a plasmid encoding a zinc-finger DNA-binding protein, intentionally designed to upregulate VEGF expression, has resulted in a clear improvement of pain and neurological symptoms in 50% of the patients with diabetic neuropathy (www.Sangamo.com), and an additional phase I/II trial with a VEGF-expressing plasmid is currently ongoing for the same disease (www.wiley.co.uk/genetherapy/clinical; trial ID: US-467).

Until now, three major classes of viral vectors, based on Adenovirus, Herpes Simplex Virus (HSV), and Adeno-Associated Virus (AAV), have been exploited to target the PNS. Although not originally neurotropic, adenoviral and AAV vectors are able to transduce spinal sensory and motor neurons after either intramuscular or intraneural injections. When injected at the site of a nerve injury, these vectors are retrogradely transported to motor neuron cell bodies, and can thus be exploited to deliver therapeutic genes along the route of the nerve. A variety of cytokines, such as GDNF (glial-derived neurotropic factor), BDNF (brain-derived neurotrophic factor), CNTF (ciliary neurotrophic factor), CT-1 (cardiotrophin-1), IGF-1 (insulin-like growth factor-1), HGF (hepatocyte growth factor), and TGFβ2 (transforming growth factor-beta 2), delivered directly after an acute nerve lesion through peripheral gene transfer, have been shown to increase neuronal survival, axonal diameter, myelination, and functional recovery in various animal models of nerve injury (Araki *et al.*, 2006; Li *et al.*, 2008; Sakamoto *et al.*, 2003). Consistently, a similar retrograde transport of neurotrophic factors has been successful in improving the outcome in a mouse model of amyotrophic lateral sclerosis (ALS) (Kaspar *et al.*, 2003). However, the strong absorption of both adenoviral and AAV vectors to skeletal muscle fibers might represent a limitation for efficient neuronal transduction and retrograde transport. Thus, novel strategies are being developed in order to enhance viral neurotropism upon intramuscular gene transfer. Muscle injection with botulinum neurotoxin significantly enhanced retrograde transfer of adenoviral vectors, likely as a consequence of toxin-induced nerve sprouting at the end plates (Millecamps *et al.*, 2002). Alternatively, enhanced neurotropism and axonal uptake could be achieved through the manipulation of the vector capsid coat. The insertion of neurotrophic peptides into the AAV-2 capsid successfully increased the neurotropism of the recombinant vector (Federici *et al.*, 2007; Xu *et al.*, 2005). In this context, however, it is worth remarking that any modification of the AAV capsid, while conferring targeting specificity, usually determines a marked loss of virion infectivity (Buning *et al.*, 2003).

HSV vectors have been recently recognized as truly promising tools to target nerves by peripheral inoculation. HSV-1 is a double-stranded DNA virus able to establish life-long latency in neurons. As part of its natural life cycle, this virus has evolved to spread from the cutaneous or mucosal cells to nerve terminals innervating the infection site, and finally to the respective neuronal cell bodies., thus resulting particularly efficient at transducing dorsal root ganglia (DRG) neurons, following injection into a peripheral nerve. HSV is a large virus with a genome of 152 kb and much of the genetic material in this genome can be replaced by transgenes without significantly impairing essential viral functions, thus allowing the insertion of up to 30 kb of exogenous genetic material, for instance, large or multiple transgenes (Palmer *et al.*, 2000). Once transported to the cell body, HSV-based vectors do not integrate into the host cell genome, but persist as nuclear extrachromosomal entities, therefore avoiding insertional activation or inactivation of host cell genes and ensuring long-term transgene expression. The subcutaneous inoculation of an HSV vector expressing nerve growth factor (NGF) preserved nerve function in streptozotocin diabetic mice (Goss *et al.*, 2002) as well as in rodent models of toxic neuropathy (Chattopadhyay *et al.*, 2003, 2004, 2005). Based on their specific neurotropism, HSV vectors are beginning to be largely exploited to promote nerve repair after injury. The HSV-mediated expression of the antiapoptotic protein BCL2 or the neurotrophic factor GDNF, upon direct injection into the ventral horn of the spinal cord, was shown to preserve motor neuron viability and motor function following ventral root avulsion (Natsume *et al.*, 2003).

B. Design and Optimization of Novel Vectors to Target the Peripheral Nervous System

1. *Design of Expression Cassettes with Optimized Promoter Choice and Regulatory Elements*

Targeting gene expression to a specific cell population by use of cell-selective promoters is an attractive approach, limiting undesirable side effects, such as activation of the host-cell defence machinery or promoter inactivation, as compared to viral ubiquitous promoters (Liu *et al.*, 2004). A general limitation of the applicability of cellular promoters to drive the expression of exogenous genes is their relatively weak transcriptional activity. The transcriptional amplification strategy (TAS) has been exploited to enhance the activity of such promoters without loss of cell type specificity. The original idea of TAS entailed the use of a cell-specific promoter to drive the simultaneous expression of the desired transgene and a strong artificial transcriptional activator to potentiate transcription by binding to a specific binding site artificially introduced into the promoter. To conform to the space limitations often imposed by conventional viral gene expression vectors, a novel, more compact TAS-amplified cell-specific promoter was incorporated into a lentiviral vector, and was shown to be effective in driving gene expression in both neuronal and glial cells (Liu *et al.*, 2008), the two important cell targets of the PNS.

2. *Problems and Promises in the Use of Herpes Simplex Virus Vectors*

Despite the striking potential of HSV-based vectors to target the PNS, several *in vivo* experiments in rodents have clearly shown that even strong promoters, such as the CMV IE gene promoter, only sustain transgene expression for a few weeks. The same mechanisms repressing wild-type (wt) HSV gene expression during latency is likely to be involved in the shutoff of the transgene carried by the vector. This problem of promoter suppression could be overcome by exploiting the unique region of the HSV-1 genome that remains transcriptionally active during HSV-1 latency. This region is named as the latency-associated transcript (LAT) and produces a messenger RNA species during latency utilizing two promoter regions (LAT P1 and LAT P2) (Perez *et al.*, 2004). These promoter regions have been recognized as transposable elements and can confer long-term activity onto neighboring exogenous promoters, allowing prolonged transgene expression in both the central nervous system (CNS) and the PNS (Chattopadhyay *et al.*, 2005; Goins *et al.*, 1994).

A recent, promising application of HSV vectors aims at delivering short interfering RNAs (siRNAs) to DRG neurons *in vivo*, resulting in specific silencing of the targeted gene. For instance, HSV-mediated silencing of the *trpv1* gene, which is one of the most important players in nociception, is emerging as an interesting potential approach for neuropathic pain relief (Anesti *et al.*, 2008).

3. *Novel Lentiviral Vectors*

A recent class of vectors exploitable to target the PNS is based on a rabies-pseudotyped lentivirus (equine infectious anemia virus). These vectors have been recently used to express the retinoic acid receptor $\beta 2$ into the spinal cord after a dorsal root nerve crush; this treatment promoted the regeneration of DRG axons through their entry zone, and resulted in improved sensory and motor functions (Wong *et al.*, 2006). More recently, novel VSV-G-pseudotyped human immunodeficiency virus type 1 (HIV-1)-based lentiviral vectors have been proven to exhibit exquisite neurotropism and to promote retrograde axonal transport *in vitro* and *in vivo* (Federici *et al.*, 2009).

III. Emerging Concepts in Gene Therapy for Nerve Repair

A. Neurovascular Cross-Talk in the PNS

The striking similarities and parallelisms between the vascular and nervous systems have recently aroused much excitement and opened new intervention strategies for both neurodegenerative and vascular disorders (Carmeliet and Tessier-Lavigne, 2005; Zacchigna *et al.*, 2008a). The nervous and vascular systems share remarkable conservation in their overall anatomical architecture among individuals of the same species and across the vertebrate phyla (Zacchigna *et al.*, 2008b). In addition, they often pattern along parallel routes in peripheral tissues, indicating the occurrence of extensive developmental links. To follow the same path, axons and vessels often take advantage of each other: vascular cells produce soluble signals to track axons alongside pioneer vessels (Honma *et al.*, 2002; Kuruvilla *et al.*, 2004) and, conversely, nerves secrete molecules to guide blood vessels (Mukouyama *et al.*, 2002).

The biological and molecular similarities between the nervous and vascular systems have drawn much attention since several families of guidance cues, first discovered in the nervous system, have also been shown to guide blood vessels to their targets using similar mechanisms. This neurovascular cross-talk has been largely documented and investigated in the CNS. For instance, neural crest cells give rise to smooth muscle cells of certain brain vessels or to sympathetic nerves, which innervate and control blood flow of peripheral resistance vessels, while neurogenesis occurs in vascular niches in the adult CNS. Common signals also determine the cell fate of both neuronal and endothelial cells (i.e., the Notch signaling pathway), as well as the sorting, segregation, and establishment of boundaries between neuronal and vascular cell populations with different identities (i.e., ephrin signaling) (Zacchigna *et al.*, 2008a). This interplay of signals between the two systems is highly consistent and confirmed by recent genetic studies indicating that the insufficient production of angiogenic signals can cause neurodegeneration (Zacchigna *et al.*, 2008a). Remarkably, such angiogenic factors, which have been recently named "angioneurins" (and include, among others, VEGF, PDGF, TFGβ, and HGF), can also stimulate neuroregeneration, and have direct neuroprotective and other pleiotropic effects on various neural cell types, thus paving the way for novel therapeutic opportunities.

The similarities between nerves and vessels appear even more striking at the periphery, where vessels and nerves adopt similar morphological and molecular strategies to locate their targets and find their way toward their final destination, particularly after damage. Growing axons send out a highly motile and sensitive structure, the growth cone, which dynamically extends numerous filopodia and lamellipodia to "feel" the appropriate guidance cues and reassess its trajectory among the warren of possible routes. Similarly, a specialized endothelial cell type in the leading edge of capillaries, referred to as the "tip cell," continuously explores its environment through the extension and retraction of filopodia to define the direction in which the new vascular sprout has to grow. These tip cells do not proliferate and basically "pave the path" for subjacent "stalk" endothelial cells, which in contrast replicate extensively, allowing the oriented extension of the nascent vessel (Eichmann *et al.*, 2005).

On top of these striking similarities, even more unexpected is the recent increasing evidence that classical angiogenic factors, such as VEGF, may not only preserve nerve viability, but also promote distal axonal regeneration, thus becoming attractive targets for peripheral nerve repair. Indeed, the administration of recombinant VEGF facilitated nerve regeneration, sprouting, and functional recovery in a rat model of sciatic neurotomy (Fu *et al.*, 2007), as well as in acellular nerve grafts (Rovak *et al.*, 2004).

B. AAV-Transduced Muscle Scaffolds

Whenever a large gap is produced at the site of a nerve lesion, an autologous nerve graft represents the gold standard therapy. However, this strategy is often hampered by important concerns, since (i) the availability of donor tissue for bridging extensive nerve gaps is often limiting, (ii) the procedure requires the sacrifice of one or more sensory nerves, and (iii) the quality of donor grafts may be insufficient to guide and support axonal regeneration. In these difficult cases, the use of biological conduits made of collagen, arteries, veins, muscle, or even synthetic materials can be a good alternative to nerve grafting. In particular, the insertion of a muscle scaffold inside a vein offers several advantages over other techniques, including ease of access to all graft material in the lesion area without the need for additional incisions, the filling of large gaps (up to 5 cm), complete preservation of healthy sensory nerves, and possible spatial orientation of fiber regeneration. This approach has been successful for both sensory and mixed nerves, providing a good functional recovery in 85% of cases (Battiston *et al.*, 2007).

Besides their "structural" support in guiding nerve repair, biological and synthetic conduits might also provide a useful platform for the production of neurotrophic factors by gene therapy. Different tissues can be introduced within the conduit and genetically engineered to secrete useful proteins. For instance, the filling of the nerve graft with Schwann cells genetically modified to overexpress GDNF (Li *et al.*, 2006), FGF-2 (fibroblast growth factor-2) (Haastert *et al.*, 2006), or the enzyme sialyl transferase-X (STX, which modifies cell adhesion molecules, thus impacting Schwann cell motility) (Haile *et al.*, 2007), provided a significant advantage over empty conduits in terms of morphological and functional regeneration.

C. Scwhann Cells: An Overlooked Target in the PNS

Schwann cells play a critical role in peripheral nerve regeneration. Within autologous nerve grafts, these cells perform two primary functions, namely, guiding regenerating axons along longitudinal tubes and producing trophic factors needed to support the regeneration process. Even when a biological conduit is used to bridge a nerve defect, the conduit is soon colonized by a number of Schwann cells that make a pathway for regrowing axons. Therefore, Schwann cells may represent an ideal target to introduce therapeutic genes for nerve repair.

A promising option to foster nerve repair entails the genetic modification of autologous Schwann cells either *in vivo* or *ex vivo*, followed by their reimplantation into the same individual. Both electroporation and viral transduction have been used to modify Schwann cells *ex vivo*, either as single cells or as an intact graft (Aspalter *et al.*, 2009; Haastert *et al.*, 2007). *In vivo* manipulation of Schwann cell gene expression has so far been limited to viral transduction. Increased neurotrophin secretion by Schwann cells was shown to lead to a significant increase in the number of axons and myelination at the contusion site, associated with functional recovery, in different models of spinal cord and nerve injury (Golden *et al.*, 2007; Guo *et al.*, 2007). Alternatively, the axon itself can be genetically modified by using viral vectors to overexpress neurotrophic factors, such as BDNF, able to act on Schwann cells, thereby promoting myelination. Very few studies have directly assessed the efficiency of viral vector-mediated transduction of Schwann cells *in vivo*. Among these studies, efficient transgene expression by Schwann cells could be obtained by either central or peripheral delivery of adenoviral vectors (Watanabe *et al.*, 2006), as well as by lentiviral delivery to the basal forebrain of aged rats (Nagahara *et al.*, 2009), whereas the actual capacity of AAV vectors to transduce Schwann cells is still unsettled.

Emerging evidence indicates that a subpopulation of Schwann cells (called perisynaptic Schwann cells, PSCs) actively participates in the formation, function, maintenance, and repair of the chemical synapse at the neuromuscular junction (Feng and Ko, 2008a). In particular, PSCs cells express agrin and neuregulins, which might help postsynaptic differentiation and synaptic repair after injury (Feng and Ko, 2008b). Thus, gene transfer strategies aimed at the overexpression of synaptogenic molecules by PSCs is foreseen as a promising approach to foster functional recovery of nerve function (Ng *et al.*, 2007).

IV. Conclusions

The possibility to use gene therapy to sustain structural and functional repairs in the PNS is becoming an attractive reality. Here, we have briefly reviewed the main approaches attempted so far and discussed a few emerging concepts, which might help design and optimize novel therapeutic strategies. Much experimental research is still needed, however, in order to translate these strategies into actual therapies for patients. On one side, a deeper understanding of the molecular and cellular mechanisms underlying axonal repair and functional recovery will be required to define novel candidate genes. On the other side, the improvement of current gene delivery systems, together with a combination of gene transfer and tissue engineering, will offer new tools to increase both efficiency and safety of peripheral nerve gene therapy.

References

Abrams, M., and Widenfalk, J. (2005). Emerging strategies to promote improved functional outcome after peripheral nerve injury. *Restor. Neurol. Neurosci.* **23**(5–6), 367–382.

Anesti, A. M., Peeters, P. J., Royaux, I., and Coffin, R. S. (2008). Efficient delivery of RNA interference to peripheral neurons *in vivo* using herpes simplex virus. *Nucleic Acids Res.* **36**(14), e86.

Araki, K., Shiotani, A., Watabe, K., Saito, K., Moro, K., and Ogawa, K. (2006). Adenoviral GDNF gene transfer enhances neurofunctional recovery after recurrent laryngeal nerve injury. *Gene Ther.* **13**(4), 296–303.

Aspalter, M., Vyas, A., Feiner, J., Griffin, J., Brushart, T., and Redett, R. (2009). Modification of Schwann cell gene expression by electroporation *in vivo*. *J. Neurosci. Methods* **176**(2), 96–103.

Battiston, B., Tos, P., Conforti, L. G., and Geuna, S. (2007). Alternative techniques for peripheral nerve repair: Conduits and end-to-side neurorrhaphy. *Acta Neurochir. Suppl.* **100,** 43–50.

Buning, H., Nicklin, S. A., Perabo, L., Hallek, M., and Baker, A. H. (2003). AAV-based gene transfer. *Curr. Opin. Mol. Ther.* **5**(4), 367–375.

Carmeliet, P., and Tessier-Lavigne, M. (2005). Common mechanisms of nerve and blood vessel wiring. *Nature* **436**(7048), 193–200.

Chattopadhyay, M., Goss, J., Lacomis, D., Goins, W. C., Glorioso, J. C., Mata, M., and Fink, D. J. (2003). Protective effect of HSV-mediated gene transfer of nerve growth factor in pyridoxine neuropathy demonstrates functional activity of trkA receptors in large sensory neurons of adult animals. *Eur. J. Neurosci.* **17**(4), 732–740.

Chattopadhyay, M., Goss, J., Wolfe, D., Goins, W. C., Huang, S., Glorioso, J. C., Mata, M., and Fink, D. J. (2004). Protective effect of herpes simplex virus-mediated neurotrophin gene transfer in cisplatin neuropathy. *Brain* **127**(Pt. 4), 929–939.

Chattopadhyay, M., Wolfe, D., Mata, M., Huang, S., Glorioso, J. C., and Fink, D. J. (2005). Long-term neuroprotection achieved with latency-associated promoter-driven herpes simplex virus gene transfer to the peripheral nervous system. *Mol. Ther.* **12**(2), 307–313.

Eichmann, A., Le Noble, F., Autiero, M., and Carmeliet, P. (2005). Guidance of vascular and neural network formation. *Curr. Opin. Neurobiol.* **15**(1), 108–115.

Federici, T., Kutner, R., Zhang, X. Y., Kuroda, H., Tordo, N., Boulis, N. M., and Reiser, J. (2009). Comparative analysis of HIV-1-based lentiviral vectors bearing lyssavirus glycoproteins for neuronal gene transfer. *Genet. Vaccines Ther.* **7**(1), 1.

Federici, T., Liu, J. K., Teng, Q., Yang, J., and Boulis, N. M. (2007). A means for targeting therapeutics to peripheral nervous system neurons with axonal damage. *Neurosurgery* **60**(5), 911–918. discussion 911–918.

Feng, Z., and Ko, C. P. (2008a). Schwann cells promote synaptogenesis at the neuromuscular junction via transforming growth factor-beta1. *J. Neurosci.* **28**(39), 9599–9609.

Feng, Z., and Ko, C. P. (2008b). The role of glial cells in the formation and maintenance of the neuromuscular junction. *Ann. N.Y. Acad. Sci.* **1132,** 19–28.

Fu, C., Hong, G., and Wang, F. (2007). Favorable effect of local VEGF gene injection on axonal regeneration in the rat sciatic nerve. *J. Huazhong Univ. Sci. Technol. Med. Sci.* **27**(2), 186–189.

Goins, W. F., Sternberg, L. R., Croen, K. D., Krause, P. R., Hendricks, R. L., Fink, D. J., Straus, S. E., Levine, M., and Glorioso, J. C. (1994). A novel latency-active promoter is contained within the herpes simplex virus type 1 UL flanking repeats. *J. Virol.* **68**(4), 2239–2252.

Golden, K. L., Pearse, D. D., Blits, B., Garg, M. S., Oudega, M., Wood, P. M., and Bunge, M. B. (2007). Transduced Schwann cells promote axon growth and myelination after spinal cord injury. *Exp. Neurol.* **207**(2), 203–217.

Goss, J. R., Goins, W. F., Lacomis, D., Mata, M., Glorioso, J. C., and Fink, D. J. (2002). Herpes simplex-mediated gene transfer of nerve growth factor protects against peripheral neuropathy in streptozotocin-induced diabetes in the mouse. *Diabetes* **51**(7), 2227–2232.

Guo, J. S., Zeng, Y. S., Li, H. B., Huang, W. L., Liu, R. Y., Li, X. B., Ding, Y., Wu, L. Z., and Cai, D. Z. (2007). Cotransplant of neural stem cells and NT-3 gene modified Schwann cells promote the recovery of transected spinal cord injury. *Spinal Cord* **45**(1), 15–24.

Haastert, K., and Grothe, C. (2007). Gene therapy in peripheral nerve reconstruction approaches. *Curr. Gene Ther.* **7**(3), 221–228.

Haastert, K., Lipokatic, E., Fischer, M., Timmer, M., and Grothe, C. (2006). Differentially promoted peripheral nerve regeneration by grafted Schwann cells over-expressing different FGF-2 isoforms. *Neurobiol. Dis.* **21**(1), 138–153.

Haastert, K., Mauritz, C., Chaturvedi, S., and Grothe, C. (2007). Human and rat adult Schwann cell cultures: Fast and efficient enrichment and highly effective non-viral transfection protocol. *Nat. Protoc.* **2**(1), 99–104.

Haile, Y., Haastert, K., Cesnulevicius, K., Stummeyer, K., Timmer, M., Berski, S., Drager, G., Gerardy-Schahn, R., and Grothe, C. (2007). Culturing of glial and neuronal cells on polysialic acid. *Biomaterials* **28**(6), 1163–1173.

Honma, Y., Araki, T., Gianino, S., Bruce, A., Heuckeroth, R., Johnson, E., and Milbrandt, J. (2002). Artemin is a vascular-derived neurotropic factor for developing sympathetic neurons. *Neuron* **35**(2), 267–282.

Kaspar, B. K., Llado, J., Sherkat, N., Rothstein, J. D., and Gage, F. H. (2003). Retrograde viral delivery of IGF-1 prolongs survival in a mouse ALS model. *Science* **301**(5634), 839–842.

Kirchmair, R., Tietz, A. B., Panagiotou, E., Walter, D. H., Silver, M., Yoon, Y. S., Schratzberger, P., Weber, A., Kusano, K., Weinberg, D. H., Ropper, A. H., Isner, J. M., *et al.* (2007). Therapeutic angiogenesis inhibits or rescues chemotherapy-induced peripheral neuropathy: Taxol- and thalidomide-induced injury of vasa nervorum is ameliorated by VEGF. *Mol. Ther.* **15**(1), 69–75.

Kuruvilla, R., Zweifel, L. S., Glebova, N. O., Lonze, B. E., Valdez, G., Ye, H., and Ginty, D. D. (2004). A neurotrophin signaling cascade coordinates sympathetic neuron development through differential control of TrkA trafficking and retrograde signaling. *Cell* **118**(2), 243–255.

Li, Q., Ping, P., Jiang, H., and Liu, K. (2006). Nerve conduit filled with GDNF gene-modified Schwann cells enhances regeneration of the peripheral nerve. *Microsurgery* **26**(2), 116–121.

Li, Z., Peng, J., Wang, G., Yang, Q., Yu, H., Guo, Q., Wang, A., Zhao, B., and Lu, S. (2008). Effects of local release of hepatocyte growth factor on peripheral nerve regeneration in acellular nerve grafts. *Exp. Neurol.* **214**(1), 47–54.

Liu, B., Paton, J. F., and Kasparov, S. (2008). Viral vectors based on bidirectional cell-specific mammalian promoters and transcriptional amplification strategy for use *in vitro* and *in vivo*. *BMC Biotechnol.* **8,** 49.

Liu, B. H., Wang, X., Ma, Y. X., and Wang, S. (2004). CMV enhancer/human PDGF-beta promoter for neuron-specific transgene expression. *Gene Ther.* **11**(1), 52–60.

Makwana, M., and Raivich, G. (2005). Molecular mechanisms in successful peripheral regeneration. *FEBS J.* **272**(11), 2628–2638.

Millecamps, S., Mallet, J., and Barkats, M. (2002). Adenoviral retrograde gene transfer in motoneurons is greatly enhanced by prior intramuscular inoculation with botulinum toxin. *Hum. Gene Ther.* **13**(2), 225–232.

Mukouyama, Y. S., Shin, D., Britsch, S., Taniguchi, M., and Anderson, D. J. (2002). Sensory nerves determine the pattern of arterial differentiation and blood vessel branching in the skin. *Cell* **109**(6), 693–705.

Murakami, T., Arai, M., Sunada, Y., and Nakamura, A. (2006). VEGF 164 gene transfer by electroporation improves diabetic sensory neuropathy in mice. *J. Gene Med.* **8**(6), 773–781.

Nagahara, A. H., Bernot, T., Moseanko, R., Brignolo, L., Blesch, A., Conner, J. M., Ramirez, A., Gasmi, M., and Tuszynski, M. H. (2009). Long-term reversal of cholinergic neuronal decline in aged non-human primates by lentiviral NGF gene delivery. *Exp. Neurol.* **215**(1), 153–159.

Natsume, A., Wolfe, D., Hu, J., Huang, S., Puskovic, V., Glorioso, J. C., Fink, D. J., and Mata, M. (2003). Enhanced functional recovery after proximal nerve root injury by vector-mediated gene transfer. *Exp. Neurol.* **184**(2), 878–886.

Ng, B. K., Chen, L., Mandemakers, W., Cosgaya, J. M., and Chan, J. R. (2007). Anterograde transport and secretion of brain-derived neurotrophic factor along sensory axons promote Schwann cell myelination. *J. Neurosci.* **27**(28), 7597–7603.

Palmer, J. A., Branston, R. H., Lilley, C. E., Robinson, M. J., Groutsi, F., Smith, J., Latchman, D. S., and Coffin, R. S. (2000). Development and optimization of herpes simplex virus vectors for multiple long-term gene delivery to the peripheral nervous system. *J. Virol.* **74**(12), 5604–5618.

Perez, M. C., Hunt, S. P., Coffin, R. S., and Palmer, J. A. (2004). Comparative analysis of genomic HSV vectors for gene delivery to motor neurons following peripheral inoculation *in vivo*. *Gene Ther.* **11**(13), 1023–1032.

Rovak, J. M., Mungara, A. K., Aydin, M. A., and Cederna, P. S. (2004). Effects of vascular endothelial growth factor on nerve regeneration in acellular nerve grafts. *J. Reconstr. Microsurg.* **20**(1), 53–58.

Sakamoto, T., Kawazoe, Y., Shen, J. S., Takeda, Y., Arakawa, Y., Ogawa, J., Oyanagi, K., Ohashi, T., Watanabe, K., Inoue, K., Eto, Y., and Watabe, K. (2003). Adenoviral gene transfer of GDNF, BDNF and TGF beta 2, but not CNTF, cardiotrophin-1 or IGF1, protects injured adult motoneurons after facial nerve avulsion. *J. Neurosci. Res.* **72**(1), 54–64.

Schmidt, C. E., and Leach, J. B. (2003). Neural tissue engineering: Strategies for repair and regeneration. *Ann. Rev. Biomed. Eng.* **5,** 293–347.

Schratzberger, P., Schratzberger, G., Silver, M., Curry, C., Kearney, M., Magner, M., Alroy, J., Adelman, L. S., Weinberg, D. H., Ropper, A. H., and Isner, J. M. (2000). Favorable effect of VEGF gene transfer on ischemic peripheral neuropathy. *Nat. Med.* **6**(4), 405–413.

Simovic, D., Isner, J. M., Ropper, A. H., Pieczek, A., and Weinberg, D. H. (2001). Improvement in chronic ischemic neuropathy after intramuscular *phVEGF165* gene transfer in patients with critical limb ischemia. *Arch. Neurol.* **58**(5), 761–768.

Streppel, M., Heiser, T., and Stennert, E. (2000). Historical development of facial nerve surgery with special reference to hypoglossal-facial nerve anastomosis. *HNO* **48**(11), 801–808.

Watanabe, T. S., Ohtori, S., Koda, M., Aoki, Y., Doya, H., Shirasawa, H., Yamazaki, M., Moriya, H., Takahashi, K., and Yamashita, T. (2006). Adenoviral gene transfer in the peripheral nervous system. *J. Orthop. Sci.* **11**(1), 64–69.

Wong, L. F., Yip, P. K., Battaglia, A., Grist, J., Corcoran, J., Maden, M., Azzouz, M., Kingsman, S. M., Kingsman, A. J., Mazarakis, N. D., and McMahon, S. B. (2006). Retinoic acid receptor beta2 promotes functional regeneration of sensory axons in the spinal cord. *Nat. Neurosci.* **9**(2), 243–250.

Xu, J., Ma, C., Bass, C., and Terwilliger, E. F. (2005). A combination of mutations enhances the neurotropism of AAV-2. *Virology* **341**(2), 203–214.

Zacchigna, S., Lambrechts, D., and Carmeliet, P. (2008a). Neurovascular signalling defects in neurodegeneration. *Nat. Rev. Neurosci.* **9**(3), 169–181.

Zacchigna, S., Ruiz de Almodovar, C., and Carmeliet, P. (2008b). Similarities between angiogenesis and neural development: What small animal models can tell us. *Curr. Top. Dev. Biol.* **80,** 1–55.

USE OF STEM CELLS FOR IMPROVING NERVE REGENERATION

Giorgio Terenghi,* Mikael Wiberg,[†,‡] and Paul J. Kingham*[,†]

*Blond McIndoe Laboratories, Tissue Injury and Repair Group, School of Medicine, University of Manchester, Manchester, United Kingdom

[†]Section of Anatomy, Department of Integrative Medical Biology, Umeå University, Umeå, Sweden

[‡]Section of Hand and Plastic Surgery, Department of Surgical and Perioperative Science, University Hospital, Umeå, Sweden

A clear need exists for new surgical approaches to enhance the recuperation of functions after peripheral nerve injury and repair. At present, advances in the regenerative medicine fields of biomaterials, cellular engineering, and molecular biology are all contributing to the development of a bioengineered nerve implant, which could be used clinically as an alternative to nerve autograft. In this review we examine the recent progress in this field, looking in particular at the applicability of Schwann cells and stem cell transplantation to enhance nerve regeneration.

I. Nerve Repair and Regeneration

Peripheral nerve repair, either by direct suturing of the nerve ends or by grafting, remains the standard treatment for nerve injuries. In the most severe cases peripheral nerve injury is associated with significant loss of nervous tissue. This causes a long nerve gap, which is necessary to bridge by using an autologous nerve graft, generally the sural nerve. This bridging strategy is essential to provide a physical substrate for axonal growth, as the grafted nerve contains the cellular and molecular elements essential for axonal regeneration. However, there is a limitation of the amount of nerve that can be harvested, with additional donor site

INTERNATIONAL REVIEW OF
NEUROBIOLOGY, VOL. 87
DOI: 10.1016/S0074-7742(09)87021-9

morbidity for the patients, such as sensory loss, scarring, and possible neuroma formation (Evans, 2001; Wu and Chiu, 1999).

An alternative to nerve autografting is the use of conduits. This technique has several advantages as it provides a guidance channel and mechanical support for the regenerating axons, reduces invasion by connective tissue, and could be used to bridge any gap length. The search for an alternative method to bridge a gap is nothing new, and a number of biological and synthetic material have been shown to be useful in the formation of nerve conduits for gap repair (Ciardelli and Chiono, 2006; Valentini and Aebischer, 1997), some of which have been tested in clinical trials with some degree of success (Aberg *et al.*, 2009; Schlosshauer *et al.*, 2006). Despite these advances, it has become clear that the number of regenerating neurons following injury and repair is suboptimal, and experiments have calculated that only 50–60% of sensory and motor neurons regenerate after primary repair or nerve grafting (Welin *et al.*, 2008). It has become evident that a more complex strategy is needed to improve regeneration. In particular, the addition to the conduit of molecular and cellular elements is needed in order to mimic a nerve microenvironment and to promote an improved axonal regrowth, in essence creating an "artificial nerve."

II. Schwann Cells for Nerve Regeneration

Schwann cells are the main glial cells of the peripheral nervous system, forming the myelin surrounding the nerve fibers, and creating a structural support to the axons. Schwann cells also secrete growth factors essential for the maintenance of the neuronal cells, and importantly are essential for the regeneration process (Hall, 2001). Their ability to align themselves to provide directional cues to the regrowing axons (Thompson and Buettner, 2006) and the possibility to enhance their capability to synthesize and secrete increased levels of growth factors by using genetic modification (Li *et al.*, 2006) make Schwann cells an attractive choice for addition to nerve conduits.

Schwann cells can be cultured *in vitro*, and their transplantation within a nerve conduit has been shown to improve both the quality and the rate of the regenerating axons (Hadlock *et al.*, 2000; Mosahebi *et al.*, 2001a). By using stable genetic labeling methodology for the identification of the transplanted cells, it has been possible to demonstrate their active participation in the regeneration process (Mosahebi *et al.*, 2002; Tohill *et al.*, 2004b). However, the use of autologous cultured Schwann cells for the treatment of acute injuries may be impractical due to the technical difficulties and the time required in harvesting and expanding this slow growing type of cell, and 10 weeks of culture may be needed to obtain a sufficient number of Schwann cells for transplantation into a conduit (Mosahebi *et al.*, 2001a). Such a delay in carrying out the nerve repair following injury would be deleterious to the clinical outcome, as a longer time between injury and repair results in a greater neuronal cell death, and therefore a reduced potential for functional recovery (Hart *et al.*, 2002). Although neuronal cell death can be prevented by neuroprotective pharmacological intervention (Hart *et al.*, 2008), the problem remains as how to improve the regenerating potential of the neuronal population.

Allogeneic Schwann cell transplantation is an attractive option, as it would allow the immediate use of cells stored in tissue banks following a screening for immunological tissue match, as it is done for organ transplant. Some initial success has been obtained experimentally with transplantation of genetically marked allogeneic Schwann cells without any concomitant immune suppression (Mosahebi *et al.*, 2002). However, the results demonstrated that the rejection process was too rapid for the cells to have a significant influence on the regeneration process, although the immune reaction did not have a deleterious effect on the nerve regeneration. With further improvement and limited immune suppression, in the future it may be possible to use allogeneic Schwann cells for transplantation. A similar approach has been used for allogeneic nerve graft both experimentally (Midha *et al.*, 2001) and clinically (Mackinnon *et al.*, 2001), where immunosuppression has been shown to avoid rejection and improve axonal regeneration (Kvist *et al.*, 2003; Sulaiman *et al.*, 2002).

One problem encountered with transplantation of cells is their dispersion in a homogeneous way within the conduit. This can be achieved by the use of a matrix, which has the potential to provide similar properties as the extracellular matrix (Lutolf and Hubbell, 2005; Novikova *et al.*, 2006). The matrix should permit cell proliferation while lasting long enough to entrap the transplanted cells in the conduit for the duration of axonal regeneration, and synthetic or natural hydrogels appear to be the favored choice. Natural components of the extracellular matrix such as collagen, fibrin, and hyaluronic acid have been used in a hydrogel form with encouraging results (Kalbermatten *et al.*, 2008; Labrador *et al.*, 1998; Madison *et al.*, 1985; Seckel *et al.*, 1995), although the bovine origin of the collagen has somehow discouraged any further development of this type of matrix because of the risk of prion transmission. Alginate hydrogel, extracted from brown algae, has been used for Schwann cell transplantation (Mosahebi *et al.*, 2001b) and growth factor release (Mohanna *et al.*, 2003, 2005), but in the long term its slow degradation rate may cause a physical obstruction to the regenerating axons. A more sophisticated system of longitudinally aligned micro- (Chew *et al.*, 2008; Ngo *et al.*, 2003) and nanofibers (Wang *et al.*, 2008) within the conduit has also been successfully applied, and it was found that the axonal regeneration was dependent on the diameter and grouping of the fibers (Wen and Tresco, 2006). Constructing or coating of the fibers with extracellular matrix molecule such as collagen or laminin was also suggested (Matsumoto *et al.*, 2000;

Tong *et al.*, 1994), and in our experience the coating of PHB fibers with extracellular matrix molecules improves both Schwann cells proliferation and axon elongation *in vitro* (Armstrong *et al.*, 2007). Despite these clear indications, more research is needed to find a matrix which satisfies all the requirements needed for a successful three-dimensional support to the regrowing axons and the proliferating Schwann cells.

III. Stem Cells for Regenerative Medicine

There is a large body of literature describing embryonic stem cells, hematopoietic stem cells, and stem cells in epithelia that display a rapid turnover, that is, in the skin and intestinal epithelium, as previously reviewed (Tohill and Terenghi, 2004). Discrete stem cell populations have now been identified in other adult tissues such as bone marrow, adipose tissue, and neural tissue. These are defined as multipotent stem cells, which are naturally capable of developing into a restricted number of cell types specific to that tissue (Alison *et al.*, 2002). Unlike embryonic stem cells, organ-specific stem cells were believed to have lost the capacity to generate other somatic lineage. Until recently, dogma dictated that organ-specific stem cells were restricted to differentiate only into cell types of the tissue from which they originate. However, many recent reports have shown that stem cells from one tissue can cross lineage boundaries to differentiate into cells of other lineages either *in vitro* or *in vivo* after transplantation. This plasticity, or ability for cells to trans-differentiate, has created great interest for its therapeutic potential in tissue engineering, particularly in the area of nerve regeneration.

Bone marrow contains two distinct populations of progenitor cells: hematopoietic and bone marrow stromal progenitors. Orthodox differentiation of marrow stromal cells, also called mesenchymal stem cells, leads to mesenchymal progenitors for bone, cartilage, tendon, adipose tissue, and muscle (Muraglia *et al.*, 2000). Mesenchymal progenitors are clinically advantageous in that they are readily accessible by bone marrow biopsy, and their renewable population represents an abundant source. It is now evident that mesenchymal stem cells have the ability to cross lineage boundaries and differentiate into diverse cell types. Recently mesenchymal stem cells have been shown to trans-differentiate into cardiomyocytes, endothelial cells, and smooth muscle cells following direct injection into adult heart (Barbash *et al.*, 2003; Gojo *et al.*, 2003).

Thus, stem cells represent an ideal candidate for regenerative medicine, and several aspects of their biological activity have been investigated, including their ultrastructural phenotypic characterization (Raimondo *et al.*, 2006). It has been found that both the age of the donor and the number of passages during culture *in vitro* can influence negatively the proliferation and differentiation potential

(Galderisi *et al.*, 2008; Kretlow *et al.*, 2008). There are some concerns about the possible oncogenic transformation of transplanted stem cells. Interestingly, a recent paper has shown that human stem cells do not seem to have such potential (Vilalta *et al.*, 2008), even if this study has to be interpreted with caution given that the testing was carried out inter-species rather than within-species. The culture of stem cells in 3-D structure rather than in 2-D culture dish may also influence the proliferation and differentiation potential (Burdick and Vunjak-Novakovic, 2009; Chu *et al.*, 2009; Chung and Burdick, 2009). Stem cells have also been shown to proliferate without addition of animal serum (Schallmoser *et al.*, 2008), a useful precursor study for the clinical application of cells grown without animal derived products.

IV. Stem Cells for Nerve Regeneration

More relevant to nerve regeneration is the ability of mesenchymal stem cells to differentiate into both neurons and glial cells. Both rat and human mesenchymal stem cells can differentiate *in vitro* into neuronal cells (Woodbury *et al.*, 2000), and express markers for both neuronal and glial cells (Bossolasco *et al.*, 2005; Kim *et al.*, 2002). There are also reports describing the use of bone marrow mesenchymal stem cell transplantation in models of peripheral nerve injury. Expression of glial cell markers found typically on Schwann cells was described *in vitro* in rat mesenchymal stem cells following exposure to a cocktail of growth factors, and these cells integrate in the regenerating growth cone upon transplantation into a conduits grafted into the rat sciatic nerve (Dezawa *et al.*, 2001; Tohill *et al.*, 2004a). Migration and differentiation of mesenchymal stem cells was found at the site of sciatic nerve injury following the injection of cultured undifferentiated stem cells (Cuevas *et al.*, 2002; Zhang *et al.*, 2004). Other studies have shown that the transplantation of undifferentiated mesenchymal stem cells in nerve conduits stimulates axonal regrowth and motor function recovery (Chen *et al.*, 2006; Pereira *et al.*, 2006; Wang *et al.*, 2008). Different types of nerve conduits have also been used for the grafting of differentiated mesenchymal stem cells, resulting in increased nerve regeneration in rat (Choi *et al.*, 2005; Hou *et al.*, 2006; Kalbermatten *et al.*, 2009; Yamakawa *et al.*, 2007; Zhang *et al.*, 2005), also using human mesenchymal stem cells in combination with immunosuppression (Shimizu *et al.*, 2007), and in primates using autologous stem cells (Hu *et al.*, 2007; Wang *et al.*, 2008).

In our laboratories, we have shown that mesenchymal stem cells can differentiate into Schwann cell-like phenotype, which can be transplanted in nerve conduits and actively participate in the nerve regeneration process (Tohill *et al.*, 2004a). We have also demonstrated that the differentiated mesenchymal stem cells have all the morphological, molecular, and functional characteristics of Schwann cells, both in rat (Caddick *et al.*, 2006; Mahay *et al.*, 2008a) and in human (Brohlin *et al.*, 2009). Importantly, differentiated mesenchymal stem cells synthesize and secrete neurotrophins (Mahay *et al.*, 2008b) and *in vitro* they upregulate myelin genes and proteins expression when cocultured with neuronal cells (Mantovani *et al.*, 2009). Similarly, other studies have shown the potential of differentiated mesenchymal stem cells to express various levels of growth factors (Chen *et al.*, 2007; Crigler *et al.*, 2006) and to myelinate axons (Keilhoff *et al.*, 2006).

Recently, it has also been shown the existence of multipotent stem cells in adipose tissue (Gimble and Guilak, 2003, Strem *et al.*, 2005), which represents an easy and accessible source for obtaining such cells, for example, from lipoaspirate, with minimal discomfort to the donor. Adipose-derived stem cells are similar to bone-marrow-derived mesenchymal stem cells in both phenotypic and gene expression profile (De Ugarte *et al.*, 2003; Strem *et al.*, 2005), and in their differentiation into osteoblasts, chondrocytes, and adipocytes (Tholpady *et al.*, 2003). More importantly, the frequency of stem cells in adipose tissue is 100- to 1000-fold higher than in bone marrow (Banfi *et al.*, 2001; Muschler *et al.*, 2001; Strem *et al.*, 2005). This is a considerable advantage as it reduces the period of expansion of the stem cells prior to differentiation, further reducing the possible delay between injury and cell transplantation in nerve conduits.

It has been proposed that stem cells within the adipose tissue are derived from vascular precursors (Lin *et al.*, 2008), which would be consistent with the finding that the yield of adipose-derived stem cells is directly correlated to the blood vessel density in the adipose tissue (Da Silva Mirelles *et al.*, 2009) and the site of harvesting (Jurgens *et al.*, 2008). We have shown that rat adipose-derived stem cells proliferate at a faster rate than bone-marrow-derived mesenchymal stem cells (Kingham *et al.*, 2007), while comparison of adipose- and bone-marrow-derived stem cells between human and nonhuman primate showed some differences in their differentiation capabilities with increasing number of passages (Izadpanah *et al.*, 2006). Our studies have also shown that adipose-derived stem cells can be stably differentiated into glial cells expressing markers found in Schwann cells (Kingham *et al.*, 2007). These differentiated cells also stimulate neurite outgrowth and myelination when cocultured with neuronal cells in a similar way as Schwann cells (Kingham *et al.*, 2007, Mantovani *et al.*, 2009) suggesting a differentiation into Schwann cell phenotype.

V. Conclusions

All these studies show strong evidence of the potential of stem cells for application in nerve engineering, and they may be promising candidates for cell therapy procedures once their features are fully assessed experimentally. In this review we have shown how regenerative medicine for nerve repair comprises the combination of biomaterials, biological mechanism, and cell therapy. The development of new biomaterials and novel approaches to cell therapy are bringing the concept of new surgical procedure for nerve repair nearer to realization. Also, the combination of pharmacological intervention to control the mechanism of neuroprotection would assure the regeneration of all involved neuronal cells, hopefully leading to enhanced functional recovery. Although many biological problems are still to be resolved and much experimental work still to be carried out, the fast development of the research in this field would indicate that the concept of an "artificial nerve" for surgical repair of nerve lesion is not an abstract concept, but a reality which will be available in the near future.

References

Aberg, M., Ljungberg, C., Edin, E., Millqvist, H., Nordh, E., Terenghi, G., and Wiberg, M. (2009). Clinical evaluation of a resorbable wrap-around implant as an alternative to nerve repair. *J. Plast. Reconstr. Aesthet. Surg.* in press.

Alison, M. R., Poulsom, R., Forbes, S., and Wright, N. A. (2002). An introduction to stem cells. *J. Pathol.* **197,** 419–423.

Armstrong, S. J., Wiberg, M., Terenghi, G., and Kingham, P. J. (2007). ECM Molecules mediate both Schwann cell proliferation and activation to enhance neurite outgrowth. *Tissue Eng.* **13**(12), 2863–2870.

Banfi, A., Bianchi, G., Galotto, M., Cancedda, R., and Quarto, R. (2001). Bone marrow stromal damage after chemo/radiotherapy: Occurrence, consequences and possibilities of treatment. *Leuk. Lymphoma* **42,** 863–870.

Barbash, I. M., Chouraqui, P., Baron, J., Feinberg, M. S., Etzion, S., Tessone, A., Miller, L., Guetta, E., Zipori, D., Kedes, L. H., Kloner, R. A., and Leor, J. (2003). Systemic delivery of bone-marrow derived mesenchymal stem cells to the infarcted myocardium. *Circulation* **108,** 863–868.

Bossolasco, P., Cova, L., Calzaross, C., Rimoldi, S. G., Borsotti, C., Deliliers, G. L., Silani, V., and Soligo, D. (2005). Polli. Neuro-glial differentiation of human bone marrow stem cells *in vitro*. *Exp. Neurol.* **193**(2), 312–325.

Brohlin, M., Mahay, D., Novikov, L. N., Terenghi, G., Wiberg, M., Shawcross, S. G., and Novikova, L. N. (2009). Characterisation of human mesenchymal stem cells following differentiation into Schwann cell-like cells. *Neurosci. Res.* **64,** 41–49.

Burdick, J. A., and Vunjak-Novakovic, G. (2009). Engineered microenvironments for controlled stem cell differentiation. *Tissue Eng.* **15,** 205–219.

Caddick, J., Kingham, P. J., Gardiner, N. J., Wiberg, M., and Terenghi, G. (2006). Phenotypic and functional characteristics of mesenchymal stem cells differentiated along a Schwann cell lineage. *Glia* **54**(8), 840–849.

Chen, C. J., Ou, Y. C., Liao, S. L., Chen, W. Y., Chen, S. Y., Wu, C. W., Wang, C. C., Wang, W. Y., Huang, Y. S., and Hsu, S. H. (2007). Transplantation of bone marrow stromal cells for peripheral nerve repair. *Exp. Neurol.* **204,** 443–453.

Chen, X., Wang, X. D., Chen, G., Liu, W. W., Yao, J., and Gu, X. S. (2006). Study of *in vivo* differentiation of rat bone marrow stromal cells into Schwann cell-like cells. *Microsurgery* **26**(2), 111–115.

Chew, S. Y., Mi, R., Hoke, A., and Leong, K. W. (2008). The effect of the alignment of electrospun fibrous scaffolds on Schwann cell maturation. *Biomat* **29**(6), 653–661.

Choi, B. H., Zhu, S. J., Kim, B. Y., Huh, Y. H., Lee, S. H., and Jung, J. H. (2005). Transplantation of cultured bone marrow stromal cells to improve peripheral nerve regeneration. *Int. J. Oral Maxillofac. Surg.* **34,** 537–542.

Chu, C., Schmidt, J. J., Carnes, K., Zhang, Z., Kong, H. J., and Hofmann, M. C. (2009). Three-dimensional synthetic niche components to control germ cell proliferation. *Tissu Eng.* **15,** 255–262.

Chung, C., and Burdick, J. A. (2009). Influence of three-dimensional hyaluronic acid microenvironments on mesenchymal stem cell chondrogenesis. *Tissue Eng.* **15,** 243–254.

Ciardelli, G., and Chiono, V. (2006). Materials for peripheral nerve regeneration. *Macromol. Biosci.* **6,** 13–26.

Crigler, L., Robey, R. C., Asawachaicharn, A., Gaupp, D., and Phinney, D. G. (2006). Human mesenchymal stem cell subpopulations express a variety of neuro-regulatory molecules and promote neuronal cell survival and neuritogenesis. *Exp. Neurol.* **198,** 54–64.

Cuevas, P., Carceller, F., Dujovny, M., Garcia-Gomez, I., Cuevas, B., Gonzalez-Corrochano, R., Diaz-Gonzalez, D., and Reimers, D. (2002). Peripheral nerve regeneration by bone marrow stromal cells. *Neurol. Res.* **24,** 634–638.

Da, L., Sand, T. T., Harman, R. J., Lennon, D. P., and Caplan, A. I. (2009). MSC frequency correlates with blood vessel density in equine adipose tissue. *Tissue Eng.* **15,** 221–229.

De Ugarte, D. A., Morizono, K., Elbarbary, A., Alfonso, Z., Zuk, P. A., Zhu, M., Dragoo, J. L., Ashjian, P., Thomas, B., Benhaim, P., Chen, I., Fraser, J., *et al.* (2003). Comparison of multi-lineage cells from human adipose tissue and bone marrow. *Cells Tissues Organs* **174,** 101–109.

Dezawa, M., Takahashi, I., Esaki, M., Takano, M., and Sawada, H. (2001). Sciatic nerve regeneration in rats induced by transplantation of *in vitro* differentiated bone-marrow stromal cells. *Eur. J. Neurosci.* **14,** 1771–1776.

Evans, G. R. D. (2001). Peripheral nerve injury: A review and approach to engineered constructs. *Anat. Rec.* **263,** 396–404.

Galderisi, U., Helmbold, H., Squillaro, T., Alessio, N., Komm, N., Khadang, B., Cipollaro, M., Bohn, W., and Giordano, A. (2008). *In vitro* senescence of rat mesenchymal stem cells is accompanied by downregulation of stemness-related and DNA damage repair genes. *Stem Cell Dev.* [e-pub ahead of print].

Gimble, J., and Guilak, F. (2003). Adipose-derived adult stem cells, Isolation, characterization, and differentiation potential. *Cytotherapy* **5,** 362–369.

Gojo, S., Gojo, N., Mori, T., Abe, H., Kyo, S., Hata, J., and Umezawa, A. (2003). *In vivo* cardiovasculogeneis by direct injection of isolated adult mesenchymal stem cells. *Exp. Cell Res.* **288,** 51–59.

Hadlock, T., Sundback, C., Hunter, D., Cheney, M., and Vacanti J, M. (2000). A polymer foam conduit seeded with Schwann cells promotes guided peripheral nerve regeneration. *Tissue Eng.* **6,** 119–127.

Hall, S. (2001). Nerve repair: A neurobiologist's view. *J. Hand Surg.* **26B,** 129–136.

Hart, A., Brannstrom, T., Wiberg, M., and Terenghi, G. (2002). Timecourse of primary sensory neuron and satellite cell death after peripheral axotomy in adult rat. *Exp. Brain Res.* **142,** 308–318.

Hart, A. M., Terenghi, G., and Wiberg, M. (2008). Neuronal death after peripheral nerve injury and experimental strategies for neuroprotection. *Neurol. Res.* **30,** 999–1011.

Hou, S. Y., Zhang, H. Y., Quan, D. P., Liu, X. L., and Zhu, J. K. (2006). Tissue-engineered peripheral nerve grafting by differentiated bone marrow stromal cells. *Neuroscience* **140,,** 101–110.

Hu, J., Zhu, Q. T., Liu, X. L., Xu, Y. B., and Zhu, J. K. (2007). Repair of extended peripheral nerve lesions in rhesus monkeys using acellular allogenic nerve grafts implanted with autologous mesenchymal stem cells. *Exp. Neurol.* **204**(2), 658–666.

Izadpanah, R., Trygg, C., Patel, B., Christopher, K., Dufour, J., Gimble, J. M., and Bunnell, B. A. (2006). Biological properties od mesenchymal stem cells derived from bone marrow and adipose tissue. *J. Cell Biochem.* **99,** 1285–1297.

Jurgens, W. J. F. M., Oedayrajsingh-Varma, M. J., Helder, M. N., Zandich Doulabi, B., Schouten, T. E., Kuik, D. J., Ritt, M. J. P. F., and van Milligen, F. J. (2008). Effect of tissue harvesting site on yield of stem cells derived from adipose tissue: Implications for cell-based therapies. *Cell Tissue Res.* **332,** 415–426.

Kalbermatten, D. F., Kingham, P. J., Mahay, D., Mantovani, C., Pettersson, J., Raffoul, W., Balchin, H., Pierer, G., and Terenghi, G. (2008). Fibrin matrix for suspension of regenerative cells in an artificial nerve conduit. *J. Plast. Reconstr. Aesthet. Surg.* **61,** 669–675.

Kalbermatten, D. F., Pettersson, J., Kingham, P. J., Mahay, D., Erba, P., Pierer, G., and Terenghi, H. (2009). New fibrin conduit for peripheral nerve repair. *J. Recontr. Microsurg.* **25,** 27–33.

Keilhoff, G., Stang, F., Goihl, A., Wolf, G., and Fansa, H. (2006). Mesenchymal stem cells as alternative therapy in supporting nerve regeneration and myelination. *Cell Mol. Neurobiol.* **26,** 1233–1250.

Kim, B. J., Seo, J. H., Bubien, J. K., and Oh, Y. S. (2002). Differentiation of adult bone marrow stem cells into neuroprogenitor cells *in vitro*. *Neuroreport* **13,** 1185–1188.

Kingham, P. J., Kalbermatten, D. F., Mahay, D., Armstrong, S. J., Wiberg, M., and Terenghi, G. (2007). Adipose-derived stem cells differentiate into a Schwann cell phenotype and promote neurite outgrowth *in vitro*. *Exp. Neurol.* **207**(2), 267–274.

Kretlow, J. D., Jin, Y. Q., Liu, W., Zhang, W. J., Hong, T. H., Zhou, G., Baggett, L. S., Mikos, A. G., and Cao, Y. (2008). Donor age and cell passage affects differentiation potential of murine bone marrow-derived stem cells. *BMC Cell Biol.* **9,** 60.

Kvist, M., Danielsen, N., and Dahlin, L. B. (2003). Effects of FK506 on regeneration and macrophages in injured rat sciatic nerve. *J. Peripher. Nerv. Syst.* **8,** 251–259.

Labrador, R. O., Buti, M., and Navarros, X. (1998). Influence of collagen and laminin gels concentration on nerve regeneration after resection and tube repair. *Exp. Neurol.* **149,** 243–252.

Li, Q., Ping, P., Jiang, H., and Liu, K. (2006). Nerve conduit filled with GDNF gene-modified Schwann cells enhances regeneration of the peripheral nerve. *Microsurgery* **26,** 116–121.

Lin, G., Garcia, M., Ning, H., Banie, L., Guo, Y. L., Lue, T. F., and Lin, C. S. (2008). Defining stem and progenitor cells within adipose tissue. *Stem Cells Dev.* **17,** 1053–1064.

Lutolf, M. P., and Hubbell, J. A. (2005). Synthetic biomaterials as instructive extracellular microenvironments for morphogenesis in engineering. *Nat. Biotechnol.* **23,** 47–55.

Mackinnon, S. E., Doolabh, V. B., Novak, C. B., and Trulock, E. P. (2001). Clinical outcome following nerve allograft transplantation. *Plast. Reconstr. Surg.* **107,** 1419–1429.

Madison, R., da Silva, C. F., Dikkes, P., Chiu, T. H., and Sidman, R. L. (1985). Increased rate of peripheral nerve regeneration using bioresorbable nerve guides and a laminin-containing gel. *Exp. Neurol.* **88,** 767–772.

Mahay, D., Terenghi, G., and Shawcross, S. (2008a). Growth factors in mesenchymal stem cells following glial cell differentiation. *Biotechnol. Appl. Biochem.* **51,** 167–176.

Mahay, D., Terenghi, G., and Shawcross, S. G. (2008b). Schwann cell mediated trophic effects by differentiated mesenchymal stem cells. *Exp. Cell Res.* **314,** 2692–2701.

Mantovani, C., Mahay, D., Shawcross, S., Terenghi, G., and Wiberg, M. (2009). Development of an *in vitro* co-culture model to investigate peripheral myelin formation. *Neurosci. Res.* (submitted for publication).

Matsumoto, K., Ohnishi, K., Kiyotani, T., Sekine, T., Ueda, H., Nakamura, T., Eo, K., and Shimizu, Y. (2000). Peripheral nerve regeneration across an 80 mm gap bridged by a polyglycolic acid (PGA)-collagen tube filled with laminin-coated fibres: A histological and electrophysiological evaluation of regenerated nerves. *Brain Res.* **868,** 325–328.

Midha, R., Nag, S., Munro, C. A., and Ang, L. C. (2001). Differential response of sensory and motor axons in nerve allografts after withdrawal of immunosuppressive therapy. *J. Neurosurg.* **94,** 102–110.

Mosahebi, A., Woodward, B., Wiberg, M., Martin, R., and Terenghi, G. (2001a). Retroviral labelling of Schwann cells: *In vitro* characterisation and *in vivo* transplantation to improve peripheral nerve regeneration. *Glia* **34,** 8–17.

Mosahebi, A., Simon, M., Wiberg, M., and Terenghi, G. (2001b). A novel use of alginate hydrogel as Schwann cell matrix. *Tissue Eng.* **7,** 525–534.

Mosahebi, A., Fuller, P., Wiberg, M., and Terenghi, G. (2002). Effect of allogeneic Schwann cell transplantation on peripheral nerve regeneration. *Exp. Neurol.* **173,** 213–223.

Mohanna, P. N., Young, R. C., Wiberg, M., and Terenghi, G. (2003). A composite poly-hydroxybutyrate-glial growth factor conduit for long nerve gap repair. *J. Anat.* **203,** 553–565.

Mohanna, P. N., Terenghi, G., and Wiberg, M. (2005). Composite PHB-GGF conduit for long nerve gap repair: A long-term evaluation. *Scan. J. Plast. Reconstr. Hand Surg.* **39,** 129–137.

Muraglia, A., Cacedda, R., and Quarto, R. (2000). Clonal mesenchymal progenitors from human bone marrow differentiate *in vitro* according to a hierarchical model. *J. Cell Sci.* **113,** 1161–1166.

Muschler, G. F., Nitto, H., Boehm, C. A., and Easley, K. A. (2001). Age- and gender-related changes in the cellularity of human bone marrow and the prevalence of osteoblastic progenitors. *J. Orthop. Res.* **19,** 117–125.

Novikova, L. N., Mosahebi, A., Wiberg, M., Terenghi, G., Kellerth, J. O., and Novikov, L. N. (2006). Alginate hydrogel and matrigel as potential cell carrier for neurotransplantation. *J. Biomed. Mat. Res.* **77A,** 242–252.

Ngo, T. T. B., Waggoner, P. J., Romero, A. A., Nelson, K. D., Eberhart, R. C., and Smith, G. M. (2003). Poly-l-lactide microfilaments enhance peripheral nerve regeneration across extended nerve lesions. *J. Neurosci. Res.* **72,** 227–238.

Pereira Lopes, F. R., de Moura Campos, L. C., Correa, J. D. Jr., Balduino, A., Lora, S., Langone, F., Borojevic, R., and Martinez, A. M. B. (2006). Bone marrow stromal cells and resorbable collagen guidance tubes enhance sciatic nerve regeneration in mice. *Exp. Neurol.* **198,** 457–468.

Raimondo, S., Penna, C., Pagliaro, P., and Geuna, S. (2006). Morphological characterisation of GFP stably transfected adult mesenchymal bone marrow stem cells. *J. Anat.* **208,** 3–12.

Schallmoser, K., Rodhe, E., Reinisch, A., Bartmann, C., Thaler, D., Drexler, C., Obenauf, A. C., Lanzer, G., Linkesch, W., and Strunk, D. (2008). Rapid large-scale expansion of functional mesenchymal stem cells from unmanipulated bone marrow without animal serum. *Tissue Eng.* **14,** 185–196.

Schlosshauer, B., Dreesmann, L., Schaller, H. E., and Sinis, N. (2006). Synthetic nerve guide implants in humans: A comprehensive survey. *Neurosurgery* **59**(4), 740–748.

Seckel, B. R., Jones, D., Hekimian, K. J., Wang, K. K., Chakalis, D. P., and Costas, P. D. (1995). Hyaluronic acid through a new injectable nerve guide delivery stem enhances peripheral nerve regeneration in the rat. *J. Neurosci. Res.* **40,** 3138–3324.

Shimizu, S., Kitada, M., Ishikawa, H., Itokazu, Y., Wakao, S., and Dezawa, M. (2007). Peripheral nerve regeneration by the *in vitro* differentiated human bone marrow stromal cells with Schwann cell property. *Biochem. Biophys. Res. Commun.* **359,** 915–920.

Strem, B. M., Hicok, K. C., Zhu, M., Wulur, I., Alfonso, Z., Schreiber, R. E., Fraser, J. K., and Hedrick, M. H. (2005). Multipotential differentiation of adipose tissue-derived stem cells. *Keio J. Med.* **54,** 132–141.

Sulaiman, O. A., Voda, J., Gold, B. G., and Gordon, T. (2002). FK506 increases peripheral nerve regeneration after chronic axotomy but not after chronic Schwann cell denervation. *Exp. Neurol.* **175,** 127–137.

Tholpady, S. S., Katz, A. J., and Ogle, R. C. (2003). Mesenchymal stem cells from rat visceral fat exhibit multipotential differentiation in vitro. *Anat. Rec. A Discov. Mol. Cell. Evol. Biol.* **272,** 398–402.

Thompson, D. M., and Buettner, H. M. (2006). Neurite outgrowth is directed by Schwann cell alignment in the absence of other guidance cues. *Ann. Biomed. Eng.* **34,** 161–168.

Tohill, M., Mantovani, C., Wiberg, M., and Terenghi, G. (2004a). Rat bone marrow mesenchymal stem cells express glial markers and stimulate nerve regeneration. *Neurosci. Lett.* **362,** 200–203.

Tohill, M. P., Mann, D. J., Mantovani, C., Wiberg, M., and Terenghi, G. (2004b). Green fluorescent protein is a stable morphological marker for Schwann cell transplants in bioengineered nerve conduits. *Tissue Eng.* **10,** 1359–1367.

Tohill, M., and Terenghi, G. (2004). Stem-cell plasticity and therapy for injuries of the peripheral nervous system. *Biotechnol. Appl. Biochem.* **40,** 17–24.

Tong, X. J., Hirai, K. I., Shimada, H., Mizutani, Y., Izumi, T., Toda, N., and Yu, P. (1994). Sciatic nerve regeneration navigated by laminin-fibronectin double coated biodegradable collagen grafts in rats. *Brain. Res.* **663,** 155–162.

Valentini, R. F., and Aebischer, P. (1997). Strategies for the engineering of peripheral nervous tissue regeneration. *In* "Principles of Tissue Engineering" (R. P. Lanza, R. Langer, and W. L. Chick, Eds.), pp. 671–684. Academic Press, Austin.

Vilalta, M., Degano, I. R., Bago, J., Gould, D., Santos, M., Garcia-Arranz, M., Ayatas, R., Fuster, C., Chernajovsky, Y., Garcia-Olmo, D., Rubio, N., and Blanco, J. (2008). Biodistribution, long term survival and safety of human adipose tissue-derived mesenchymal stem cells transplanted in nude mice by high sensitivity non-invasive bioluminescence imaging. *Stem. Cell Dev.* **17,** 993–1003.

Wang, D., Liu, X. L., Zhu, J. K., Jiang, L., Hu, J., Zhang, Y., Yang, L. M., Wang, H. G., and Yi, J. H. (2008). Bridging small-gap peripheral nerve defects using acellular nerve allograft implanted with autologous bone marrow stromal cells in primates. *Brain Res.* **1188,** 44–53.

Wang, W., Itoh, S., Konno, K., Kikkawa, T., Ichinose, S., Sakai, K., Ohkuma, T., and Watabe, K. (2008). Effects of Schwann cell allignment along the oriented electrospun chitosan nanofibres on nerve regeneration. *J. Biomed. Mater. Res.* [Epub ahead of print].

Welin, D., Novikova, L. N., Wiberg, M., Kellerth, J. O., and Novikov, L. N. (2008). Survival and regeneration of cutaneous and muscular afferent neurons after peripheral nerve injury in adult rats. *Exp. Brain. Res.* **186,** 315–323.

Wen, X., and Tresco, P. A. (2006). Effect of filament diameter and extracellular matrix molecule precoating on neurite outgrowth and Schwann cell behaviour on multifilament entubulation bridging device *in vitro*. *J. Biomed. Mat. Res.* **76,** 626–637.

Woodbury, D., Schwarz, E. J., Prockop, D. J., and Black, I. B. (2000). Adult rat and human bone marrow stromal cells differentiate into neurons. *J. Neurosci. Res.* **61,** 364–370.

Wu, J., and Chiu, D. T. (1999). Painful neuromas: a review of treatment modalities. *Ann. Plast. Surg.* **43,** 661–667.

Yamakawa, T., Kakinoki, R., Ikeguchi, R., Nakayama, K., Morimoto, Y., and Nakamura, T. (2007). Nerve regeneration promoted in a tube with vascularity containing bone marrow-derived cells. *Cell Transplant.* **16,** 811–822.

Zhang, P., He, X., Liu, K., Fu, Z., Zhang, D., Zhang, Q., and Jiang, B. (2004). Bone marrow stromal cells differentiated into functional Schwann cells in injured rat sciatic nerve. *Artif. Cells Blood Substit. Immobil. Biotechnol.* **32,** 509–518.

Zhang, P., He, X., Zhao, F., Zhang, D., Fu, Z., and Jiang, B. (2005). Bridging small-gap peripheral nerve defects using biodegradable chitin conduits with cultured Schwann and bone marrow stromal cells in rats. *J. Reconstr. Microsurg.* **21,** 565–572.

TRANSPLANTATION OF OLFACTORY ENSHEATHING CELLS FOR PERIPHERAL NERVE REGENERATION

Christine Radtke,* Jeffery D. Kocsis,†,‡ and Peter M. Vogt*

*Department of Plastic, Hand and Reconstructive Surgery, Hannover Medical School, Hannover, Germany

†Department of Neurology and Center for Neuroscience and Regeneration Research, Yale University School of Medicine, New Haven, Connecticut 06510, USA

‡Rehabilitation Research Center, Veterans Affairs Connecticut Healthcare System, West Haven, Connecticut 06516, USA

Peripheral nerve injury is a common clinical problem, and the development of novel strategies to enhance peripheral nerve regeneration is important. Traumatic events, including motor vehicle accidents, sports-related injuries, violence, and falls, lead to significant numbers of peripheral nerve lesions. Traumatic nerve injuries are often associated with life-threatening injuries, which must be treated first. During the delay in nerve repair, the transected nerves undergo Wallerian degeneration. Therefore, delay before surgical treatment is critical, but care must also be taken to ensure that nerve reapposition is performed in a manner that will result in a therapeutic benefit. Peripheral nerve repair after transection injury combined with transplantation of myelin-forming glia cells, for example, Schwann cells (SCs) or olfactory ensheathing cells (OECs), may facilitate the regenerative process. Cell-based therapies are being considered in clinical trials for a number of neurological diseases, including multiple sclerosis, spinal cord injury, Parkinson's disease, and stroke. The rationale is that transplanted cells may provide neuroprotection by production of chemokines and neurotrophins or could serve as a replacement therapy. A number of cells derived from adult peripheral tissues for cell therapies are also being actively investigated. These cells include SCs from peripheral nerve, olfactory OECs from the olfactory

INTERNATIONAL REVIEW OF NEUROBIOLOGY, VOL. 87
DOI: 10.1016/S0074-7742(09)87022-0

system, and stromal cells from bone marrow (mesenchymal stem cells, MSCs). In principle, these cells could be derived autologously, and used acutely or expanded in culture and used for cell-based therapies. Here, we review experimental work demonstrating the potential of one of these cells, the OEC, as an experimental tool for promoting recovery in peripheral nerve injury.

I. Consequences of Nerve Injury

Trauma often leads to peripheral nerve injury, which is a common and devastating complication that can cause irreversible impairment or even complete functional loss of the affected limb. Some axonal regeneration and functional recovery can occur after surgical reapposition, but the clinical outcome is not optimal and research to develop further interventional approaches to optimize functional recovery after nerve repair is ongoing and important. Peripheral nerve injury can result not only in transection of axonal fibers, but also in axonal demyelination. If the proximal end of the nerve is obstructed from the peripheral target, the sprouting axons can regenerate within the nerve and result in neuroma formation with attendant pain and dysesthesia. Thus, achieving regeneration after nerve injury is important not only for restoration of function, but also to prevent the potentially devastating effects of aberrant sensory impulse generation associated with neuroma formation. During the regenerative process, it is not only important that the transected axons regenerate and reinnervate the appropriate target, be it muscle or a sensory organ, the axons must form nodes of Ranvier with appropriate sodium organization in order to restore rapid impulse conduction. Experimental transplantation of myelin-forming cells into a nerve repair site has been shown to facilitate this complex repair process.

Surgical techniques for reapposition of nerve represent an important tool for the peripheral nerve repair. Early studies have evaluated the efficacy of appropriate nerve anastomotic techniques and application of biological materials to nerve stumps to study their role in promoting regeneration of peripheral nerves (Terzis, 1979). Our current understanding is that minimal tissue manipulation, attenuation of inflammatory responses, and minimization of tension at suture sites are important conditions for successful structural and functional anastomosis of peripheral nerves (Grant *et al.*, 1999). Nerve conduction is often blocked at the repair site where extensive scar tissue can form as mechanical obstruction and surgical neurolysis is necessary. This scarring is thought to impede the axonal regeneration.

Peripheral nerve injury is a common clinical problem, and strategies to enhance peripheral nerve regeneration are clinically important. Cell transplantation approaches are being already used experimentally to enhance regeneration of spinal cord axons. One cell of particular interest for both the central nervous system (CNS) and the peripheral nervous system (PNS) is a specialized glia cell derived from olfactory tissue, the olfactory ensheathing cell (OEC). These cells can form cellular bridges and provide trophic support in the spinal cord, which allows axons to regenerate, and can enhance nerve regeneration after suture repair (Radtke *et al.*, 2009).

II. Unique Properties of Olfactory Ensheathing Cells

OECs are specialized cells which support axons that leave the olfactory epithelium and project through the olfactory nerve system into the olfactory bulb of the CNS. They are glial cells found in the olfactory mucosa, olfactory nerve, and the outer nerve layer of olfactory bulb in the brain. The OECs are pluripotent in that they display Schwann cell (SC)- or astrocyte-like cell properties (Devon and Doucette, 1992). Most investigators have harvested OECs from the olfactory bulb, thus providing a potential autologous source of cells to reduce the risk of immunological rejection (Marshall *et al.*, 2006). Olfactory epithelial neurons in the nasal mucosa are continuously replaced and regenerate axons into the CNS (olfactory bulbs) even in the adult (Graziadei and Graziadei, 1979; Moulton, 1974). After olfactory nerve transection, cell regeneration in the olfactory epithelium is accelerated (Carr and Farbman, 1992). This unique property of olfactory receptor neurons and their close association with OECs may led to the suggestion that they guide and enhance regenerating CNS axons through a normally growth inhibitory environment (PNS–CNS boundary) (Li *et al.*, 1997; Ramon-Cueto and Valverde, 1995). In support of this hypothesis, OECs are permissive for regenerating axons to grow through inhibitory substrates such as gliotic tissue (Liuzzi and Lasek, 1987; Ramon-Cueto and Nieto-Sampedro, 1994; Rudge and Silver, 1990), though under normal circumstances neither CNS nor PNS axons are able to traverse this environment (Liuzzi and Lasek, 1987; Rudge and Silver, 1990). OECs possess the ability to remove degenerating axons via phagocytosis and can produce channels along which newly formed axons can regenerate (Li *et al.*, 2004). Thus, OECs might have the ability to navigate glial scars and remyelinate axons in damaged CNS (Ramon-Cueto and Nieto-Sampedro, 1994) by traversing the intervening mesenchyme, utilizing a mechanism employed during embryogenesis in the navigation of olfactory nerves to the olfactory bulbs (Raisman and Li, 2007). OECs are also a source of neurotrophic factors, such as nerve growth factor (NGF), brain derived-neurotrophic factor (BDNF) (Marshall *et al.*, 2006), platelet-derived growth factor, and neuropeptide Y (Ubink *et al.*, 1994), suggesting that trophic factor production by OECs might enhance the survival of damaged axons. Additionally, OECs normally do not form myelin, but can do so after transplantation into demyelinated lesions.

Following nerve injury, fibroblasts invade the lesion site and scarring occurs. This mechanical obstruction reduces axonal growth through the lesion site and, consequently, impairs functional outcome. The scar is thought to impede the migration of regenerating axons and cells. In contrast to SCs that invade the lesion, OECs are able to cross this barrier (Franklin and Barnett, 1997). Furthermore, OECs have a greater migratory potential than SCs (Li *et al.*, 1998; Ramon-Cueto and Nieto-Sampedro, 1994) and they produce more neurotrophic factors (Wewetzer *et al.*, 2002). In addition, OECs do not accumulate proteoglycans as do SCs, which cause growth cone collapse (Bovolenta *et al.*, 1992; Plant *et al.*, 2002). Therefore, OECs may be better candidates for cell-based regenerative therapies, and clinical trials have begun using OECs to repair spinal cord lesions in humans (Feron *et al.*, 2005; Huang *et al.*, 2003; Lima *et al.*, 2006).

III. OECs in Spinal Cord Injury

Several studies have demonstrated enhanced functional recovery after OEC transplantation into the injured spinal cord (García-Alías *et al.*, 2004; Li *et al.*, 1998; Plant *et al.*, 2002; Ramon-Cueto *et al.*, 2000; Sasaki *et al.*, 2004; Verdú *et al.*, 2003). While the precise mechanism of this functional recovery is not fully understood, several mechanisms have been suggested, including remyelination (Devon and Doucette, 1992; Franklin *et al.*, 1996; Imaizumi *et al.*, 2000; Sasaki *et al.*, 2004), long axon tract regeneration (Imaizumi *et al.*, 2000; Li *et al.*, 1997; Ramon-Cueto *et al.*, 2000), axonal sparing (Plant *et al.*, 2002), and plasticity associated with novel polysynaptic pathways (Bareyre *et al.*, 2004; Keyvan-Fouladi *et al.*, 2003). In addition, recruitment of endogenous SCs (Boyd *et al.*, 2004; Ramer *et al.*, 2004; Takami *et al.*, 2002) and remote inhibition of apoptosis of motor cortical neurons (Sasaki *et al.*, 2006) have been suggested to contribute to improvement in functional outcome of injured spinal cord after OEC transplantation.

IV. OECs in Peripheral Nerve Repair and Contribution of OEC Transplantation to Peripheral Nerve Repair

Nerve repair combined with myelin-forming cell transplantation is a relatively simple, rapid, and efficient means of peripheral nerve repair. However, functional nerve regeneration requires not only axonal sprouting and elongation, but also remyelination and appropriate ion-channel deployment at the node of Ranvier.

Remyelination is necessary for the restoration of normal nerve conduction and for the protection of axons from neurodegenerative processes. The combination of surgical nerve repair and transplantation of peripheral myelin-forming cells, which enhance axonal regeneration and remyelinate demyelinated fibers, is currently being investigated.

Unlike in the CNS, regeneration can occur in peripheral nerve. The SCs in the distal segment of a cut nerve dissociate from the degenerating axons and upregulate the low-affinity NGF receptor, $p75^{NGFR}$ (p75 nerve growth factor receptor), and express NGF (Johnson *et al.*, 1988). The axons in the proximal nerve stump sprout and regenerate through SC-enriched basal lamina tubes and can re-establish functional connections in peripheral targets, such as skin and muscle, and various degrees of functional recovery can occur. However, a number of issues such as the delay for SCs to express NGF, dying back of the injured axons, navigation of axons across a complex nerve injury site, and appropriate targeting to peripheral end structures are major clinical concerns.

For several reasons transplantation of OECs has been considered as an adjunct for peripheral nerve repair. The reasoning is that they may provide a scaffold for axons to regenerate as well as trophic factors and directional cues (Deumens *et al.*, 2006). OEC engraftment into axotomized facial nerve enhances axonal sprouting (Deumens *et al.*, 2006; Guntinas-Lichius *et al.*, 2001) and promotes recovery of vibrissae motor performance (Guntinas-Lichius *et al.*, 2001). The rate of eye closure was increased following OEC transplantation in a facial nerve lesion model, but aberrant nerve branching was not changed (Choi and Raisman, 2005). Detailed mophological analysis indicates that both SCs (Radtke *et al.*, 2005) and OECs (Dombrowski *et al.*, 2006) transplanted into transected sciatic nerve integrate into the injury site and form peripheral myelin on the regenerated axons. Moreover, the nodes of Ranvier of the regenerated axons myelinated by the transplanted cells express the appropriate sodium channel (voltage-gated sodium channel, NaV1.6). The question was raised whether these engrafted cells accelerate or improve functional outcome following nerve injury (Ibrahim *et al.*, 2006). To address this issue, OECs were isolated from the olfactory bulbs of adult green fluorescent protein (GFP)-expressing transgenic rats and transplanted into a sciatic nerve lesion. Five weeks after transplantation, the nerves were studied histologically. GFP-expressing OECs survived in the lesion and distributed longitudinally across the lesion zone. The internodal regions of individual teased living fibers were identified by GFP in the cytoplasmic and nuclear compartments of cells surrounding the axons. Immunoelectron microscopy for GFP indicated that the transplanted OECs formed peripheral-type myelin. Immunostaining for sodium channel and Caspr revealed a high density of NaV1.6 at the newly formed nodes of Ranvier, which were flanked by paranodal Caspr staining. Thus, the transplanted OECs extensively integrate into transected peripheral nerve, form myelin on regenerated peripheral nerve fibers, and nodes of Ranvier were formed on the regenerated axons with proper sodium channel organization. Furthermore, the regenerated fibers show reconstructed nodes of Ranvier with proper sodium channel organization, that is, sodium channel NaV1.6. Thus, transplanted GFP-expressing OECs are able to remyelinate regenerated peripheral nerve fibers and restore proper nodal structure in injured PNS, indicating that they can contribute to local nerve repair (Dombrowski *et al.*, 2006).

In a recent study, we demonstrated that OECs transplanted at the time of microsurgical nerve repair enhanced nerve regeneration and functional outcome (Radtke *et al.*, 2009). Interestingly, we found reduced die-back of the axons proximal to the transection site and an increased number of regenerated axons distal to the transection site. Moreover, the microsuture repaired nerves were thinner at the repair site as compared to the nerves that had OEC transplants. We hypothesize that the transplanted OECs are primed to produce neurotrophins and, therefore, can have an immediate effect on the injured axons. That is, they may allow for less axonal die-back and early regeneration of the injured axons. This more rapid regeneration induced by the OECs could allow them to navigate across the injury site before significant scar formation occurs. This hypothesis certainly will need further testing.

V. Challenges in Cell-Therapy Approaches for Peripheral Nerve Repair

An important concern with regard to cell transplantation therapies is the harvesting of sufficient numbers of appropriate cells for transplantation. While in the laboratory a number of cell types can be expanded in culture with trophic factors and mitogens, it is not clear if such expansion will alter the physiology of the cells in a negative way with regard to their ability to produce appropriate neurotrophins and form functional myelin. Another concern with expanded cells is the potential risk of tumor formation. It is absolutely essential to determine if experimental *in vivo* transplantation of expanded cells not only retain their ability to carry out neural repair, but also do not form tumors. This concern emphasizes the importance of Phase I clinical studies to assess safety in small numbers of patients. It cannot be assumed that observed safety in animal models will apply to humans.

Transplantation of glial cells may not only remyelinate axons, but could also produce trophic factors that may alter axonal sprouting and synapse formation. One concern is that the newly organized neural structures could be maladaptive and elicit neurological problems, such as pain or allodynia (Hofstetter *et al.*, 2005). Care must be taken so that functional properties of regenerated axons elicited by interventional approaches, such as cell transplantation, do not result in maladaptive responses. Importantly, we found that nodal sodium channels are appropriate following transplantation in injured nerve (Dombrowski *et al.*, 2006;

Radtke *et al.*, 2005) and spinal cord (Sasaki *et al.*, 2006). Appropriate design of functional experiments to assess efficacy will be essential for evaluating these approaches in clinical studies. Additionally, because patterns of remyelination and potentially partial restoration of axon and synapses may not recapitulate pre-injury structures, rehabilitation strategies may be needed to better understand and maximize the extent of achievable functional recovery.

Another challenge is the time delay that currently is required for preparation of SCs or OECs, particularly if they are to be derived autologously. OECs can be prepared from biopsy of nasal mucosa and SCs from sural nerve. However, these autologous sources of cells would require days in culture for preparation and, thus, a critical therapeutic window could be missed. Clearly, the development of immunologically compatible cells that could be produced in high quantity and banked for later use is a critical experimental challenge.

Unlike animal models where cell survival and the anatomical pattern of remyelination can be studied at the termination of the experiments with histological processing, extensive histological examination in human studies is not possible. Moreover, while magnetic resonance imaging (MRI) techniques are powerful, there are currently limitations in assaying donor cell survival *in vivo* after transplantation in humans. Techniques to label donor cells for recognition with MRI are experimental and not currently routinely employed. As these techniques become compatible with routine clinical use, more powerful methods will be available to assess donor cell survival and distribution in human clinical studies. Some information of cell survival in transplanted patients could be obtained from biopsy, but the amount of tissue is limited and safety issues are associated with biopsy procedures.

VI. Prospects of Cell-Based Clinical Approaches

In cell transplantation approaches to treat Parkinson's disease, neuronal replacement of a specific neuronal type is required, that is, mesencephalic dopaminergic neurons. This presents many challenges in terms of cell collection and expansion in culture and directing cells to a specific neuron type that must synaptically integrate into the CNS. In many ways, the use of myelin-forming cells or precursors to remyelinate CNS axons is less daunting. Endogenous remyelination can occur in both the CNS and PNS, and committed myelin-forming cells and precursors can be derived from a number of sources including autologous CNS, PNS, and bone marrow. There are strong signaling mechanisms between axons and appropriate glial cells for induction of myelin formation. Therefore, delivery of sufficient numbers of myelin-forming cells to areas of demyelination may readily result in new myelin formation.

In addition to addressing peripheral nerve injury, cell-therapy approaches may be useful in traumatic CNS injury. Contusive spinal cord injury results in necrosis of spinal cord tissue, but even in humans, there are often areas surrounding the central necrotic core that are demyelinated. OECs are being used in clinical cell transplantation studies in China (Huang *et al.*, 2003) and Australia (Feron *et al.*, 2005) for spinal cord injury patients. Transplantation of both OECs (Imaizumi *et al.*, 2000; Li *et al.*, 1997; Ramon-Cueto and Nieto-Sampedro, 1994) and SCs (Imaizumi *et al.*, 2000; Takami *et al.*, 2002) suggest that axonal regeneration with subsequent myelination can be elicited by implantation of these cells. It is thought that the peripheral myelin-forming cells can bridge a lesion zone in the spinal cord and provide a permissive environment for axonal regeneration. Significant behavioral improvement (Li *et al.*, 1997; Sasaki *et al.*, 2004) occurs after OEC transplantation. However, whether the functional improvement is the result of axonal regeneration or preservation (i.e., neuroprotection) is an important concern. Several recent studies on transplanting OECs and bone marrow cells (Chopp *et al.*, 2000; Li *et al.*, 2006) into a contusive spinal cord injury model demonstrate that the extent of tissue damage is significantly reduced. This suggests that a neuroprotective influence resulting in tissue sparing is effectuated by the cell transplantation. Transplanted cells may produce neurotrophic factors *in vivo*, and several trophic factors have been reported to have therapeutic effects following CNS trauma. Exogenous application of BDNF, GDNF (glial cell-line-derived neurotrophic factor), NGF, EGF (epidermal growth factor), and bFGF (basic fibroblast growth factor), have been reported to limit the extent of ischemic lesion volume (Ay *et al.*, 2001; Schäbitz *et al.*, 1997). The neural tissue rescue may act through several mechanisms, such as free-radical scavenging, antiapoptotic activity, anti-inflammatory activity, and antiglutamate excitotoxicity (Hirouchi and Ukai, 2002). Moreover, mesenchymal stem cells (MSCs) genetically modified to produce neurotrophins or angiogenic factors can improve the functional outcome after cerebral infarction (Onda *et al.*, 2008). Thus, in addition to the prospect of enhancing axonal regeneration and reconstructing myelin, cell-therapy approaches utilizing OECs, SCs, or any of a number of stem-cell types may provide an important neuroprotective strategy in the treatment of traumatic CNS and peripheral nerve injury.

In summary, experimental work indicates that transplantation of OECs at the time of surgical nerve repair can enhance regeneration and functional outcome. Beneficial therapeutic effects may well result from both an initial neuroprotective effect mediated by donor cell production of trophic factors to facilitate sprouting of the injured axons and reduction in axonal die-back, and later reparative effects on the regenerated axons, such as remyelination. The early effects may be critical in that they allow regeneration across the repair site before significant scar formation occurs. Although much more experimental laboratory investigation is required, progress to date holds the exciting prospect of a novel adjunct cell therapy to suture repair to improve treatment of peripheral nerve injuries.

References

Ay, I., Sugimori, H., and Finklestein, S. P. (2001). Intravenous basic fibroblast growth factor (bFGF) decreases DNA fragmentation and prevents downregulation of Bcl-2 expression in the ischemic brain following middle cerebral artery occlusion in rats. *Brain Res. Mol. Brain Res.* **87,** 71–80.

Bareyre, F. M., Kerschensteiner, M., Raineteau, O., Mettenleiter, T. C., Weinmann, O., and Schwab, M. E. (2004). The injured spinal cord spontaneously forms a new intraspinal circuit in adult rats. *Nat. Neurosci.* **7,** 269–277.

Bovolenta, P., Wandosell, F., and Nieto-Sampedro, M. (1992). CNS glial scar tissue: A source of molecules which inhibit central neurite outgrowth. *Prog. Brain Res.* **94,** 367–379.

Boyd, J. G., Lee, J., Skihar, V., Doucette, R., and Kawaja, M. D. (2004). LacZ-expressing olfactory ensheathing cells do not associate with myelinated axons after implantation into the compressed spinal cord. *Proc. Natl. Acad. Sci. USA* **101,** 2162–2166.

Carr, V. M., and Farbman, A. I. (1992). Ablation of the olfactory bulb up-regulates the rate of neurogenesis and induces precocious cell death in olfactory epithelium. *Exp. Neurol.* **115,** 55–59.

Choi, D., and Raisman, G. (2005). Disorganization of the facial nucleus after nerve lesioning and regeneration in the rat: Effects of transplanting candidate reparative cells to the site of injury. *Neurosurgery* **56,** 1093–1100.

Chopp, M., Zhang, X. H., Li, Y., Wang, L., Chen, J., Lu, D., Lu, M., and Rosenblum, M. (2000). Spinal cord injury in rat: Treatment with bone marrow stromal cell transplantation. *Neuroreport* **11,** 3001–3005.

Deumens, R., Koopmans, G. C., Lemmens, M., Mollers, S., Honig, W. M., Steinbusch, H. W., Brook, G., and Joosten, E. A. (2006). Neurite outgrowth promoting effects of enriched and mixed OEC/ONF cultures. *Neurosci. Lett.* **397,** 20–24.

Devon, R., and Doucette, R. (1992). Olfactory ensheathing cells myelinate dorsal root ganglion neurites. *Brain Res.* **589,** 175–179.

Dombrowski, M. A., Sasaki, M., Lankford, K. L., Kocsis, J. D., and Radtke, C. (2006). Myelination and nodal formation of regenerated peripheral nerve fibers following transplantation of acutely prepared olfactory ensheathing cells. *Brain Res.* **1125,** 1–8.

Feron, F., Perry, C., Cochrane, J., Licina, P., Nowitzke, A., Urquhart, S., Geraghty, T., and Mackay-Sim, A. (2005). Autologous olfactory ensheathing cell transplantation in human spinal cord injury. *Brain* **128,** 2951–2960.

Franklin, R. J., Gilson, J. M., Franceschini, I. A., and Barnett, S. C. (1996). Schwann cell-like myelination following transplantation of an olfactory bulb-ensheathing cell line into areas of demyelination in the adult CNS. *Glia* **17,** 217–224.

Franklin, R. J., and Barnett, S. C. (1997). Do olfactory glia have advantages over Schwann cells for CNS repair? *J. Neurosci. Res.* **50,** 665–672.

García-Alías, G., Lopez-Vales, R., Fores, J., Navarro, X., and Verdú, E. (2004). Acute transplantation of olfactory ensheathing cells or Schwann cells promotes recovery after spinal cord injury in the rat. *J. Neurosci. Res.* **75,** 632–641.

Grant, G. A., Goodkin, R., and Kliot, M. (1999). Evaluation and surgical management of peripheral nerve problems. *Neurosurgery* **44,** 825–840.

Graziadei, P. P., and Graziadei, G. A. (1979). Neurogenesis and neuron regeneration in the olfactory system of mammals. I. Morphological aspects of differentiation and structural organization of the olfactory sensory neurons. *J. Neurocytol.* **8,** 1–18.

Guntinas-Lichius, O., Angelov, D. N., Tomov, T. L., Dramiga, J., Neiss, W. F., and Wewetzer, K. (2001). Transplantation of olfactory ensheathing cells stimulates the collateral sprouting from axotomized adult rat facial motoneurons. *Exp. Neurol.* **172,** 70–80.

Hofstetter, C. P., Holmstrom, N. A., Lilja, J. A., Schweinhardt, P., Hao, J., Spenger, C., Wiesenfeld-Hallin, Z., Kurpad, S. N., Frisen, J., and Olson, L. (2005). Allodynia limits the usefulness of intraspinal neural stem cell grafts; directed differentiation improves outcome. *Nat. Neurosci.* **8,** 346–353.

Hirouchi, M., and Ukai, Y. (2002). Current state on development of neuroprotective agents for cerebral ischemia. *Nippon Yakurigaku Zasshi* **120,** 107–113.

Huang, H., Chen, L., Wang, H., Xiu, B., Li, B., Wang, R., Zhang, J., Zhang, F., Gu, Z., Li, Y., Song, Y., Hao, W., *et al.* (2003). Influence of patients' age on functional recovery after transplantation of olfactory ensheathing cells into injured spinal cord injury. *Chin. Med. J. (Engl.)* **116,** 1488–1491.

Ibrahim, A., Li, Y., Li, D., Raisman, G., and El Masry, W. S. (2006). Olfactory ensheathing cells: Ripples of an incoming tide? *Lancet Neurol.* **5,** 453–457.

Imaizumi, T., Lankford, K. L., and Kocsis, J. D. (2000). Transplantation of olfactory ensheathing cells or Schwann cells restores rapid and secure conduction across the transected spinal cord. *Brain Res.* **854,** 70–78.

Johnson, E. M. Jr., Taniuchi, M., and DiStefano, P. S. (1988). Expression and possible function of nerve growth factor receptors on Schwann cells. *Trends Neurosci.* **11,** 299–304.

Keyvan-Fouladi, N., Raisman, G., and Li, Y. (2003). Functional repair of the corticospinal tract by delayed transplantation of olfactory ensheathing cells in adult rats. *J. Neurosci.* **23,** 9428–9434.

Li, Y., Carlstedt, T., Berthold, C. H., and Raisman, G. (2004). Interaction of transplanted olfactory-ensheathing cells and host astrocytic processes provides a bridge for axons to regenerate across the dorsal root entry zone. *Exp. Neurol.* **188,** 300–308.

Li, Y., Field, P. M., and Raisman, G. (1997). Repair of adult rat corticospinal tract by transplants of olfactory ensheathing cells. *Science* **277,** 2000–2002.

Li, Y., Field, P. M., and Raisman, G. (1998). Regeneration of adult rat corticospinal axons induced by transplanted olfactory ensheathing cells. *J. Neurosci.* **18,** 10514–10524.

Li, Y., McIntosh, K., Chen, J., Zhang, C., Gao, Q., Borneman, J., Raginski, K., Mitchell, J., Shen, L., Zhang, J., Lu, D., and Chopp, M. (2006). Allogeneic bone marrow stromal cells promote glial-axonal remodeling without immunologic sensitization after stroke in rats. *Exp. Neurol.* **198,** 313–325.

Lima, C., Pratas-Vital, J., Escada, P., Hasse-Ferreira, A., Capucho, C., and Peduzzi, J. D. (2006). Olfactory mucosa autografts in human spinal cord injury: A pilot clinical study. *J. Spinal Cord Med.* **29,** 191–203; discussion 204–206.

Liuzzi, F. J., and Lasek, R. J. (1987). Astrocytes block axonal regeneration in mammals by activating the physiological stop pathway. *Science* **237,** 642–645.

Marshall, C. T., Lu, C., Winstead, W., Zhang, X., Xiao, M., Harding, G., Klueber, K. M., and Roisen, F. J. (2006). The therapeutic potential of human olfactory-derived stem cells. *Histol. Histopathol.* **21,** 633–643.

Moulton, D. G. (1974). Dynamics of cell populations in the olfactory epithelium. *Ann. N. Y. Acad. Sci.* **237,** 52–61.

Onda, T., Honmou, O., Harada, K., Houkin, K., Hamada, H., and Kocsis, J. D. (2008). Therapeutic benefits by human mesenchymal stem cells (hMSCs) and Ang-1 gene-modified hMSCs after cerebral ischemia. *J. Cereb. Blood Flow Metab.* **28,** 329–340.

Plant, G. W., Currier, P. F., Cuervo, E. P., Bates, M. L., Pressman, Y., Bunge, M. B., and Wood, P. M. (2002). Purified adult ensheathing glia fail to myelinate axons under culture conditions that enable Schwann cells to form myelin. *J. Neurosci.* **22,** 6083–6091.

Radtke, C., Aizer, A. A., Agulian, S. K., Lankford, K. L., Vogt, P. M., and Kocsis, J. D. (2009). Transplantation of olfactory ensheathing cells enhances peripheral nerve regeneration after microsurgical nerve repair. *Brain Res.* **1254,** 10–17.

Radtke, C., Akiyama, Y., Lankford, K. L., Vogt, P. M., Krause, D. S., and Kocsis, J. D. (2005). Integration of engrafted Schwann cells into injured peripheral nerve: Axonal association and nodal formation on regenerated axons. *Neurosci. Lett.* **387,** 85–89.

Raisman, G., and Li, Y. (2007). Repair of neural pathways by olfactory ensheathing cells. *Nat. Rev. Neurosci.* **8,** 312–319.

Ramer, L. M., Au, E., Richter, M. W., Liu, J., Tetzlaff, W., and Roskams, A. J. (2004). Peripheral olfactory ensheathing cells reduce scar and cavity formation and promote regeneration after spinal cord injury. *J. Comp. Neurol.* **473,** 1–15.

Ramon-Cueto, A., Cordero, M. I., Santos-Benito, F. F., and Avila, J. (2000). Functional recovery of paraplegic rats and motor axon regeneration in their spinal cords by olfactory ensheathing glia. *Neuron* **25,** 425–435.

Ramon-Cueto, A., and Nieto-Sampedro, M. (1994). Regeneration into the spinal cord of transected dorsal root axons is promoted by ensheathing glia transplants. *Exp. Neurol.* **127,** 232–244.

Ramon-Cueto, A., and Valverde, F. (1995). Olfactory bulb ensheathing glia: A unique cell type with axonal growth-promoting properties. *Glia* **14,** 163–173.

Rudge, J. S., and Silver, J. (1990). Inhibition of neurite outgrowth on astroglial scars *in vitro*. *J. Neurosci.* **10,** 3594–3603.

Sasaki, M., Black, J. A., Lankford, K. L., Tokuno, H. A., Waxman, S. G., and Kocsis, J. D. (2006). Molecular reconstruction of nodes of Ranvier after remyelination by transplanted olfactory ensheathing cells in the demyelinated spinal cord. *J. Neurosci.* **26,** 1803–1812.

Sasaki, M., Lankford, K. L., Zemedkun, M., and Kocsis, J. D. (2004). Identified olfactory ensheathing cells transplanted into the transected dorsal funiculus bridge the lesion and form myelin. *J. Neurosci.* **24,** 8485–8493.

Schäbitz, W. R., Schwab, S., Spranger, M., and Hacke, W. (1997). Intraventricular brain-derived neurotrophic factor reduces infarct size after focal cerebral ischemia in rats. *J. Cereb. Blood Flow Metab.* **17,** 500–506.

Takami, T., Oudega, M., Bates, M. L., Wood, P. M., Kleitman, N., and Bunge, M. B. (2002). Schwann cell but not olfactory ensheathing glia transplants improve hindlimb locomotor performance in the moderately contused adult rat thoracic spinal cord. *J. Neurosci.* **22,** 6670–6681.

Terzis, J. K. (1979). Clinical microsurgery of the peripheral nerve: The state of the art. *Clin. Plast. Surg.* **6,** 247–267.

Ubink, R., Halasz, N., Zhang, X., Dagerlind, A., and Hokfelt, T. (1994). Neuropeptide tyrosine is expressed in ensheathing cells around the olfactory nerves in the rat olfactory bulb. *Neuroscience* **60,** 709–726.

Verdú, E., García-Alías, G., Fores, J., Lopez-Vales, R., and Navarro, X. (2003). Olfactory ensheathing cells transplanted in lesioned spinal cord prevent loss of spinal cord parenchyma and promote functional recovery. *Glia* **42,** 275–286.

Wewetzer, K., Verdú, E., Angelov, D. N., and Navarro, X. (2002). Olfactory ensheathing glia and Schwann cells: Two of a kind? *Cell Tissue Res.* **309,** 337–345.

MANUAL STIMULATION OF TARGET MUSCLES HAS DIFFERENT IMPACT ON FUNCTIONAL RECOVERY AFTER INJURY OF PURE MOTOR OR MIXED NERVES

Nektarios Sinis,* Thodora Manoli,* Frank Werdin,* Armin Kraus,*
Hans E. Schaller,* Orlando Guntinas-Lichius,[†] Maria Grosheva,[‡] Andrey Irintchev,[†]
Emanouil Skouras,[§] Sarah Dunlop,[¶] and Doychin N. Angelov[||]

*Department of Hand, Plastic, Reconstructive Surgery and Burn Unit,
Eberhard-Karls-University of Tuebingen, BG Trauma Center Tuebingen, Germany
[†]ENT-Department, Friedrich-Schiller University, Jena, Germany
[‡]ENT-Department, University of Cologne, Germany
[§]Department of Trauma, Hand and Reconstructive Surgery, University of Cologne, Germany
[¶]School of Animal Biology and Western Australian Institute for Medical Research,
The University of Western Australia, Perth, Australia
[||]Anatomical Institute I, University of Cologne, Germany

Direct coaptation and interpositional nerve grafting (IPNG) of an injured peripheral nerve is still associated with poor functional recovery. Main reasons for that are thought to be an extensive collateral axonal branching at the site of transection and the polyinnervation of motor endplates due to terminal axonal and intramuscular sprouting. Moreover, severe changes occurring within the muscle after long-term denervation, like loss of muscle bulk and circulation as well as progressive fibrosis, have a negative effect on the quality of functional recovery after reinnervation. We have recently shown that manual stimulation (MS) of paralyzed vibrissal muscles in rat promotes full recovery after facial nerve coaptation. Furthermore, MS improved functional recovery after hypoglossal nerve repair, hypoglossal-facial IPNG of the facial nerve in rat. In contrary, MS did not improve recovery after injury of the median nerve in rat, which is however a mixed peripheral nerve comparing to the facial nerve. It is speculated that manually stimulated recovery of motor function requires an intact sensory input, which is affected in case of mixed peripheral nerves but not in case of pure motor

INTERNATIONAL REVIEW OF
NEUROBIOLOGY, VOL. 87
DOI: 10.1016/S0074-7742(09)87023-2

0074-7742/09 $35.00

nerves. In this article, we summarize our results of MS in several peripheral nerve injury models in order to illustrate the application potential of this method and to give insights into further investigations on that field.

I. Introduction

Despite advances in reconstructive microsurgery and the innate capacity of peripheral nerves to regenerate, functional recovery after direct nerve coaptation or interpositional nerve grafting (Hall, 2005) is usually poor. After long-term muscle denervation, severe changes occur within the muscles, which later have a considerable effect on functional recovery after reinnervation, such as loss of bulk and circulation, shrinkage of connective tissue, and fibrosis (Bardosi *et al.*, 1987). After several months of complete denervation, muscle membrane properties also change, becoming no longer responsive to electrical stimulation (ES) (Brown and Holland, 1979). Reinnervated muscles, furthermore, show the so-called "post-paralytic syndrome" involving paresis, synkinesis, dysreflexia, and a considerable disfigurement (Valls-Sole, 2002). The pathomechanism of the post-paralytic syndrome was thought to be associated with a misrouting of the original motor axons lying proximal to the transection site in terms of "collateral axonal branching" (Morris *et al.*, 1972; Shawe, 1955) and in terms of inappropriate terminal and intramuscular reinnervation (Brown *et al.*, 1981; Gordon *et al.*, 2004; Rich and Lichtman, 1989). Sensory axons show sprouting within the peripheral target tissue as well as centrally in the spinal cord (Rajan *et al.*, 2003; Tsuyoshi *et al.*, 2006).

Proximal stumps of transected motor axons give off up to 25 collateral branches in order to reach distal fascicles at the lesion site. These branches normally fail to reach their original fascicles, leading to "collateral axonal branching" and reinnervation of different muscles, often with antagonistic functions, by a single motoneuron (Ito and Kudo, 1994). However, limiting axonal sprouting using antibodies against growth factors at the transection site did not improve functional recovery after facial nerve repair in rat (Guntinas-Lichius *et al.*, 2005).

Upon reaching an endplate, axons undergo further sprouting, in terms of "terminal axonal" and "intramuscular sprouting," which often leads to reinnervation of more than one muscle fiber by a single motoneuron. This state is known as "polyinnervation" of endplates (Brown *et al.*, 1981; Rich and Lichtman, 1989). The degree of endplate polyinnervation was shown to be reciprocally associated with the degree of functional recovery after transection of facial and hypoglossal nerve in rat (Angelov *et al.*, 2007; Evgenieva *et al.*, 2008). Intramuscular sprouting, and therefore endplate polyinnervation, could be inhibited by ES of denervated soleus muscle. ES reduces the number of terminal Schwann cells (TSC) bridges that form between endplates during reinnervation of a completely denervated muscle and cause nerve terminal sprouting (Love *et al.*, 2003). On the contrary, regular ES of completely denervated muscles suppresses the production of chemical mediators required for reconnection of an axon branch with its motor endplate. It also reduces spontaneous fibrillation of orphaned muscle fibers that is thought to be a signal for sprouting of remaining intact nerve axons (Cohan and Kater, 1986). ES therefore showed no satisfactory improvement of functional recovery after complete nerve injury.

We tried a novel approach using manual stimulation (MS) in order to achieve a better functional recovery in several models of peripheral nerve transection. The idea behind that was the clinical benefit of soft tissue massage that was supposed to improve muscle blood flow and to preserve the contractile properties of it (Hovind and Nielsen, 1974). "Facial retraining" using physical rehabilitation methods was recently shown to improve outcome in several conditions after facial nerve injury, like acoustic neuroma, Bell's palsy, Ramsay Hunt syndrome, and facial nerve repair (Barbara *et al.*, 2003; VanSwearingen and Brach, 2003).

First, it was shown that MS of paralyzed vibrissal muscles following facial nerve injury results in full recovery of whisking in adult rats (Angelov *et al.*, 2007). An indirect hint for the importance of muscle stimulation on whisking recovery after facial nerve injury was the observation that blind rats with retinal dystrophy demonstrated a better functional outcome than normal rats, presumably due to forced whisker use (Tomov *et al.*, 2002). Positive results were obtained by applying MS on the orbicularis oculi muscle in adult rats after facial nerve injury as well (Bischoff *et al.*, 2008). Furthermore, MS of facial muscles improved functional recovery after hypoglossal-facial nerve coaptation and interpositional nerve grafting (IPNG) of facial nerve in rats (Guntinas-Lichius *et al.*, 2007). As next step, we decided to test a different nerve transection model in rats, the one of hypoglossal nerve. MS of the suprahyoid-sublingual region improved tongue function (Evgenieva *et al.*, 2008).

In both models of facial and hypoglossal nerves, the sensory input of the affected area came from a different, noninjured nerve. The sensory input of the facial innervated area is being mediated via the trigeminal nerve (V) and the one of the hypoglossal innervated area via the trigeminal (V), the glossopharyngeal (IX), and the vagus (X) nerves. Indeed, after extirpation of the ipsilateral sensory input, mediated by the infraorbital nerve, MS of vibrissal muscles had a rather negative effect on whisking recovery during facial nerve regeneration in rats (Pavlov *et al.*, 2008). This hypothesis was supported by the observation that MS of forearm muscles did not improve recovery of motor function after injury of the mixed median nerve (Sinis *et al.*, 2008). In that case, nerve transection led also to an interruption of the sensory feedback to the brain cortex, which seems to be an important stimulus for functional recovery after reinnervation.

In this review, we will summarize our observations on the role of MS in functional recovery and in several pathomorphological changes occurring after muscle reinnervation.

II. Manual Stimulation

Manual stimulation is thought to be applied as regular stroking or massage of denervated muscles during peripheral nerve regeneration until reinnervation occurs. The background of this treatment is to keep the muscle in a basic metabolic state in order to avoid muscle spindle inactivation and progressive fibrosis. It is important to keep muscle architecture movable so that sufficient muscle contraction after reacquisition of contractile proteins during reinnervation can take place (Lomo *et al.*, 1974). Moreover, MS maintains the sensory input to the cortex or the spinal cord, if this is not affected by the nerve transection. Sufficient sensory input is thought to be an important stimulus for specific target reinnervation, for example, by stimulating plasticity in the cortex or the spinal cord (Galtrey *et al.*, 2007; Guntinas-Lichius *et al.*, 2005) and reducing the degree of polyinnervated endplates (Pavlov *et al.*, 2008).

Very few studies have defined specific stimulation protocols for enhancing function after nerve injury. Though ES was proven to be not useful for complete nerve lesions in long term (Cohan and Kater, 1986), in some models MS of target muscles of peripheral nerve injury really showed satisfactory results. In the examined models, MS was initiated 1 day after surgery and was applied for 5–10 min per day, 5 days a week. The stimulation was applied in a way mimicking the natural movement of the affected muscles (Fig. 1). Whiskers were stimulated mimicking protraction and retraction, by stroking them forward and then letting them passively return backward (Angelov *et al.*, 2007). The orbicularis oculi muscle was subjected to gentle rhythmic manual closure (Bischoff *et al.*, 2008). In the case of hypoglossal nerve injury model, the suprahyoid-sublingual region was stimulated by gentle stroking of the lower jaw and upper neck to stimulate all three extrinsic muscles of the tongue (styloglossus, genoioglossus, and hyoglossus) involved in swallowing (Evgenieva *et al.*, 2008). In the case of median nerve lesion model, volar forearm muscles (palmaris longus, flexor digitorum sublimis, and pronator quadratus) were manually stimulated by stroking (Sinis *et al.*, 2008).

A. Effects of MS on Functional Recovery After Facial Nerve Injury

Manual stimulation of the ipsilateral whiskers after facial nerve injury had a dramatic effect in comparison to untreated rats, resulting in a complete recovery of normal whisking as indicated by the amplitude of movement and the speed during protraction. Recovery under only enriched environmental conditions

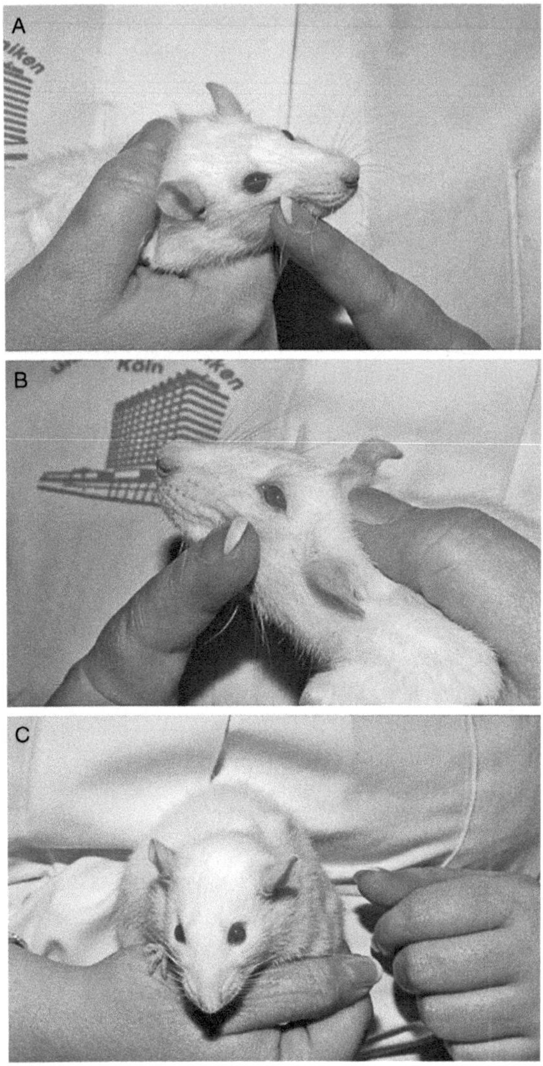

FIG. 1. Postoperative treatment of rats after facial nerve transection and suture. (A) Manual mechanical stimulation of the right, that is, ipsilateral to the nerve transection and suture (FFA) vibrissae and whiskerpad muscles. (B) Manual mechanical stimulation of the left, that is, contralateral to FFA vibrissae and whiskerpad muscles. (C) Handling of the animals. Adopted from Angelov *et al.* (2007).

stimulating the use of face muscles, in particular those controlling the vibrissae, was completely ineffective. Triple retrograde labeling of the zygomatic, buccal, and mandibular branch of facial nerve was applied to assess the projection patterns of motor axons from the facial nucleus to its different motor branches and showed that the degree of collateral axonal branching at the lesion site remains elevated regardless stimulation (Dohm *et al.*, 2000). Moreover, MS reduced the degree of motor endplate polyinnervation in fibers of the vibrissal muscle levator labii superioris (Angelov *et al.*, 2007).

After hypoglossal-facial IPNG of the facial nerve, MS resulted in a significant increase, but not complete restoration, of the whisking amplitude. Also in that case, collateral axonal branching remained elevated, while the degree of endplate polyinnervation was significantly diminished (Guntinas-Lichius *et al.*, 2007). MS after extirpation of the trigeminal sensory input of the vibrissal area (infraorbital nerve) worsened the functional outcome after ipsilater facial nerve injury, illustrating the importance of intact afferent input for motor axon regeneration. In the same study, it was shown that MS counteracts posttraumatic loss of synaptic input in the facial nucleus (Pavlov *et al.*, 2008). This could be achieved by quantifying the levels of synaptophysin in the facial nucleus, an established marker for presynaptic terminals, using immunocytochemistry and fluorescent image analysis (Calhoun *et al.*, 1996).

Application of MS following facial nerve injury on a different facial muscle, the orbicularis oculi, improved dramatically the blink capacity of the eye (Fig. 2). Also, the fraction of polyinnervated neuromuscular junctions was significantly reduced (Bischoff *et al.*, 2008).

B. Effects of MS on Functional Recovery After Hypoglossal Nerve Injury

In the model of hypoglossal nerve repair MS of the suprahyoid-sublingual region gave similar results as in the case of the facial nerve injury models mentioned above. There was a beneficial effect on the restoration of tongue position, measured as the deviation of the tongue tip from the middle line (Fig. 3), no effect on the degree of collateral axonal branching at the lesion site, and a decrease in the proportion of polyinnervated endplates. MS remarkably restored normal levels of total synaptic input (synaptophysin levels) in the hypoglossal nucleus as in intact animals. Examining the size of motoneurons 2 months after hypoglossal nerve repair, it was shown that back-labeled perikarya in rats receiving MS were significantly larger than in nonstimulated animals. Finally, the comparison of left and right cortical tongue muscle volume 2 months after coaptation of the right hypoglossal nerve gave no differences between stimulated and nonstimulated animals (Evgenieva *et al.*, 2008).

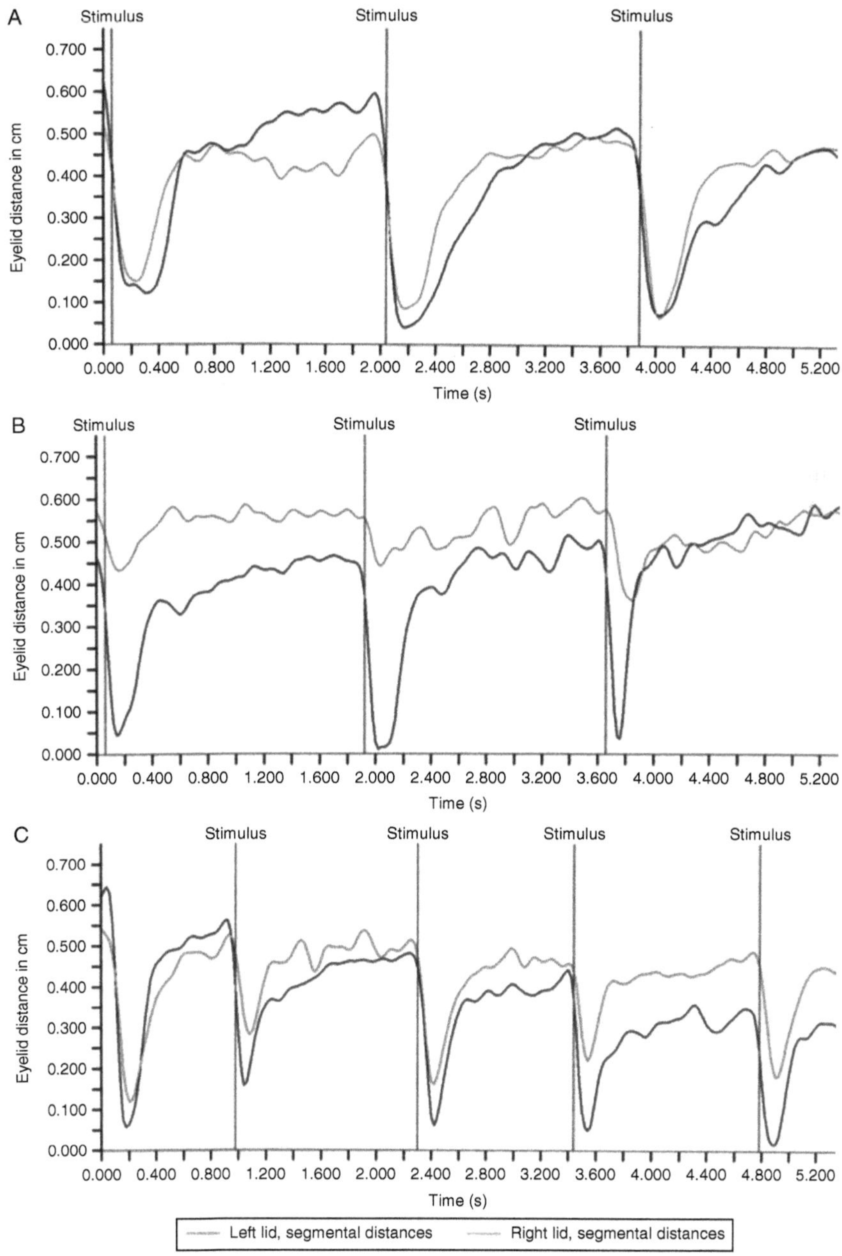

FIG. 2. Biometric analysis of eye closure (representative curves). (A) In unoperated rats, both curves, indicating the closure of the eyelids of the left (in blue) and right (in red) eye display a parallel course with a very good closure (minimum intereyelid distance) after each air puff stimulus. (B) This is

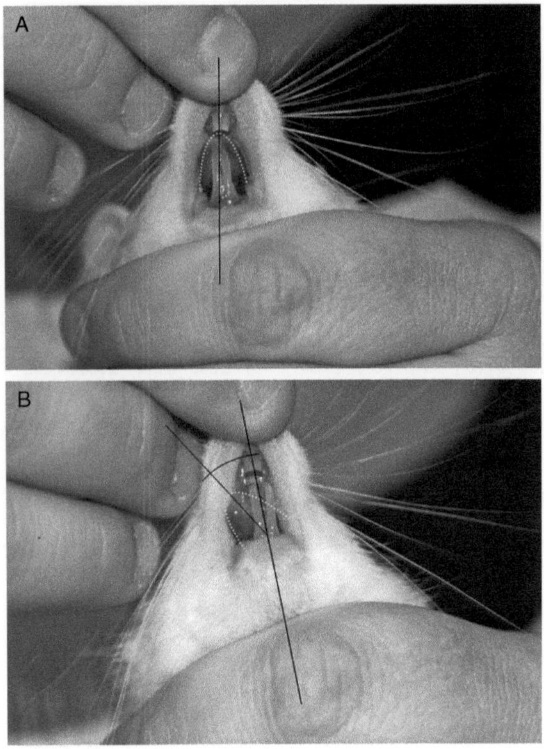

FIG. 3. Measurement of tongue-tip deviation from the midline, that is, of the angle between the long axis of the organ and the median line of the body running between the incisor teeth, in representative animals. The edges of the tongue are outlined by a dotted line. (A) In intact rats, the identical tonus of the right and left protruders and transverse muscles situated the tip of the tongue exactly in the middle behind the lower incisors, that is, the deviation from the midline was 0°. (B) In operated animals, the left protruder dominated and displaced the tongue-tip to the right, that is, the long axis of the organ was no more covering the median line and the angle between them was proportional to the recovery of function. Adopted from Evgenieva *et al.* (2008).

---

in sharp contrast with the situation in operated rats in which the orbicularis oculi muscle on the operated side is, as indicated by the lack of blink reflex responses, paretic (red curve) even 2 months after FFA and no manual stimulation. (C) Definite improvement of the eye closure on the right side (red curve almost parallel to the blue one) 2 months after facial nerve coaptation and manual stimulation of the orbicularis oculi muscle. Adopted from Bischoff *et al.* (2008).

C. Effects of MS on Functional Recovery After Injury of a Mixed Peripheral Nerve

No beneficial effects were in contrary obtained by MS of the forearm muscles after median nerve repair (Fig. 4). There was no significant increase in the grasping force, no effect on the collateral axonal branching at the lesion site, no effect on the quality of endplate reinnervation, and no increase in size of the medianus somata in manual stimulated rats. The main difference between the median and the hypoglossal or facial nerve model however is the depletion of sensory input in the case of median nerve. Mixed peripheral nerves, like the median nerve, consist of both afferent and efferent axons, while motor peripheral nerves, like the hypoglossal and facial nerve, consist mainly of efferent axons. The sensory input of hypoglossal and facial area is being carried by different nerves (Sinis *et al.*, 2008).

III. Discussion

Efficient function restoration after peripheral nerve injury is a great challenge for reconstructive surgeons and rehabilitation medicine. Prompt microsurgical reconstruction is necessary to reduce denervation time, maximize reinnervation, and optimize function of target muscles. Depending on the extent of the nerve lesion, different reconstructive techniques are available, namely, primary or secondary nerve suture, nerve grafting, and end-to-side coaptation (Bontioti *et al.*, 2005). However, even the most proximally located injuries show poor recovery (Kelly *et al.*, 2007; Sinis *et al.*, 2005; Terzis and Konofaos, 2008). Our recent experimental findings have indicated that a simply applicable rehabilitation method, namely, MS of denervated muscles, improves or even restores function and reduces the proportion of polyinnervated endplates after injury of motor peripheral nerves, like the facial and the hypoglossal nerve. MS had also a positive effect after hypoglossal-facial IPNG of the facial nerve. On the contrary, MS provided no benefits after injury of a mixed peripheral nerve, like the median nerve, illustrating the importance of an intact sensory input.

The pathomechanisms limiting functional recovery after nerve injury are poorly understood. It was believed that collateral axonal branching at the lesion site and the subsequent misdirected reinnervation were the major factors limiting recovery (Choi and Raisman, 2005; Ito and Kudo, 1994). Despite the good functional outcome after applying MS on target muscles after denervation, the degree of axonal branching was surprisingly not affected at all, and therefore it seems to be not as necessary as it was thought to be. This conclusion is supported by the lack of improved function after reduction of collateral axonal branching

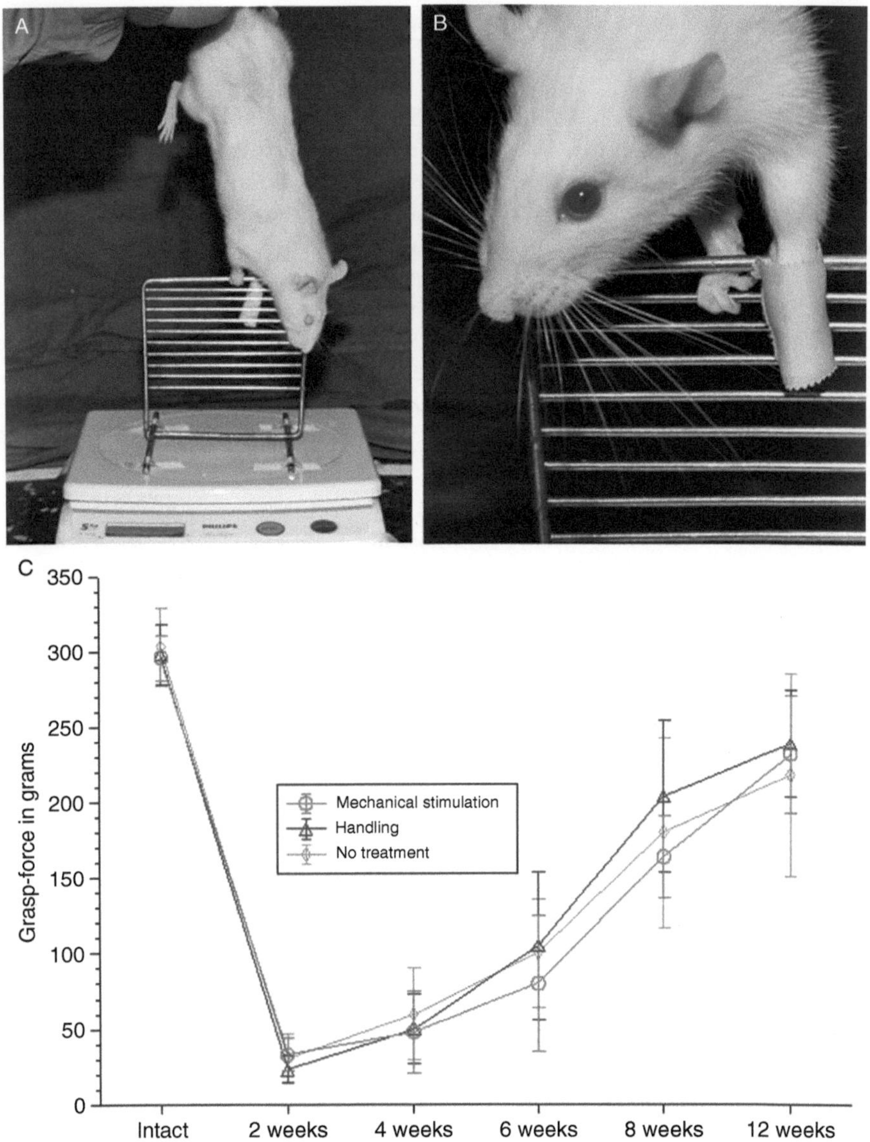

FIG. 4. Grip force test. (A) Measuring of the grip force by means of a wire grid taped onto an electronic balance. (B) To avoid undesired participation in grasping, the fingers of the left extremity were covered by a piece of adhesive textile tape. (C) Time course of the recovering grip force in grams. No significant differences among the three groups of operated animals are detectable. Adopted from Sinis *et al.* (2008).

using antibodies to a variety of nerve growth factors (Guntinas-Lichius *et al.*, 2005). Similarly, use of conduits to reconstruct the sciatic nerve after injury and improve accuracy of reinnervation did not improve locomotor performance in rats (Valero-Cabre and Navarro, 2002). It is speculated that, recruitment of aberrantly innervating motoneurons may be reduced or modified by supra-spinal control circuits and spinal reflex mechanisms, as a result of use-dependent plasticity in the central nervous system (CNS). In that context it was shown that, cells in motor nuclei can be "re-educated" to subserve new functional use after tendon transfer (Illert *et al.*, 1986). Another evidence for CNS plasticity during reinnervation was the modification of the recruitment of single motor units via descending control mechanisms, as observed by feedback EMG studies (Guntinas-Lichius, 2004).

A correlation between the degree of endplate polyinnervation and the functional recovery after peripheral nerve lesion could be constantly observed, and is therefore considered as one of the mandatory factors for successful motor reinnervation. The phenomenon of polyinnervation was described to be a limiting factor for function restoration already years ago (Barry and Ribchester, 1995; Gorio *et al.*, 1983; Guntinas-Lichius *et al.*, 2005; Tam and Gordon, 2003). MS significantly reduced the proportion of polyinnervated endplates in target muscles after peripheral nerve injury. This result can be explained by previous studies showing that muscle activity during the phase of synaptic formation and consolidation leads to reduction of intramuscular sprouting (Brown *et al.*, 1980; Deschenes *et al.*, 2006; Tam *et al.*, 2001).

An earlier study suggested that intramuscular axonal sprouting in response to muscle paralysis comes from short-range sprouting stimuli generated by the inactive muscle fibers (Brown and Ironton, 1977; Pockett and Slack, 1982). Various growth factors have been identified as possible candidates for this role. Their amount should be reciprocally proportional to muscle activity (Brown *et al.*, 1981). These sprouting stimuli are thought to be counteracted or neutralized by interaxonal competition when MS takes place (Brown and Holland, 1979; Love *et al.*, 2003).

An important factor triggering reinnervation, and hence polyinnervation of endplates, is the bridges built by TSC to adjacent innervated motor endplates. Using these bridges TSC reach, attract, and direct intramuscular axonal sprouts toward denervated endplates (Son *et al.*, 1996). It has been shown that the outgrowth of TSC bridges precedes the outgrowth of sprouts from intact intramuscular axons, which means that TSC bridges are responsible for the initiation of intramuscular axonal sprouting (Dickens *et al.*, 2003). Thus, the beneficial effect of MS on the quality of muscle reinnervation may be partially due to a reduction of the number of TSC bridges (Tam *et al.*, 2001). A perturbed formation of TSC bridges could be achieved by running exercise (Tam and Gordon, 2003) or ES (Love *et al.*, 2003). It can be concluded that any form of artificial muscular activity could inhibit the TSC bridge formation and consequently reduce intramuscular axonal sprouting.

Unaffected sensory input seems to be another essential factor influencing motor recovery. In the paradigms of facial nerve repair (Angelov *et al.*, 2007; Bischoff *et al.*, 2008) or hypoglossal nerve repair (Evgenieva *et al.*, 2008) motor axons were lesioned, but the circuity conveying sensory information from the facial or suprahyoid-sublingual region to the facial or hypoglossal nuclei were intact. Depletion of the sensory trigeminal input after facial nerve repair even worsened the functional outcome of whisking in manual stimulated, than in untreated rats (Pavlov *et al.*, 2008). Moreover, no benefits in functional recovery were obtained after MS following injury of a mixed peripheral nerve (after median nerve injury and repair), where both motor and sensory axons were depleted (Sinis *et al.*, 2008). One reason for that could be the excessive collateral branching of sensory dorsal root ganglia (DRG) cells, which prevents them conveying accurate sensory information to their target motor neurons in the ventral horns (Sulaiman *et al.*, 2002). It can be speculated that an increased sensory input supports the regenerative response of injured motoneurons via stimulating plasticity in the spinal cord (Galtrey *et al.*, 2007).

Afferent axons may exert direct trophic effects on the motoneuron somata and dendrites leading to enhanced production of regeneration associated molecules, like GAP-43, synapsin I, cAMP, and BDNF, which are known to stimulate dendrite growth and synaptic remodeling (Al-Majed *et al.*, 2000, 2004; Pearse *et al.*, 2004). The importance of sensory stimulation, achieved also by MS in case of injury models of pure motor peripheral nerves, is supported also by the observation that stimulation and neuromuscular re-education programs of paralyzed facial muscles are very effective for increasing facial muscle control and function, even after prolonged paralysis (Barbara *et al.*, 2003; Cronin and Steenerson, 2003; VanSwearingen and Brach, 2003). However, sensory re-training paradigms after peripheral mixed nerve injury have not yet been developed in rats, and it is therefore unknown whether sensory re-training alone, or sensory re-training combined with MS would yield to a decrease of endplate polyinnervation and a better functional outcome.

Central events are also critical during peripheral nerve regeneration (Horvath *et al.*, 2005) and plasticity within cortical and subcortical networks, and are thought to play an important role in muscle reanimation (Sanes and Donoghue, 2000). A profound functional reorganization in the somatosensory cortex was observed after injuries, especially of the sensory inputs (Chen *et al.*, 2002; Florence *et al.*, 1994; Wall *et al.*, 2002). Therefore, several sensory re-training protocols to induce central plasticity after peripheral mixed nerve injury have been successful in improving functional outcome (Bjorkman *et al.*, 2005; Rosen and Lundborg, 2007). In the case of the hypoglossal nerve injury model, no changes in cortical representation of tongue muscles were observed in manually stimulated rats 2 months after hypoglossal nerve repair. One reason could be that target reinnervation was completed by the time of examination. A second reason could be that sensory innervation of the tongue musculature remained intact with no alteration in somatosensory cortical representation. This observation implies that cortical plasticity may be mainly important for restoration of sensory, but not motor, function.

Following peripheral nerve injury, synaptic terminals detach from motoneurons, a phenomenon known as "synaptic stripping" (Blinzinger and Kreutzberg, 1968). This posttraumatic deafferentation can be reversible if target reinnervation occurs (Guntinas-Lichius *et al.*, 1994; Mader *et al.*, 2004; Neiss *et al.*, 1992). Quantitative electron microscopic analysis of regenerated cat gastrocnemius motoneurons has, however, revealed that restoration of synaptic input is incomplete (Brannstrom and Kellerth, 1999). Synaptic input, measured by synaptophysin levels in facial or hypoglossal nucleus after facial nerve or hypoglossal nerve repair, respectively, could be restored to normal levels or at least counteracted by MS. These observations suggest that MS may prevent loss of facial- or hypoglossal-nucleus afferents. Moreover, it seems that synaptic input is another parameter, apart from the degree of muscle polyinnervation that influences the degree of functional recovery after peripheral nerve lesion (Pavlov *et al.*, 2008).

The fact that motoneuron perikarya were larger in manually stimulated rats than in untreated animals in case of pure motor nerve injury (hypoglossal nerve), but not in case of mixed nerve injury (median nerve), illustrates the importance of an intact sensory input. It is possible that MS inhibits the axotomy-induced atrophy during nerve recovery by influencing positively regenerating motoneurons via enhanced sensory input. Larger cell bodies of stimulated motoneurons are supposed to be an indication for a better functional state. Indeed, there was a correlation between degree of recovery and soma size of retrogradely labeled motoneurons reinnervating the quadriceps muscle 3 months after femoral nerve injury (Simova *et al.*, 2006).

References

Al-Majed, A. A., *et al.* (2000). Electrical stimulation accelerates and increases expression of BDNF and trkB mRNA in regenerating rat femoral motoneurons. *Eur. J. Neurosci.* **12,** 4381–4390.

Al-Majed, A. A., *et al.* (2004). Electrical stimulation accelerates and enhances expression of regeneration-associated genes in regenerating rat femoral motoneurons. *Cell Mol. Neurobiol.* **24,** 379–402.

Angelov, D. N., *et al.* (2007). Mechanical stimulation of paralyzed vibrissal muscles following facial nerve injury in adult rat promotes full recovery of whisking. *Neurobiol. Dis.* **26,** 229–242.

Barbara, M., *et al.* (2003). Early rehabilitation of facial nerve deficit after acoustic neuroma surgery. *Acta Otolaryngol.* **123,** 932–935.

Bardosi, A., *et al.* (1987). The ultrastructure of normal and denervated human facial muscle. *Plast. Reconstr. Surg.* **79,** 171–176.

Barry, J. A., and Ribchester, R. R. (1995). Persistent polyneuronal innervation in partially denervated rat muscle after reinnervation and recovery from prolonged nerve conduction block. *J. Neurosci.* **15,** 6327–6339.

Bischoff, A., *et al.* (2008). Manual stimulation of the orbicularis oculi muscle improves eyelid closure after facial nerve injury in adult rats. *Muscle Nerve.* **39,** 197–205.

Bjorkman, A., *et al.* (2005). Anaesthesia of the axillary plexus induces rapid improvement of sensory function in the contralateral hand: An effect of interhemispheric plasticity. *Scand. J. Plast. Reconstr. Surg. Hand Surg.* **39,** 234–237.

Blinzinger, K., and Kreutzberg, G. (1968). Displacement of synaptic terminals from regenerating motoneurons by microglial cells. *Z. Zellforsch Mikrosk Anat.* **85,** 145–157.

Bontioti, E., *et al.* (2005). End-to-side nerve repair in the upper extremity of rat. *J. Peripher. Nerv. Syst.* **10,** 58–68.

Brannstrom, T., and Kellerth, J. O. (1999). Recovery of synapses in axotomized adult cat spinal motoneurons after reinnervation into muscle. *Exp. Brain Res.* **125,** 19–27.

Brown, M. C., and Holland, R. L. (1979). A central role for denervated tissues in causing nerve sprouting. *Nature* **282,** 724–726.

Brown, M. C., *et al.* (1981). An assessment of the spread of the signal for terminal sprouting within and between muscles. *Brain Res.* **210,** 145–151.

Brown, M. C., *et al.* (1980). Nodal and terminal sprouting from motor nerves in fast and slow muscles of the mouse. *J. Physiol.* **306,** 493–510.

Brown, M. C., and Ironton, R. (1977). Motor neurone sprouting induced by prolonged tetrodotoxin block of nerve action potentials. *Nature* **265,** 459–461.

Calhoun, M. E., *et al.* (1996). Comparative evaluation of synaptophysin-based methods for quantification of synapses. *J. Neurocytol.* **25,** 821–828.

Chen, R., *et al.* (2002). Nervous system reorganization following injury. *Neuroscience* **111,** 761–773.

Choi, D., and Raisman, G. (2005). Disorganization of the facial nucleus after nerve lesioning and regeneration in the rat: Effects of transplanting candidate reparative cells to the site of injury. *Neurosurgery* **56,** 1093–1100. (discussion).

Cohan, C. S., and Kater, S. B. (1986). Suppression of neurite elongation and growth cone motility by electrical activity. *Science* **232,** 1638–1640.

Cronin, G. W., and Steenerson, R. L. (2003). The effectiveness of neuromuscular facial retraining combined with electromyography in facial paralysis rehabilitation. *Otolaryngol. Head Neck Surg.* **128,** 534–538.

Deschenes, M. R., *et al.* (2006). Increased and decreased activity elicits specific morphological adaptations of the neuromuscular junction. *Neuroscience* **137,** 1277–1283.

Dickens, P., *et al.* (2003). Schwann cell dynamics with respect to newly formed motor-nerve terminal branches on mature (Bufo marinus) muscle fibers. *J. Neurocytol.* **32,** 381–392.

Dohm, S., *et al.* (2000). Local application of extracellular matrix proteins fails to reduce the number of axonal branches after varying reconstructive surgery on rat facial nerve. *Restor. Neurol. Neurosci.* **16,** 117–126.

Evgenieva, E., *et al.* (2008). Manual stimulation of the suprahyoid-sublingual region diminishes polynnervation of the motor endplates and improves recovery of function after hypoglossal nerve injury in rats. *Neurorehabil. Neural Repair* **22,** 754–768.

Florence, S. L., *et al.* (1994). Sensory afferent projections and area 3b somatotopy following median nerve cut and repair in macaque monkeys. *Cereb. Cortex* **4,** 391–407.

Galtrey, C. M., *et al.* (2007). Promoting plasticity in the spinal cord with chondroitinase improves functional recovery after peripheral nerve repair. *Brain* **130,** 926–939.

Gordon, T., *et al.* (2004). Adaptive and maladaptive motor axonal sprouting in aging and motoneuron disease. *Neurol Res.* **26,** 174–185.

Gorio, A., *et al.* (1983). Muscle reinnervation–II. Sprouting, synapse formation and repression. *Neuroscience* **8,** 403–416.

Guntinas-Lichius, O. (2004). The facial nerve in the presence of a head and neck neoplasm: Assessment and outcome after surgical management. *Curr. Opin. Otolaryngol. Head Neck Surg.* **12,** 133–141.

Guntinas-Lichius, O., *et al.* (2007). Manual stimulation of facial muscles improves functional recovery after hypoglossal-facial anastomosis and interpositional nerve grafting of the facial nerve in adult rats. *Neurobiol. Dis.* **28,** 101–112.

Guntinas-Lichius, O., *et al.* (2005). Factors limiting motor recovery after facial nerve transection in the rat: Combined structural and functional analyses. *Eur. J. Neurosci.* **21,** 391–402.

Guntinas-Lichius, O., *et al.* (1994). Differences in glial, synaptic and motoneuron responses in the facial nucleus of the rat brainstem following facial nerve resection and nerve suture reanastomosis. *Eur. Arch. Otorhinolaryngol.* **251,** 410–417.

Hall, S. (2005). The response to injury in the peripheral nervous system. *J. Bone Joint Surg. Br.* **87,** 1309–1319.

Horvath, S., *et al.* (2005). Use of a recombinant pseudorabies virus to analyze motor cortical reorganization after unilateral facial denervation. *Cereb. Cortex* **15,** 378–384.

Hovind, H., and Nielsen, S. L. (1974). Effect of massage on blood flow in skeletal muscle. *Scand. J. Rehabil. Med.* **6,** 74–77.

Illert, M., *et al.* (1986). Forearm muscles of man can reverse their function after tendon transfers: An electromyographic study. *Neurosci. Lett.* **67,** 129–134.

Ito, M., and Kudo, M. (1994). Reinnervation by axon collaterals from single facial motoneurons to multiple muscle targets following axotomy in the adult guinea pig. *Acta Anat. (Basel.)* **151,** 124–130.

Kelly, E. J., *et al.* (2007). End-to-side nerve coaptation: A qualitative and quantitative assessment in the primate. *J. Plast. Reconstr. Aesthet. Surg.* **60,** 1–12.

Lomo, T., *et al.* (1974). Contractile properties of muscle: Control by pattern of muscle activity in the rat. *Proc. R. Soc. Lond. B. Biol. Sci.* **187,** 99–103.

Love, F. M., *et al.* (2003). Activity alters muscle reinnervation and terminal sprouting by reducing the number of Schwann cell pathways that grow to link synaptic sites. *J. Neurobiol.* **54,** 566–576.

Mader, K., *et al.* (2004). Dual mode of signalling of the axotomy reaction: Retrograde electric stimulation or block of retrograde transport differently mimic the reaction of motoneurons to nerve transection in the rat brainstem. *J. Neurotrauma* **21,** 956–968.

Morris, J. H., *et al.* (1972). A study of degeneration and regeneration in the divided rat sciatic nerve based on electron microscopy. II. The development of the "regenerating unit". *Z. Zellforsch. Mikrosk. Anat.* **124,** 103–130.

Neiss, W. F., *et al.* (1992). The hypoglossal-facial anastomosis as model of neuronal plasticity in the rat. *Ann. Anat.* **174,** 419–433.

Pavlov, S. P., *et al.* (2008). Manually-stimulated recovery of motor function after facial nerve injury requires intact sensory input. *Exp. Neurol.* **211,** 292–300.

Pearse, D. D., *et al.* (2004). cAMP and Schwann cells promote axonal growth and functional recovery after spinal cord injury. *Nat. Med* **10,** 610–616.

Pockett, S., and Slack, J. R. (1982). Source of the stimulus for nerve terminal sprouting in partially denervated muscle. *Neuroscience* **7,** 3173–3176.

Rajan, B., *et al.* (2003). Epidermal reinnervation after intracutaneous axotomy in man. *J. Comp. Neurol.* **457,** 24–36.

Rich, M. M., and Lichtman, J. W. (1989). *In vivo* visualization of pre- and postsynaptic changes during synapse elimination in reinnervated mouse muscle. *J. Neurosci.* **9,** 1781–1805.

Rosen, B., and Lundborg, G. (2007). Enhanced sensory recovery after median nerve repair using cortical audio-tactile interaction. A randomised multicentre study. *J. Hand Surg. Eur.* **32,** 31–37.

Sanes, J. N., and Donoghue, J. P. (2000). Plasticity and primary motor cortex. *Ann. Rev. Neurosci.* **23,** 393–415.

Shawe, G. D. (1955). On the number of branches formed by regenerating nerve-fibres. *Br. J. Surg.* **42,** 474–488.

Simova, O., *et al.* (2006). Carbohydrate mimics promote functional recovery after peripheral nerve repair. *Ann. Neurol.* **60,** 430–437.

Sinis, N., *et al.* (2008). Manual stimulation of forearm muscles does not improve recovery of motor function after injury to a mixed peripheral nerve. *Exp. Brain Res.* **185,** 469–483.

Sinis, N., *et al.* (2005). Nerve regeneration across a 2-cm gap in the rat median nerve using a resorbable nerve conduit filled with Schwann cells. *J. Neurosurg.* **103,** 1067–1076.

Son, Y. J., *et al.* (1996). Schwann cells induce and guide sprouting and reinnervation of neuromuscular junctions. *Trends Neurosci.* **19,** 280–285.

Sulaiman, O. A., *et al.* (2002). Chronic Schwann cell denervation and the presence of a sensory nerve reduce motor axonal regeneration. *Exp. Neurol.* **176,** 342–354.

Tam, S. L., *et al.* (2001). Increased neuromuscular activity reduces sprouting in partially denervated muscles. *J. Neurosci.* **21,** 654–667.

Tam, S. L., and Gordon, T. (2003). Mechanisms controlling axonal sprouting at the neuromuscular junction. *J. Neurocytol.* **32,** 961–974.

Terzis, J. K., and Konofaos, P. (2008). Nerve transfers in facial palsy. *Facial Plast. Surg.* **24,** 177–193.

Tomov, T. L., *et al.* (2002). An example of neural plasticity evoked by putative behavioral demand and early use of vibrissal hairs after facial nerve transection. *Exp. Neurol.* **178,** 207–218.

Tsuyoshi, H., *et al.* (2006). Sprouting of sensory neurons in dorsal root ganglia after transection of peripheral nerves. *Arch. Histol. Cytol.* **69,** 173–179.

Valero-Cabre, A., and Navarro, X. (2002). Changes in crossed spinal reflexes after peripheral nerve injury and repair. *J. Neurophysiol.* **87,** 1763–1771.

Valls-Sole, J. (2002). Facial palsy, postparalytic facial syndrome, and hemifacial spasm. *Mov. Disord.* **17**(2), S49–S52.

VanSwearingen, J. M., and Brach, J. S. (2003). Changes in facial movement and synkinesis with facial neuromuscular reeducation. *Plast. Reconstr. Surg.* **111,** 2370–2375.

Wall, J. T., *et al.* (2002). Human brain plasticity: An emerging view of the multiple substrates and mechanisms that cause cortical changes and related sensory dysfunctions after injuries of sensory inputs from the body. *Brain Res. Brain Res. Rev.* **39,** 181–215.

ELECTRICAL STIMULATION FOR IMPROVING NERVE REGENERATION: WHERE DO WE STAND?

Tessa Gordon,* Olewale A. R. Sulaiman,† and Adil Ladak‡

*Division of Neuroscience, University of Alberta, Edmonton, Alberta, Canada T6G 2S2
†Department of Neurosurgery, Ochsner Clinic Foundation, New Orleans, Louisiana 70131, USA
‡Division of Plastic Surgery, University of Alberta, Edmonton, Alberta, Canada T6G 2S2

While injured neurons regenerate their axons in the peripheral nervous system, it is well recognized that functional recovery is frequently poor. Animal experiments in which injured motoneurons remain without peripheral targets (chronic axotomy) and Schwann cells in distal nerve stumps remain without innervation (chronic denervation) revealed that it is the duration of chronic axotomy and Schwann cell denervation that accounts for this poor functional recovery and *not* irreversible muscle atrophy that has been so commonly thought to be the reason. More recently, we demonstrated that axon outgrowth across lesion sites is a major contributing factor to the long delays incurred between the injury and the reinnervation of denervated targets. In the rat, a period of 1 month transpires before all motoneurons regenerate their axons across a lesion site. We have developed a technique of 1 h low-frequency electrical stimulation (ES) of the proximal nerve stump just after surgical repair of a transected peripheral nerve that greatly accelerates axon outgrowth. This technique has been applied in patients after carpal tunnel release surgery where the ES promoted the regeneration of all median nerves to reinnervate thenar muscles within 6–8 months, which contrasted with failure of any injured nerves to reinnervate muscles in the same time frame without ES. These findings are very promising such that the ES method could become a clinically viable tool for accelerating axon regeneration and muscle reinnervation.

INTERNATIONAL REVIEW OF
NEUROBIOLOGY, VOL. 87
DOI: 10.1016/S0074-7742(09)87024-4

I. Introduction

Recovery following peripheral nerve injury, particularly proximal nerve injury, remains a significant clinical problem despite the known capacity for axon regeneration in the peripheral nervous system. Poor functional outcomes following repair of peripheral nerve lesions have been attributed to two main factors impeding axonal regeneration: (1) slow growth across coaptation sites and/or nerve gaps and (2) a relatively short time frame when the injured neurons and the denervated Schwann cells can support regenerating axons following injury. Expedition of regenerating axons across repair sites in order to optimize functional regeneration is a major focus of current research. This review will consider why functional recovery in humans remains an extensive clinical problem and will discuss the developing role of electrical stimulation (ES) as a method of accelerating the outgrowth of axons across a site of nerve repair, both in animals and humans.

II. Basis for Poor Functional Recovery After Nerve Injury and Repair

Despite the permissive growth environment of the peripheral nervous system, functional recovery after surgical repair of injured peripheral nerves is often suboptimal (Kim *et al.*, 2003; Sunderland, 1978; Terzis and Smith, 1990). This is true particularly when nerves regenerate over long distances and/or long delays occur between injury and target reinnervation. Experimentally, when axon regeneration from the proximal nerve stump and through the distal the nerve stump is delayed in rat hindlimbs, the numbers of motoneurons that regenerate their axons through the distal nerve stump and the number that reinnervate denervated muscle targets are reduced to 33%, and 10%, respectively (Fu and Gordon, 1995a,b; Sulaiman and Gordon, 2000, 2008). While the poor regeneration after prolonged target denervation has been attributed to the deterioration and dissolution of the denervated target tissues (Gutmann *et al.*, 1942; Kline and Hudson, 1995; Politis *et al.*, 1982), experimental evidence indicates that the deterioration of the capacity of the neurons for regeneration and of the distal stumps to support regenerating axons rather than the inability of denervated muscles to recover from denervation atrophy after reinnervation, accounts for the reduced regenerative capacity (Fu and Gordon, 1995b; Gordon *et al.*, 2003; Sulaiman and Gordon, 2000, 2008; Sulaiman *et al.*, 2002).

Most injuries to peripheral nerves are contusive in nature and tend to leave the nerves in continuity, thereby making it difficult to determine the degree/completeness of nerve injuries and whether urgent repair would be beneficial (reviewed by Sulaiman *et al.*, 2008). Spontaneous recovery of these type of injuries provide patients the best prognosis for recovery compared to surgical repair, which may require excision of scarred nerve segments and graft repair (Kim *et al.*, 2003). However, the delay incurred in waiting to observe any clinical or electrophysiological evidence of functional recovery may be associated with loss of the regenerative ability of the neurons and the growth-supportive environment in the distal nerve stump with consequent reduced success of axon regeneration and, in turn, poor functional recovery. Hence, delayed repair of injured nerve with or without graft repair as well as the naturally slow rate of axonal regeneration results in the progressive loss of the capacity of chronically axotomized neurons to sustain their regenerative capacity and for Schwann cells to support regenerating axons (Boyd and Gordon, 2002; Fu and Gordon, 1995a,b; Gordon *et al.*, 2003; Sulaiman and Gordon, 2000, 2008; Sulaiman *et al.*, 2002). Current clinical practice recommends a period of expectant observation to determine the degree of nerve injury prior to surgical intervention, therefore accelerating axonal regeneration may be a vital way to counteract the deleterious effect of delayed and graft nerve repairs.

III. The Potential of Brief Electrical Stimulation for Accelerating Axon Regeneration

A. Wallerian Degeneration and Staggered Axon Regeneration into the Distal Nerve Stump

Axons do not regenerate across surgical sites within the time frames predicted by well-established rates of regeneration of 1–3 mm/day. Using retrograde dyes that are taken up by regenerated axon tips and transported to the cell bodies of motoneurons and sensory neurons, we observed that the regeneration of axons across a site of nerve repair is very slow when compared with the 1–3 mm/day rate of axon regeneration (Fig. 1) (Al-Majed *et al.*, 2000b; Brushart *et al.*, 2002; Gordon *et al.*, 2008). Cajal (1928) had observed in silver-stained sections of a repair site that axon sprouts emerge from the proximal nerve stump and progress in a disordered fashion into the repair site (Cajal, 1928). This disarray of regenerating axons was observed when green fluorescent protein (GFP)-labeled regenerating nerves proceeded across a suture site amongst the initially disorganized extracellular matrix (ECM) and Schwann cells in the distal nerve stump (Brushart *et al.*, 2002; Witzel *et al.*, 2005). The ordering of Schwann cells and the ECM at the surgical site proceeds over a 10-day period with an increasing number of motoneurons regenerating their axons across the suture site. Thus, regeneration of axons from the proximal nerve stump is not synchronous with axons progressing across a nerve lesion and suture site slowly, entering the distal stump in a staggered fashion.

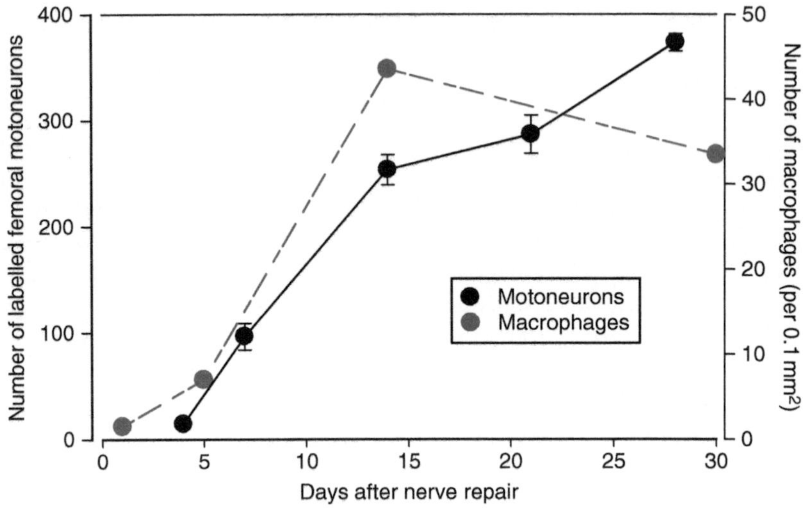

FIG. 1. Mean (±S.E.) number of motoneurons that regenerate their axons 1.5 mm into the distal nerve stump of the femoral nerve after transection and surgical repair (●) and number of macrophages that enter the distal nerve stump at the same distance from the transection site, visualized with the ED1 antibody (○).

Many axon sprouts emerge from the proximal nerve stump with injury, but they are not guided through the distal stump unless there has been organization of ECM elements, including laminin and fibronectin (Platt *et al.*, 2003). The organization of ECM is of inferior quality when nerve conduits are used to bridge a nerve gap, under which circumstance, the first process is the formation of a fibrin bridge joining the proximal and distal nerve stumps, followed by some ECM organization and, only then do Schwann cells migrate into the conduit to form Bands of Bungner, along which sprouting axons migrate (Lundborg, 1988; Williams *et al.*, 1983). In the case of a nerve transection and primary coaptation where there is little gap between the proximal and distal nerve stumps, the same process must proceed although the formation of the initial fibrin bridge is likely to be minimal. After nerve injuries, the proximal and distal stumps of the injured nerve undergo structural and molecular changes, in preparation for the process of axonal regeneration. The proximal stump undergoes *die back* degeneration up to the first node of Ranvier and the growth cones elaborate multiple "regenerating units"; the distal nerve stump undergoes *Wallerian* degeneration during which molecules that could be inhibitory to regeneration, such as myelin-derived proteins, are eliminated (Fu and Gordon, 1997; Gordon *et al.*, 2008). Schwann cells that proliferate in the distal nerve stump upregulate growth factor receptors that include the p75 receptor for neurotrophins, epidermal growth factor receptors

(EGFr) and receptors for IGF-1 (Fig. 2) (Johnson *et al.*, 1988; Taniuchi *et al.*, 1988; Toma *et al.*, 1992) and downregulate the ciliary neurotrophic factor (Sendtner *et al.*, 1992; Smith *et al.*, 1993). Schwann cells play a major role in Wallerian degeneration

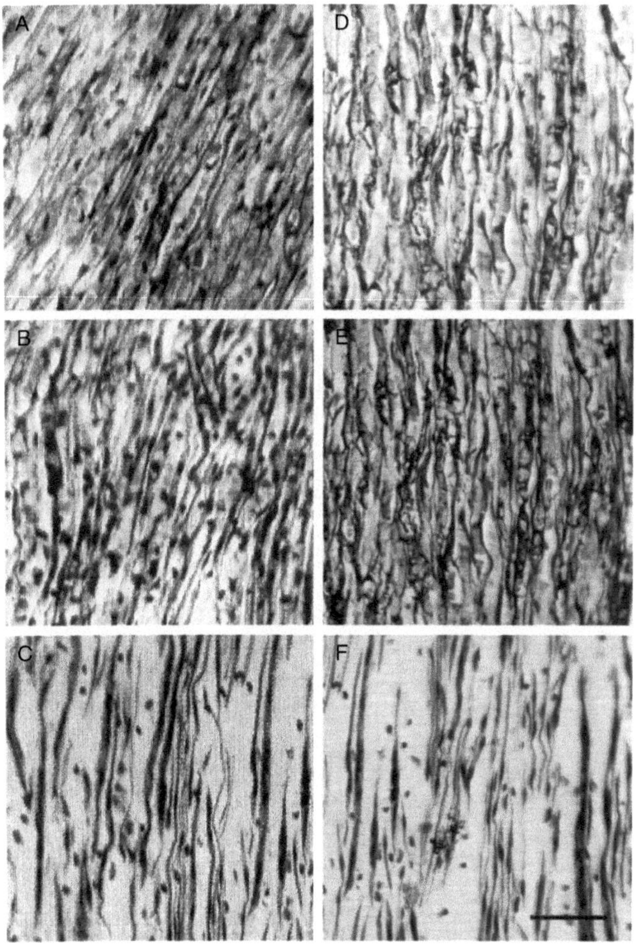

FIG. 2. DAB immunocytochemical visualization (in brown) of the neurotrophin receptor, p75 (A–C) and the receptor for epidermal growth factor (EGFr) (D–F) in longitudinal 14 μm cryosections of denervated distal nerve stumps, 7 days (A, D), 14 days (B, E) and 28 days (C, F) after sciatic nerve section. The sections are counterstained with hemotoxylin (purple cells). Degeneration of axons and myelin was evidence from the visibly swollen axons and the myelin debris 7 days after denervation of the nerve stumps (A, D) but much less evident at 14 days after transection. Note the decreased tissue density in C and F as a result of the removal of most of the degenerated debris following Wallerian degeneration; almost all the myelin debris had been removed. Note the rapid onset of p75 upregulation in cells in the distal nerve stump with later expression of EGFr. The scale bar is 50 μm.

through: (i) phagocytosis of the axonal and myelin debris, (ii) proliferation and lining of the endoneurial pathways as a scaffold for regenerating axons—"bands of Bungner"; and (iii) active interaction with infiltrating macrophages to secrete the required adhesion molecules and neurotrophic factors (Hoke and Mi, 2007; Sulaiman and Gordon, 2003, 2008; Sulaiman *et al.*, 2005). The phagocytosis of axon and myelin debris during Wallerian degeneration is not completed until weeks after the injury following which Schwann cells form bands of Bungner (Fig. 2) (You *et al.*, 1997). In this context, it is not surprising that few axons regenerate across the suture site within the first 4 days (Brushart *et al.*, 2002). Regenerating axons progress slowly with all the motoneurons regenerating their axons across the suture site and into the distal nerve stump by 4 weeks (Fig. 1). Interestingly, we found that the progression of axonal regeneration across the suture site is in direct correlation with both the infiltration of hematogenous macrophages at the suture site and in the distal nerve stumps (Fig. 1), and clearance of myelin and axonal debris (Fig. 2) confirming the essential role that the process of Wallerian degeneration plays in the process of axonal regeneration. The number of macrophages is the highest at the injury site but the relative increase in numbers is approximately equal throughout the distal stump (Avellino *et al.*, 1995, 2004).

B. Electrical Stimulation Accelerates Axon Outgrowth in Nerve Injury

Electrical stimulation of denervated muscle to prevent muscle atrophy has been a long-studied modality in peripheral nerve injury. More recently, the use of brief ES of a transected nerve to promote axonal outgrowth has emerged as a potential therapeutic measure to enhance functional reinnervation. Delivery of an electrical stimulus directly to a peripheral nerve initially stemmed from studies by Hoffman (Hoffman, 1952), where delivery of an electrical stimulus to the spinal cord or nerve roots immediately following partial denervation of the sciatic nerve resulted in accelerated axonal sprouting. The application of ES directly to an injured peripheral nerve was first investigated by Nix and Hopf (1983) who stimulated the soleus nerve of a rabbit following crush injury and reported accelerated recovery of twitch force, tetanic tension, and muscle action potential in the soleus muscle. Thereafter, Pockett and Gavin (1985) reported that after a sciatic nerve crush injury, the recovery of the toe spread reflex improved significantly with ES proximal to the crush site. The findings of Nix and Hopf (1983) and Pockett and Gavin (1985) provided evidence for improved functional reinnervation following brief ES. However, the mechanism by which this improved regeneration was achieved was not elucidated. Was the observed improvement in reinnervation due to stimulation of enhanced axon outgrowth of parent axons in the proximal stump through the repair site or an increase in the overall rate of regeneration?

Our subsequent investigation of the use of ES in peripheral nerve injury has provided insights to the mechanism of action. From previous studies, it was unclear how ES accelerated regeneration. To answer this fundamental question, we electrically stimulated the rat femoral nerve following transection and primary repair. We found that ES significantly increased the number of motoneurons that regenerated their axons into the nerve branches of the femoral nerve (Fig. 3). This acceleration was found to be due to accelerated sprouting of axons across the nerve repair site and not due to an accelerated rate of regeneration (Al-Majed *et al.*, 2000b; Brushart *et al.*, 2002). For motoneurons, low-frequency ES at 20 Hz for a period of 1 h was found to be as effective as continuous stimulation over 2 weeks. Similar results were found for the sensory neurons with ES for 1 h; however, ES extending beyond 1 h did *not* accelerate regeneration (Geremia *et al.*, 2007). The molecular basis for this difference in stimulation time for motor and sensory neurons is the difference in the sensitivity of trkB receptors to released neurotrophic factors. ES upregulates neurotrophic factor and trk receptor expression in motoneurons (Al-Majed *et al.*, 2000a). These receptors remain receptive to neurotrophic factors on motoneurons in contrast to upregulated trkB receptors on sensory neurons which are rapidly down regulated when ES exceeds 1h. (Geremia *et al.*, 2007). Consistent with these findings, English *et al.* (2005, 2007) have demonstrated a critical role of brain-derived neurotrophic factor (BDNF) in mediating the effectiveness of ES in accelerating sensory axon regeneration, where ES results in increased immunoreactivity of trkB receptor and BDNF. In addition to these findings, we have recently demonstrated an upregulation of BDNF and trkB mRNA expression in motoneurons following ES using *in situ* hybridization (Al-Majed *et al.*, 2000a). Furthermore, delivery of a functional blocking antibody against BDNF administered during the first 3 days after nerve section, repair, and 1 h ES completely blocks the accelerating effect of the ES on axon regeneration (Tyreman *et al.*, 2008). Additional genes, termed regeneration associated genes (RAGs), have been found to be upregulated as a result of increased BDNF expression, most notably Tα1-tubulin and GAP-43. We have demonstrated that delivery of ES at 20 Hz for a period of 1 h following nerve transection and primary repair results in upregulation of Tα1-tubulin and downregulation of neurofilament (NF) (Al-Majed *et al.*, 2004). This gene expression profile is normally observed in peripheral nerve injury allowing for more rapid transport of tubulin and faster axon elongation (Hoffman and Lasek, 1980). With ES, the relative increase in Tα1-tubulin and decrease in NF is enhanced, presumably accounting for increased axonal growth (Al-Majed *et al.*, 2004). Finally, to confirm that ES following peripheral nerve injury exerts its effect through gene expression at the cell body, we have demonstrated complete negation of any beneficial effect of ES following nerve injury and repair with the application of tetrodotoxin (which functions to block the retrograde conduction of action potential) in both motor and sensory neurons (Fig. 3) (Al-Majed *et al.*, 2000a; Geremia *et al.*, 2007).

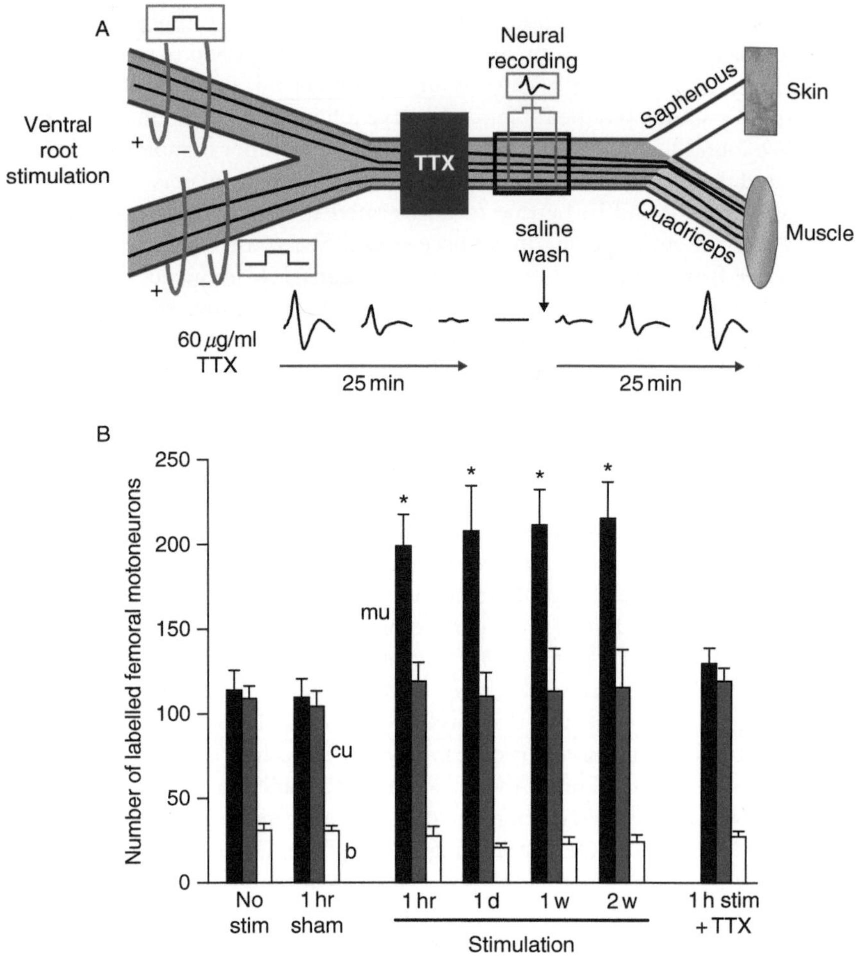

FIG. 3. (A) The dose of tetrodotoxin (TTX) that blocked the conduction of action potentials in response to supramaximal stimulation of the L3 and L4 ventral roots that supply the femoral nerve was established in an acute experiment. The dose of 60 μg/ml produced full blockade within 25 min. This block was completely reversible within the same time frame. (B) The number of femoral motoneurons that was retrogradely labeled by application of fluororuby and fluorogold to the motor branch to the quadriceps muscle and to the sensory saphenous nerve branch, 3 weeks after femoral nerve transection and nerve repair. The number of motoneurons that regenerated their axons appropriately into the muscle branch (mu) and inappropriately into the sensory (cu) branch was the same at 3 weeks for the nerves that were not stimulated and those that were sham stimulated for 1 h. After 1 h, 1 day, 1 week, and 2 weeks electrical stimulation (20 Hz continuous alternating current) significantly more motoneurons regenerated their axons into the appropriate muscle nerve branch with those regenerating into the inappropriate cu branch remaining the same as with no stimulation. Indeed the electrical stimulation promoted axon regeneration of *all* the motoneurons. However, in the presence of TTX, the stimulation effect was completely blocked. Hence the accelerating effect of the axon regeneration was attributed to the electrical activation of the motoneurons (Modified from Al-Majed *et al.*, 2000a,b).

To achieve functional recovery, accurate motor and sensory specific reinnervation is required in addition to timely regeneration. Following transection of the femoral nerve in a rat, there is an initial misdirection of regeneration resulting in motor-specific neurons regenerating through the cutaneous branch of the femoral nerve (Fig. 3B). At 3–4 weeks following transection injury, the proportion of motoneurons regenerating axons into the appropriate motor pathway increases while the number of "misdirected" motoneurons remains unchanged, a phenomenon that has been coined preferential motor reinnervation (PMR) (Al-Majed *et al.*, 2000b). With the delivery of 1 h of ES at 20 Hz, we demonstrated a significantly increased proportion of motoneurons regenerating their axons into the appropriate motor pathway following transection and repair compared to sham controls (Fig. 3B) (Al-Majed *et al.*, 2000b). Similarly ES promoted specific reinnervation of sensory pathways by the axotomized dorsal root ganglion sensory neurons (Brushart *et al.*, 2005; Geremia *et al.*, 2007). Thus, in addition to accelerating axon outgrowth, ES promotes appropriate pathway selection by regenerating axons of motor and sensory neurons.

These findings taken together demonstrate that the effect of the ES is mediated at the cell body and that the effect is to accelerate axon outgrowth and pathway specific regeneration. The findings, moreover, confirm the conclusions that were made in the 1970s as to the promise of the effectiveness of ES for accelerating functional recovery after nerve injury by promoting earlier target reinnervation (Nix and Hopf, 1983). Recent findings that ES accelerates the reinnervation of hand muscles and sensory organs in humans after carpal tunnel surgical release and 1 h of ES at 20 Hz are promising for extension of the technique of brief ES for accelerating axon regeneration in human patients (Gordon *et al.*, 2007, 2008).

IV. Conclusions

Despite continuous advancements in knowledge and technique, regeneration in the peripheral nervous system remains a significant source of morbidity. The recent application of ES to promote axonal sprouting in nerve injury has resulted in promising functional recovery in animal models. With the limitation of ES to 1 h, the translational potential of this modality is significant. To date, one study has made the translation to human trial where post-surgical ES of the median nerve following carpal tunnel release accelerated and completely restored muscle innervation after 6 months in contrast to no significant improvement in patients whose median nerve was not stimulated. (Gordon *et al.*, 2007, 2008). Although the mechanism by which ES exerts its effect is not completely understood, the emerging utility and effectiveness of ES in peripheral nerve injury is exciting, and we anticipate further advancement and application of this modality in the near future.

Acknowledgments

We appreciate the financial support by the Canadian Institutes for Health Research (CIHR) for operating grants to TG from CIHR and a CIHR group grant that includes Dr. Zochodne, Dr. Sayed, and Dr. Midha from University of Calgary, VMKV from University of Saskatchewan, and Dr. Chan and T.G. from University of Alberta. T.G. is an Alberta Heritage Foundation for Medical Research (AHFMR) Senior Investigator.

References

Al-Majed, A. A., Brushart, T. M., and Gordon, T. (2000a). Electrical stimulation accelerates and increases expression of BDNF and trkB mRNA in regenerating rat femoral motoneurons. *Eur. J. Neurosci.* **12,** 4381–4390.

Al-Majed, A. A., Neumann, C. M., Brushart, T. M., and Gordon, T. (2000b). Brief electrical stimulation promotes the speed and accuracy of motor axonal regeneration. *J. Neurosci.* **20,** 2602–2608.

Al-Majed, A. A., Tam, S. L., and Gordon, T. (2004). Electrical stimulation accelerates and enhances expression of regeneration-associated genes in regenerating rat femoral motoneurons. *Cell Mol Neurobiol.* **24,** 379–402.

Avellino, A. M., Dailey, A. T., Harlan, J. M., Sharar, S. R., Winn, R. K., McNutt, L. D., and Kliot, M. (2004). Blocking of up-regulated ICAM-1 does not prevent macrophage infiltration during Wallerian degeneration of peripheral nerve. *Exp. Neurol.* **187,** 430–444.

Avellino, A. M., Hart, D., Dailey, A. T., MacKinnon, M., Ellegala, D., and Kliot, M. (1995). Differential macrophage responses in the peripheral and central nervous system during wallerian degeneration of axons. *Exp. Neurol.* **136,** 183–198.

Boyd, J. G., and Gordon, T. (2002). A dose-dependent facilitation and inhibition of peripheral nerve regeneration by brain-derived neurotrophic factor. *Eur. J. Neurosci.* **15,** 613–626.

Brushart, T. M., Hoffman, P. N., Royall, R. M., Murinson, B. B., Witzel, C., and Gordon, T. (2002). Electrical stimulation promotes motoneuron regeneration without increasing its speed or conditioning the neuron. *J. Neurosci.* **22,** 6631–6638.

Brushart, T. M., Jari, R., Verge, V., Rohde, C., and Gordon, T. (2005). Electrical stimulation restores the specificity of sensory axon regeneration. *Exp. Neurol.* **194,** 221–229.

Cajal, S. R. (1928). "Degeneration and Regeneration of the Nervous System." Oxford University Press, New York, Oxford (translated by R. M. May).

English, A. W., Meador, W., and Carrasco, D. I. (2005). Neurotrophin-4/5 is required for the early growth of regenerating axons in peripheral nerves. *Eur. J. Neurosci.* **21,** 2624–2634.

English, A. W., Schwartz, G., Meador, W., Sabatier, M. J., and Mulligan, A. (2007). Electrical stimulation promotes peripheral axon regeneration by enhanced neuronal neurotrophin signaling. *Dev. Neurobiol.* **67,** 158–172.

Fu, S. Y., and Gordon, T. (1995a). Contributing factors to poor functional recovery after delayed nerve repair: Prolonged axotomy. *J. Neurosci.* **15,** 3876–3885.

Fu, S. Y., and Gordon, T. (1995b). Contributing factors to poor functional recovery after delayed nerve repair: Prolonged denervation. *J. Neurosci.* **15,** 3886–3895.

Fu, S. Y., and Gordon, T. (1997). The cellular and molecular basis of peripheral nerve regeneration. *Mol. Neurobiol.* **14,** 67–116.

Geremia, N. M., Gordon, T., Brushart, T. M., Al-Majed, A. A., and Verge, V. M. (2007). Electrical stimulation promotes sensory neuron regeneration and growth-associated gene expression. *Exp. Neurol.* **205,** 347–359.

Gordon, T., Brushart, T. M., Amirjani, N., and Chan, K. M. (2007). The potential of electrical stimulation to promote functional recovery after peripheral nerve injury–comparisons between rats and humans. *Acta Neurochir. Suppl.* **100,** 3–11.

Gordon, T., Chan, K. M., Sulaiman, O. A. R., Udina, E., Amirjani, N., and Brushart, T. M. (2008). Accelerating axon growth to overcome limitations in functional recovery after peripheral nerve injury. *Neurosurgery* (in press).

Gordon, T., Sulaiman, O. A. R., and Boyd, J. G. (2003). Experimental strategies to promote functional recovery after peripheral nerve injuries. *J. Peripher. Nerv. Syst.* **8,** 236–250.

Gutmann, E., Guttmann, L., Medawar, P. B., and Young, J. Z. (1942). The rate of regeneration of nerve. *J. Exp. Biol.* **19,** 14–44.

Hoffman, H. (1952). Acceleration and retardation of the process of axon-sprouting in partially denervated muscles. *Aust. J. Exp. Biol. Med. Sci.* **30,** 541–566.

Hoffman, P. N., and Lasek, R. J. (1980). Axonal transport of the cytoskeleton in regenerating motor neurons: Constancy and change. *Brain Res.* **202,** 317–333.

Hoke, A., and Mi, R. (2007). In search of novel treatments for peripheral neuropathies and nerve regeneration. *Discov. Med.* **7,** 109–112.

Johnson, E. M., Jr., Taniuchi, M., and DiStefano, P. S. (1988). Expression and possible function of nerve growth factor receptors on Schwann cells. *Trends Neurosci.* **11,** 299–304.

Kim, D. H., Cho, Y. J., Tiel, R. L., and Kline, D. G. (2003). Outcomes of surgery in 1019 brachial plexus lesions treated at Louisiana State University Health Sciences Center. *J. Neurosurg.* **98,** 1005–1016.

Kline, D. G., and Hudson, A. R. (1995). "Nerve Injuries: Operative Results for Major Nerve Injuries, Entrapments and Tumors." W.B. Saunders, Philadelphia.

Lundborg, G. (1988). Nerve regeneration. *In* "Nerve Injury and Repair" (G. Lundborg, Ed.), pp. 149–195. Churchill Livingstone, London.

Nix, W. A., and Hopf, H. C. (1983). Electrical stimulation of regenerating nerve and its effect on motor recovery. *Brain Res.* **272,** 21–25.

Platt, C. I., Krekoski, C. A., Ward, R. V., Edwards, D. R., and Gavrilovic, J. (2003). Extracellular matrix and matrix metalloproteinases in sciatic nerve. *J. Neurosci. Res.* **74,** 417–429.

Pockett, S., and Gavin, R. M. (1985). Acceleration of peripheral nerve regeneration after crush injury in rat. *Neurosci. Lett.* **59,** 221–224.

Politis, M. J., Ederle, K., and Spencer, P. S. (1982). Tropism in nerve regeneration *in vivo*. Attraction of regenerating axons by diffusible factors derived from cells in distal nerve stumps of transected peripheral nerves. *Brain Res.* **253,** 1–12.

Sendtner, M., Stockli, K. A., and Thoenen, H. (1992). Synthesis and localization of ciliary neurotrophic factor in the sciatic nerve of the adult rat after lesion and during regeneration. *J. Cell Biol.* **118,** 139–148.

Smith, G. M., Rabinovsky, E. D., McManaman, J. L., and Shine, H. D. (1993). Temporal and spatial expression of ciliary neurotrophic factor after peripheral nerve injury. *Exp Neurol.* **121,** 239–247.

Sulaiman, O. A. R., Midha, R., and Gordon, T. (2008). Pathophysiology of surgical nerve disorders. *In* "Chapter 2: Youmans' Neurological Surgery" (H. R. Winn, ed.), 6th edn.

Sulaiman, O. A. R., Boyd, J. G., and Gordon, T. (2005). Axonal regeneration in the peripheral system of mammals. *In* "Neuroglia" (H. Kettenmann and B. R. Ransom, Eds.), pp. 454–466. Oxford University Press, Oxford.

Sulaiman, O. A. R., and Gordon, T. (2003). Cellular and molecular interactions after peripheral and central injury. *Biomed. Rev.* **14,** 51–62.

Sulaiman, O. A. R., and Gordon, T. (2000). Effects of short- and long-term Schwann cell denervation on peripheral nerve regeneration, myelination, and size. *GLIA* **32,** 234–246.

Sulaiman, O. A. R., and Gordon, T. (2008). Role of chronic Schwann cell denervation in poor functional recovery after peripheral nerve injuries and strategies to combat it. *Neurosurgery* (in press).

Sulaiman, O. A. R., Voda, J., Gold, B. G., and Gordon, T. (2002). FK506 increases peripheral nerve regeneration after chronic axotomy but not after chronic schwann cell denervation. *Exp. Neurol.* **175,** 127–137.

Sunderland, S. (1978). "Nerve and Nerve Injuries." Livingstone, Edinburgh.

Taniuchi, M., Clark, H. B., Schweitzer, J. B., and Johnson, E. M. Jr. (1988). Expression of nerve growth factor receptors by Schwann cells of axotomized peripheral nerves: Ultrastructural location, suppression by axonal contact, and binding properties. *J. Neurosci.* **8,** 664–681.

Terzis, J. K., and Smith, K. L. (1990). "The Peripheral Nerve: Structure, Function and Reconstruction." Raven, New York.

Toma, J. G., Pareek, S., Barker, P., Mathew, T. C., Murphy, R. A., Acheson, A., and Miller, F. D. (1992). Spatiotemporal increases in epidermal growth factor receptors following peripheral nerve injury. *J. Neurosci.* **12,** 2504–2515.

Tyreman, N., Pettersson, L. M. E., Verge, V. M., and Gordon, T. (2008). BDNF-mediated acceleration of motor axonal regeneration by brief low frequency electrical stimulation (ES). *Soc. Neurosci.* **33,** 752–759.

Williams, L. R., Longo, F. M., Powell, H. C., Lundborg, G., and Varon, S. (1983). Spatial-temporal progress of peripheral nerve regeneration within a silicone chamber: Parameters for a bioassay. *J. Comp. Neurol.* **218,** 460–470.

Witzel, C., Rohde, C., and Brushart, T. M. (2005). Pathway sampling by regenerating peripheral axons. *J. Comp. Neurol.* **485,** 183–190.

You, S., Petrov, T., Chung, P. H., and Gordon, T. (1997). The expression of the low affinity nerve growth factor receptor in long-term denervated Schwann cells. *GLIA* **20,** 87–100.

PHOTOTHERAPY IN PERIPHERAL NERVE INJURY: EFFECTS ON MUSCLE PRESERVATION AND NERVE REGENERATION

Shimon Rochkind,* Stefano Geuna,[†] and Asher Shainberg[‡]

*Division of Peripheral Nerve Reconstruction, Department of Neurosurgery, Tel Aviv Sourasky Medical Center, Tel Aviv University, Israel

[†]Department of Clinical and Biological Sciences, San Luigi Gonzaga School of Medicine, University of Turin, Turin 10043, Italy

[‡]Faculty of Life Science, Bar-Ilan University, Israel

Posttraumatic nerve repair and prevention of muscle atrophy represent a major challenge of restorative medicine. Considerable interest exists in the potential therapeutic value of laser phototherapy for restoring or temporarily preventing denervated muscle atrophy as well as enhancing regeneration of severely injured peripheral nerves.

Low-power laser irradiation (laser phototherapy) was applied for treatment of rat denervated muscle in order to estimate biochemical transformation on cellular and tissue levels, as well as on rat sciatic nerve model after crush injury, direct or side-to-end anastomosis, and neurotube reconstruction. Nerve cells' growth and axonal sprouting were investigated in embryonic rat brain cultures. The animal outcome allowed clinical double-blind, placebo-controlled randomized study that measured the effectiveness of 780-nm laser phototherapy on patients suffering from incomplete peripheral nerve injuries for 6 months up to several years.

In denervated muscles, animal study suggests that the function of denervated muscles can be partially preserved by temporary prevention of denervation-induced

DOI: 10.1016/S0074-7742(09)87025-6

0074-7742/09 $35.00

biochemical changes. The function of denervated muscles can be restored, not completely but to a very substantial degree, by laser treatment initiated at the earliest possible stage post injury.

In peripheral nerve injury, laser phototherapy has an immediate protective effect. It maintains functional activity of the injured nerve for a long period, decreases scar tissue formation at the injury site, decreases degeneration in corresponding motor neurons of the spinal cord, and significantly increases axonal growth and myelinization.

In cell cultures, laser irradiation accelerates migration, nerve cell growth, and fiber sprouting.

In a pilot, clinical, double-blind, placebo-controlled randomized study in patients with incomplete long-term peripheral nerve injury, 780-nm laser irradiation can progressively improve peripheral nerve function, which leads to significant functional recovery.

A 780-nm laser phototherapy temporarily preserves the function of a denervated muscle, and accelerates and enhances axonal growth and regeneration after peripheral nerve injury or reconstructive procedures. Laser activation of nerve cells, their growth, and axonal sprouting can be considered as potential treatment for neural injury. Animal and clinical studies show the promoting action of phototherapy on peripheral nerve regeneration, which makes it possible to suggest that the time for broader clinical trials has come.

I. Introduction

When muscles are denervated, in cases of complete peripheral nerve injury, they deteriorate progressively. Although some muscle regeneration does occur (Carraro *et al.*, 2005), it is at a le Carraro vel insufficient to replace the degenerative loss. There is a need to find effective methods for muscle preservation and nerve regeneration enhancement, especially after surgical nerve repair (Dvali and Mackinnon, 2003; Lundborg, 2002). Surgical repair is the preferred modality of treatment for complete or severe peripheral nerve injury (Belzberg *et al.*, 2004; MacKinnon and Dellon, 1988; Midha, 2008; Noble *et al.*, 1998; Spinner 2008; Sunderland, 1978; Terzis and Smith, 1990). In most cases, the results can be successful if the surgery is performed in the first 6 months after injury, in comparison to long-term cases where surgical management is less successful. Nonetheless, in related literature, there are several publications of surgical treatment of long-term injuries (most of which were severe, incomplete, and with minimal or partial preservation of muscle activity) of the brachial plexus and peripheral nerve (Kline and Hackett, 1975; Narakas, 1978; Rochkind and Alon, 2000; Rochkind *et al.*, 2007b). The reason for early surgical intervention has to do with the fact that between 1 and 3 years post injury, denervated muscles undergo progressive degeneration, which leads to loss of muscle fibers and their replacement with fat and fibrous connective tissue. For most patients who suffer from long-term peripheral nerve injuries, spontaneous recovery is often unsatisfactory. The usual results after such an injury are degeneration of the distal axons and retrograde degeneration of the corresponding neurons of the spinal cord, followed by a very slow regeneration. Recovery may eventually occur, but it is slow and frequently incomplete. The secondary effects of peripheral nerve injury are wasted muscles and a high incidence of pressure sores. Therefore, numerous attempts have been made to enhance and/or accelerate the recovery of injured peripheral nerves and decrease or prevent atrophy of the corresponding muscles. Among the various proposed methods for enhancing nerve repair, phototherapy has received increasing attention over the last two decades. The term phototherapy refers to the use of light for producing a therapeutic effect on living tissues. Although a pioneering report on the effects of laser phototherapy on the regeneration of traumatically injured peripheral nerves was published in the late 1970s (Rochkind, 1978), it is only since the late 1980s that scientific interest was kindled in this therapeutic approach for neural rehabilitation, leading to the publication of a number of studies that have shown positive effects of phototherapy on peripheral nerve regeneration (Gigo-Benato *et al.*, 2005; Rochkind 2009).

The possible mechanism of action of phototherapy on the nervous tissue with respect to peripheral nerve regeneration has been provided by the *in vitro* studies, which showed that phototherapy induces massive neurite sprouting and outgrowth in cultured neuronal cells (Wollman *et al.*, 1996), as well as Schwann cell proliferation (Van Breugel and Bar, 1993). Also, it has been suggested that phototherapy may enhance recovery of neurons from injury by altering mitochondrial oxidative metabolism (Elles *et al.*, 2003) and guide neuronal growth cones *in vitro*, perhaps due to the interaction with cytoplasmic proteins and, particularly, due to the enhancement of actin polymerization at the leading axon edge (Ehrlicher *et al.*, 2002). Phototherapy alters nerve cell activity, including upregulation of a number of neurotrophic growth factors and extracellular matrix proteins known to support neurite outgrowth (Byrnes *et al.*, 2005). A possible molecular explanation was provided by demonstrating an increase in growth-associated protein-43 (GAP-43) immunoreactivity in early stages of rat sciatic nerve regeneration after phototherapy (Shin *et al.*, 2003). Another study (Snyder *et al.*, 2002) showed that application of phototherapy upregulates calcitonin gene-related peptide (CGRP) mRNA expression in facial motor nuclei after axotomy. By altering the intensity or temporal pattern of injury-induced CGRP expression, phototherapy may thus optimize the rate of regeneration and target innervation and neuronal survival of axotomized neurons.

In this chapter, we report the results of an experimental study aimed at investigating how laser phototherapy affects long-term denervated muscles by examining acetylcholine receptors (AChR), which play a special role in neuromuscular transmission, and creatine kinase (CK) content, which is an important enzyme for supplying a source of energy to the muscle. The results of this investigation supplement our previous studies (Rochkind, 2009) pertaining to the effectiveness of laser phototherapy in treating severely injured peripheral nerve after crush injury, neurorraphy, side-to-end anastomosis, or neurotube reconstruction, based on our 30 years of research.

II. Phototherapy in Denervated Muscle Preservation

Using the denervated rat gastrocnemius muscle (*in vivo*) as a model of study, we investigated the influence of low-power laser irradiation on CK activity and the level of AChR in denervated muscle in order to estimate biochemical transformation on cellular and tissue levels. Much of the literature on the effects of long-term denervation of mammalian skeletal muscle has focused on experimental studies of total sciatic section in rats (Borisov *et al.*, 2001; Dow *et al.*, 2004). In our study (Rochkind *et al.*, in preparation), rats were chosen for investigation in the vast majority of cases due to their availability, good survival record, and ease of treatment. For the surgical procedure Wister rats were anesthetized and complete denervation of the gastrocnemius muscle was done (cut and remove 1 cm segment of the sciatic nerve). After operation, the rats were divided into four groups: group I—denervated nonirradiated group (15 rats); group II—denervated laser-treated group (15 rats); group III—intact nonirradiated group (15 rats); and group IV—intact irradiated group (15 rats). The rats underwent laser treatment (HeNe laser, 35 mW, 30 min) every day, for 14 days. Low-power laser irradiation was delivered transcutaneously to the gastrocnemius muscle of denervated group II and intact group IV. Under general anesthesia, the rats were sacrificed and the gastrocnemius muscle was homogenized.

CK activity was measured by the specific spectrophotometrical method using spectrophotometer at 340 nm and a Sigma kit (Rosaki, 1967; Shainberg and Isac, 1984) 7, 30, 60, and 120 days after denervation in both denervated and intact muscles.

Internal and membrane-inserted AChR was quantitated by ^{125}I-alpha-bungarotoxin on the same homogenates (Almon *et al.*, 1974; Chin and Almon, 1980) 7, 30, 60, and 120 days after denervation in both denervated and intact muscles. The data obtained was evaluated as cpm of bound ^{125}I-a-BuTX/mg protein. Radioactivity was assessed with Auto-Gamma Counter in denervated and intact muscles.

A. Creatine Kinase (CK) Activity in Intact and Denervated Rat Gastrocnemius Muscle

Muscle contraction and relaxation require the action of CK. Phosphocreatine, formed by the reaction of this enzyme, constitutes a reservoir of high-energy phosphate, which is available for quick resynthesis of ATP. This high concentration of ATP is then accessible for muscle contraction. Following muscle denervation, the level of CK and muscle weight decreases (Goldspink, 1976). Like others (Kloosterboer *et al.*, 1979), we found (Rochkind *et al.*, in preparation) that in the control nonirradiated group, denervation of the gastrocnemius muscle reduces CK activity. The decrease of CK activity in both groups (nonirradiated and laser-treated) progresses to a similar value for 7 days after denervation and is followed by a sharp fall in the non-laser-treated group in comparison to the delayed and attenuated decrease of the CK activity in the laser-irradiated group. Thus, in the control nonirradiated group, 30 days after denervation, the amount of CK decreased markedly to 41% of the normal value (intact muscle). In the same time, delayed and attenuated decrease of the CK activity was observed in the laser-treated group. The CK activity of the laser-treated denervated muscle decreased only by 17% of the normal value. The analysis of CK activity in the denervated laser-treated group compared to the control denervated group showed a statistically significant difference ($p = 0.008$). After the 30-day period, the CK activity gradually began to decrease in both groups and 4 months after denervation it reached similar levels (Fig. 1). It is known that in denervated muscle, the protein degradation rate is accelerated (Goldspink, 1976). The temporary prevention of denervation-induced biochemical changes may be prompted by a trophic signal for increased synthesis of CK, thus preserving a reservoir of high-energy phosphate available for quick resynthesis of ATP. This data supports Bolognani and Volpi (1991), which shows that laser irradiation increased ATP production in the mitochondria.

B. Acetylcholine Receptors Synthesis in Intact and Denervated Rat Gastrocnemius Muscle

Acetylcholine receptors (AChR), which play a special role in neuromuscular transmission, are concentrated at the neuromuscular junction of the adult muscle. A nerve impulse triggers the release of acetylcholine, producing a much larger end-plate potential, which excites the muscle membrane and leads to muscle contraction. The amount of AChR in neuromuscular junction appears to increase their number and to cover the entire extrajunctional area following muscle denervation. In the denervated muscle, the amount of AChR increases prior to muscle degeneration (Lomo and Westgaard, 1975).

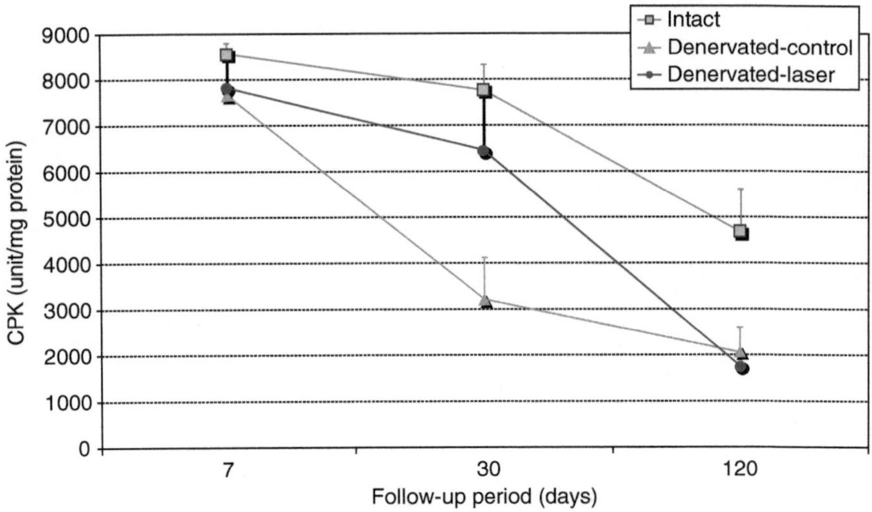

FIG. 1. Graph illustrating the results of study evaluating creatine kinase (CK) activity (unit/mg protein) in intact and denervated rat gastrocnemius muscle. Graph showing content of CK (unit/mg protein) during 7, 30, and 120 days in intact and denervated muscles with and without laser treatmet.

In the control nonirradiated group, 7 days after muscle denervation, as expected, the amount of AChR increased to 161% of the normal value (intact muscle). In contrast, the amount of AChR of the laser-irradiated denervated muscle remained near normal value. Thirty days after denervation in the laser-treated group, the amount of AChR increased to 180% as compared to 278% in the nonlaser group. It is interesting that 4 months after denervation, in spite of progressive muscle atrophy, the amount of AChR in laser-treated group remains at 53% of normal value compared to only 27% in the nonirradiated group. Statistical analysis showed borderline significance ($p = 0.056$) between denervated laser-treated and nonirradiated denervated muscles (Fig. 2).

Our findings suggest that in early stage of muscle degeneration, laser treatment may temporarily preserve the denervated muscle close to its physiological status before injury, and during progressive stages of muscle degeneration, partially maintain the amount of AChR in the denervated muscle compared to the non-laser-treated muscle.

C. Is Laser Phototherapy Damaging to the Muscle?

During 4 months of follow-up period, we found no evidence of laser-induced damage after irradiation. Moreover, in the laser-irradiated intact muscle group, we found a significant increase in CK activity 60 days into the follow-up

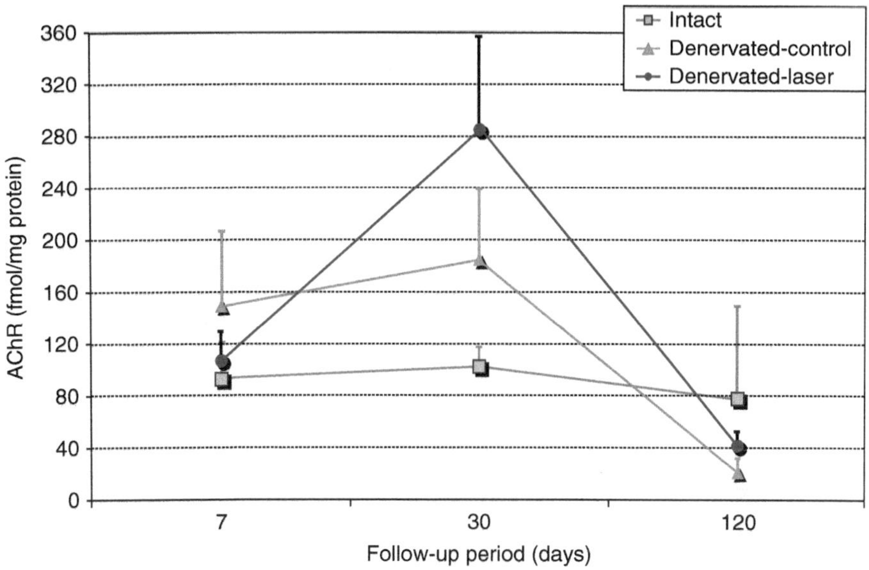

FIG. 2. Graph illustrating the results of study evaluating the level of acetylcholine receptors (AChR) in intact and denervated rat gastrocnemius muscle. Graph showing content of AChR (fmol/mg protein) during 7, 30, and 120 days in intact and denervated muscles with and without laser treatmet.

period ($p = 0.008$) and an increasing amount of AChR ($p = 0.0008$) compared to nonirradiated intact muscle. These findings suggest a possible positive therapeutic effect of laser phototherapy on the muscle.

D. Can Laser Phototherapy Prevent Denervation Muscle Atrophy?

Late denervation has been widely studied in animal models. In rats, it has been shown that for the first 7 months after denervation, myofibers exhibit a net loss of nuclear domains followed by nuclear groupings (Viguie *et al.*, 1997). If not reinnervated, the regenerating myofibers undergo atrophy and degeneration (Missini *et al.*, 1987). For decrease or temporary prevention of this process, especially in cases of complete peripheral nerve injury, where affected nerve is reconstructed by grafts, tube, or primary anastomosis, laser phototherapy can be an effective tool that preserves denervated muscle until nerve sprouting into the muscle occurs. This experimental study suggests that the function of denervated muscles can be restored, not completely but to a very substantial degree, by laser treatment initiated at the earliest possible stage post injury. These findings could have direct therapeutic applications of possible treatment of denervated muscles.

III. Phototherapy in Peripheral Nerve Regeneration

Posttraumatic nerve repair continues to be a major challenge of restorative medicine. Although enormous progress has been made in surgical techniques over the past three decades, functional recovery after severe lesion of a major nerve trunk is often incomplete and sometimes unsatisfactory.

A. Incomplete Peripheral Nerve Injury

1. *Experimental Peripheral Nerve Crush Injury*

Under general anesthesia, the rat sciatic nerve was exposed and crushed with applied pressure of 6.3 ± 0.7 MPa of an ordinary closed hemostat for 30 s. Studies investigating the effects of low-power laser irradiation on injured peripheral nerves of rats have found that it provides the following: (1) immediate protective effects which increase the functional activity of the injured peripheral nerve (Rochkind *et al.*, 1988); (2) maintenance of functional activity of the injured nerve over time (Rochkind *et al.*, 1987a); (3) decrease or prevention of scar tissue formation at the site of injury (Fig. 3) (Rochkind *et al.*, 1987b); (4) prevention or decreased degeneration in corresponding motor neurons of the spinal cord (Fig. 4) (Rochkind *et al.*, 1990); and (5) increase in the rate of axonal growth and myelinization (Fig. 5) (Rochkind *et al.*, 1987a). Moreover, direct laser irradiation of the spinal cord improves recovery of the corresponding injured peripheral nerve (Rochkind, *et al.*, 2001). These results, as those of Andres *et al.*, (1993), suggest that laser phototherapy accelerates and improves the regeneration of the incomplete injured peripheral nerve.

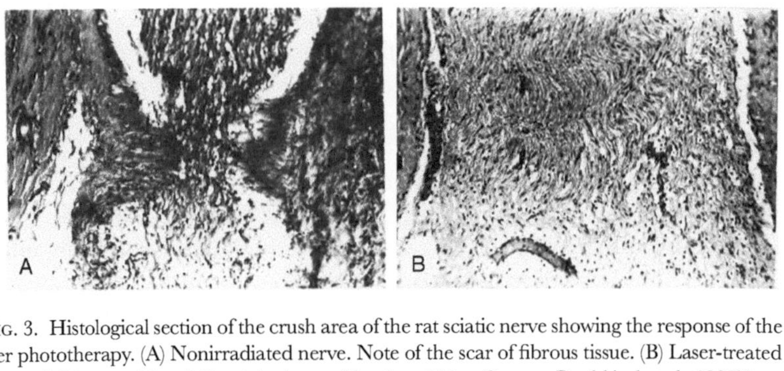

FIG. 3. Histological section of the crush area of the rat sciatic nerve showing the response of the nerve to laser phototherapy. (A) Nonirradiated nerve. Note of the scar of fibrous tissue. (B) Laser-treated nerve shows no visible scar. H and E, original magnification: 150× (Source: Rochkind *et al.*, 1987b).

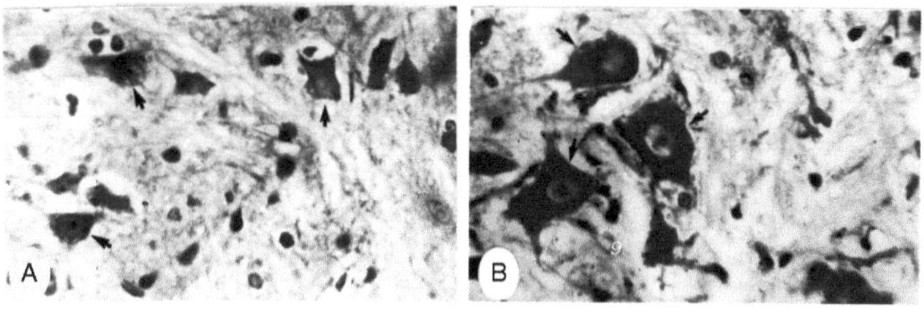

FIG. 4. Paraffin section from the anterior horn of corresponding segments of the rat spinal cord 14 days after crush injury to the sciatic nerve, showing the spinal cord response to laser treatment of the injured peripheral nerve. (A) Section from a control animal shows extensive chromatolysis and cytoplasmic atrophy found in 40% of the motor neurons (arrows). (B) Section from a laser-treated animal shows minimal degenerative changes found in 20% of the motor neurons (arrows). Stained by cresyl fast violet, magnification: 800× (Source: Rochkind *et al.*, 1990).

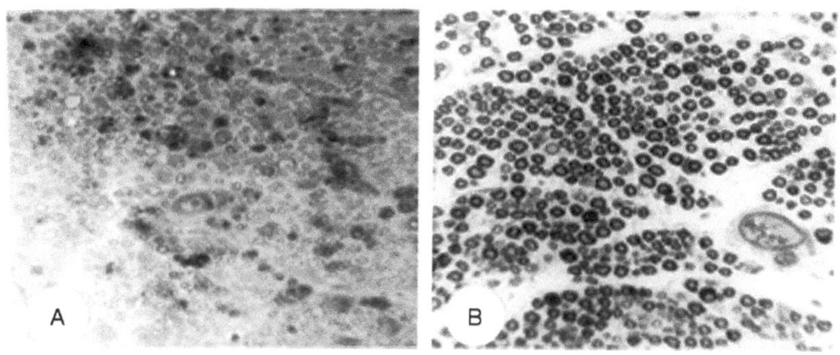

FIG. 5. Photomicrographs of semithin sections stained with toluidine blue showing the axonal response to laser treatment of the injured (crushed) peripheral nerve in rat. One group of rats was treated using laser phototherapy for 20 consecutive days after injury. Twenty-one days after injury, the nerves were excised and stained. (A) Site of crush injury in an untreated nerve showing nerve fibers that appear to be smaller and mostly nonmyelinated. Numerous macrophages and phagocytes are observed. (B) Site of crush injury in a laser-treated nerve demonstrating that most axons are ensheathed with myelin and a very few infiltrating macrophages are observed. Magnification: 300× (Source: Rochkind *et al.*, 1987a).

B. Complete Peripheral Nerve Injury

1. *Regeneration of the Transected Sciatic Nerve in Rat After Primary Anastomosis*

In acute cases where a peripheral nerve is completely transected, the treatment of choice is direct anastomosis. Means of enhancing regeneration are essential, since degeneration is always inevitable in severely damaged peripheral nerves.

The therapeutic effect of 780-nm laser irradiation on peripheral nerve regeneration after complete transection and direct anastomosis of rat sciatic nerve was evaluated in a double-blind randomized study (Shamir *et al.*, 2001). After surgery, 13 of 24 rats received postoperative laser treatment applied transcutaneously for 30 min on a daily basis for 21 consecutive days—15 min to the injured sciatic nerve and 15 min to the corresponding segments of the spinal cord. Positive somatosensory evoked responses were found in 69.2% of the irradiated rats ($p = 0.019$), compared to 18.2% of the nonirradiated rats. Immunohistochemical staining in the laser-treated group showed an increased total number of axons ($p = 0.026$) and better quality of regeneration process, which became evident by an increased number of large diameter axons ($p = 0.021$), compared to the nonirradiated control group (Fig. 6). The study suggests that postoperative laser phototherapy enhances the regenerative processes of peripheral nerves after complete transection and anastomosis.

2. *Median Nerve Regeneration in the Rat After End-to-Side Anastomosis*

A double-blind randomized study in the rat median nerve model (Gigo-Benato *et al.*, 2004) investigated the effects of low-power laser irradiation after the employment of an innovative technique in nerve surgery—namely, end-to-side anastomosis that can be used in case of a particularly severe nerve lesion characterized by complete loss of the proximal nerve stump. In such cases, when grafting is impossible to be done, it has been shown that regeneration along the severed nerve can be obtained by inducing collateral axonal sprouting from a neighbor intact nerve (Bontioti and Dahlin, 2009; Rovak *et al.*, 2001). Rat median nerves were repaired by end-to-side anastomosis on the ulnar intact nerve and then laser irradiated. Results showed that in laser-treated groups, compared to untreated controls, phototherapy induced a significantly faster recovery of the motor function (measured by means of the grasping test) and of target muscle mass, and a significantly faster myelinization of the regenerated nerve axons. Figure 7 shows the gross appearance of the repaired median nerve, 16 weeks postoperatively, in a non-laser-treated animal versus the laser-irradiated animal.

3. *Regeneration of the Sciatic Nerve in the Rat After Complete Segmental Loss and Neurotube Reconstruction*

In cases where peripheral nerve is injured and complete segmental loss exists, the treatment of choice is nerve reconstruction using an autogenous nerve graft. The use of a regenerating guiding tube for the reconstruction of segmental loss of a peripheral nerve has some advantages over the regular nerve grafting procedure.

This double-blind randomized study was done to evaluate the efficacy of 780-nm laser phototherapy on the acceleration of axonal growth and regeneration after experimental peripheral nerve reconstruction by guiding tube

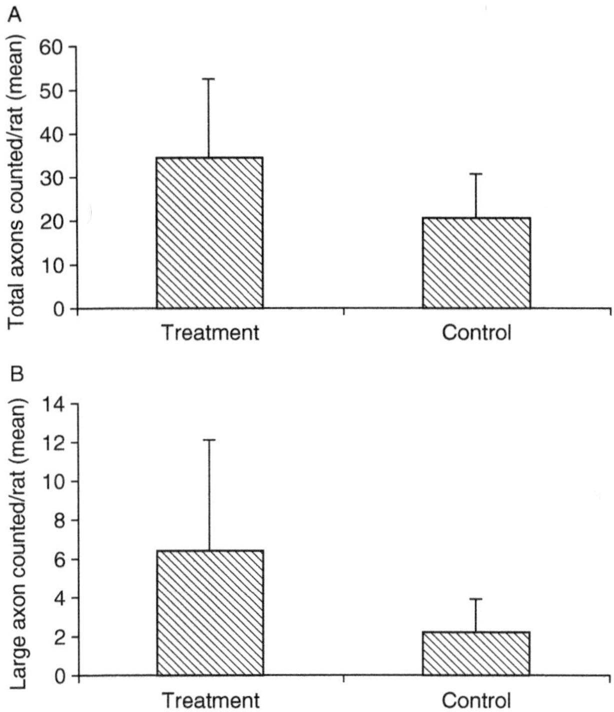

FIG. 6. Bar graphs illustrating the results of double-blind randomized study evaluating regeneration of the transected rat sciatic nerve after suturing and postoperative low-power laser treatment. (A) Graph showing a statistically significant increase in the total number of axons in the laser-treated group ($p = 0.026$), compared to the nontreated control group. (B) Graph showing a statistically significant increase in large diameter axons in the laser-irradiated group ($p = 0.021$), compared to the nonirradiated control group (Source: Shamir *et al.*, 2001).

(Rochkind *et al.*, 2007c). The 5-mm segment of the right sciatic nerve was removed and proximal and distal parts were inserted into an artificial neurotube (Fig. 8). The rats were divided into two groups, laser-treated and non-laser-treated. Postoperative low-power laser irradiation was applied transcutaneously for 30 min: 15 min on the transplanted peripheral nerve area and 15 min on corresponding segments of the spinal cord during 14 consecutive days. Conductivity of the sciatic nerve was studied by stimulating the sciatic nerve and recording the somatosensory-evoked potentials (SSEP) from the scalp. Three months after surgery, SSEP were found in 70% of the rats in the laser-treated group in comparison with 40% of the rats in the nonirradiated group. Morphologically, the transected nerve had good reconnection in both groups and the neurotube had dissolved (Fig. 9). The growth of myelinated axons, which crossed through the

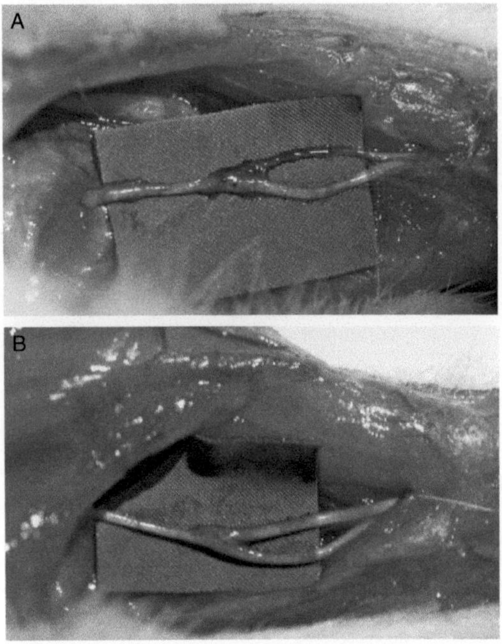

FIG. 7. Intraoperative photograph of end-to-side anastomosis between affected median nerve and intact rat ulnar nerve. Macroscopic appearance at week 16 postoperatively of the regenerated median nerve (arrow) in a non-laser-treated animal (A) and laser-treated animal (B). The better recovery of nerve trophism in the laser-treated animal is clearly evident (Source: Gigo-Benato *et al.*, 2004).

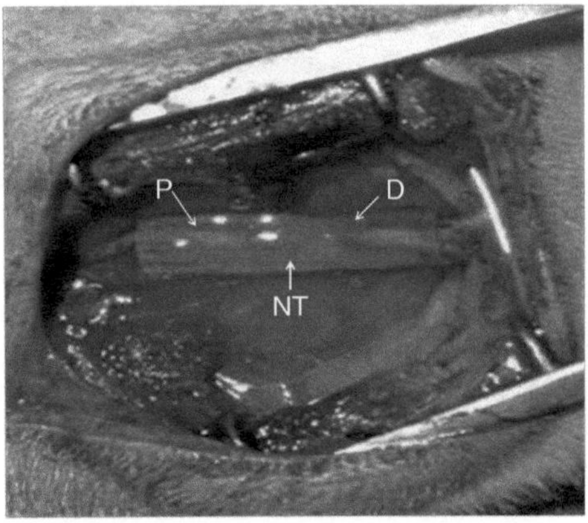

FIG. 8. Intraoperative photograph of the neurotube (NT) reconstruction procedure. An NT placed between the proximal (P) and the distal (D) parts of the rat sciatic nerve for the reconnection of 0.5-cm nerve defect.

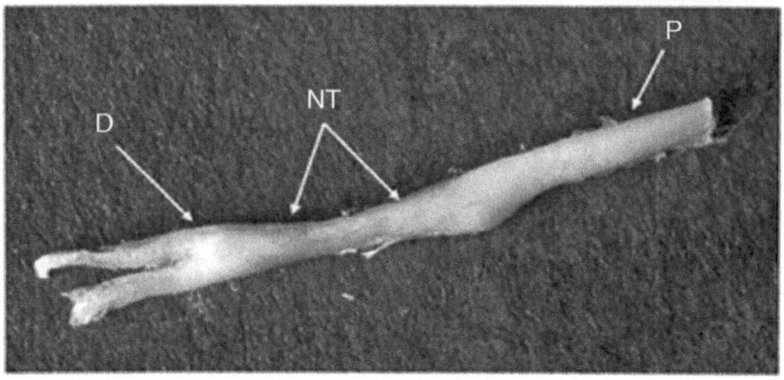

FIG. 9. Photograph of the sciatic nerve of an adult rat 3 months after neurotube (NT) reconstruction. The NT recreated the anatomical connection of the previously transected and divided nerve, and a distance of 0.5 cm was recreated (Source: Rochkind *et al.*, 2007c).

composite neurotube, was found and the continuation of axonal sprouting through the area of the tube to the distal part of the nerve was recognized. The laser-treated group showed more intensive axonal growth compared to the nonirradiated control group.

IV. Phototherapy on Nerve Cell Growth *In Vitro* as a Potential Procedure for Cell Therapy

Neuronal loss and degeneration resulting from peripheral nerve injuries has led us to explore the possibility of using laser phototherapy on cells as a method of preventing or decreasing this phenomena. Rochkind *et al.*, (2009) investigated the effect of 780-nm laser phototherapy on sprouting and cell size of embryonic rat brain cells, which were grown on microcarriers (MC) and embedded in neurogel.

Cell cultures: Whole brains were dissected from 16-day-old rat embryos (Sprague Dawley). After mechanical dissociation, cells were seeded directly in neurogel or suspended in positively charged cylindrical MC. Single-cell MC aggregates were either 780-nm laser irradiated within 1 h after seeding or cultured without irradiation.

Neurogel (hyaluronic acid and laminin) was enriched with growth factors BDNF (brain-derived neurotrophic factor) and IGF-1 (insulin-like growth factor-1) (Rochkind *et al.*, 2006).

780-nm low-power laser irradiation of 10, 30, 50, 110, 160, 200, and 250 mw were used to optimize energy density for activation of nerve cell cultures. Dissociated cells or cell–MC aggregates embedded in neurogel were irradiated for 1, 3, 4, or 7 min.

Fluorescent staining: Cultures were fixed with 4% paraformaldehyde and incubated with antibodies against neural cell marker: Mouse anti-Rat-microtubule-associated protein. Cells were then washed and incubated with Texas Red-conjugated goat antimouse IgG. A rapid sprouting of nerve processes from the irradiated cell–MC aggregates was detected already within 24 h after seeding. The extension of nerve fibers was followed by active neuronal migration. Differences between controls and irradiated stationary dissociated brain cultures became evident at about the end of the first week of cultivation—several neurons in the irradiated cultures exhibited large perikarya and thick elongated processes (Fig. 10). Furthermore, during the next 2–3 weeks of cultivation, neurons in the irradiated cultures developed a dense branched interconnected network of neuronal fibers. The sprouting of long processes from large cell body was mainly observed in immunofluorescent MAP-2 staining (Fig. 11).

This study suggests that laser phototherapy may play a role in prevention of neuronal loss and accelerate axonal regeneration.

V. 780-nm Laser Phototherapy in Clinical Trial

Based on the outcome of animal studies, a clinical double-blind, placebo-controlled randomized study was performed to measure the effectiveness of 780-nm low-power laser irradiation on patients who had been suffering from incomplete peripheral nerve and brachial plexus injuries from 6 months up to several years (Rochkind *et al.*, 2007a). Most of these patients were discharged by

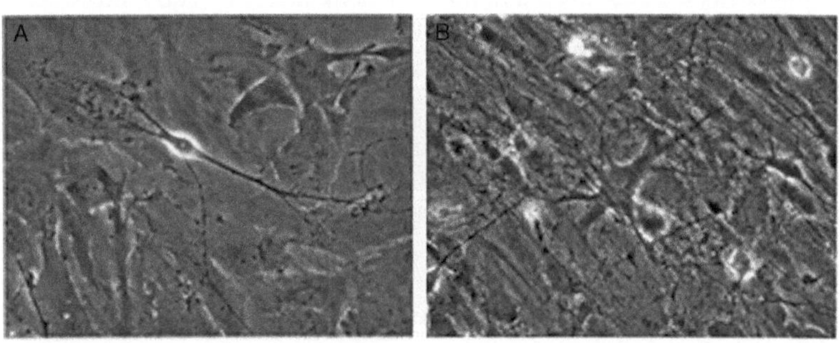

FIG. 10. Photographs of the effect of 780-nm laser irradiation treatment on perikarya and fibers of nerve cells derived from rat embryonic brain. Dissociated brain cells were embedded in neurogel and were either exposed to single radiation dose of 50 mW for 3 min (B) or were nonirradiated controls (A). Large neural cells exhibiting thick fibers were observed after 8 days *in vitro* irradiated cultures (B). Original magnification: 200× (Source: Rochkind *et al.*, 2009).

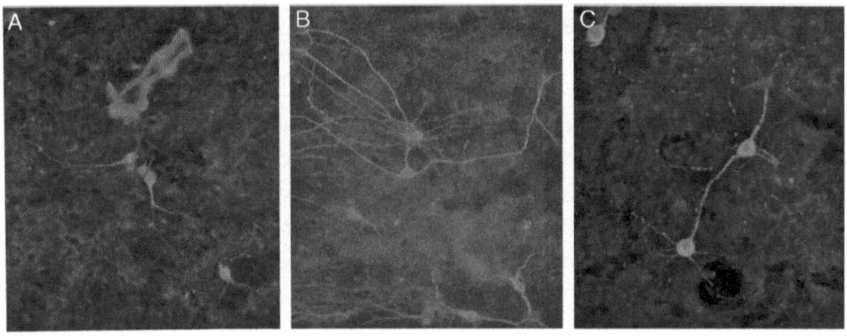

FIG. 11. Immunofluorescent staining of controls and laser-irradiated neuronal brain cells migrated from cell–MC aggregates in neurogel. (A) nonirradiated control. (B) 50-mW, 1-min irradiation. (C) 50-mW, 4-min irradiation. In the irradiated cultures, note large perikarya-bearing long interconnected fibers are positively stained for the neuronal marker MAP-2. Original magnification: 200× (Source: Rochkind *et al.*, 2009). (See Color Insert.)

orthopedics, neurosurgeons, and plastic surgeons without further treatment. In this study, 18 patients with a history of traumatic peripheral nerve/brachial plexus injury (mean duration from injury to treatment, 7 months in laser-treated group and 11.5 months in placebo group) with a stable neurological deficit and a significant weakness were randomly divided to receive either 780-nm laser or placebo (nonactive light) irradiation.

The laser or placebo (nonactive light) treatment was applied transcutaneously; each day for 21 consecutive days, 5 h daily (3 h to the injured area of the peripheral nerve and 2 h to the corresponding segments of the spinal cord). The laser or placebo device were placed approximately 40 cm from the skin-treated point, focused on the injured area of the peripheral nerve or corresponding level of the spine (area of corresponding segments of the spinal cord).

A. Laser Dosage

In the spinal cord area, laser irradiation was performed transcutaneously directly above the projection of the corresponding segments of the spinal cord, which was divided into two intravertebral levels. Each level was irradiated for 60 min a day (150 J/mm^2), totaling 120 min a day (300 J/mm^2).

In the peripheral nerve area, laser irradiation was performed transcutaneously directly above the projection of the injured nerve, which was divided into three parts: proximal, injured area, and distal. Each section was irradiated for 60 min a day (150 J/mm^2), totaling 180 min a day (450 J/mm^2).

The irradiating spot size was 3 × 2 mm (6 mm^2). The penetration of the near-infrared 780-nm wavelength was approximately 4 cm. Analysis of results of this trial in the laser-irradiated group showed statistically significant improvement in motor function in the previously partially paralyzed limbs, compared to the placebo group, where no statistical significance in neurological status was found (Fig. 12). Mean motor function of the most influential (functionally dominant) muscle for movement of the affected limb showed a statistically significant improvement in the laser-irradiated group compared to the placebo group. The function was improved mostly by increasing power of the dominant muscles and not intrinsic muscles. Electrophysiological observation during the trial supplied us with important diagnostic information and helped to determine the degree of functional recovery in nerve-injured patients. The electrophysiological analysis also showed statistically significant improvement in recruitment of voluntary muscle activity in the laser-irradiated group, compared to the placebo group (Fig. 13).

This study is not the ultimate word regarding 780-nm laser phototherapy in peripheral nerve injured patients. The disadvantages of this study are the small amount of patients, different nerves, and etiology of injury. Nevertheless, this pilot

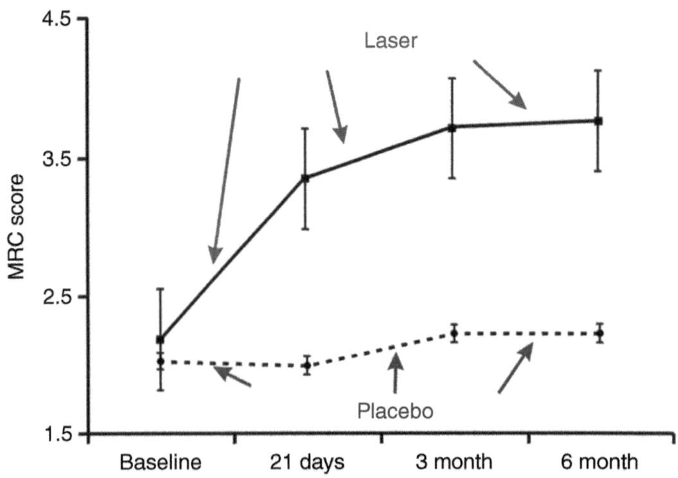

FIG. 12. Graph of the motor function follow-up in injured patients who underwent 780-nm laser phototherapy or placebo treatment. Mean motor function (±S.D.) of all affected muscles was examined in injured patients using the Medical Research Council (MRC) Grading System. The analysis of the results showed that at baseline, the 780-nm laser-treated and placebo groups were in clinically similar conditions ($p = 0.887$). The analysis of motor function during the 6-month follow-up period compared to baseline showed statistically significant improvement ($p = 0.0001$) in the laser-treated group compared with the placebo group (Source: Rochkind *et al.*, 2007a).

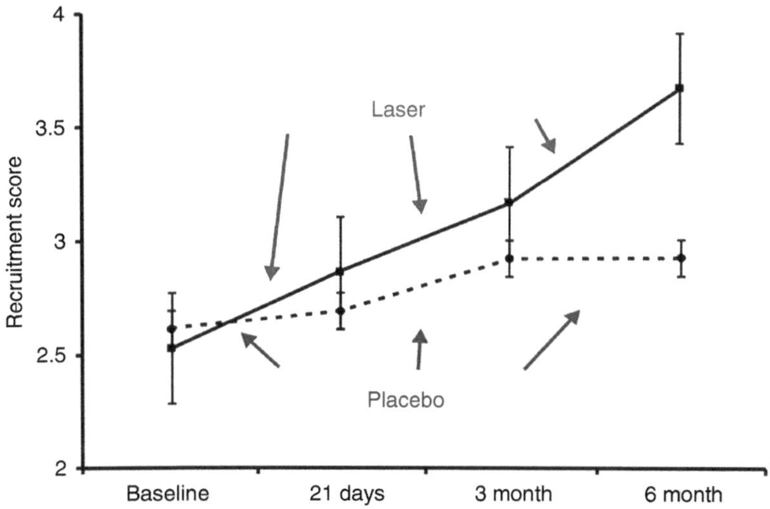

FIG. 13. Graph of the motor unit recruitment in injured patients who underwent either 780-nm laser phototherapy or placebo treatment. Motor unit recruitment, the mean of all examined muscles (±S.D.), was monitored in injured patients. The 780-nm laser-treated and placebo groups were in similar conditions at baseline ($p = 0.934$). In the laser-treated group, statistically significant improvement ($p = 0.0006$) was found in motor unit recruitment during the 6-month follow-up period, compared with the placebo group (Source: Rochkind *et al.*, 2007a).

study suggests that in peripheral nerve injured patients, 780-nm low-power laser irradiation can progressively improve peripheral nerve function. That leads us to continue this study with the perspective to multicenter trial.

VI. Conclusions

Results of the experimental study on denervated muscles suggest that laser treatment can restore its function to a substantial degree when initiated at the earliest possible postinjury stage. These findings could have direct therapeutic applications on preserving the function of denervated muscle after peripheral nerve injury.

The extensive review of published articles reported in this chapter as well as in previous ones published in *Muscle and Nerve* (Gigo-Benato *et al.*, 2005) and *Neurosurgical Focus* (Rochkind, 2009), revealed that most of the experimental studies showed phototherapy to promote the recovery of the severely injured peripheral nerve. This review makes it possible to suggest that a time for broader clinical trials has arrived. The significance of the experimental and clinical studies is the provision of new laser technology for treatment of severe nerve injury.

References

Almon, R. R., Andrew, C. G., and Appel, S. H. (1974). Acetylcholine receptor in normal and denervated slow and fast muscle. *Biochemistry* **13,** 5522–5528.

Andres, J. J., Borke, R. C., Woolery, S. K., and Van de Merwe, W. P. (1993). Low-power laser irradiation alters the rate of regeneration of the rat facial nerve. *Lasers Surg. Med.* **13,** 72–82.

Belzberg, A. J., Dorsi, M. J., Storm, P. B., and Moriarity, J. L. (2004). Surgical repair of brachial plexus injury: A multinational survey of experienced peripheral nerve surgeons. *J. Neurosurg.* **101,** 365–376.

Bolognani, L., and Volpi, N. (1991). Low power laser enzymology: Reactivation of myosin ATPase by GaAs an HeNe lasers. *In* "Basic and Applied Research in Photobiology and Photomedicine" (S. Passarella and E. Quagliariello, Eds.), pp. 315–338. University of Bari, Italy.

Bontioti, E., and Dahlin, L. (2009). Mechanisms underlying the end-to-side nerve regeneration. *Int. Rev. Neurobiol.* **87,** In Press.

Borisov, A. B., Dedkov, E. I., and Carlson, B. M. (2001). Interrelations of myogenic response, progressive atrophy of muscle fibers, and cell death in denervated skeletal muscle. *Anat. Rec.* **264,** 203–218.

Byrnes, K. R., Wu, X., Waynant, R. W., and Anders, J. J. (2005). Low-power laser irradiation alters gene expression of olfactory ensheathing cells *in vitro*. *Lasers Surg. Med.* **37,** 161–171.

Carraro, U., Rossini, K., Mayr, W., and Kern, H. (2005). Muscle fiber regeneration in human permanent lower motoneuron denervation: Relevance to safety and effectiveness of FES-training, which induces muscle recovery in SCI subjects. *Artif. Organs* **29,** 187–191.

Chin, H., and Almon, R. R. (1980). Fiber-type effects of castration on the cholinergic receptor population in skeletal muscle. *J. Pharmacol. Exp. Ther.* **212,** 553–559.

Dow, D. E., Cederna, P. S., Hassett, C. A., Kostrominova, T. Y., Faulkner, J. A., and Dennis, R. G. (2004). Number of contractions to maintain mass and force of a denervated rat muscle. *Muscle Nerve* **30,** 77–86.

Dvali, L., and Mackinnon, S. (2003). Nerve repair, grafting, and nerve transfers. *Clin. Plast. Surg.* **30,** 203–221.

Ehrlicher, A., Betz, T., Stuhrmann, B., Koch, D., Milner, V., Raizen, M. G., and Kas, J. (2002). Guiding neuronal growth with light. *Proc. Natl. Acad. Sci. USA* **99,** 16024–16028.

Elles, J. T., Henry, M. M., Summerfelt, P., *et al.* (2003). Therapeutic photobiomodulation for methanol-induced retinal toxicity. *Proc. Natl. Acad. Sci. USA* **100,** 3439–3444.

Gigo-Benato, D., Geuna, S., de Castro Rodrigues, A., Tos, P., Fornaro, M., Boux, E., *et al.* (2004). Low-power laser biostimulation enhances nerve repair after end-to-side neurorrhaphy: A double-blind randomized study in the rat median nerve model. *Laser Med. Sci.* **19,** 57–65.

Gigo-Benato, D., Geuna, S., and Rochkind, S. (2005). Phototherapy for enhancing peripheral nerve repair: A review of the literature. *Muscle Nerve* **31,** 694–701.

Goldspink, D. F. (1976). The effects of denervation on protein turnover of rat skeletal muscle. *Biochem. J.* **156,** 71–80.

Kline, D. G., and Hackett, E. R. (1975). Reappraisal of timing for exploration of civilian peripheral nerve injuries. *Surgery* **78,** 54–65.

Kloosterboer, H., Faassen, H., Stroker-de Vries, S. A., and Hommes, F. A. (1979). Effect of cyclic nucleotides on weight of gastrocnemius and creatine kinase activity after denervation of muscle in young rats. *Biol. Neonate* **36,** 160–167.

Lomo, T., and Westgaard, R. H. (1975). Further studies on the control of ACH sensitivity by muscle activity in the rat. *J. Physiol.* **252,** 603–626.

Lundborg, G. (2002). Enhancing posttraumatic nerve regeneration. *J. Peripher. Nerv. Syst.* **7,** 139–140.

MacKinnon, S. E., and Dellon, A. L. (1988). *In* "Surgery of the Peripheral Nerve". Thieme Medical publishers, Inc., New York.

Midha, R. (2008). Mechanism and pathology of injury. *In* "Nerve Injury" (D. G. Kline and A. R. Hudson, Eds.), pp. 23–42. Saunders Elsevier, Philadelphia.

Missini, I., Favaro, G., and Carraro, U. (1987). Maturation: Dystrophic changes and the continuous production of fibers in skeletal muscle regenerating in the absence of nerve. *J. Neuropathol. Exp. Neurol.* **46,** 315–331.

Narakas, A. (1978). Surgical treatment of traction injuries of the brachial plexus. *Clin. Orth. Rel. Res.* **133,** 71–90.

Noble, J., Munro, C. A., Prasad, V. S., and Midha, R. (1998). Analysis of upper and lower extremity peripheral nerve injuries in a population of patients with multiple injuries. *J. Trauma* **45,** 6–122.

Rochkind, S. (1978). Stimulation effect of laser energy on the regeneration of traumatically injured peripheral nerves. *Morphogen Regen.* **83,** 25–27.

Rochkind, S. (2009). Phototherapy in peripheral nerve regeneration: From basic science to clinical study. *Neurosurg. Focus* **26**(2), E8, 1–6.

Rochkind, S., and Alon, M. (2000). Microsurgical management of old injuries of the peripheral nerve and brachial plexus. *J. Reconstr. Microsurg.* **16,** 541–546.

Rochkind, S., Barr-Nea, L., Razon, N., Bartal, A., and Schwartz, M. (1987a). Stimulatory effect of HeNe laser low-dose laser on injured sciatic nerves of rats. *Neurosurgery* **20,** 843–847.

Rochkind, S., Nissan, M., Barr-Nea, L., Schwartz, M., and Bartal, A. (1987b). Response of peripheral nerve to HeNe Laser: Experimental studies. *Lasers Surg. Med.* **7,** 441–443.

Rochkind, S., Nissan, M., Lubart, R., Avram, J., and Bartal, A. (1988). The *in vivo* nerve response to direct low-energy laser irradiation. *Acta Neurochir. (Wien)* **94,** 74–77.

Rochkind, S., Barr-Nea, L., and Volger, I. (1990). Spinal cord response to laser treatment of injured peripheral nerve. *Spine* **15,** 6–10.

Rochkind, S., Nissan, M., Alon, M., Shamir, M., and Salame, K. (2001). Effects of laser irradiation on the spinal cord for the regeneration of crushed peripheral nerves in rats. *Lasers Surg. Med.* **28,** 216–219.

Rochkind, S., Shahar, A., Fliss, D., El-Ani, D., Astachov, L., Hayon, T., Alon, M., Zamostiano, R., Ayalon, O., Biton, I. E., Cohen, Y., Halperin, R., *et al.* (2006). Development of a tissue-engineered composite implant for treating traumatic paraplegia in rats. *Eur. Spine J.* **15,** 234–245.

Rochkind, S., Drory, V., Alon, M., Nissan, M., and Ouaknine, G. E. (2007a). The treatment of incomplete peripheral nerve injuries using a new modality—laser phototherapy (780 nm). *Photomed. Laser Surg.* **25,** 436–442.

Rochkind, S., Filmar, G., Kluger, Y., and Alon, M. (2007b). Delayed microsurgery of penetrating peripheral nerve injuries: Pre, intra- and postoperative analysis and results. *Acta Neurochir. Suppl.* **100,** 21–24.

Rochkind, S., Leider-Trejo, L., Nissan, M., Shamir, M., Kharenko, O., and Alon, M. (2007c). Efficacy of 780-nm laser phototherapy on peripheral nerve regeneration after neurotube reconstruction procedure (double-blind randomized study). *Photomed. Laser Surg.* **25,** 137–143.

Rochkind, S., El-Ani, D., Hayun, T., Nevo, Z., and Shahar, A. (2009). Increase of neuronal sprouting and migration using 780-nm laser phototherapy as procedure for cell therapy. *Lasers Surg. Med.* in press.

Rosaki, S. B. (1967). Creatine phosphokinase—A spectrophotometric method with improved sensitivity. *Am. J. Clin. Pathol.* **50,** 89–97.

Rovak, J. M., Cederna, P. S., and Kuzon, W. M., Jr. (2001). Terminolateral neurorrhaphy: A review of the literature. *J. Reconstr. Microsurg.* **17,** 615–624.

Shainberg, A., and Isac, A. (1984). Inhibition of acetylcholine receptor synthesis by conditioned medium of electrically stimulated muscle cultures. *Brain Res.* **13,** 373–376.

Shamir, M. H., Rochkind, S., Sandbank, J., and Alon, M. (2001). Double-blind randomized study evaluating regeneration of the rat transected sciatic nerve after suturing and postoperative low power laser treatment. *J. Reconstruct. Microsurg.* **17,** 133–138.

Shin, D. H., Lee, E., Hyun, J. K., Lee, S. J., Chang, Y. P., Kim, J. W., *et al.* (2003). Growth-associated protein-43 is elevated in the injured rat sciatic nerve after low-power laser irradiation. *Neurosci. Lett.* **344,** 71–74.

Snyder, S. K., Byrnes, K. R., Borke, R. C., Sanchez, A., and Anders, J. J. (2002). Quantification of calcitonin gene-related peptide mRNA and neuronal cell death in facial motor nuclei following axotomy and 633 nm low power laser treatment. *Lasers Surg. Med.* **31,** 216–222.

Spinner, R. J. (2008). Operative care and techniques. *In* "Nerve Injury" (D. G. Kline and A. R. Hudson, Eds.), pp. 87–105. Saunders Elsevier, Philadelphia.

Sunderland, S. (1978). "Nerves and Nerve Injuries", 2nd ed. Churchill Livingstone, Edinburgh.

Terzis, J. K., and Smith, K. L. (1990). The Peripheral Nerve: Structure, Function and Reconstruction A Hampton Press Publication, Raven Press, Hampton.

Van Breugel, H. H., and Bar, P. R. (1993). HeNe laser irradiation affects proliferation of cultured rat Schwann cells in a dose-dependent manner. *J. Neurocytol.* **22,** 185–190.

Viguie, C. A., Lu, D. X., Huang, S. K., Rengen, H., and Carlson, B. M. (1997). Quantitative study of the effects of long-term denervation on the extensor f digitorum longus muscle of the rat. *Anat. Rec.* **248,** 346–354.

Wollman, Y., Rochkind, S., and Simantov, R. (1996). Low-power laser irradiation enhances migration and neurite sprouting of cultured rat embryonal brain cells. *Neurol. Res.* **18,** 467–470.

AGE-RELATED DIFFERENCES IN THE REINNERVATION AFTER PERIPHERAL NERVE INJURY

Uroš Kovačič, Janez Sketelj, and Fajko F. Bajrović

Institute of Pathophysiology, Faculty of Medicine, University of Ljubljana, SI-1000 Ljubljana, Slovenia

Numerous and extensive functional, structural, and biochemical changes characterize intact aged peripheral nervous system. Functional recovery after peripheral nerve injury depends on survival of injured neurons and functional reinnervation of target tissue by regeneration of injured axons and collateral sprouting of uninjured (intact) adjacent axons. The rate of axonal regeneration becomes slower and its extent (density and number of regenerating axons) decreases in aged animals. Aging also impairs terminal sprouting of regenerated axons and collateral sprouting of intact adjacent axons, thus further limiting target reinnervation and its functional recovery. Decreased survival of aged noninjured and injured neurons, limited intrinsic growth potential of neuron, alteration in its responsiveness to stimulatory or inhibitory environmental factors, and changes in the peripheral neural pathways and target tissues are possible reasons for impaired reinnervation after peripheral nerve injury in old age. The review of present data suggests that this impairment is mostly due to the age-related changes in the peripheral neural pathways and target tissues, and not due to the limited intrinsic growth capacity of neurons or their reduced responsiveness to trophic factors. Age-related alterations in the soluble

INTERNATIONAL REVIEW OF NEUROBIOLOGY, VOL. 87
DOI: 10.1016/S0074-7742(09)87026-8

target derived neurotrophic factors, like nerve growth factor, and nonsoluble extracellular matrix components of neural pathways, like laminin, might be important in this respect.

I. Introduction

An increasing number of clinical reports and experimental studies in recent years indicates that the influence of systemic factors, like metabolic disorders, sex, and age, on the outcome of the injuries of the peripheral nervous system (PNS) might be as crucial as the nerve lesion size, its location, and the surgical techniques used for its repair (Jones *et al.*, 2001; Kovačič *et al.*, 2003, 2004; Verdú *et al.*, 2000). Since the population of the industrialized countries is progressively aging, the question of age-related differences regarding the outcome of PNS injuries has become a particularly important issue in clinical practice. Indeed, clinical observations indicating that ageing is associated with less favorable prognosis after peripheral nerve injury (Barrios and de Pablos, 1991; Frykman, 1976) are now supported by a series of experimental studies showing that both regeneration of injured axons and collateral sprouting of uninjured axons after peripheral nerve injury decline with aging (Cowen and Gavazzi, 1998; Verdú *et al.*, 2000). Here we review the changes that occur in the aging PNS, with a particular focus on reinnervation after peripheral nerve injury. Recent results indicating the reasons and possible mechanisms of age-related impairment in peripheral nerve regeneration and collateral sprouting will be summarized.

II. Age-Related Changes in the PNS

Functional, structural, and biochemical changes taking place in intact peripheral nerves during aging are well known and have been extensively reviewed recently (Melcangi *et al.*, 2003; Peters, 2002; Ulfhake *et al.*, 2000; Verdú *et al.*, 2000). Therefore, only the most important age-related changes in the PNS will be described in the following section.

With respect to peripheral organ function, several clinical and experimental investigations demonstrated that normal aging is accompanied by reduction in maximum muscle tetanic force (Einsiedel and Luff, 1992), impairment of thermal, tactile, and vibration sensitivity (Mitchell and Schady, 1988; Navarro and Kenedy, 1991), and autonomic dysfunction (for review see Verdú *et al.*, 2000). However, it should be noted that not all nerve-related functional modalities are reduced in senescence. Namely, aged rats have a lower basal mechanical nociceptive threshold (Kitagawa *et al.*, 2005) and they exhibit more severe tactile allodynia after the partial nerve injury (Crisp *et al.*, 2003) than young rats. In addition, although several electrophysiologic studies demonstrated significantly lower amplitudes of compound action potentials and nerve conduction velocities in older subjects of various species when compared to young ones (see Verdú *et al.*, 1996, 2000; Hort-Legrand *et al.*, 2001), the conduction velocity of unmyelinated fibers was relatively unaffected by the aging process (Sato *et al.*, 1985).

Age-related impairment in peripheral organ function may be due to the loss of target organ tissue, functional reduction in response of the target organs to neural control, decrease of target organ innervation and/or intrinsic impairment of nerve function. In addition to degenerative and regressive changes, as pointed out by Ulfhake *et al.* (2000), sensorimotor impairment during aging includes also phenotypic switches among neurons as well as remodeling of sensorimotor innervation. The pattern of changes suggests that the underlying mechanism is sustained dependence of PNS neurons on target tissues, which, at least in part, appears to be mediated through signaling by target-derived trophic factors.

The age related changes in functional parameters of aged PNS have been attributed to numerous morphological alterations like nerve fiber density, atrophy and other changes (Hort-Legrand *et al.*, 2001; Sato *et al.*, 1985; Verdú *et al.*, 1996). At older ages, histologic and cellular changes may lead to fiber degeneration (Ceballos *et al.*, 1999; Verdú *et al.*, 1996). A variable reduction in the number and density of myelinated und unmyelinated axons has been reported with aging in different types of peripheral nerves of several animal species (Azcoitia *et al.*, 2003; Jeronimo *et al.*, 2008; Knox *et al.*, 1989; Verdú *et al.*, 2000; but see Kovačič *et al.*, 2008). It should be noted that the unaltered number of myelinated axons in the aged sensory nerves, as observed in some studies (Jeronimo *et al.*, 2008; Knox *et al.*, 1989; Kovačič *et al.*, 2008), does not necessarily indicate the absence of axon impairment since striking changes in myelinated axons' size distribution, pathologic alterations of fibers and the presence of regeneration clusters suggest extensive age-related degenerative and regenerative events (Knox *et al.*, 1989).

It is well known that aged myelinated axons undergo atrophy and shape alterations, while myelin sheaths show various irregularities (e.g., ballooning, splitting, infolding, reduplication, and remyelination) (see Peters, 2002; Melcangi *et al.*, 2003). The reduction of the myelin thickness observed in peripheral nerves of aged animals, together with changes in myelin structure, may be related to decrease in myelin protein expression with aging (see Melcangi *et al.*, 2003). In addition, the *g* ratio distribution of old animals' myelinated axons was shifted to the left (Jeronimo *et al.*, 2008), which suggests marked axonal atrophy. Atrophy and changes in the shape of the axons of the neurons of all sizes have been related to a decrease in neurofilament expression with aging (Parhad *et al.*, 1995).

It has been reported that an aging peripheral nerve has reduced energy requirements and progressive reduction in energy stores (Low *et al.*, 1986; Martinelli *et al.*, 2003). The maintenance of nerve cell integrity at long distances from the cell body directly involves the axonal transport, which depends on energy metabolism. All materials transported by the anterograde or retrograde transport use carrier proteins that rely on ATP as the energy source. Accordingly, a decrease of the axonal transport with aging has been demonstrated in several studies (Fernandez and Hodges-Savola, 1994; McQuarrie *et al.*, 1989).

III. Ageing and Reinnervation After Peripheral Nerve Injury

In principle, recovery of motor, sensory, and autonomic function after peripheral nerve injuries is possible only if adequate and functional reinnervation of target tissues is achieved by regeneration of the injured axons and/or by collateral sprouting of intact adjacent axons. Generally, regeneration of injured axons is more vigorous than collateral sprouting of intact axons and is, therefore, the primary process of reinnervation after peripheral nerve injury.

A. Aging and Axon Regeneration After Peripheral Nerve Injury

Several clinical investigations reported that both the rate and the degree of functional recovery after peripheral nerve lesions decline with aging (see Verdú *et al.*, 2000). These observations are in agreement with experimental studies clearly demonstrating age-related reduction in the rate of axon regeneration. Guttmann *et al.* (1942) found that motor axon regeneration, as assessed by toe spreading in response to pinching the regenerating peroneal nerve, was more rapid in young than in old rabbits. This observation was later supported and extended by several morphological and functional studies which demonstrated that the motor, sensory, and sudomotor reinnervation by regenerating axons occurred earlier in young than in old mice and rats (Choi *et al.*, 1995, 1996; Jacob and Croes, 1998; Navarro *et al.*, 1988; Pestronk *et al.*, 1980; Streppel *et al.*, 1998; Verdú *et al.*, 2000).

The impaired functional reinnervation with aging may be also due to reduction of the number of regenerating axons that succeed to reach the target, and due to a limited capability for terminal regenerative axon sprouting in the target tissues. This possibilities are supported by the histological studies which showed the age-related decrease in the number of regenerating axons after crush injury of the sciatic nerve in mice (Tanaka and Webster, 1991; Tanaka *et al.*, 1992) and motor buccal branch of the facial nerve in rat (Vaughan, 1992). In addition, a more detailed morphological analysis demonstrated that the axonal diameter and myelin sheath thickness in regenerated nerves were smaller in aged than in young adult mice (Tanaka *et al.*, 1992). These changes are in agreement with reduced density of cytoskeletal elements in regenerating aged axons (Vaughan, 1992) and significantly lower levels of myelin proteins in the normal aged PNS (Melcangi *et al.*, 2003). They provide a possible explanation for the decreased nerve conduction velocity of regenerated nerves in aged animals (Choi *et al.*, 1995, 1996; Verdú *et al.*, 2000). At late stages of regeneration after nerve injury, the neuromuscular junctions in reinnervated muscles are less stable in aged than in young adult rats, as demonstrated by increased fragmentation, degradation, and loss of endplate area by confocal microscopy (Apel *et al.*, 2009; Wang *et al.*, 2007). Other age-related pathological changes, including reduced numbers of terminal axonal sprouts, greater incidence of abnormal nerve bundles, immature nerve terminals, and multiple innervations of single terminal endplate in the reinnervated muscles of rats were detected after nerve crush injury (Kawabuchi *et al.*, 1998; Wang *et al.*, 2007) and after the treatment of muscle with botulinum toxin, which pharmacologically denervates the muscle fibers (Pestronk *et al.*, 1980). These changes in aged rats are in correlation with impaired terminal Schwann cell (SC) extensions in the reinnervated muscles, which act as a guide for regenerating-sprouting axons during reinnervation (Kawabuchi *et al.*, 1998; Reynolds and Woolf, 1992). Taken together, these results suggest that despite the capability of the nerve fibers from old animals to regenerate and extend sprouts in the muscle, re-establishment of normal single-fiber motor innervation is reduced due to the impaired SC–axon interactions (Kawabuchi *et al.*, 1998; Streppel *et al.*, 1998). Notably, reduction with aging in terminal sprouting of regenerating sympathetic neurons in the iris has also been reported (Gavazzi, 1995).

B. Ageing and Collateral Sprouting of Uninjured Axons

Peripheral nerve injury is usually followed by regeneration of the injured axons. However, if the injured axons fail to reinnervate the denervated area, its functional recovery could only be achieved by collateral sprouting of adjacent intact axons. This leads to expansion of the innervation territory of an uninjured neurons or even nerves into a foreign territory abandoned by the injured neurons or nerve. In the PNS of adult mammals, collateral sprouting of motor, sympathetic and sensory axons has been demonstrated after injury of the adjacent axons (Brown *et al.*, 1981; Devor *et al.*, 1979; Navarro and Kennedy, 1988). Navarro and Kennedy (1988) reported that the rate and percentage of functional reinnervation of sweat glands in the mouse due to collateral sprouting of intact postganglionic sympathetic axons is reduced with aging. Aging also impaired collateral sprouting of uninjured motor axons after partial denervation (transection of the L5 root) of the soleus muscle in mice which resulted in less extensive expansion of motor units when compared to mature young animals (Jacob and Robbins, 1990). In aged rats, collateral sprouting-related expansion of the cutaneous nociceptive territory of the uninjured sural nerve into the adjacent denervated skin, as detected by the pinch test, was slower than in young adult rats, and the absolute reinnervated area after 8 weeks of recovery was about 50% smaller in aged than in young adult rats (Kovačič *et al.*, 2008). This result is in agreement with an early coincidental finding that collateral sensory reinnervation in the skin was more rapid in very young than in older rabbits (Weddell *et al.*, 1941). In addition, the age-related difference in the extension of the collateral sprouting of cutaneous nociceptive axons remained significant even after a twice longer recovery period (Kovačič *et al.*, 2009). Our long-term experiments showed that the recovery of skin nociception in the dorsum of the rat foot due to collateral sprouting of the spared sural nerve reached a plateau about 4 months after denervation of the surrounding territory (see Kovačič *et al.*, 2007). Therefore, we suggested that the age-related difference in the recovery of nociception by collateral sprouting was not due to a slower rate of reinnervation in the aged rats, but due to the reduced extent of collateral sprouting or lower density of the collateral sprouts in the skin.

In the nerve injuries, in which the proximal nerve stump is not available for the classical end-to-end nerve repair, the end-to-side nerve repair technique may have considerable therapeutic potential (Geuna *et al.*, 2006; Pannucci *et al.*, 2007). By using this technique, functional motor and sensory reinnervation of a previously denervated tissue is achieved by sprouting of the uninjured axons of the donor nerve through the end-to-side coapted distal stump of the injured peripheral nerve without causing long-term functional impairment to the donor nerve (see Kovačič *et al.*, 2007). Histomorphometry of the recipient nerves has revealed limited sprouting response after end-to-side nerve repair in older rats (Hess *et al.*, 2006). However, in this study a mixed nerve was used as a donor nerve and, therefore, the results do not allow any conclusion regarding specifically the sprouting of either sensory or motor axons. In our recent studies, we demonstrated that sprouting of nociceptive afferents into the end-to-side coapted peripheral nerve is less abundant in aged than in young adult rats (Kovačič *et al.*, 2008, 2009). This conclusion is supported by the hystomorphometric analysis, which showed that the total number of thin myelinated axons in the recipient nerve 8 and 19 weeks after the end-to-side nerve repair was about 6-fold and 7.6-fold, respectively, lower in aged than in young adult rats. Furthermore, age-related effects on functional recovery of cutaneous mechano-nociception after end-to-side nerve repair were similar to those previously described in functional recovery by collateral axon sprouting after peripheral nerve injury without end-to-side nerve repair.

C. Are all Subpopulations of Peripheral Neurons Equally Affected by Aging?

Pestronk *et al.* (1980) reported that the average rate of regeneration after sciatic nerve crush in the rat, as demonstrated with radiotracer method, slowed with increasing age, although a small population of rapidly regenerating motor and sensory axons continued to grow at fast rate regardless of the age. This is in accordance with an earlier study (Guttmann *et al.*, 1942), which failed to demonstrate age-related difference in the rate of sensory axon regeneration in rabbits. Notably, in this study the sensitive functional nerve pinch test was used to assess the sensory nerve outgrowth, which selectively detects the position of the fastest-growing nociceptive fibers (see also McQuarrie, 1978).

In our recent studies of the end-to-side nerve repair in aging rats, it was found that only the sprouting of myelinated nociceptive axons, but not the unmyelinated nociceptive and/or sympathetic axons, was significantly reduced (Kovačič *et al.*, 2008, 2009). This is in agreement with earlier results showing that the total number of regenerating myelinated axons was significantly reduced 2 weeks after the sciatic nerve crush in aged mice, but the numbers of regenerating unmyelinated axons in old and young adults did not differ significantly (Tanaka and Webster, 1991). The lack of age-related difference in sprouting and regeneration of the unmyelinated axons raises the question of whether the findings about the sprouting and regeneration of aged myelinated axons can be extrapolated to axons of other neuronal populations. Notably, some sympathetic neuron subpopulations in mammals showed little evidence of age-related decline in target innervation (Cowen, 2002; Kudwa *et al.*, 2002). Taken together, these findings suggest that there is no general age-related reduction in the regenerative and sprouting axon growth response of neuronal subpopulations in the PNS.

IV. Possible Reasons for Impaired Reinnervation with Aging

Which differences between the mature and aged PNS determine how vigorously their axons will reinnervate the target tissues after injury? In principal, both regeneration of injured axons and collateral sprouting of uninjured axons can be impaired with aging due to at least four age-dependent changes: (1) decreased survival of aged noninjured and injured neurons, (2) decreased regenerating and sprouting capacity of aged neurons, (3) reduced responsiveness of aged regenerating and sprouting axons to the pathway- or target-derived neurotrophic factors, and (4) alterations in the aged peripheral neural pathways guiding the regenerating and sprouting axons, or changes in the target tissues which they supply.

A. Survival of Aged Noninjured and Injured Neurons

Ageing reduces facial motoneuron number and increases their vulnerability to axotomy in Wistar rats, but not in Fischer 344 rats (Johnson and Duberley, 1998; see also Streppel *et al.*, 1998). To our knowledge, there are no published data on the vulnerability of primary sensory neurons after peripheral nerve injury. Bergman and Ulfhake (1998) found only a small decrease (approximately 12%) in the number of cervical and lumbar dorsal root ganglion (DRG) neurons in intact aged Sprague-Dawley rats of both sexes. On the contrary, in our histomorphometric analysis of myelinated axons in the intact sensory (sural) nerves in Wistar rats of different ages we demonstrated that the total number of myelinated axons in the sural nerves of the young adult rats was minimally (about 10%), but statistically significantly, smaller than that in middle-aged and aged rats (Kovačič *et al.*, 2008). However, there was no significant difference between the groups in the number of thin myelinated axons (diameter less than 5 m) in the sural nerves. In addition, except for minimal changes in the peroneal nerve, no statistically significant alteration in the total number of myelinated axons in the intact sural nerve was associated with aging in Fisher rats (Knox *et al.*, 1989) and Wistar rats (Jeronimo *et al.*, 2008).

There was no significant correlation between the degree of the DRG neuron loss and the extent of behavioral deficits among the aged individuals (Bergman and Ulfhake, 1998). Therefore, the authors concluded that the loss of primary sensory neurons alone cannot explain the functional deficits in sensory perception among senescent individuals. Nevertheless, these results suggest that, at least in some species and in some neuronal populations, less extensive functional reinnervation after PNS injury could at least partly depend on increased susceptibility for the neuronal loss following axonal damage in old age.

B. Regenerating and Sprouting Capacity of Aged Neurons

Axon growth is a product of the combined properties of the neuron and its environment. Therefore, it is difficult to separate the two processes and assign the roles to the individual participants; in particular, it is difficult to design *in vivo* experiments that specifically reveal intrinsic neuron-related differences in aging. However, experimental situations can be created in which axons growing from young and aged neurons can be observed in the same environment, so that differences between these neurons and their causes can be assessed. In order to demonstrate this, experiments have to be designed in which neurons of different age can be transplanted to the same site in a young adult host or allowed to grow along neural pathways from animals of the same ages. In this way it has been demonstrated that there is no general decrease in the vigor of axon growth with increasing age. Accordingly, Gavazzi and Cowen (1993) did not observe any difference in the pattern and time course of reinnervation of young host irises between superior cervical ganglia taken from young and old donors, which were implanted *in oculo* for 4 or 6 weeks. We found that the average number of myelinated sensory axons in the nerve grafts from young rats coapted to the side of the sural nerves of aged rats (host) was (a) 2.4-fold higher than that in the aged nerve grafts coapted to the side of the young host sural nerves, and (b) not statistically significantly smaller than that in the end-to-side coapted grafts in the control group of young rats (Kovačič *et al.*, 2008). These results indicate that the intrinsic growth capacity of aged sensory and sympathetic neurons is not a limiting factor for less abundant growth response of axons during aging. *In vitro*, it is relatively easy to arrange for DRG neurons of different ages to interact with the same type of environment. The results of these experiments (Hall *et al.*, 2001; Niwa *et al.*, 2002) are consistent with the *in vivo* experiments, suggesting that the intrinsic capacity of sensory neurons for outgrowth of new axonal sprouts is not considerably reduced with aging (see also Gavazzi *et al.*, 2001). It should be notified that the conditioning lesion of the sciatic nerves in Fisher rats stimulated aged motor axons to regenerate at rates comparable to those seen in younger animals (Jacob and Croes, 1998). However, increased growth capacity of neurons is suggested to be one of the stimulative effects of conditioning lesion on nerve regeneration (Dahlin and Kanje, 1992; Lankford *et al.*, 1998; McQuarrie, 1978). These observations suggest that essentially reduced intrinsic growth capacity of neurons may have some, though minor, contribution to the decreased axon regeneration with aging.

C. Responsiveness of Aged Regenerating and Sprouting Axons to the Pathway- or Target-Derived Neurotrophic Factors

Axonal atrophy in aging is associated with reduced expression of neurofilament (Parhad *et al.*, 1995), a member of the intermediate filament (IF) network, which play a major role in the regulation of axon caliber (Hoffman *et al.*, 1987). Accordingly, axonal diameter in regenerated nerves is smaller in old than in young adult mice (Tanaka *et al.*, 1992). This could partially explain the decreased conduction velocity of regenerated nerves in aged animals (see above), but not the reduced number of regenerating axons or slower growth rate. Increased levels of peripherin, which is another member of IF proteins, in regenerating DRG neurons and spinal motor neurons shortly after axotomy suggest that peripherin is involved in axonal growth and regeneration (Troy *et al.*, 1990). In addition, intact peripherin is required for collateral sprouting of nociceptive axons (Belecky-Adams *et al.*, 2003). It should be noted that in the DRG peripherin is specifically expressed in small-sized neurons with unmyelinated fibers (Fornaro *et al.*, 2008).

However, the levels of peripherin were even higher in aged than in young adult peripheral nerves and DRGs, suggesting that this filament network is not disrupted in the PNS of aged rats (Kovačič *et al.*, 2008). This could explain the lack of age-related difference in sprouting and regeneration of the unmyelinated fibers into the end-to-side coapted recipient nerve in rats (Kovačič *et al.*, 2008, 2009) and after sciatic nerve crush in mice (Tanaka and Webster, 1991), respectively. The synthesis of other growth-associated proteins, such as GAP-43 and tubulin, which is increased in regenerating and collaterally sprouting DRG neurons (Mearow *et al.*, 1994), is induced in DRGs of aged rats to the same level as in young adult rats during peripheral nerve regeneration (Willcox and Scott, 2004). Taken together, these results suggest that the production of growth-associated proteins is well maintained during aging.

Nevertheless, the elongation of regenerating axons depends on the incorporation of new biologic materials transported from the soma to the growth cone (Oblinger and Lasek, 1988; Tashiro and Komiya, 1992). It was reported that some metabolic changes, that affect the axonal transport of cytoskeletal components and possibly trophic factors, occur in neurons during aging (Fernandez and Hodges-Savola, 1994; McQuarrie *et al.*, 1989). The retrograde transport of nerve growth factor (NGF) is a particularly sensitive marker of normal trophic state of a sensory neuron (Hendry *et al.*, 1974). In addition, NGF regulates collateral sprouting of sensory axons (Diamond *et al.*, 1987) by binding to TrkA and p75 NGF receptors (Gallo *et al.*, 1997). Accordingly, expression of both neurotrophin receptors increased in the DRG sensory neurons during collateral sprouting (Mearow *et al.*, 1994). Downregulation of NGF receptor expression in intact aged DRG (Bergman *et al.*, 1996; Parhad *et al.*, 1995) suggests that reduced responsiveness of sensory neurons to NGF might contribute to the reduced collateral sprouting in aged rats. However, after transection/ligation of either the peroneal nerve (Kovačič *et al.*, 2008) or the sciatic nerve (Bergman *et al.*, 1999) in the rat, no age-related difference in the protein levels of NGF and TrkA receptor, or in the neurotrophin receptors' mRNA levels, was found in the DRGs that contained injured neurons and uninjured sprouting sensory neurons. This argues against the above-mentioned possibility of reduced growth responsiveness of sensory neurons in aged rats. In particular, considering the lack of age-related difference in the collateral sprouting of the unmyelinated axons (Kovačič *et al.*, 2008), we can conclude that there is no general age-related reduction in the axon growth response. Notably, NGF treatment increased the sprouting response of aged rat iridial regenerating nerves on *in oculo* grafts of young and old peripheral tissues (middle cerebral arteries) (Gavazzi, 1995). However, the increase was statistically significant only on young cerebral arteries, suggesting that other target tissue factors might also be important for age-related impairment of PNS plasticity.

D. Alterations in the Aged Peripheral Neural Pathways and Target Tissues

When attempting to explain why aged neurons regenerate less vigorously than young ones, age-related changes in the environment surrounding the axon growth cone are clearly relevant. Delay in Wallerian degeneration in the distal stump of regenerating nerve in aged animals is one characteristic of these changes. Ten days after facial nerve crush, the nerves of young animals (3 months) were characterized by signs of axonal regeneration, whereas the nerves of aged rats (15 months) were still undergoing extensive degeneration of myelin and axons, myelin clearance, whereas axonal sprouts were sparsely found (Vaughan, 1992). A similar pattern was observed at 15 days after crush in the sciatic nerves of aged mice (Tanaka and Webster, 1991; Tanaka *et al.*, 1992) and 4–8 weeks after transection and resuturing the peroneal nerves of aged rats (Choi *et al.*, 1995). A smaller SC cytoplasm and myelin at all time intervals (2–8 weeks) after sciatic nerve crush in aged mice than in young ones (Tanaka *et al.*, 1992) suggests that proliferative response of SCs in the injured nerve is reduced with aging (see also Santos *et al.*, 2000). Taken together, the delay in Wallerian degeneration in aged injured nerves may be most likely attributed to a reduced reactivity of nonneuronal cells (SCs and infiltrating macrophages). The deficit to re-establish normal single motor innervation due to the impaired SC-axon interactions in the reinnervated aged muscle has already been mentioned above (see Kawabuchi *et al.*, 1998). In addition, the peripheral nerves of aged mice are unable to locally upregulate vascular endothelial growth factor, a prototypical angiogenic cytokine, after injury, and have substantial deficits in mounting an appropriate intraneural angiogenic response during nerve regeneration (Pola *et al.*, 2004). By using cross-transplantation procedures in the rat end-to-side nerve coaptation model, we demonstrated that the sprouting of sensory axons from intact young sural nerves into the aged nerve grafts was significantly decreased when compared to young adult controls (Kovačič *et al.*, 2008). In contrast, we found much smaller and statistically not significant reduction of axon sprouting from the aged sural nerves into the young nerve grafts when compared to young adult controls. Therefore, reduced collateral sprouting of sensory axons in aged rats seems to be largely due to age-induced alterations in the peripheral neural pathways, and not due to a decreased intrinsic capability of aged neurons for sprouting. Axonal sprouting into the end-to-side coaptated nerve graft is primarily dependent on SCs (Bajrović *et al.*, 2002). Therefore, a decreased proliferative response of SCs during aging (Santos *et al.*, 2000; Tanaka *et al.*, 1992) might be important in limiting the sprouting of sensory axons from young nerves into aged nerve grafts. In addition, SCs are the main source of axonal growth-promoting substances, such as soluble neurotrophic factors and nonsoluble extracellular matrix (ECM) components. Collateral sprouting of cutaneous nociceptive axons after denervation of the adjacent skin critically depends upon the availability of the NGF (Diamond *et al.*, 1987; see also Bajrović *et al.*, 2002). On the other hand, regeneration of sensory and sympathetic neurons is, at least at the beginning, an NGF-independent process (Bajrović *et al.*, 2001; Diamond *et al.*, 1987; Gloster and Diamond, 1992), even though NGF administration can prevent some of the degenerative responses to axonal damage (Otten *et al.*, 1983; Rich *et al.*, 1987). At present, there is no clear evidence that peripheral tissues synthesize less NGF in aged than in young animals. By using Western blot method, we found that only the levels of the 32 and 53 kDa NGF isoforms (but not of other isoforms) were significantly higher in the denervated skin in young adult than in aged rats (Kovačič *et al.*, 2008). However, there were no differences between the two age groups of rats regarding the content of NGF isoforms in the intact and degenerated peroneal nerves. By using more sensitive methods, the levels of NGF mRNA and protein in the intact and degenerated sciatic nerves were found to be even higher in aged than in young adult rats and mice (Date *et al.*, 1994; Ming *et al.*, 1999). Mature NGF (13.5 kDa), 25 and 32 kDa precursor proteins (proNGF), and high-molecular-weight glycosylated precursor NGF isoforms (170, 70, 60, and 53 kDa) have been described earlier in the rat DRG, peripheral nerve and skin (see Kovačič *et al.*, 2008). It should be pointed out that the biological roles of precursor NGF isoforms are not fully understood but might exhibit proapoptotic activity (Lee *et al.*, 2001), in particular in the peripheral sympathetic neurons of aged but not of young adult rodents (Al-Shawi *et al.*, 2008). One obvious alteration is a dramatic increase (up to 50-fold) in NGF protein isoforms, corresponding to proNGF, in the superior cervical ganglia and targets (pineal gland and extracerebral blood vessels) in which the sympathetic innervation shows age-related atrophy in rats. In the iris, where sympathetic innervation is protected into the old age, proNGF was decreased (see Randolph *et al.*, 2007; Al-Shawi *et al.*, 2008). In addition, sympathectomy in young adult animals was followed by the same alterations in proNGF levels (Randolph *et al.*, 2007). Notably, survival of aged sympathetic neurons was rescued by neurotensin, an alternative sortilin ligand that blocks the sortilin-mediated effects of proNGF (Al-Shawi *et al.*, 2008). Because of these data it was proposed that selective age-related neuronal atrophy and neurodegeneration may be mediated by elevated levels of proNGF expression in some targets of sympathetic axons.

On the other hand, we found that the levels of laminin polypeptides in the peroneal nerves were 50–100% higher in young adult than in aged rats. Non-soluble ECM components, such as laminin, promote axonal outgrowth of sensory neurons *in vitro* (Millaruelo *et al.*, 1988; Tucker *et al.*, 2006) and regeneration of the rat sciatic nerve *in vivo* (Wang *et al.*, 1992). In addition, age-dependent decrease of laminin content in cerebral vessels from old rats reduced their sensory reinnervation after transplantation *in oculo* into the young host when compared with young rat cerebral vessels (Gavazzi *et al.*, 1995, 1996). These findings suggest that age-induced alterations of the ECM components in the peripheral neural pathways probably play an important role in age-related reduction of nerve regeneration (Wang *et al.*, 1992) and collateral sprouting of sensory axons into end-to-side coapted nerve grafts and into the skin (Kovačič *et al.*, 2008). In summary, the results of all presented studies argue that the aged axon growth pathways and target tissues are largely responsible for the observed reduced ability of PNS for recovery after nerve injury in old age.

V. Conclusions

The data demonstrating the age-related reduction in the capacity of functional reinnervation of the PNS after injury have increased over the last years. And yet, as human longevity increases and brings about an increase in the number of elderly citizens, we need to understand what is happening to the PNS during normal aging, and in particular, what mechanisms underlie the impaired reinnervation after PNS injuries in old age. The present data suggest that this is mostly due to the age-induced alterations in the peripheral neural pathways and target tissues, and not due to the essentially decreased intrinsic growth capacity of neurons or their reduced responsiveness to trophic factors. Age-related alterations in the soluble neurotrophic factors and nonsoluble growth-promoting ECM components of neural pathways, like laminin, might be important in this respect. Notably, studies on age-related changes in the expression of inhibitory environmental factors in the PNS are lacking at present. Although the translation of animal studies to human applications should be performed with caution, understanding the relative contributions of neurons and their peripheral targets to age-related differences in the plasticity of the PNS after injury might be important for the future treatment of PNS injuries in the old age.

References

Al-Shawi, R., Hafner, A., Olsen, J., Chun, S., Raza, S., Thrasivoulou, C., Lovestone, S., Killick, R., Simons, P., and Cowen, T. (2008). Neurotoxic and neurotrophic roles of proNGF and the receptor sortilin in the adult and ageing nervous system. *Eur. J. Neurosci.* **27,** 2103–2114.

Apel, P. J., Alton, T., Northam, C., Ma, J., Callahan, M., Sonntag, W. E., and Li, Z. (2009). How age impairs the response of the neuromuscular junction to nerve transection and repair: An experimental study in rats. *J. Orthop. Res.* **27,** 385–393.

Azcoitia, I., Leonelli, E., Magnaghi, V., Veiga, S., Garcia-Segura, L. M., and Melcangi, R. C. (2003). Progesterone and its derivatives dihydroprogesterone and tetrahydroprogesterone reduce myelin fiber morphological abnormalities and myelin fiber loss in the sciatic nerve of aged rats. *Neurobiol. Aging* **24,** 853–860.

Bajrović, F., Srpčić, M., and Sketelj, J. (2001). Schwann cell dependence of regenerating rat sensory neurons is inversely related to the quality of axon growth substratum. *J Neuropathol. Exp. Neurol.* **60,** 65–74.

Bajrović, F., Kovačič, U., Pavčnik, M., and Sketelj, J. (2002). Interneuronal signalling is involved in induction of collateral sprouting of nociceptive axons. *Neuroscience* **111,** 587–596.

Barrios, C., and de Pablos, J. (1991). Surgical management of nerve injuries of the upper extremity in children: A 15-year survey. *J. Pediatr. Orthop.* **11,** 641–645.

Belecky-Adams, T., Holmes, M., Shan, Y., Tedesco, C. S., Mascari, C., Kaul, A., Wight, D. C., Morris, R. E., Sussman, M., Diamond, J., and Parysek, L. M. (2003). An intact intermediate filament network is required for collateral sprouting of small diameter nerve fibers. *J. Neurosci.* **23,** 9312–9319.

Bergman, E., and Ulfhake, B. (1998). Loss of primary sensory neurons in the very old rat: Neuron number estimates using the disector method and confocal optical sectioning. *J. Comp. Neurol.* **396,** 211–222.

Bergman, E., Johnson, H., Zhang, X., Hokfelt, T., and Ulfhake, B. (1996). Neuropeptides and neurotrophin receptor mRNAs in primary sensory neurons of aged rats. *J. Comp. Neurol.* **375,** 303–319.

Bergman, E., Fundin, B. T., and Ulfhake, B. (1999). Effects of aging and axotomy on the expression of neurotrophin receptors in primary sensory neurons. *J. Comp. Neurol.* **410,** 368–386.

Brown, M. C., Holland, R. L., and Hopkins, W. G. (1981). Motor nerve sprouting. *Annu. Rev. Neurosci.* **4,** 17–42.

Ceballos, D., Cuadras, J., Verdú, E., and Navarro, X. (1999). Morphometric and ultrastructural changes with ageing in mouse peripheral nerve. *J. Anat.* **195,** 563–576.

Choi, S. J., Harii, K., Lee, M. J., Furuya, F., and Ueda, K. (1995). Electrophysiological, morphological, and morphometric effects of aging on nerve regeneration in rats. *Scand. J. Plast. Reconstr. Surg. Hand Surg.* **29,** 133–140.

Choi, S. J., Harii, K., Asato, H., and Ueda, K. (1996). Aging effects in skeletal muscle recovery after reinnervation. *Scand. J. Plast. Reconstr. Surg. Hand Surg.* **30,** 89–98.

Cowen, T. (2002). Selective vulnerability in adult and ageing mammalian neurons. *Auton. Neurosci.* **96,** 20–24.

Cowen, T., and Gavazzi, I. (1998). Plasticity in adult and ageing sympathetic neurons. *Prog. Neurobiol.* **54,** 249–288.

Crisp, T., Giles, J. R., Cruce, W. L., McBurney, D. L., and Stuesse, S. L. (2003). The effects of aging on thermal hyperalgesia and tactile-evoked allodynia using two models of peripheral mononeuropathy in the rat. *Neurosci. Lett.* **339,** 103–106.

Dahlin, L. B., and Kanje, M. (1992). Conditioning effect induced by chronic nerve compression. An experimental study of the sciatic and tibial nerves of rats. *Scand. J. Plast. Reconstr. Surg. Hand Surg.* **26,** 37–41.

Date, I., Furukawa, S., and Ohmoto, T. (1994). Increased nerve growth factor level in the distal stump of transected sciatic nerve in relation to aging and its application for neural grafting. *Exp. Neurol.* **130,** 168–171.

Devor, M., Schonfeld, D., Seltzer, Z., and Wall, P. D. (1979). Two modes of cutaneous reinnervation following peripheral nerve injury. *J. Comp. Neurol.* **185,** 211–220.

Diamond, J., Coughlin, M., MacIntyre, L., Holmes, M., and Visheau, B. (1987). Evidence that endogenous nerve growth factor is responsible for collateral sprouting, but not the regeneration of nociceptive axons in adult rats. *Proc. Natl. Acad. Sci. USA* **84,** 6596–6600.

Einsiedel, L. J., and Luff, A. R. (1992). Alterations in the contractile properties of motor units within the ageing rat medial gastrocnemius. *J. Neurol. Sci.* **112,** 170–177.

Fernandez, H. L., and Hodges-Savola, C. A. (1994). Axoplasmic transport of calcitonin gene-related peptide in rat peripheral nerve as a function of age. *Neurochem. Res.* **19,** 1369–1377.

Fornaro, M., Lee, J. M., Raimondo, S., Nicolino, S., Geuna, S., and Giacobini-Robecchi, M. (2008). Neuronal intermediate filament expression in rat dorsal root ganglia sensory neurons: An *in vivo* and *in vitro* study. *Neuroscience* **153,** 1153–1163.

Frykman, G. K. (1976). Peripheral nerve injuries in children. *Orthop. Clin. North Am.* **7,** 701–716.

Gallo, G., Lefcort, F. B., and Letourneau, P. C. (1997). The trkA receptor mediates growth cone turning toward a localized source of nerve growth factor. *J. Neurosci.* **17,** 5445–5454.

Gavazzi, I. (1995). Collateral sprouting and responsiveness to nerve growth factor of ageing neurons. *Neurosci. Lett.* **189,** 47–50.

Gavazzi, I., and Cowen, T. (1993). Axonal regeneration from transplanted sympathetic ganglia is not impaired by age. *Exp. Neurol.* **122,** 57–64.

Gavazzi, I., Boyle, K. S., Edgar, D., and Cowen, T. (1995). Reduced laminin immunoreactivity in the blood vessel wall of ageing rats corelates with reduced innervation *in vivo* and following transplantation. *Cell Tissue Res.* **281,** 23–32.

Gavazzi, I., Boyle, K. S., and Cowen, T. (1996). Extracellular matrix molecules influence innervation density in rat cerebral blood vessels. *Brain Res.* **734,** 167–174.

Gavazzi, I., Railton, K. L., Ong, E., and Cowen, T. (2001). Responsiveness of sympathetic and sensory iridial nerves to NGF treatment in young and aged rats. *Neurobiol. Aging* **22,** 287–296.

Geuna, S., Papalia, I., and Tos, P. (2006). End-to-side (terminolateral) nerve regeneration: A challenge for neuroscientists coming from an intriguing nerve repair concept. *Brain Res. Rev.* **52,** 381–388.

Gloster, A., and Diamond, J. (1992). Sympathetic nerves in adult rats regenerate normally and restore pilomotor function during an anti-NGF treatment that prevents their collateral sprouting. *J. Comp. Neurol.* **326,** 363–374.

Guttmann, E., Guttmann, L., Medawar, P. B., and Young, J. Z. (1942). The rate of regeneration of nerve. *J. Exp. Biol.* **19,** 14–44.

Hall, K. E., Sheng, H. C., Srinivasan, S., Spitsbergen, J. M., Tuttle, J. B., Steers, W. D., and Wiley, J. W. (2001). Treatment of aged rat sensory neurons in short-term, serum-free culture with nerve growth factor reverses the effect of aging on neurite outgrowth, calcium currents, and neuronal survival. *Brain Res.* **888,** 128–137.

Hendry, I. A., Stockel, K., Thoenen, H., and Iversen, L. L. (1974). The retrograde axonal transport of nerve growth factor. *Brain Res.* **68,** 103–121.

Hess, J. R., Brenner, M. J., Myckatyn, T. M., Hunter, D. A., and Mackinnon, S. E. (2006). Influence of aging on regeneration in end-to-side neurorrhaphy. *Ann. Plast. Surg.* **57,** 217–222.

Hoffman, P. N., Cleveland, D. W., Griffin, J. W., Landes, P. W., Cowan, N. J., and Price, D. L. (1987). Neurofilament gene expression: A major determinant of axonal caliber. *Proc. Natl. Acad. Sci. USA* **84,** 3472–3476.

Hort-Legrand, C., Lestrade, R., and Behar, A. (2001). Motor and sensory conduction velocities and amplitude of nerve or muscle potentials in the normal rat according to age. *Acta Physiol. Hung.* **88,** 239–249.

Jacob, J. M., and Croes, S. A. (1998). Acceleration of axonal outgrowth in motor axons from mature and old F344 rats after a conditioning lesion. *Exp. Neurol.* **152,** 231–237.

Jacob, J. M., and Robbins, N. (1990). Age differences in morphology of reinnervation of partially denervated mouse muscle. *J. Neurosci.* **10,** 1530–1540.

Jeronimo, A., Jeronimo, C. A., Rodrigues Filho, O. A., Sanada, L. S., and Fazan, V. P. (2008). A morphometric study on the longitudinal and lateral symmetry of the sural nerve in mature and aging female rats. *Brain Res.* **1222,** 51–60.

Johnson, I. P., and Duberley, R. M. (1998). Motoneuron survival and expression of neuropeptides and neurotrophic factor receptors following axotomy in adult and ageing rats. *Neuroscience* **84,** 141–150.

Jones, K. J., Brown, T. J., and Damaser, M. (2001). Neuroprotective effects of gonadal steroids on regenerating peripheral motoneurons. *Brain Res. Rev.* **37,** 372–382.

Kawabuchi, M., Chongjian, Z., Islam, A. T., Hirata, K., and Nada, O. (1998). The effect of aging on the morphological nerve changes during muscle reinnervation after nerve crush. *Restor. Neurol. Neurosci.* **13,** 117–127.

Kitagawa, J., Kanda, K., Sugiura, M., Tsuboi, Y., Ogawa, A., Shimizu, K., Koyama, N., Kamo, H., Watanabe, T., Ren, K., and Iwata, K. (2005). Effect of chronic inflammation on dorsal horn nociceptive neurons in aged rats. *J. Neurophysiol.* **93,** 3594–3604.

Knox, C. A., Kokmen, E., and Dyck, P. J. (1989). Morphometric alteration of rat myelinated fibers with aging. *J. Neuropathol. Exp. Neurol.* **48,** 119–139.

Kovačič, U., Sketelj, J., and Bajrović, F. F. (2003). Sex-related difference in collateral sprouting of nociceptive axons after peripheral nerve injury in the rat. *Exp. Neurol.* **184,** 479–488.

Kovačič, U., Žele, T., Osredkar, J., Sketelj, J., and Bajrović, F. F. (2004). Sex-related differences in the regeneration of sensory axons and recovery of nociception after peripheral nerve crush in the rat. *Exp. Neurol.* **189,** 94–104.

Kovačič, U., Tomšič, M., Sketelj, J., and Bajrović, F. F. (2007). Collateral sprouting of sensory axons after end-to-side nerve coaptation: A longitudinal study in the rat. *Exp. Neurol.* **203,** 358–369.

Kovačič, U., Žele, T., Marš, T., Sketelj, J., and Bajrović, F. F. (2008). Aging impairs collateral sprouting of nociceptive axons in the rat. *Neurobiol Aging* doi:10.1016/j.neurobiolaging.2008.03 (Epub ahead of print)..

Kovačič, U., Sketelj, J., and Bajrović, F. F. (2009). Effect of aging on recovery of cutaneous nociception after end-to-side nerve repair in the rat. *Ann. Plast. Surg.* doi: 10.1097/SAP.0b013e318180c8cb (in press).

Kudwa, A. E., Shoemaker, S. E., Crutcher, K. A., and Isaacson, L. G. (2002). Evidence for reduced accumulation of exogenous neurotrophin by aged sympathetic neurons. *Brain Res.* **948,** 24–32.

Lankford, K. L., Waxman, S. G., and Kocsis, J. D. (1998). Mechanisms of enhancement of neurite regeneration *in vitro* following a conditioning sciatic nerve lesion. *J. Comp. Neurol.* **391,** 11–29.

Lee, R., Kermani, P., Teng, K. K., and Hempstead, B. L. (2001). Regulation of cell survival by secreted proneurotrophins. *Science* **294,** 1945–1948.

Low, P. A., Schmelzer, J. D., and Ward, K. K. (1986). The effect of age on energy metabolism and resistance to ischaemic conduction failure in rat peripheral nerve. *J. Physiol.* **374,** 263–271.

Martinelli, C., Sartori, P., Ledda, M., and Pannese, E. (2003). Age-related quantitative changes in mitochondria of satellite cell sheaths enveloping spinal ganglion neurons in the rabbit. *Brain Res. Bull.* **61,** 147–151.

McQuarrie, I. G. (1978). The effect of a conditioning lesion on the regeneration of motor axons. *Brain Res.* **152,** 597–602.

McQuarrie, I. G., Brady, S. T., and Lasek, R. J. (1989). Retardation in the slow axonal transport of cytoskeletal elements during maturation and aging. *Neurobiol. Aging* **10,** 359–365.

Mearow, K. M., Kril, Y., Gloster, A., and Dimond, J. (1994). Expression of NGF receptor and GAP-43 mRNA in DRG neurons during collateral sprouting and regeneration of dorsal cutaneous nerves. *J. Neurobiol.* **25,** 127–142.

Melcangi, R. C., Azcoitia, I., Ballabio, M., Cavarretta, I., Gonzalez, L. C., Leonelli, E., Magnaghi, V., Veiga, S., and Garcia-Segura, L. M. (2003). Neuroactive steroids influence peripheral myelination: A promising opportunity for preventing or treating age-dependent dysfunctions of peripheral nerves. *Prog. Neurobiol.* **71,** 57–66.

Millaruelo, A. I., Nieto-Sampedro, M., and Cotman, C. W. (1988). Cooperation between nerve growth factor and laminin or fibronectin in promoting sensory neuron survival and neurite outgrowth. *Brain Res.* **466,** 219–228.

Ming, Y., Bergman, E., Edstrom, E., and Ulfhake, B. (1999). Reciprocal changes in the expression of neurotrophin mRNAs in target tissues and peripheral nerves of aged rats. *Neurosci. Lett.* **273,** 187–190.

Mitchell, S., and Schady, W. (1988). Somatic sensation in the elderly. *In* "New Issues in Neurosciences: Peripheral Nerve Changes in the Elderly" (P. K. Thomas, Ed.), pp. 157–164. John Wiley and Sons, Chichester.

Navarro, X., Kamei, H., and Kennedy, W. R. (1988). Effect of age and maturation on sudomotor nerve regeneration in mice. *Brain Res.* **447,** 133–140.

Navarro, X., and Kennedy, W. R. (1988). Effect of age on collateral reinnervation of sweat glands in the mouse. *Brain Res.* **463,** 174–181.

Navarro, X., and Kennedy, W. R. (1991). Evaluation of thermal and pain sensitivity in type I diabetic patients. *J. Neurol. Neurosurg. Psychiatr.* **54,** 60–64.

Niwa, H., Hayakawa, K., Yamamoto, M., Itoh, T., Mitsuma, T., and Sobue, G. (2002). Differential age-dependent trophic responses of nodose, sensory, and sympathetic neurons to neurotrophins and GDNF: Potencies for neurite extension in explant culture. *Neurochem. Res.* **27,** 485–496.

Oblinger, M. M., and Lasek, R. J. (1988). Axotomy-induced alterations in the synthesis and transport of neurofilaments and microtubules in dorsal root ganglion cells. *J. Neurosci.* **8,** 1747–1758.

Otten, U., Lorez, H. P., and Businger, F. (1983). Nerve growth factor antagonizes the neurotoxic action of capsaicin on primary sensory neurones. *Nature* **301,** 515–517.

Pannucci, C., Myckatyn, T. M., Mackinnon, S. E., and Hayashi, A. (2007). End-to-side nerve repair: Review of the literature. *Restor. Neurol. Neurosci.* **25,** 45–63.

Parhad, I. M., Scott, J. N., Cellars, L. A., Bains, J. S., Krekoski, C. A., and Clark, A. W. (1995). Axonal atrophy in aging is associated with a decline in neurofilament gene expression. *J. Neurosci. Res.* **41,** 355–366.

Pestronk, A., Drachman, D. B., and Griffin, J. W. (1980). Effects of aging on nerve sprouting and regeneration. *Exp. Neurol.* **70,** 65–82.

Peters, A. (2002). The effects of normal aging on myelin and nerve fibers: A review. *J. Neurocytol.* **31,** 581–593.

Pola, R., Aprahamian, T. R., Bosch-Marcé, M., Curry, C., Gaetani, E., Flex, A., Smith, R. C., Isner, J. M., and Losordo, D. W. (2004). Age-dependent VEGF expression and intraneural neovascularization during regeneration of peripheral nerves. *Neurobiol. Aging* **25,** 1361–1368.

Randolph, C. L., Bierl, M. A., and Isaacson, L. G. (2007). Regulation of NGF and NT-3 protein expression in peripheral targets by sympathetic input. *Brain Res.* **1144,** 59–69.

Reynolds, M. L., and Woolf, C. J. (1992). Terminal Schwann cells elaborate extensive processes following denervation of the motor endplate. *J. Neurocytol.* **21,** 50–66.

Rich, K. M., Luszczynski, J. R., Osborne, P. A., and Johnson, E. M., Jr. (1987). Nerve growth factor protects adult sensory neurons from cell death and atrophy caused by nerve injury. *J. Neurocytol.* **16,** 261–268.

Santos, A. R. Jr., Wada, M. L., Langone, F., and Oliveira, A. L. (2000). Differential schwann cell migration in adult and old mice: An *in vitro* study. *Brain Res.* **881,** 73–76.

Sato, A., Sato, Y., and Suzuki, H. (1985). Aging effects on conduction velocities of myelinated and unmyelinated fibers of peripheral nerves. *Neurosci. Lett.* **53,** 15–20.

Streppel, M., Angelov, D. N., Guntinas-Lichius, O., Hilgers, R. D., Rosenblatt, J. D., Stennert, E., and Neiss, W. F. (1998). Slow axonal regrowth but extreme hyperinnervation of target muscle after suture of the facial nerve in aged rats. *Neurobiol. Aging* **19,** 83–88.

Tanaka, K., and Webster, H. D. (1991). Myelinated fiber regeneration after crush injury is retarded in sciatic nerves of aging mice. *J. Comp. Neurol.* **308,** 180–187.

Tanaka, K., Zhang, Q. L., and Webster, H. D. (1992). Myelinated fiber regeneration after sciatic nerve crush: Morphometric observations in young adult and aging mice and the effects of macrophage suppression and conditioning lesions. *Exp. Neurol.* **118,** 53–61.

Tashiro, T., and Komiya, Y. (1992). Organization and slow axonal transport of cytoskeletal proteins under normal and regenerating conditions. *Mol. Neurobiol.* **6,** 301–311.

Troy, C. M., Muma, N. A., Greene, L. A., Price, D. L., and Shelanski, M. L. (1990). Regulation of peripherin and neurofilament expression in regenerating rat motor neurons. *Brain Res.* **529,** 232–238.

Tucker, B. A., Rahimtula, M., and Mearow, K. M. (2006). Laminin and growth factor receptor activation stimulates differential growth responses in subpopulations of adult DRG neurons. *Eur. J. Neurosci.* **24,** 676–690.

Ulfhake, B., Bergman, E., Edstrom, E., Fundin, B. T., Johnson, H., Kullberg, S., and Ming, Y. (2000). Regulation of neurotrophin signaling in aging sensory and motoneurons: Dissipation of target support? *Mol Neurobiol.* **21,** 109–135.

Vaughan, D. W. (1992). Effects of advancing age on peripheral nerve regeneration. *J. Comp. Neurol.* **323,** 219–237.

Verdú, E., Butí, M., and Navarro, X. (1996). Functional changes of the peripheral nervous system with aging in the mouse. *Neurobiol. Aging* **17,** 73–77.

Verdú, E., Ceballos, D., Vilches, J. J., and Navarro, X. (2000). Influence of aging on peripheral nerve function and regeneration. *J. Peripher. Nerv. Syst.* **5,** 191–208.

Wang, G. Y., Hirai, K., Shimada, H., Taji, S., and Zhong, S. Z. (1992). Behavior of axons, Schwann cells and perineurial cells in nerve regeneration within transplanted nerve grafts: Effects of anti-laminin and anti-fibronectin antisera. *Brain Res.* **583,** 216–226.

Wang, Y. J., Zhou, C. J., Shi, Q., Smith, N., and Li, T. F. (2007). Aging delays the regeneration process following sciatic nerve injury in rats. *J. Neurotrauma* **24,** 885–894.

Weddell, G., Guttmann, L., and Guttmann, E. (1941). The local extension of nerve fibers into denervated areas of skin. *J. Neurol. Psych.* **4,** 206–225.

Willcox, B. J., and Scott, J. N. (2004). Growth-associated proteins and regeneration-induced gene expression in the aging neuron. *Mech. Ageing Dev.* **125,** 513–516.

NEURAL PLASTICITY AFTER NERVE INJURY AND REGENERATION

Xavier Navarro*,†

*Institute of Neurosciences and Department of Cell Biology, Physiology and Immunology, Universitat Autònoma de Barcelona, E-08193 Bellaterra, Spain

†Centro de Investigación en Red sobre Enfermedades Neurodegenerativas (CIBERNED), Spain

Injuries to the peripheral nerves result in partial or total loss of motor, sensory, and autonomic functions in the denervated segments of the body due to the interruption of axons, degeneration of distal nerve fibers, and eventual death of axotomized neurons. Functional deficits caused by nerve injuries can be compensated by reinnervation of denervated targets by regenerating injured axons or by collateral branching of undamaged axons, and remodeling of nervous system circuitry related to the lost functions. Plasticity of central connections may compensate functionally for the lack of adequate target reinnervation; however, plasticity has limited effects on disturbed sensory localization or fine motor control after injuries, and may even result in maladaptive changes, such as neuropathic pain and hyperreflexia. After axotomy, neurons shift from a transmitter to a regenerative phenotype, activating molecular pathways that promote neuronal survival and axonal regeneration. Peripheral nerve injuries also induce a cascade of events, at the molecular, cellular, and system levels, initiated by the injury and progressing throughout plastic changes at the spinal cord, brainstem nuclei, thalamus, and brain cortex. Mechanisms involved in these changes include neurochemical changes, functional alterations of excitatory and inhibitory

DOI: 10.1016/S0074-7742(09)87027-X

0074-7742/09 $35.00

synaptic connections, sprouting of new connections, and reorganization of sensory and motor central maps. An important direction for research is the development of therapeutic strategies that enhance axonal regeneration, promote selective target reinnervation, and are also able to modulate central nervous system reorganization, amplifying positive adaptive changes that improve functional recovery and also reducing undesirable effects.

I. Introduction

Injuries to the peripheral nerves result in a partial or total loss of motor, sensory, and autonomic functions mediated by the lesioned nerves to the denervated segments of the body due to the interruption of axons, degeneration of nerve fibers distal to the lesion, and eventual death of axotomized neurons. Injuries to the peripheral nervous system can result in substantial functional loss and decreased quality of life because of permanently impaired sensory and motor functions and secondary problems, such as neuropathic pain.

Functional deficits caused by nerve injuries can be compensated by three neural mechanisms: the reinnervation of denervated targets by regeneration of injured axons, the reinnervation by collateral sprouting of undamaged axons in the vicinity, and the remodeling of nervous system circuits. However, clinical and experimental evidences demonstrate that such mechanisms do not usually allow for a satisfactory functional recovery after severe nerve injuries (Kline and Hudson, 1995; Lundborg, 2004; Sunderland, 1991). The capability of transected nerves to regenerate and recover functional connections is dependent on the age of the subject, the site and type of lesion, the type and delay of surgical repair, and the distance over which axons must regrow to span the injury. Despite that fact that peripheral nerve fibers are able to regenerate across the injury site and along the distal nerve, reinnervation of target organs does not always lead to adequate recovery of motor and sensory functions. The selectivity of axon-target reconnection plays an important role in the impairment of function after nerve injury and regeneration. Misdirection of regenerated axons leads to inappropriate reinnervation of end organs (Bodine-Fowler *et al.*, 1997; Molander and Aldskogius, 1992; Valero-Cabré *et al.*, 2004), although preferential motor reinnervation has been observed as the result of progressive withdrawal or pruning of misdirected axons (Brushart, 1993). On the other hand, collateral reinnervation has temporal and spatial constraints, especially for large sensory and motor axons (Brown *et al.*, 1980; Jackson and Diamond, 1984), although it is helpful to recover cutaneous pain sensibility and motor strength in partially denervated muscles.

Factors contributing to poor functional recovery after peripheral nerve injuries include the following: (1) damage to the neuronal cell body due to axotomy and retrograde degeneration, excluding the possibility of regeneration; (2) inability for axonal growth due to the nerve lesion or underlying diseases; (3) poor specificity of reinnervation by regenerating axons, when target organs become reinnervated by nerve fibers of different function; (4) changes in the central circuits in which the injured neurons participate due to plasticity of neural connections. Nowadays, there are no repair techniques that ensure the recovery of normal sensorimotor functions following severe nerve trauma, and it is generally agreed that a plateau has been reached for the refinement of surgical repair techniques (Lundborg, 2000a). New therapeutic strategies are needed to potentiate axonal regeneration, promote selective target reinnervation, and modulate central reorganization.

The peripheral and central nervous systems are functionally integrated regarding the consequences of a nerve injury: a peripheral nerve lesion always results in profound and long-lasting central modifications and reorganization (Kaas and Collins, 2003; Navarro *et al.*, 2007; Wall *et al.*, 2002). The mechanisms of plasticity and reorganization of spinal and brain circuits linked with the axotomized neurons are complex; they may result in either beneficial adaptative functional changes or maladaptive changes resulting in positive symptoms, such as pain, disesthesia, hyperreflexia, and dystonia. Plastic changes occur at the molecular, cellular, and circuit levels and affect injured neurons as well as neurons that interconnect with them and glial cells (Fig. 1).

II. Neuronal Survival and Reaction to Axotomy

After nerve injuries, axons distal to the lesion site are disconnected from the neuronal body and degenerate. The axotomized neurons undergo a series of phenotypic changes, known as neuronal reaction and chromatolysis, which represent the changes necessary for survival and axonal regeneration (Fu and Gordon, 1997; Verdú and Navarro, 1998). The success of nerve regeneration depends at a first instance on the capacity of axotomized neurons to survive and shift toward a regenerative phenotype. The availability of several neurotrophic factors, provided by autocrine and paracrine sources, influences the response of axotomized neurons (Boyd and Gordon, 2003; Terenghi, 1999).

The most obvious morphological changes in the neuronal body after axotomy are dissolution of the Nissl bodies (chromatolysis), nuclear eccentricity, nuclear and nucleolar enlargement, cell swelling, and retraction of dendrites (Kreutzberg, 1995; Lieberman, 1971). The intensity and time course of the neuronal response are mainly influenced by the severity of the injury, distance of lesion to cell body,

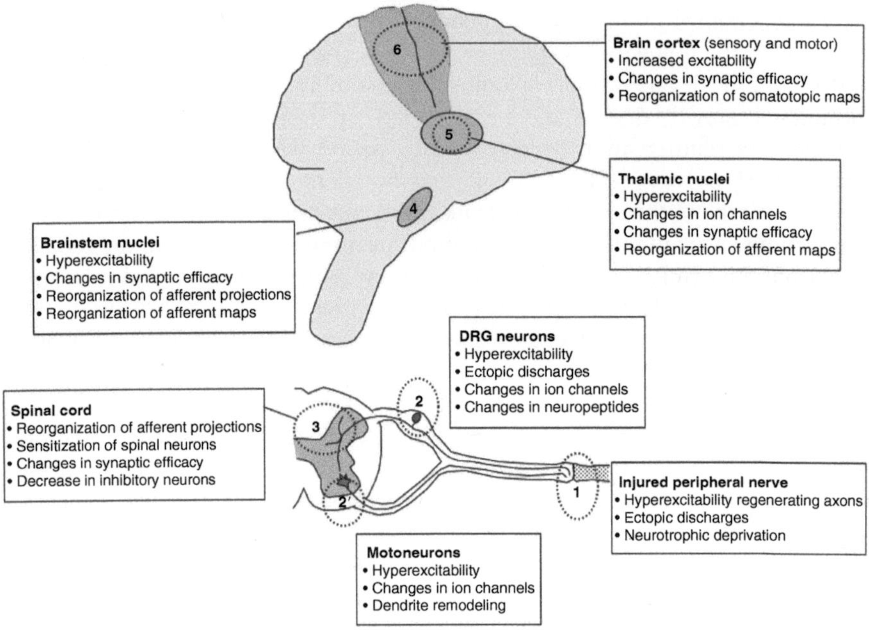

FIG. 1. Schematic summary of the main plastic changes that develop after nerve injury at the different levels of the somatic sensorimotor nervous system. (Reprinted from Navarro *et al.*, 2007 with permission from Elsevier.)

type of neuron, and age. Chromatolysis starts in hours following nerve section and is not fully reversed by 3 months after axotomy. The reaction is more intense and long lasting if distal reinnervation is impeded. A neuronal response continuum appears to exist, beginning as chromatolysis and evolving into either survival or apoptosis of neurons after axotomy (Martin *et al.*, 1999).

The proportion of neuronal death among sensory neurons of the dorsal root ganglia (DRG) after sciatic nerve injury in rodents has been reported between 10% and 30%, affecting more small than large neurons (Groves *et al.*, 1997; Ygge, 1989). In contrast, only a nonsignificant loss of motoneurons has been found after sciatic nerve injury (Valero-Cabré *et al.*, 2001; Vanden-Noven *et al.*, 1993), but 50–80% dye following ventral root lesions in adults (Koliatsos *et al.*, 1994; Martin *et al.*, 1999). Adult neurons are less susceptible to die than immature neurons, and lesions near the cell body cause a higher proportion of neuronal death than distal lesions.

The rapid arrival of injury-related signals (Hanz and Fainzilber, 2006) induces the surviving neurons to shift from a "transmitting" state to a "regenerative" state, underlied by prominent changes in gene expression, which lead to a decrease in the synthesis of neurotransmission-related products and an increased synthesis of growth-associated proteins and structural components of the membrane. The modifications in the activity of transcription factors result in characteristic changes of gene expression in the injured and regenerating neurons, as well as in associated glial cells following peripheral nerve injuries (Bosse *et al.*, 2006; Cameron *et al.*, 2003; Costigan *et al.*, 2002; Kubo *et al.*, 2002; Yang *et al.*, 2006). Hundreds of genes have been found either up- or downregulated by axotomy, but the function of the majority of them remains unknown. Gene expression changes affect the encoding of transcription factors, cytoskeletal proteins, cell adhesion and guidance molecules, trophic factors and receptors, cytokines, neuropeptides and neurotransmitter synthesizing enzymes, ion channels, receptors, and membrane transporters (Fig. 2).

One of the most relevant phenotypic changes in the neurons after axotomy is the downregulation of neurotransmitters and genes encoding transmitter-related proteins. However, they also show marked changes in the expression of neuropeptides. Axotomized motoneurons show a reduction in choline acetyltransferase (ChAT) and an increase of calcitonin gene-related peptide (CGRP) expression (Borke *et al.*, 1993; Calderó *et al.*, 1992). Sympathetic neurons that normally express neuropeptide Y (NPY) show a reduced expression of NPY and tyrosine hydroxylase after axotomy, while they overexpress VIP, galanin, and substance

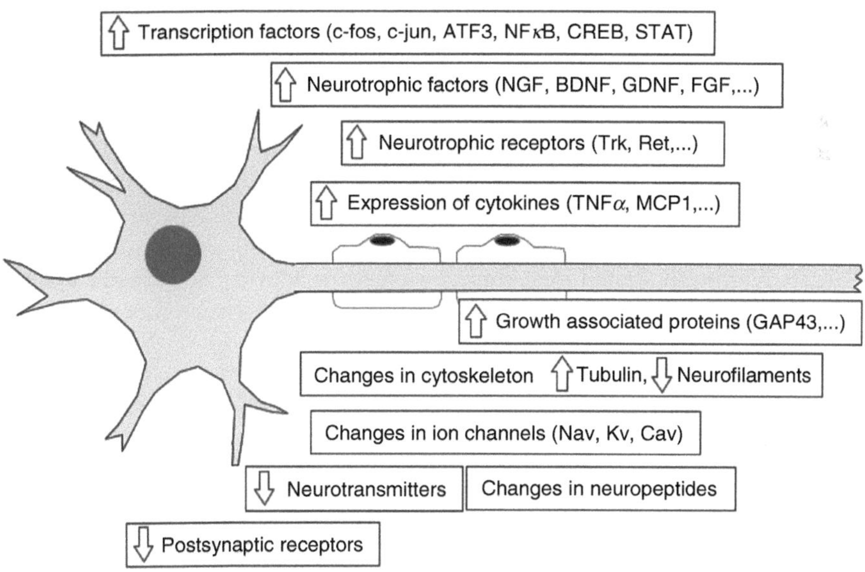

FIG. 2. Summary of the changes in the molecular expression in axotomized neurons after peripheral nerve injury. The arrows represent ⇑: increase, ⇓: decrease.

P (Zigmond, 1997). Primary sensory neurons also change their phenotype with regard to messengers after peripheral injury. Particularly, substance P and CGRP are markedly downregulated in peptidergic small DRG neurons after axotomy, whereas other neuropeptides such as VIP, galanin, and NPY are increased (McGregor *et al.*, 1984; Villar *et al.*, 1991; Wakisaka *et al.*, 1991).

A. Changes in Ion Channels and Excitability in Injured Neurons

Axotomy triggers plastic changes in the expression of several ion channels, transducers, and receptors in the injured neurons, which are related with the modifications of neuronal phenotype, including electrical hyperexcitability.

Following injury to their peripheral axons, DRG neurons downregulate some ion channel genes and upregulate others. As a result, a different repertoire of sodium channels is inserted into the neuron membrane following injury, a molecular change that contributes to hyperexcitability in these cells. Following nerve injury, the expression of tetrodotoxin (TTX)-sensitive sodium channels $Na_v1.3$, normally expressed only during development, is upregulated in primary sensory neurons, whereas other sodium channels are downregulated (Waxman *et al.*, 2000; Wood *et al.*, 2004). The accelerated recovery from inactivation of the $Na_v1.3$ channels produces a decrease in refractory period that contributes to hyperexcitability of DRG neurons. The expression of potassium channels in DRG cells also changes following axotomy; voltage-gated potassium channels $K_v1.2$ and 2.1 largely decrease, whereas $K_v1.1$ and 1.3 show smaller decreases (Ishikawa *et al.*, 1999). Modifications of potassium channels may represent an adaptive mechanism to stabilize the injured axon membrane, which has been made hyperexcitable by the changes in sodium channels. Increases in excitability in all types of DRG sensory neurons occur within weeks of axotomy, with significant reduction of excitation threshold and increase of spontaneous activity (Abdulla and Smith, 2001; Zhang *et al.*, 1997). These electrophysiological changes are more intense in animals that exhibit signs of neuropathic pain.

Ion-channel expression is also modified in motoneurons after axonal lesion, with reappearance of $Na_v1.3$ mRNA and decrease of $Na_v1.1$ mRNA (Iwahashi *et al.*, 1994; Patko *et al.*, 2003). Axotomized motoneurons exhibit changes in membrane electrical properties (Kuno *et al.*, 1974; Titmus and Faber, 1990), consistent with dedifferentiation of motoneuronal properties following axotomy. When muscle reinnervation is not allowed, these changes are maintained chronically, whereas following reinnervation, normal electrical properties recover, suggesting that functional contact with muscle is required for the normal expression of motoneuron electrical properties.

B. Synaptic Plasticity of Axotomized Neurons

In parallel to the neuronal body reaction, there is a retraction of the dendritic tree and a reduction of the synapses received by axotomized neurons—morphological changes that account for a functional isolation of the injured, nonfunctional neurons from the remaining neural circuits (Purves, 1975; Sumner and Sunderland, 1973). The reduction in dendritic size is due to loss of preterminal and terminal dendritic segments. Following reinnervation of target muscles, the dendritic volume of axotomized motoneurons is normalized due to an increase in diameter and number of dendrites per neuron, although the number of dendritic branches remains reduced (Brannstrom *et al.*, 1992a,b). The distal dendritic arborizations of these motoneurons exhibit axon-like characteristics, supporting the claim that axotomy leads to a remodeling of the neuronal polarity (Rose *et al.*, 2001). This process might be generated as replacement for the injured axon, but have also the potential to make inappropriate connections within the central nervous system (CNS) that may be detrimental for functional recovery.

In parallel to dendrite retraction, the number and area of synaptic terminals on chromatolyzed motoneurons is reduced shortly after axotomy. The reduction in number and percentage of boutons on the surface membrane is due to detachment of synapses associated to the presence of reactive astrocytes and microglia (Aldskogius *et al.*, 1999; Chen, 1978; Sumner and Sunderland, 1973). Synaptic restoration takes place in parallel to peripheral target reinnervation. In contrast, after permanent axotomy, synaptic cover remains reduced for months. The amount of synaptic loss and recovery is variable and affects various types of synapses to different degrees. For example, in motoneurons, excitatory terminals are eliminated to a larger degree than inhibitory ones, reflecting a reorganization of the synaptic input to diminish the excitotoxic influence on the neurons.

The changes in synaptic functions have been extensively studied for the synapse between Ia muscle spindle afferents and alpha motoneurons (Ia–MN synapse) in the spinal cord. After nerve lesion, three distinct phases in the Ia–MN synaptic function were described: an early increase in the postsynaptic responses, followed by a longer lasting decline, and a phase of recovery when the nerve reinnervates target muscles (Mendell, 1984; Seburn and Cope, 1998). The early increase in synaptic transmission strength is attributable to the lack of activation of the synapse that results when Ia afferents are transected, to reduced activity of homonymous motoneurons, and likely to an increase in the probability of transmitter release. The second phase of progressive decline in the excitatory postsynaptic potentials (EPSP) amplitude at the Ia–MN synapse develops by 1 week after axotomy, in relation with stripping of spinal cord synapses. The EPSPs are restored to normal size when the nerve reinnervates the muscle, even after long delay. Neuronal deprivation or exposition to neurotrophic factors might be the signals for regenerating neurons to modify their phenotype, thus enhancing or depressing synaptic contacts (Mendell and Munson, 1999).

III. Plastic Changes and Remodeling at the Spinal Cord

Nerve injury may sensitize and induce remodeling of central neural structures. Experimental evidence of these changes is illustrated by the development of wind-up, classic central sensitization, long-term potentiation (LTP), distortion of receptive fields of CNS neurons, as well as the enhancement of spinal reflexes and the persistence of pain or hyperalgesia (Melzack *et al.*, 2001; Woolf and Salter, 2000).

A. Changes in Spinal Reflexes After Nerve Lesions

Marked plastic changes in the connections and function of spinal reflexes occur after nerve injuries in parallel to axonal regeneration and reinnervation. Such changes may play important effects on movement control and sensory processing, especially when reinnervation is incomplete or defective. The monosynaptic spinal stretch reflex shows complete recovery after nerve crush lesion (Scott and Panesar, 1995), but in contrast, is permanently abolished after nerve section, probably due to either stripping of central synapses or poor muscle spindle reinnervation (Cope *et al.*, 1994). In fact, the structure of reinnervated muscle spindles is altered after nerve lesions (Verdú and Navarro, 1997), probably determining deficits in stimulus transduction.

Recovery of the H-wave reflex, the electrical counterpart of the monosynaptic stretch reflex, induced by stimulation of afferent fibers proximal to the injury, indicates that spinal reflex circuits are functional as soon as the muscle is reinnervated. The H reflex is highly facilitated at early stages of reinnervation, resulting in an increase of the H/M amplitude ratio, thus indicating a facilitated response of the Ia–MN synapse. Muscle reinnervation tends to revert this facilitatory effect, but the H/M ratio remains higher than the control during months following nerve section and repair (Valero-Cabré and Navarro, 2001, 2002a; Vivó *et al.*, 2008). Interestingly, the reflex facilitation affects also the adjacent spinal cord segments, where intact motoneurons undergo compensatory synaptic rearrangements, receiving more excitatory boutons (Havton and Kellerth, 2004; Holmberg and Kellerth, 2000). After sciatic nerve injury, crossed extensor reflex responses conveyed by myelinated afferent fibers show a marked increase in amplitude during the first stages of regeneration, and tend to decline with reinnervation. In contrast, the slow component mediated by unmyelinated afferents is normal or sometimes absent (Valero-Cabré and Navarro, 2002b). Therefore, after nerve injuries, reflexes mediated by myelinated afferents exhibit a higher degree of facilitation than those mediated by unmyelinated fibers.

Different mechanisms may account for the facilitation of reflex responses following nerve injuries, including increase of neuronal and synaptic excitability, decrease of the effectiveness of propiospinal and descending inhibitory systems, and remodeling of sensory afferent projections within the spinal cord.

B. Remodeling of Spinal Cord Projections and Circuits

Peripheral regeneration following nerve section results in the mismatch of the connections between sensory receptors and afferent fibers with second-order neurons in the dorsal horn of the spinal cord and alters the somatotopy of the body representation at the spinal cord level. These alterations result in loss or decrease of tactile acuity and discrimination, and underlie in part positive symptoms, such as dysesthesia and neuropathic pain. The lack of specificity in peripheral reinnervation after complete nerve section followed by surgical repair mainly arises from topographic intrafascicular changes generated after the injury, the random growth of regenerative sprouts, and misrouting of regenerated axons (Abernethy *et al.*, 1992; Madison *et al.*, 1996). Regenerating sensory axons can reinnervate foreign target territories and inappropriate receptors, including reinnervation of muscle receptors by cutaneous afferents and vice versa, or reinnervation of tactile receptors by nociceptive afferents.

Peripheral nerve lesions produce changes in the spinal cord receptive fields, the functional efficacy of central connections, and the laminar projections of afferent fibers, so that the properties of the central projections of regenerated sensory neurons are not well-matched with the peripheral receptor properties. After nerve section and repair, the receptive fields of dorsal horn cells are larger than normal and distorted, in agreement with the random and disperse nature of axonal regeneration. At long intervals the receptive fields shrink, approaching normal sizes, but the somatotopic organization remains partially scrambled. This plastic process implies changes in synaptic efficacy, with some synaptic inputs being strengthened while others weaken or disappear, and new synaptic inputs may be established in order to make the receptive fields more continuous (Koerber *et al.*, 2006).

After a peripheral nerve lesion, DRG neurons show variable responses regarding sprouting or growth of their central axonal branch. Peptidergic (CGRP+) sensory neurons are able to sprout into segments denervated by rhizotomy in a nonsomatotopic manner. In contrast, nonpeptidergic (IB4+) sensory neurons maintain the somatotopic distribution centrally, but their spinal projections are reduced after nerve injury (Molander *et al.*, 1996). Therefore, considerably heterogeneous responses to injury and regeneration occur among different types of sensory neurons, particularly between peptidergic (NGF-dependent) and nonpeptidergic (GDNF-dependent) neurons.

There has been extensive controversy regarding the rearrangement of central projections of injured and intact sensory afferents in the spinal cord following nerve lesions, and its potential contribution to neuropathic pain and mechanical hypersensitivity. After peripheral nerve lesions, the central axons of injured myelinated Aβ fibers were found to sprout from their normal termination territory within deep laminae III and V of the dorsal horn to partially invade laminae II and I, which normally receive inputs from C and Aδ sensory fibers (Shortland and Woolf, 1993; Woolf *et al.*, 1995). The death or atrophy of a number of small sensory DRG cells and degeneration of their terminations within the dorsal horn after nerve injury would induce sprouting of Aβ and Aδ afferents into the empty laminae I and II. Nevertheless, more recent studies found only a very limited number of large A fiber sprouts into inner lamina II, normally innervated by C-fibers, but not into outer lamina II or lamina I, and no obvious changes in the laminar locations of axonal collaterals (Hughes *et al.*, 2003; Koerber *et al.*, 2006; Shehab *et al.*, 2003). Sprouting of the central projections of undamaged small C and Aδ fibers into the region occupied by central projection of fibers whose peripheral projections had been interrupted has been demonstrated (Hu *et al.*, 2004). The somatotopic reorganization following nerve lesions in adult animals might be better explained by an increased synaptic efficacy and formation of new synaptic boutons of existing silent projections, that can activate second-order neurons that were originally outside the corresponding receptive field, without notable structural changes or collateral sprouting of central fibers (Koerber *et al.*, 2006). This functional reorganization of the sensory circuitry may constitute an underlying mechanism for sensory abnormalities following peripheral nerve injuries.

Peripheral nerve injuries also alter spinal pathways, mainly those involving inhibitory interneurons. An early and progressive decrease of dorsal horn GABAergic interneurons (Castro-Lopes *et al.*, 1993; Moore *et al.*, 2002) and of inhibitory Renshaw interneurons (Sanna *et al.*, 1993) has been reported after nerve lesions. The reduction of activity of inhibitory interneurons in the spinal cord may therefore lead to increased excitation of ascending sensory pathways and motoneurons.

IV. Plastic Changes and Reorganization at Cortical and Subcortical Levels

Functional reorganization of sensory and motor systems following peripheral nerve damage affects neural networks including spinal cord, brainstem, thalamus, and cortical regions directly or indirectly involved in the processing of the impacted functions (for review see Chen *et al.*, 2002; Kaas, 1991; Kaas and Collins, 2003; Lundborg, 2000b; Navarro *et al.*, 2007; Wall *et al.*, 2002). Injuries in a nerve or root induce a cascade of functional changes progressing to reorganization of the entire pathway, from peripheral neurons to cortex (Fig. 1). The higher the level in the neuraxis, the larger the degree of functional plasticity that such a region might have potential to undergo after nerve injury.

A. Reorganization at Subcortical Levels

Fast plastic changes have been demonstrated in the dorsal column nuclei and thalamus, in which anesthesia-induced peripheral block of cutaneous afferents caused an immediate and reversible reorganization, suggesting that new responses arise from unmasking of pre-existing inputs (Nicolelis *et al.*, 1993).

Mapping of the hand representation in the brainstem cuneate nucleus and in somatosensory cortex after section of both median and ulnar nerves in primates showed substantial changes at both locations, starting within minutes post injury and continuing over several days (Xu and Wall, 1997, 1999). Acute changes, characterized by relatively larger reorganization in the cuneate nucleus than in the cortex, progress to a later stage, whereby cortical and thalamic changes achieve the levels of brainstem reorganization (Churchill *et al.*, 2001; Xu and Wall, 1999). The more extensive thalamic and cortical reorganizations emerging weeks to months after sensory loss may also depend on the growth of new connections. Studies on primates, years after hand amputation, demonstrated that neurons in the thalamic nucleus hand map were activated by inputs from the upper limb stump and also from the face (Florence *et al.*, 2000). This expansion of thalamic receptive fields tended to be larger than in the somatosensory cortex. Feed-forward and feed-back projections seem to play a role in the induction and maintenance of central plasticity after nerve lesions. Cortical reorganization depends on similar processes in subcortical structures, but subcortical plasticity is at least in part determined by corticosubcortical projections (Kaas, 1999; Krupa *et al.*, 1999).

B. Reorganization of Somatosensory Cortex

Peripheral nerve lesion implies an acute deafferentation with immediate and evolving plastic changes on the corresponding cortical representation areas. Early evidences for reorganization in the somatosensory cortex were obtained in primate experiments (Garraghty and Kaas, 1991; Merzenich *et al.*, 1983a,b; Wall *et al.*, 1986), but it has been shown to occur throughout species (Wall and Cusick, 1984). Peripheral nerve injuries result in loss of evoked activity in the corresponding cortical map in response to stimulation of the skin covered by the injured nerve. Short thereafter, the same region may become responsive to inputs originated in other body parts with adjacent cortical representations. The result is shrinkage in size of the cortical field deprived from its original inputs, and its takeover by adjacent representations for the processing of information from intact sources. Initially, receptive fields in the reorganized cortex are large, with hardly discernible somatotopy. Over time, the receptive fields become smaller and a new somatotopy representing the adjacent nondenervated parts of the body emerges. Cortical reorganizations of a larger scale occur after extensive deafferentation induced by multiple dorsal rhizotomy. For example, in monkeys studied after more than 10 years of complete deafferentation of the arm, the deprived hand and arm cortical representation showed evoked activity in response to inputs from the face—an expansion of more than 10 mm in length.

If peripheral nerves are allowed to regenerate, the cortical remodeling may be partially restored depending on the severity of the injury and the accuracy of reinnervation. Following a crush injury, the corresponding cortical map is re-established to a quite normal pattern (Wall *et al.*, 1983). However, after nerve transection and repair, there is a large degree of misdirection in axonal growth, resulting in an abnormal pattern of input or output activity to the cortical maps. Therefore, it results in reorganization of the somatosensory cortex with a distorted map of the skin areas originally innervated by the damaged nerve (Florence *et al.*, 1994; Wall *et al.*, 1986). In humans who suffer nerve section and repair, sensory mislocalizations persist for many years, indicating that the errors in target reinnervation cannot be corrected by brain plasticity.

The changes in cortical maps follow a fast time course. Sensory deafferentation by anesthetic blockade of a finger results in minutes in an expansion of the cortical representations of intact fingers into the map of anesthetic finger, which returns to normal in a matter of minutes to hours (Rossini *et al.*, 1994). This rapid time course suggests that there is an extended network of connectivity across cortical sensory areas, so that sensory inputs from one finger normally inhibit existing inputs from adjacent fingers (Wall *et al.*, 2002). On the other hand, large deafferentation such as in amputations induces a slow cortical reorganization with progressive shrinkage of the affected territory accompanied by the expansion of adjacent cortical regions. This new organization emerges within weeks after injury and becomes permanent on chronic amputations (Pearson *et al.*, 2001, 2003). Studies in humans after chronic upper limb amputation indicate that areas of primary somatosensory cortex may become functionally reactivated by inputs originated in the limb stump or even the face. Nevertheless, parts of the deprived cortex may remain unresponsive after the lesion. Consequently, abnormal identification of touch stimuli to the face is referred in the missing hand (Flor *et al.*, 1998). The basis of phantom sensations secondary to amputation seems to include reorganization phenomena at the cortical and subcortical levels. Furthermore, a sensory map corresponding to the different portions of the missing limb can be traced in the stump or in the face of such subjects (Ramachandran and Hirstein, 1998). The extent of shifts in cortical representation has been found to correlate with neuropathic pain in the phantom limb, thus becoming an example that cortical reorganization does not always result in adaptive solutions.

C. Reorganization of Motor Cortex

Peripheral nerve injuries block the output flow from the motor cortex to the denervated muscles, resulting not only in paralysis, but also in motor cortex plasticity (Donoghue *et al.*, 1990; Sanes *et al.*, 1988, 1990). Cortical motor representations corresponding to denervated muscles become initially silent, so that their electrical stimulation is unable to evoke muscle activity. In a few hours, stimulation of the same region may elicit activity in muscles with motor representations adjacent to the paralyzed muscles. Stimulation thresholds required to elicit movements in expanded representations are at or below normal levels, suggesting the withdrawal of inhibitory local circuits.

In humans, the pattern resulting from peripheral nerve lesions consists of enlarged representation of cortical areas and large motor evoked potentials for muscles immediately proximal to the nerve lesion, in parallel to the reduction or disappearance of the motor map of denervated muscles (Chen *et al.*, 2002). Some of these changes can be apparent minutes after nerve block, weeks after spinal cord injury, and a few months after amputation (Mano *et al.*, 2003). The fast changes apparent after anesthetic nerve block suggest that motor cortex excitability can be rapidly increased during deafferentation (Brasil-Neto *et al.*, 1992). Intrathalamic connectivity between somatosensory nuclei and motor nuclei projecting into premotor or supplementary motor areas could also indirectly mediate the motor cortical effects.

Profound long-standing reorganization changes occur within the primary motor cortex after chronic amputation. Studies in primates subjected to forelimb amputation showed that stimulation of the motor cortex originally devoted to the missing hand evoked movements of the stump, shoulder, and trunk (Qi *et al.*, 2000; Wu and Kaas, 1999). Similar changes have been reported in humans with upper- or lower-limb amputations (Chen *et al.*, 1998; Cohen *et al.*, 1991). Several observations point out that motor plastic changes following amputation occur predominantly at the cortical level without the intervention of the spinal cord or the brainstem. Cortical reorganization following a limb amputation can be reversed if the severed body part is reimplanted and functionally used (Giraux *et al.*, 2001; Röricht *et al.*, 2001). New sensory inputs combined with viable motor activity provided by the implanted hand reversed the functional reorganization that emerged after amputation, so that the hand and arm motor and somatosensory cortical representations return to their original location and extent.

Such cases suggest that plastic reorganization can be reversed even at chronic stages, thus indicating that an already remodeled cortex retains the ability to reorganize its pattern of representations regardless of the time post injury.

D. Mechanisms of Cortical and Subcortical Plasticity

Different mechanisms are involved in the development of plastic changes at cortical and subcortical levels after nerve injury. The fast changes, occurring within minutes after deafferentation, are probably based on the unmasking of previously present, but functionally inactive, synaptic connections. Unmasking of connections can occur as a result of increased excitatory neurotransmitter release, increased density of postsynaptic receptors, changes in membrane conductance, or removal of inhibitory projections. Removal of inhibition of excitatory synapses due to reduction of GABAergic inhibition has been suggested as the most important cause of short-term plastic changes, contributing to maintain projection neurons to the deafferented nuclei and cortex in a tonic state of increased excitation (Chen *et al.*, 2002).

Injuries to the peripheral nerve and the spinal cord produce an increase in the excitability of neurons in the sensory pathways, contributing to the physiopathology of neuropathic pain and also to subcortical and cortical reorganizations. Recently it has been demonstrated that after peripheral nerve injury, third-order nociceptive neurons in the thalamus undergo changes in expression of sodium channels, with upregulation of $Na_v1.3$ (Zhao *et al.*, 2006). Such a change in channel phenotype is accompanied by neuronal hyperexcitability and expanded peripheral receptive fields.

CNS plastic reorganization occurring over long periods likely involves more stable functional or structural mechanisms, including LTP and long-term depression (LTD) phenomena, or the formation of *de novo* connections by axonal sprouting and synaptogenesis (Kaas, 1991). Administration of NMDA receptor antagonists during the first month after nerve injury reduces the changes of sensory cortex maps, an effect that does not occur after this critical time (Garraghty and Muja, 1996; Myers *et al.*, 2000). Thus, it was proposed that LTP/LTD NMDA-mediated mechanisms contribute to cortical map reorganization during the first month after injury, but is not needed at later times. The consolidation of longer lasting plastic reorganization requires structural processes consisting of new projections by sprouting of axon collaterals, dendritic branching, and formation of new synaptic connections. The extent and relevance of these mechanisms, however, remain still poorly understood, particularly given the highly inhibitory CNS environment, which normally prevents axonal growth after injuries. Nevertheless, afferents in the spinal cord and in the cuneate nucleus display abnormally large termination fields after limb amputation (Florence *et al.*, 1998), which is indicative of new growth.

Subcortical levels may play a major role in providing large-scale cortical reorganization. Within the subcortical nuclei, a few millimeters shortcut may be sufficient to induce the rapid shift of cortical representation areas, which if developed at cortical level would require considerably longer intracortical projections. Tracing studies demonstrated aberrant sprouting of gracile nucleus neurons, normally relaying hind limb sensory information into the cuneate nucleus, which relays forelimbs' information, following dorsal rhizotomies in the rat (Sengelaub *et al.*, 1997). Also in monkeys, many years after dorsal rhizotomies resulting in extensive thalamic and cortical reorganization, there was evidence of neuronal atrophy and loss of axons of the cuneate and the thalamic nuclei (Woods *et al.*, 2000). Therefore, transneuronal atrophy associated with retraction of axons and compensatory axonal sprouting seems to play a significant influence on the reorganization of somatotopic maps in the brain cortex.

Structural changes in dendritic arborization within the cortex have been recently related to plastic reorganization after nerve injuries (Hickmott and Steen, 2005). After median and ulnar nerve ligation in monkeys, the deprived cortical area 3b, investigated months later, showed a progressive expansion in distal regions of the dendritic arbor of both pyramidal and spiny stellate neurons as compared to unaffected cortical areas (Churchill *et al.*, 2004). This expansion correlated with the degree of functional reorganization, supporting the notion that latent inputs gain expression after nerve injury via their influence onto distally located dendritic sites, thus implicating intracortical connections in sustaining cortical reorganization. Adult reorganization seems to occur primarily through changes in the strength and efficacy of existing synapses, rather than implicate active remodeling of connections.

V. Remodeling CNS Plasticity

Peripheral nerve injury induces dramatic processes of reorganization in structures across the neuraxis. These changes consist of decrease of excitability, metabolism, and surface extension of the disconnected central substrates with compensatory enhancement of neighboring representations. Plastic reorganization changes are reversible provided that adequate patterns of activity conveyed by regenerated peripheral axons are reinstated. However, when nerve regeneration is hampered or when profound misrouting of regenerated axons to mismatched targets occurs, plasticity of central connections in adult mammals has limited effects and may be even detrimental for the recovery of fine sensory processing and motor control.

It has been argued that expanded cortical and subcortical representations might help in optimizing the use of central resources for motor action or sensory processing of intact muscles and sensory receptors. Nevertheless, plastic changes following peripheral injury do not always result in better adaptation. On the contrary, the reorganization of somatosensory systems is related to the development of neuropathic pain and phantom pain in amputees. In order to therapeutically address the effects of maladaptive plasticity, recent research has studied the possibilities to modulate the abnormal enlargement or shrinkage of cortical representations by manipulating the flow of sensory inputs or motor outputs. Electrical stimulation of selective nerve fascicles in patients with chronic nerve injuries or amputations demonstrated that somatosensory localization remains accurate despite the presumed central reorganization that takes place after nerve injury (Dhillon *et al.*, 2005; Schady *et al.*, 1994). These observations open the basis for treatments attempting to restore normal neural processing and cortical maps by acting on the injured nerve. Different strategies have been implemented, including peripheral nerve electrical stimulation (Field-Fote, 2004; Vivó *et al.*, 2008), transcraneal magnetic stimulation on cortical regions (Rosenkranz and Rothwell, 2006), specific sensory training (Florence *et al.*, 2001; Shieh *et al.*, 1998), delivery of alternate modality stimulation (Rosén and Lundborg, 2003), or deprivation of inputs from intact nerve territories (Bjorkman *et al.*, 2005; Rosén *et al.*, 2006). Such interventions might eventually be effective for increasing sensory discrimination, improving motor control, reducing hyperreflexia, or ameliorating neuropathic pain, although more definite studies are needed to assess the magnitude and durability of effects and the physiopathological mechanisms involved.

Acknowledgments

The author's research was supported by grants from the the Ministerio de Sanidad y Consumo (PI060201) of Spain, the European Commission (NEUROBOTICS project, IST-001917; TIME project, ICT- 224012), and FEDER funds.

References

Abdulla, F. A., and Smith, P. A. (2001). Axotomy- and autotomy-induced changes in the excitability of rat dorsal root ganglion neurons. *J. Neurophysiol.* **85,** 630–643.

Abernethy, D. A., Rud, A., and Thomas, P. K. (1992). Neurotrophic influence of the distal stump of transected peripheral nerve on axonal regeneration: Absence of topographic specificity in adult nerve. *J. Anat.* **180,** 395–400.

Aldskogius, H., Liu, L., and Svensson, M. (1999). Glial responses to synaptic damage and plasticity. *J. Neurosci. Res.* **58,** 33–41.

Bjorkman, A., Rosén, B., and Lundborg, G. (2005). Enhanced function in nerve-injured hands after contralateral deafferentation. *Neuroreport* **16,** 517–519.

Bodine-Fowler, S. C., Meyer, R. S., Moscovitz, A., Abrams, R., and Botte, J. (1997). Inaccurate projection of rat soleus motoneurons: A comparison of nerve repair techniques. *Muscle Nerve* **20,** 29–37.

Borke, R. C., Curtis, M., and Ginsberg, C. (1993). Choline acetyltransferase and calcitonin gene-related peptide immunoreactivity in motoneurons after different types of nerve injury. *J. Neurocytol.* **22,** 141–153.

Bosse, F., Hasenpusch-Theil, K., Kury, P., and Muller, H. W. (2006). Gene expression profiling reveals that peripheral nerve regeneration is a consequence of both novel injury-dependent and reactivated developmental processes. *J. Neurochem.* **96,** 1441–1457.

Boyd, J. G., and Gordon, T. (2003). Neurotrophic factors and their receptors in axonal regeneration and functional recovery after peripheral nerve injury. *Mol. Neurobiol.* **27,** 277–324.

Brannstrom, T., Havton, L., and Kellerth, J. O. (1992a). Changes in size and dendritic arborization patterns of adult cat spinal alpha-motoneurons following permanent axotomy. *J. Comp. Neurol.* **318,** 439–451.

Brannstrom, T., Havton, L., and Kellerth, J. O. (1992b). Restorative effects of reinnervation on the size and dendritic arborization patterns of axotomized cat spinal alpha-motoneurons. *J. Comp. Neurol.* **318,** 452–461.

Brasil-Neto, J. P., Cohen, L. G., Pascual-Leone, A., Jabir, F. K., Wall, R. T., and Hallett, M. (1992). Rapid reversible modulation of human motor outputs after transient deafferentation of the forearm: A study with transcranial magnetic stimulation. *Neurology* **42,** 1302–1306.

Brown, M. C., Holland, R. L., and Ironton, R. (1980). Nodal and terminal sprouting from motor nerves in fast and slow muscles in the mouse. *J. Physiol.* **306,** 493–510.

Brushart, T. M. (1993). Motor axons preferentially reinnervate motor pathways. *J. Neurosci.* **13,** 2730–2738.

Calderó, J., Casanovas, A., Sorribas, A., and Esquerda, J. E. (1992). Calcitonin gene-related peptide in rat spinal cord motoneurons: Subcellular distribution and changes induced by axotomy. *Neuroscience* **48,** 449–461.

Cameron, A. A., Vansant, G., Wu, W., Carlo, D. J., and Ill, C. R. (2003). Identification of reciprocally regulated gene modules in regenerating dorsal root ganglion neurons and activated peripheral or central nervous system glia. *J. Cell. Biochem.* **88,** 970–985.

Castro-Lopes, J. M., Tavara, I., and Coimbra, A. (1993). GABA decreases in the spinal cord dorsal horn after peripheral neurectomy. *Brain Res.* **620,** 287–291.

Chen, D. H. (1978). Qualitative and quantitative study of synaptic displacement in chromatolyzed spinal motoneurons of the cat. *J. Comp. Neurol.* **177,** 635–664.

Chen, R., Corwell, B., Yaseen, Z., Hallet, M., and Cohen, L. (1998). Mechanisms of cortical reorganization in lower-limb amputees. *J. Neurosci.* **18,** 3443–3450.

Chen, R., Cohen, L. G., and Hallett, M. (2002). Nervous system reorganization following injury. *Neuroscience* **111,** 761–773.

Churchill, J. D., Arnold, L. L., and Garraghty, P. E. (2001). Somatotopic reorganization in the brainstem and thalamus following peripheral nerve injury in adult primates. *Brain Res.* **910,** 142–152.

Churchill, J. D., Tharp, J. A., Wellman, C. L., Sengelaub, D. R., and Garraghty, P. E. (2004). Morphological correlates of injury-induced reorganization in primate somatosensory cortex. *BMC Neurosci.* **5,** 43.

Cohen, L. G., Bandinelli, S., Findley, T. W., and Hallett, M. (1991). Motor reorganization after upper limb amputation in man. *Brain* **114,** 615–627.

Cope, T. C., Bonasera, J., and Nichols, T. R. (1994). Reinnervated muscles fail to produce stretch reflexes. *J. Neurophysiol.* **71,** 817–820.

Costigan, M., Befort, K., Karchewski, L., Griffin, R. S., D'Urso, D., Allchorne, A., Sitarski, J., Mannion, J. W., Pratt, R. E., and Woolf, C. J. (2002). Replicate high-density rat genome oligonucleotide microarrays reveal hundreds of regulated genes in the dorsal root ganglion after peripheral nerve injury. *BMC Neurosci.* **3,** 16.

Dhillon, G. S., Kruger, T. B., Sandhu, J. S., and Horch, K. W. (2005). Effects of short-term training on sensory and motor function in severed nerves of long-term human amputees. *J. Neurophysiol.* **93,** 2625–2633.

Donoghue, J. P., Suner, S., and Sanes, J. N. (1990). Dynamic organization of primary motor cortex output to target muscles in adult rats. II. Rapid reorganization following motor nerve lesions. *Exp. Brain. Res.* **79,** 492–503.

Field-Fote, E. C. (2004). Electrical stimulation modifies spinal and cortical neural circuitry. *Exerc. Sports Sci. Rev.* **32,** 155–160.

Flor, H., Elbert, T., Muhlnickel, W., Pantev, C., Wienbruch, C., and Taub, E. (1998). Cortical reorganization and phantom phenomena in congenital and traumatic upper-extremity amputees. *Exp. Brain Res.* **119,** 205–212.

Florence, S. L., Garraghty, P. E., Wall, J. T., and Kaas, J. H. (1994). Sensory afferent projections and area 3b somatotopy following median nerve cut and repair in macaque monkeys. *Cereb. Cortex* **4,** 391–407.

Florence, S. L., Taub, H. B., and Kaas, J. H. (1998). Large-scale sprouting of cortical connections after peripheral injury in adult macaque monkeys. *Science* **282,** 1117–1121.

Florence, S. L., Hackett, T. A., and Strata, F. (2000). Thalamic and cortical contributions to neural plasticity after limb amputation. *J. Neurophysiol.* **83,** 3154–3159.

Florence, S. L., Boydston, L. A., Hackett, T. A., Lachoff, H. T., Strata, F., and Niblock, M. M. (2001). Sensory enrichment after peripheral nerve injury restores cortical, not thalamic, receptive field organization. *Eur. J. Neurosci.* **13,** 1755–1766.

Fu, S. Y., and Gordon, T. (1997). The cellular and molecular basis of peripheral nerve regeneration. *Mol. Neurobiol.* **14,** 67–116.

Garraghty, P. E., and Kaas, J. H. (1991). Large-scale functional reorganization in adult monkey cortex after peripheral nerve injury. *Proc. Natl. Acad. Sci. USA* **88,** 6976–6980.

Garraghty, P. E., and Muja, N. (1996). NMDA receptors and plasticity in adult primate somatosensory cortex. *J. Comp. Neurol.* **367,** 19–26.

Giraux, P., Sirigu, A., Schneider, F., and Dubernard, J. M. (2001). Cortical reorganization in motor cortex after graft of both hands. *Nat. Neurosci.* **4,** 691–692.

Groves, M. J., Christopherson, T., Giometto, B., and Scaravilli, F. (1997). Axotomy-induced apoptosis in adult rat primary sensory neurons. *J. Neurocytol.* **26,** 615–624.

Hanz, S., and Fainzilber, M. (2006). Retrograde signalling in injured nerve: The axon reaction revisited. *J. Neurochem.* **99,** 13–19.

Havton, L. A., and Kellerth, J. O. (2004). Plasticity of lumbosacral monosynaptic reflexes after a ventral root transection injury in the adult cat. *Exp. Brain Res.* **155,** 111–114.

Hickmott, P. W., and Steen, P. A. (2005). Large-scale changes in dendritic structure during reorganization of adult somatosensory cortex. *Nat. Neurosci.* **8,** 140–142.

Holmberg, P., and Kellerth, J. O. (2000). Do synaptic rearrangements underlie compensatory reflex enhancement in spinal motoneurons after partial cell loss?. *Synapse* **38,** 384–391.

Hu, J., Mata, M., Hao, S., Zhang, G., and Fink, D. J. (2004). Central sprouting of uninjured small fiber afferents in the adult rat spinal cord following spinal nerve ligation. *Eur. J. Neurosci.* **20,** 1705–1712.

Hughes, D. I., Scott, D. T., Todd, A. J., and Riddell, J. S. (2003). Lack of evidence for sprouting of Abeta afferents into the superficial laminas of the spinal cord dorsal horn after nerve section. *J. Neurosci.* **23,** 9491–9499.

Ishikawa, K., Tanaka, M., Black, J. A., and Waxman, S. G. (1999). Changes in expression of voltage-gated potassium channels in dorsal root ganglion neurons following axotomy. *Muscle Nerve* **22,** 502–507.

Iwahashi, Y., Furuyama, T., Inagaki, S., Morita, Y., and Takagi, H. (1994). Distinct regulation of sodium channel types I, II and III following nerve transection. *Brain Res. Mol. Brain Res.* **22,** 341–345.

Jackson, P. C., and Diamond, J. (1984). Temporal and spatial constraints on the collateral sprouting of low-threshold mechanosensory nerves in the skin of rats. *J. Comp. Neurol.* **226,** 336–345.

Kaas, J. H. (1991). Plasticity of sensory and motor maps in adult mammals. *Annu. Rev. Neurosci.* **14,** 137–167.

Kaas, J. H. (1999). Is most of neural plasticity in the thalamus cortical?. *Proc. Natl. Acad. Sci. USA* **96,** 7622–7623.

Kaas, J. H., and Collins, C. E. (2003). Anatomic and functional reorganization of somatosensory cortex in mature primates after peripheral nerve and spinal cord injury. *Adv. Neurol.* **93,** 87–95.

Kline, D. G., and Hudson, A. R. (1995). "Nerve Injuries: Operative Results for Major Nerve Injuries, Entrapments, and Tumors." Saunders, Philadelphia.

Koerber, H. R., Mirnics, K., and Lawson, J. J. (2006). Synaptic plasticity in the adult spinal dorsal horn: The appearance of new functional connections following peripheral nerve regeneration. *Exp. Neurol.* **200,** 468–479.

Koliatsos, V. E., Price, W. L., Pardo, C. A., and Price, D. L. (1994). Ventral root avulsion: an experimental model of death of adult motor neurons. *J. Comp. Neurol.* **342,** 35–44.

Kreutzberg, G. W. (1995). Reaction of the neuronal cell body to axonal damage. *In* "The Axon: Structure, Function and Pathophysiology." (S. G. Waxman, J. D. Kocsis, and P. K. Stys, Eds.), pp. 355–374. Oxford University Press, Oxford.

Krupa, D. J., Ghazanfar, A. A., and Nicolelis, M. A. (1999). Immediate thalamic sensory plasticity depends on corticothalamic feedback. *Proc. Natl. Acad. Sci. USA* **96,** 8200–8205.

Kubo, T., Yamashita, T., Yamaguchi, A., Hosokawa, K., and Tohyama, M. (2002). Analysis of genes induced in peripheral nerve after axotomy using cDNA microarrays. *J. Neurochem.* **82,** 1129–1136.

Kuno, M., Miyata, Y., and Muñoz-Martinez, E. J. (1974). Differential reaction of fast and slow alpha-motoneurones to axotomy. *J. Physiol.* **240,** 725–739.

Lieberman, A. R. (1971). The axon reaction: A review of the principal features of perikarial responses to axon injury. *Int. Rev. Neurobiol.* **14,** 49–124.

Lundborg, G. (2000a). A 25-year perspective of peripheral nerve surgery: Evolving neuroscientific concepts and clinical significance. *J. Hand Surg.* **25A,** 391–414.

Lundborg, G. (2000b). Brain plasticity and hand surgery: An overview. *J. Hand Surg.* **25B,** 242–252.

Lundborg, G. (2004). "Nerve Injury and Repair: Regeneration, Reconstruction, and Cortical Remodeling" 2nd ed. Elsevier, Philadelphia.

Madison, R. D., Archibald, S. J., and Brushart, T. M. E. (1996). Reinnervation accuracy of the rat femoral nerve by motor and sensory neurons. *J. Neurosci.* **16,** 5698–5703.

Mano, Y., Chuma, T., and Watanabe, I. (2003). Cortical reorganization in training. *J. Electromyogr. Kinesiol.* **13,** 57–62.

Martin, L. J., Kaiser, A., and Price, A. C. (1999). Motor neuron degeneration after sciatic nerve avulsion in adult rat evolves with oxidative stress and is apoptosis. *J. Neurobiol.* **40,** 185–201.

McGregor, G. P., Gibson, S. J., Sabate, I. M., Blank, M. A., Christofides, N. D., Wall, P. D., Polak, J. M., and Bloom, S. R. (1984). Effect of peripheral nerve section and nerve crush on spinal cord neuropeptides in the rat; increased VIP and PHI in the dorsal horn. *Neuroscience* **13,** 207–216.

Melzack, R., Coderre, T. J., Katz, J., and Vaccarino, A. L. (2001). Central neuroplasticity and pathological pain. *Ann. NY Acad. Sci.* **933,** 157–174.

Mendell, L. (1984). Modifiability of spinal synapses. *Physiol. Rev.* **64,** 260–324.

Mendell, L. M., and Munson, J. B. (1999). Retrograde effects on synaptic transmission at the Ia/motoneuron connection. *J. Physiol. (Paris)* **93,** 297–304.

Merzenich, M. M., Kaas, J. H., Wall, R. J., Nelson, M., Sur, D., and Felleman, D. (1983a). Topographic reorganization of somatosensory cortical areas 3B and 1 in adult monkeys following restricted deafferentation. *Neuroscience* **8,** 33–55.

Merzenich, M. M., Kaas, J. H., Wall, R. J., Sur, M., Nelson, R. J., and Felleman, D. J. (1983b). Progression of change following median nerve section in the cortical representation of the hand in areas 3b and 1 in adult owl and squirrel monkeys. *Neuroscience* **10,** 639–665.

Molander, C., and Aldskogius, H. (1992). Directional specificity of regenerating primary sensory neurons after peripheral nerve crush or transection and epineurial suture. A sequential double-labelling study in the rat. *Restor. Neurol. Neurosci.* **4,** 339–344.

Molander, C., Wang, H. F., Rivero-Melián, C., and Grant, G. (1996). Early decline and late restoration of spinal cord binding and transganglionic transport of isolectin B4 from *Griffonia simplicifolia* I after peripheral nerve transection or crush. *Restor. Neurol. Neurosci.* **10,** 123–133.

Moore, K. A., Kohno, T., Karchewski, L. A., Scholz, J., Baba, H., and Woolf, C. J. (2002). Partial peripheral nerve injury promotes a selective loss of GABAergic inhibition in the superficial dorsal horn of the spinal cord. *J. Neurosci.* **22,** 6724–6731.

Myers, W. A., Churchill, J. D., Muja, N., and Garraghty, P. E. (2000). Role of NMDA receptors in adult primate cortical somatosensory plasticity. *J. Comp. Neurol.* **418,** 373–382.

Navarro, X., Vivó, M., and Valero-Cabré, A. (2007). Neural plasticity after peripheral nerve injury and regeneration. *Progr. Neurobiol.* **82,** 163–201.

Nicolelis, M. A., Lin, R. C., Woodward, D. J., and Chapin, J. K. (1993). Induction of immediate spatiotemporal changes in thalamic networks by peripheral block of ascending cutaneous information. *Nature* **361,** 533–536.

Patko, T., Vassias, I., Vidal, P. P., and De Waele, C. (2003). Modulation of the voltage-gated sodium- and calcium-dependent potassium channels in rat vestibular and facial nuclei after unilateral labyrinthectomy and facial nerve transection: An *in situ* hybridization study. *Neuroscience* **117,** 265–280.

Pearson, P. P., Arnold, P. B., Oladehin, A., Li, C. X., and Waters, R. S. (2001). Large-scale cortical reorganization following forelimb deafferentation in rat does not involve plasticity of intracortical connections. *Exp. Brain Res.* **138,** 8–25.

Pearson, P. P., Li, C. X., Chappell, T. D., and Waters, R. S. (2003). Delayed reorganization of the shoulder representation in forepaw barrel subfield (FBS) in first somatosensory cortex (SI) following forelimb deafferentation in adult rats. *Exp. Brain Res.* **153,** 100–112.

Purves, D. (1975). Functional and structural changes in mammalian sympathetic neurones following interruption of their axons. *J. Physiol.* **252,** 429–463.

Qi, H. X., Stepniewska, I., and Kaas, J. H. (2000). Reorganization of primary motor cortex in adult macaque monkeys with long-standing amputations. *J. Neurophysiol.* **84,** 2133–2147.

Ramachandran, V. S., and Hirstein, W. (1998). The perception of phantom limbs. The D. O. Hebb lecture. *Brain* **121,** 1603–1630.

Röricht, S., Machetanz, J., Irlbacher, K., Niehaus, L., Biemer, E., and Meyer, B. U. (2001). Reorganization of human motor cortex after hand replantation. *Ann. Neurol.* **50,** 240–249.

Rose, P. K., MacDermid, V., and Neuber-Hess, M. (2001). Structural plasticity of motoneuron dendrites caused by axotomy. *In* "Motor Neurobiology of the Spinal Cord" (T. C. Cope, Ed.), pp. 251–270. CRC Press, Boca Raton, FL.

Rosén, B., and Lundborg, G. (2003). Early use of artificial sensibility to improve sensory recovery after repair of the median and ulnar nerve. *Scand. J. Plast. Reconstr. Surg. Hand Surg.* **37,** 54–57.

Rosén, B., Bjorkman, A., and Lundborg, G. (2006). Improved sensory relearning after nerve repair induced by selective temporary anaesthesia: A new concept in hand rehabilitation. *J. Hand Surg. [Br.]* **31,** 126–132.

Rosenkranz, K., and Rothwell, J. C. (2006). Differences between the effects of three plasticity inducing protocols on the organization of the human motor cortex. *Eur. J. Neurosci.* **23,** 822–829.

Rossini, P. M., Martino, G., Narici, L., Pasquarelli, A., Peresson, M., Pizzela, V., Tecchio, F., Torrioli, G., and Romani, G. L. (1994). Short term brain plasticity in humans: Transient finger representation changes in sensory cortex somatotopy following ischemic anesthesia. *Brain Res.* **642,** 169–177.

Sanes, J. N., Suner, S., Lando, J. F., and Donoghue, J. P. (1988). Rapid reorganization of adult rat motor cortex somatic representation patterns after motor nerve injury. *Proc. Natl. Acad. Sci. USA* **85,** 2003–2007.

Sanes, J. N., Suner, S., and Donoghue, J. P. (1990). Dynamic organization of primary motor cortex output to target muscles in adult rats. I. Long-term patterns of reorganization following motor or mixed peripheral nerve lesions. *Exp. Brain. Res.* **79,** 479–491.

Sanna, P. P., Celio, M., Bloom, F., and Rende, M. (1993). Presumptive Renshaw cells contain decreased calbindin during recovery from sciatic nerve lesions. *Proc. Natl. Acad. Sci. USA* **90,** 3048–3052.

Schady, W., Braune, S., Watson, S., Torebjork, H. E., and Schmidt, R. (1994). Responsiveness of the somatosensory system after nerve injury and amputation in the human hand. *Ann. Neurol.* **36,** 68–75.

Scott, J. J., and Panesar, G. (1995). Small-signal sensitivity of muscle spindles reinnervated after long-term denervation. *Brain Res.* **671,** 325–328.

Seburn, K. L., and Cope, T. C. (1998). Short-term afferent axotomy increases both strength and depression at Ia–motoneuron synapses in rat. *J. Neurosci.* **18,** 1142–1147.

Sengelaub, D. R., Muja, N., Mills, A. C., Myers, W. A., Churchill, J. D., and Garraghty, P. E. (1997). Denervation-induced sprouting of intact peripheral afferents into the cuneate nucleus of adult rats. *Brain Res.* **769,** 256–262.

Shehab, S. A., Spike, R. C., and Todd, A. J. (2003). Evidence against cholera toxin B subunit as a reliable tracer for sprouting of primary afferents following peripheral nerve injury. *Brain Res.* **964,** 218–227.

Shieh, S. J., Chiu, H. Y., and Hsu, H. Y. (1998). Long-term effects of sensory reeducation following digital replantation and revascularization. *Microsurgery* **18,** 334–336.

Shortland, P., and Woolf, C. J. (1993). Chronic peripheral nerve section results in a rearrangement of the central axonal arborizations of axotomized A-beta primary afferent neurons in the rat spinal cord. *J. Comp, Neurol.* **330,** 65–82.

Sumner, B. E. H., and Sunderland, S. (1973). Quantitative electron microscopy on the injured hypoglossal nucleus in the rat. *J. Neurocytol.* **2,** 315–328.

Sunderland, S. (1991). "Nerve Injuries and Their Repair. A Critical Appraisal." Churchill Livingstone, Edinburgh.

Terenghi, G. (1999). Peripheral nerve regeneration and neurotrophic factors. *J. Anat.* **194,** 1–14.

Titmus, M. J., and Faber, D. S. (1990). Axotomy-induced alterations in the electrophysiological characteristics of neurons. *Prog. Neurobiol.* **35,** 1–51.

Valero-Cabré, A., and Navarro, X. (2001). H reflex restitution and facilitation after different types of peripheral nerve injury and repair. *Brain Res.* **919,** 302–312.

Valero-Cabré, A., and Navarro, X. (2002a). Functional impact of axonal misdirection after peripheral nerve injuries followed by graft or tube repair. *J. Neurotrauma* **19,** 1475–1485.

Valero-Cabré, A., and Navarro, X. (2002b). Changes in crossed spinal reflexes after peripheral nerve injury and repair. *J. Neurophysiol.* **87,** 1763–1771.

Valero-Cabré, A., Tsironis, K., Skouras, E., Perego, G., Navarro, X., and Neiss, W. F. (2001). Superior muscle reinnervation after autologous nerve graft or poly-l-lactide-ε-caprolactone (PLC) tube implantation in comparison to silicone tube repair. *J. Neurosci. Res.* **63,** 214–223.

Valero-Cabré, A., Tsironis, K., Skouras, E., Navarro, X., and Neiss, W. F. (2004). Peripheral and spinal motor reorganization after nerve injury and repair. *J. Neurotrauma* **21,** 95–108.

Vanden-Noven, S., Wallace, N., Muccio, D., Turtz, A., and Pinter, M. J. (1993). Adult spinal motoneurons remain viable despite prolonged absence of functional synaptic contact with muscle. *Exp. Neurol.* **123,** 147–156.

Verdú, E., and Navarro, X. (1997). Comparison of immunohistochemical and functional reinervation of skin and muscle after peripheral nerve injury. *Exp. Neurol.* **146,** 187–198.

Verdú, E., and Navarro, X. (1998). The role of Schwann cell in nerve regeneration. *In* "Understanding Glial Cells" (B. Castellano, B. González, and M. Nieto-Sampedro, Eds.), pp. 319–359. Kluwer Academic, Boston.

Villar, M. J., Wiesenfeld-Hallin, Z., Xu, X. J., Theodorsson, E., Emson, P. C., and Hökfelt, T. (1991). Further studies on galanin-, substance P-, and CGRP-like immunoreactivities in primary sensory neurons and spinal cord: Effects of dorsal rhizotomies and sciatic nerve lesions. *Exp. Neurol.* **112,** 29–39.

Vivó, M., Puigdemasa, A., Casals, L., Asensio, E., Udina, E., and Navarro, X. (2008). Immediate electrical stimulation enhances regeneration and reinnervation and modulates spinal plastic changes after sciatic nerve injury and repair. *Exp. Neurol.* **211,** 180–193.

Wakisaka, S., Kajander, K. C., and Bennett, G. J. (1991). Increased neuropeptide (NPY)-like immunoreactivity in rat sensory neurons following peripheral axotomy. *Neurosci. Lett.* **124,** 200–203.

Wall, J. T., and Cusick, C. G. (1984). Cutaneous responsiveness in primary somatosensory (S-I) hindpaw cortex before and after partial hindpaw deafferentation in adult rats. *J. Neurosci.* **4,** 1499–1515.

Wall, J. T., Felleman, D. J., and Kaas, J. H. (1983). Recovery of normal topography in the somatosensory cortex of monkeys after nerve crush and regeneration. *Science* **221,** 771–773.

Wall, J. T., Kaas, J. H., Sur, M., Nelson, R. J., Fellman, D. J., and Merzenich, M. M. (1986). Functional reorganization in somatosensory cortical areas 3b and 1 of adult monkeys after median nerve repair: Possible relationships to sensory recovery in humans. *J. Neurosci.* **6,** 218–233.

Wall, J. T., Xu, J., and Wang, X. (2002). Human brain plasticity: an emerging view of the multiple substrates and mechanisms that cause cortical changes and related sensory dysfunctions after injuries of sensory inputs from the body. *Brain Res. Rev.* **39,** 181–215.

Waxman, S. G., Cummins, T. R., Dib-Hajj, S. D., and Black, J. A. (2000). Voltage-gated sodium channels and the molecular pathogenesis of pain: A review. *J. Rehabil. Res. Dev.* **37,** 517–528.

Wood, J. N., Boorman, J. P., Okuse, K., and Baker, M. D. (2004). Voltage-gated sodium channels and pain pathways. *J. Neurobiol.* **61,** 55–71.

Woods, T. M., Cusick, C. G., Pons, T. P., Taub, E., and Jones, E. G. (2000). Progressive transneuronal changes in the brainstem and thalamus after long-term dorsal rhizotomies in adult macaque monkeys. *J. Neurosci.* **20,** 3884–3899.

Woolf, C. J., and Salter, M. W. (2000). Neuronal plasticity: Increasing the gain in pain. *Science* **288,** 1765–1768.

Woolf, C. J., Shortland, P., Reynolds, M., Ridings, J., Doubell, T., and Coggeshall, R. E. (1995). Reorganization of central terminals of myelinated primary afferents in the rat dorsal horn following peripheral axotomy. *J. Comp. Neurol.* **360,** 121–131.

Wu, C. W. H., and Kaas, J. H. (1999). Reorganization in primary motor cortex of primates with long-standing therapeutic amputations. *J. Neurosci.* **19,** 7679–7697.

Xu, J., and Wall, J. T. (1997). Rapid changes in brainstem maps of adult primates after peripheral injury. *Brain Res.* **774,** 211–215.

Xu, J., and Wall, J. T. (1999). Functional organization of tactile inputs from the hand in the cuneate nucleus and its relationship to organization in the somatosensory cortex. *J. Comp. Neurol.* **411,** 369–389.

Yang, Y., Xie, Y., Chai, H., Fan, M., Liu, S., Liu, H., Bruce, I., and Wu, W. (2006). Microarray analysis of gene expression patterns in adult spinal motoneurons after different types of axonal injuries. *Brain Res.* **1075,** 1–12.

Ygge, J. (1989). Neuronal loss in lumbar dorsal root ganglia after proximal compared to distal sciatic nerve resection: A quantitative study in the rat. *Brain Res.* **478,** 193–195.

Zhang, J. M., Donnelly, D. F., Song, X. J., and Lamotte, R. H. (1997). Axotomy increases the excitability of dorsal root ganglion cells with unmyelinated axons. *J. Neurophysiol.* **78,** 2790–2794.
Zhao, P., Waxman, S. G., and Hains, B. C. (2006). Sodium channel expression in the ventral posterolateral nucleus of the thalamus after peripheral nerve injury. *Mol. Pain* **2,** 27.
Zigmond, R. E. (1997). LIF, NGF, and the cell body response to axotomy. *Neuroscientist* **3,** 176–185.

FUTURE PERSPECTIVE IN PERIPHERAL NERVE RECONSTRUCTION

Lars Dahlin,* Fredrik Johansson,[†] Charlotta Lindwall,[‡] and Martin Kanje[†]

*Department of Hand Surgery, Malmö University Hospital, SE-205 02 Malmö, Sweden
[†]Department of Cell and Organism Biology, Lund University, SE-223 62 Lund, Sweden
[‡]Institute of Neuroscience and Physiology, Gothenburg University, SE-413 90, Gothenburg, Sweden

Nerve injuries induce severe disability and suffering for patients. Profound alterations in nerve trunks, neurons, and the central nervous system are induced rapidly after injury. This includes activation of intracellular signal transduction mechanisms aiming at the transfer of the cells into a regenerative state through the induction of the appropriate gene programs. The understanding of the neurobiological mechanisms that occur after injury can be used to design modern strategies for reconstruction after nerve injuries. Signal transduction mechanisms for instance may be targets for pharmacological intervention to stimulate nerve regeneration. Nerve injuries, particularly where there is a defect between the severed nerve trunks like in brachial plexus lesions, remain a challenge for the surgeon. Reconstruction of nerve injuries with a defect requires utilization of graft material, which can be of various designs. Application of autologous nerve grafts and use of nerve transfers are the most common clinical solutions to overcome problems with nerve defects. In this chapter we discuss the future perspective of nerve reconstruction with focus on signal transduction mechanisms and new avenues to bridge nerve defects using nanomodified graft surfaces.

DOI: 10.1016/S0074-7742(09)87028-1

0074-7742/09 $35.00

I. Introduction

In spite of extensive research on nerve regeneration, with focus on both clarifications of the delicate molecular and cellular mechanisms, as well as direct clinically applied projects, results after nerve injury and repair are generally still insufficient. The outcome is particularly troublesome when larger nerve defects have to be reconstructed. A variety of factors influence the results after nerve injury and repair and reconstruction (Dahlin, 2008a,b,c, 2009). A poor result may often be observed when a nerve injury occurs in an adult, when a mixed nerve is injured proximally, such as the median and ulnar nerves, or when a nerve trunk or a spinal nerve root is severely lacerated, such as in the brachial plexus. Hence, such nerve injuries cause frustration not only for the patient, but also for the surgeon that has to deal with poor outcome. In contrast, injuries in young children, injuries to a pure motor nerve, such as the posterior interosseous nerve, and injuries where there is a short distance to the target, may sometimes have a favorable outcome, particularly in recovery of motor function. The present issue of *The International Review of Neurobiology* focuses on repair and reconstruction of nerve injuries today as well as future possibilities. Here we focus on future perspectives of nerve repair and in particular the possibility of targeting signal transduction to improve regeneration and secondly the possibility to use nanomodified surfaces for nerve reconstruction.

II. Intracellular Signaling

The reactions of neurons and nonneuronal cells after nerve injury are very complex processes that consist of temporally and spatially orchestrated mechanisms aimed at cellular repair. After an injury in larger and mature organisms, the axons have to grow over long distances, usually along basal lamina tubes. Sometimes the basal lamina tubes have to be recreated. An important question is if these pathways are optimal and if the Schwann cells along the paths are not optimally receptive for the outgrowing axons. A possible strategy to explore for improvements would be to try to mimic some of the developmental mechanisms of axonal growth. In such conditions there is a growth of axons over a limited distance with very receptive cells in a perfect environment with appropriate tropic and trophic signals to guide the axons to their target. Finally, the brain is very well adaptive to the new signals received from the periphery in young subjects and during development. Intensive research on mechanisms of nerve regeneration is necessary to clarify the events induced in neurons, Schwann cells, and other cells, like endothelial and inflammatory cells, in the regeneration process (Dahlin, 2008b).

It can be anticipated that stimulation or inhibition of these events with proper timing and positioning can promote both pathfinding and axonal outgrowth. In analogy with a symphony orchestra, a large number of molecular instruments are played in the cells, each with their own specific and optimal function, but in contrast to an orchestra, in the cellular context, the conductor—the surgeon—may have very limited influence on the cellular orchestra in a mature subject. New findings have been presented on the subcellular mechanisms, particularly when it comes to intracellular signaling—signal transduction, which are initiated rapidly after a cellular injury and then continuous during the entire regeneration process through autocrine and paracrine cellular signaling. Such injury-induced signal transduction will be one of the topics discussed in the present review.

III. Development of Nerve Repair and Reconstruction

Although nerve repair and reconstruction is problematic, some success has yet been achieved with the aim to improve outcome for nerve repair following injury. This progress was initiated by researchers in the beginning of the last century, such as by Ramon y Cajal (1928), who described, in meticulous studies, the biology after nerve injuries. With focus on clinical nerve repair and reconstruction, Bunnell (1944) presented an impressive description of the problem of nerve injury and repair. He described factors that influenced the outcome, techniques for how to repair nerves after injury, and also results of repair in individual patients. Other important contributions to the understanding of nerve injuries have been made by Sunderland, Seddon, Moberg, Narakas, Gilbert, Birch and many others. During the last 50 years, new strategies for nerve repair and reconstruction have evolved (Lundborg, 2000). Nerve graft techniques have been introduced by the pioneering work of particularly Millesi *et al.* (1972), and such procedures are now routine in the clinic. Thus, extensive reconstruction of brachial plexus lesions is possible by the use of nerve grafts performed at specific centers in the world, where also various nerve transfers are utilized to improve function after particularly nerve root avulsions. Recently, nerve transfers were described, which were made more distally in the arm and hand, and this technique is now frequently used (Brown and Mackinnon, 2008). For shorter nerve defects, various conduits have been, or will be, introduced in the clinic based on extensive experimental research in our laboratories (Dahlin *et al.*, 2007; Lundborg *et al.*, 2004; Nilsson *et al.*, 2005a; Scherman *et al.*, 2001, 2004). Specific rehabilitation strategies with sensory re-education are routine in the clinic. Finally, better tools to treat pain and allodynia in the injured patients are also available.

However, in spite of these improvements the outcome of nerve repair is still poor, particularly when sensory functions are considered.

IV. Nerve Reconstruction: Technique and Alternatives

The clinical routine to bridge defects between injured nerve trunks is the use of autologous nerve grafts. Many problems remain after such a procedure, such as the likely discrepancy in caliber between the graft and repaired nerve. In addition, there is probably a limit of how long a graft can be, which will still permit the growth of axons to reinnervate the target. Furthermore, there may also be a lack of availability of graft material. Alternatives to nerve grafts have been experimentally developed, but few of them are clinically applied. For short gaps, and in specific circumstances as an alternative to nerve repair, various conduits are available (Lundborg *et al.*, 2004; Weber *et al.*, 2000). Other alternatives are the simple technique by the use of longitudinal sutures to bridge short defects, which is developed in our laboratories (Scherman *et al.*, 2001). Acellular nerve grafts have been developed experimentally and are becoming more popular in the clinic. The presumption is that acellular nerve grafts are less prone to be attacked by the immune system than cell-containing nerve grafts (Hudson *et al.*, 2004; Kvist *et al.*, 2008; Sondell *et al.*, 1997). Making grafts acellular may for this reason even allow xenografting making the problem of shortage of graft material and the sacrifice of healthy donor nerve void. Recently, acellular nerve allografts, additionally treated with chondrotinase A (Krekoski *et al.*, 2001), have been used to bridge short defects in digital nerves (Karabekmez *et al.*, 2009). By making nerve grafts acellular with different techniques (Hudson *et al.*, 2004; Krekoski *et al.*, 2001; Sondell *et al.*, 1999), three-dimensional structures are obtained which still contain growth-stimulating substances like laminin. To improve the regeneration process through such acellular nerve grafts and other matrices used for bridging (e.g., tendon autografts), Schwann cells from the recipient, cultured or acutely dissociated from the injured nerve segment (Brandt *et al.*, 2005; Nilsson *et al.*, 2005a), have been added to such structures. Although axonal outgrowth can be improved initially in such Schwann cell enriched structures (Nishiura *et al.*, 2004), long-term functional recovery may be disappointing (Arino *et al.*, 2008), although a "blow-through" effect in experimental studies has been suggested (Keune *et al.*, 2006). However, further clinical studies are required to elucidate long-term outcomes. In the future, we may expect that stem cells from the recipient may be used. These may have advantageous influence on axonal outgrowth through different structures, where the stem cells can get the same characteristics and function as Schwann cells after differentiation (see Terenghi *et al.*, Chapter 21, this issue).

V. Signal Transduction in Peripheral Nerve Regeneration

Some decades ago the research on effects of different growth factors showed promising results with hopes to apply these factors to improve peripheral nerve regeneration. However, the use of neurotrophic factors to improve regeneration in the peripheral nervous system has not come to clinical use, either since the mechanisms of nerve regeneration are much more complex than first anticipated or because the treatment has drawbacks, like the induction of allodynia by NGF treatment.

Harvey *et al.* (2006) cites the neurobiologist Larry Benowitz, who described the regenerative response in neurons in the central nervous system with the analogy of the "break and the gas pedal." Growth-inhibitory substrates are the breaks, whereas the growth and trophic factors that provide growth enhancing signals is the accelerator or the gas pedal. To continue the analogy of a moving vehicle, it is not enough to take the foot from the break or to push harder on the accelerator if the handbrake is still engaged. In addition, we also have to add the "steering wheel" that directs the extension of the growth cones with their filopodia. Finally, a vehicle also has a clutch, which may represent so-far unrecognized signaling pathways, or specific switches in the cell machinery that have to be engaged for an optimal and directed growth (Harvey *et al.*, 2006).

A. The Injury Signal and the Cell Body Reaction

Cells respond to signals in their environment by translating them into intracellular messengers, which through their actions induce the appropriate stimuli-specific response. The neural response to an axonal injury is not simply localized to the site of the damage, but profound changes also occur in the cell body, sometimes long distances away from the injury site. These changes are collectively known as the cell body reaction, and involves alterations in transcription, translation, and posttranslational processes. How these changes are induced and orchestrated, both spatially and temporally, and how the information of the injury is conveyed from the injury site to the cell body still remains an enigma, but this matter is the focus of intense research (Ambron and Walters, 1996; Ambron *et al.*, 1995, 1996; Befort *et al.*, 2003; Boeshore *et al.*, 2004; Brindle and Montminy, 1992; Bussmann and Sofroniew, 1999; Chen and Strickland, 2003; Chen *et al.*, 1996; Costigan *et al.*, 2002).

Microarray analysis of injured neurons has revealed injury-induced regulation of hundreds of genes (Curtis *et al.*, 1998; Drysdale *et al.*, 1996; Gunstream *et al.*, 1995; Gupta *et al.*, 1996), including those encoding neurotrophin receptors, transcription factors and cytoskeletal components. It is hypothesized that such transcriptional changes come about following injury as a response to both negative and positive signals, that is, lack of signals from target tissue and injury-induced signals from the damaged axon, respectively. The cell body reaction can, for instance, be initiated by the disruption of retrograde trophic support from the target tissue. There is ample experimental support for such negative control of the cell body response in injured sensory neurons, and it has been shown that pharmacological inhibition of retrograde axonal transport mimics several aspects of the cell body reactions associated with axotomy (Hai *et al.*, 1989, 1999; Hibi *et al.*, 1993). Also, injury-induced downregulation of the neurotransmitters substance P and neuropeptide Y in sensory neurons can be mitigated by distal application of nerve growth factor (NGF) or acidic fibroblast growth factor (aFGF), respectively, to the nerve (Ji *et al.*, 1996; Kallunki *et al.*, 1996). However, the cell body may also be triggered by positive signaling, that is, retrograde transport of proteins from the site of injury. The mollusk *Aplysia californica* has been extensively used as a model system to illustrate such positive regulatory mechanisms (Ambron *et al.*, 1995; Costigan *et al.*, 2002; Karin, 1995; Leah *et al.*, 1991).

Currently, there is no established treatment other than surgery for peripheral nerve injuries. However, the molecular mechanisms that regulate the neuronal injury response could in the future be used as the basis for developing new clinical therapies. Ultimately, the goal would be to modify how a peripheral axonal lesion activates the intrinsic growth capacity of the injured neuron, which in turn would be aimed to promote the speed and accuracy of regeneration.

The intrinsic growth capacity of peripheral neurons has been suggested to be mediated through the actions of cyclic adenosine monophosphate (cAMP) (Liang *et al.*, 1996; Lindwall and Kanje, 2005; Lindwall *et al.*, 2004), which ultimately regulates organization of the cytoskeleton (Snider *et al.*, 2002). Also, among others, two transcription factors that have been demonstrated to be rapidly induced by peripheral nerve injury are c-Jun (Raivich *et al.*, 2004) and activating transcription factor 3 (ATF3) (Lindwall *et al.*, 2004) (Fig. 1). c-Jun is the fundamental component of the activating protein 1 (AP-1) complex (Karin, 1995), and is one of the targets of the stress activated c-Jun *N*-terminal kinase (JNK), which catalyzes its phosphorylation (Neumann *et al.*, 2002; Nilsson *et al.*, 2005b; Perlson *et al.*, 2004a). Conditional knockout of c-Jun (Raivich *et al.*, 2004) as well as pharmacological inhibition of JNK (Lindwall *et al.*, 2004) has been demonstrated to inhibit nerve regeneration. Thus, the JNK family of kinases is required for successful regeneration of peripheral sensory neurons (Middlemas *et al.*, 2003; Perlson *et al.*, 2004b; Qiu *et al.*, 2002a), and as such represents a target for future clinical therapies. ATF3 is a member of the ATF/CREB transcription factor family (Qiu *et al.*, 2002b; Raivich *et al.*, 2004), and is rapidly induced by a variety of signals, including agents that induce the JNK signaling pathway (Seijffers *et al.*, 2006, 2007; Snider *et al.*, 2002; Sung *et al.*, 2001; Tanabe *et al.*, 2003). Inhibition of JNK reduces ATF3 protein levels (Lindwall *et al.*, 2004), which in turn hamper regeneration. Importantly, ectopic expression of ATF3 can actually promote neurite outgrowth of peripheral neurons, possibly through an increase in the intrinsic

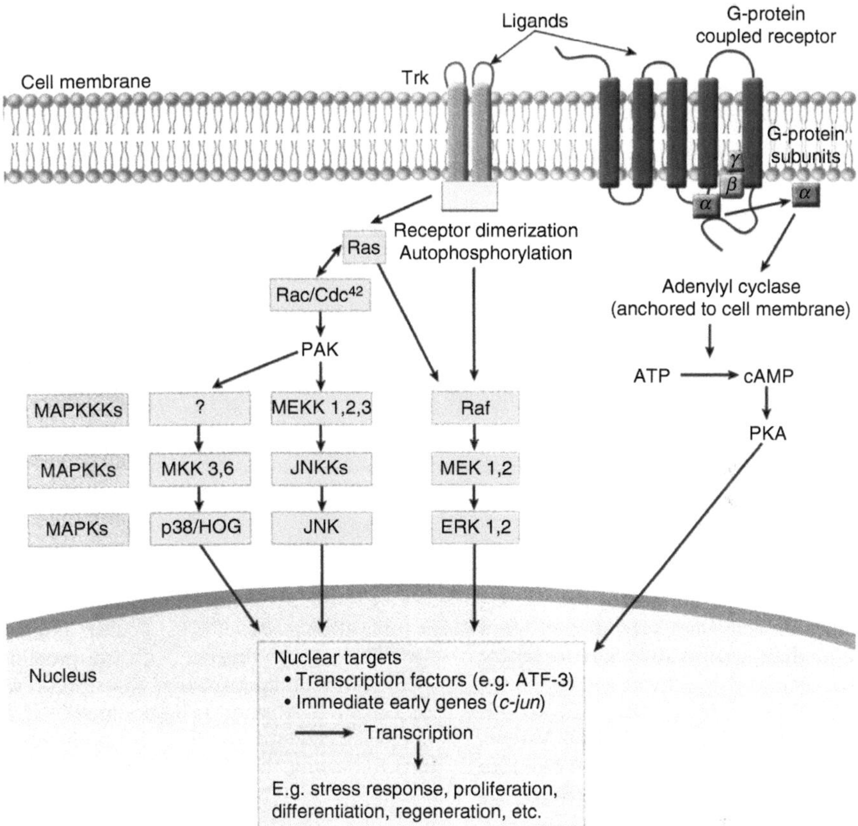

FIG. 1. Signal transduction steps in neurons and nonneuronal cells that occur after nerve injury. The schematic drawing shows the various steps that are needed to activate cells in response to a trauma. Different intracellular steps include phosphorylation steps by the MAPK modules. (Reproduced by kind permission of Elsevier.)

growth state of the neurons (Wong and Oblinger, 1991; Woolf *et al.*, 1990). The presence of these factors over time does probably influence the efficiency of axonal outgrowth; knowledge of outmost importance when considering timing of nerve repair (see below). Thus, if the levels of injury induced molecules, such as JNK, c-Jun, and ATF3, can be selectively modified following a peripheral nerve injury, augmentation of neuritogenesis can be obtained. Such treatment strategies for severe nerve injuries must, however, await a better understanding of the intrinsic molecular mechanisms initiating, and underlying, the regeneration process.

B. Extrinsic Properties Required for Axon Growth and Target Finding

The intrinsic growth capacity of peripheral nerve regeneration has to be combined with a proper environment to encourage axonal growth. Normally, peripheral axons are ensheathed and myelinated by Schwann cells. These cells also provide a basal lamina surrounding bundles of axons. Following an injury, Schwann cells de-differentiate and aid in the clearing of damaged debris, while during regeneration they act as guides for sprouting axons. During the regenerative process they upregulate several genes; the protein products of which may be involved in the guidance of axonal sprouts by Schwann cell-axon attachment (Martini *et al.*, 1994). For instance, the previously mentioned extracellular matrix (ECM) molecule laminin, which is produced by Schwann cells, plays a significant role during regeneration. Laminin receptors, such as integrins, are expressed on the growing axons, which supports regeneration. On the other hand, in mouse knockout models of laminin, axonal regeneration is significantly impaired (Zhang and Ambron, 2000). Thus, regeneration depends on a complex interplay and signals between several cell types within the nerve.

During regeneration the axonal sprouts grow down the distal nerve segment and, if successful, reinnervate their correct targets. Axonal outgrowth is, however, slow in humans, and occur at a rate of around 1 mm per day. Success of regeneration can only be judged following reinnervation of the target tissue, a process which, depending on where the damage was done, may take weeks or months after the initial insult, although the regeneration process can be followed by advancement of the Tinel sign. Unfortunately, at the time of reinnervation the window of successful regeneration may already have passed. Axons must also make correct discriminatory choices in order to reinnervate the correct target tissue, and during this process they are often misrouted. In order to develop therapeutic strategies to improve both rate and accuracy of target reinnervation we need to clarify the molecular events that influence the intrinsic growth capacity as well as axonal discrimination of the extrinsic cues, both substrate bound and diffusible, that is encountered by the axon along the regenerative pathway (Fig. 2). However, it should be stressed that these signal transduction mechanisms occur not only in neurons and its axons but also in all types of cells in a temporal and spatial resolution.

VI. Nanotechnology and Nerve Regeneration

The complex treatment of peripheral nerve injuries but also of the injured spinal cord, which involves a variety of strategies, has been emphasized by many authors (see, e.g., Garbossa *et al.*, 2006). To enhance axonal regeneration with a possible application in spinal cord repair a new generation of tissue compatible

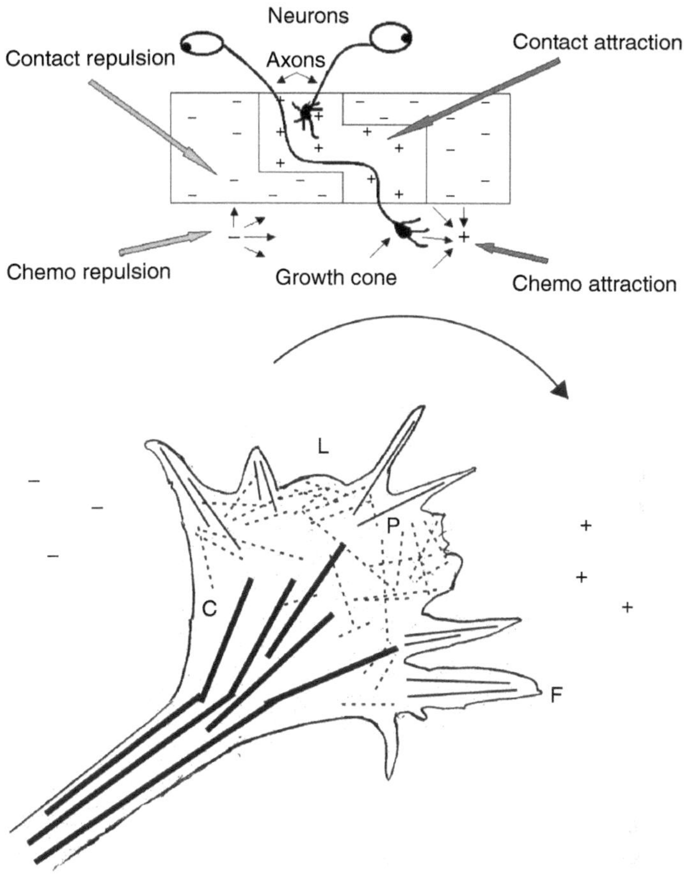

FIG. 2. Schematic drawing showing four different mechanisms that direct axonal outgrowth after nerve injury involving both attraction and repulsion of the growth cone (upper drawing). The lower drawing show details of a growth cone formed on the tip of each of numerous sprouts, which originate from the proximal end of the transected axon. C = domain where microtubules are located. P = domain where F-actin monomers and actin filaments are found. F = finger-like filopodia that palpates the surroundings directing growth. L = veil-like lamellipodia. (Reproduced by kind permission of American Society for Surgery of the Hand.)

matrices are currently developed using recent developments in nanotechnology. Nanostructures can be a suitable environment for outgrowing axons in different situations; not only to create bridges for nerve defects but also for other reasons with focus on brain machine interface (BMI) issues. Such reasons are directionality of axonal growth and sorting of nerve fibers (Johansson *et al.*, 2006; Prinz *et al.*, 2008). Nanotechnology and nerve regeneration is a future exciting field which also will be covered in the present chapter. Development of biologically compatible scaffolds that can serve as permissive substrates for growth of neurons, migration of Schwann cells, influence differentiation, and minimize scar and inhibitory environment is a challenge where specific criteria have been outlined for application in the central nervous system (Ellis-Behnke *et al.*, 2007). In the CNS, the four important P's of regeneration as a framework has been stressed (Ellis-Behnke *et al.*, 2007): *Preservation* of neurons (no cell death); Growth-*permissive* environment; *Promotion* of growth through the permissive environment of preserved neurons, axons and their sprouts and growth cones, and, finally, utilized and improved *plasticity* after reconnection. Nanotechnology may offer solutions to several of these criteria. In combination with genomic and proteomic revolution this nanomic one will help to understand pathophysiological events and to improve results as has been suggested for the visual system (Harvey *et al.*, 2006).

A. Nanostructures for Neurite Regeneration

For the development of a new generation of artificial scaffold implants, which are tissue compatible, which smoothly integrate with the host, and which also enhance axonal regeneration, the influence of the implant surface, that is, the topography and chemophysical properties on which the cells/neurites will grow, is of paramount importance. Such tissue engineering principles can also be adopted for the study of cellular behavior associated with regenerating nerve tissue *in vitro*.

Nanotechnology has provided us with new tools that allow the design of structures with dimensions of only a few nanometers that may interact with cells and subcellular processes on a suitable cellular scale (Yim *et al.*, 2005). Extensive research on cell reactions to nanostructures *in vitro*, as well as on cell and cell extensions—neurites—has been performed during the last decade. It is conceivable that nanostructured implant surfaces can be tuned to interact smoothly with the tissue on the implant site and evoke less of an immune response than would nonstructured surfaces. Furthermore, such surfaces can be modified for the organization of the attached cells and thereby tissue formation, resulting in enhanced regeneration. We are presently pursuing these ideas for the repair and reconstruction of peripheral nerves.

B. Neurites and Topography: From Micro to Nano

During nerve regeneration the outgrowth of axons is influenced by a variety of factors, local or distant, and by the cues in the surrounding (Fig. 2). Almost a hundred years ago, R. G. Harrison reported that cells and neurites grown on threads from a spider's web followed the fibers (Harrison, 1911). In 1945, the American developmental neurobiologist Paul Weiss named this behavior "contact guidance" (Weiss, 1945), but it was not until the early 1970s that biologists seriously tested the idea of contact guidance again, starting with growing cells on grooved substrata and on spheres (Maroudas, 1972; Rovensky *et al.*, 1971). Since then, the cellular responses to topographical cues of many different kinds have been tested, including curved surfaces, single steps, angled planes, pillars, pits, pores, cylinders, spheres, and last, but not least, the most studied structure, parallel grooves and ridges (Flemming *et al.*, 1999). The explosion of research on such artificial topographical cues was mainly due to the rapid development of techniques in the computer industry. Hence, structures with micrometer, and in the last 10–15 years even submicrometer, sized objects are possible to produce and have become available for biomedical research.

Although the exact cell reaction to a specific topography may not easily be predicted, since it is cell type dependent, a great pool of structures and cell types have been tested. Today, it is clear that structures as small as 5–10 nm can change the morphology of some cells, that is, macrophages (Wojciak-Stothard *et al.*, 1996) and that axons may follow grooves and fibers with widths of around 100 nm (Johansson *et al.*, 2006). The latter is perhaps not too surprising considering the fasciculation (minifasciculation) that occurs once a pioneering axon (that in mice may be as thin as 100 nm) has found a path during embryogenesis or regeneration. A simple, although elegant, model for neurite guidance on fibers with different diameters has been presented (Smeal *et al.*, 2005), and shows an enhanced neurite alignment along thin fibers as compared with thicker ones. The basic idea for this guidance phenomenon appears to be the stiffness of the cytoskeleton of the extending neurites. These extensions can simply not curve around a fiber with too small a curvature radius and therefore extend along the fiber in an aligned manner.

For tissue engineering applications in general, and nerve grafts in particular, ordered outgrowth/morphology is often requested. Even though 100 years have passed, the old finding of Harrison is still applicable, not only in the micrometer domain, but also in the nanometer range: parallel structures of grooves with ridges and fibers will orient cells and cell extensions, for example, axons along the structures. Such guided axons often display a simplified growth cone and a higher outgrowth rate as compared to a similar smooth or irregular surface (Corey *et al.*, 2007), a desirable feature in clinical applications. We have found excellent axonal guidance on substrates with rows of 2.5 μm long, vertically standing nanowires separated by 400 nm. Axons from dorsal root ganglia (mouse) were found to be unable to cross between, or climb the nanowires when the distance between two standing wires was small enough (sufficient with 400 nm separation but not with 1 μm) (Prinz *et al.*, 2008). Again, the explanation of this behavior is probably due to the rigidity of the cytoskeleton, a model that may explain several contact guidance phenomena.

On nonordered substrates, such as porous silicon, which has a sponge-like appearance, attachment and proliferation have been shown to be dependent on pore size *in vitro* (Bayliss *et al.*, 1999a,b, 2000; Sapelkin *et al.*, 2006). In this way, we have demonstrated ordered axonal outgrowth from mouse dorsal root ganglia on porous stripes in otherwise smooth silicon (Johansson *et al.*, 2005, 2008). Such porous silicon has also been shown to induce less encapsulation than smooth silicon *in vivo*, indicating a more biocompatible structure (Rosengren *et al.*, 2002). On random meshes of polymer fibers produced by electrospinning (see below), axonal outgrowth is hampered as compared to aligned fibers of the same material, probably due to irregular contact guidance cues (Corey *et al.*, 2007; Wang *et al.*, 2008). For clinical applications, such as nerve grafts where fast Schwann cell migration and axonal regeneration is crucial, ordered linear structures are obviously of essence.

Some of the intracellular molecular components of the guiding system have been identified (Nobes and Hall, 1999; Patel and Van Vactor, 2002), although still many pieces in the puzzle are missing. All cell reactions to the topographies described above depend on highly coordinated assembly and disassembly of the cytoskeleton and in particular microfilaments. The intracellular signaling pathways arising from the extracellular cues, and leading to the rearrangement of the cytoskeleton, involve signal transduction described previously.

For migrating cells the small GTPases, Rac, Cdc42, Ras, and Rho have been shown to be important for organizing the cytoskeleton during migration. Rac is essential for the protrusion of lamellipodia and thereby forward movement, Cdc42 is necessary for maintaining cell polarity, while Ras regulates focal adhesions and associated actin fibers (Nobes and Hall, 1999). The last one, Rho, has been reported as necessary for cell adhesion during movement and thereby contact guidance (Nobes and Hall, 1999; Rajnicek *et al.*, 2008). The guidance of axons thus depends on the same molecules, Cdc42 and Rac, that mediate growth cone attraction and elongation, while Rho mediates repulsion and growth cone collapse (Patel and Van Vactor, 2002). The alternating activation GTPases of by the external cues via the membrane receptors can thus guide the axon in a stop and go fashion. These mechanisms are examples of the intracellular signal transduction pathways.

C. Why Nanostructures?

So, if the results of an experiment performed 100 years ago on spider silk gave the same result as the most advanced structures today, why is there still an interest in nanotechnology in tissue engineering? Artificial nanopatterns can be controlled with respect to size, chemical composition, and physical properties. From an engineering point of view, the spatial resolution is extremely high using nanotechnology enabling influence and guidance on single axons (Figs. 3 and 4). This may be very important for high-resolution neural interfaces (BMI) that may support axons bypassing injuries, controlling artificial limbs or restore other functions including hearing and vision (Donoghue, 2002). From a more biological/clinical point of view, the ECM is composed of fibers and fibrils

FIG. 3. Upper and lower: Axons grown *in vitro* are highly sensitive for topographies such as grooves and ridges. Here, DRG axons grown on grooves and ridges as small as 100 nm wide, display contact guidance and follow the patterned areas (squares with orthogonal grooves/ridges in an otherwise plane polymer surface produced with Nanoimprint Lithography). (Scanning electron microscope images.)

Fig. 4. *Top*: A DRG mounted in Matrigel sends out axons in a random fashion on a flat polymer surface. Once entering squares with grooves and ridges, most axons are guided along the imprinted topography. The top row holds horizontal grooves, while the lower row holds vertical grooves. Note that the thicker axons appears to be less guided than the thinner on those nanometer sized grooves. (Flourescence microscope image.) *Bottom*: On stripes (light green) of porous silicon, pore sizes of 500 nm, in an otherwise flat substrate (black), both axons and Schwann cells from an explanted DRG *in vitro*, prefer to grow and elongate on the porous stripes rather than on the flat areas. This behavior could be utilized for guidance of such cells and extensions in many different applications. (Flourescence microscope image.)

ranging from nanometers to micrometers. Hence, the addition of nanostructures for tissue engineered implants can mimic the ECM structure—the natural environment of cells and growing axons *in vivo* (Ma *et al.*, 2005).

In many other bio-nano applications, such as quantum dots, the effect of quantum physics is taken advantage of. These quantum effects and the potential for such nanostructures are so far not included in tissue engineering, although the extreme area to volume ratio can be employed for built-in drug delivery systems. The substance that should be delivered can either be adsorbed on the artificial substrate, and then the area-to-volume ratio is critical for how much substance an implant can hold. For biodegradable implants (usually polymers), where substances can be incorporated in the implant material, the delivery rate depends on the area rather than the volume when the substrate is degraded. Nanostructured/porous substrates will therefore represent faster delivery systems than bulk substrates of the same volume. The possibility to include substances which promotes survival of neurons in nanostructured nerve implants should be explored.

D. From Cell Reactions to Nanostructures *In Vitro* to Nerve Regeneration Applications

Most *in vitro* studies on nanostructures have been made on flat (rather 2.5-D than true 3-D), hard substrates such as silicon, glass, and plastics. The reason is the limitation of the patterning techniques inherited from the computer industry, that is, photolithography and electron beam lithography that usually must be used at some stage in processing a structured surface. "Soft lithography" is an overall description of many techniques where rubber molds from templates, created by the techniques mentioned above, can produce new topographical or chemical patterns. The use of such techniques can transit such structured surfaces onto irregular shapes to some degree, but only within certain limits. Although very different from an *in vivo* situation, such flat test structures have supplied us with most of the basic knowledge concerning cell adhesion, migration, and alignment etc.

From the clinical perspective, a technique called electrospinning may be better suited for nerve repair. In short, this technique is based on a polymer that is pushed out of a thin syringe. At the syringe tip, the polymer is surface charged and forms a jet stream toward an electrically grounded target, where the polymer is collected when the solvent evaporates. The thickness of such polymer fibers can be tuned from some nanometers to micrometers and the fiber alignment can be manipulated by rapid movement of the target. In this way, fabrics of nanometer polymer fibers resembling the ECM have been produced (Kumbar *et al.*, 2008; Ma *et al.*, 2005; Murugan and Ramakrishna, 2006) and aligned such fibers guide extending neurites and migrating Schwann cells (Kim *et al.*, 2008; Schnell *et al.*, 2007; Yang *et al.*, 2005). The layers of ECM-like, biodegradable polymers, for example, poly-L-lactic acid (PLLA), can be added onto many macroscopic surfaces as a way to enhance biofunctionality, or work on its own as an artificial ECM scaffold. Besides the obvious resemblance of the ECM structure, the polymer itself can be blended with axon promoting factors, such as laminin (Koh *et al.*, 2008). By using highly aligned structures, contact guidance may also help to enhance axonal outgrowth and nerve regeneration *in vitro* and *in vivo* (Kim *et al.*, 2008; Wang *et al.*, 2008). The use of biodegradable polymers, such as PLLA with the opportunity to blend in other substances, together with the porous structure of the fabric that enables diffusion of nutrition and oxygen prior to vascularization, fulfills many clinical requirements of a nerve graft.

The use of a tissue-engineered nerve graft that performs as good as, or even better, than a standard autologous graft may minimize costs and trauma after nerve injuries. Nanostructures assist in mimicking actual tissue, enable designs on a subcellular level, and may thus be used in future nerve grafts.

VII. Clinical Development: Future Perspectives

To improve nerve regeneration and the outcome after various injuries there is a requirement for the exploration of new research avenues. Such avenues can be signal transduction and nanotechnology as discussed above. There are several other aspects which require attention from the clinical perspective. One is the problem of comparing new with conventional repair and reconstruction techniques. Another is the timing of repair and reconstruction. A third is the problem of neuronal cell death which may be a target for pharmacological intervention. Finally, focus is now also directed towards brain plasticity and the patient's ability to utilize coping strategies to adjust to the impaired function.

In the short perspective, in clinical studies, and particularly multicenter studies, we can investigate the effectiveness of different nerve reconstruction techniques, such as the new alternatives to nerve grafts. However, in more extensive nerve injuries, like in brachial plexus injuries, there are difficulties to evaluate used repair and reconstruction techniques since no lesion is similar to the other, that is, there are differences in the individual extent of injury and thereby the need for different reconstruction procedures. Thus, it is difficult to collect an appropriate number of patients with similar injuries, where such injuries are reconstructed with well-defined techniques. Previous findings have revealed an impaired functional recovery if nerve reconstruction of the brachial plexus lesion is done 6 months or later following the injury. The timing for nerve reconstruction has been emphasized based on neurobiological alterations in neurons and Schwann cells (Saito and Dahlin, 2008). Interestingly, Kay and co-workers (Jivan *et al.*, 2009) have recently presented data indicating that functional outcome after brachial plexus lesions involving C5-C6 is better if reconstruction is done within 2 weeks after injury. Such notion is supported by the neurobiological data indicating that cellular alterations in both neurons and Schwann cells are time dependent. In Schwann cells, signal transduction mechanisms are rapidly initiated, even within 30 min, which are important for the proliferation of the Schwann cells after a nerve injury (Martensson *et al.*, 2007) and thereby the outgrowth of axons. Schwann cells can also modify the growth environment in the distal nerve segment after injury (Danielsen *et al.*, 1995). Transcription factors, upregulated rapidly in Schwann cells and neurons, subside over time with a subsequent impaired activation of Schwann cell in the distal nerve segment and decreased axonal outgrowth (Saito and Dahlin, 2008). Similarly, a rapid upregulation of the transcription factor ATF3 in neurons is also deteriorated over time. The diminution of that ATF3 response in neurons seems to correlate to impaired nerve regeneration (Saito and Dahlin, 2008). Interestingly, the decline of ATF3-containing neurons is more rapid in motor neurons than in sensory neurons (Kataoka *et al.*, 2007; Saito and Dahlin, 2008). However, the cell death of neurons is more pronounced in sensory neurons than in motor neurons (Hart *et al.*, 2004; McKay Hart *et al.*, 2002). Neuronal cell death can also be diminished if nerve trunks are repaired early after injury (Ma *et al.*, 2003). To prevent or decrease neuronal cell death, particularly among motor neurons, a pharmacological intervention can be considered. Experimental data indicate that early treatment, perhaps within the first 24 h after injury, with *N*-acetylcystein (Hart *et al.*, 2002, 2004) can reduce the number of neurons that go through programmed cell death, apoptosis. However, such treatment has to be tested clinically, preferably in multicenter studies, utilizing the specific protocols for evaluation of function after nerve injury and repair.

A problem after repair and reconstruction of proximal nerve injuries is the extended time before reinnervation of the target can be expected. The Schwann cell response to injury deteriorates over time leading to impaired axonal regeneration after proximal nerve injuries. In this respect, the described nerve transfers in the hand and distal forearm (Brown and Mackinnon, 2008) is an alternative since the surgeon can transfer the nerve injury from a proximal to a distal one. The growing axons from the transferred nerve are thus allowed to grow into an environment that is still permissive in the originally injured distal nerve segment. Nerve transfers can also be applied for a distal nerve segment when there is a lack of a proximal nerve trunk as a source of axon—end-to-side nerve repair (see Bontioti and Dahlin; Chapter 12, this issue). In addition, there are also a large number of other clinically potentially exciting additional treatments as an adjunct to conventional nerve repair and nerve grafting techniques, which can improve nerve regeneration covered in the present issue of *International Review of Neurobiology*.

Following nerve reconstruction the surgeon also has to plan and initiate the rehabilitation phase and focus on the central nervous system, where extensive reorganization in the cerebral cortex and other levels occur after injury. Early, before reinnervation (phase I), new concepts for rehabilitation have to be considered followed by novel rehabilitation techniques when reinnervation of the hand and arm has occurred (phase II). Several new strategies have been introduced in recent years including the use of EMLA® (local anesthetics) cream application to the forearm leading to improved sensibility in the hand after repaired median and ulnar nerve injuries (Lundborg *et al.*, 2007; Rosen *et al.*, 2006). Thus, brain plasticity is a central issue in nerve reconstruction. Furthermore, individual care of the patients is crucial to direct them along their inborn strategies to cope with such an injury (Cederlund *et al.*, 2008), strategies which they may not immediately be aware of.

Taken together, although conventional nerve reconstruction techniques are used frequently in clinical practice, the outcome is generally still insufficient. Thus, new treatment strategies have to be introduced based on new avenues of research. The utilization of knowledge of intracellular signal transduction mechanisms, and the use of nanotechnologies are exciting perspectives in nerve reconstruction in the future.

Acknowledgments

The authors are supported by grants from the Swedish Research Council (Medicine and Natural Sciences), Crafoord's Fund for Medical Research, Konsul Thure Carlsson Fund for Medical Research, Region Skåne, and Funds from the University Hospital Malmö, Sweden.

References

Ambron, R. T., and Walters, E. T. (1996). Priming events and retrograde injury signals. A new perspective on the cellular and molecular biology of nerve regeneration. *Mol. Neurobiol.* **13,** 61–79.

Ambron, R. T., Dulin, M. F., Zhang, X. P., Schmied, R., and Walters, E. T. (1995). Axoplasm enriched in a protein mobilized by nerve injury induces memory-like alterations in Aplysia neurons. *J. Neurosci.* **15,** 3440–3446.

Ambron, R. T., Zhang, X. P., Gunstream, J. D., Povelones, M., and Walters, E. T. (1996). Intrinsic injury signals enhance growth, survival, and excitability of Aplysia neurons. *J. Neurosci.* **16,** 7469–7774.

Arino, H., Brandt, J., and Dahlin, L. B. (2008). Implantation of Schwann cells in rat tendon autografts as a model for peripheral nerve repair: Long term effects on functional recovery. *Scand. J. Plast. Reconstr. Surg. Hand Surg.* **42,** 281–285.

Bayliss, S. C., Buckberry, L. D., Fletcher, I., and Tobin, M. J. (1999a). The culture of neurons on silicon. *Sens. Actuators a-Phys.* **74,** 139–142.
Bayliss, S. C., Heald, R., Fletcher, D. I., and Buckberry, L. D. (1999b). The culture of mammalian cells on nanostructured silicon. *Adv. Mater.* **11,** 318–321.
Bayliss, S. C., Buckberry, L. D., Harris, P. J., and Tobin, M. (2000). Nature of the silicon-animal cell interface. *J. Porous Mater.* **7,** 191–195.
Befort, K., Karchewski, L., Lanoue, C., and Woolf, C. J. (2003). Selective up-regulation of the growth arrest DNA damage-inducible gene Gadd45 alpha in sensory and motor neurons after peripheral nerve injury. *Eur. J. Neurosci.* **18,** 911–922.
Boeshore, K. L., Schreiber, R. C., Vaccariello, S. A., Sachs, H. H., Salazar, R., Lee, J., Ratan, R. R., Leahy, P., and Zigmond, R. E. (2004). Novel changes in gene expression following axotomy of a sympathetic ganglion: A microarray analysis. *J. Neurobiol.* **59,** 216–235.
Brandt, J., Nilsson, A., Kanje, M., Lundborg, G., and Dahlin, L. B. (2005). Acutely-dissociated Schwann cells used in tendon autografts for bridging nerve defects in rats: A new principle for tissue engineering in nerve reconstruction. *Scand. J. Plast. Reconstr. Surg. Hand Surg.* **39,** 321–325.
Brindle, P. K., and Montminy, M. R. (1992). The CREB family of transcription activators. *Curr. Opin. Genet. Dev.* **2,** 199–204.
Brown, J. M., and Mackinnon, S. E. (2008). Nerve transfers in the forearm and hand. *Hand Clin.* **24,** 319–340.
Bunnell, S. (1944)."Surgery of the Hand." J.B. Lippincott Company, Philadelphia.
Bussmann, K. A., and Sofroniew, M. V. (1999). Re-expression of p75NTR by adult motor neurons after axotomy is triggered by retrograde transport of a positive signal from axons regrowing through damaged or denervated peripheral nerve tissue. *Neuroscience* **91,** 273–281.
Cederlund, R., Thorén-Jönsson, A. L., and Dahlin, L. B. (2008). Adaptation in daily activities three months after a major traumatic hand injury. *In* "XIIIth Congress of the FESSH, IXth Congress of the EFSHT," Lausanne, 19-21 June.
Chen, Z. L., and Strickland, S. (2003). Laminin gamma1 is critical for Schwann cell differentiation, axon myelination, and regeneration in the peripheral nerve. *J. Cell Biol.* **163,** 889–899.
Chen, B. P., Wolfgang, C. D., and Hai, T. (1996). Analysis of ATF3, a transcription factor induced by physiological stresses and modulated by gadd153/Chop10. *Mol. Cell. Biol.* **16,** 1157–1168.
Corey, J. M., Lin, D. Y., Mycek, K. B., Chen, Q., Samuel, S., Feldman, E. L., and Martin, D. C. (2007). Aligned electrospun nanofibers specify the direction of dorsal root ganglia neurite growth. *J. Biomed. Mater. Res. A* **83,** 636–645.
Costigan, M., Befort, K., Karchewski, L., Griffin, R. S., D'Urso, D., Allchorne, A., Sitarski, J., Mannion, J. W., Pratt, R. E., and Woolf, C. J. (2002). Replicate high-density rat genome oligonucleotide microarrays reveal hundreds of regulated genes in the dorsal root ganglion after peripheral nerve injury. *BMC Neurosci.* **3,** 16.
Curtis, R., Tonra, J. R., Stark, J. L., Adryan, K. M., Park, J. S., Cliffer, K. D., Lindsay, R. M., and DiStefano, P. S. (1998). Neuronal injury increases retrograde axonal transport of the neurotrophins to spinal sensory neurons and motor neurons via multiple receptor mechanisms. *Mol. Cell Neurosci.* **12,** 105–118.
Dahlin, L. B. (2008a). Nerve injuries. *Curr. Orthop.* **22,** 9–16.
Dahlin, L. B. (2008b). The nerve response to injury. *In* "Upper Extremity Nerve Repair: Tips and Techniques—A Master Skills Publication Volume," (D. Slutsky, Ed.), pp. 15–28. ASSH.
Dahlin, L. B. (2008c). Techniques of peripheral nerve repair. *Scand. J. Surg.* **97,** 310–316.
Dahlin, L., Brandt, J., Nilsson, A., Lundborg, G., and Kanje, M. (2007). Schwann cells, acutely dissociated from a predegenerated nerve trunk, can be applied into a matrix used to bridge nerve defects in rats. *Acta Neurochir. Suppl.* **100,** 57–59.
Danielsen, N., Kerns, J. M., Holmquist, B., Zhao, Q., Lundborg, G., and Kanje, M. (1995). Predegeneration enhances regeneration into acellular nerve grafts. *Brain Res.* **681,** 105–108.

Donoghue, J. P. (2002). Connecting cortex to machines: Recent advances in brain interfaces. *Nat. Neurosci.* **5**(Suppl), 1085–1088.

Drysdale, B. E., Howard, D. L., and Johnson, R. J. (1996). Identification of a lipopolysaccharide inducible transcription factor in murine macrophages. *Mol. Immunol.* **33,** 989–998.

Ellis-Behnke, R. G., Teather, L. A., Schneider, G. E., and So, K. F. (2007). Using nanotechnology to design potential therapies for CNS regeneration. *Curr. Pharm. Des.* **13,** 2519–2528.

Flemming, R. G., Murphy, C. J., Abrams, G. A., Goodman, S. L., and Nealey, P. F. (1999). Effects of synthetic micro- and nano-structured surfaces on cell behavior. *Biomaterials* **20,** 573–588.

Garbossa, D., Fontanella, M., Fronda, C., Benevello, C., Muraca, G., Ducati, A., and Vercelli, A. (2006). New strategies for repairing the injured spinal cord: The role of stem cells. *Neurol. Res.* **28,** 500–504.

Gunstream, J. D., Castro, G. A., and Walters, E. T. (1995). Retrograde transport of plasticity signals in Aplysia sensory neurons following axonal injury. *J. Neurosci.* **15,** 439–448.

Gupta, S., Barrett, T., Whitmarsh, A. J., Cavanagh, J., Sluss, H. K., Derijard, B., and Davis, R. J. (1996). Selective interaction of JNK protein kinase isoforms with transcription factors. *EMBO J.* **15,** 2760–2770.

Hai, T. W., Liu, F., Coukos, W. J., and Green, M. R. (1989). Transcription factor ATF cDNA clones: An extensive family of leucine zipper proteins able to selectively form DNA-binding heterodimers. *Genes Dev.* **3,** 2083–2090.

Hai, T., Wolfgang, C. D., Marsee, D. K., Allen, A. E., and Sivaprasad, U. (1999). ATF3 and stress responses. *Gene Exp.* **7,** 321–335.

Harrison, R. G. (1911). On the stereotropism of embryonic cells. *Science* **34,** 279–281.

Hart, A. M., Wiberg, M., Youle, M., and Terenghi, G. (2002). Systemic acetyl-L-carnitine eliminates sensory neuronal loss after peripheral axotomy: A new clinical approach in the management of peripheral nerve trauma. *Exp. Brain Res.* **145,** 182–189.

Hart, A. M., Terenghi, G., Kellerth, J. O., and Wiberg, M. (2004). Sensory neuroprotection, mitochondrial preservation, and therapeutic potential of n-acetyl-cysteine after nerve injury. *Neuroscience* **125,** 91–101.

Harvey, A. R., Hu, Y., Leaver, S. G., Mellough, C. B., Park, K., Verhaagen, J., Plant, G. W., and Cui, Q. (2006). Gene therapy and transplantation in CNS repair: The visual system. *Prog. Retin. Eye Res.* **25,** 449–489.

Hibi, M., Lin, A., Smeal, T., Minden, A., and Karin, M. (1993). Identification of an oncoprotein- and UV-responsive protein kinase that binds and potentiates the c-Jun activation domain. *Genes Dev.* **7,** 2135–2148.

Hudson, T. W., Zawko, S., Deister, C., Lundy, S., Hu, C. Y., Lee, K., and Schmidt, C. E. (2004). Optimized acellular nerve graft is immunologically tolerated and supports regeneration. *Tissue Eng.* **10,** 1641–1651.

Ji, R. R., Zhang, Q., Pettersson, R. F., and Hokfelt, T. (1996). aFGF, bFGF and NGF differentially regulate neuropeptide expression in dorsal root ganglia after axotomy and induce autotomy. *Regul. Pept.* **66,** 179–189.

Jivan, S., Kumar, N., Wiberg, M., and Kay, S. (2009). The influence of pre-surgical delay on functional outcome after reconstruction of brachial plexus injuries. *J. Plast. Reconstr. Aesthet. Surg.* **62,** 472–479

Johansson, F., Kanje, M., Eriksson, C., and Wallman, L. (2005). Guidance of neurons on porous patterned silicon: Is pore size important? *Phys. Status Solidi c- current topics in solid state physics* **2**(9), 3258–3262.

Johansson, F., Carlberg, P., Danielsen, N., Montelius, L., and Kanje, M. (2006). Axonal outgrowth on nano-imprinted patterns. *Biomaterials* **27,** 1251–1258.

Johansson, F., Kanje, M., Linsmeier, C. E., and Wallman, L. (2008). The influence of porous silicon on axonal outgrowth *in vitro*. *IEEE Trans. Biomed. Eng.* **55,** 1447–1449.

Kallunki, T., Deng, T., Hibi, M., and Karin, M. (1996). c-Jun can recruit JNK to phosphorylate dimerization partners via specific docking interactions. *Cell* **87,** 929–939.

Karabekmez, F. E., Duymaz, A., Mardini, A., and Moran, S. L. (2009). Early clinical outcomes with the use of decellularized nerve allograft for repair of sensory defects within the upper extremity. *In* "American Society for Peripheral Nerve Annual Meeting." Maui, Hawaii.

Karin, M. (1995). The regulation of AP-1 activity by mitogen-activated protein kinases. *J. Biol. Chem.* **270,** 16483–16486.

Kataoka, K., Kanje, M., and Dahlin, L. B. (2007). Induction of activating transcription factor 3 after different sciatic nerve injuries in adult rats. *Scand. J. Plast. Reconstr. Surg. Hand Surg.* **41,** 158–166.

Keune, J. D., Brenner, M. J., Schwetye, K. E., Yu, J. W., Fox, I. K., Hunter, D. A., and Mackinnon, S. E. (2006). Temporal factors in peripheral nerve reconstruction with suture scaffolds: An experimental study in rodents. *Restor. Neurol. Neurosci.* **24,** 181–190.

Kim, Y. T., Haftel, V. K., Kumar, S., and Bellamkonda, R. V. (2008). The role of aligned polymer fiber-based constructs in the bridging of long peripheral nerve gaps. *Biomaterials* **29,** 3117–3127.

Koh, H. S., Yong, T., Chan, C. K., and Ramakrishna, S. (2008). Enhancement of neurite outgrowth using nano-structured scaffolds coupled with laminin. *Biomaterials* **29,** 3574–3582.

Krekoski, C. A., Neubauer, D., Zuo, J., and Muir, D. (2001). Axonal regeneration into acellular nerve grafts is enhanced by degradation of chondroitin sulfate proteoglycan. *J. Neurosci.* **21,** 6206–6213.

Kumbar, S. G., James, R., Nukavarapu, S. P., and Laurencin, C. T. (2008). Electrospun nanofiber scaffolds: Engineering soft tissues. *Biomed. Mater.* **3,** 034002.

Kvist, M., Kanje, M., Ekberg, H., Corbascio, M., and Dahlin, L. B. (2008). Costimulation blockade in transplantation of nerve allografts: Long-term effects. *J. Peripher. Nerv. Syst.* **13,** 200–207.

Leah, J. D., Herdegen, T., and Bravo, R. (1991). Selective expression of jun proteins following axotomy and axonal transport block in peripheral nerves in the rat: Evidence for a role in the regeneration process. *Brain Res.* **66,** 198–207.

Liang, G., Wolfgang, C. D., Chen, B. P., Chen, T. H., and Hai, T. (1996). ATF3 gene. Genomic organization, promoter, and regulation. *J. Biol. Chem.* **271,** 1695–1701.

Lindwall, C., and Kanje, M. (2005). The role of p-c-Jun in survival and outgrowth of developing sensory neurons. *Neuroreport* **16,** 1655–1659.

Lindwall, C., Dahlin, L., Lundborg, G., and Kanje, M. (2004). Inhibition of c-Jun phosphorylation reduces axonal outgrowth of adult rat nodose ganglia and dorsal root ganglia sensory neurons. *Mol. Cell Neurosci.* **27,** 267–279.

Lundborg, G. (2000). A 25-year perspective of peripheral nerve surgery: Evolving neuroscientific concepts and clinical significance. *J. Hand Surg.* **25A,** 391–414.

Lundborg, G., Rosen, B., Dahlin, L., Holmberg, J., and Rosen, I. (2004). Tubular repair of the median or ulnar nerve in the human forearm: A 5-year follow-up. *J. Hand Surg. Br.* **29,** 100–107.

Lundborg, G., Bjorkman, A., and Rosen, B. (2007). Enhanced sensory relearning after nerve repair by using repeated forearm anaesthesia: Aspects on time dynamics of treatment. *Acta Neurochir. Suppl.* **100,** 121–126.

Ma, J., Novikov, L. N., Kellerth, J. O., and Wiberg, M. (2003). Early nerve repair after injury to the postganglionic plexus: An experimental study of sensory and motor neuronal survival in adult rats. *Scand. J. Plast. Reconstr. Surg. Hand Surg.* **37,** 1–9.

Ma, Z., Kotaki, M., Inai, R., and Ramakrishna, S. (2005). Potential of nanofiber matrix as tissue-engineering scaffolds. *Tissue Eng.* **11,** 101–109.

Maroudas, N. G. (1972). Anchorage dependence: Correlation between amount of growth and diameter of bead, for single cells grown on individual glass beads. *Exp. Cell Res.* **74,** 337–342.

Martensson, L., Gustavsson, P., Dahlin, L. B., and Kanje, M. (2007). Activation of extracellular-signal-regulated kinase-1/2 precedes and is required for injury-induced Schwann cell proliferation. *Neuroreport* **18,** 957–961.

Martini, R., Schachner, M., and Brushart, T. M. (1994). The L2/HNK-1 carbohydrate is preferentially expressed by previously motor axon-associated Schwann cells in reinnervated peripheral nerves. *J. Neurosci.* **14,** 7180–7191.

McKay Hart, A., Brannstrom, T., Wiberg, M., and Terenghi, G. (2002). Primary sensory neurons and satellite cells after peripheral axotomy in the adult rat: Timecourse of cell death and elimination. *Exp. Brain Res.* **142,** 308–318.

Middlemas, A., Delcroix, J. D., Sayers, N. M., Tomlinson, D. R., and Fernyhough, P. (2003). Enhanced activation of axonally transported stress-activated protein kinases in peripheral nerve in diabetic neuropathy is prevented by neurotrophin-3. *Brain* **126,** 1671–1682.

Millesi, H., Meissl, G., and Berger, A. (1972). The interfascicular nerve grafting of the median and ulnar nerve. *J. Bone Joint Surg.* **54A,** 727–750.

Murugan, R., and Ramakrishna, S. (2006). Nano-featured scaffolds for tissue engineering: A review of spinning methodologies. *Tissue Eng.* **12,** 435–447.

Neumann, S., Bradke, F., Tessier-Lavigne, M., and Basbaum, A. I. (2002). Regeneration of sensory axons within the injured spinal cord induced by intraganglionic cAMP elevation. *Neuron* **34,** 885–893.

Nilsson, A., Dahlin, L., Lundborg, G., and Kanje, M. (2005a). Graft repair of a peripheral nerve without the sacrifice of a healthy donor nerve by the use of acutely dissociated autologous Schwann cells. *Scand. J. Plast. Reconstr. Surg. Hand Surg.* **39,** 1–6.

Nilsson, A., Moller, K., Dahlin, L., Lundborg, G., and Kanje, M. (2005b). Early changes in gene expression in the dorsal root ganglia after transection of the sciatic nerve; effects of amphiregulin and PAI-1 on regeneration. *Brain Res. Mol. Brain Res.* **136,** 65–74.

Nishiura, Y., Brandt, J., Nilsson, A., Kanje, M., and Dahlin, L. B. (2004). Addition of cultured schwann cells to tendon autografts and freeze-thawed muscle grafts improves peripheral nerve regeneration. *Tissue Eng.* **10,** 157–164.

Nobes, C. D., and Hall, A. (1999). Rho GTPases control polarity, protrusion, and adhesion during cell movement. *J. Cell Biol.* **144,** 1235–1244.

Patel, B. N., and Van Vactor, D. L. (2002). Axon guidance: The cytoplasmic tail. *Curr. Opin. Cell Biol.* **14,** 221–229.

Perlson, E., Hanz, S., Medzihradszky, K. F., Burlingame, A. L., and Fainzilber, M. (2004a). From snails to sciatic nerve: Retrograde injury signaling from axon to soma in lesioned neurons. *J. Neurobiol.* **58,** 287–294.

Perlson, E., Medzihradszky, K. F., Darula, Z., Munno, D. W., Syed, N. I., Burlingame, A. L., and Fainzilber, M. (2004b). Differential proteomics reveals multiple components in retrogradely transported axoplasm after nerve injury. *Mol. Cell. Proteomics* **3,** 510–520.

Prinz, C., Hallstrom, W., Martensson, T., Samuelson, L., Montelius, L., and Kanje, M. (2008). Axonal guidance on patterned free-standing nanowire surfaces. *Nanotechnology* **19,** 345101.

Qiu, J., Cai, D., Dai, H., McAtee, M., Hoffman, P. N., Bregman, B. S., and Filbin, M. T. (2002a). Spinal axon regeneration induced by elevation of cyclic AMP. *Neuron* **34,** 895–903.

Qiu, J., Cai, D., and Filbin, M. T. (2002b). A role for cAMP in regeneration during development and after injury. *Prog. Brain Res.* **137,** 381–387.

Raivich, G., Bohatschek, M., Da Costa, C., Iwata, O., Galiano, M., Hristova, M., Nateri, A. S., Makwana, M., Riera-Sans, L., Wolfer, D. P., Lipp, H. P., Aguzzi, A., *et al.* (2004). The AP-1 transcription factor c-Jun is required for efficient axonal regeneration. *Neuron* **43,** 57–67.

Rajnicek, A. M., Foubister, L. E., and McCaig, C. D. (2008). Alignment of corneal and lens epithelial cells by co-operative effects of substratum topography and DC electric fields. *Biomaterials* **29,** 2082–2095.

Ramon y Cajal, S. (1928)."Degeneration and Regeneration of the Nervous System." Oxford University Press, London.

Rosen, B., Bjorkman, A., and Lundborg, G. (2006). Improved sensory relearning after nerve repair induced by selective temporary anaesthesia - a new concept in hand rehabilitation. *J. Hand Surg. Br.* **31,** 126–132.

Rosengren, A., Wallman, L., Danielsen, N., Laurell, T., and Bjursten, L. M. (2002). Tissue reactions evoked by porous and plane surfaces made out of silicon and titanium. *IEEE Trans. Biomed. Eng.* **49,** 392–399.

Rovensky, Y. A., Slavnaja, I. L., and Vasiliev, J. M. (1971). Behaviour of fibroblast-like cells on grooved surfaces. *Exp. Cell Res.* **65,** 193–201.

Saito, H., and Dahlin, L. B. (2008). Expression of ATF3 and axonal outgrowth are impaired after delayed nerve repair. *BMC Neurosci.* **9,** 88.

Sapelkin, A. V., Bayliss, S. C., Unal, B., and Charalambou, A. (2006). Interaction of B50 rat hippocampal cells with stain-etched porous silicon. *Biomaterials* **27,** 842–846.

Scherman, P., Lundborg, G., Kanje, M., and Dahlin, L. B. (2001). Neural regeneration along longitudinal polyglactin sutures across short and extended defects in the rat sciatic nerve. *J. Neurosurg.* **95,** 316–323.

Scherman, P., Kanje, M., and Dahlin, L. B. (2004). Local effects on triiodothyronine-treated polyglactin sutures on regeneration across peripheral nerve defects. *Tissue Eng.* **10,** 455–464.

Schnell, E., Klinkhammer, K., Balzer, S., Brook, G., Klee, D., Dalton, P., and Mey, J. (2007). Guidance of glial cell migration and axonal growth on electrospun nanofibers of poly-epsilon-caprolactone and a collagen/poly-epsilon-caprolactone blend. *Biomaterials* **28,** 3012–3025.

Seijffers, R., Allchorne, A. J., and Woolf, C. J. (2006). The transcription factor ATF-3 promotes neurite outgrowth. *Mol. Cell Neurosci.* **32,** 143–154.

Seijffers, R., Mills, C. D., and Woolf, C. J. (2007). ATF3 increases the intrinsic growth state of DRG neurons to enhance peripheral nerve regeneration. *J. Neurosci.* **27,** 7911–7920.

Smeal, R. M., Rabbitt, R., Biran, R., and Tresco, P. A. (2005). Substrate curvature influences the direction of nerve outgrowth. *Ann. Biomed. Eng.* **33,** 376–382.

Snider, W. D., Zhou, F. Q., Zhong, J., and Markus, A. (2002). Signaling the pathway to regeneration. *Neuron* **35,** 13–16.

Sondell, M., Fex-Svenningsen, A., and Kanje, M. (1997). The insulin-like growth factors I and II stimulate proliferation of different types of Schwann cells. *Neuroreport* **8,** 2871–2876.

Sondell, M., Lundborg, G., and Kanje, M. (1999). Vascular endothelial growth factor stimulates Schwann cell invasion and neovascularization of acellular nerve grafts. *Brain Res.* **846,** 219–228.

Sung, Y. J., Povelones, M., and Ambron, R. T. (2001). RISK-1: A novel MAPK homologue in axoplasm that is activated and retrogradely transported after nerve injury. *J. Neurobiol.* **47,** 67–79.

Tanabe, K., Bonilla, I., Winkles, J. A., and Strittmatter, S. M. (2003). Fibroblast growth factor-inducible-14 is induced in axotomized neurons and promotes neurite outgrowth. *J. Neurosci.* **23,** 9675–9686.

Wang, H. B., Mullins, M. E., Cregg, J. M., Hurtado, A., Oudega, M., Trombley, M. T., and Gilbert, R. J. (2008). Creation of highly aligned electrospun poly-L-lactic acid fibers for nerve regeneration applications. *J. Neural. Eng.* **6,** 16001.

Weber, A., Breidenback, W. C., Brown, R. E., Jabaley, M. E., and Mass, D. P. (2000). A randomized prospective study of polyglycolic acid conduits for digital nerve reconstraction in humans. *Plast. Reconstr. Surg.* **106,** 1036–1045.

Weiss, P. (1945). Experiments on cell and axon orientation *in vitro*: The role of colloidal exudates in tissue organization. *J. Exp. Zool.* **100,** 353–386.

Wojciak-Stothard, B., Curtis, A., Monaghan, W., MacDonald, K., and Wilkinson, C. (1996). Guidance and activation of murine macrophages by nanometric scale topography. *Exp. Cell Res.* **223,** 426–435.

Wong, J., and Oblinger, M. M. (1991). NGF rescues substance P expression but not neurofilament or tubulin gene expression in axotomized sensory neurons. *J. Neurosci.* **11,** 543–552.

Woolf, C. J., Reynolds, M. L., Molander, C., O'Brien, C., Lindsay, R. M., and Benowitz, L. I. (1990). The growth-associated protein GAP-43 appears in dorsal root ganglion cells and in the dorsal horn of the rat spinal cord following peripheral nerve injury. *Neuroscience* **34,** 465–478.

Yang, F., Murugan, R., Wang, S., and Ramakrishna, S. (2005). Electrospinning of nano/micro scale poly(L-lactic acid) aligned fibers and their potential in neural tissue engineering. *Biomaterials* **26,** 2603–2610.

Yim, E. K., Reano, R. M., Pang, S. W., Yee, A. F., Chen, C. S., and Leong, K. W. (2005). Nanopattern-induced changes in morphology and motility of smooth muscle cells. *Biomaterials* **26,** 5405–5413.

Zhang, X. P., and Ambron, R. T. (2000). Positive injury signals induce growth and prolong survival in Aplysia neurons. *J. Neurobiol.* **45,** 84–94.

INDEX

D

E

O

P

R

CONTENTS OF RECENT VOLUMES

Volume 39

Volume 40

Volume 41

Volume 42

Volume 43

Volume 44

Volume 45

Volume 46

Volume 47

Volume 48

Volume 49

Volume 50

Volume 51

Volume 52

Volume 53

Volume 54

Volume 55

Volume 56

Volume 57

Volume 58

Volume 59

Volume 60

Volume 61

Volume 62

Volume 63

Volume 64

Volume 65

Volume 66

Volume 67

Volume 68

Volume 69

Volume 70

Volume 71

Volume 72

Volume 73

Volume 74

Volume 75

Volume 76

Volume 77

Volume 78

Volume 79

Volume 79

Volume 81

Volume 82

Volume 83

Volume 84

Volume 85

Volume 86

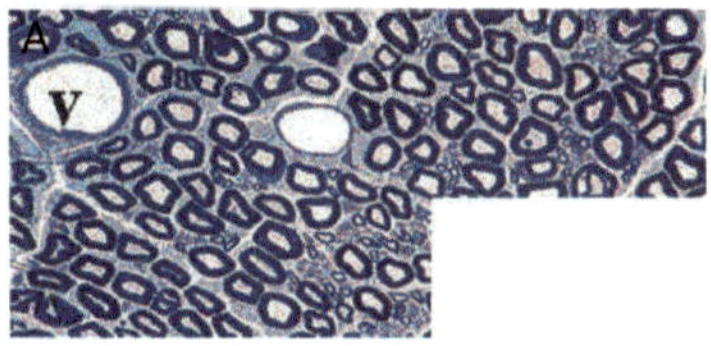

KAPLAN ET AL., CHAPTER 2, FIG. 1. (See Legend in Text.)

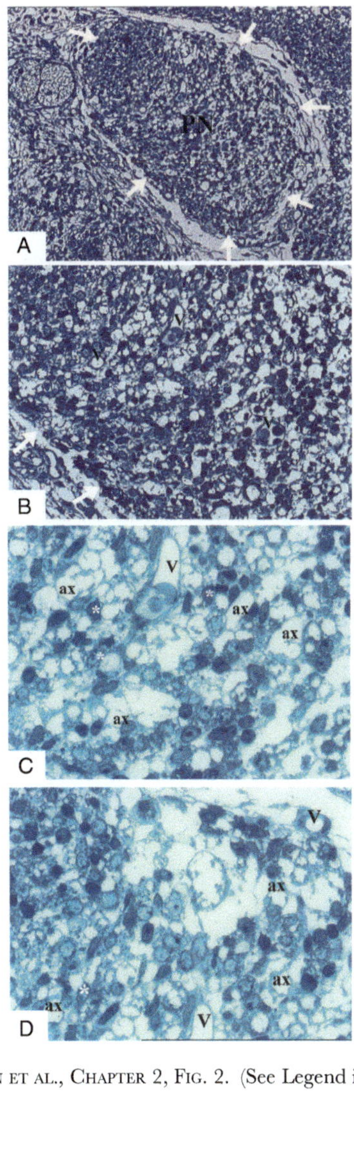

KAPLAN ET AL., CHAPTER 2, FIG. 2. (See Legend in Text.)

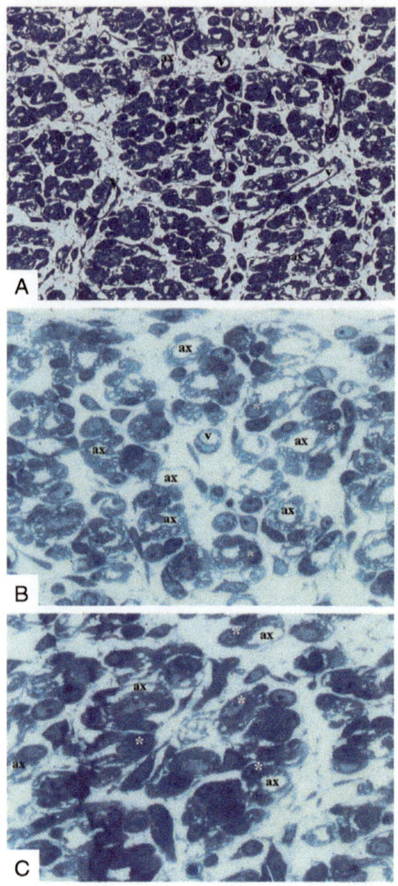

KAPLAN ET AL., CHAPTER 2, FIG. 3. (See Legend in Text.)

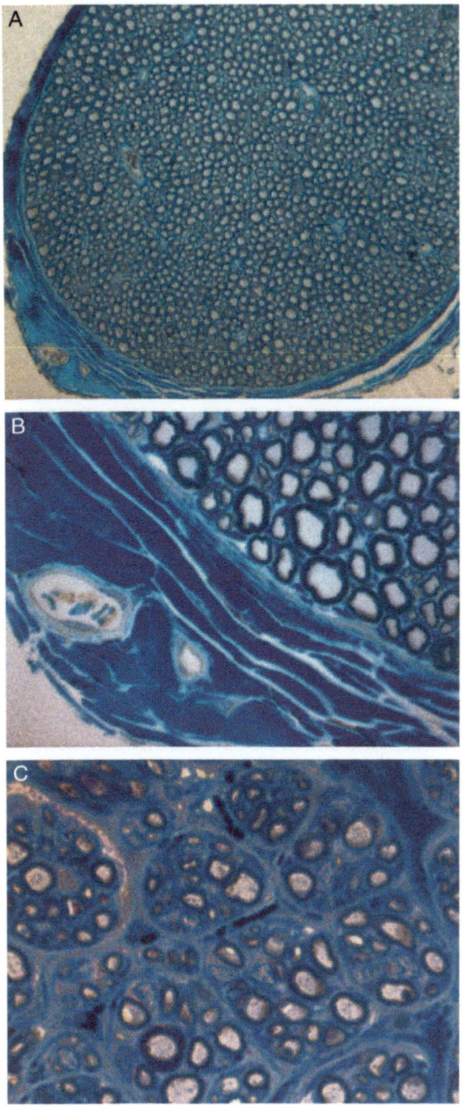

Geuna et al., Chapter 3, Fig. 3. (See Legend in Text.)

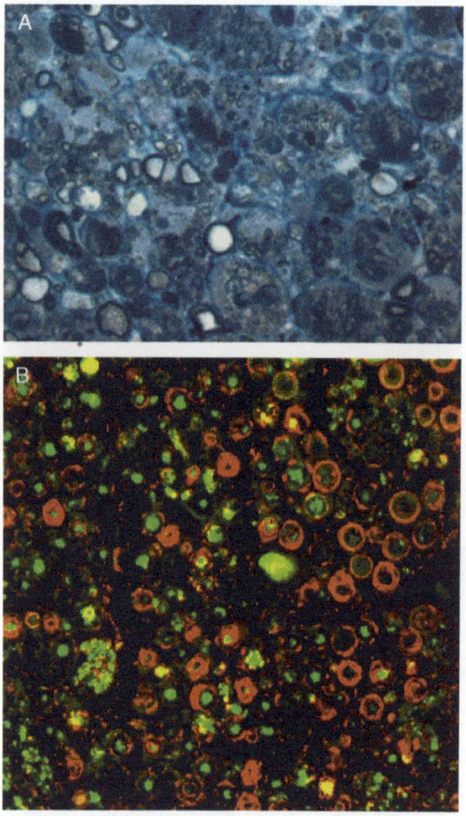

GEUNA ET AL., CHAPTER 3, FIG. 4. (See Legend in Text.)

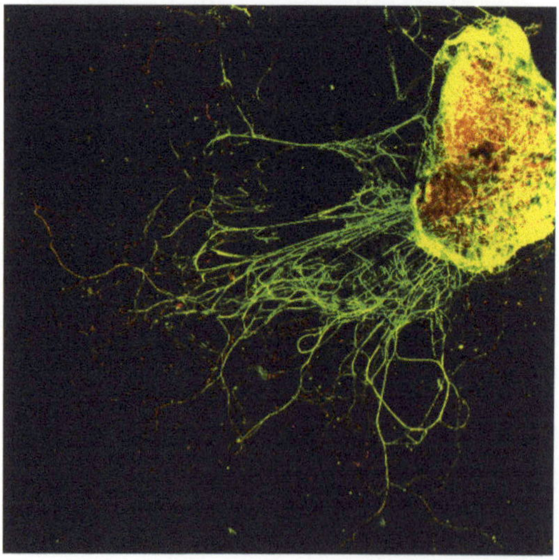

TOS ET AL., CHAPTER 4, FIG. 1. (See Legend in Text.)

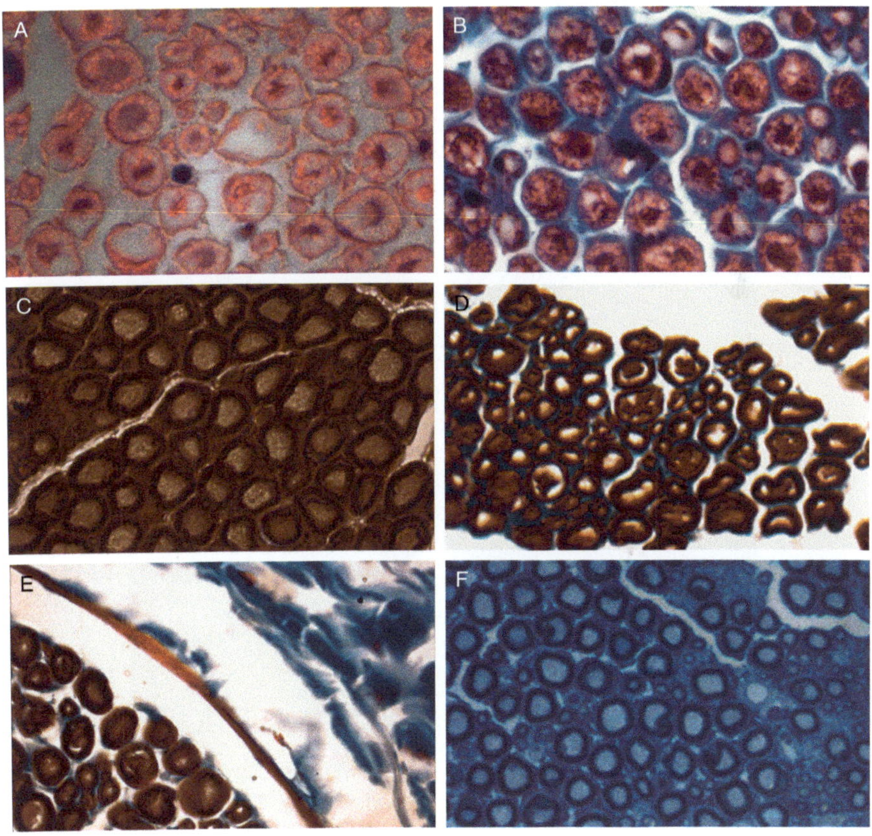

RAIMONDO ET AL., CHAPTER 5, FIG. 1. (See Legend in Text.)

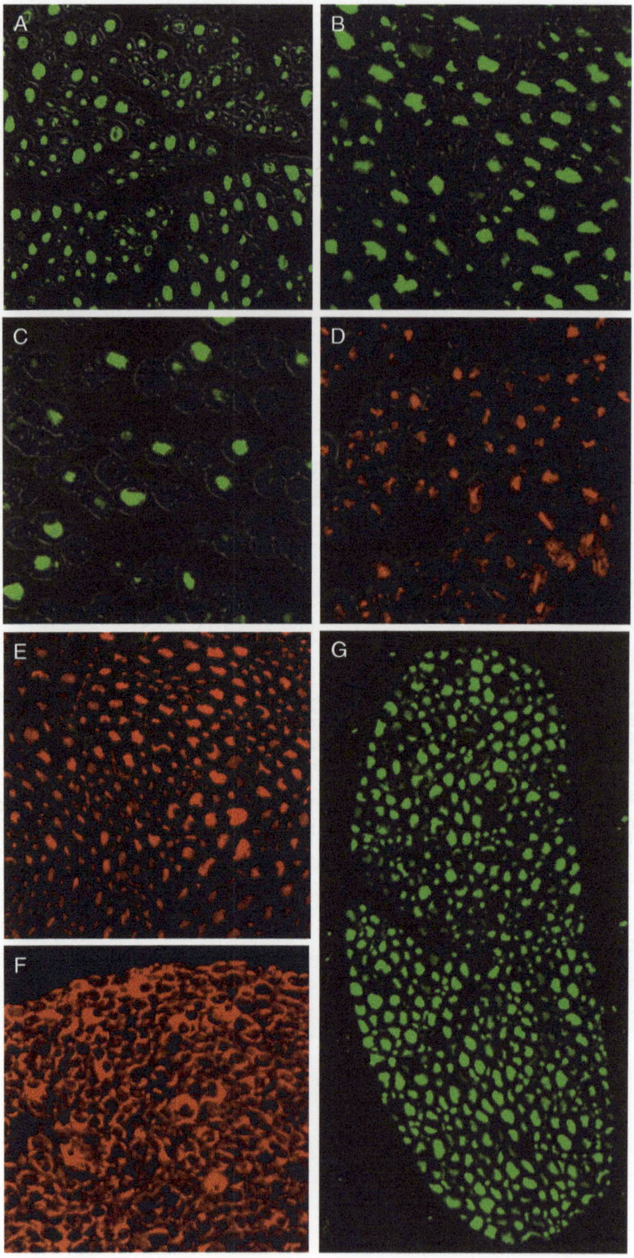

RAIMONDO ET AL., CHAPTER 5, FIG. 2. (See Legend in Text.)

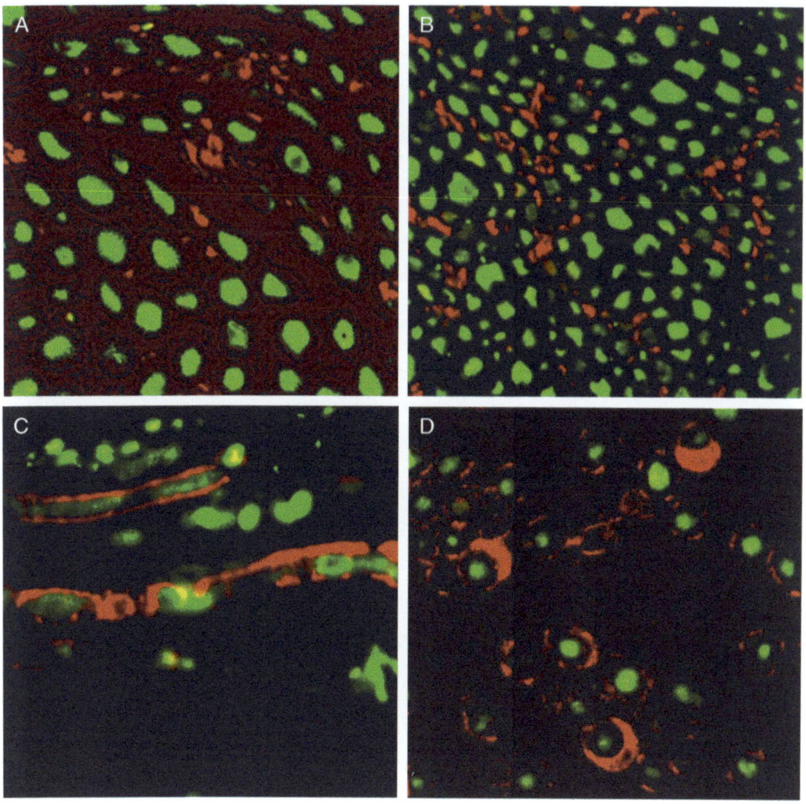

RAIMONDO ET AL., CHAPTER 5, FIG. 3. (See Legend in Text.)

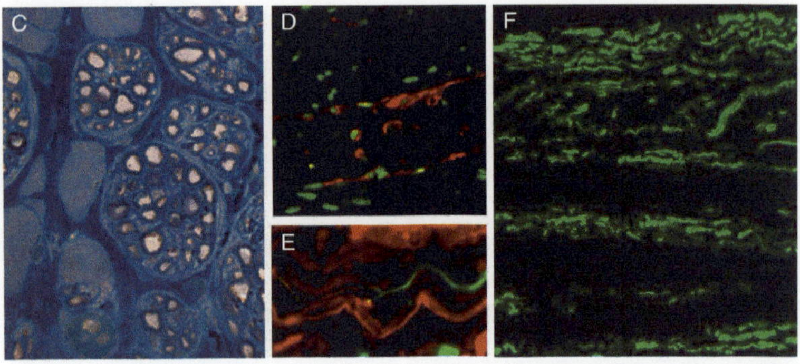

BATTISTON ET AL., CHAPTER 11, FIG. 4. (See Legend in Text.)

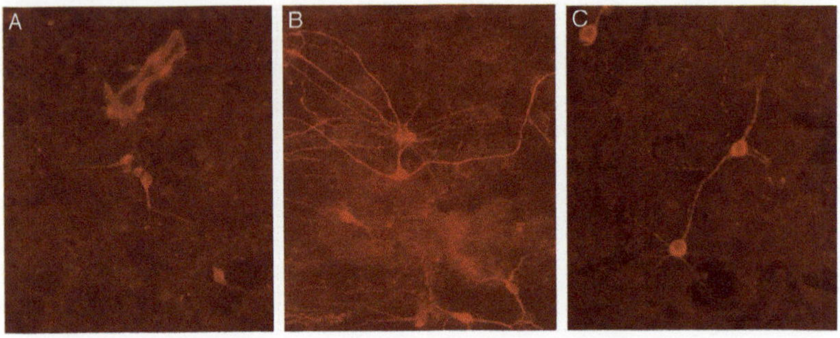

ROCHKIND ET AL., CHAPTER 25, FIG. 11. (See Legend in Text.)

FSC
www.fsc.org
MIX
Papier | Fördert gute Waldnutzung
FSC® C083411

Kolibri 360 GmbH
Ferdinand-Jühlke-Straße 7
99095 Erfurt, Deutschland
produktsicherheit@kolibri360.de